Atomic and Nuclear Clusters

Sponsors

American Express Bank Ltd.

Aristotle University of Thessaloniki, Greece

Boutari Brothers Wineries, S. A.

Commercial Bank of Greece

European Physical Society

Folkloristic Group "Krystallis"

Greek Ministry of Culture and Sciences

Greek Ministry of Industry, Research and Technology

Greek National Commission for UNESCO

Greek National Tourist Organization

Hahn-Meitner-Institut Berlin GmbH, Germany

Hellenic Nuclear Physics Society

Hellenic Physical Society

Institute of Nuclear Physics, N.C.S.R. "Demokritos"

International Science Foundation

INTRACOM (Hellenic Telecommunications & Industry)

National Bank of Greece

Olympic Airways

Olympic Aviation

University of Aegeon

Venturis Sea Lines

Chairmen

G. S. Anagnostatos, N.C.S.R. "Demokritos", Athens, Greece
W. von Oertzen, Hahn-Meitner-Institut Berlin GmbH, Germany

Secretaries

W. Ekardt, Fritz-Haber-Institut, Berlin, Germany
M. E. Grypeos, Aristotle University of Thessaloniki, Greece

International Organizing Committee

J. A. Alonso, Universidad de Valladolid, Spain
M. Brenner, Abo Akademi, Turku, Finland
R. A. Broglia, Instituto Nazionale di Fisica Nucleare, Milano, Italy
C. Detraz, IN2P3, Paris, France
K. A. Gridnev, St. Petersburg University, Russia
F. B. Malik, Southern Illinois University at Carbondale, USA
T. P. Martin, Max-Planck-Institut, Stuttgart, Germany
B. R. Mottelson, Nordisk Institut for Theoretisk Fysik, Copenhagen, Denmark
A. C. Xenoulis, N.C.S.R. "Demokritos", Athens, Greece

International Advisory Committee

R. S. Berry, The University of Chicago, USA
S. Bjornholm, University of Copenhagen, Denmark
C. Brechignac, Laboratoire Aime Cotton, Orsay, France
D. M. Brink, University of Oxford, U.K.
M. Broyer, Université Claude Bernard-Lyon I, France
R. Coussement, University of Leuven, Belgium
O. Echt, University of New Hampshire, Durham, USA
A. Faessler, Eberhard-Karls-Universität Tubingen, Germany
C. Fotakis, Foundation for Research and Technology, Heraklion, Greece
W. Greiner, Universität Frankfurt a. M., Germany
P. E. Hodgson, University of Oxford, U.K.
H. Horiuchi, Kyoto University, Japan
P. Jena, Virginia Commonwealth University, Richmond, USA
J. Jortner, Tel Aviv University, Israel
A. A. Katsanos, Technical University of Crete, Chania, Greece
J. Koutecky, Freie Universität Berlin, Germany
H. O. Lutz, Universität Bielefeld, Germany
E. Merzbacher, University of North Carolina at Chapel Hill, USA
J. W. Negele, Massachusetts Institute of Technology, Cambridge, USA
A. Oglobin, Kurchatov Institute, Moscow, Russia
S. Rosati, Universita die Pisa, Italy

Local Organizing Committee

G. S. Anagnostatos, N.C.S. R. "Demokritos", Athens, Greece
D. Bonatsos, N.C.S.R. "Demokritos", Athens, Greece
A. Clouvas, Aristotle University of Thessaloniki, Greece
M. E. Grypeos, Aristotle University of Thessaloniki, Greece
C. P. Panos, Aristotle University of Thessaloniki, Greece
L. Trouposkiadis, N.C.S.R. "Demokritos", Athens, Greece
M. Velegrakis, Foundation for Research and Technology, Heraklion, Greece
A. C. Xenoulis, N.C.S.R. "Demokritos", Athens, Greece

Technical Head

Ph. Trouposkiadis, N.C.S.R. "Demokritos", Athens, Greece

G. S. Anagnostatos W. von Oertzen (Eds.)

Atomic and Nuclear Clusters

Proceedings of the Second International Conference
at Santorini, Greece
June 28 – July 2, 1993

With 225 Figures

Springer

Professor G. S. Anagnostatos
NCSR "Demokritos"
Institute of Nuclear Physics
GR-15310 Aghia Paraskevi, Attiki
Greece

Professor Dr. W. von Oertzen
Hahn-Meitner-Institut Berlin
Glienicker Straße 100
D-14109 Berlin
Germany

Some contributions to this book originally appeared in the journals
Zeitschrift für Physik A – Hadrons and Nuclei, Volume 349, Number 3/4
(ISSN 0939-7922)
Zeitschrift für Physik D – Atoms, Molecules and Clusters, Volume 31, Numbers 3 and 4
(ISSN 0178-7638)

ISBN 3-540-59160-5 Springer-Verlag Berlin Heidelberg New York

CIP data applied for

Printed in Germany

SPIN 10490061 55/3144-5 4 3 2 1 0 – Printed on acid-free paper

Preface

The Second International Conference on Atomic and Nuclear Clusters '93 was organized in a joint effort by the 'Demokritos' National Center for Scientific Research, G.S. Anagnostatos (representing the atomic physics) and the Hahn-Meitner-Institut, W. von Oertzen (representing the nuclear physics).

The subject of clusters – small aggregates of particles – is a topic of primary interest in both atomic and nuclear physics, and also in other fields like in the case of quark-structure of baryons and in cosmology. The interplay between atomic and nuclear physics is a particularly fascinating one because many concepts are common to both fields (quantal effects, shells, geometric structures, collective modes, fission etc.)

This conference was the second after the first one organized by Professor M. Brenner in Åbo (Finland) in 1991. The general atmosphere of a joint forum for atomic and nuclear physicists was very fruitful and thus the decision to have a sequence of such meetings has been taken. A third one is planned in St. Petersburg (Russia) with Professor K. Gridnev (St. Petersburg) and Mme. Professor C. Bréchignac (Orsay) as Chairpersons.

The conference site, Firá on Santorini island (Greece), was a wonderful choice for a conference. It is small, which helps to keep people concentrated in a smaller community, it has a perfect convention center, the P. Nomikos Conference Center, and a very beautiful landscape formed by a large volcanic crater.

Apart from the sessions with oral presentations mostly composed of invited papers, poster sessions were held early in the afternoons. The posters and a limited number of selected contributions presented orally, are also printed in these proceedings as two-page contributions. The order of presentation in these proceedings follows that of the official program. The poster papers are collected together at the end of the proceedings in alphabetical order. As apparent from the present volume, during the conference the relevant topics of atomic and nuclear clusters were grouped into common sessions.

The papers accepted in these proceedings are the result of a refereeing procedure directed by the chairmen of the conference. **Originality and quality were the main criteria of acceptance for both invited and contributed papers.** After this procedure, the accepted papers on atomic clusters were sent to Zeitschrift für Physik D and those on nuclear clusters to Zeitschrift für Physik A. The editors-in-chief of these journals, by employing their own criteria, accepted certain papers to appear as regular articles in their respective journals. These articles are indicated by a citation line at the top of the first page of the article.

The general atmosphere of the conference was very lively and international. The overall feeling that the subject of the conference needs common efforts in both atomic and nuclear fields was definitely confirmed.

Athens – Berlin,
March 1995

G.S. Anagnostatos
W. von Oertzen

Acknowledgements

We are grateful to Professor T.R. Motteleson for his continuous support and encouragement of a joint study of atomic and nuclear clusters which constitutes the main theme of this series of conferences.

We deeply appreciate the suggestions provided by the members of the International Organizing Committee and the members of the International Advisory Committee which contributed to the balanced composition of topics and speakers. Of course, the conference could not have taken place without the untiring efforts of the members of the Local Organizing Committee during the long period of preparation. We also thank the European Physical Society under whose auspices the conference was held.

Finally, we are indebted to the conference sponsors listed on the first page of these proceedings.

Contents

List of Participants

Aberg Sven
Dept of Mathematical Physics
Lund Institute of Technology
P.O. Box 118
22100 Lund
Sweden
Tel (46)-46109633
Fax (46)-46104416
e-MAIL SVEN.ABERG@MAT.FYS.LTH.SE

Adamides Eleftherios
N.T.U., Athens
T.E.I. Pireaus
Greece
Tel (30)-1-770.7842

Alonso Julio
Dept Fisica Teorica
Universidad de Valladolid
Valladolid 47011
Spain
Tel (34)-83-423142
FAX (34)-83-423013
e-MAIL FISATO@CPD.UVA.ES

Anagnostatos Gerasimos S.
Institute of Nuclear Physics
NCSR "Demokritos"
Aghia Paraskevi
Athens 15310
Greece
Tel (30)-1-6518770
FAX (30)-1-6511215
Telex 601(21)6199
e-MAIL ANAGNOS@CYCLADES.NRCPS.ARIADNE-T.GR

Athanassenas Kalliopi
Steacie Institute for Molecular Sciences
National Research Council of Canada
100 Sussex Drive
Ottawa, Ontario KIA OR6
Canada
Tel (1)-613-9900947
FAX (1)-613-9548902

Balbas Luis Carlos
Depto Fisica Teorica
Universidad de Valladolid
Valladolid 47011
Spain
Tel (34)-83-423144
FAX (34)-83-423114, 423013
e-MAIL BALBAS@CPD.UVA.ES

Benfield Robert Edward
Chemical Lab.
University of Kent
Canterbury CT2 7NH
U.K.
Tel (44)-227-764000 ext. 3558
FAX (44)-227-475475
e-MAIL reb1@UKc.ac.UK

Bettiolo Massimo
Instituto Fisica Nucleare
Laboratori Nazionali di Legnaro
Via Romea 4
35020 Legnaro
Italy
Tel (39)-49-8292311
FAX (39)-49-641925
e-MAIL YANNUCCI@LNL.INFN.IT

Bezrukov Oleg
St. Petersburg 196 105
Yakvlevskii pet cl 6 KV 88
Russia
Tel (7)-812-2942926
FAX (7)-812-4287240
e-MAIL BEZRUKOV@INII.PHYS.LGU.CPB.SU.

Bilwes Brigitte
C.R.N.
BP 20-CRO
67037 Strasbourg Cedex
France
Tel (33)-88286526
Fax (33)-88286794
e-MAIL BILWESB@FRCPN11.IN2P3.FR

Bonatsos Dionissis
Institute of Nuclear Physics
NCSR Demokritos
Aghia Paraskevi-Attiki 153 10
Greece
Tel (30)-1-6518770
FAX (30)-1-6511215

Borgenstrand Hakan
Manne Siegbahn Institute for Physics, MSI
Frescativaegen 24
S-10405 Stockholm
Sweden
Tel (46)-8-161023 (1000)
FAX (46)-8-158674
e-MAIL BORGENSTRAND@MSI.SUNET.SE

Borggreen Jorn
Niels Bohr Institut
University of Copenhagen
Riso 4000 Roskilde
Denmark
Tel (45)-46775600, 42371616
FAX (45)-42373516
e-MAIL JOBO@NBITAL.DK

Bracco Angela
Dept of Physics
Universita Degli Studi di Milano
Via Celoria 16
20133 Milano
Italy
Tel (39)-2-2392300, 2392248, 2392231
FAX (39)-2-2392487

Brechignac Catherine
CNRS
Laboratoire Aime Cotton
Bat 505 Campus Orsay
91405 Orsay Cedex
France
Tel (33)-1-69824050
FAX (33)-1-69853616

Brenner Marten
Abo Akademi University
Porthansgatan 3-5
SF-20500 Abo
Finland
Tel (358)-21-654861
FAX (358)-21-654776
Telex AABIB 62301 SF
e-MAIL MBRENNER@FINABO

Bressani Tullio
I.N.F.N. Sezione di Torino
Via P. Giuria 1
Torino 10125
Italy
Tel (39)-11-6707322
FAX (39)-11-6707323
e-MAIL VAXTO::BRESSANI

Broglia Ricardo A.
Dept of Physics
Universita Degli Studi di Milano
Via Celoria 16
20133 Milano
Tel (39)-2-2392300, 2392248, 2392231
FAX (39)-2-2392487

Broyer Michel
Laboratoire de Spectrometrie Ionique et Moleculaire (LASIM)
Universite Lyon I
Bat 205
43 Bd du 11 Novembre 1918
69622 Villeurbanne Cedex
France
Tel (33)-72448260
FAX (33)-78894415

Borve Knut
Dept of Chemistry
University of Bergen
Allegt. 41
N-5007 Bergen
Norway
Tel (47)-5-213365
FAX (47)-5-329058
e-MAIL: Knut.Borve@Kj.uib.no

Busso Luigi
INFN-Sezione di Torino
Via P. Giuria 1
Torino 10125
Italy
Tel (39)-11-6707417
FAX (39)-11-6699579
e-MAIL VAXTO::BUSSO

Carjan Nicolae
University of Bordeaux
CENBG
33175 Gradignan Cedex
France
Tel (33)-56891800
FAX (33)-56751180
e-MAIL CARJAN@FRCPN11

Castleman A.Welford Jr.
Dept of Chemistry
Pennsylvania State University
152 Davey Laboratory
University Park
PA 16802
U.S.A.
Tel (1)-814-8657242
FAX (1)-814-8653314
e-MAIL AWC@PSUVM.EDU

Chbihi Abdelouahad
GANIL
B.P. 5027
14000 Caen
France
Tel (33)-31454708
FAX (33)-31454665
e-MAIL CHBIHI@FRCPN11

Chermette Henry
Instit. Phys. Nucleaire
43 Bd 11 Novembre 1918
F-69622 Villeurbanne Cedex
France
Tel (33)-72448427
FAX (33)-72448004
e-MAIL CHERM@FRCPN11.in2p3.fr

Clarke Norman
University of Birmingham
School of Physics and Space Research
Birmingham B15 2TT
U.K.
Tel (44)-21-4144680
FAX (44)-21-4144719, 4146709
e-MAIL nmc@np.ph.bham.ac.uk.

Clouvas Alexander
Dept of Electrical Engineering
University of Thessaloniki
54006 Thessaloniki
Greece
Tel (30)-31-996353
FAX (30)-31-274868
e-MAIL clouvas@vergina.eng.auth.Gr

Coleman Valerie
International Technology Gateway
155 Sansome St. # 1200
San Francisco, CA 94 104
U.S.A.
Tel (1)-415-4217233
FAX (1)-415-4217238
e-MAIL 71732.3253@Compuserve.Com

Corradi Lorenzo
INFN-Laboratori Nazionali di Legnaro
Via Romea 4
35020 Legnaro Padova
Italy
Tel (39)-49-8292409, 8292311
FAX (39)-49-641200, 641927
e-MAIL VAXLNL::CORRADI

Cseh Jozsef
Institute of Nuclear Research
Hungarian Academy of Sciences
Debrecen BEMT. 18/c, Pf 51
4001 Debrecen
Hungary
Tel (36)-52-317266
FAX (36)-52-316181
e-MAIL H908CSE@ELLA.HU

Da Providencia Joao
Depto de Fisica Da Universidade De Coimbra
Universidade de Coimbra
P-3000 Coimbra
Portugal
FAX (351)-39-29158
Telex 52601 DEFIUC-P

Demetriou Vivian
Graduate Student Nuclear Physics (DPHR)
Particle & Nuclear Physics
Nuclear Physics Laboratory
Keble Road
Oxford OX1 3RH
U.K.
Tel (44)-865-273371
FAX (44)-865-273418
e-MAIL DEMETRIOU@V1.PH.OX.AC.UK

Ehlich Rudolf
Max-Born-Institut
Rudower Chaussee 6
12489 Berlin
Germany
Tel (49)-30-63921221
FAX (49)-30-63921229
e-MAIL ehlich@mbi.wtza-berlin.de

Ekardt Walter
Max-Planck-Institut
Faradayweg 4-6
1000 Berlin 33
Germany
Tel (49)-30-8305338
FAX (49)-30-8305333
telex 185676 thimp-d

Erba Gadioli Enrica
Milan University
Via Celoria 16
20133 Milano
Italy
Tel (39)-2-2392241, 2392298
FAX (39)-2-2392487
Telex 02/334687
e-MAIL VAXMI::GADIOLI

Farantos Stavros
Dept of Chemistry
Universiity of Crete and FO.R.T.H-IESL
Heraklion, Crete 71110
Greece
Tel (30)-81-239757
FAX (30)-81-239735
e-MAIL farantos@chaos.iesl.forth.gr
or farantos@grearn.bitnet

Farizon Bernadette
Institut de Physique Nucleaire
43 Bvd du 11 Nov. 1918
69622 Villeurbanne
France
Tel (33)-72448389
FAX (33)-72448004
e-MAIL MAZUY@FRCPN11

Farizon Michel
Institut de Physique Nucleaire
Universite Lyon I
43 Bvd du 11 Novembre 1918
69622 Villeurbanne Cedex
France
Tel (33)-72448389
FAX (33)-72448004
e-MAIL FARIZON@FRCPN11

Fotakis Costas
FO.R.T.H.
P.O. Box 1527
71110 Heraklion, Crete
Greece
Tel (30)-81-210035
FAX (30)-81-239735

Fulton Brian
Physics Dept.
University of Birmingham
Birmingham B15 2TT
U.K.
Tel (44)-21-4144683
FAX (44)-21-4144719
e-MAIL bw@np.ph.bham.ac.uk

Furman Walter
Frank Lab. Neutron Physics
Joint Institute for Nuclear Research
Mira St. 7-15
Dubna, Moscow Region
141980 Russia
Tel (7)-96-2165085, 9243914
FAX (7)-96-2165085, 2002283
Telex 911621 Dubna SU
e-MAIL FURMAN@INp20.JINR.DUBNA.SU

Gadioli Ettore
Physics Dept.
Milan University
Via Celoria 16,
20133 Milano
Italy
Tel (39)-2-2392293, 2392298
FAX (39)-2-2392487
e-MAIL VAXMI::GADIOLI or
GADIOLI@MILANO.INFN.IT

Galmarini Marina
Physics Dept.
Milan University
Via Celoria 16
20133 Milano
Italy
Tel (39)-2-2392298
FAX (39)-2-2392487
e-MAIL VAXMI::GADIOLI

Garcia Martin E.
Institut fur Theoretische Physik
Freie Universitat Berlin
Arnimallee 14,
1000 Berlin 33
Germany
Tel (49)-30-8385503, 8383798
FAX (49)-30-8386799
e-MAIL GARCIA@DEC1.PHYSIK.FU-
-BERLIN.DE

Garcias Francesca
Dept of Physics
University of the Balearic Islands
07071 Palma de Mallorca
Spain
Tel (34)-71-173218
FAX (34)-71-173426
e-MAIL DFSFGG8@PS.UIB.ES

Gerlic Evelyne
Institut de Physique Nucleaire de Lyon
CNRS (IN2P3)
Universite Claude Bernard
43 Bd du 11 Novembre 1918
69622 Villeurbanne Cedex
France
Tel (33)-72431090
FAX (33)-72448004
e-MAIL GEREVE@FRCPN11

Gherghescu Radu Alexandru
Institute of Atomic Physics
P.O. Box MG-6
76900 Bucharest
Romania
Tel (40)-1-7807040, 7802190
FAX (40)-1-3122247
e-MAIL RGHERG@ROIFA.BITNET

Gingerich Karl A.
Dept of Chemistry
Texas A+M University
College Station
Texas 77843-3250
U.S.A.
Tel (1)-409-8452431
FAX (1)-409-8454719
e-MAIL E404IS@TAMVM

Goldberg V.Z.
I.V. Kurchatov Institute of Atomic Energy
Kurchatov Sq. 1
123182 Moscow
Russia

Gonnenwein Friedrich
University of Tubingen
Auf Der Morgenstelle 14
Tubingen D-72076
Germany
Tel (49)-7071-292424
FAX (49)-7071-296296
e-MAIL goennenwein@pit.physik.uni-
-tuebingen.de

Greiner Walter
Institut fur Theoretische Physik
University of Frankfurt
Robert Mayer Str. 8-10
6000 Frankfurt 11
Germany
Tel (49)-69-7982332
FAX (49)-69-7988350
e-MAIL UF00@DDAGSI3

Gridnev Konstantin
Nuclear Physics Dept.
Institut of Physics
Saint Petersburg University
Uljanova 1
St. Petersburg 198904
Russia
Tel (7)-812-4287409
FAX (7)-812-4286649
e-MAIL gridnev@onti.phys.lgu.spb.su

Gross Dieter
Hahn Meitner Institut-Berlin
Glienickerstr. 100
D-14109 Berlin
Germany
Tel (49)-30-80092677
FAX (49)-30-80092098
e-MAIL gross@VAX.HMI.DBP.DE

Grypeos Michael
Aristotle University of Thessaloniki
54006 Thessaloniki
Greece
Tel (30)-31-998031
FAX (30)-31-206138
Telex 412181 AUTHGR

Gugliemetti Alessandra
Instituto di Fisica Generale Applicata
Universita di Milano
Via Celoria 16
Milano 20133
Italy
Tel (39)-2-70630854
FAX (39)-2-2665922
e-MAIL RBONETTI@MILANO.INFN.IT

Haberland Hellmut
Fakultat fur Physik
Albert-Ludwigs-Universitat
H. Herderstrasse 3
7800 Freiburg
Germany
Tel (49)-761-2032435
FAX (49)-761-2034527
e-MAIL HHAB@ibm.ruf.uni-freiburg.de

Hill David L.
Patent Enforcement Fund, Inc.
1095 Sasco Hill Road
Fairfield CT 06430
U.S.A.
Tel (1)-203-2597789, 2556624
FAX (1)-203-2543307

Hirschmann Thomas
University of Regensburg
Institut fur Theoretische Physik
Universitatstr. 31
Regensburg W-9340
Germany
Tel (49)-9419432018
e-MAIL hirschma@rphnw2.ngate.uni-
-regensburg.de

Hodgson Peter E.
Nuclear Physics Lab.
Dept of Physics
University of Oxford
Keble Road
Oxford OXI 3RH
U.K.
Tel (44)-865-273333
HORIUCHI@JPNYITP
FAX (44)-865-273418
Telex 83295 Nuclox D

Horiuchi Hisashi
Dept of Physics
Kyoto University
Kyoto 606-01
Japan
Tel (81)-75-7533823
FAX (81)-75-7533886
e-MAIL

Hussonnois Michel
Institut de Physique Nucleaire
BP no 1
Orsay Cedex 91406
France
Tel (33)-1-69415174
FAX (33)-1-69416470

Iwamoto Akira
Japan Atomic Energy Research Institute
Tokai-mura,
Ibaraki-Ken 319-11
Japan
Tel. (81)-292-825453
FAX (81)-292-825939
e-MAIL J2935@JPNJAERI or
iwamoto@hadvon01.tokai.jaeri.qo.jp

Jena Purusottam
Dept of Physics
Virginia Commonwealth University
1020 West Main Street
P.O. Box 2000, Richmond
Virginia 23284-2000
U.S.A.
Tel (1)-804-3678991
FAX (1)-804-3677073
e-MAIL jena@ruby.vcu.edu.

Kadmensky Stanislav
Dept of Nuclear Physics
Voronegh State University
Plekhanovskaya 22
Voronegh 394 030
Russia
Tel (8)-732-524654

Kaldor Uzi
School of Chemistry
Tel Aviv University
69978 Tel Aviv
Israel
Tel (972)-3-6408590
FAX (972)-3-6409293
e-MAIL kaldor@chemib2.tau.ac.il

Karpeshin Feodor
Institute of Physics
St. Petersburg University
Uljanovskaya 1, Petrodvorets
SU-198904 St. Petersburg
Russia
Tel (7)-812-4286348
FAX (7)-812-4287240
e-MAIL FFKARP@ONTI.Phys.LGU.SPB.SU

Kartavenko Vladimir
Lab. of Theoretical Physics
Joint Institute for Nuclear Research
Head P.O. Box 79
Moscow
Russia
FAX (7)-97-52381
Telex 911621 dubna SU
e-MAIL KARTAVEN@THEOR.JINRC.DUBNA.SU

Katsanos A.
Dept of Sciences
Technical University of Crete
34, El. Venizelou St.
73100 Chania
Greece

Knospe Olaf
TU Dresden
Institut fur Theoretische Physik
Mommsenstr. 13
01062 Dresden
Germany
Tel (49)-351-4636151
FAX (49)-351-4637079
e-MAIL Knospe@urzdfn.physik.tu-dresden.dbp.de

Kondow Tamotsu
Dept of Chemistry
School of Science
The University of Tokyo
Bunkyo-ku, Tokyo 113
Japan
Tel (81)-3-38122111 ext 4334, 56897347
FAX (81)-3-56897347
e-MAIL Kondow@utsc.chem.s-tokyo.ac.jp

Koskinen Matti
Dept of Physics
University of Jyvaskyla
P.O. Box 35
40351 Jyvaskyla
Finland
Tel (358)-41-602373
FAX (358)-41-602351
e-MAIL MKOSKINEN@JYEL.JYV.FI

Kotsos Vasilios
Dept of Theoretical Physics
Aristotle University of Thessaloniki
54006 Thessaloniki
Greece
Tel (30)-31-998031

Kubono Shigeru
Institute for Nuclear Study
University of Tokyo
3-2-1 Midori-cho, Tanashi-shi
Tokyo 188
Japan
Tel (81)-424-614131
FAX (81)-424-685844
e-MAIL KUBONO@JPNUTINS.bitnet

Last Isidore
Soreq NRC
Yavne 70600
Israel
Tel (972)-8-434106
FAX (972)-8-434315

Lenis Dimitrios
Institute of Nuclear Physics
NCSR "Demokritos"
Aghia Paraskevi
15310 Athens
Greece
Tel (30)-1-6518770
FAX (30)-1-6511215

Li Chaochen
Chinese Embassy to Greece
2A Krinon Str. P. Psychico
Athens 154 42
Greece
Tel (30)-1-6876743
FAX (30)-1-6876743

Lievens Peter
Laboratorium Voor Vaste Stof-Fysyka
En Magnetisme
Katholieke Universiteit Leuven
Celestijnenlaan 200D
Leuven 3001
Belgium
Tel (32)-16-201015
FAX (32)-16-201368
e-MAIL FGBEA07@BLEKUL11

Liolios Theodore
Dept of Theoretical Physics
Aristotle University of Thessaloniki
540 06 Thessaloniki
Greece
Tel (30)-31-991436

Lodi Rizzini Evandro
Ingegneria
INFN-Brescia University
Via Branze
Brescia 25100
Italy
Tel (39)-30-3715710
FAX (39)-30-380014
e-MAIL BSVAX::LODI
LODI@BSVAX.UNIBS.IT

Luder Christian
FO.R.T.H.-IESL Laser and Applications
P.O. Box 1527
71110 Heraklion, Crete
Greece
Tel (30)-81-238430
FAX (30)-81-239735
e-MAIL CHR@IESL.FORTH.GR

Luetzenkirchen Klaus
Institut fur Kernchemie
University of Mainz
D-55099 Mainz
Germany
Tel (49)-6131-395974
FAX (49)-6131-395253
e-MAIL LUTZ@VKCMZD.Chemie.Uni-Mainz.DE

Martin T. Patrick
Max Plank Institut FKF
Heisenberg str. 1
7000 Stuttgart 80
Germany
Tel (49)-711-6891346
FAX (49)-711-6891010
e-MAIL Martin@Vaxff3.MPI-Stuttgart.
.MPG.DE

Malik Bary F.
Physics Dept.
Southern Illinois University
Carbondale 62901
U.S.A.
Tel (1)-618-4532020
FAX (1)-618-4531056
e-MAIL GA0378@SIUCVMB

Manngard Peter
Dept of Physics
Abo Akademi
Porthansgatan 3-5
20500 Turku
Finland
Tel (358)-21-654630
FAX (358)-21-654776
e-MAIL PMANNGARD@FINABO.FI

Mansikka-aho Juha
Dept of Physics
University of Jyvaskyla
P.O. Box 35
40351 Jyvaskyla
Finland
Tel (358)-41-602373
FAX (358)-41-602351
e-MAIL MANSIKKA-AHO@JYFL.JYV.FI

Merchant Alan Clive
Nuclear Physics Lab.
Oxford University
Keble Road
Oxford OX1 3RH
England
Tel (44)-865-273312
FAX (44)-865-273418
Telex 83295 Nuclox G
e-MAIL merchant@V1.PH.OX.AC.UK

Mohring Klaus
Hahn Meitner Institut Berlin
Glienicker Str. 100
14091 Berlin
Germany
Tel (49)-30-80092300
FAX (49)-30-80092098
e-MAIL MOEHRING@VAX.HMI.DBP.DE

Muller Hans
Friedrich Schiller University
Institut fur Physicalische Chemie
Am Steiger 3, Haus 3
07743 Jena
Germany
Tel (49)-3641-8225054
FAX (49)-3641-8225062

Naher Ulrich
Max-Planck-Institut
fur Festkorperforschung
Heisenberg str. 1
7000 Stuttgart 80
Germany
Tel (49)-711-6891341
FAX (49)-711-6891010
e-MAIL naeher@vaxff3.mpi-stuttgart.mpg.de

Nesterenko Valentin
Joint Institut for Nuclear Research
Laboratory of Theoretical Physics
141980 Dubna
Moscow region
Russia
Tel (7)-9621-63383
FAX (7)-9621-65084
Telex 911621 DUBNA SU
e-MAIL NESTER@LTP03.JINR.DUBNA.SU

Nielsen Mek Buhl
The Niels Bohr Institute
DK-4000 Roskilde
Denmark
Tel (45)-46-775605
FAX (45)-42-373516
e-MAIL MBN@NBITAL.NBI.DK

Noto Hirochi
Hokusei Gakuen University
West 2-3-1 Ohyachi Atsubetsu-ku,
Sapporo 004
Japan
Tel (81)-11-8912731
FAX (81)-11-8943690

Novoselsky Akiva
The Racah Institute of Physics
The Hebrew University of Jerusalem
Jerusalem 91904
Israel
Tel (972)-2-585674
FAX (972)-2-584497
e-MAIL AKIVANOV@HUJIVMS

Oberhummer Heinz
TU Wien
Institut fur Kernphysik
Wiedner Hauptstrasse 8-10
1040 Wien
Austria
Tel (43)-1-58801 ext. 5574
FAX (43)-1-564203
e-MAIL OBERHUMMER@EVAX.
TUWIEN.AC.AT

Osaki Toshihiko
Solar Research Lab.
Government Industrial Research Institute, Nagoya
1-1 Hirate-cho Kita-ku
Nagoya 462
Japan
Tel (81)-52-9112111
FAX (81)-52-9162802

Ohkubo Shigeo
Dept of Applied Science
Kochi Women's University
Kochi 780
Japan
FAX (81)-888-738551
e-MAIL 950121@JPNKUDPC.BITNET

Pacheco Jorge M.
Depto de Fisica da Universidade
University of Coimbra
3000 Coimbra
Portugal
Tel (351)-39-23671
FAX (351)-39-29158
Telex 52601 DEFIUC P
e-MAIL PACHECO@CIUC.UC.PT

Palmer Richard E.
University of Cambridge
Cavendish Laboratory
Madingley Road
CB3 OHE Cambridge
U.K.
Tel (44)-223-337410
FAX (44)-223-350266
e-MAIL REP10@PHX.CAM.AC.UK

Panos Christos
Dept of Theoretical Physics
Aristotle University of Thessaloniki
Thessaloniki 54006
Greece
Tel (30)-31-992842
e-MAIL CABZ0800@GRTHEUN1

Pedersen Jorgen
Niels Bohr Institute
4000 Roskilde
Denmark
Tel (45)-46-775600, 371616
FAX (45)-46-373516
e-MAIL JP@NBITAL.DK

Petridou Theodora
Dept of Theoretical Physics
Aristotle University of Thessaloniki
Thessaloniki 54006
Greece
Tel (30)-31-998031
FAX (30)-31-206138
Telex 412181 AUTHGR

Pilar Iniguez De La Torre
Dpto Fisica Teorica
Universidad de Valladolid
Valladolid 47011
Spain
Tel (34)-83-423000, ext.4197
FAX (34)-83-423013
e-MAIL CLUSTER@CPD.UVA.ES

Plonski Ileana Hamia
Institute of Atomic Physics
P.O. Box 196-6
RO-76 900 Bucharest
Romania
Tel (40)-1-7807040, 2190
FAX (40)-1-3122247
e-MAIL RGHERG@ROIFA.BITNET

Poenaru Dorin
Institute of Atomic Physics
P.O. Box MG-6
RO-76 900 Bucharest
Romania
Tel (40)-1-7807040, 2190
FAX (40)-1-3122247
e-MAIL POENARU@ROIFA.BITNET

Politis Constantin
School of Engineering
Engineering Science Dept.
University of Patras
26110 Patras
Greece
Tel (30)-61-997222
FAX (30)-61-997255

Pougheon Francoise
Institut de Physique Nucleaire
91406 Orsay Cedex
France
Tel (33)-69417944
FAX (33)-69416470
e-MAIL POUGHEON@FRCPN11

Puente Antonio
Dpto de Fisica
Universidad de las Islas Baleares
07071 Palma de Mallorca (Baleares)
Spain
Tel (34)-71-172538
FAX (34)-71-173426
e-MAIL DFSAPF8@PS.UIB.ES

Pugatch Valery
Institut for Nuclear Research
Prospekt Nauki 47
252 028 Kiev-28
Ukraine
Tel (044)-265-3350
FAX (044)-265-4463
e-MAIL kinr@sovamsu.sovusa.com.

Rastikerdar Samad
Isfahan University
P.O. Box 166
Isfahan
Iran
Tel (98)-31-275402, 247654
FAX (98)-31-277396
Telex (98)-31-2312295

Reinhard Dietrich
Institut de Physique Experimentale EPFL
PHB-Ecublens
CH-1015 Lausanne
Switzerland
Tel (41)-21-6934404
FAX (41)-21-6933604
e-MAIL REINHARD@ELDPA.EPFL.CH

Reinhard Paul-Gerhard
Institut fur Theoretische Physik
University Erlangen
Staudtstr. 7
8520 Erlangen
Germany
Tel (49)-91-31858458, 31858462
FAX (49)-91-31858444
e-MAIL reinhard@theorie2.physik.uni-erlangen.de

Ren Hai-Cang
The Rockefeller University
Box 254
New York 10021-6399
U.S.A.
Tel (1)-212-3278838
FAX (1)-212-3277517
e-MAIL REN@ROCKVAX

Roussel Francois
CEA
DRECAM SPAM Bt 522
C.E. Saclay
91191 Gif sur Yvette Cedex
France
Tel (33)-1-69084145
FAX (33)-1-69088446
Telex ENERGX 604641F PAM+

Samanta Chhanda
Saha Institut of Nuclear Physics
1/AF Bidhan Nagar
Calcutta 700 064
India
Tel (91)-33-375345, 375346
FAX (91)-33-374637
Telex 21-4103 SINP IN
e-MAIL Chhanda@Saha.ernet.in

Sassemberg Ulf
Dept of Physics
Stockholm University
Vamadisvagen 9
11346 Stockholm
Sweden
Tel (46)-8-164695
FAX (46)-8-347817
e-MAIL US@VAND.PHYSTO.SE

Satoh Eiji
Chiba-Keizai University
3-59-5 Todoroki-cho, Inage-ku
Chiba-Shi, Chiba 263
Japan
Tel (81)-43-2539111
FAX (81)-43-2546600

Scheier Paul
Institut fur Ionenphysik
Technikerstr. 25
Innsbruck 6020
Austria
Tel (43)-512-2185145
FAX (43)-5122185032
e-MAIL IONEN@UNIIBK.AC.AT

Schmidt Rudiger
Fak. fur Naturwissenschaften und Mathematic.
Institut fur Theor. Physik
Technische Universitat Dresden
Leichtbau Haus 3
Zeumerstrasse 1C
O-8027 Dresden
Germany
Tel (49)-351-4636150
FAX (49)-351-4632826

Simmons Peter
Nuclear Physics Lab.
Dept of Physics
Oxford University
Keble Road
Oxford OX1 3RH
U.K.
Tel (44)-865-273349
FAX (44)-865-273418
e-MAIL Simmons@ph.ox.ac.uk

Stampfli Peter
Physics Dept.
Freie Universitat Berlin
Arnimallee 14
1000 Berlin 33
Germany
Tel (49)-30-8386124, 7919787
FAX (49)-30-8311355

Staudt Guenter
Physikalische Institut
University of Tuebingen
Auf der Morgenstelle 14
72076 Tuebingen
Germany
Tel (49)-7071-296075
FAX (49)-7071-296296
e-MAIL staudt@pit.physik.uni tuebingen.de

Stert Volker
Max-Born-Institut Berlin
Rudower Chaussee 6
12489 Berlin
Germany
Tel (49)-30-63921240
FAX (49)-30-63921259

Stienkemeier Frank
Fak. fur Physik
Universitaet Bielefeld
Universitaets Str.
33615 Bielefeld
Germany
Tel (49)-521-1065362
FAX (49)-521-1066002
e-MAIL STIENKEMEIER@PHYSIK.UNI-
-BIELEFELD.DBP.DE

Tai Yutaka
Radiation Research Dept.
Solar Research Lab.
Government Industrial Research Institute,
Nagoya
1-Hirate, Kita
Nagoya 462
Japan
Tel (81)-52-9112111
FAX (81)-52-9162802

Takami Michio
The Institute of Physical and Chemical Research
(RIKEN)
Wako
Saitama 351-01
Japan
Tel (81)-48-4621111 ext. 3611
FAX (81)-48-4624661
e-MAIL mtakami@rkna50.riken.go.jp

Teghil Roberto
Dpto di Chimica
Universita di Basilicata
Via N. Sauro 85
85 100 Potenza
Italy
Tel (39)-97-1474238
FAX (39)-97-1474223

Temmer George M.
Dept of Physics
Rutgers University
Piscata Way
NJ 08855-0849
U.S.A.
Tel (1)-908-9322402
FAX (1)-908-9324343
Telex 703528

Tretiakova Svetlana
Flevor Lab. of Nuclear Reaction
Joint Institute for Nuclear Research
141 980 Dubna
Russia
Tel (7)-95-9262261
FAX (7)-95-9752381
Telex 911621 Dubna SU
e-MAIL oyuts@ljar.jinr.dubna.su or
LSME2.JINR.DUBNA.SU

Trouposkiadis Leonidas
Tandem Lab. of Nuclear Physics
NCSR "Demokritos"
Aghia Paraskevi-Attiki 15310
Greece
Tel (30)-1-6518770, 6513111 ext. 454
FAX (30)-1-6511215

Urban Joachim
Fritz-Haber-Inst. der MPG
Abt. Electronenmikroskopie
Faraday Weg 4-6
14195 Berlin (Dahlem)
Germany
Tel (49)-30-8305322
FAX (49)-30-8305333

Vaagen Jan S.
SENTEF
Dept Physics
University of Bergen
Allegaten 55
5007 Bergen
Norway
Tel (47)-5-212744, 2806
FAX (47)-5-318334
e-MAIL JanS.Vaagen@fi.uib.no

Vannucci Luigi
Laboratori Nazionali di Legnaro INFN
Via Romea 4
35020 Legnaro (Padova)
Italy
Tel (39)-49-8292311
FAX (39)-49-641925
e-MAIL VANNUCCI@VAXLNL.LNL.INFN.IT

Varga Kalman
Institute of Nuclear Research
Hellas
Hungarian Academy of Sciences (ATOMKI)
P.O. Box 51
Debrecen 4001
Hungary
Tel (36)-52-317266
FAX (36)-52-316181
e-MAIL H906VAR@ELLA.HU
or VARGAK@HUKLTES1.BITNET

Velegrakis Michalis
Foundation for Research and Technology
Institute of Electronic Structure and Laser
P.O. Box 1527
711 10 Heraklion
Greece
Tel (30)-81-239757, 210035
FAX (30)-81-239735
Telex 262860 RCC GR
e-MAIL VELE@IESL.FORTH.GR

Vialle Jean-Louis
Laboratoire de Spectrometrie Ionique et
Moleculaire (LASIM)
Universite Claude Bernard-Lyon I
43 Bd. du 11 Novembre 1918
69 622 Villeurbanne Cedex
France
Tel (33)-72431136
FAX (33)-78894415

Vostrikov Anatoli
Institute of Thermophysics
Lavrentleva 1
630090 Novosibirsk
Russia
Tel (7)-3832-351551, 355552
FAX (7)-3832-357880
e-MAIL root@otani.thermo.nsk.su

Von Oertzen Wolfram
Hahn-Meitner-Institut Berlin
1000 Berlin 39
Germany
Tel (49)-30-80092421
FAX (49)-30-80092097
e-MAIL OERTZEN@VAX.HMI.DBP.DE

Wuosmaa Alan H.
Argonne National Laboratory
9700 South Cass Ave.
Argonne IL 604 39
U.S.A.
Tel (1)-708-2526545, 2523663
FAX (1)-708-2526210
e-MAIL WUOSMAA@ANLPHY

Xenoulis Alexandros
Institute of Nuclear Physics
Tandem Lab.
NCSR "Demokritos"
Aghia Paraskevi-Attiki 15310
Greece
Tel (30)-1-6518770
FAX (30)-1-6511215

Yamaya Takashi
Dept of Physics
Faculty of Science
Tohoku University
Aramaki Aoba-ku
Sendai 980
Japan
Tel (81)-22-2221800 ext. 3260
FAX (81)-22-2251891
e-MAIL YAMAYA@JPNTOHOK.BITNET

Zagrebaev Valeri
Chuvash State University
Cheboksary 428000
Russia
Tel (7)-8350-426540
FAX (7)-95-9313084
Telex 658127 OPYT SU
e-MAIL cher@ljar.jinr.dubna.su

Zurmuhle Robert W.
Dept of Physics
University of Pennsylvania
Philadelphia PA 19104
U.S.A.
Tel (1)-215-8986022
FAX (1)-215-8982010
e-MAIL zurmuhle@tandem.physics.edu

Zamick Larry
Rutgers University
P.O.Box 849 Piscataway
NJ 08855-0849
U.S.A.
Tel (1)-908-9323874
FAX (1)-908-9324343
e-MAIL ZAMICK@RUTHEP

Opening Address: Please, Keep Beauty Alive

"Oh! beautiful earth
As dawn rises
And you appear before me
With the birds singing the miracle of this world"[1]

I want to continue to see "the miracle of this world", in the BEAUTY of the sudden incredible light that comes from the new transparent cicada that has just changed into its new dress leaving the old one on the branch of the almond tree in my island garden

in the BEAUTY of the brand new courageous leaf that has won the fight against the aged dry trunk of the spring tree that welcomes me at my study window

in the BEAUTY of the good hands of the fisherman carved with salt, effort and night air

in the BEAUTY of the red and gold morning sky that announces the gift of a generous god: one more day on the beautiful planet Earth

in the BEAUTY of the confidence of a child's hand that this apple, this orange on the tree heavy with fruit above its head contains life not death

in the BEAUTY of my young pupil's skin, my young pupil's hair, incredibly alive, a being in itself, my young pupil's new and wondering eyes

in the BEAUTY of my body alive and confident falling blissfully into a clear radiant blue-green eternal sea

I do not want to see the dead fish lying like live guilt on the shore of a poisoned sea, a poisoned lake, a poisoned river

I do not want to see the fruits of the earth looking beautiful but holding death the way I saw them lying in their boxes in the market on Chernobyl day

I do not want to see those eternal ghosts that will keep visiting our dreams wearing the destroyed backs, the destroyed faces of Hiroshima's amazed – still amazed in their death – victims

I do not want to see the faces of the Karyatides on the Acropolis eaten up by the rude, ruthless, deadly modern air, hiding in some skyless room of some museum. They were born to inhabit the sky

I do not want to see PROGRESS THAT KILLS. I want to see PROGRESS THAT HELPS LIFE.

As a Greek, a teacher and a poet, I do not want to see KNOWLEDGE and SCIENCE serve DEATH, serve and save MATTER and kill human life, animal life, plant life, the way the damned, criminal neutron bomb is very able to do: stop life, dissolve life and maintain buildings, factories, bridges and roads. To be used by whom? Ants? They do not need them. How much KNOWLEDGE and SCIENCE did it take to create that anti-God?

As a Greek, a teacher and a poet, I want to continue to see that BEAUTY which is the last and highest stage of love of which Socrates talks in Plato's "Symposium":

"Ενταύθα του βίου, ω φίλε Σώκρατες,
έφη η Μαντινική ξένη, είπερ που άλλοθι,
βιωτόν ανθρώπω, θεωμένω αυτό το καλόν."

"And if, my dear Socrates," Diotima went on,
"man's life is ever worth the living, it is when he
has attained this vision of the very soul of beauty."

This BEAUTY and its continuation lies in the hands of you scientists today more than ever before, in a way that has never been before. This BEAUTY and its live manifestations.

This BEAUTY, this COSMOS (and COSMOS in Greek means JEWEL, ADORNMENT) we wish and demand of you scientists of the world to PLEASE, KEEP ALIVE.

[1] From the the record "The Songs of the Sun and the Fog" for a children's choir by Olympia Karayiorga.

IF NOT, Robert Oppenheimer's fate may be yours. This fate that our vastly humane poet NIKIFOROS VRETTAKOS describes in his poem: "To Robert Oppenheimer":

> [...]
> "What do you want, friend Oppenheimer
> What do you want now? There isn't, there isn't.
> We know everything: that you have
> built yourself a prison made of guilt
> and have shut yourself up
> that you spend your days weeping.
> [...]
> As from a sense of divine duty
> We have come to torture you,
> because the world is beautiful,
> the sky drips light,
> and you,
> you have shot the day of the world,
> right in the heart
> I do not speak to you from another star,
> I talk to you from the window of your brother
> [...]
> Which wind will carry of its own accord
> your voice?
> Which animal will graze by you?
> Who will dare risk pronounce your name
> in time of prayer?
> The sky and the heart have rejected you.
> [...]
> Friend Oppenheimer, I will torture you
> no more.
> If you can get any sleep, well, sleep".

In the name of life's precious BEAUTY, do not make this guilt come true for you, for any of you.

Instead side up with that surgeon of London who, after operating successfully on a Greek poetess, whose hands sweated without stop ever since she could remember, depriving her of the complete happiness of touch, received, a few days ago, this letter:

"Thank you dear Doctor,
You've spent years of your life working on a rare problem which happened to be my own torturing problem. You finally found the way to solve it.
I knew what science could do, but now I see the effect on me. Thank you, I'm ready to enter the full kingdom of my new sense: TOUCH."

So, dear scientists of the world, let GRATITUDE rise to you like a prayer. AMEN.

Olympia Karayiorga, Poetess

8 Pindou St.
N. Psychico
15451 Athens

Part I

Papers Presented Orally

Z. Phys. A 349, 197–204 (1994)

A survey of alpha-clustering

P.E. Hodgson

Department of Physics, University of Oxford, UK

Received: 27 October 1993

Abstract. The concept of alpha-clustering has found many applications to nuclear reactions and nuclear structure. Calculations of the properties of nuclear matter indicate that condensation into alpha-particles occurs when the density is somewhat less than the central nuclear density, suggesting that alpha clustering is favoured in the nuclear surface. This provides an explanation of alpha-decay which is now quantitatively successful. The concept of the alpha-particle mean field is now widely used to unify the alpha-particle bound and unbound states found by alpha-transfer reactions, and the scattering of alpha-particles by nuclei. The analysis of the structure of ^{44}Ti using the alpha-^{40}Ca potential has been particularly successful. There are now several phenomenological alpha-alpha interactions that fit the low energy phase shifts, and these are used to calculate the structure of light nuclei, particularly those composed of 2N protons and 2N neutrons. Many excited states of such nuclei can be studied in this way, and a particularly interesting class of states is those composed of a linear chain of alpha-particles, some of which have been identified experimentally.

PACS: 21.60.Gx

1. Introduction

The concept of clustering is very widely used in physics. The essential idea is that for many purposes it is possible and useful to consider a complicated object as a simple one, ignoring many features of its internal structure. This has many applications in nuclear physics. Very often we are first aware of the simple object, and only later do we find that it has a structure; thus the alpha-particle was known through studies of radioactivity long before it was found to consist of two neutrons and two protons, and these nucleons themselves were later found to have a quark substructure.

Even when we are well aware of the substructure it is still useful for many purposes to treat the complex structure as a single entity. What we are concerned with here is the usefulness of recognising that sometimes the nucleons inside the nucleus form transient clusters. Many such clusters are possible in principle, but the formation probability depends on the stability of the cluster, and of all possible clusters the alpha particle is the most stable due to its high symmetry and binding energy. Thus a discussion of clustering in nuclei is mainly confined to alpha-particle clustering.

We can learn about alpha clustering in nuclei both theoretically and experimentally. Nuclear structure calculations show that alpha-clustering is more likely to take place in the nuclear surface where the density is lower, and this leads us to expect that light nuclei will have a cluster structure (see Sect. 2).

The presence of alpha-particles near the nuclear surface makes it likely that they will escape, if this is energetically possible. Alpha decay, known since the beginning of nuclear physics, is of great importance for clustering, and is discussed in Sect. 3.

Another consequence of the presence of alpha-particles on the nuclear surface is the enhancement of reactions involving the emission or capture of an alpha-particle. The study of such reactions enables the alpha-particle bound and unbound states of nuclei to be found, as described in Sect. 4. These data, together with optical model analysis of alpha-particle scattering by nuclei, enable the alpha-particle mean field to be defined for both negative and positive energies.

Light nuclei are essentially all surface, and so it is to be expected that they will show a pronounced alpha-structure, especially if they contain 2N neutrons and 2N protons and so can be formed entirely of alpha-particles. Using a phenomenological alpha-alpha interaction it is then possible to calculate the structure of such nuclei, and some results are discussed in Sect. 5. A particularly remarkable phenomenon is the formation of alpha-chain states in light nuclei, and these are described in Sect. 6.

Over the years there have been many theoretical and experimental studies of alpha-clustering in nuclei, and several reviews have been published (Wildermuth and Tang [1]; Hodgson [2], [3]). In addition, there is a series of international conferences on clustering aspects in nuclear and

Correspondence to: Dr P.E. Hodgson, Department of Physics, Nuclear Physics Laboratory, Keble Road, Oxford OX1 3RH, UK

subnuclear systems, and the proceedings of these conferences contain much information on alpha-clustering. A vast amount of data has been obtained in recent years, and in this brief survey it is possible to select only a few highlights.

2. Alpha matter

The likelihood of alpha-clustering can be studied by considering an infinite medium consisting initially of nucleons in the absence of Coulomb forces, and seeing how its stability depends on its density. This was done by Brink and Castro [4] using several different nucleon-nucleon forces and spatial configurations of nucleons. They found that the nucleons condense into alpha-particles when the density falls to about one-third of the central nuclear density. This suggests that alpha-particle formation may be energetically favoured in the region of the nuclear surface. In the nuclear interior it is less favoured, but this does not prevent the transient formation of alpha clusters in that region.

Further studies of the effect of alpha clustering were made by Akaishi and Bandō [5] using a variational wave function and the Brink-Boeker [6] and Serber forces. A detailed study of the properties of alpha-cluster matter has been made by Tohsaki-Suzuki [7] using the generator coordinate method (GCM). He considered cubic lattices of alpha-particles in one, two and three dimensions with a Brink-Boeker effective interaction. The nucleon spatial wave functions at $\mathbf{R}_1$, $\mathbf{R}_2 \ldots \mathbf{R}_n$ in GCM space are

$$\phi_i(r) = (\pi b^2)^{-3/4} \exp\left(-\frac{1}{2b^2}(\mathbf{r}-\mathbf{R}_i)^2\right) \qquad (i = 1, 2 \ldots N) \tag{2.1}$$

where b is the size parameter of the four-nucleon or alpha cluster.

The configuration of N alpha-clusters is then given by a Slater determinant of the 4N nucleon wave functions characterised by the distance parameter s giving the separations of the alpha-particles in the lattice in the x, y and z directions. The microscopic Hamiltonian is expressed as the sum of kinetic and potential energy terms with direct and exchange components. The total energy per particle is then calculated as a function of the size parameter b and the distance parameter s. It was found that the energy rapidly converges to limiting values as the number of alpha-particles increases, and these limits, in one, two and three dimensions, characterise infinite alpha matter. For a three-dimensional lattice, the energy has a sharp minimum at $s = 2.8\,\text{fm}$, but is relatively insensitive to the size parameter b with optimum value $1.3\,\text{fm}$ (Fig.1). The minimum in the energy curve is not so sharp in one and two dimensions. The value of the binding energy per nucleon in three dimensions is $14.5\,\text{MeV}$, very similar to that in nuclear matter.

The surface energy was studied by considering a semi-infinite slab of alpha matter, and it was found that this rapidly converges to 12 MeV, corresponding to a value of 28 MeV for alpha matter. The calculation also gives the incompressibility of alpha matter. These properties are compared with those for nuclear matter in Table 1.

These results depend on the character of the force used. Additional calculations showed that alpha matter collapses

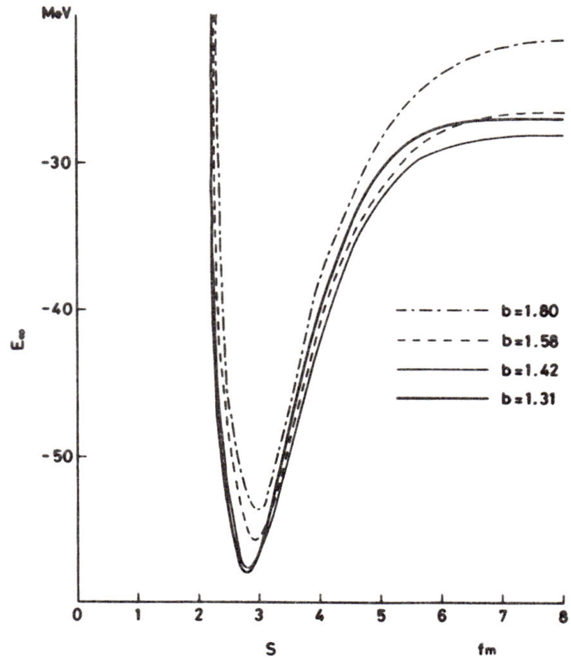

Fig. 1. The energy of three-dimensional alpha matter as a function of the size parameter b of the alpha-clusters and of their mean separation s (Tohsaki-Suzuki [7])

Table 1. Characteristics of alpha matter and nuclear matter calculated with the Brink-Boeker force

	Alpha Matter	Nuclear Matter	Empirical (Nuclear)
Volume Energy	14.5	15.7	16
Surface Energy	28.0		17
Incompressibility	169	184	210

for the Volkov 2 force, although it gives the saturation properties of the individual alpha-particle quite well. This collapse can be prevented by using a density-dependent force such as that of Skyrme or by increasing the Majorana exchange component.

The microscopic formulation of the theory of alpha matter has been further extended by Tohsaki [8].

3. Alpha decay

Many nuclei in their ground states are unstable to particle emission, and this also applies to cluster emission. Indeed, cluster emission is frequently favoured compared with nucleon emission due to the high binding energy of the cluster. This is particularly the case for alpha-decay which has been familiar since the earliest days of nuclear physics. Much more recently, it has been found that heavier clusters can be emitted from some heavy nuclei.

The probability of the spontaneous emission of an alpha-particle from a nucleus depends on two factors: firstly the probability of the formation of the alpha-particle inside the nucleus and secondly the probability that it will escape from

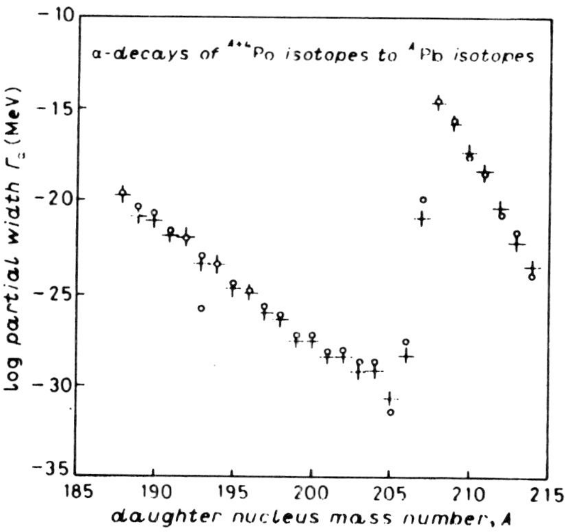

Fig. 2. The alpha-decay widths of the polonium isotopes (*crosses*) compared with the values calculated by Merchant and Buck [12] (*points*). The discontinuity occurs at the magic number $N = 126$

the nucleus. To effect this escape, the alpha-particle must free itself from the attractive forces that hold the nucleus together. It is usual to model these forces by a one-body nuclear potential that is attractive throughout the nuclear volume but repulsive at larger distances from the nucleus where the long-range Coulomb repulsion exceeds the short range nucleon-nucleon attraction. The probability of the alpha-particle escaping is then that of it penetrating the potential barrier due to the combined action of the nuclear and Coulomb forces.

The large variation in alpha decay rates is due to their extreme sensitivity to the width of this potential barrier. This can be quite well reproduced by some semi-empirical formulae (Poenaru and Ivascu [9]; Brown [10]) and by a simple potential model (Buck and Merchant [11]; Merchant and Buck [12]; Buck, Merchant and Perez [13]). Assuming pre-formation factors of unity, they showed that the decay rates are obtained to within about a factor of three, and thus are largely determined by the Q-value of the reaction. Some of their results are shown in Fig.2.

It is much more difficult to calculate the absolute alpha decay width. Many shell model calculations of increasing complexity have given good values for the relative widths but the absolute values were too low by two or three orders of magnitude. Essentially the reason for this is that even a very large shell-model basis is often insufficient to describe the particular spatial configuration of nucleons forming an alpha particle as it leaves the nucleus. Progress has recently been made in two ways: by applying a very sophisticated shell model analysis to light nuclei and by supplementing the shell model wave function with a cluster component.

The former method has been applied by Grigorescu, Brown and Dumitrescu [14] to calculate the alpha decay of excited states of ^{16}O. They considered ^{16}O to consist of a ^{12}C nucleus with four additional nucleons (forming the alpha particle) in the $1p_{1/2}$, $1d_{5/2}$ and $2s_{1/2}$ orbits. All possible shell-model configurations in this model space were included, and the interior wave functions were generated by several different interactions. The exterior wave functions were calculated from an alpha-particle folded potential obtained from the M3Y interaction. The one-level R-matrix theory was used to calculate the alpha-decay widths for several low-lying 0^+, 2^+ and 4^+ states, and it was found that these are very sensitive to the channel radius. If this is chosen in a consistent way, the results agree with experiment usually to within an order of magnitude, where the experimental values vary over several orders of magnitude. This suggests that the method of calculation is basically sound, and that more accurate results can be obtained by a further extension of the model space.

More rapid convergence can be obtained by adding a cluster model component to the shell model wavefunction, so that the total wave function becomes (Varga, Lovas and Liotta [15])

$$\Psi = \Psi(\text{shell}) + \Psi(\text{cluster}) \tag{3.1}$$

The cluster-model wavefunction is an antisymmetrised product of the intrinsic wave functions $\Phi_c(\xi_c)$ and $\Phi_\alpha(\xi_\alpha)$ of the core nucleus and the emitted alpha-particle and the wavefunction of their relative motion $\phi(\mathbf{R})$, where $\mathbf{R}$ is the separation of their centres of mass

$$\Psi(\text{cluster}) = \mathcal{A}\{\Phi_c(\xi_c)\Phi_\alpha(\xi_\alpha)\phi(\mathbf{R})\} \tag{3.2}$$

The decay is described by a Gamow wavefunction, which at large alpha-core distances becomes an outgoing Coulomb wave. The alpha-particle itself is described by 1s harmonic oscillator wavefunctions $\varphi(\mathbf{x})$ and the intrinsic state Φ_α is related to the 4×4 Slater determinant by

$$\Psi_\alpha(\{x_i\}) = \varphi^{(4\alpha)}\left(\frac{1}{4}\sum_{i=1}^{4}\mathbf{x}_i\right)\Phi_\alpha \tag{3.3}$$

The shell-model basis states are obtained by diagonalising the Hamiltonian

$$\mathcal{H} = \sum_{i=1}^{4}(T_i + U_i) + \sum_{i>j}^{4} V(i,j) \tag{3.4}$$

where T_i is the nucleon kinetic energy, U_i the average single-particle potential and $V(i,j)$ the residual interaction.

The calculation of the alpha decay width was carried out avoiding the approximations inherent in previous work, and all the parameters were fixed from independent experimental data. Thus the parameters of the single-particle and harmonic oscillator potentials were chosen to fit the experimental energies of the single-particle states and the binding energies of the initial and final nuclei.

This formalism was applied to calculate the absolute decay width of ^{212}Po, which decays to ^{208}Pb. This is a particularly favourable case, as the final nucleus has a double-closed shell. The decay width was calculated from the R-matrix formula

$$\Gamma = 2P_L\left(\frac{\hbar^2}{2M_\alpha a_L}\right)g_L^2(a_L) \tag{3.5}$$

where M_α is the reduced mass, P_L the Coulomb penetration factor, a_L the channel radius, and $g_L(a_L)$ the clustering amplitude. The result, $\Gamma = 1.45 \times 10^{-15}$ MeV agrees well with the experimental value of 1.5×10^{-15} MeV. The corresponding spectroscopic factor $S = 0.025$ is about twenty times the best shell-model estimates, and the probability of formation of an alpha-particle inside the nucleus is 0.302.

A careful study was made of the sensitivity of the final result to the input parameters and to the details of the calculation. Providing the experimental data used as input is optimally fitted, they estimate that their final result is reliable to within a factor of two or thee.

In many studies, harmonic oscillator wavefunctions were used because the Brody-Moshinsky transformation then greatly simplifies the calculation. This is however likely to give rise to difficulties in the nuclear surface, where the harmonic oscillator wavefunctions are too small. It would be preferable to use Saxon-Woods wavefunctions that do not have this defect. The advantage of the Brody-Moshinsky transformation can still be retained by expanding the Saxon-Woods wavefunctions in harmonic oscillator wavefunctions. Calculations along these lines are being made by Brown [16].

It is important to extend these calculations to a range of nuclei, in particular to see if they give the discontinuity at the neutron shell closure apparent in Fig.2 without the assumption of an increase in the principal quantum number that enabled Merchant and Buck to fit the data so well. It would also be interesting to see if the apparent reduction of discontinuities at the proton shell closures can also be explained.

4. The alpha-particle mean field

The concept of the nucleon mean field has proved to be very successful in unifying the single-nucleon bound states and the scattering of nucleons by nuclei. For negative energies, the energies of the bound states are the eigenvalues of an optical model potential and the widths of their fragment distribution can be related to the imaginary part of the potential. For positive energies, the elastic scattering of nucleons is well described by a complex optical model potential. The potential is continuous over the whole energy region, with the real and imaginary potentials depending on the energy. A further unification is achieved by connecting the real and imaginary parts of the potential by the dispersion relations and this explains the apparently anomalous energy variation around the Fermi surface (see Hodgson [17] and references therein).

The same concept of a mean field can be applied to unify our knowledge of alpha-particle bound states and elastic scattering. The real part of the alpha-particle potential can be obtained by a double folding of the nuclear and alpha-particle densities with the nucleon-nucleon interaction, giving

$$V_\alpha(r) = \iint \rho(\mathbf{r}_1)\rho(\mathbf{r}_2)v(|\mathbf{r} + \mathbf{r}_2 - \mathbf{r}_1|)d\mathbf{r}_1 d\mathbf{r}_2. \tag{4.1}$$

A good approximation to the radial variation may be obtained by using a delta function for the nucleon-nucleon interaction and analytical expressions for the nuclear densities. For many purposes it is more convenient to use an analytic expression for the potential, such as the Saxon-Woods potential or the cosh potential (Buck and Merchant [11], Buck, Merchant and Perez [13])

$$V_\alpha(r) = \frac{V(1 + \cosh(R/a))}{\cosh(r/a) + \cosh(R/a)} \tag{4.2}$$

The Saxon-Woods potential is often improved by raising it to a power of about two, and then it gives results similar to those obtained with the folded and cosh potentials (Brenner [18]).

The energies,widths and B(E2) transition rates for the alpha-particle states in ^{16}O were successfully analysed using an alpha-^{12}C folded potential by Buck et al. [19] and subsequently they extended the analysis to ^{20}Ne, which has a pronounced alpha-cluster structure (Katō and Bandō [20]; Chung et al. [21]). More recently, there has been much interest in applying the model to heavier nuclei, especially to the nuclei analogous to ^{16}O and ^{20}Ne in the fp shell, namely ^{40}Ca and ^{44}Ti.

The nucleus ^{44}Ti (= ^{40}Ca + alpha) is a particularly favourable case, and it has been extensively studied (Horiuchi [22]; Michel Reidemeister and Ohkubo [23]; Ohkubo [24]; Wada and Horiuchi [25], Merchant, Pál and Hodgson [26]). The earlier analyses were made of the cross-section of the ^{40}Ca(^{6}Li,d) reaction at 28 and 32 MeV (Fulbright et al. [27]) by applying the distorted wave theory to determine the quantum numbers of many of the bound and unbound states of ^{44}Ti. As shown in Fig.3, these states are quite well given by the alpha-cluster model, using the alpha-particle optical potential that fits very well the extensive data on the elastic scattering of alpha-particles by ^{40}Ca (Delbar et al. [28]). It is notable that the model predicts a $N = 13$, $K = 0^-$ negative parity band of alpha-particle states in ^{44}Ti, which at that time had not been observed. The 1^-, 3^- and 5^- members of this band were subsequently found by Yamaya et al. [29] using an incident energy of 50 MeV (see Fig.4). This provides striking confirmation of the alpha-cluster model.

The absolute alpha-particle spectroscopic factors obtained in this work were however very low, ranging from 0.04 for the ground state to even lower values for the excited states. Such low values raised a serious question about the validity of the alpha-cluster model. This was further investigated by Yamaya et al. [30], who found that the spectroscopic factors are very sensitive to the values of the alpha-particle wave functions in the region of the nuclear surface. The previous analysis used a wave function obtained from a Saxon-Woods potential which was not obtained from alpha-particle elastic scattering, whereas later analyses such as that of Michel et al. [23] used the more realistic squared Saxon-Woods potential which gives smaller values of the wave function in the surface region, and hence larger spectroscopic factors. Re-analysis of their data using this potential gave spectroscopic factors between 0.1 and 0.2 for the $K = 0^+$ and $K = 0^-$ bands.

The experimental data at 28 and 50 MeV have also been analysed by Kim and Udagawa [31] using the breakup-fusion model of Udagawa et al. [32]. More recent data at 60 MeV has also been analysed with this model (Guazzoni et al. [33]). They identified the 12^+ member of the $N = 12$ band and probably also the 0^+ member of the $N = 14$ band.

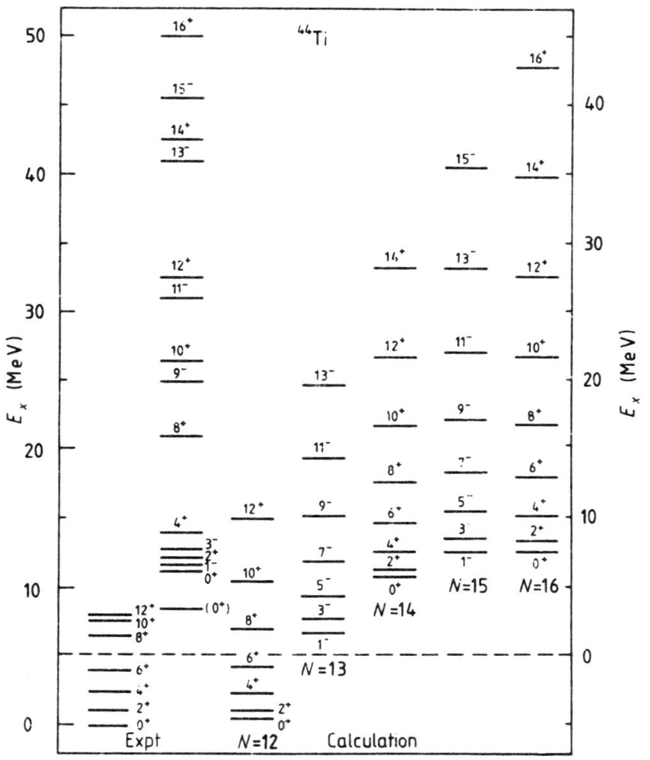

Fig. 3. A comparison of the calculated energies of the bands with principal quantum numbers N = 12 to N = 16 of alpha-cluster states in ^{44}Ti generated by a finite-range folded potential with the experimental data (Merchant, Pál and Hodgson [26])

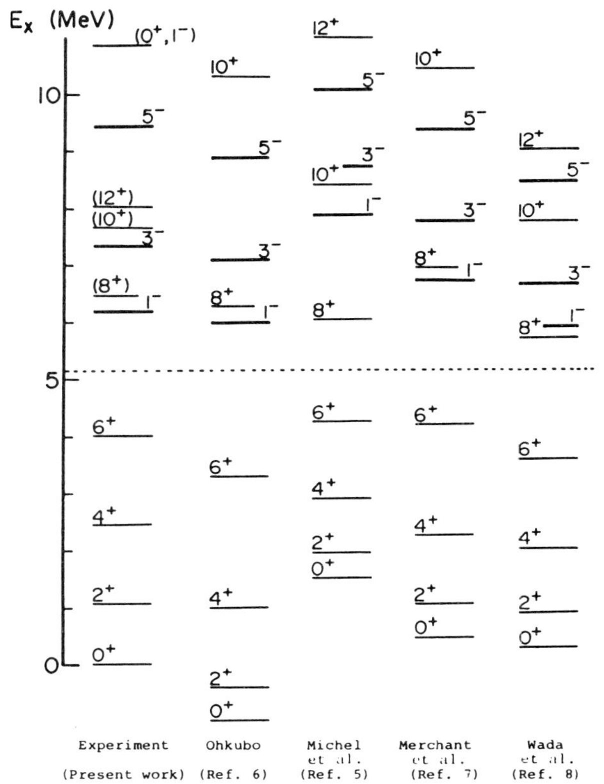

Fig. 4. Observed parity doublet band in comparison with theoretical predictions of α-cluster states in ^{44}Ti. The negative parity states observed by Yamaya et al. are indicated by thick lines. The 10.86-MeV state is a candidate for the head of the higher nodal state of N = 14 or 15 (Yamaya et al. [29])

The spectroscopic factors of the 0^+–12^+ states agree quite well with the theoretical values of Tomoda and Arima [34]. They also found that the cluster strengths of the N = 12 and N = 13 states is spread among many fragments.

Alpha-cluster states in ^{40}Ca have similarly been studied by the ^{36}Ar(^{6}Li,d) and other reactions. Analysis of data obtained at 50 MeV by Yamaya et al. [35] gave five members of the $K = 0^+$ band and fragments of four members of the $K = 0^-$ band, as shown in Fig.5. The summed alpha-particle spectroscopic factors are about three times as large as those found for ^{44}Ti.

It is notable that the fragmentation of the alpha-cluster states in ^{40}Ca seems to be much greater than for those in ^{44}Ti. Yamaya et al. suggest that this may be an effect of the open shell nature of the ^{36}Ar core. As a result, the alpha clusters in ^{40}Ca are formed from nucleons distributed over a larger number of shell model orbits than in the case of ^{44}Ti. In addition, the higher level density above the alpha-particle threshold energy in ^{40}Ca compared with ^{44}Ti is probably due to the softness of the ^{36}Ar core.

The structure of ^{40}Ca has recently been studied by Sakuda and Ohkubo [36] using the orthogonality-condition cluster model for α+^{36}Ar. This gave a unified account of the $K = 0^+$ and $K = 0^-$ band shell model states and also the fragmentation of the alpha-particle spectroscopic strength and B(E2) values, all in good accord with experiment. This shows that the co-existence of the alpha-cluster and shell models is important for an understanding of the structure of ^{40}Ca.

All these analyses were made using standard optical potentials. However experience with the nucleon mean field indicates that it would be useful to apply the dispersion relations to connect the real and imaginary parts of the potential. This also provides a means of studying the fragmentation of the alpha-particle states. The available data on the elastic scattering of alpha-particles by ^{40}Ca were therefore analysed by Hodgson and Haider [37] using the dispersion relations. This gave a dispersive alpha-particle potential that fits the elastic scattering data as well as the previous analyses. It would be interesting to see whether the imaginary part of this potential at negative energies can be connected to the observed fragmentation of the bound and unbound states.

5. The alpha-particle structure of nuclei

There are other aspects of nuclear structure that are not conveniently analysed using the alpha-particle mean field described in the previous section. In particular, the structures of many light nuclei can be usefully treated as alpha-particle structures, with some additional nucleons.

The ground state nuclear wave functions often have a large component consisting of the maximum number of complete alpha-particles coupled to the remaining nucleons.

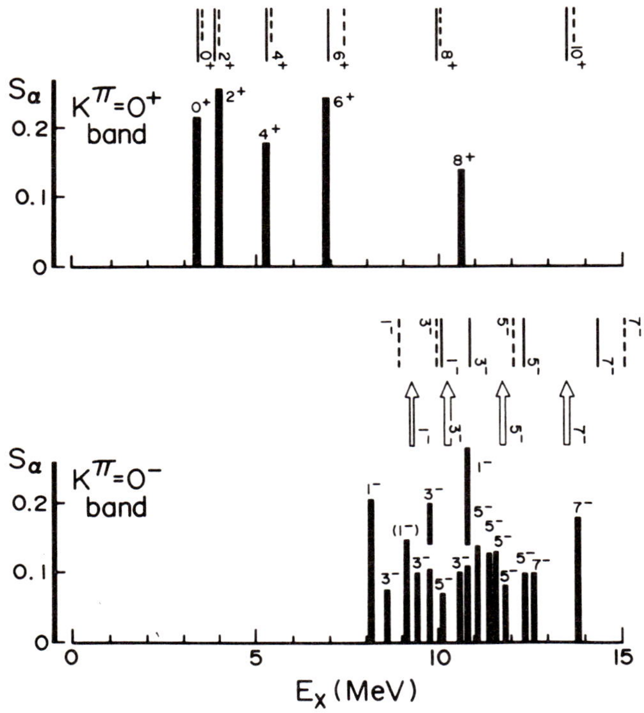

Fig. 5. The alpha-particle spectroscopic factors for the members of the 0^+ and 0^- bands, compared with calculations with a squared Saxon-Woods potential (*full lines*) and with the spline potential of Reidemeister et al. [38] (*dashed lines*). The centroids of the fragmented $K^\pi = 0^-$ states are shown by arrows (Yamaya et al. [35])

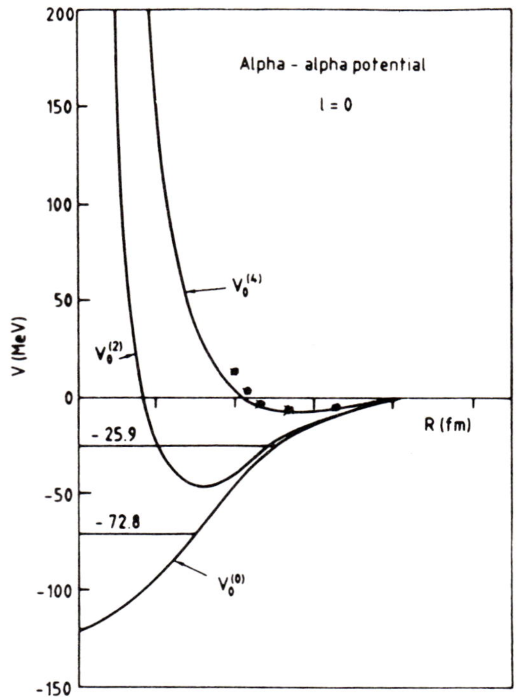

Fig. 6. The alpha-alpha potential of Buck, Friedrich and Wheatley [43] for $\ell = 0$ together with its supersymmetric partners $V_0^{(2)}$ and $V_0^{(4)}$. The solid circles show the phenomenological potential of Shimodaya et al. [44] which has a hard core at 2 fm. The Marchenko potential obtained by the resonating group method is almost identical to $V_0^{(4)}$. All potentials include the Coulomb term (Sofianos, Panda and Hodgson [40])

Thus ^{6}Li is largely $\alpha + d$. Excited states show a similar alpha-structure, and this ensures large cross-sections for alpha-transfer reactions. At higher energies, more excited states have configurations in which all the alpha-particles are broken up, and these show correspondingly smaller alpha-transfer cross-sections.

Many investigations have been made of nuclei composed entirely of alpha-particles. By postulating an alpha-alpha interaction it is possible to describe many of their properties, both in the ground state and in excited states in which the alpha-particles remain intact.

The alpha-particle model of light nuclei can be tested by using a phenomenological alpha-alpha interaction to calculate the properties of alpha-particle nuclei. Such an investigation was carried out by Tohsaki-Suzuki, Matsuse and Arima [39]. They used an effective nuclear force composed of finite range two-body and three-body operators

$$V_{ij}^{(2)} = \sum_n v_n^{(2)} \exp[-(r_{ij}/r_n)^2](W_n^{(2)} + M_n^{(2)} P_{ij}) \tag{5.1}$$

and

$$V_{ijk}^{(3)} = \sum_n v_n^{(3)} \exp[-(r_{ij}/r_n)^2 - (r_{jk}/r_n)^2] \times (W_n^{(3)} + M_n^{(3)} P_{ij})(W_n^{(3)} + M_n^{(3)} P_{jk}) \tag{5.2}$$

The parameters of this force were optimised to fit the saturation properties of light double-closed shell nuclei. The binding energies of the alpha-particle, ^{16}O, ^{40}Ca and nuclear matter are reproduced rather better than they are by the Brink-Boeker I or Skyrme II potentials, and the incompressibility $K \sim 220$. The phase shifts for alpha-alpha scattering are also well fitted.

The alpha-alpha interaction has been extensively studied in recent years, and a whole series of phase-equivalent potentials of increasing depth has been found. More recently, it has been realised that successive applications of the supersymmetric transformation and Marchenko inversion to the deeper potentials transform them to phase-equivalent shallower potentials without unphysical Pauli states. This essentially removes the ambiguities in the potential, as shown in Fig.6 (Sofianos, Panda and Hodgson [40]).

The coherent density fluctuation model has been extensively used to obtain the momentum distribution of nucleons in nuclei, which is particularly important for the accurate calculation of nucleon knock-out reactions such as $(p, 2p)$ and $(e, e'p)$ (Antonov, Hodgson and Petkov [41]). Recently, the model has been extended to calculate the alpha-particle momentum distribution, which is important for alpha-particle knock-out reactions such as $(\alpha, 2\alpha)$, $(p, p'\alpha)$ and $(e, e'\alpha)$. The calculations require the evaluation of a twelve-dimensional integral that is first reduced to a nine-dimensional integral and then evaluated by the Monte Carlo method. Some typical results are shown in Fig.7. These alpha-particle momentum distributions are being used to calculate the cross-sections of some of the reactions just mentioned, for comparison with experimental data (Antonov et al. [42]).

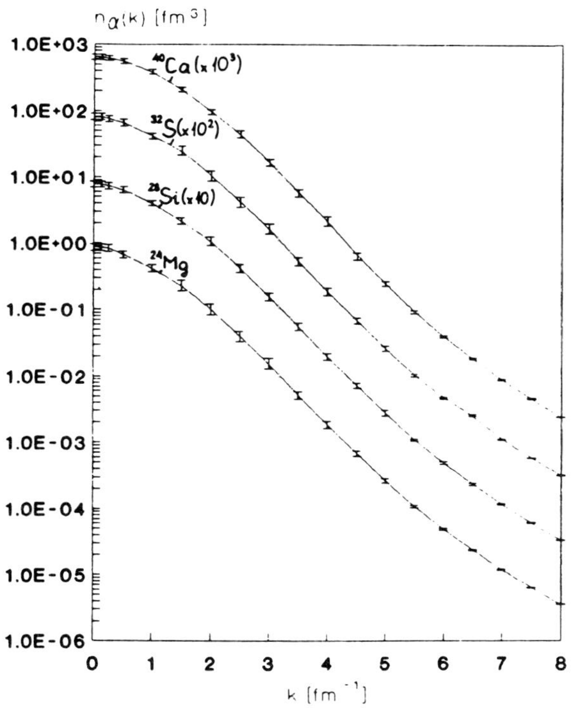

Fig. 7. The momentum distributions of alpha-particles in the ^{24}Mg, ^{28}Si, ^{32}S and ^{40}Ca nuclei obtained with the coherent density fluctuation model. The error bars indicate the uncertainties in the Monte Carlo calculation, and the normalisation is $\int_0^\infty n_{cm}^{(\alpha)}(P)P^2dP = 1$ (Antonov et al. [42])

6. Alpha-chain states

A particularly interesting class of alpha-particle states is provided by the alpha chain states. These are usually excited states of nuclei consisting of a linear chain of alpha-particles. Such states are more stable than might be thought, because, in the case of ^{24}Mg for example, the energy of an oscillator quantum is about 24 MeV in a direction perpendicular to the chain but only about 4 MeV in the direction of the chain. It thus requires about 20 MeV to bend the chain.

The first of these is ^{8}Be, which has a well-known rotational band consisting of 0^+, 2^+ and 4^+ states at 0, 3 and 11 MeV. The ground state is unstable by 92 KeV, with a width of 6 eV, and the two higher states have widths of 1 MeV and 3 MeV.

The next alpha-particle nucleus, ^{12}C, has a 0_2^+ state at 7.65 MeV that is believed to have the structure of a linear chain of three alpha-particles (Brink [6]). This ^{12}C state preferentially breaks up into ^{8}Be+α, and is known from elastic scattering to have a large charge radius.

There are no known 0^+ states of ^{16}O that have ^{8}Be, ^{12}C(0_2^+) or four alpha-particles among their decay products. There are however three 2^+ states around 17 MeV that decay into ^{8}Be and at least one alpha-particle, and they may have a chain-structure. There are several other possible candidates, and so chain states in ^{16}O may still be found. Calculations of alpha-chain states in ^{12}C and ^{16}O have been made by Zamick and Zheng [45].

In 1986, Marsh and Rae [46] made calculations of the states of ^{24}Mg using a cranked version of the Bloch-Brink cluster model and predicted a band of states around 45 MeV corresponding to a linear chain of six alpha-particles. Experimental evidence for this band has been found by Wuosmaa et al. [47] at 48 MeV, approximately the energy predicted by Marsh and Rae. This state breaks up into two ^{12}C alpha-chain states.

One of the remarkable features of the data of Wuosmaa et al. is that on resonance the cross-section has a very strong peak at 90° in the centre-of-mass system. Rae, Merchant and Buck [48] find that this can be explained only if many partial waves contribute with particular relative phases. They suggest that the data are consistent with the resonance being due to an alpha-particle chain with a very large moment of inertia, as that the peak is due to the superposition of resonances that are essentially degenerate, their spacings being less than their widths. The angular distribution can thus be considered to be an approximate shape or angle eigenstate.

Rae and Merchant are able to account for the data in terms of alpha-particle resonances in a hyperdeformed potential. A semi-classical analysis gives the relatively small widths (Rae and Merchant [49]) and fully quantum-mechanical calculations are also able to account for the observations.

More recently, Merchant and Rae [50] have used the same cranking model, with the Brink-Boeker B1 nucleon-nucleon force, to calculate the energies of the alpha-chain states in fourteen nuclei from ^{8}Be to ^{60}Zn. The binding energies of the lowest energy chain state in each nucleus were found to be given quite accurately by

$$BE_{\text{Chain}}^{\text{Head}} = 11.042 + 17.484N \text{ MeV} \tag{6.1}$$

where N is the number of alpha-particles in the chain. Subtracting the ground state binding energy gives the energies E^* of the alpha-chain states and these are above the threshold energies for breaking into alpha-particles.

Evidence for a chain of seven alpha-particles was found last year by the Charissa group at the Daresbury laboratory (Rae and Merchant [49]; Simmons et al. [51]; Rae et al. [48]). The energy of this state is found to be at 60 MeV in ^{28}Si, within a few MeV of the predictions of Merchant and Rae [50].

Much work is still needed to confirm this identification. The angular distribution of the ^{12}C(^{16}O,7α) reaction should show features similar to the ^{24}Mg data of Wuosmaa et al. Experiments to test this are planned using the accelerator in Canberra.

7. Conclusion

The last few years have seen many advances in our knowledge of alpha-clustering in nuclei. Very detailed calculations of alpha-matter have given a fundamental insight into alpha-clustering. Relative alpha-decay rates can now be quite accurately calculated by a potential model, and meticulous shell and cluster model studies now give good values for the absolute decay rates. The alpha mean field is able to unify our knowledge of bound and unbound alpha-particle states using a potential that also fits alpha-particle elastic scattering. The momentum distribution of alpha-particles in the nucleus has been calculated, and this provides the essential basis for realistic calculations of alpha knock-out reactions. The alpha-alpha interaction is now much better known; and the alpha-particle structure of light nuclei is a fruitful field of study. Exotic features such as the alpha-chain states have now been found in several nuclei, reminding us that our intuition may be insufficiently bold, as has been recently demonstrated by the discovery of buckminster fullerene. Perhaps we should think more carefully about the possibility of more exotic alpha states such as bubbles and toroids (Wilkinson [52]). Is it possible that rings of alpha-particles are produced like smoke rings in heavy ion reactions, and that these subsequently decay into alpha-chain states?

Acknowledgement. I thank all colleagues, in particular Prof. B.A. Brown, Dr A.C. Merchant, Dr W.D.M. Rae, Prof. S. Ohkubo, Prof. A. Tohsaki, Prof. T. Yamaya, Prof. P. Guazzoni and Prof. L. Zetta for valuable discussions and for information on their unpublished work.

References

Cluster Conferences

Chester Conference. Clustering Aspects of Nuclear Structure: The 4th International Conference on Clustering Aspects of Nuclear Structure and Nuclear Reactions. Chester 23–27 July, 1984. J.S. Lilley, M.A. Nagarajan (eds.) Dordrecht: Reidel 1985)

Kyoto 1988. Proceedings of the Fifth International Conference on Clustering Aspects in Nuclear and Subnuclear Systems. Kyoto, 25–29 July 1988. K. Ikeda, K. Katori and Y. Suzuki (eds.)(Supplement) Journal of the Physical Society of Japan, Vol.58, 1989

Turku, 1991. Clustering Phenomena in Atoms and Nuclei. Proceedings of the International Conference on Nuclear and Atomic Clusters. Turku, Finland June 1991. M. Brenner, T. Lönnroth, F.B. Malik (eds.) Berlin, Heidelberg, New York: Springer 1992

Santorini, 1993. Proceedings of the Second International Conference on Atomic and Nuclear Clusters

Articles

1. Wildermuth, K., Tang, Y.C.: A Unified Theory of the Nucleus. Academic New York: Academic Press 1977
2. Hodgson, P.E.: The Uncertainty Principle and Foundations of Quantum Mechanics. W.C. Price, S. Chissick (eds.) New York Wiley 1977, Ch.23
3. Hodgson, P.E.: Proceedings of the Fifth International Conference on Clustering Aspects in Nuclear and Subnuclear Systems, July 1988, Kyoto Japan. J. Phys. Soc. Jpn **58**, 755 (1989)
4. Brink, D., Castro, J.J.: Nucl. Phys. **A216**, 109 (1973)
5. Akaishi, Y., Bandō, H.: Prog. Theor. Phys. **41**, 1594 (1969)
6. Brink, D.M.: Proc. Int. School of Physics, Enrico Fermi, Course XXXVI, Varenna, 1966. C. Bloch (ed.) New York: Academic Press 1966;
   Brink, D.M., Boeker E.: Nucl. Phys. **A91**, 1 (1967)
7. Tohsaki-Suzuki, A.: Prog. Theor. Phys. **81**, 370 (1989)
8. Tohsaki, A.: Prog. Theor. Phys. **88**, 1119 (1992)
9. Poenaru, D.N., Ivascu, M.: J. Phys. (Paris) **44**, 791 (1983)
10. Brown, B.Alex: Phys. Rev. **C46**, 811 (1992)
11. Buck, B., Merchant, A.C.: J. Phys. **G15**, 615 (1989)
12. Merchant, A.C., Buck, B.: Europhys. Lett. **8**, 409 (1989)
13. Buck, B., Merchant, A.C., Perez, S.M.: Phys. Rev. Lett. **65**, 2975 (1990); J. Phys. G**17**, 1223 (1991a); Mod. Phys. Lett. A**6**, 2453 (1991b); J. Phys. **G17**, 1223 (1991c); Phys. Rev. C**45**, 2247 (1992)
14. Grigorescu, M., Brown, B.Alex, Dumitrescu, O.: Phys. Rev. C**47**, 2666 (1993)
15. Varga, K., Lovas, R.G., Liotta, R.J.: Phys. Rev. Lett. **69**, 37 (1992;) Nucl. Phys. A**550**, 421 (1992)
16. Brown, B.A.: Private Communication (1993)
17. Hodgson, P.E.: Proc. Int. Conf. on Nuclear Data for Science and Technology, Julich, 1992, 768
18. M. Brenner, Turku Conference, 327 (1992)
19. Buck, B., Dover, C., Vary, J.P.: Phys. Rev. **C11**, 1803 (1975)
20. Katō, K., Bandō, H.: Prog. Theor. Phys. **59**, 774 (1978)
21. Chung, W., van Hienen, J., Wildenthal, B.H., Bennett, C.L.: Phys. Lett. **79B**, 381 (1978)
22. Horiuchi, H.: Prog. Theor. Phys. **73**, 1172 (1985)
23. Michel, F., Reidemeister, G., Ohkubo, S.: Phys. Rev. Lett. **57**, 1215 (1986a); Phys. Rev. C**34**, 1248 (1986b); C**37**, 292 (1988)
24. Ohkubo, S.: Phys. Rev. C**38**, 2377 (1988)
25. Wada, T., Horiuchi, H.: Phys. Rev. C**38**, 2063 (1988)
26. Merchant, A.C., Pál, K.F., Hodgson, P.E.: J. Phys. G**15**, 601 (1989)
27. Fulbright, H.W., Bennet, C.L., Lindgren, R.A., Markham, R.G., McGuire, S.C., Morrison, G.C., Strohbusch, U., Tōke, J.: Nucl. Phys. A**84**, 329 (1977)
28. Delbar, Th., Grégoire, Gh., Paic, G., Ceuleneer, R., Michel, F., Vanderpoorten, R., Budzanowski, A., Dixon, W.R., Storey, R.S., Simpson, J.J.: Nucl. Phys. **202**, 579 (1973)
29. Yamaya, T., Oh-ami, S., Fujiwara, M., Itahashi, T., Katori, K., Tosaki, M., Kato, S., Hatori, S., Ohkubo, S.: Phys. Rev. C**42**, 1935 (1990)
30. Yamaya, T., Ohkubo, S., Okabe, S., Fujiwara, M.: Phys. Rev. C**47**, 2389 (1993a)
31. Kim, Chong-Yeal, Udagawa, T.: Phys. Rev. C**46**, 532 (1992)
32. Udagawa, T., Lee, Y.J., Tamura, T.: Phys. Lett. B**196**, 291 (1987); Phys. Rev. C**39**, 47 (1989)
33. Guazzoni, P., Jaskola, M., Zetta, L., Kim, Chong-Yeal, Udagawa, T., Bohlen, G.: Nucl. Phys. (in press)
34. Tomoda, T., Arima, A.: Nucl. Phys. A**303**, 217 (1978)
35. Yamaya, T., Saito, M., Fujiwara, M., Itahashi, T., Katori, K., Suehiro, T., Kato, S., Hatori, S., Ohkubo, S.: Phys. Lett B**306**, 1 (1993b)
36. Sakuda, T., Ohkubo, S.: Santorini Conference, 1993.
37. Hodgson, P.E., Haider, W.: Proceedings of the International Symposium on Microscopic Cluster Models of Light Nuclei, Yukawa Institute for Theoretical Phzsics, Kyoto B 222 (1993)
38. Reidemeister, G., Ohkubo, S., Michel, F.: Phys. Rev. C**41**, 63 (1990)
39. Tohsaki-Suzuki, A., Matsuse, T., Arima, A.: Florence Conference **I**, 36 (1983)
40. Sofianos, S.A., Panda, K.C., Hodgson, P.E.: J. Phys. G. **19**, 1929 (1993)
41. Antonov, A.N., Hodgson, P.E., Petkov, I.Zh.: Nucleon Momentum and Density Distributions in Nuclei. Oxford: Clarendon 1988; Nucleon Correlations in Nuclei. Berlin Heidelberg New York: Springer 1993
42. Antonov, A.N., Nikolov, E.N., Petkov, I.Zh., Hodgson, P.E., Lalazissis, G.A.: Bulg. J. Phys. **19**, 11 (1992)
43. Buck, B., Friedrich, H., Wheatley, C.: Nucl. Phys. **A275**, 241 (1977)
44. Shimodaya, I., Tamagaki, R., Tanaka, H.: Prog. Theor. Phys. **27**, 793 (1962)
45. Zamick, L., Zheng, D.C.: Santorini Conference, 1993
46. Marsh, S., Rae, W.D.M.: Phys. Lett. B**180**, 185 (1986)
47. Wuosmaa, A.H., Betts, R.R., Back, B.B., Freer, M., Glagola, B.G., Happ, Th., Henderson, D.J., Wilt, P., Bearden, I.G.: Phys. Rev. Lett. **68**, 1295 (1992)
48. Rae, W.D.M., Merchant, A.C., Buck, B.: Phys. Rev. Lett. **69**, 3709 (1992)
49. Rae, W.D.M., Merchant, A.C.: Institute of Physics Conference, Glasgow, 1993. J. Phys. G (to be published)
50. Merchant, A.C., Rae, W.D.M.: Nucl. Phys. A**549**, 431 (1992); J. Phys. G**19**, L89 (1993); Santorini Conference 1993b
51. Simmons, P.M., Rae, W.D.M., Davinson, T., Fox, S., Gyapong, G.J., Jones, C.D., Watson, D.L., Clarke, N.M., Fulton, B.R., Leddy, M., Murgatroyd, J.T., Murphy, A., Pople, J., Catford, W.N., Lilley, J.: Institute of Physics Conference, Glasgow, 1993
52. Wilkinson, D.H.: Nucl. Phys. A**452**, 296 (1986)

This article was processed using Springer-Verlag TEX Z.Physik A macro package 1.0 and the AMS fonts, developed by the American Mathematical Society.

Z. Phys. D 31, 181–185 (1994)

Oscillations of finite quantal Fermi systems

R.A. Broglia [1,3], **F. Alasia** [2], **G. Colo** [1,2], **H.E. Roman** [1,2,4], **Ll. Serra** [1]

[1] Dipartimento di Fisica, Università di Milano, via Celoria 16, I-20133 Milano, Italy
[2] INFN, Sezione di Milano, via Celoria 16, I-20133 Milano, Italy
[3] The Niels Bohr Institute, University of Copenhagen, DK-2100 Copenhagen Ø, Denmark
[4] Institut für Theoretische Physik, Universität Hamburg, D-20355 Hamburg, Germany

Received: 5 January 1994

Abstract. Mean field theory and linear response methods as used in the description of nuclei and clusters are reviewed. These methods are applied to the electromagnetic response of C_{60}.

PACS: 21.00; 36.40; 33.80.Eh; 71.45.−d

1. Introduction

Fourty years ago, the paper which contains the foundations of modern nuclear structure, namely the collective model, was published [1]. The organizing element in this model is the mean field, where nucleons move independently of each other, in quantized orbits. The existence of a long mean free path for fermions reveals an extreme collective response of the whole system. In fact, for a nucleon to move undisturbed and to be able to close its trajectory bouncing elastically off the nuclear surface, requires that the other nucleons display a very ordered, coherent motion. A main element making this possible is the fact that fermions obey the exclusion principle. In fact, the way they interact depend not only on the residual interaction, but also on the availability of final states.

The mean field of a finite system can hardly be static but is expected to be highly dynamic [2]. As a consequence, nuclei react elastically to a sudden external field, and plastically over long periods of time as compared to single-particle frequencies [3]. Giant resonances, namely the nuclear zero sound, are excited in the first case. The second motion is associated with large amplitude fluctuations of which fission and exotic decay are well known examples. The concepts mentioned above have proved to carry on well into the field of simple metal clusters and, more recently, also to provide the basis for a microscopic description of the fullerenes (cf. e.g. [4]).

In the present paper we shall briefly review the theoretical concepts of the model with special emphasis to its suitability to describe cluster physics, and discuss applications to the optical response of C_{60}.

2. Mean field theory

Mean field theory (MFT) is inspired by the Hartree-Fock approximation of atomic physics. This theory finds the optimum determinant of single-particle wavefunctions, in the sense that it minimizes the expectation of the N-particle Hamiltonian,

$$H=-\sum_{i=1}^{N}\frac{\hbar^2}{2m}\nabla_i^2+\sum_{i<j}^{N}v(r_{ij}), \tag{1}$$

sum of a kinetic energy term and of a two-body interaction. The optimum many-particle wavefunction is obtained by using single-particle wavefunctions ϕ_i that satisfy

$$-\frac{\hbar^2}{2m}\nabla^2\phi_i(\mathbf{r})+\int d^3r'\sum_{j=1}^{N}|\phi_j(\mathbf{r}')|^2v(\mathbf{r}-\mathbf{r}')\phi_i(\mathbf{r})$$
$$-\int d^3r'\sum_{j=1}^{N}\phi_j^*(\mathbf{r}')\phi_i(\mathbf{r}')\phi_j(\mathbf{r})v(\mathbf{r}-\mathbf{r}')$$
$$=\varepsilon_i\phi_i(\mathbf{r}). \tag{2}$$

The quantities ε_i play the role of Lagrange multipliers for minimizing $\langle H\rangle$, and they are interpreted as the single-particle energies. The first interaction term may be rewritten as an ordinary potential term $V_{ee}\phi_i$ with the potential given by

$$V_{ee}(\mathbf{r})=\int d^3r'\,n(\mathbf{r}')v(\mathbf{r}-\mathbf{r}') \tag{3}$$

This paper was originally submitted in connection with the 2nd. Int. Conference on Atomic and Nuclear Clusters held in Santorini from 28. June–2. July 1993 and is published here as a regular article after an independent refereeing procedure according to the standards of Z. Phys. D

* On leave from Department de Física, Universitat de les Illes Balears, E-07071 Palma de Mallorca, Spain

where

$$n(\mathbf{r})=\sum_{j=1}^{N} |\phi_j(\mathbf{r})|^2 \tag{4}$$

is the electron density.

The third term in (2), the exchange interaction, cannot be expressed so simply. As a practical matter, the Schrödinger equation is only easy to solve for such simple potential fields. For this reason one attempts to approximate the exchange interaction by a form that would make it fit into a potential field. In atomic physics, the following approximation was introduced by Slater:

$$-\int d^3r' \frac{e^2}{|\mathbf{r}-\mathbf{r}'|} \sum_j \phi_j^*(\mathbf{r}')\phi_i(\mathbf{r}')\phi_j(\mathbf{r})$$
$$\cong c(\mathbf{r})\phi_i(\mathbf{r}), \tag{5}$$

where the coefficient $c(\mathbf{r})$ is calculated assuming the occupied wavefunction can be represented by an infinite Fermi gas in the vicinity of the point $\mathbf{r}$. This yields

$$c(\mathbf{r})=-e^2\left(\frac{3n(\mathbf{r})}{\pi}\right)^{1/3}=-0.984\,e^2[n(\mathbf{r})]^{1/3}. \tag{6}$$

Once the Slater approximation is made to simplify the Hartree-Fock equation, it is tempting to make further changes in the interaction in order that the final energies converge better to the actual solution of the many-particle Schrödinger equation. This approach is known as the local-density approximation (LDA) and the Schrödinger equation with the modified potential field is called the Kohn-Sham equation [5].

Besides the direct and exchange Coulomb interaction between the electrons, the potential includes a term $V_{\text{el-ion}}=-e^2\sum_\alpha 1/|\mathbf{r}-\mathbf{R}_\alpha|$ describing the attractive interaction between the electrons and the ions (at positions $\mathbf{R}_\alpha$), and the density-dependent contact interaction chosen to reproduce the correlation energy of an interacting electron gas. Thus, the Kohn-Sham equation attains the form

$$-\frac{\hbar^2}{2m}\nabla^2\phi_i(\mathbf{r})+V_{ee}\phi_i+V_{\text{el-ion}}\phi_i$$
$$-0.984\,e^2 n^{1/3}\phi_i+\frac{\partial E_{\text{corr}}}{\partial n}\phi_i=\varepsilon_i\phi_i, \tag{7}$$

where

$$E_{\text{corr}}=-0.91\log\left[1+\frac{11.4}{(4\pi n/3)^{1/3}}\right], \tag{8}$$

is the correlation energy in the interacting electron gas. Here $n\equiv n(\mathbf{r})$. The LDA has been very succesful in solid state physics [6]. For example, the total energy derived from LDA gives a good indication of the most stable state, when a choice of states is possible. However, it has also definite limitations. In particular, LDA can underestimate band gaps in semiconductors and insulators by as much as 50%, testifying to the importance of the non locality of the exchange interaction in describing single-particle energies.

In nuclear physics, the corresponding local density approximation has also been applied with success [7]. However, the nuclear interaction is rather complicated and for many purposes a more crude approach is adequate, in particular, the one which makes use of the Skyrme Hamiltonian. The single-particle Hamiltonian has the form

$$H=-\frac{\hbar^2}{2m_k}\nabla^2+V, \tag{9}$$

the sum of a kinetic energy operator (with an effective mass m_k) and a potential functional. The essential feature of V is that it is local, a function of $\mathbf{r}$ that depends only on quantities calculated from the density matrix in the vicinity of $\mathbf{r}$. In the usual parametrization, V is expressed as

$$V(\mathbf{r})=\tfrac{3}{4}t_0\rho+\tfrac{3}{16}t_3\rho^2+\tfrac{1}{16}(3t_1+5t_2)\tau$$
$$+\tfrac{1}{32}(5t_2-9t_1)\nabla^2\rho, \tag{10}$$

where $\rho\equiv\rho(\mathbf{r})$ and $\tau\equiv\tau(\mathbf{r})$ are the nuclear density and the kinetic energy density, respectively. The full expression has been simplified by assuming equal numbers of neutrons and protons and completely filled shells. The first term proportional to t_0 is associated with the basic short-ranged attraction between nucleons. This term alone will lead to a collapse of the nucleus in Hartree-Fock theory. The next two terms are repulsive and serve to offset the attraction, producing nuclear saturation at some finite density. The second term with t_3 depends quadratically on ρ, and the third depends on the kinetic energy density τ which varies as the 5/3 power of the density in the Fermi gas limit. The kinetic term in the single-particle Hamiltonian has an effective mass that depends explicitely on the density,

$$\frac{1}{m_k}=\frac{1}{m}[1+(3t_1+5t_2)\rho], \tag{11}$$

which is not appreciably more difficult to treat than the usual Schrödinger equation with constant mass. This effective mass arise from momentum-dependent contact interactions. The momentum dependence is present in the bare nucleon-nucleon interaction and is a consequence of the meson exchange origin of it. However, quantitatively the coefficients t_i cannot be calculated very well from the basic interactions.

Like LDA in solid state physics, the Skyrme model has been succesful in describing the structure of nuclei and their binding energies. Since the entire interaction is phenomenological, it is not surprising that the average systematics of energies is reproduced. It also gives the basic shell physics, that is, magic numbers, deformations, the quantum numbers of single-particle orbitals near the Fermi energy. Also the charge distributions within nuclei are reasonably well reproduced by mean field theory.

However, at a closer look one also observes cospicuous discrepancies. In particular, the density of levels around the Fermi surface is much too low and the accupancy of the associated levels is either 0 or 1, at variance with the observations indicating strong deviations from a sharp Fermi surface. At the basis of this discrepancy is the fact that in the present model, the couplings of the nucleons with collective fluctuations of the mean field are not considered. These couplings renormalize in an important way the single-particle levels leading to an ω-dependent effective mass m_ω, arising from the non-locality in time of the couplings (cf. e.g. [2]). As a result, the effective mass becomes $m^*=m_k m_\omega/m$, and the occupancies turn out to be inversely proportional to m_ω. Because typical values of m_k/m and m_ω/m are 0.7 and 1.4 respectively, the final result corresponds to nucleons moving in an average potential and carrying the bare nucleon mass, but displaying occupancies around the Fermi surface which can differ considerably from 0 and 1. Much of these couplings are contained in the correlation term within the LDA. However, in this approximation the Fermi surface is sharp, and the fermion occupancies are still either 0 or 1 (cf. e.g. [8]).

3. Linear response theory

Mean field theory is one of the most useful approximations in many branches of physics and chemistry. In a mean field theory we replace the many-particle Schrödinger equation by a single-particle Schrödinger equation.

The most familiar applications of MFT are those describing stationary states, but the extension to time-dependent states is straightforward and leads directly to linear response theory or random phase approximation (RPA). There are many derivations of RPA and we shall approach it from the solution of the time-dependent Schrödinger equation in an external field. Let

$$H=H_0+V_x(t) \tag{12}$$

be the total Hamiltonian, sum of an unperturbed, time independent term, $H_0=-\hbar^2/2m\nabla^2+V$, and of an external field $V_x(\mathbf{r},t)=V_x(\mathbf{r})\cos(\omega t)$, displaying a periodic time dependence. One can take the perturbed wavefunction to have the form

$$\phi_i'=\left[\phi_i+\sum_j(x_{ij}\,\mathrm{e}^{-\mathrm{i}\omega t}+y_{ij}\,\mathrm{e}^{\mathrm{i}\omega t})\phi_j\right]\mathrm{e}^{-\mathrm{i}\varepsilon_i t/\hbar}, \tag{13}$$

where $H_0\phi_i=\varepsilon_i\phi_i$ is the solution of the unperturbed Hamiltonian. The coefficients x_{ij} and y_{ij} are obtained by substituting the wavefunction in the perturbed Schrödinger equation,

$$i\hbar\frac{\partial\phi_i'}{\partial t}=-\frac{\hbar^2}{2m}\nabla^2\phi_i'+V\phi_i'+V_x(t)\phi_i'. \tag{14}$$

If the perturbation is turned on slowly at $t=-\infty$, the perturbed wavefunctions are found to be

$$\phi_i'(t)=(\phi_i+i\phi_i^R+\phi_i^I)\,\mathrm{e}^{\mathrm{i}\varepsilon_i t/\hbar}, \tag{15}$$

where

$$\phi_i^R=\sum_j\frac{(\varepsilon_i-\varepsilon_j)\langle j|V_x|i\rangle}{(\varepsilon_i-\varepsilon_j)^2-\omega^2}\phi_j\cos(\omega t),$$

and

$$\phi_i^I=\sum_j\frac{\omega\langle j|V_x|i\rangle}{(\varepsilon_i-\varepsilon_j)^2-\omega^2}\phi_j\sin(\omega t).$$

The time-dependent density $n(\mathbf{r},t)$ (and current) may easily be found from the wave-functions, i.e.

$$n(\mathbf{r},t)=\sum_{i=1}^N|\phi_i'(t)|2=n_0(\mathbf{r})+\delta n_{\mathrm{ip}}(\mathbf{r})\cos(\omega t), \tag{16}$$

where $n_0(\mathbf{r},t)=\sum_i^N|\phi_i|^2$ is the equilibrium density and

$$\delta n_{\mathrm{ip}}=\int\mathrm{d}^3r'\,\Pi^0(\mathbf{r},\mathbf{r}',\omega)\,V_x(\mathbf{r}').$$

The subindex ip in the above expressions indicates that, at this level, we are dealing with the independent-particle response. The polarization operator is defined as

$$\Pi^0(\mathbf{r},\mathbf{r}',\omega)=\sum_{i,j}\langle i|n(\mathbf{r})|j\rangle\langle j|n(\mathbf{r}')|i\rangle\times\left(\frac{1}{\varepsilon_i-\varepsilon_j-\hbar\omega}+\frac{1}{\varepsilon_i-\varepsilon_j+\hbar\omega}\right), \tag{17}$$

and describes the response function for the density charge at $\mathbf{r}$ induced by a potential field at $\mathbf{r}'$. The quantity

$$\langle i|n(\mathbf{r})|j\rangle=\sum_{k=1}^N\int\phi_i^*(\mathbf{r}_1,\ldots,\mathbf{r}_N)\times\phi_j(\mathbf{r}_1,\ldots,\mathbf{r}_N)\,\Pi_{\ell\neq k}\,\mathrm{d}^3r_\ell \tag{18}$$

is the density of the system, which in the case of a single particle reads $\langle i|n(\mathbf{r})|j\rangle=\phi_i^*(\mathbf{r})\phi_j(\mathbf{r})$.

One more thing has to be added to arrive at the RPA theory of response. So far, the only perturbation in the single-particle Schrödinger equation was the external field. However, the induced density oscillations will cause the self-consistent field to oscillate at the same frequency as well. Given the dependence of the potential field on particle density, one can evaluate the change with respect to small density changes. Let us define the functional derivative $\delta V(\mathbf{r})/\delta n(\mathbf{r}')$ which gives this dependence. The time-varying mean field is then

$$\delta V(\mathbf{r})=\int\mathrm{d}^3r'\,\frac{\delta V(\mathbf{r})}{\delta n(\mathbf{r}')}\,\delta n(\mathbf{r}'). \tag{19}$$

Adding this potential to the external potential, one obtains an implicit equation for the self-consistent density,

$$\delta n_{\mathrm{RPA}}(\mathbf{r})=\int\mathrm{d}^3r_1\,\Pi^0(\mathbf{r},\mathbf{r}_1)\times\left(V_x(\mathbf{r}_1)+\int\mathrm{d}^3r'\,\frac{\delta V(\mathbf{r}_1)}{\delta n(\mathbf{r}')}\,\delta n_{\mathrm{RPA}}(\mathbf{r}')\right).$$

This is the response of the system to the total field, sum of the external and internal fields. We must define an RPA polarization in terms of the external field according to

$$\delta n_{\mathrm{RPA}} = \int d^3r' \, \Pi^{\mathrm{RPA}}(\mathbf{r}, \mathbf{r}', \omega) \, V_x(\mathbf{r}') .$$

The two equations above imply an integral equation which can be compactly written as an operator equation

$$\Pi^{\mathrm{RPA}} = \Pi^0 + \Pi^0 \frac{\delta V}{\delta n} \Pi^{\mathrm{RPA}} .$$

This operator has the formal solution

$$\Pi^{\mathrm{RPA}} = \left(1 - \Pi^0 \frac{\delta V}{\delta n}\right)^{-1} \Pi^0 ,$$

which is the basic equation for the RPA response.

4. Electromagnetic response of C_{60}

The molecule C_{60}, the third form of carbon after graphite and diamond, has remarkable physical and chemical properties. In particular its electromagnetic response displays a series of peaks at energies in the interval 4–7 eV and a broad resonance peaked around 20 eV (cf. [9–13]).

In what follows we apply the theory developed in the previous sections to this system. Two main differences are to be noted. One is associated with the fact that in order to obtain a correct description of the motion of the four valence electrons corresponding to each *C*-atom, it is necessary to make use of pseudopotentials (cf. e.g. [14]). The second is connected with the fact that C_{60} is a truncated icosahedron. Consequently, the position of the atoms have to be properly dealt with. This has been done by carrying a multipole expansion of the associated field.

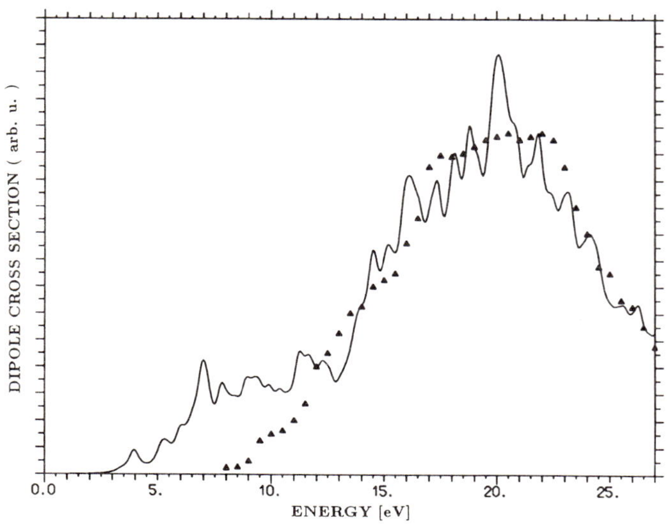

Fig. 1. Calculated photoabsorption cross section of C_{60} *(continuous line)* in comparison with experimental results [13]. The theoretical cross-section exausts ≅90% of the oscillator strength within the energy interval shown

The RPA response is shown in Fig. 1 in comparison with the beautiful experimental data of [13]. They were obtained from photoemission reactions on a beam of C_{60} molecules. With this technique, information is only available above the ionization threshold at ≈7.5 eV. The theory provides an overall description of the Mie resonance with a centroid at ≈20 eV. These results imply that the breaking of the strength associated with the decay of the giant dipole resonance on single-particle excitations (Landau damping) is the essential relaxation mechanism of the plasmon. This result is gratifying in view of the fact that tight-binding approximation calculations [15], which were able to predict correctly the centroid of the Mie resonance, lead to damping widths which are considerably narrower than experimentally observed. Furthermore, the coupling of the Mie resonance to the thermal fluctuations of the cluster surface gives a negligible contribution to the width [16]. This is because C_{60}, in spite of displaying zero point fluctuations [17], is rather rigid [18].

5. Conclusions

Mean field theory and RPA provide a powerful tool for describing the electromagnetic response of nuclei and clusters, including fullerenes. It does not escape the attention of the authors the fact that the method employed in the present paper may be used at profit to calculate the electromagnetic response of molecules of very large atomic weight in a compact way.

We have benefitted from many discussions with G. Benedek, G.F. Bertsch and K. Yabana. H.E.R. gratefully acknowledges financial support from the Alexander von Humboldt Stiftung (Feodor Lynen program). Ll. S. gratefully acknowledges financial support from the Commission of the European Communities (Community training project 920027).

References

1. Bohr, A., Mottelson, B.R.: Mat. Fys. Medd. Dan. Vid. Selsk. **27**, 16 (1953)
2. Mahaux, C., Bortignon, P.F., Broglia, R.A., Dasso, C.H.: Phys. Rep. **120**, 1 (1985)
3. Bertsch, G.F.: Prog. Theor. Phys. **74–75**, 115 (1983)
4. Bertsch, G.F., Broglia, R.A.: Oscillations in finite quantum systems. London: Cambridge University Press (in press)
5. Kohn, W., Vashita, P., In: Theory of the inhomogeneous electron gas, p. 79. Lundquist, S., March, N.H. (eds.). New York: Plenum Press (1983)
6. Williams, R., von Barth, U., In: Theory of the inhomogeneous electron gas, p. 189. Lundquist, S., March, N.H. (eds.). New York: Plenum Press (1983)
7. Negele, J.: Rev. Mod. Phys. **54**, 913 (1986)
8. Bernath, M., Hansen, M., Broglia, R.A.: Phys. Rev. A**49**, 1115 (1994)
9. Keller, J.W., Coplan, M.A.: Chem. Phys. Lett. **193**, 89 (1992)
10. Gensterblum, G., Pireaux, J.J., Thiry, P.A., Caudano, R., Vigneron, J.P., Lambin, Ph., Lucas, A.A., Krätschmer, W.: Phys. Rev. Lett. **67**, 2171 (1991)
11. Ren, S.L. et al.: Appl. Phys. Lett. **68**, 784 (1992)
12. Lorosso, G.: Private communication

13. Hertel, I.V., Steger, H., de Vries, J., Weisser, B., Menzel, C., Kamke, B., Kamke, W.: Phys. Rev. Lett. **68**, 784 (1992)
14. Bachelet, G.B., Hamann, D.R., Schlüter, M.: Phys. Rev. B**26**, 4199 (1982)
15. Bertsch, G.F., Bulgac, A., Tomanek, D., Wang, Y.: Phys. Rev. Lett. **67**, 2690 (1991)
16. Hansen, M.: Private communication
17. Kohanoff, J., Andreoni, W., Parrinello, M.: Phys. Rev. B**46**, 4371 (1992)
18. Onida, G., Benedek, G.: Europhys. Lett. **18**, 403 (1992)

Z. Phys. A 349, 205–211 (1994)

Clustering aspects of nuclei with octupole and superdeformation

Sven Åberg, Lars-Ove Jönsson

Department of Mathematical Physics, Lund Institute of Technology, PO Box 118, S-221 00 Lund, Sweden

Received: 27 October 1993

Abstract. Extreme nuclear shapes are considered from the clustering point of view. General aspects of light nuclei which possess very exotic, elongated shapes ("α-strings") are given. The relation between superdeformation and di-molecules is considered, and it is shown that nuclear di-molecules have the same magic numbers and orbits filled as the ellipsoidal superdeformations for light nuclei, while for heavier nuclei they are expected in different mass-regions. The structure of possible octupole-unstable 3:1 nuclear shapes ("hyperdeformations") is discussed. Indications and consequences of a four-fold symmetry of the nuclear hamiltonian are briefly discussed.

PACS: 21.10.−k; 21.60−n

1. Introduction

Nuclear mean field theories have been found to quite well describe a rich variety of nuclear shapes all over the periodic table [1]. In this paper we shall try to relate some of the extreme shapes nuclei may possess to a possible underlying clustering of nuclear fragments. The most well-known clustering effects appear for very light nuclei where the natural building block is the α-particle. Although the concept of a collective deformation is less well founded for very light nuclei [2], several properties may be understood from the simplest Nilsson-Strutinsky type of calculation, also for nuclei in the p-shell and the sd-shell, see e.g. [3, 4]. For light nuclei protons and neutrons move in more or less the same single-particle orbitals, and the Nilsson diagram shown in Fig. 1 is valid for protons as well as for neutrons for p-shell and sd-shell nuclei. Each single-particle level may thus be filled by two protons and two neutrons coupled to spin and isospin zero. Consequently, the Nilsson diagram for light nuclei may be effectively considered as a single-particle diagram valid for α-particles, where despite their boson character at most one α-particle can fill each single-particle level. This procedure gives of course only a subset of states in even-even $N=Z$ nuclei; e.g. negative-parity states are excluded. On the other hand we obtain a nice geometrical picture of each configuration which also can be related to states obtained in α-cluster models.

As an example we shall consider a special class of nuclear shapes related to strings of α-particles. Harvey has suggested a phenomenological way to describe the (diabatic) configuration in a fusion reaction [5]. The configuration of the fused system is obtained by fixing one of the ions, and filling the particles from the other ion into free orbitals under the condition that oscillator quanta may change only in the direction of the reaction axis, e.g. the z-axis. In the quantum number notation

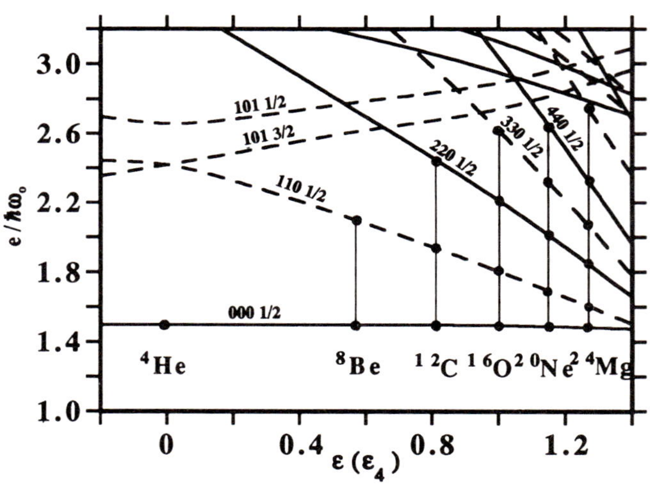

Fig. 1. Nilsson diagram valid for protons as well as for neutrons at prolate deformations. The shell filling of n:1 α-string configurations are marked out at relevent deformations for ^{4}He, ^{8}Be, ^{12}C, ^{16}O, ^{20}Ne and ^{24}Mg. Single-particle states are denoted by the quantum numbers $N n_z \Lambda \Omega$

This work was supported by the Swedish Natural Science Research Council (NFR)

used in Fig. 1, $[N n_Z \Lambda]\Omega$, this means that n_Z may change while $n_\perp(=N-n_Z)$, Λ and Ω must remain constant. The α-particle is in an s-state, i.e. filling the orbit [000]1/2. Adding α-particles along the z-axis implies that the second α-particle must occupy [110]1/2, the third [220]1/2, etc, up to $[n-1\,n-1\,0]1/2$ for n α-particles on a string.

These orbits all carry large single-particle quadrupole moments; at prolate deformations ($\varepsilon>0$) they are strongly down-sloping in the Nilsson diagram (Fig. 1). The filling of the $[NN0]1/2$ orbits with α-particles is thus strongly driving the corresponding nucleus towards prolate shapes. The approximate equilibrium deformations, resulting from a minimization of the shell energy plus the liquid-drop energy (Nilsson-Strutinsky procedure), are marked out in Fig. 1 for $n=1$, 2, 3, 4, 5 and 6, i.e. for ^{4}He, ^{8}Be, ^{12}C, ^{16}O, ^{20}Ne and for ^{24}Mg, respectively [3, 4]. These configurations correspond to excited states in the respective nuclei (except for ^{4}He and ^{8}Be). As the intuitive picture of α-strings suggests, the corresponding shapes are prolate with an axis ratio between the z-axis and the perpendicular axes being very close to n:1. We thus have a nice relation between the α-cluster description and the deformed shell model, see also [6, 7]. Experimental evidence for these exotic nuclear shapes exist for ^{8}Be, ^{12}C, ^{16}O [8, 9], and recently also for the 6:1 configuration in ^{24}Mg [10].

Thus, very exotic nuclear shapes do indeed exist for light nuclei. In this paper we want to discuss exotic shapes also for heavier nuclei, and the possible relation to clustering. We shall particularly consider 2:1 and 3:1 shapes.

Superdeformed (SD) nuclei have s shape corresponding to an axis ratio of about 2:1. Such exotic shapes have now been observed in several mass regions. The question whether strongly necked-in SD nuclei (nuclear di-molecules) may exist will be discussed in Sect. 2. In Sect. 3 the instability of SD shapes towards reflection asymmetry will be briefly discussed. Also more exotic shapes are discussed, namely 3:1 "hyperdeformed" nuclei. The instability towards octupole shapes seems to be much stronger for 3:1 shapes than for 2:1 shapes, and we find that several hyperdeformed nuclei are indeed expected to be octupole deformed. Indications of C_4-symmetry (90° symmetry of the hamiltonian) from observed rotational spectra are discussed in Sect. 4, and in Sect. 5 a short summary is given.

2. Superdeformations versus nuclear di-molecules

Superdeformed nuclei have been observed in (see e.g. [11] and references quoted therein):

(*i*) light nuclei (as discussed above),
(*ii*) fission isomers around ^{232}Th; with angular momenta about $0 \leqq I \leqq 8$,
(*iii*) high-spin rotational bands in nuclei around ^{152}Dy; $25 \leqq I \leqq 65$,
(*iv*) rotation bands in nuclei around ^{192}Hg; $8 \leqq I \leqq 25$.

The nuclear shapes for the cases (*ii*), (*iii*) and (*iv*) are all believed to be close to prolate ellipsoidals with the larger axis approximately double as long as the perpendicular axes. In case (*i*) the shapes are less well defined but may be described in terms of α-clusters, or alternatively, in terms of mean field nuclear shapes. For example, ^{8}Be may be considered as either a nuclear di-molecule with two α-particles placed side by side, or as an approximate 2:1 deformed ellipsoid. The same is valid for the SD states in ^{32}S which also may be described as two (spherical) ^{16}O nuclei placed side by side. How do these two descriptions relate? Is a clustering picture valid also for heavier nuclei? Obviously, it is not a good idea to try to describe the superdeformed nucleus ^{152}Dy as a clustering of two ^{76}As nuclei. But can nuclear di-molecular shapes exist as a new class of SD nuclei?

Effects caused by the formation of di-molecular (quasi)states have indeed been seen in heavy nuclei. In studies by Möller et al. [12] two paths to fission were found in the nuclear potential-energy surface for nuclei around ^{264}Fm. One path corresponds to the usual fission path where scission occurs at a quite elongated shape. The other path, "the fusion path", corresponds to shapes where the scission occurs at a much smaller elongation, roughly corresponding to the shape of a di-molecule. The latter path was found to be more preferred the closer the neutron number is to $N=164$, which can be understood as an effect of a very favorable formation of two double magic (spherical) ^{132}Sn clusters in the ^{264}Fm compound system. The scenario could give a nice explanation of observed [13] peaks in the total kinetic energy distribution curves [12].

We shall first study effects of necking in the pure harmonic oscillator potential (Subsect. 2.1), and then in Subsect. 2.2 utilize the more realistic Woods-Saxon potential.

2.1. *Harmonic oscillator potential*

In the anisotropic harmonic oscillator potential, single-particle energy degeneracies occur when the oscillator frequencies, ω_x, ω_y and ω_z, relate as small integer numbers [14]. Particularly large gaps in the single-particle spectrum show up at ratio 1:1:1 (spherical shape) and at 2:2:1, i.e. when the z-axis is double the size of the perpendicular axes (SD prolate shape). An interesting observation is that every second SD magic number corresponds to twice a spherical magic number. This may be understood by considering the shape and configuration of the system obtained through the fusion of two equal (double-magic, $N=Z$) spherical nuclei. Through the Harvey prescription [5] it is then found that the resulting configurations exactly correpond to the closed-shell configurations at superdeformation, see also Refs. [15–17]. In the left-hand part of Fig. 2 we show how the single-particle spectra of two spherical $A/2$ nuclei add up and give rise to a single-particle spectrum of one nucleus with the shape of a double-sphere containing A nucleons. Volume conservation implies that the oscillator frequency of the double-sphere is $2^{1/3}$ times larger than that of a sphere.

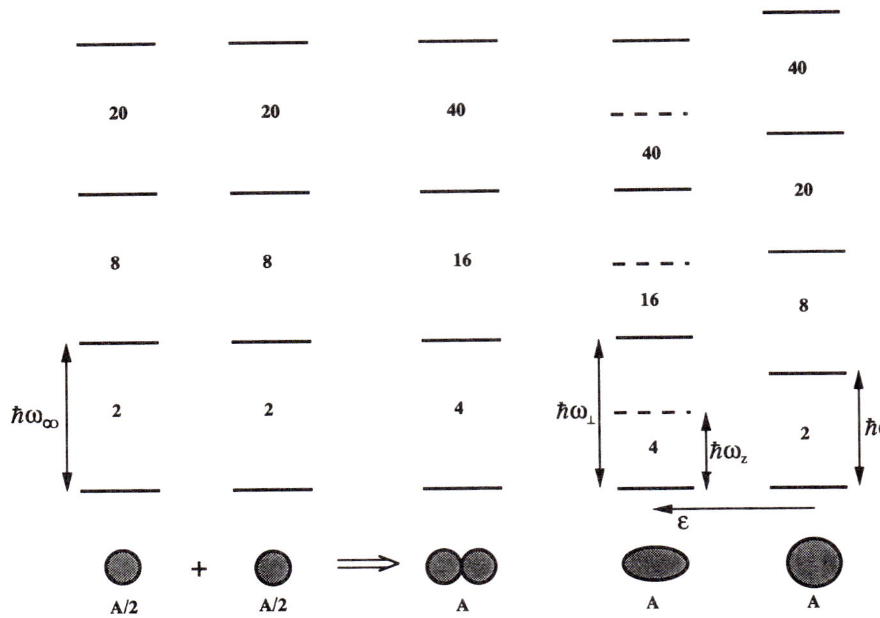

Fig. 2. Single-particle spectra for different versions of the pure harmonic oscillator potential. The three spectra on the left-hand side illustrate the fusion of two spheres with $A/2$ particles into a double-sphere. The two spectra on the right-hand side show how the sphere with A particles changes versus quadrupole deformation, ε, into a 2:1 deformed spheroid

In the right-hand part of Fig. 2 it is shown how the spherical magic numbers change as the sphere is deformed into a superdeformed spheroidal nucleus, with the ratio between the oscillator frequencies 2:2:1. The same particle numbers for closed shells appear for the spheroidal nucleus as for the double-sphere. Also the single-particle energy distances are the same ($\hbar\omega_\perp = \hbar\omega_\infty = 2^{1/3}\hbar\omega_0$). The double-sphere and the 2:2:1 deformed oscillator are thus expected to have very similar properties. But although the filled orbitals are the same for the two types of shape, there are also differences. For the spheroidal nucleus additional states appear in the middle of the gaps in the single-particle spectrum of the double-sphere, as shown by dashed lines in Fig. 2. These states correspond to the excitation of oscillator quanta in the z-direction with the oscillator frequency $\hbar\omega_Z(\omega_\perp:\omega_Z = 2:1)$. The degeneracies are thus smaller for spheroidal SD nuclei than for nuclei with a dumb-bell shape which implies a smaller amplitude of the shell energy. (The shell energy of a double-sphere with A particles is expected to be approximately twice the shell energy of one sphere with $A/2$ particles.)

Another general difference between SD spheroids and double-spheres is that the giant dipole resonance (GDR) should look quite different in the two cases. In the superdeformed spheroidal nucleus two peaks are expected to appear, one at about $2\hbar\omega_Z$ and one at about $2\hbar\omega_\perp$, while due to the lack of the $\hbar\omega_Z$ states in the system of a double-sphere (see Fig. 2), the lower GDR-peak disappears and the higher-lying peak becomes correspondingly larger. An increase of the necking degree of freedom is thus expected to decrease the lower component of the GDR peak, and in fact increase the centroid energy.

Superdeformed nuclei are generally expected to be populated by statistical $E1$-transitions, where the $B(E1)$ strength becomes strongly enhanced through the low-lying GDR component [18]. In that scenario the population of states which are strongly necked-in should thus be less favorable than of "normal" SD shapes.

2.2. *Woods-Saxon potential*

In order to study varies types of shapes in a realistic mean-field potential, including 2:1 deformed spheroids and double-spheres, we have introduced [19] a shape parametrization based on the usual expansion into spherical harmonics (β-parametrization). Compared to other shape parametrizations aimed for describing double-spheres (which are usually based on two-center shell models, see e.g. [20]), the present parametrization has the advantage of including nuclear ground states as well as superdeformed shapes which are well described in the β-parametrization.

In the β-parametrization axially symmetric shapes are described by

$$R(\theta, \hat{\beta}) = f(\hat{\beta})\, R_0 \Big[1 + \sum_{i=2,4,\ldots n} \beta_i Y_i^0(\theta)\Big], \tag{1}$$

where $f(\hat{\beta})$ takes care of the volume conservation. Two new deformation parameters, c and a, are defined. If β_i^∞ denotes β-parameters fitted to a double-sphere configuration, and $\beta_i^{2:1}$ denotes β-parameters fitted to a 2:1 spheroid, then for arbitrary values of the parameters c and a in the interval [0, 1], the β-parameters are given by

$$\beta_i = (1-a)\cdot c^{i/2}\cdot \beta_i^{2:1} + a\cdot c\cdot \beta_i^\infty . \tag{2}$$

In order to get a good description of the double-sphere it was found necessary to include multipoles up to the order $n = 10$ [19]. The family of shapes obtained by varying c and a can be seen in Fig. 3. Here a varies from 0 to 1 in the vertical direction. With $a = 0$ we get a succes-

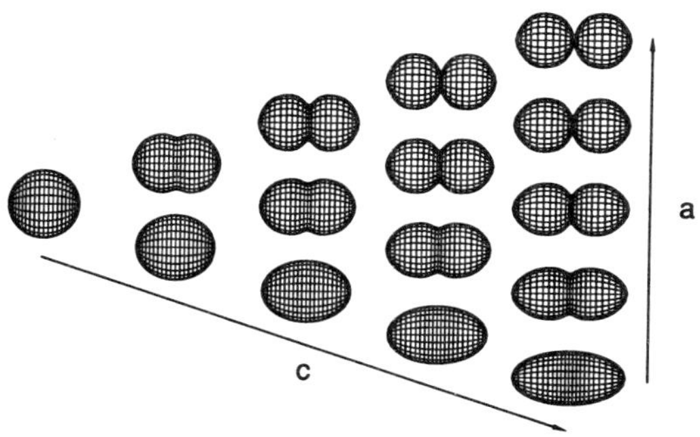

Fig. 3. Family of shapes obtained by the $c-a$ parametrization described in the text. Parameters c and a are varied from 0 to 1

sion of spheroidal shapes with varying c, and with $a=1$ we get a smooth transition from a sphere to a double-sphere, maintaining the sphericity of the end fragments. By varying a from 0 to 1 at $c=1$ a successive change in shape from a 2:1 deformed spheroid to a double sphere is obtained.

In Fig. 4 we show the Woods-Saxon neutron single-particle energies valid around ^{96}Zr as functions of the deformation parameter a at $c=1$, i.e. the transition from the 2:1 spheroid ($a=0$) to the double-sphere ($a=1$). Note the restoration of the spherical quantum numbers at $a=1$, but with a twofold degeneracy as compared to the spherical configuration. Due to the diffuse surface of the Woods-Saxon potential the restoration is only approximative as compared to two independent spheres, since the two spheres still interact. Nice single-particle gaps are seen at $N=4$, 16, 28, 40, 56, ... corresponding to the spherical magic numbers at $N=2$, 8, 14, 20, 28, ... At the spheroidal 2:1 deformation ($a=0$) closed shells appear for $N=4$, (10), 16, 24, 46, 66, ..., i.e. the same as for the double-sphere for the lowest neutron numbers only. In Fig. 2 we saw that in the pure harmonic oscillator potential closed shells appear at the same particle numbers for the spheroid and for the double-sphere. Since the pure harmonic oscillator model gives a fairly good approximation of the single-particle spectrum of very light nuclei, this explains the agreement between the two types of shapes for low particle numbers. This also explains the appearance of a semi-closed shell at $N=10$ for the spheroidal shape. For heavier nuclei the two kinds of shapes are favored at quite different particle numbers, and the occupied orbits are quite different. For example, for $N=66$ and 86 (^{152}Dy) there are (as expected) no favoring at all for double-spherical configurations, while for $N=40$ and 56 the shell structure strongly favors double-spherical shapes.

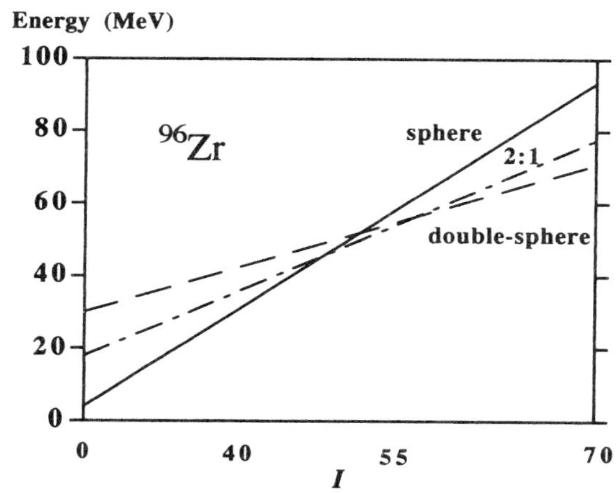

Fig. 5. The approximate (rigid-body moments of inertia have been assumed) rotational behavior of ^{96}Zr at spherical, 2:1 spheroidal and double-spherical shapes. The shell energies were approximated by their spin-zero values

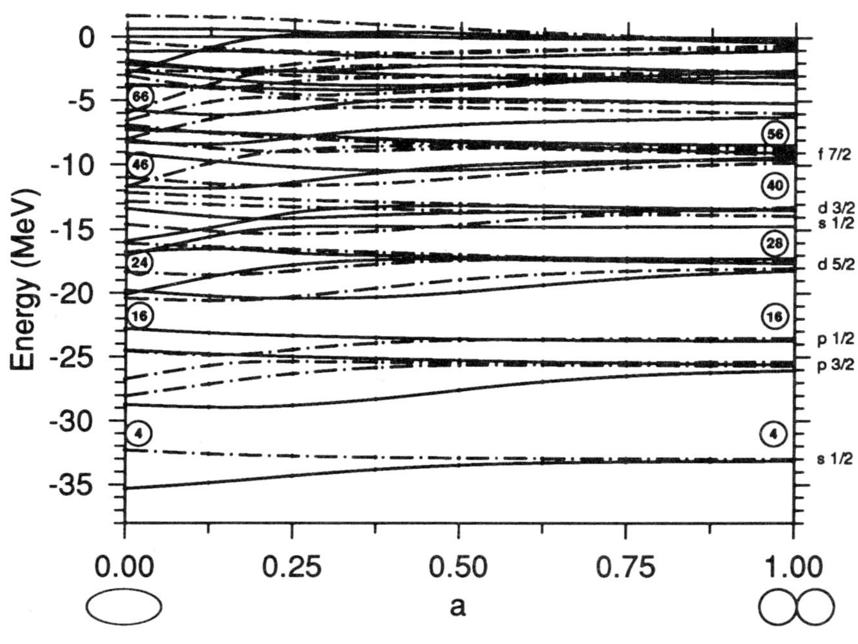

Fig. 4. S.p. energies for Woods-Saxon neutron levels as functions of the shape variable a, see (2). The two limits $a=0$ and $a=1$ correspond to spheroidal 2:1 deformation and double-sphere, respectively. Particle numbers at closed shells are encircled. At the double-sphere deformation j-shell configurations are given. Each j-shell has double degeneracy as compared to one sphere. Woods-Saxon parameters from [21] are used

For such exotic shapes as the double-sphere it is quite necessary to include a finite-range surface energy term [22]. For example, for ^{96}Zr the double-sphere thereby gains almost 30 MeV while the gain in the spheroidal 2:1 nucleus is only about 4 MeV. Still, the surface energy disfavors the double-spherical shapes, and for ^{96}Zr the total energy comes about 10 MeV higher for the double-sphere than for the 2:1 deformed spheroidal. However, the moment of inertia of the double-sphere is about 2.2 times larger than the value for the rigid sphere, while the corresponding value for the 2:1 deformed spheroid is about 1.5. Since the rotational energy is inversely proportional to the moment of inertia (which at high spin should take the rigid-body values), this implies that the double-sphere would easily get more favored at high spins. This is illustrated in Fig. 5, where the double-spherical shape of ^{96}Zr is seen to become more favored than the spherical as well as the 2:1 spheroidal shapes for spin values above $I \approx 50$.

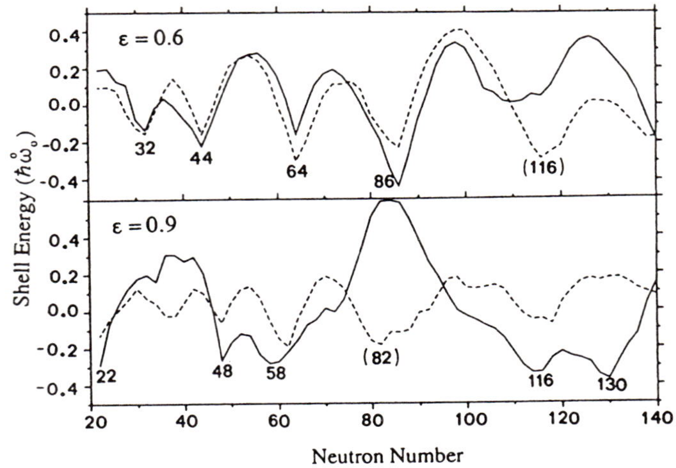

Fig. 6. Shell energy obtained from the Nilsson hamiltonian vs neutron number for 2:1 and 3:1 shapes at $\varepsilon_3=0$ (*solid lines*) and $\varepsilon_3=0.15$ (*dashed lines*). The shell energy is expressed in units of $\hbar\omega_0=41\,A^{-1/3}$ MeV and calculated at the fixed rotational frequency $\omega=0.06\mathring{\omega}_0$. From [25]

3. Instability towards reflection asymmetry – hyperdeformations

In order to understand the importance of the octupole degree of freedom at very elongated shapes we shall once again consider to pure harmonic oscillator potential [15–17]. The double-sphere and the spheroidal 2:1 shape both have closed shells at $N=4$, 16, 40, ..., see Fig. 2. This could be understood as the sum of two equal spherical magic numbers, or the fusion of two spheres with equal size. But, in addition, the spheroid has closed shells at $N=10$, 28, 60, ... These numbers can be viewed as the sum of two *neighboring* spherical magical numbers; or as the fusion of two spheres with *different* sizes. Intuitively, nuclei at such magic numbers should be soft or unstable against reflection asymmetric changes of the nuclear mean field. Indeed, by considering octupole vibrations on a quadrupole deformed H.O. mean field (studied in the RPA based on selfconsistently derived octupole operators, i.e. four different combinations of $\{Y_3^\mu\}$, and coupling constant), it was found for all four modes that a lower vibrational energy arises for the asymmetric combination of spherical magic numbers than for the symmetric combination [23, 24]. Similar result have been obtained in Nilsson-Strutinsky type of calculations (based on the Nilsson potential) of the nuclear shell energy as a function of the particle number at different ($Y_{3^-}^0$) octupole deformations [15, 25]. At the spheroidal 2:1 deformation particle numbers corresponding to asymmetric combinations of spherical magic numbers were found to have a more negative shell energy (more favorable) for a static octupole deformation (reflection asymmetric shape) than without, see the upper part of Fig. 6.

In the pure H.O. magic numbers can be described in terms of spherical magic numbers also at the spheroidal 3:1 deformation. In this case there are three different types of magic numbers. One corresponds to three times the spherical magic numbers (symmetric combination), one to combining two bigger spherical numbers with one smaller, and finally, the third way is to combine one bigger with two smaller, see [17, 23]. Asymmetric combinations correspond to a lower octupole vibrational energy in the RPA solutions, and the symmetric combination corresponds to a higher energy [23]. However, the shell energy calculated in the modified oscillator potential behaves much more irregular and does not show a corresponding staggering, see the lower part of Fig. 6. In fact, octupole correlations are in general expected to be very strong at such large quadrupole deformations since groups of single-particle levels with opposite parity come very close at these deformations [26]. For example, the so called 3:rd minimum, calculated in the potential-energy surfaces for nuclei around ^{232}Th, was found to be strongly octupole deformed [27].

An interesting clustering aspect of the hyperdeformed ^{232}Pu ("3rd minimum deformation") has been suggested by Nazarewicz et al. [28]. The shape of the hyperdeformed ^{232}Pu ($\beta_2=0.84$, $\beta_3=0.59$) overlaps very nicely with the shapes of the spherical nucleus ^{132}Sn and the superdeformed nucleus ^{100}Ru. The clustering would show up in the fission channel as an enhanced production of ^{132}Sn and ^{100}Ru [28].

In the rare-earth region the hyperdeformed minimum is calculated to come more than 20 MeV above the ground state. But in a similar way as for superdeformed nuclei, rotation is the tool by which we can study hyperdeformed nuclei. Due to the very large moment of inertia such exotic shapes may become yrast at high spins, and could possibly be populated in heavy-ion fusion reactions. One especially favorable case for hyperdeformation was calculated to be ^{146}Gd [29]. The lowest hyperdeformed configuration in this nucleus was found to become yrast at about 80 $\hbar$, and despite the extreme deformation rather large gaps (almost 1 MeV) were found in the single-particle spectrum for the proton as well as for the neutron system. The population of such hyperdeformed states would thus be favored by a low density of states, and also by a quite low-lying GDR component. Recently it was reported that hyperdeformed states were formed in ^{152}Dy through a (HI, xn)-reaction [30].

4. C_4-symmetry in nuclei [1]?

The well-known 0^+ state at 6.05 MeV in ^{16}O is mainly built by the excitation of two protons and two neutrons from the p-shell to the sd-shell. This gives rise to a very large triaxial deformation with $\varepsilon \approx 0.8$ and $\gamma \approx 40°$ ($\varepsilon_4 \approx 0.16$), and a rotational band up to spin 8 may be built by rotation around the axis with the largest moment of inertia (the x-axis) [31]. In the α-cluster model this band corresponds to a configuration where all four α-particles are placed in a plane forming a tetrahedron, see Fig. 7a, and the rotation axis is perpendicular to the plane. This configuration is "almost" symmetric against 90° rotation around the x-axis, i.e. is approximately invariant under the point group C_4. A shape corresponding to an exact C_4 symmetry is illustrated in Fig. 7b. At spin 8^+ the $4p-4h$ band terminates and the corresponding nuclear shape is oblate with the rotation axis coinciding with the symmetry axis. This oblate shape is similar to the C_4-symmetric shape shown in Fig. 7b, and the $4p-4h$ rotational band in ^{16}O should be a good candidate for showing evidence for C_4-symmetry. The corresponding hamiltonian would thus be invariant to the $\Re_x(\pi/2)\,(=\exp(iI_x\pi/2))$ operator, and a new quantum number, 4-signature (r_4), would be good:

$$\Re_x(\pi/2)\,\Psi = r_4\,\Psi.$$

If we write this new quantum number as

$$r_4 \equiv \exp(i\alpha_4\pi/2)$$

the quantum number α_4 may take the values 0, 1, 2 and 3 in an even-even nucleus, corresponding to four different sequences of rotation states with the spin values $I = \alpha_4 \bmod 4$. Most deformed nuclei have a shape with C_2-symmetry, and (2-)signature is a good quantum number corresponding to the appearance of two different sequences of $\Delta I = 2$ rotational states. In the $4p-4h$ rotational band in ^{16}O a splitting between the bands with $\alpha_4 = 0$ and 2 ($\Delta I = 4$ staggering) can indeed be seen, see Fig. 8, which may imply the existence of an approximate C_4-symmetry. The separation between the two bands with the spin values $I = 0$ and 4, and with $I = 2$ and 6, respectively, i.e. the 4-signature splitting, is about 300 keV.

[1] This section is based on a collaboration with the authors of [32]

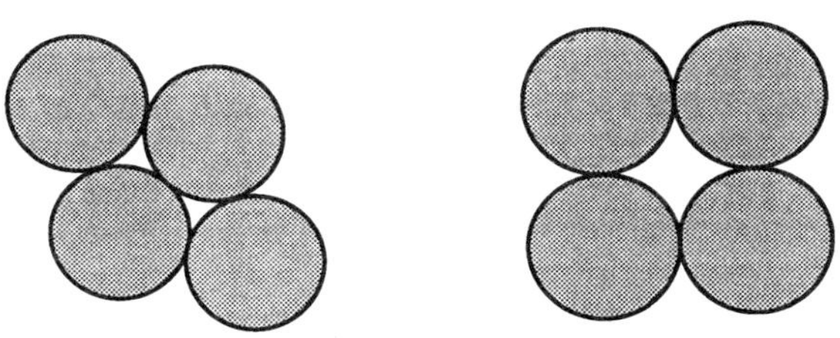

Fig. 7a, b. Schematic α-particle configurations in ^{16}O, as seen in the plane perpendicular to the rotation axis

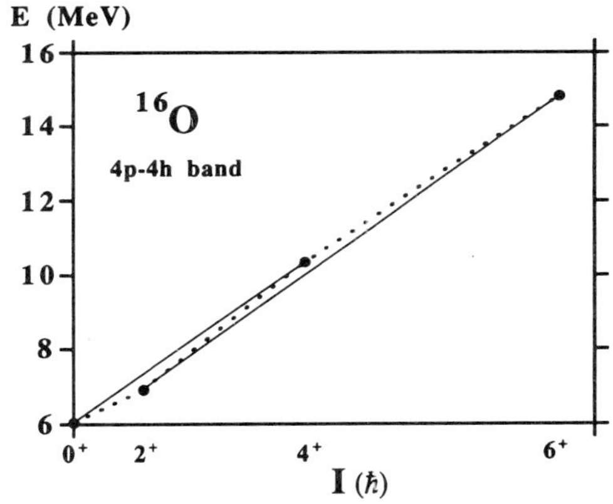

Fig. 8. The experimental energies of the excited $4p-4h$ band starting at 6.05 MeV in ^{16}O plotted vs $I(I+1)$. The 0^+ and 4^+, and 2^+ and 6^+ states, respectively seem to form two different bands, separated by about 300 keV

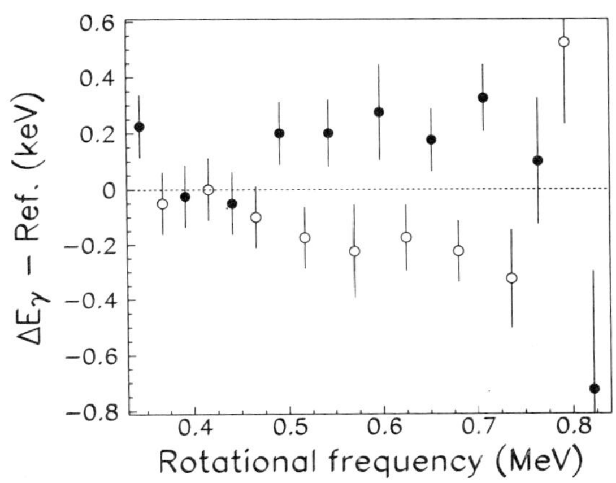

Fig. 9. Observed transition energies in the superdeformed nucleus ^{149}Gd plotted versus the rotational frequency. A smooth average has been subtracted. Filled and empty symbols refer to different values of α_4. From [32]

It is certainly of very great interest that recently a rotational band in the superdeformed nucleus ^{149}Gd was observed [32] to show such a $\Delta I = 4$ staggering, see Fig. 9. In this case the staggering between the two sequences of $\Delta I = 4$ bands is found to be several orders of magnitude smaller than the ^{16}O case, about 120 eV, suggesting that the C_4-symmetry has a dynamical rather than a statical origin. It should also be pointed out the C_4-symmetry of the nuclear hamiltonian does not necessarily imply a C_4-symmetry of the corresponding shape, see e.g. [33].

5. Summary

We have discussed several types of exotic shapes where the shape may be described as consisting of two or several clusters. In light nuclei the clusters may consist of α-particles building up very elongated strings. In the mean field description such α-strings are described as n:1 deformed prolate shapes.

Another possibility is that the α-particles are arranged in a symmetric way around the axis of rotation, giving rise to interesting properties of the corresponding rotational band. The example of the excited rotational band in ^{16}O was discussed in terms of C_4-symmetry, and the rotational band shows clear signs of a $\Delta I = 4$ staggering. In a completely different context a similar rotational behavior has recently been observed in the superdeformed nucleus ^{149}Gd.

Hamiltonians which are invariant under the symmetry group C_3 would also show very interesting rotational properties. The simplest example would be ^{12}C which can be considered as consisting of three α-particles in a triangular plane. However, the rotation takes place around an axis within the triangular plane, and not along the axis which has the C_3-symmetry. Consequently no signs of C_3-symmetry are expected in the rotational spectrum.

The clustering aspect was also considered for heavier nuclei, and the possible existence of dumb-bell shaped nuclei was discussed. Light nuclei with such shapes are similar to (spheroidal) superdeformed nuclei, while for heavier nuclei they are quite different and appear for other particle numbers. The possible existence of a "new class" of SD nuclei, where the necking is very large, is certainly very interesting. It was found that the lower-lying peak of the giant dipole resonance should be considerably smaller for dumb-bell shaped nuclei than for SD's with a spheroidal shape.

Also very alongated nuclei with a strong octupole deformation was discussed. The octupole deformation is much more important at hyperdeformations than at superdeformations, and the favorable cases found for hyperdeformations with a strong degree of reflection asymmetry are ^{146}Gd (or nuclei around) at very high spins, and nuclei around ^{232}Th at very low spin. While the first class of hyperdeformations possibly might form nice rotational sequences, fission is so dominating for the heavier nuclei (around ^{232}Th) that the hyperdeformed states only show up as resonances in the fission cross section.

References

1. Åberg, S., Flocard, H., Nazarewicz, W.: Ann. Rev. Nucl. Part. Sci. **40**, 439 (1990)
2. Brack, M.: In: Physics and chemistry of fission, Vol. 1, p. 227, Vienna: IAEA 1980
3. Leander, G., Larsson, S.E.: Nucl. Phys. A **239**, 93 (1975)
4. Ragnarsson, I., Åberg, S., Sheline, R.K.: Phys. Scr. **24**, 215 (1981)
5. Harvey, M.: Proc. 2nd Int. Cinf. on Clustering Phenomena in Nuclei, College Park 1975, USDERA Report ORO-4856-26, p. 549
6. Rae, W.D.M.: Proc. 5th Int. Conf. on Clustering Aspects in Nuclear and Subatomic Systems p. 77, Kyoto, Japan, 1988 Ikeda, K., Katori, K., Suzuki, Y. (eds.)
7. Cseh, J., Scheid, W.: J. Phys. G **18**, 1419 (1992)
8. Morinaga, H.: Phys. Rev. **101**, 245 (1956)
9. Chevallier, P., et al.: Phys. Rev. **160**, 827 (1967)
10. Wuosmaa, A.H., et al.: Phys. Rev. Lett. **68**, 1295 (1992)
11. Proc. 21st Int. Symp. on Rapidly Rotating Nuclei, Tokyo, Japan, Oct. 26–30, 1992, Nucl. Phys. **A557** (1993)
12. Möller, P., Nix, J.R., Swiatecki, W.J.: Nucl. Phys. A **469**, 1 (1987)
13. Hulet, E.K., et al.: Phys. Rev. Lett. **56**, 313 (1986)
14. Bohr, A. Mottelson, B.R.: Nuclear structure, Vol. 2. New York: Benjamin 1975
15. Bengtsson, T., et al.: Phys. Scr. **24**, 200 (1981)
16. Åberg, S.: In: Nuclear structure and heavy-ion reaction dynamics (Notre Dame 1990), Inst. Phys. Conf. Ser. No 109, Chap 6, p. 143 Betts, R.R., Kolata, J.J. (eds.) 1991
17. Nazarewicz, W., Dobaczewski, J.: Phys. Rev. Lett. **68**, 154 (1992)
18. Schiffer, K., Herskind, B., Gascon, J.: Z. Phys. A **332**, 17 (1989); Schiffer, K., Herskind, B.: Nucl. Phys. A **520**, 521c (1990)
19. Jönsson, L-O, Nazarewicz, W., Ragnarsson, I., Åberg, S.: (in preparation)
20. Holzer, P., Mosel, U., Greiner, W.: Nucl. Phys. A **138**, 241 (1969)
21. Dudek, J., et al.: Phys. Rev. C **26**, 1712 (1982)
22. Krappe, H.J., Nix, J.R.: In: Physics and chemistry of fission, Vol. 1, p. 159. Vienna: IAEA 1974
23. Nazmitdinov, R., Åberg, S.: Phys. Lett. B **289**, 238 (1992)
24. Nakatsukasa, T., Mizutori, S., Matsuyanagi, K.: Prog. Theor. Phys. **87**, 607 (1992)
25. Höller, J., Åberg, S.: Z. Phys. **336**, 363 (1990)
26. Ragnarsson, I., Nilsson, S.G., Sheline, R.K.: Phys. Rep. **45**, 1 (1978)
27. Möller, P.: Nucl. Phys. A **192**, 529 (1972); Bengtsson, R., et al.: Nucl. Phys. A **473**, 77 (1987)
28. Cwiok, S., et al.: Phys. Lett. B **322**, 304 (1994)
29. Åberg: Nucl. Phys. A **557**, 17c (1993)
30. Galindo-Uribarri, A., et al.: Phys. Rev. Lett. **71**, 231 (1993)
31. Åberg, S., et al.: Nucl. Phys. A **391**, 327 (1982)
32. Flibotte, S., et al.: Phys. Rev. Lett. **71**, 4299 (1993)
33. Hamamoto, I.: Phys. Lett. B **193**, 399 (1987)

Electronic shells and supershells in large trivalent metal clusters

M. Pellarin, B. Baguenard, C. Bordas, J. Lermé, J.L. Vialle and M. Broyer

Laboratoire de Spectrométrie Ionique et Moléculaire (URA n°171), CNRS et Université Lyon I, Bât. 205, 43 Bd du 11 Novembre 1918, 69622 Villeurbanne Cedex, France

Received 28 June 1993

Abstract. Electronic shell structures are observed for the first time in large gallium and aluminum clusters. In gallium, the electronic shells extend up to 7000 electrons and a nice supershell structure is obtained around 2500 electrons. This number is considerably larger than expected from standard jellium calculations and is consistent with the introduction of a small surface softness. In aluminum, the clusters are more rigid and in usual experimental conditions they have specific geometric arrangement. However by heating the source nozzle, we obtain liquid-like aluminum clusters and observe the electronic shell structure up to 1800 electrons.

PACS: 36.40+d, 31.20.Sy, 82.80.Ms

1. Introduction

The electronic properties of alkali clusters are well understood through the electronic shell model introduced by Knight et al. [1]. In the spherical symmetry, the electronic energy levels nl have high degeneracy $2(2l+1)$ and high values of the cohesive energy and of the Ionization Potentials are obtained for closed shells. This electronic shell structure leads to "magic" numbers in mass spectra [1] and to energy gaps in Ionization Potential measurements [2].

For large sizes the nl levels tend to condense into highly degenerate shells. This bunching of quantum levels is then modulated by a supershell structure [3]. These quantum shells and supershells may be interpreted in a semi-classical approach [4,5] and are in fact related to quantization conditions for the closed classical orbits of the elevtron in spherical potential. At non zero temperature, the triangle and square orbits are dominant and the interference between the contribution of both orbits (with lengths close to each other) leads to a beating structure which corresponds to the quantum supershell. The group of Copenhagen [3,6] was the first to point out that such phenomena could be observed in alkali clusters and they succeeded to make the experiment in 1991 [7], as other groups did later [8,9].

In trivalent metal clusters, three valence electrons per atom are expected to be delocalized and the situation is in principle more favorable for the observation of the same effects because the energy gap between electronic shells are higher. The first attempt to observe the electronic shell structure in trivalent metal clusters was made by R.L. Whetten [10,11] and his group on aluminum (Al_N). They werre able to explain some features obtained in the size range $20<N<400$ by the electronic shell structure but not all the structures observed in the mass spectra and Ionization Potentials. Lermé et al. showed [10,11] that a clear shell structure emerges at N=250 and extended up to 900 atoms. These results were tentatively interpreted in terms of electronic shell structure within the semi-classical theory [10] (classical star orbit). However, this model predicts that this shell structure must change for larger clusters. T.P. Martin [12] observed the same effects up to 10000 atoms and gave a quite convincing explanation: this shell structure may be correlated with the coverage of triangular faces on octahedral clusters. This means that in usual experimental conditions, aluminum clusters are rigid and dominated by specific geometric arrangement. This is not surprising because the electronic shell structure is related to the spherical symmetry of small droplets of liquid-metal. Since the aluminum melting point is rather high (930 K), the quite cold aluminum clusters (T≈300 K) produced in molecular beams are expected to be rigid or solid-like. In this respect, gallium clusters are much better candidates for the observation of electronic shell structure: the gallium melting point is 300 K and moreover this material is known to undergo a strong supercooling. We report in this paper near threshold photoionization experiments in large gallium clusters. Clear electronic shells and supershell structures are obtained and extend up to 7000 electrons. The observed quantum supershell is shifted as compared to the prediction of the standerd self-consistent jellium model and this point is discussed. Moreover, we have performed new experiments in aluminum clusters by varying the nozzle temperature and we have observed the transition between the octahedral geometric arrangement and the electronic shell structure.

2. Gallium clusters

The clusters are produced by the laser vaporization technique. The design of our source is similar to the de Heer design [15]. The block source is cooled to about 100 K by liquid nitrogen, to prevent the gallium from melting. The clusters are photoionized by a U.V. laser having a photon energy very close to the ionization threshold. They are then analysed through a time-of-flight (TOF) mass spectrometer perpendicular to the beam. The initial transverse velocity of the clusters in the TOF is compensated by an electric field applied between deflector plates. With this geometry, the mass resolution is good (450 in our case [13]) but the mass range of the spectrometer is limited. However, it allows us to observe the oscillations in mass spectra, and it may be increased by applying a linear ramp voltage to the deflector plates [15].

The mass spectra of gallium clusters are very different from those of aluminum presented in references [12],[13]. Figure 1 shows two mass spectra of gallium clusters in two different size ranges: $100<N<800$ and $800<N<2400$. Clear oscillation may be seen and at least for small sizes they are rather step-like as expected from photoionization threshold effects. Since the number of valence electrons in the clusters is $N_e=3N$, we may plot the $N_e^{1/3}$ values corresponding to the observed dips in experimental spectra as a function of the dip index K. We obtain $N_e^{1/3} = 0.603\ K + c$ and the slope (0.603) is very similar to the corresponding slope for sodium clusters (0.61). Moreover, if for gallium clusters we plot $N^{1/3}$ values, the slope (0.418) is different from the one obtained for any known atomic shell or subshell structure [16,17]. Therefore, we have indeed observed electronic shell structures in gallium clusters.

Since our results extend up to 7200 electrons, we expect to observe supershell structures. Self consistent standard jellium calculations predict this supershell [18] around $N_e \approx 1150$ electrons for a Wigner-Seitz radius $r_s=3.93$ a.u. (sodium case). In sodium, the supershell is observed experimentally around N_e=900-1000 electrons depending on the authors [7,8]. Since the node region of the beating structure extends over about 300 electrons, the agreement between theory and experiment may be considered as reasonable. However, in gallium clusters no beating structure occurs around N_e=1150-1200 electrons ($N\approx400$) (Fig. 1). In fact the supershell structure of gallium clusters is shown in Fig. 2 and the node region is observed around $N_e=2500$ electrons ($N=830$ atoms). Strong oscillations are obtained in the region $400<N<700$ and again in the region $N>1000$. For $700<N<1000$, the oscillations are much less intense and

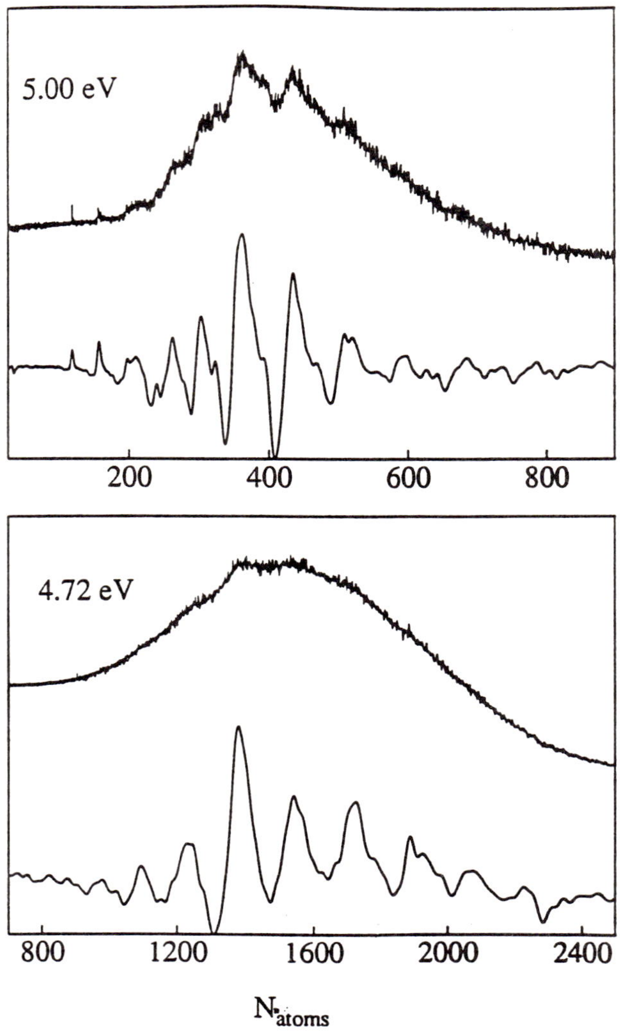

Fig. 1. Mass spectra of Ga_N clusters: top: $200<N<800$, bottom: $800<N<2400$. The photon energy is different in the two spectra, in order to keep the ionization energy close to the threshold. Clear oscillations emerge in the two mass spectra but no beating structure is observed around N_e=1200 electrons (N=400 atoms).

two times faster on a $N^{1/3}$ (or $N_e^{1/3}$) scale. This quantum beat mode results from the interference between classical triangular and square orbits. The doubling of the oscillation periodicity in the node region corresponds to contributions of the turn triangular and square orbits [9] which are in phase when the single-turn ones are in opposite phase (π phase shift).

The position of the beating region in gallium clusters at 2500 electrons is in complete disagreement with self-consistent standard jellium calculations which predict this beating at 1150 electrons. However, the beating region is very sensitive to the surface potential profiles because the square orbits are localized closer to the cluster surface than the triangular orbits. It may be shown [19] that at first approximation a slight modification of the surface potential introduces a shift between the contribution of the two orbits, leading to a shift of the beating location.

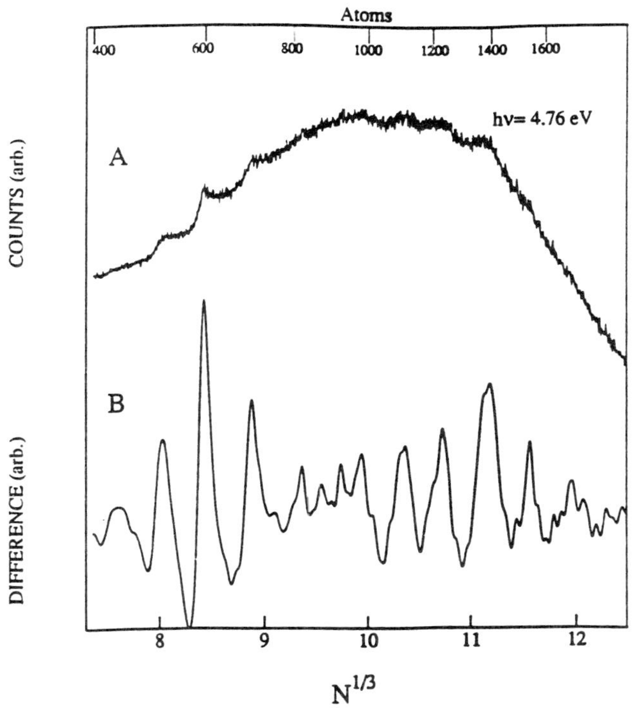

Fig. 2. A: photoionization mass spectrum of Ga_N clusters for 400<N<1600 atoms. B: Slightly smoothed difference between A and its highly smoothed envelope. The beating region is obtained for N=830 ($N^{1/3}$=9.4).

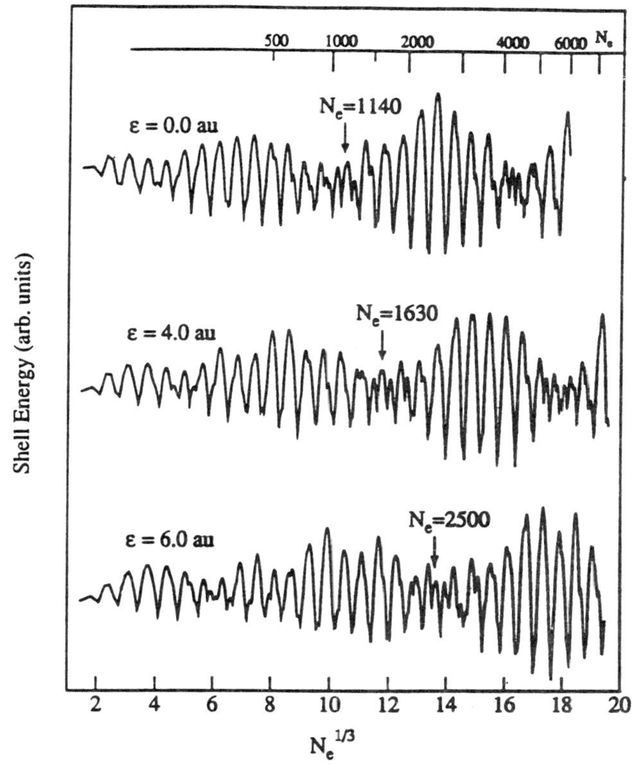

Fig. 3. Calculated shell energy of Ga_N clusters as a function of softness parameter ε. The beating region shifts as ε increases.

In order to perform calculations taking into account a possible softness of the surface potential profile, we have slightly modified the square shape of the jellium density profile. We have used a trapezoidal density for the positive ionic sea: the ionic density is assumed constant except close to the surface where it varies linearly over a thickness ε between the bulk value and zero. For a given ε we perform self-consistent jellium calculations for some selected sizes between 300 and 6600 electrons and we extract for each size the effective potential experienced by the electron. This effective potential does not vary significantly with the cluster size except for a scaling factor. We then use this optimized averaged effective potential to obtain the shell energy for all sizes by an independent-electrons quantum calculation. This procedure allows us a fast exploration of the beating shift as a function of ε (Fig. 3). Finally, complete self-consistent calculations are performed for the best value of ε.

We find the best agreement with the experiment for ε=6 a.u. This value is comparable to the average interatomic distance (6.3 a.u. in gallium) and may be considered as reasonable: in a liquid, the surface fluctuations as well as the instantaneous surface irregularities are of the order of an interatomic distance. Our semi-empirical approach must be confirmed by pseudopotential calculations or by molecular dynamics in a liquid cluster. However our very simple model gives a quite good agreement with the experiment.

3. Aluminum clusters

For aluminum, the experimental set-up is the same than for gallium. However the temperature of the block source may be heated up to 600 K. Figure 4A shows a Al_N mass spectrum recorded for a source at room temperature (300 K). Regular oscillations appear in the mass spectrum (for N>350), as already observed in the previous experiments [12,13]. The periodicity of these oscillations ($N^{1/3}$=0.22 K) corresponds quite well to the coverage of triangular faces on octahedral structures.

By increasing the source temperature (Figs. 4B and 4C), the mass spectrum gradually changes. At 473 K, a new structure emerges on the mass spectrum (Fig. 4C) while in Fig. 4B (373 K) there is a competition between these two structures.

In fact, the shell structure of Fig. 4C is the same than the electronic shell structure observed in gallium clusters. This is illustrated in Fig. 5 where aluminum mass spectra are compared to galium mass spectra. Therefore we can conclude that by heating the nozzle we are able to obtain the electronic shell structure in large aluminum clusters. The transition between the octahedral geometry and the electronic shell structure is similar to a liquid-solid transition in clusters. In our experiment, the clusters temperature is close to

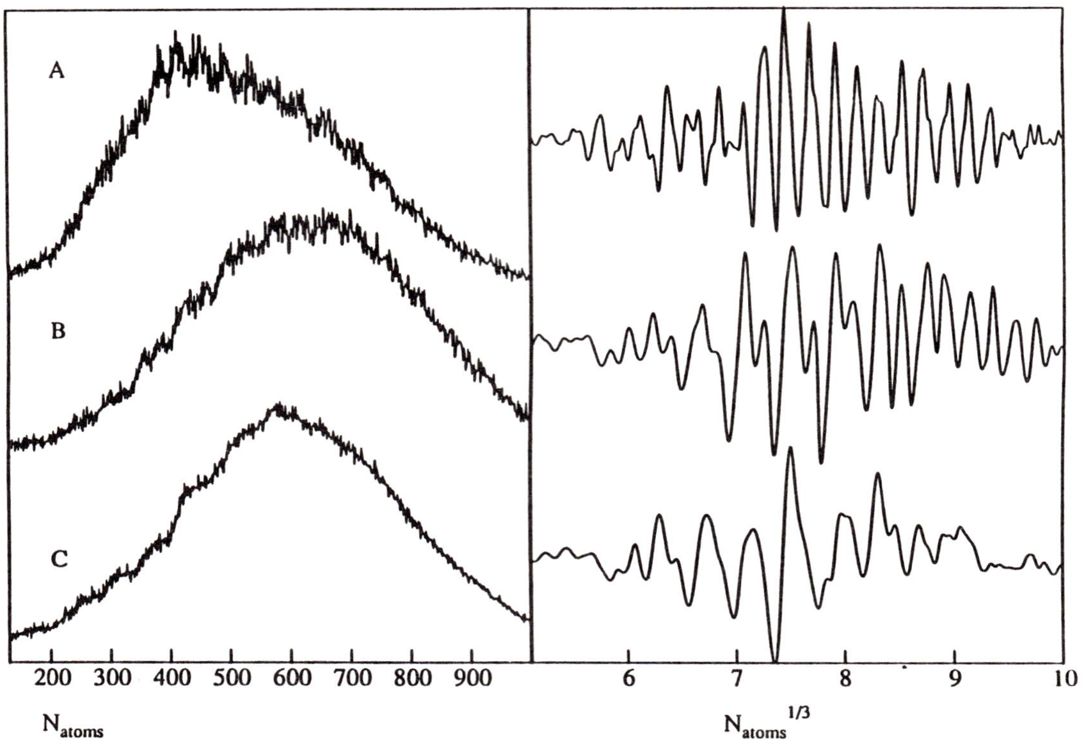

Fig. 4. On the right, aluminum mass spectra for different nozzle temperatures: A: T=473 K. B: T=373 K and C: T=273 K. On the left, some mass spectra but the bell shape envelope has been removed. Spectrum A: the fast oscillations correspond to the octahedral structure. Spectrum C: the oscillations correspond to electronic shell structure. Spectrum B: competition between the two structures.

or slightly below the nozzle temperature. The nozzle temperature in the experiment of Fig. 4C (473 K) is far below the aluminum bulk melting point (930 K). Even if we know that the melting temperature of small particles decreases when their size decreases, we cannot exclude that in our experiment we observe supercooled clusters or surface melting predicted in molecular dynamics calculation [20]. Further experiments are necessary to bring definitive statements about this problem.

Moreover as shown in Figs. 4 and 5, the electronic shell structure of aluminum clusters extendes up to 600 atoms (1800 electrons) without supershell. The beating region is also shifted in aluminum liquid droplets as in gallium. From the present stage of the experiment we can merely assume that the beating region is above 1800 electrons.

4. Conclusion

We have observed for the first time electronic shell structure in large trivalent metal clusters. In gallium, these electronic effects associated to metallic droplet model occur in usual beam conditions because the gallium melting point is quite low (300 K). In aluminum, the observation of such effects are less trivial due to the high melting point of aluminum bulk (930 K). By varying the source temperature, we are able to observe the transition between well defined geometric structure and electronic shell structure. Moreover, the quantum

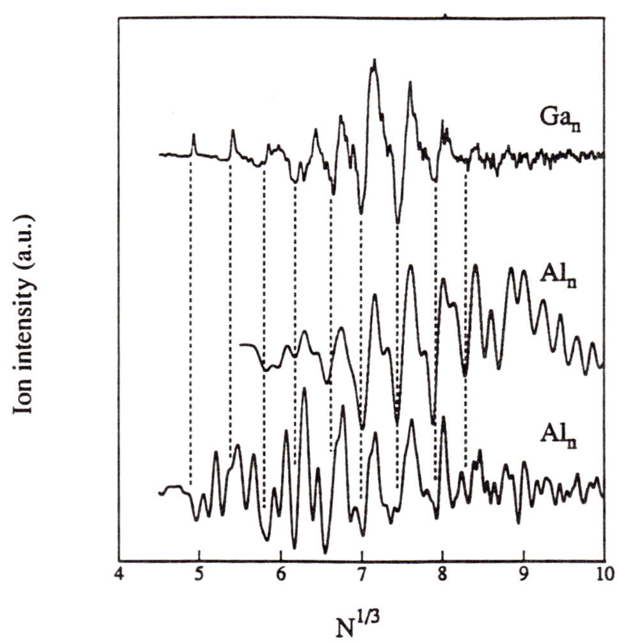

Fig. 5. Comparison between processed mass spectra of gallium and aluminum clusters (the bell shape envelope has been removed from these spectra). The same main oscillations are observed in aluminum and gallium.c

supershell corresponding to the interference between

the contributions of triangular and square classical electronic orbits is obtained in gallium clusters at about 2500 electrons. This number is considerably larger than expected from self-consistent jellium calculations. This shift may be interpreted by introducing a surface softness.

References

1 Knight, W.D., Clemenger, K., de Heer, W.A., Saunders, W.A., Chou, M.Y., and Cohen, M.L.: Phys. Rev. Lett. **52**, 2141 (1984)

2 Saunders, W.A., Clemenger, K., de Heer, W.A., and Knight, W.D.: Phys. Rev. B **32**, 1366 (1985)

3 Nishioka, H., Hansen, K., and Mottelson, B.R.: Phys. Rev. B **42**, 9377 (1990)

4 Gutzwiller, M.C.: J. Math. Phys. **12**, 343 (1971)

5 Balian, R., Bloch, C.: Ann. Phys. **69**, 76 (1971); Berry, M.V., and Tabor, M.: Proc. R. Soc. London A **356**, 375 (1977)

6 Bjørnholm, S., Borggreen, J., Echt, O., Hansen, K., Pedersen, J., and Rasmussen, H.D.: Phys. Rev. Lett. **65**, 1627 (1990)

7 Pedersen, J., Bjørnholm, S., Borggreen, J., Hansen, K., Martin, T.P., and Rasmussen, H.D.: Nature **353**, 733 (1991)

8 Martin, T.P., Bjørnholm, S., Borggreen, J., Bréchignac, C., Cahuzac, Ph., Hansen, K., and Pedersen, J.: Chem. Phys. Lett. **186**, 53 (1991)

9 Bréchignac, C., Cahuzac, Ph., Carlier, F., de Frutos, M., and Roux, J.Ph.: Phys. Rev. B **47**, 2271 (1993)

10 Schriver, K.E., Persson, J.L., Honea, E.C., and Whetten, R.L.: Phys. Rev. Lett. **64** 2539 (1990)

11 Persson, J.L., Whetten, R.L., Cheng, H.P., and Berry, R.S.: Chem. Phys. Lett. **186**, 215 (1991)

12 Lermé, J., Pellarin, M., Vialle, J.L., Baguenard, B., and Broyer, M.: Phys. Rev. Lett. **68**, 2818 (1992)

13 Pellarin, M., Baguenard, B., Broyer, M., Lermé, J., Vialle, J.L., and Perez, A.: J. Chem. Phys. **98**, 944 (1993)

14 Martin, T.P., Näher, U., and Schaber, H.: Chem. Phys. Lett. **199**, 470 (1992)

15 de Heer, W.A., and Milani, P.: Rev. Sci. Instrum. **62**, 670 (1991)

16 Martin, T.P., Bergmann, T., Göhlich, H., and Lange, T.: J. Phys. Chem. **95**, 6421 (1991)

17 Van Hardeveld, R., Hartog, F.: Surface Science **15**, 189 (1969)

18 Genzken. O., Brack, M., Chabanat, E., and Meyer, J.: Ber. Bunsenges. Phys. Chem. **96**, 1217 (1992)

19 Clemenger, K.: Phys. Rev. B **44**, 12991 (1991); Lermé, J., Bordas, C., Pellarin, M., Baguenard, B., Vialle, J.L., and Broyer, M.: submitted to Phys. Rev. B

20 Güvenç, Z.B., and Jellinck, J.: Z. Phys. D (in press)

This article was processed using Springer-Verlag TEX Z.Physik D macro package 1.0 and the AMS fonts, developed by the American Mathematical Society.

Optical response and shapes of charged sodium clusters; an analogue of the nuclear giant dipole response

J. Pedersen[1], J. Borggreen[1], P. Chowdhury[2], N. Kebaïli[3], L. Lundsberg-Nielsen[4], K. Lützenkirchen[5], M.B. Nielsen[1] and H.D. Rasmussen[1]

[1] The Niels Bohr Institute, University of Copenhagen, DK-4000 Roskilde, Denmark
[2] on leave from Yale University, New Haven, CT 06511, U.S.A.
[3] on leave from Laboratorie Aimé Cotton, CNRS II, F-91405 Orsay, France
[4] present address: Danish Inst. of Fundamental Metrology, DK-2800 Lyngby, Denmark
[5] on leave from Institut für Kernchemie, Universität Mainz, D-6500 Mainz, Germany

Received 28 June 1993

Abstract. Collective vibrations of de-localized electrons against the positive charged ionic background in sodium clusters (plasmon resonances) and the collective vibrations of protons against neutrons in nuclei (giant dipole resonances) have several features in common. However, fundamental differences exist due to differences in the two media; the nucleus is a quantum liquid whereas the metallic medium is more like a plasma with classical positive ions and quantized electrons. The similarities and differences are illustrated by results from optical response measurements for charged sodium clusters with 14 to 48 atoms.

PACS: 32.80.Fb, 36.40.+d, 33.20Kf

1. Introduction

Metal clusters are a form of matter intermediate between the atom and the world of bulk solid. For metallic droplets smaller than the mean free path of electrons (typically around a few hundred Ångstrøms, equivalent to $\approx 10^5$ atoms) the conduction electrons will move freely and independently inside a metal, but be confined by the surface of the drop. This picture of independent fermions moving in a common mean field is analogous to the shell model description of the nucleus.

This was strikingly demonstrated in 1984 by W.D. Knight et al. [1], who found energy shells in sodium clusters with 8,20,40,... valence electrons, similar to the occurrence of the noble gas atoms and magic numbers in nuclei. The existence of quantum shells has later been extended to clusters with several thousand valence electrons [2,3,4].

The independent particle motion is, however, only one aspect of a many-body fermion system. Both nuclei and clusters show pronounced collective features. Important examples are the vibration of protons against neutrons and electrons against the positive ions.

In daily life, collective oscillation of the electron gas in metal particles is experienced through spectacular optical effects seen in stained glasses where ultra fine metal particles are embedded in the glass. The absorption in the visible frequence range, due to excitation of the plasmon oscillation, gives rise to different colours for different metal particles.

Giant dipole resonances in nuclei have been known for more than forty years, and with the recent measurements of plasmon resonances [5-11] it is now possible to compare the characteristic response of the two media when absorbing photons:

A) degree of collectivity
B) resonance energy and its dependence on the number of constituent particles
C) deformation splitting; onset and relative occurrence of prolate-oblate shapes
D) damping, width of resonances.

2. Experimental technique

In our experiment, the photo absorption cross section of a cluster is determined by measuring the photo fragmentation cross section. Details of the experiment have been presented elsewhere [12], and only the most basic aspects will be repeated here (see fig. 1).

The sodium clusters are formed in a seeded expansion source and ionized by broad-band UV-light (3.0 to 4.0 eV). After acceleration to 2 keV energy, the clusters are mass selected by a Wien filter. The charged sodium cluster beam is then crossed by monochromatic light (spectral resolution $\approx$ 10nm) obtained by filtering the continuous spectrum from a 1 kW Xe-arc lamp with interference filters. In total 15 filters, ranging from 366 to 605 nm (3.4 to 2.0 eV), mounted on two wheels controlled by stepping motors, are rotated such that each filter is exposed to light for five seconds. For each wavelength the corresponding spectrum of evaporation residues (see below) is stored in its own part of the computer memory. A full cycle of rotation takes $\approx$ 100 seconds, but

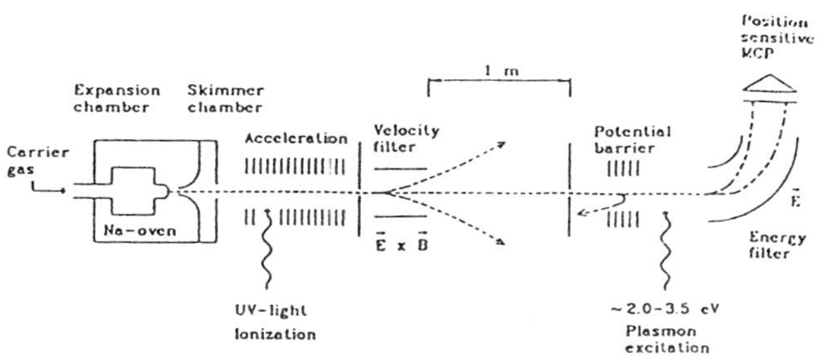

Fig. 1. Schematic drawing of the experiment

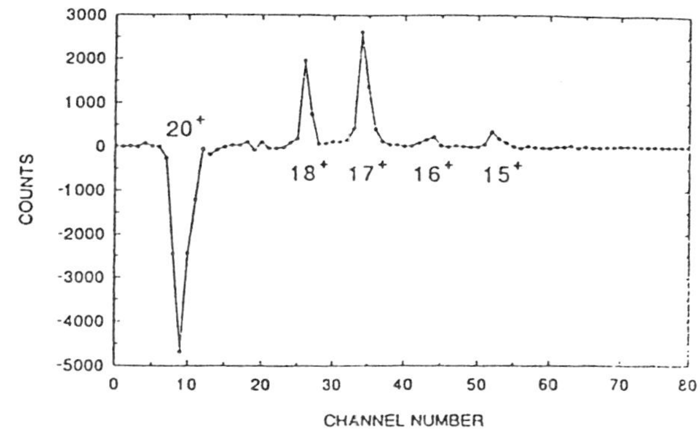

Fig. 2. Photo fragmentation spectrum of a mass selected Na_{20}^+ beam irradiated with photons of energy, $\approx$ 2.7 eV, minus the corresponding spectrum without irradiation. The energy of 2.7 eV is only enough for three evaporations, but due to monomer-dimer competition a mass loss up to 5 atoms is observed

to obtain sufficient statistical accuracy, typically more than one hundred full rotations are required. The measurements for different wavelengths are thus obtained under essentially identical conditions and the only normalization required to obtain the optical response is the overlap with the beam and the yield of transmitted light through each filter.

Photons are initially absorbed by the electron cloud of the cluster. Subsequently a rapid damping of the electron excitation into ionic motion i.e. heating, takes place. As a result the cluster dissociates by monomer (or dimer) evaporation. Thus, the kinetic energy of the daughter clusters is reduced to $E(1-x/N)$ where E and N are the energy and number of atoms of the parent cluster, respectively, and x is the number of evaporated atoms (we neglect the energy smearing due to recoil contributions).

The daughter clusters can now be identified by measuring their kinetic energies. The energy analysis is done in a cylindrical electrostatic field placed 15 cm downstream from the irradiation zone. The clusters traverse the field with different bending radii according to their energies and are detected in a position sensitive microchannel-plate.

Since some evaporation takes place in the flight path from acceleration to the plasmon excitation region, mainly due to energy acquired in the ionization process or by collisions with rest-gas molecules, one of the mass spectra is recorded without light irradiation (a non-transmitting metal plate). This spectrum is representative of the parent cluster intensity and of the background in the region of daughter clusters and is subtracted from each of the spectra obtained with light. An example of such a subtraction is shown in fig. 2.

3. Data analysis

The measured optical response for selected charged sodium clusters is shown in fig. 3. For each wavelength it is obtained by integrating the measured, background corrected intensity of all daughter clusters. It is normalized to the integrated parent intensity and to the number of filtered photons intersecting the cluster beam.

The general impression, looking at the photon absorption profiles, is that they consist of one or two broad peaks. We have performed an analysis of the data by fitting one or two Lorentz curves to the resonance shapes (neither the number of data points nor their statistical accuracy justify fits based on three or more independent components). Furthermore, we have assumed a common width in cases with two components. (Allowing for two independent widths does not increase the information gained from the fit, nor is there an improvement in terms of χ^2. It should be emphasized, however, that there is no obvious physical reason for choosing equal widths).

The fitting function has the form:

$$f(E) = \sum_{i=1}^{j} \frac{H_i W_i^2 E^2}{(E^2 - E_i^2)^2 + W_i^2 E^2} \qquad (1)$$

where H_i, W_i and E_i are the height, width and resonance energy of each component, respectively. For a single peak j=1 and for double peaks j=2 with $W_1 = W_2$. The peak resonance energies are determined to within $\pm 2\%$ (and are not affected beyond the uncertainties when six free parameters are used). The uncertainties in the widths are considerably larger, $10 - 20\%$. The result of the analysis is shown in fig.3.

4. Comparison of plasmon resonances in clusters and giant dipole resonances in nuclei

A. Collectivity.

A first indication for the collective nature of the resonances in fig.3 is given by the resonance energy. The energy is pushed to higher values than the typical single-electron excitation energy which is of the order of the shell spacing ($\approx$ 1 eV) [13]. A better way of estimating

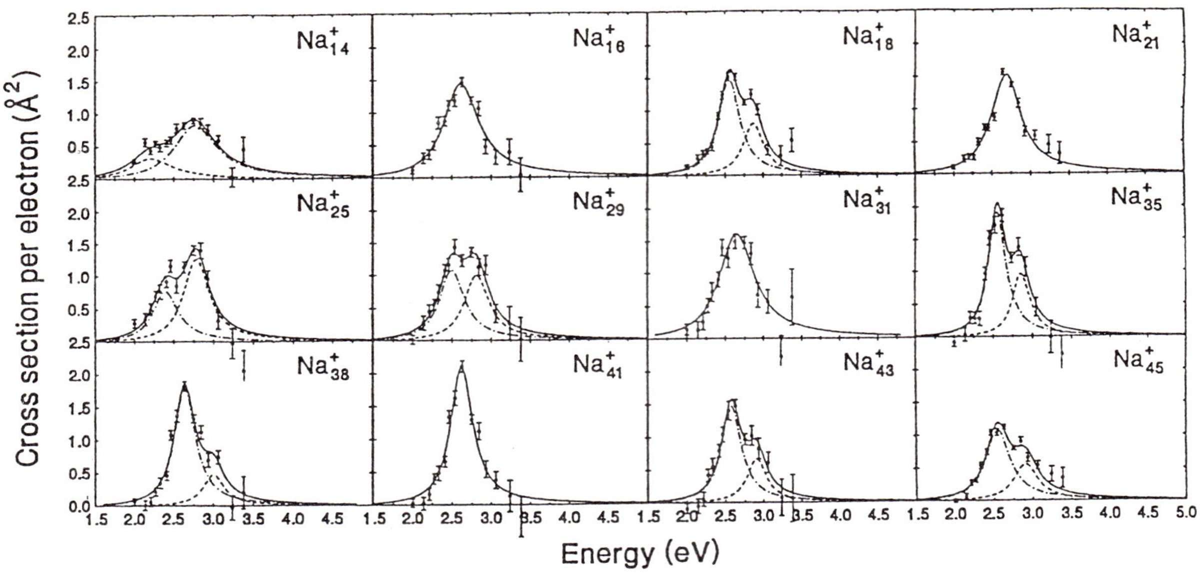

Fig. 3. Examples of the measured plasmon profiles, in absolute cross section per valence electron.The curves are the Lorentzian fits

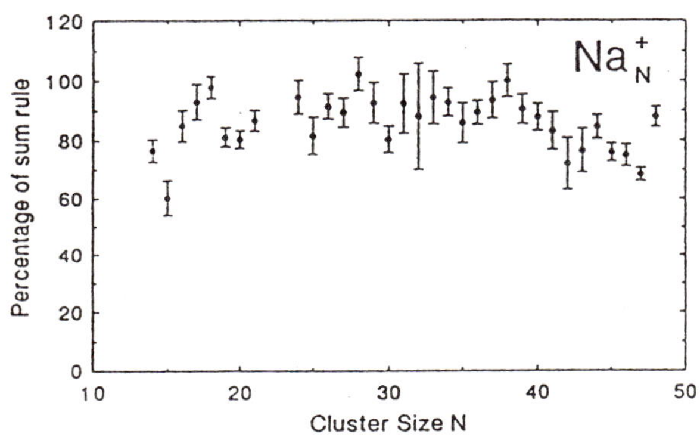

Fig. 4. The integrated cross section in percentage of the theoretical energy weighted dipole sum rule (1.0975 eV Å^2 /electron, ref.[14])

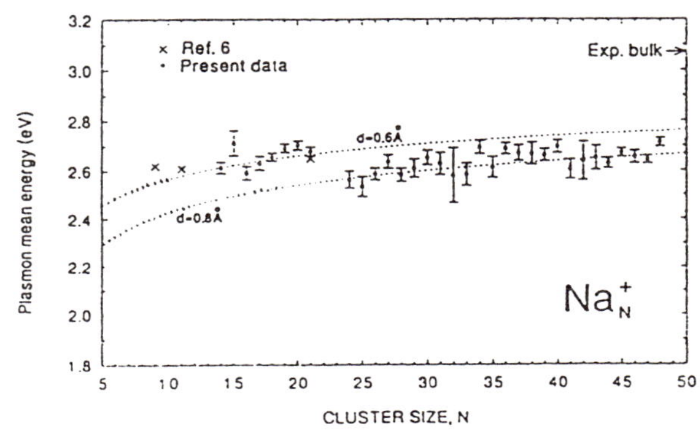

Fig. 5. Mean energy of the resonances

the collectivity is to express the measured photo absorption cross section relative to the energy weighted dipole sum rule prediction [14]. Both for nuclei and clusters the sum-rule is found, theoretically, to be proportional to the total number of relevant charged particles (protons and valence electrons,respectively). Figure 4 shows the integrated cross sections in percentage of the sum rule. Like in the nuclear case [15] the measured oscillator strength almost exhausts the theoretical value, i. e. in both cases we are dealing with collective dipole excitations.

B. Mean resonance energy.

The centroid energy $< E >$ of the plasmon absorption profile is extracted according to:

$$< E > = \frac{\sum_{i=1}^{15} E_i \sigma(E_i) \Delta E_i}{\sum_{i=1}^{15} \sigma(E_i) \Delta E_i} \tag{2}$$

where $\sigma(E_i)$ is the cross section at the energy E_i and ΔE_i is the energy interval from $(E_i + E_{i-1})\ /2$ to $(E_i + E_{i+1})\ /2$. Fig.5 shows $< E >$ as a function of cluster size. It turns out to be rather constant, $\approx$ 2.6 eV, over the measured size interval. The energy of the plasmon resonance for a sphere, with the assumption that all the oscillator strength is concentrated in a single peak, is given by [16]:

$$\omega_0^2 = \frac{Ne^2}{m\alpha} \tag{3}$$

where N is the number of delocalized electrons, m and e the mass and charge of the electron and α the electric polarizability. For a classical metal sphere, $\alpha = (N^{1/3} r_{ws})^3$, where r_{ws} is the Wigner-Seitz radius of the metal. Thus, the plasmon frequency is independent of cluster size, as opposed to nuclei. However, it is known [17] that the electron density extends beyond the jellium edge (spill-out). This implies a lowering of the resonance energy for smaller clusters because $(N^{1/3}r_{ws})^3$, should be replaced by $(N^{1/3}r_{ws} + d)^3$, where d is the spill-out. Expected variations of the resonance energy assuming two different values for the spill-out of 0.6 Å and 0.8 Å are shown as dotted curves in fig.5. Both curves converge towards the experimentally measured planar surface plasmon energy [18], transformed to bulk value for a sphere of temperature 500 K.

The energy of ≈2.6 eV and its constancy as a function of cluster size is contrasted by the results for nuclei [15]. Here the mean energy (typically 10-15 MeV) varies inversely proportional to the nuclear radius (at least for larger nuclei). The different behaviour for the two media can be related to the different restoring forces involved; in clusters it is the long range Coulomb force acting on the electrons whereas in nuclei it is the short range attractive nuclear force between neutrons and protons.

C. Deformation splitting; onset and relative occurrence of prolate - oblate shapes.

One essential property of a quantal many-body system is its shape. Fermi droplets with partly filled, highly degenerate states, break the spherical symmetry. This is well known for non-magic nuclei and is also expected for clusters. One way of measuring deformation is through the giant dipole resonance. The resulting resonance profile in nuclei shows a splitting with frequencies inversely proportional to the different lengths of the principal axes.

In clusters the analogous splitting of the resonance, due to the different polarizabilities along the principal axes, is only 3/5 of what is found for nuclei (again a demonstration of the different restoring forces in the two media).

In axially symmetric systems the resonance splits up into two components. The signature for a prolate deformation is a doubly degenerate more intense high frequency peak and conversely for the oblate deformation. The result from the analysis of the plasmon profiles is shown in fig.6. A very pronounced systematic variation between prolate and oblate shapes is observed for the open shell clusters. Furthermore, the onset of deformation occurs with just one or two particles/holes away from the filled shell. These findings are different from what is observed in nuclei. Here more particles are required in order to break the spherical symmetry and an analogous variation between prolate and oblate shapes is not found. The nuclear deformations as a function of mass are more complicated since both protons and neutrons are quantized. They do not always collaborate towards the same preferred shape.

The immediate onset of deformation for clusters versus the stability for spherical shapes of nuclei can be ascribed to pairing interaction present in nuclei but not in

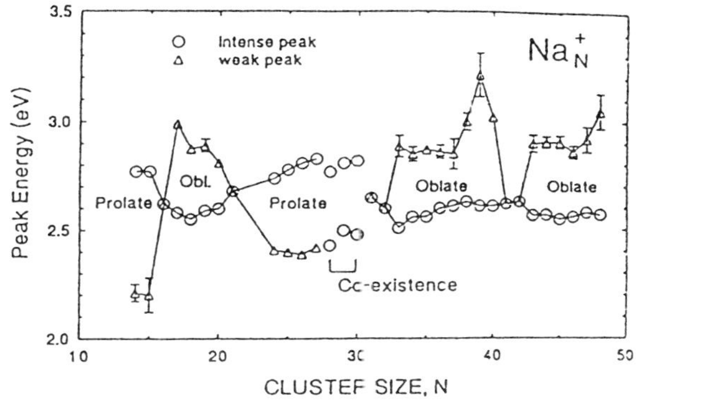

Fig. 6. Mean resonance energies as a function of cluster size. The intense (doubly degenerate) peaks shown as circles, the weaker peaks as triangles

clusters; the independent particle aspect favours shapes with a low level density above the Fermi level (Jahn - Teller effect) whereas the pairing force (scattering of paired particles into empty levels) leads to an energetic preference for a high level density around the Fermi level. Thus, in nuclei, the pairing stabilizes the spherical shape around the shell closings. The preference for prolate or oblate shape as a function of the degree of major shell filling, is an interesting and non-trivial problem. In a harmonic oscillator potential, prolate shapes are favoured at the beginning of a major shell and oblate shapes at the end. For heavy clusters numerical calculations [19,20] predict that this prolate-oblate competition is reversed. This is a feature of a single isolated l-shell [20]. The oblate shapes observed for $N > 40$ are possibly a result of the $l = 4$, g- orbital, that fills between $N = 40$ and $N = 58$, lying rather isolated. The competition between prolate-oblate shapes is furthermore affected by octupole deformations [20].

In the transition region between prolate and oblate deformations one might expect triaxial shapes. The plasmon profile for Na_{16}^+ shows a very broad unresolved resonance curve, that may be fitted with three, almost equally intense peaks. In the transition region around $N = 30$ we observe profiles with two equally intense peaks; we interprete this feature as co-existence of both prolate and oblate shapes. (This is also found theoretically [21]). The clusters Na_{31}^+ and Na_{32}^+ are again candidates for triaxiality before a more regular sequence of clusters with oblate deformation follows.

D. Damping, width of resonances.

The widths of the plasmon resonances are observed to be ≈ 15% of the resonance frequency and do not vary significantly as a function of cluster size (fig. 7).

A recent measurement [22] with a femtosecond-laser, two-photon ionization, shows a lifetime of the intermediate plasmon excitation in small clusters, corresponding to a damping width of ≈ 1 meV. This is the "true" damping width for transfer of electronic excitation en-

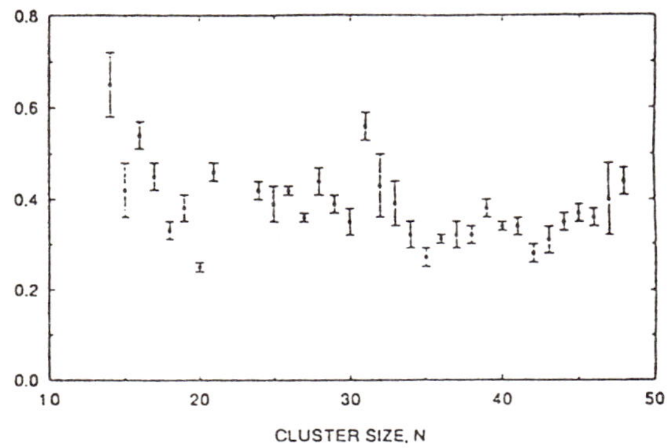

Fig. 7. The full width half maximum of the fits as a function of cluster size

ergy to the ions (i.e. the resonance oscillates $\sim 10^3$ times before it is damped out).

Theoretically two broadening mechanisms are believed to be important. The collective resonance may be fragmented as a consequence of interaction with near-lying single electron hole levels. For large numbers of valence electrons, this Landau fragmentation leads to a 1/R behaviour of the width [23] (R is the radius of the cluster). For small clusters the Landau fragmentation may appear as close-lying peaks. The Landau spreading is almost independent of temperature [23].

Another broadening mechanism is the coupling of the plasmon to surface fluctuations, which leads to a spread in frequency as the cluster changes deformation around its equilibrium shape.

The measured collective excitations, fig. 3, refer to an ensemble of rather hot clusters ≈ 450 K, i.e. above the melting point. Hence one can expect a broadening of the absorption spectrum due to contributions from several form isomers with different resonance energies. This line broadening is proportional to the square root of the temperature [13,24], but will be less significant for larger clusters. Measurements at low temperature for Na_{21}^+ and Na_{41}^+ [25] are in qualitative agreement with a $\sqrt{T}$ dependence.

In nuclei the giant dipole resonance width (for closed shell nuclei) is $\sim 15-20\%$ of the resonance frequency implying that the resonance performs few oscillations before its energy (and spin) are shared among the nucleons. For non-magic nuclei the width increases to $30-40\%$ due to extra broadening from deformation splitting [15].

Acknowledgement. The authors thank S.Bjørnholm, T.Døssing and J.J. Gaardhøje for valuable comments and stimulating discussions. This work is supported in part by USDOE Contract No DE-FG0291ER-40609, the Danish Natural Science Foundation the Carlsberg Foundation and the NOVO Foundation.

References

1. Knight, W.D., Clemenger, K., de Heer, W.A., Saunders, W.A., Chou,M- .Y., Cohen, M.L.: Phys. Rev. Lett. **52** 2141 (1984)
2. Martin, T.P., Bjørnholm, S., Borggreen, J., Bréchignac, C., Cahuzac, Ph., Hansen, K., J.Pedersen: Chem. Phys. Lett. **186** 53 (1991)
3. Pedersen, J., Bjørnholm, S., Borggreen, J., Hansen, K., Martin, T.P. Rasmussen, H.D.: Nature **353** 733 (1991)
4. Pellarin, M., Baguenard, B., Bordas, C., Royer, M., Lerme, J., Vialle, J.L.: Preprint 1992
5. Selby, K., Kresin, V., Masui, J., Vollmer, M., de Heer, W.A., Scheidemann, A., Knight, W.D.: Phys. Rev. **B43** 4565 (1991)
6. Bréchignac, C., Cahuzac, Ph., Carlier, F., de Frutos, M., Leygnier, J.: Chem. Phys. Lett. **189** 28 (1992)
7. Tiggesbäumker, J., Köller, L., Lutz, H.O., Meiwes-Broer, K.H.: Chem.Phys.Lett. **190** 42 (1992)
8. Pollack, S., Wang, C.R.C., Kappes, M.M.: J. Chem. Phys. **94** 2496 (1991)
9. Bréchignac, C., Cahuzac, Ph., Carlier, F., Leygnier, J.: Chem. Phys. Lett. **164** 433 (1989)
10. Dam, N., Saunders, W.A.: Z.Phys. **D19** 85 (1991)
11. Fallgren, H., Brown, K.M., Martin, T.P.: Z. Phys. **D19** 81 (1991)
12. Borggreen, J., Chowdhury, P., Kebaili, N., Lundsberg-Nielsen, L., Lützenkirchen, K., Nielsen, M.B., Pedersen, J. Rasmussen, H.D.: Phys.Rev. **B48** 17507 (1993)
13. Nishioka, H., Hansen, K., Mottelson, B.R.: Phys. Rev. **B42** 9377 (1990)
14. Yannouleas, C., Broglia, R.A.: Phys. Rev **A44** 5793(1991)
15. Gaardhøje, J.J.: Ann. Rev. of Nuclear and Particle Science **42** 483 (1992)
16. Mie, G.: Ann. der Phys. **25** 377 (1908)
17. de Heer, W.A., Rev. Mod. Phys. **65** 611 (1993)
18. Tsuei, K.D., Plummer, E.W., Feibelman, P.J.: Phys. Rev. Lett **63** 2256 (1989)
19. Reimann, S.M.: Thesis, Univ. of Regensburg Aug.1992
20. Hamamoto, I., Mottelson, B., Xie, Hong-xing, Zhang, Xi zhen: Z. Phys. **D21** 163 (1991)
21. Frauendorf, S., Pashkevich, V.V.: Z. Phys. D (1993) in press
22. Gerber, G.: Europhys. News **23** 154 (1992)
23. Yannouleas, C., Broglia, R.A.: Ann. of Phys. **217** 105 (1992)
24. Bertsch, G.F., Tománek, D.: Phys. Rev. **B40** 2749 (1989)
25. Ellert, C., Haberland, H., Orlik, W., Reiners, Th., Schmidt, M.: preprint (1993)

This article was processed using Springer-Verlag TEX Z.Physik D macro package 1.0 and the AMS fonts, developed by the American Mathematical Society.

Z. Phys. A 349, 213–218 (1994)

The giant dipole resonance photons and the shapes of hot Er and Sn nuclei

A. Bracco, F. Camera

Dipartimento di Fisica, Universitá di Milano and INFN sez. Milano, via Celoria, 16, I-20133 Milano, Italy

Received: 27 October 1993

Abstract. A study of the spectral and angular distributions of the high energy γ rays emitted by the giant dipole resonance (GDR) in the compound nuclei ^{165,167}Er and ^{110}Sn at a variety of excitation energies ($E^* = 50$–90 MeV) is presented. A comparison of the data with model predictions at the equilibrium deformation and with thermal fluctuations in the adiabatic limit is discussed. In general, the adiabatic calculations give a fair account of the data exception made for Sn at very high spins.

PACS: 24.30.Cz; 23.20.−g; 25.70.Gh; 27.60.+j

1. Introduction

The spectroscopy of the high energy photons emitted by the giant dipole resonance (GDR) has, in the course of the past decade, established itself as a powerful tool to study the properties of hot rotating nuclei offering the capability to probe the entire phase space available to the nuclear decay. The coupling of the giant dipole resonance to the quadrupole deformation is strongly affecting the dipole response and therefore the investigation of the GDR lineshape and angular distribution allows to obtain information on the changing of nuclear shapes as the excitation energy and angular momentum of the emitting nuclei are varied.

A topic of current interest is the detailed study of the region of excitation energy ranging from close to the yrast line, dominated by quantal effects, to the region where nuclei can be described by models based on classical and macroscopic concepts as in the case of the liquid drop model. The transition between these two regimes is thought to be associated to a transition of shapes from the ones determined by shell effects (prolate or spherical) to the ones of deformable classical objects (oblate). However, this transition region cannot be easily mapped due to the important role of thermal fluctuations [1–6] in describing hot systems with a finite number of particles.

Because of fluctuations, the experimental GDR observables can be interpreted only through the comparison with model predictions computed from a proper average over all possible nuclear shapes and orientations [7–10]. In addition, the coupling of the GDR to the quadrupole nuclear deformation is controlled by the relative time scales associated to the adjusting of the GDR response to fluctuations and to the fluctuations themselves. The times controlling fluctuations are unknown and are expected to be dependent on temperature and rotational frequency.

The experimental studies addressing the problem of the determination of the nuclear shapes at finite temperature and rotational frequency and of their fluctuations are based on the measurements of the photon spectra emitted in heavy ion fusion reactions leading to compound nucleus formation. The compound nucleus is formed with a distribution of spins that can be rather wide and this can mask spin effects in the dipole response.

In this paper we present some of the most recent studies of the strength function and of the angular distribution of the GDR associated to selected and rather narrow intervals of spin. The data were obtained with the experimental setup described in references [11, 12] and the compound nucleus reactions used are given in Table 1.

It should be noted that angular distributions are of particular interest since so far their measurements were very scarce and provide a stringent test to the fluctuation

Table 1. In this table the incident laboratory energy E_k, the excitation energy E^* and the average angular momentum of the compound nucleus associated to the fusion cross section $\langle J\rangle_{\rm fus}$ are given for all the reactions used in the present study

Reaction	E_k (MeV)	E^* (MeV)	$\langle J\rangle(\hbar)_{\rm fus}$
^{17}O + ^{150}Nd → ^{167}Er	85	61	~16
^{17}O + ^{148}Nd → ^{165}Er	70	45	~32
^{48}Ti + ^{62}Ni → ^{110}Sn	223	92	~40

models. Two subjects are discussed, namely *i)* the different sensitivity of the GDR strength function and angular distributions to fluctuations at low rotational frequency and *ii)* the rotational frequency dependence of the GDR width and angular anisotropy at higher rotational frequencies. For the first point we refer to data obtained for the ^{165,167}Er nucleus while for the second to data on ^{110}Sn. The possibility of using the two combined experimental observables to learn also on the spin and temperature dependence of the intrinsic spreading width is also discussed.

2. Fluctuation models for hot nuclei

The thermal fluctuation models, extensively discussed in [7–10], make use, among their basic ingredients, of the Boltzmann factor P describing the probability for a nuclear shape in terms of the deformation and orientation parameters (β, γ, Ω),

$$P(\beta, \gamma, \Omega, \omega, T) = Z(T, \omega)^{-1} \exp[-F(\beta, \gamma, \Omega, \omega, T)/T] \tag{1}$$

In the above expression F is the free energy, T the nuclear temperature, ω the rotational frequency and $Z(T, \omega)$ the partition function.

In the limiting case of adiabatic coupling the stretched components, $f_{\pm 1}(E_\gamma)$, (associated to dipole vibrations orthogonal to the spin axis) and the unstretched component, $f_0(E_\gamma)$, (associated to dipole vibration parallel to the spin axis) of the giant dipole resonance are calculated as follows (see (2)). For each value of the deformation (β, γ) and orientation (Ω) parameters one constructs the Lorentzian functions $L_k(E_\gamma, \beta, \gamma, \Omega)$ $(k=-1, 0, 1)$, describing a dipole vibration along the k axis, with centroid E_{GDR}, determined with the Hill-Wheeler parameterization, and width given by $\Gamma_{GDR} = \Gamma_0 (E_{GDR}/E_0)^\delta$. E_0 and Γ_0 are the GDR centroid energy and the GDR width before the thermal coupling to the shape and $\delta = 1.9$. The width Γ_0 represents the damping due to the coupling of the correlated $1p-1h$ state, describing microscopically the GDR state, to more complicated $np-nh$, states, the doorway states being the $2p-2h$ states. Small amplitude quantal fluctuations are taken into account to obtain the value of Γ_0 [13]. In the calculations, Γ_0 was kept constant at the different spins. The averaging in the adiabatic limit is made integrating over the whole shape and orientation distributions:

$$f_k(E_\gamma) = \int L_k(E_\gamma, \beta, \gamma, \Omega)\, P(\beta, \gamma, \Omega, \omega, T)\, d\tau \quad k=-1, 0, 1 \tag{2}$$

where $d\tau$ is $\beta^4 |\sin(3\gamma)|\, d\gamma\, d\beta\, d\Omega$ [7]. From these components the total strength function $(\Sigma f_k(E_\gamma))$, is constructed together with the quadrupole coefficient a_2 in terms of the unstretched (f_0) and stretched $(f_{\pm 1})$ cross sections:

$$a_2(E_\gamma) = \frac{1}{2} \frac{f_0(E_\gamma) - 0.5(f_1(E_\gamma) + f_{-1}(E_\gamma))}{f_0(E_\gamma) + f_1(E_\gamma) + f_{-1}(E_\gamma)} \tag{3}$$

Contrary to the strength function the $a_2(E_\gamma)$ depends on orientation fluctiations which produce an attenuation of the angular anisotropy that is particularly large at low spins. Therefore, measurements of the a_2 at low rotational frequencies constitute an ideal testing ground for the model of orientation fluctuations.

3. The angular distributions of Er

Some of the measured angular distributions of the gamma ray in the GDR region for Er are shown in Fig. 1 in comparison with model predictions at the equilibrium deformation and in two adiabatic scenarios, one including only shape fluctuations and the other accounting for shape and orientation fluctuations. In doing this comparison, the calculations have been shifted by 0.5 MeV to account for the effect of the response function of the high energy detectors.

The present experimental results are characterized by the rather small quadrupole coefficient $a_2(E_\gamma)$ that does not change even when the average value of the spin distribution of the decaying compound nucleus is doubled (from $16\hbar$ to $32\hbar$).

A similar rather stable behaviour with spin can be also found for both calculations at the equilibrium deformation[1] (see Fig. 1) and including shape average. In the case of the equilibrium deformation [14] a possible explanation is that at $T \leqq 1.3$ MeV shell effects are

[1] To be noted that calculations at the equilibrium deformation provide an approximate estimate of the effect of time dependence of the fluctuations in the limit where the hopping frequency in the orientation and shape fluctuations is infinite

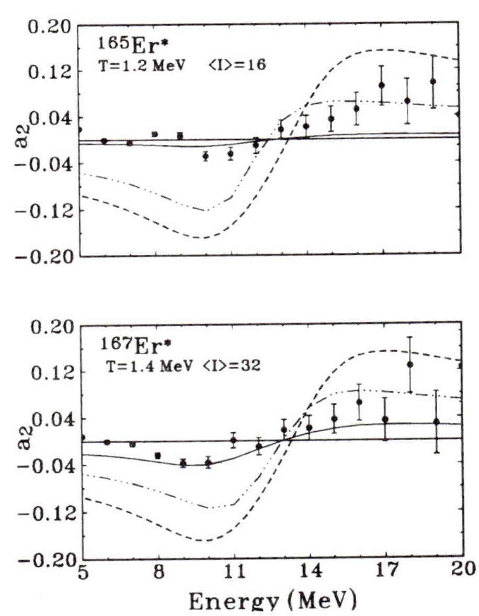

Fig. 1. The a_2 coefficient as a function of the transition energy of the gamma rays emitted by ^{165,167}Er nuclei, as obtained in the present study. The temperatures were deduced from the excitation energies given in Table 1 while the average spin is associated to the chosen intervals of coincidence fold measured with the multiplicity filter. The curves correspond to: equilibrium deformation (*dotted-dashed lines*), adiabatical model with shape fluctuations (*dashed lines*) and with shape and orientation fluctuations (*full drawn lines*)

still important and rotation is not able to induce drastic shape changes as it does at higher temperatures when shell effects are negligible. The adiabatic calculations accounting only for shape average suggest that also the average shape to which the GDR couples is not affected by spin in the spin and temperature intervals sampled by the present study. The calculated anisotropies including adiabatic shape fluctuations are larger than the predictions for the equilibrium because the term β^4 contained in the integration volume element gives larger weight to large deformations.

The experimental results show pronounced differences as far as the size of the anisotropy is concerned, with respect to both predictions for the equilibrium deformation and including only shape averaging but can be rather well reproduced with the additional inclusion of orientation fluctuations (cfr. Fig. 1, continuous line). Only in the case of the measurements at $I = 16\hbar$ the absolute values of the calculation slightly underestimate the data. The present result suggests that the wobbling of the nucleus around the spin axis is causing that the dipole resonance sees an almost spherical nucleus when I is in the interval 16–32$\hbar$ resulting in very small angular anisotropies.

The effective deformations extracted from the angular distribution of the Er data are shown in Fig. 2 together with the deformations calculated from the centroid energies deduced from the statistical model analyses reported in [15, 16]. The nuclear deformation can be calculated from the centroid energies of the two fitted components, $\beta = 1.05(d-1)\,d^{1/3}$ with $E_b/E_a = 0.911\,d + 0.089$ where b stands for the symmetry axis and a for the perpendicular axes [17]. The deformations (β) obtained from the spectral distribution analyses (filled diamonds) are in rather good agreement with the equilibrium deformations (continuous line) and this indicates that the values of the β parameter is not strongly affected by thermal fluctuations in contrast with the deformations deduced from the angular distributions (filled circles) that are a factor of five smaller than the equilibrium ones.

The comparison of the experimental effective deformations deduced from the spectral and $a_2(E_\gamma)$ distributions confirms the model predictions (see discussion in section 2): the sensitivity to thermal fluctuations, particularly orientation, is different for the two observables. Due to the fact that orientation fluctuations are not changing the dipole strength function, one couldn't find consistency among the deformations deduced from the total integrated spectrum and from the angular distribution.

4. The strength function and angular distribution for Sn

In the case of the Sn nucleus we present new data, recently obtained, that have allowed to study the spin dependence of both the width and angular distribution. Several studies [1, 6, 17, 21] were made in this mass region by different groups and although the measurements were associated to the entire spin distribution of the fusion reaction, spin effects were indicated as responsible of the found width increase. The measurements here discussed give the GDR properties at the same temperature and different spins and represent the first experimental attempt to disentangle spin and temperature effects. In Fig. 3 some of data and statistical model calculations are shown in a linearized form that helps to judge the quality of the fits (full drawn lines). The calculation was obtained with a version of the CASCADE [19] program modified to include the gamma decay in competition with particle evaporation. The values of the width from the present analyses are plotted in Fig. 3 as a function of the average spin of the distribution associated to a given coincidence fold in the multiplicity filter. The data, shown as filled circles, were obtained fitting, with a χ^2 minimization method, the measured spectra in the γ transition energy region, relevant to the GDR study, from 9 to 19 MeV with a Lorentzian function assuming 100% of the EWSR strength. It was also assumed that the initial compound nucleus had only one value of spin that was chosen to be equal to the average spin associated to each coincidence fold. The value of the centroid energy for all the measurements was 15.0 MeV ± 0.5. For the level density parameter the value $a = A/8$ was chosen since it gives the best value of the centroid energy and because it has been widely used in the previous statistical model analyses of the existing experiments. The calcula-

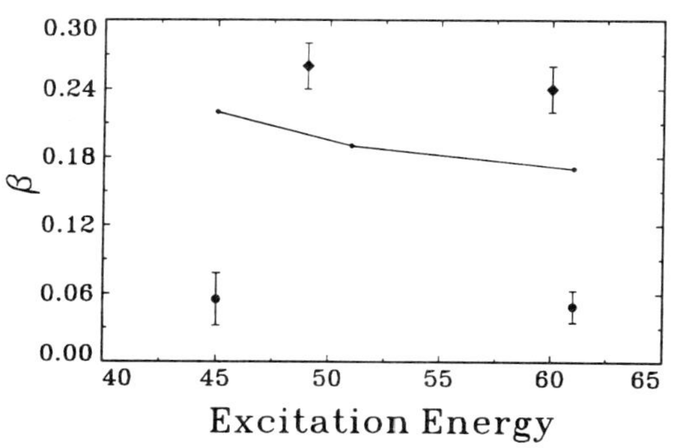

Fig. 2. Values of the β deformation parameter deduced from the present angular distributions (*filled circles*) and from the strength functions of [15] and [16] (*filled diamonds*). The equilibrium shape in ^{166}Er is shown with the *full drawn line*

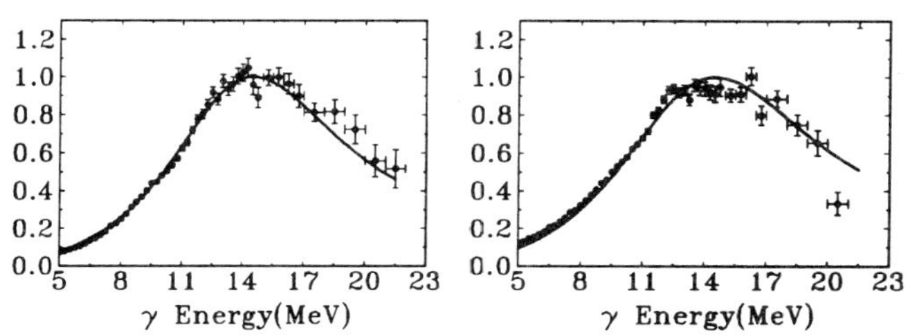

Fig. 3. Measured spectra multiplied by $(Y_\gamma^{exp}(E_\gamma)/Y_\gamma^{cal}(E_\gamma))$ $f(E_\gamma, E_{GDR}, \Gamma_{GDR})$ where $Y_\gamma^{cal}(E_\gamma)$ and $Y_\gamma^{exp}(E_\gamma)$ are the calculated, with a statistical model, and measured γ yields at transition energies E_γ and $f(E_\gamma)$ is the Lotentzian function associated to the values of E_{GDR} and Γ_{GDR} giving the best fit (*full drawn curve*). Data and calculations are for the decay of the ^{110}Sn compound nucleus at $E^* = 92$ MeV for different spin windows, the top one associated to the average value of 42$\hbar$ and the bottom one to 50$\hbar$

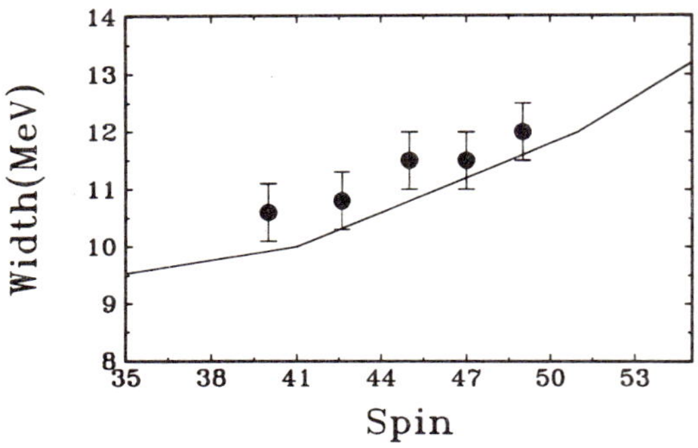

Fig. 4. Experimental widths of the GDR as a function of the compound nucleus spin at a temperature $T \sim 2$ MeV. The *full drawn line* curve gives the width predicted by the adiabatic model of thermal fluctuations

tions were folded with the detector response function calculated with the program GEANT.

With the data of the present experiment, that are the only ones so far measured at a fixed nuclear temperature, the expected increase of the width due to spin effects is now finally confirmed. Moreover, this increase is rather well reproduced by calculations of thermal fluctuations of shape and orientation in the adiabatic regime (see Fig. 4). However, it should be pointed that [20] the calculated lineshapes and the ones obtained from statistical model analyses have approximately the same FWHM but different tails and it is a difficult task to measure the tails with accuracy.

Simultaneous measurements at finite T and ω of both the width of the GDR line-shape and the $a_2(E_\gamma)$ coefficient should help to disentangle the effects due to large amplitude thermal fluctuations and to small amplitude quantal fluctuations, the latter being predicted to be temperature independent [13]. This point is explained with the help of the pedagogical example shown in Fig. 5. Note that an increase of the FWHM from 5.5 to 6.5 MeV can be achieved by increasing either the intrinsic width Γ_0 or the effective deformation of the nucleus. While the difference in the line shape associated to the two cases is such that it would be very difficult to determine it precisely with experiments, the difference in the $a_2(E_\gamma)$ is rather large.

The comparison of the expermimental widths with calculations made at a fixed Γ_0 indicates that the increase of the total width is due to an increase of the effective deformation. Examining Fig. 6, where some of the measured $a_2(E_\gamma)$ are shown (top part) together with the minimum a_2 value (bottom part), one finds indeed a strong

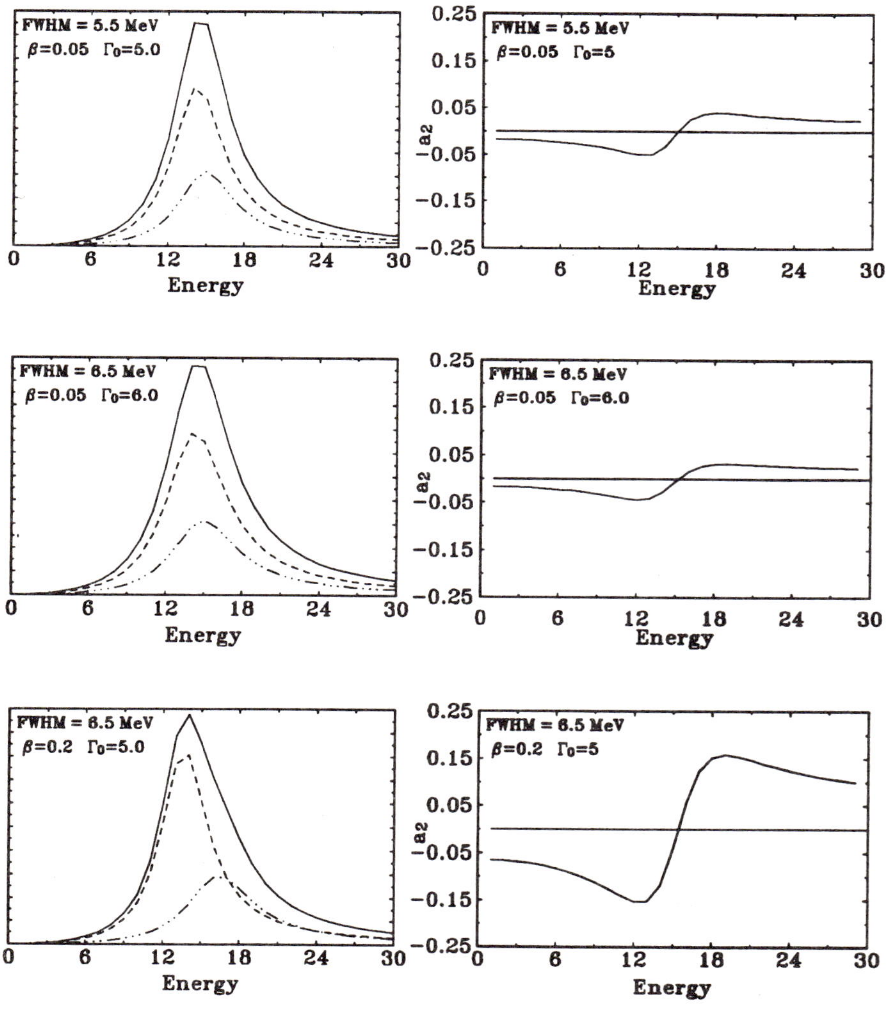

Fig. 5. Strength function (*left hand side*) and angular distribution coefficient $A_2(E_\gamma)$ (*right hand side*) of the GDR for different values of the FWHM (Γ_{GDR}), of the deformation parameter β and of the intrinsic spreading width (Γ_0). The *dashed* and *dash-dotted lines* show the stretched and unstretched components of the strength function, respectively

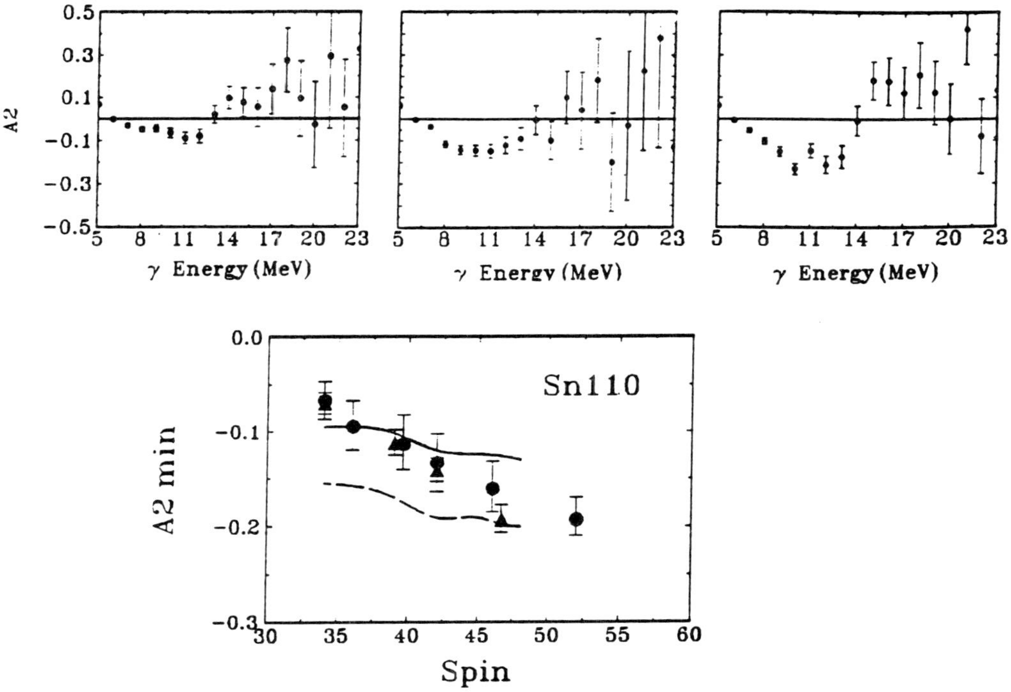

Fig. 6. The *upper part* of the figure shows the experimental angular distribution coefficient $a_2(E_\gamma)$ for three different coincidence folds (10,14 and >17, respectively) measured for ^{110}Sn. The *lower part* shows the minimum a_2 from experiments and theory. The *full circles* display the new data while the full triangles the data of [21]. The *dashed curve* was obtained from calculations at the equilibrium deformations [11] while the *full line* takes into account shape and orientation thermal fluctuations

increase of the angular anisotropy with spin again in agreement with increased effective deformations. Furthermore, the sensitivity of the $a_2(E_\gamma)$ observable to the nuclear deformation is, at this temperature and in this spin region, much larger than that of the strength function (crf. Figs. 4 and 6).

Another interesting conclusion that we have obtained with the angular distribution data is that the strong increase of the minimum value of $a_2(E_\gamma)$ is not fully reproduced by the adiabatic model (crf. Fig. 6). The fact that this was not so evident in the case of ^{162}Yb at the same spin value [11] suggests that, in addition to temperature, the physical quantity affecting the dipole response is the rotational frequency ω and not the spin.

Since the discrepancy of the adiabatic predictions with the data could indicate the presence of new effects like motional narrowing it is important to investigate the $a_2(E_\gamma)$ at these high rotational frequencies also in lighter systems.

Conclusion

The angular distribution of the γ rays emitted by the giant dipole resonance in hot ^{165,167}Er, and ^{110}Sn compound nuclei at moderate excitation energies ($E^* \leqq 100$ MeV) has been studied to probe nuclear shapes and orientations and their fluctuations.

In all cases fluctuations of shape and orientation were found to play an important role. In the case of Er, studied at moderate rotational frequency ($I \leqq 30\hbar$), the strong attenuation of the measured angular anisotropy is consistent with large effects of the orientation fluctuations as predicted by the adiabatical model.

In the case of Sn the strength function and the angular distribution at the fixed nuclear temperature $T = 2$ MeV were measured and studied as a function of the spin of the compound nucleus. The increase of the width and of the angular anisotropy of the GDR with spin in the 30–60$\hbar$ region is consistent with larger nuclear effective deformations. However, while in the case of the GDR width the increase is well reproduced by the adiabatic model the same is not true for the $a_2(E_\gamma)$. This point should be investigated with more detail and measurements of the $a_2(E_\gamma)$ at the same and higher rotational frequencies should be made. The exploration of the extreme limits of both the temperature and rotational frequency domains with the strength function and the angular distribution of the GDR will help the understanding of the models of quantal and thermal fluctuations and of the time scales involved in such processes.

We like to thank our collaborators for their very productive activity: J.J. Gaardhøje (NBI), A. Maj (Kracow), B. Million (Milano), M. Mattiuzzi (Milano), M. Pignanelli (Milano), C. Volpe (Milano), Z. Zelazny (NBI), and T. Tveter (NBI).

References

1. Gaardhøje, J.J., Ellegaard, C., Herskind, B., Diamond, R.M., Delaplanque, M.A., Dines, G., Macchiavelli, A.O., Stephens, F.S.: Phys. Rev. Lett. **56**, 1783 (1986)
2. Kicinska-Habior, M., Snover, K., Behr, J.A., Gosset, C.A., Gundlach, J.H., Feldman, G.: Phys. Rev. C **45**, 569 (1992)

3. Thirolf, M., Habs, D., Schwalm, D., Fischer, R.D., Metag, V.: Nucl. Phys. A **482**, 93c (1988)
4. Stolk, A., Balanda, A., Harakeh, M.N., Hesselink, W.H.A., Penninga, J., Rijneveld, H.: Nucl. Phys. A **505**, 241 (1989)
5. Bruce, A.M., Gaardhøje, J.J., Herskind, B., Chapman, R., Lisle, J.C., Khazaie, F., Mo, J.N., Twin, P.J.: Phys. Lett. B **215**, 237 (1988)
6. Chakrabarty, D.R., Sen, S., Thoenessen, M., Alamanos, N., Paul, P., Schicker, R., Stachel, J., Gaardhøje, J.J.: Phys. Rev. C **36**, 1886 (1987)
7. Ormand, E.W., Bortignon, P.F., Broglia, R.A., Dossing, T., Lauritzen, B.: Phys. Rev. Lett. **64**, 2254 (1990)
8. Ormand, E.W., Camera, F., Bracco, A., Maj, A., Million, B., Bortignon, P.F., Broglia, R.A.: Phys. Rev. Lett. **69**, 2905 (1992)
9. Alhassid, Y., Bush, B.: Phys. Rev. Lett. **65**, 2527 (1990)
10. Alhassid, Y., Bush, B.: Nucl. Phys. A **514**, 434 (1990)
11. Maj, A., Gaardhøje, J.J., Atac, A., Mitarai, S., Nyberg, J., Virtanen, A., Bracco, A., Camera, V., Million, B., Pignanelli, M.: Nucl. Phys. A **571**, 185 (1994)
12. Camera, F.: Ph. D. Thesis, University of Milano 1992
13. Bortignon, P.F., Broglia, R.A., Bertsch, G., Pacheco, G.: Nucl. Phys. A **460**, 149 (1986)
14. Bracco, A., Camera, F., Million, B., Pignanelli, M., Maj, A., Gaardhøje, J.J.: Nucl. Phys. A **557**, 237c (1993)
15. Gaardhøje, J.J., Ellegaard, C., Herskind, B., Steadman, S.G.: Phys. Rev. Lett. **53**, 148 (1984)
16. Gosset, C.A., Snover, K.A., Behr, J.A., Feldman, J., Osborne, J.L.: Phys. Rev. Lett. **54**, 1496 (1985)
17. Danos, M.: Nucl. Phys. **5**, 23 (1958)
18. Bracco, A., Gaardhøje, J.J., Bruce, A.M., Garret, J.D., Herskind, B., Pignanelli, M., Barneoud, D., Nifenecker, H., Pinston, J.A., Ristori, C., Schussler, F., Bacelar, J., Hofmann, H.: Phys. Rev. Lett. **62**, 2080 (1989)
19. Puhlhofer, F.: Nucl. Phys. A **280**, 267 (1977)
20. Bracco, A., Camera, F., Mattiuzzi, M., Million, B., Pignanelli, M., Volpe, C., Gaardhøje, J.J., Maj, A., Zelazny, Z., Tveter, T.: Proceeding of the Second European Workshop in Nuclear Physics, Megeve 1993 (in press)
21. Camera, F., Bracco, A., Million, B., Pignanelli, M., Maj, A., Gaardhøje, J.J.: Phys. Lett. **293** B, 18 (1992)

Z. Phys. A 349, 219–222 (1994)

RPA in nuclei and metal clusters

P.-G. Reinhard[1], **S. Weisgerber**[1], **O. Genzken**[2], **M. Brack**[3]

[1] Universität Erlangen, Institut für Theoretische Physik, Staudtstrasse 7, D-91058 Erlangen, Germany
[2] C.E.N.G., F-38041 Grenoble, France
[3] Universität Regensburg, D-93053 Regensburg, Germany

Received: 27 October 1993

Abstract. We discuss and compare the gross features of resonance excitations in nuclei and metal clusters. We point out the phenomenon of "jellium scaling" which means that different materials for metal clusters all give similar resonance spectra and we discuss the various effects which determine the exact position of the resonance.

PACS: 21.60.Jz; 36.40.+d

The excitation spectra of many-Fermion systems usually display strong resonances, e.g. the giant resonances in nuclei, the plasmons in metal clusters, or zero-sound in liquid ^{3}He. These resonances are key signatures for understanding the dynamics of a systems. They are to a good approximation harmonic vibrations and thus can be described very well in the Random-Phase-Approximation (RPA), the microscopic theory of small-amplitude vibrations of a many-Fermion system. It is the aim of this contribution to discuss and compare the gross features of the RPA excitation spectrum of nuclei and metal clusters.

The RPA belongs to the class of mean-field models which becomes obvious by the fact that it can be derived as the small-amplitude limit of time-dependent Hartree-Fock. It represents a vibration on top of a Hartree-Fock ground state. Almost all many-Fermion systems have strong correlations which invalidate a mean-field treatment as ab initio approach. The mean-field treatment requires effective Hamiltonians, a procedure which is known in atomic physics as the Kohn-Sham scheme. We use for the electrons of metal clusters the well known functional of Gunnarson and Lundquist [1]. The ions are treated in the jellium approximation yielding a homogenous positive background which is distributed over a sphere of radius $r_s N^{1/3}$. For nuclei we employ the Skyrme energy functional, here in particular the parametrization "Skyrme M*" [2].

There is a wide variety of schemes to derive and compute RPA. We prefer a formulation in terms of an algebra of $1ph$ operators, coordinate-like $\hat{Q}_\alpha$ and momentum-like $\hat{P}_\alpha$, which can be optimized in any given subspace or operators. This allows to embedd all conceivable approximations, as e.g. sum rule approach or fluid dynamics, into one and the same scheme by an appropriate choice of the basis operators. For example, the choice of simply one basis state $\hat{Q}_\alpha = r^L Y_{LM}$ yields the simple sum rule approach, free variation of the radial dependence in $\hat{Q}_\alpha = f(r) Y_{LM}$ yields the so called local RPA [3] which is equivalent to an irrotational fluid dynamics, and a choice amongst all possible $1ph$ excitations a^+a restores the full RPA. Besides this flexible switching between various levels of approximation, the scheme provides a very efficient treatment of full RPA in large configuration spaces by a mixing of detailed $1ph$ states in the region of interest with local excitations over a wide range of excitation energies. For details of the basic formulation see [4], an application to nuclei is discussed in [5], droplets of liquid ^{3}He are considered in [6], and metal clusters will be presented in [7]. We summarize here quickly the results from a comparision of different approaches: the sum rule approximation yields generally an excitation energy which is to high because it forces all strength into one peak, the local RPA provides an appropriate distribution of the overall strength distribution, and fluid dynamics with vorticity gives no improvement compared to local RPA ($\equiv$ irrotational fluid dynamcis) and produces sometimes even rather misleading details. Altogether, the local RPA is the method of choice; it is inexpensive and it gives the right global pattern of the distribution of strength. For example, the local RPA displays clearly the distinction between a surface Plasmon mode (the so called Mie plasmon) and a volume plasmon mode. It does not reproduce, of course, the detailed splitting over the nearby $1ph$ spectrum. But at least the fragmentation width can be estimated consistently within the scheme delivering strength functions which are in the average very close to those of full RPA.

In order to get an overview, one needs a compressed way to visualize the spectral properties. We do this by drawing the most important spectral information together in one plot, the energies of the resonance modes in local RPA, the unperturbed $1ph$ energies, and the particle continuum threshold. A comparison of such spectral properties for nuclei and neutral Na clusters is shown in Fig. 1. The thick vertical line in case of nuclei is to indicate that we have only access to a limited sample of nuclei whereas metal clusters allow, in principle, to study any size up the bulk limit. This

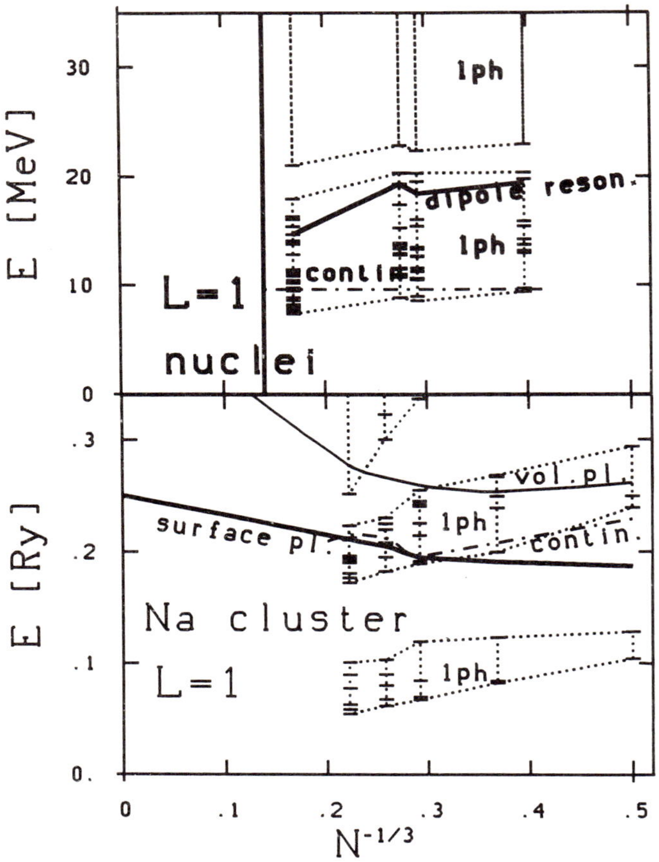

Fig. 1. Spectral properties for L=1 modes in nuclei (*upper part*) and neutral Na clusters (*lower part*) versus $N^{-1/3}$. The thick line is the dominant resonance, the *full line* show the second resonance (the volume plasmon), the horizontal bars indicate the $1ph$ energies, and the *dotted lines* embrace the various bands of $1ph$ energies. The lowest band consists out of transitions over one shell and the next over 3 shells. The particle continuum threshold (neutrons in case of nuclei, electrons for Na clusters) is shown by a *dash-dotted line*

is particularly important for semiclassical expansion which improve with increasing particle number and are usually not converged for the particle numbers provided by nuclei. Both cases display a strong resonance, the dipole giant resonance in nuclei and the surface plasmon in Na clusters. This resonance has a similar structure of collective flow, beeing close to the picture of two charged spheres which oscillate against each other (the so called Goldhaber-Teller mode). But the trends in the resonance energies differ, deacreasing with N for nuclei and slightly increasing for Na clusters. This is due to the range of the force. The short range forces in nuclei deliver collective modes as zero sound with an excitation energy

$$\hbar\omega \propto q \tag{1}$$

where q is the transferred momentum. The dominant momentum in nuclei is $q \propto 1/R$ and $R \propto N^{1/3}$ yielding the observed decrease approximately as $\hbar\omega \propto N^{-1/3}$. The Coulomb force in Na clusters, on the other hand, is of long range delivering the collective excitations as surface plasmon modes with

$$\hbar\omega = \hbar\omega_{surf} + cq^2 \tag{2}$$

where $\omega_{surf} = \sqrt{e^2\rho_{el}/3m}$ is the Mie frequency of the surface plasmon and c is some small constant. Thus the plasmon energy should be essentially constant over the systems which is approximately the case. The bend down observed in Fig. 1 is due to surface effects not contained in the estimate (2). The $1ph$ bands look similar in both cases. They decrease with N as they should. But the bands are noticeably broader for nuclei. This is due to the large spin-orbit splitting in nuclei. The behaviour of the resonances relative to the bands is much different. The nuclear dipole resonance shares the trend of the $1ph$ bands and stays always within the bands near the upper bound. This explains that we observe a fragmented resonance in every nucleus, a visible $1ph$ splitting in the light nuclei and a broad Landau damping in heavy nuclei. The resonance in Na clusters moves against the direction of the $1ph$ bands. This gives rise to strong changes of the resonance pattern with the particle number. There is a clean collective peak in Na_8 where the resonance falls nicely into a band gap, there is a transitional region between Na_{20} and Na_{40} where we see indeed isolated splittings due to interference with one or few $1ph$ styates at the lower bound of the $1ph$ band, and there is a broad Landau damped peak in heavier clusters. The width of Landau damping will increase with increasing level density. But it should shrink to zero eventually in the limit $N \to \infty$ if the $1ph$ states close to the resonance decouple due to momentum conservation. Thus there should be a transitional region of maximal Landau damping which yet needs to be explored. A final look at the continuum thresholds in Fig. 1 shows that the nuclear resonances are always far above the threshold and thus experience an additional broadening due to neutron escape. The situation is a bit more stable in neutral Na clusters where the resonance is just below the electron threshold.

There is a most crucial difference which cannot be visualized in one figure: there is only one sort of nuclei but there as a wide choice of materials, characterized by its Wigner-Seitz radius r_s, and charge states for metal clusters. At first glance, one would like to ask now for spectral plots as in Fig. 1 for a variety of materials and charges. But it turns out that this is not necessary. We find a property which one could call superficially *jellium scaling*: within the jellium approximation, all plots of spectral properties versus particle number have essentially the same structure for all reasonable r_s and charge states if the energies are properly rescaled. There is always a clean resonance for light clusters, a transitional region at $N = 20 - 40$ and a strong Landau damping for heavy systems. The only exception is the electron continuum threshold which is lifted dramatically for ionized clusters. The consequence is that one should prefer to study positively charged clusters because the essential features of the resonance dynamics are similar as in neutral systems but everything is better bound there and thus more stable. We try to visualize the jellium scaling by a plot of spectral properties versus the Wigner-Seitz radius r_s. This is done in Fig. 2 for a system in the critical transitional region with $N = 20$ electrons. This shows that the spectral relations remain essentially the same throughout all r_s, except for some tiny differences. This "jellium scaling" holds for the gross properties of the spectra in the resonance region.

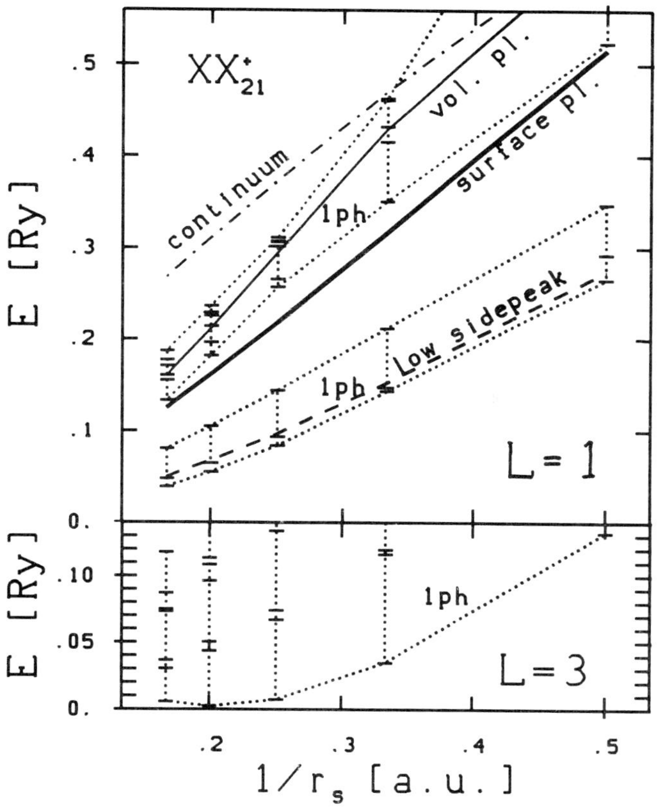

Fig. 2. *Upper part*: spectral properties for L=1 modes positively charged metal clusters with N = 21 versus r_s; explanations as in Fig. 1. *Lower part*: the lower end of the low $1ph$ band for L=3 modes

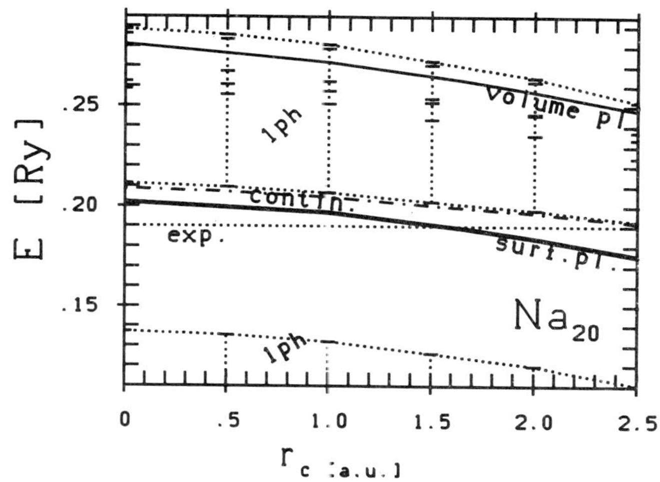

Fig. 3. Spectral properties for L=1 modes in neutral Na_{20} versus the pseudopotential core-radius r_c. Explanations as for Fig. 1. In addition comes the *dotted line* at $E \approx 0.19Ry$ to indicate the position of the experimental peak

It persists also for resonances with higher angular momenta and for higher charge states. However, there are details in the low-energy spectra which depend sensitively on the model parameters. An example is given in the lower part of Fig. 2 where we show the low energy modes with L = 3. There is a sensitive dependence on r_s with a most dramatic development near $r_s = 5a.u.$. The energy of the lowest $1ph$ state becomes nearly degenerate with the ground state. One should thus expect an extremely soft octupole vibration or even a stable octupole deformed ground state. It is to be noted, however, that this behaviour is extremely sensitive to changes in the model, as e.g. surface thickness in the jellium background. The interference of those low-energy electronic states with the phonon vibrations of the cluster will also play a crucial role. This, in turn, hints that experimental information on the low-energy spectra is highly desirable because it challenges theoretical models.

The gross features and the approximate positions of the resonances are well predicted by the Skyrme force for nuclei and the Gunnarson-Lundquist functional together with the jellium approach for metal clusters. The problems come, as usual, with the details. In nuclear, physics we have presently no working Skyrme parametrization which delivers a precise description of the nuclear ground-state properties together with appropriate resonance energies and low-energy spectra. The problem is most probably the zero-range of the Skyrme functionals and one may resolve them with finite-range functionals similar as it has been done in liquid ^3He [8, 6]. The energy-density functionals are on much better grounds in electron systems [9]. And yet, the position of the surface plasmon comes somewhat too high in calculations within the jellium approach with steep jellium surface. There is much debate about which approximation is to be blamed for that disease. There are strong arguments for a Self-Interaction-Correction (SIC) which aims to eliminate the double counting of one electron in the effective Hartree approximation of the Kohn-Sham scheme [10]. But that is not a unique explanation. There is, on the other hand, the observation that a finite surface width of the jellium can easily bring the resonance peak into the right position [11]. We show in Fig. 3 the spectral properties for Na_{20} as function of the surface width, given here from a folding of the jellium background with a pseudopotential of radius r_c. We see again a smooth dependence of all spectral features which confirm once more the notion of "jellium scaling". However, there is a clear trend of the energies which leads to a good agreement with the experimental peak energy at $r_c \approx 1.7$a.u. and that is just a typical value of the pseudopotential core-radius for Na. This shows that the concept of a finite surface is a useful, and probably necessary, completion of the jellium approach. But it would be premature to sell this as the unique explanation for the proper positioning of the resonance energy. There are several more effects to be considered, as e.g. ionic structure or core polarization, which all will have their contribution to the resonance. The various strains on the resonance energy can be shown in the sum rule model of the surface plasmon. The resonance energy becomes in the sumrule model $\omega^2 = \omega_{surf} \int d^3r \rho_{el}\rho_I / N\rho_0$ where ρ_{el} is the electron density and ρ_J the ionic jellium density (both positive definite). The bulk value ω_{sruf} for the surface plasmon is obtained in the limit of steep densities $\rho_{el} = \rho_I = \rho_0\vartheta_{Vol}$ where ϑ_{Vol} is a step function on the systems volume. We expand the actual densities as $\rho = \rho_0\vartheta_{Vol} + \delta\rho$ for ρ_{el} and ρ_I and obtain

$$\omega^2 = \omega_{surf}^2 \left[1 + \int d^3r\, \vartheta_{Vol}(\delta\rho_{el} + \delta\rho_I) + \int d^3r \frac{\delta\rho_{el}\delta\rho_I}{\rho_0}\right] \quad .(3)$$

The leading term is indeed the energy of the surface plasmon. The first order correction is determined by the portion of density which is missing inside the volume. It will be negative as both $\delta\rho$ are negative inside the volume (and positive outside for compensations). For $\delta\rho_{el}$, this is the effect from the "electron spill-out" as discussed in [12]. There is obviously an additional effect from $\delta\rho_I$, exploring the ionic spill-out. But note that an ionic surface width counts twice as it increases also the electron spill-out. This explains why a soft jellium surface can lower the resonance energy so efficiently. The consideration also explains the success of the Self-Interaction-Correction: a softer Coulomb potential will equally well enhance the electron spill-out and thus lower the resonance energy. But both (competing) explanations are at stake if we consider ionic structure. For then the $\delta\rho_I$ is more than a small correction at the surface and provides wild oscillations throughout the whole volume. The problem is still awaiting a thorough investigation.

Finally, we want to remark on the splitting of the resonances due to deformation of the cluster or nucleus. The resonance is split in a deformed nucleus according to elongation or sqeezing of the principle axes changing the freqency $\propto 1/R_{axis}$ which can be derived from scaling arguments employing a zero-range residual interaction. A similar splitting occurs for metal clusters, see e.g. [13]. But the size of the splitting is a factor of two smaller than in case of nuclei yielding a trend in frequency more like $\propto 1/\sqrt{R_{axis}}$. This is due to the long range Coulomb force which weakens the scaling effects. We hope that this last example has shown again the interesting mix of similarities and differences in the dynamics of nuclei compared to metal clusters. One thing remains yet similar: both cases deserve further theoretical fine-tuning in the mean-field models to reach a reliable quantitative agreement with the data.

References

1. Gunnarson, O., Lundquist, B.I.: Phys.Rev. **B13**, 4274 (1976)
2. Bartel, J., Quentin, P., Brack, M., Guet, C., Hakanssson, H.B.: Nucl.Phys. **A386**, 79 (1982)
3. Reinhard, P.-G., Brack, M., Genzken, O.L: Phys.Rev. **A41**, 5568 (1990)
4. Reinhard, P.-G., Gambhir, Y.K.: Ann.Phys. (Leipzig) **1**, 598 (1992)
5. Reinhard, P.-G.: Ann.Phys. (Leipzig) **1**, 632 (1992)
6. Weisgerber, S., Reinhard, P.-G.: Ann.Phys. (Leipzig) (submitted)
7. Reinhard, P.-G., Genzken, O., Brack, M.: (in preparation)
8. Weisgerber, S., Reinhard, P.-G.: Phys.Lett. **A158**, 407 (1991)
9. Reinhard, P.-G.: Phys.Lett. **A169**, 281 (1992)
10. see the contributions of W. Ekardt and J.M. Pacheco in these proceedings
11. Rubio, A., Balbas, L.C., Alonso, J.A.: Z.Phys. **D19**, 93 (1991)
12. Brack, M.: Rev.Mod.Phys. (submitted)
13. Lauritsch, G., Reinhard, P.-G., Meyer, J., Brack, M.: Phys.Lett. **A160**, 179 (1991)

This article was processed by the author using the LaTeX style file *pljour2* from Springer-Verlag.

Z. Phys. D 31, 269–273 (1994)

Optical response of bimetallic $Li_xNa_{8-x}(0 \leqq x \leqq 8)$ and of doped Na_8Zn clusters

L.C. Balbás[1], **A. Rubio**[1], **M.B. Torres**[2]

[1] Departamento de Física Teórica, Universidad de Valladolid, E-47011 Valladolid, Spain
[2] Departamento de Matemática Aplicada y Computación, Universidad de Valladolid, E-47011 Valladolid, Spain

Received: 28 June 1993 / Final version: 29 April 1994

Abstract. Using the time dependent local density approximation we calculate the photoabsorption spectra of bimetallic $Li_xNa_{8-x}(0 \leqq x \leqq 8)$ and of doped Na_8Zn clusters. Ground state cluster geometry and electronic structure are obtained by minimization of total energy from Kohn-Sham self-consistent calculations within the spherically averaged pseudopotential method. For Li_xNa_{8-x} clusters we obtain two nearly isomeric geometries. In both cases the corresponding photoabsorption spectra show a single collective excitation (surface plasmon) at increasing frequency as the number of lithium atoms increases. For a given number of lithium atoms, the plasmon frequency of the cluster in the ground state geometry is ~ 0.1 eV blueshifted with respect to that in the isomeric state. The equilibrium geometry of doped Na_8Zn cluster is formed by a nearly spherical shell of eight Na atoms enclosing the Zn impurity at the cluster center. The corresponding optical spectrum shows two peaks at higher frequencies than the surface plasmon of pure Na_8 cluster, in agreement with recent photodepletion spectroscopy experiments.

PACS: 36.40; 31.15; 31.20

1. Introduction

In this paper we use the time-dependet-local-density-approximation (TDLDA) [1, 2] to investigate the photoabsorption spectra of bimetallic clusters Li_xNa_{8-x} $(0 \leqq x \leqq 8)$, and of doped cluster Na_8Zn within the ground state structure obtained by the spherical-average-pseudopotential (SAPS) method [3], which takes into account, to some extent, the granular character of the ionic background of the clusters. This allows us to investigate the influence of the cluster geometry on its photoabsorption spectrum [4]. For the clusters considered in this work, comparison with photodepletion spectroscopy experiments is only possible for Na_8 [5], Li_8 [6], and Na_8Zn [7]. Although SAPS is not an ab-initio method it can be used to rationalize the response to light of mixed and doped alkali clusters in relation to their atomic geometry, electronic structure and chemical composition [4]. In Sect. 2 we present a brief account of the SAPS model and of the TDLDA method. In Sect. 3 we present and discuss our results for the photoabsorption cross sections. Conclusions are given in Sect. 4.

This paper was originally submitted in connection with the 2nd. Int. Conference on Atomic and Nuclear Clusters held in Santorini from 28. June – 2. July 1993 and is published here as a regular article after an independent refereeing procedure according to the standards of Z. Phys. D

2. Theory

In the SAPS model the ground state ionic and electronic structures of a cluster are obtained by minimizing the total energy, E_T, with respect to the positions of all the ions. E_T can be written as $E_T = E[n] + E_{\text{ion}-\text{ion}}$, where $E[n]$ is the energy of the valence electrons in the field created by the ions, and $E_{\text{ion}-\text{ion}}$ gives the ion-ion interaction energy. For a given configuration $(\mathbf{R}_1, \mathbf{R}_2, \dots \mathbf{R}_N)$ of the ions, $E[n]$ is obtained in the standard way from the electronic density $n(\mathbf{r})$ calculated by the Kohn-Sham equations of the density functional formalism [8] within the effective potential $V_{\text{eff}}(\mathbf{r}) = V_H(\mathbf{r}) + V_{xc}(\mathbf{r}) + V_{\text{ion}}^{\text{SAPS}}(r)$. Here V_H is the electrostatic Hartree potential, V_{xc} is the exchange-correlation potential (in the local-density-approximation), and $V_{\text{ion}}^{\text{SAPS}}$ is constructed by first forming the total ionic potential $V_{\text{ion}}(\mathbf{r}) = \sum_j^N v_{ps}(\mathbf{r} - \mathbf{R}_j)$ and then calculating the spherical average $V_{\text{ion}}^{\text{SAPS}}(r)$ of $V_{\text{ion}}(\mathbf{r})$ around the center of mass of the cluster. On the other hand, $E_{\text{ion}-\text{ion}}$ is evaluated as a pointlike interaction between the ions and consequently it takes precise account of the geometry. For the electron-ion pseudopotential, v_{ps}, we have taken the empty core model of Ashcroft [9] which is purely coulombic $(-z/r)$ outside the empty-core radius r_c and zero for $r < r_c$. The valence z is 1 for Na and Li, and 2 for Zn, and the respective r_c are (in a.u.) 1.74, 1.63 and 1.15. For Na_8Zn, we have also considered the impurity atom at the center of the cluster including all its electrons.

The fundamental ingredient in the calculation of the response properties in the time dependent local density approximation (TDLDA) [1] is the dynamical susceptibility $\chi(\mathbf{r}_1, \mathbf{r}_2, \omega)$, obtained from a Dyson-type integral equation

$$\begin{aligned}\chi(\mathbf{r}_1, \mathbf{r}_2, \omega) &= \chi_0(\mathbf{r}_1, \mathbf{r}_2, \omega) + \iint \chi_0(\mathbf{r}_1, \mathbf{r}_3, \omega) K(\mathbf{r}_3, \mathbf{r}_4) \\ &\times \chi(\mathbf{r}_4, \mathbf{r}_2, \omega)\, d\mathbf{r}_3\, d\mathbf{r}_4, \qquad (1)\end{aligned}$$

where the kernel is given by

$$K(\mathbf{r}_1, \mathbf{r}_2) = \frac{1}{r_{12}} + \frac{\partial V_{xc}(\mathbf{r}_1)}{\partial n(\mathbf{r}_1)} \delta(\mathbf{r}_1 - \mathbf{r}_2) \qquad (2)$$

and χ_0 is the independent-particle response function. χ_0 is constructed via the eigenvalues ε_i, the wave functions ϕ_i and the retarded one-electron Green functions corresponding to the self-consistent effective potential for the electronic ground state of the system

$$\begin{aligned}\chi_0(\mathbf{r}_1, \mathbf{r}_2, \omega) = \sum_{i=1}^{\mathrm{occ}} \{&\phi_i^*(\mathbf{r}_1)\phi_i(\mathbf{r}_2) G(\mathbf{r}_1, \mathbf{r}_2, \varepsilon_i + \omega) \\ &+ \phi_i(\mathbf{r}_1)\phi_i^*(\mathbf{r}_2) G^*(\mathbf{r}_1, \mathbf{r}_2, \varepsilon_i - \omega)\}. \qquad (3)\end{aligned}$$

The dynamical polarizability is obtained through the relation

$$\begin{aligned}\alpha(\omega) = \int &V_{\mathrm{ext}}^*(\mathbf{r}_1, \omega)\chi(\mathbf{r}_1, \mathbf{r}_2, \omega) \\ &\times V_{\mathrm{ext}}(\mathbf{r}_2, \omega)\, d\mathbf{r}_1\, d\mathbf{r}_2, \qquad (4)\end{aligned}$$

where V_{ext}, the external applied field, has the multipolar form $V_{\mathrm{ext}}(\mathbf{r}, \omega) = r^L Y_L^0(\hat{r}) e^{-i\omega t}$. For $L=1$ we obtain the dipole polarizability. The value of $\alpha(\omega)$ for $\omega = 0$ gives the static polarizability. Finally the photoabsorption cross section $\sigma(\omega)$ can be obtained from $\alpha(\omega)$ using Fermi's golden rule:

$$\sigma(\omega) = \frac{4\pi\omega}{c} \operatorname{Im}\alpha(\omega). \qquad (5)$$

3. Results

3.1. *Bimetallic* $Na_{8-x}Li_x$ $(0 \leq x \leq 8)$ *clusters*

The equilibrium SAPS geometries for all clusters in this isoelectronic series are of the D_{4d} form (square antiprism). We obtain also another atomic configuration with a total binding energy slightly smaller than the corresponding to D_{4d} symmetry. Such an isomeric geometry is formed by a spherical shell composed of seven sodium atoms around a central atom (lithium for $x \geq 1$). In both symmetries the average radius of a cluster, R_{av}, defined as the average distance from the atomic cores to the cluster center, decreases as the number of lithium atoms increases; conversely the total energy per atom increases. These trends are represented in Fig. 1a.

The electronic configuration for all these eight-electron clusters is $1s^2 1p^6$. The highest occupied molecular orbital (HOMO) eigenvalue, $-\varepsilon_p$, rise from Na_8 to Li_8 as well as the vertical ionization potential (VIP) (obtained as the energy difference between the neutral and cation clusters, both with the geometry of the neutral one). Note in Fig. 1b the large difference between the HOMO and VIP values, which is a well known deficiency of the local density-approximation (see e.g. [4]).

In both isomeric states the corresponding TDLDA photoabsorption spectra show only one dominant peak (surface plasmon) appearing at increasing energy as the content of lithium increases (see Fig. 1c). The plasmon peak of D_{4d} clusters is about 0.1 eV blueshifted with respect to the peak in the centered geometry. Observe in Fig. 1c that the surface plasmons of the $(x \leq 4)$ clusters with centered geometry are redshifted with respect to the plasmon in Na_8 with D_{4d} geometry. Only for Na_8 and Li_8 are there photodepletion experiments available, showing a single collective feature at 2.53 eV [5] and 2.5–2.6 eV [6] respectively. In order to obtain the observed energy resonance of Na_8 Rubio et al. [4] performed a TDLDA calculation assuming the SAPS D_{4d} symmetry but enlarging the average radius ~10%. Although the interatomic distances in clusters are expected to be reduced relative to the bulk matter distances, the SAPS method leads to larger reductions than ab initio methods (see [4] for details). Independent of this failure, which affects all the clusters in the series similarly, the trends despicted in Fig. 1 are expected to be qualitatively correct.

As long as the dipole oscillator strength is mainly concentrated in the collective resonance (about 80%), we can rationalize our predictions for the plasmon frequencies and static dipole polarizabilities in terms of the semi-classical-plasmon-pole-approximation (SPPA) [10]. In this approximation only delocalized valence electrons are allowed to contribute to the resonance, exhausting the dipolar Thomas-Reiche-Kuhn sum rule. In the SPPA the dipole resonance frequency ω_0 is related to the static polarizability $\alpha(0)$ and the number of valence electrons N_e by the expresion

$$\omega_0^2 = (e^2 N_e)/(m_e \alpha(0)). \qquad (6)$$

Using our calculated values of ω_0 and $\alpha(0)$ we represent in Fig. 1c the effective number $N_e/8$ of valence electrons per atom as a function of the number of lithium atoms. The behaviour observed in Fig. 1c can be understood qualitatively assuming that the effective number of valence electrons is smaller for lithium than for sodium, following roughly a stoichiometric weighting law, as suggested by Pollack et al. [11] in the case of $Na_{4-x}Li_x$ clusters. Alternatively, this effect can be traced back to a larger effective electron mass of lithium, as stressed in

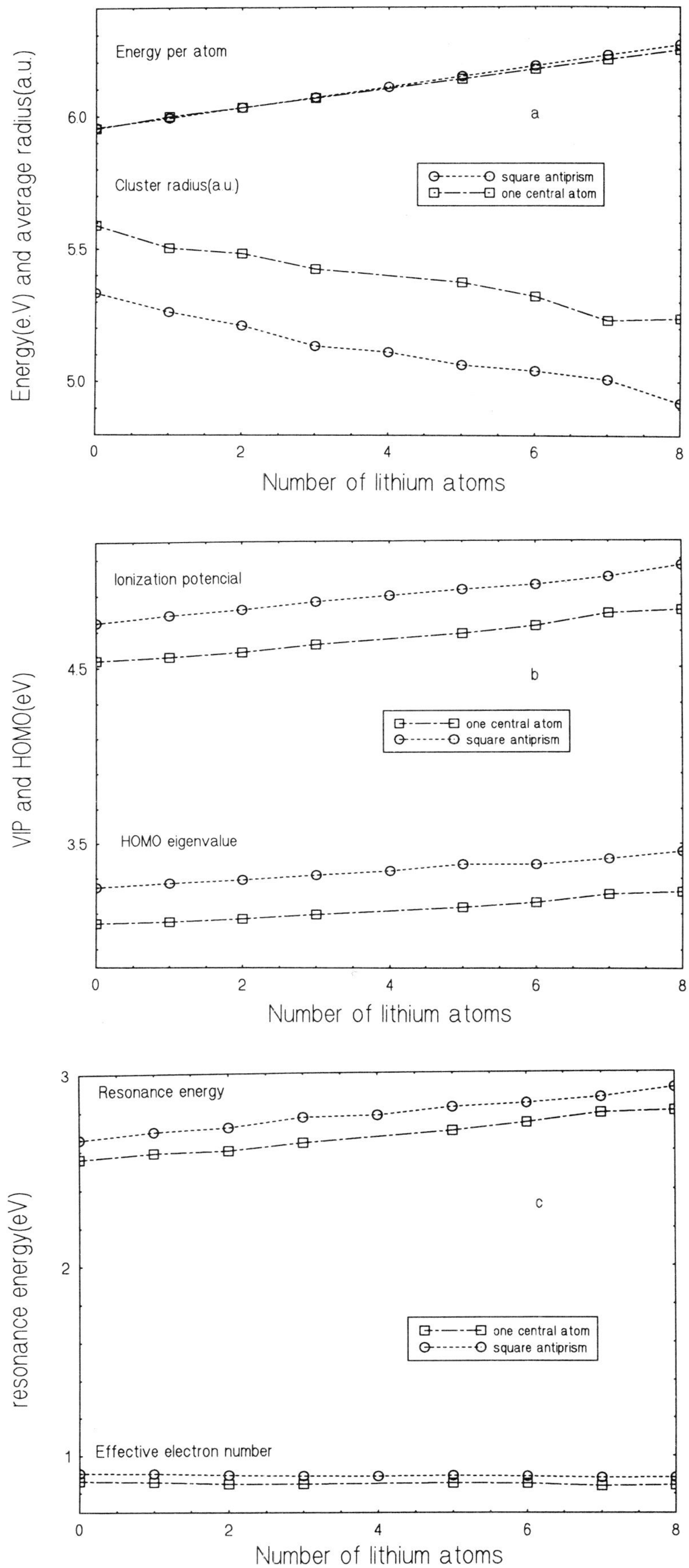

Fig. 1a–c. Evolution with lithium content of several calculated properties of $Na_{8-x}Li_x$ clusters within two isomeric geometries: D_{4d} square antiprism (----) and a seven atom shell around a central atom (lithium for $x > 0$) (–·–·–). **a** Total energy per atom (eV) and average cluster radius (a.u.); **b** vertical ionization potential (VIP) and highest occupied orbital eigenvalue (HOMO), both in eV; **c** resonance energy (eV) and effective valence electron number per atom

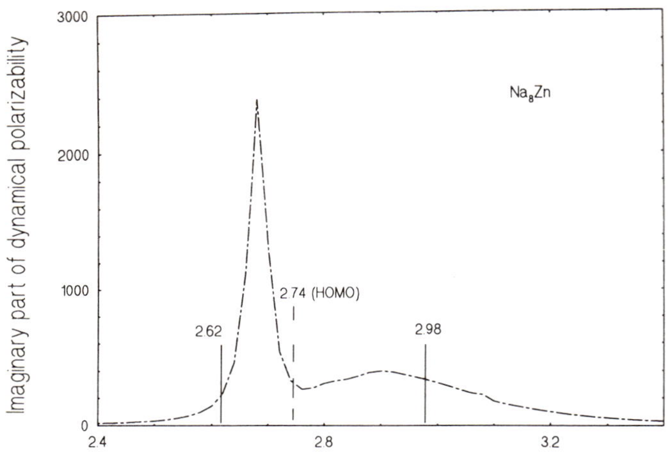

Fig. 2. Imaginary part of the TDLDA dynamical polarizability $\alpha(\omega)$ for Na_8Zn cluster with the Zn atom at the cluster center including all its electrons. The *dashed vertical line* indicates the theoretical ionization threshold (HOMO eigenvalue). *Solid vertical lines* indicates the position of experimental peaks from [7]

recent experimental [12] and theoretical [13] works. Note in Fig. 1c that the effective number of valence electrons per atom is smaller for the centered geometry than for the empty cage D_{4d} geometry, which is due to the different effective potential (different screening) for the two ionic configurations.

3.2. Photoabsorption cross section of doped Na_8Zn *cluster*

The photodepletion experiment for Na_9Zn [7] shows a broad surface plasmon centered at ~ 2.62 eV and a fragmentation peak at ~ 2.98 eV. For this cluster we adopt the equilibrium SAPS geometry obtained by J.M. López et al. [14], which consists of a spherical shell with 8 sodium atoms enclosing a central Zn divalent atom. For this ten electron cluster the electronic configuration obtained by SAPS is $1s^2\,1p^6\,2s^2$, with eigenvalues -6.89 eV, -3.60 eV and -2.88 eV for the orbitals $1s$, $1p$, and $2s$ respectively. The TDLDA result for $\mathrm{Im}\,\alpha(\omega)$ shows two peaks at 2.74 eV and 3.04 eV respectively. The first peak is due to the collective dipolar excitation (surface plasmon) and the second is due to the interaction with the theoretical ionization threshold at $-\varepsilon_{2s}=2.88$ eV (Landau Damping). The particle-hole transitions $1p\rightarrow 1d$ and $2s\rightarrow 2p$ appear at 1.74 eV and 1.88 eV respectively. By considering the central Zn atom with all its 30 electrons and the same SAPS geometry, the electronic configuration comes out to be (within the atomic notation for orbitals) $1s^2\,2s^2\,2p^6\,3s^2\,3p^6\,3d^{10}\,4s^2\,4p^6\,5s^2$, which is also a closed shell configuration. The eigenvalues of the orbitals $4s$, $4p$, and $5s$ (HOMO) are now -6.74 eV, -3.66 eV and -2.74 eV respectively. In Fig. 2 we represent the corresponding imaginary part of the TDLDA dynamical polarizability $\mathrm{Im}\,\alpha(\omega)$.

The peak at 2.68 eV in Fig. 2 is the dipolar collective excitation (surface plasmon) which appears in the photodepletion experiments [7] at ~ 2.62 eV (marked in Fig. 2 by a vertical line). The peak at 2.9 eV in Fig. 2, corresponding to the peak observed at ~ 2.98 eV in the experiment, is a fragmentation of the surface plasmon induced by its proximity to the ionization threshold (i.e., $-\varepsilon_{5s}=2.74$ eV, also shown in Fig. 2 as HOMO). The particle-hole transitions $4p\rightarrow 4d$ and $5s\rightarrow 5p$, appear at 1.83 eV and 1.62 eV respectively. The results despicted in Fig. 2 are closer to experiment than those obtained by Yannouleas et al. [15] by means of a RPA response calculation of the Na_8Zn cluster described within the two-step jellium model. In such a model, divalent Zn atom is assumed to be a jellium drop with radius per electron $r_s=1.15$ a.u. (bulk Zn has however $r_s=2.31$ a.u.), and monovalent Na atoms are described by a surrounding jellium with $r_s=4.0$ a.u. For this model cluster Yannouleas et al. [15] obtained the ground state electronic configuration $1s^2\,2s^2\,2p^6$, and the corresponding RPA spectra shows a collective feature at 2.57 eV (with 42% of the strength) and two nearly degenerate particle-hole transitions at ~ 2.87 eV ($2s\rightarrow 2p$ and $1p\rightarrow 3s$, sharing $\sim 26\%$ of the total strength). In our TDLDA calculation within the SAPS model we find, however, that the particle-hole transitions appear ~ 1 eV shifted downwards with respect to the those in the RPA calculation of Yannouleas et al. [15]. Consequently, the second peak observed in the experimental spectrum of Na_8Zn is interpreted in our calculation as a fragmentation due to the proximity of the ionization potential threshold.

4. Conclusions

We have obtained the TDLDA optical spectra of bimetallic $Na_{8-x}Li_x$ and doped Na_8Zn alkali clusters within the atomic and electronic structures given by the SAPS method. For the $Na_{8-x}Li_x$ clusters we obtain two isomeric atomic configurations, the equilibrium one with D_{4d} symmetry and the other with one central atom inside of a nearly spherical shell of seven sodium atoms. Both cases results in one single surface plasmon at increasing energy as the number of lithium atoms increases. For a given concentration of lithium the plasmon of the centered geometry is ~ 0.1 eV redshifted with respect to the one of the empty-cage geometry. Comparison with predictions from the semiclassical surface plasmon pole approximation (SSPA) indicates a smaller effetive valence electron number (equivalently a larger effective electron mass) for lithium than for sodium.

For the doped Na_8Zn cluster the geometry is formed by a nearly spherical shell with eight sodium atoms around a central Zn atom. Taking into account all the electrons of the Zn atom, we obtain a TDLDA photoabsorption spectrum in good agreement with photodepletion experiment [7]. Despite the crude SAPS approximations involved in the prediction of atomic geometries, the corresponding TDLDA optical spectra are in qualitative agreement with available photodepletion spectroscopic data for Na_8 [5], Li_8 [6] and Na_8Zn [7].

This work has been supported by the DGICYT (Spain), grant PB89-0352-C02-01. M.B.T. acknowledged financial support from a FPI grant of the M.E.C. (Spain).

References

1. Stott, H.J., Zaremba, E.: Phys. Rev. A**21**, 12 (1980)
2. Ekardt, W.: Phys. Rev. Lett. **52**, 1925 (1984); Phys. Rev. B**31**, 6360 (1985)
3. Iñiguez, M.P., López, M J., Alonso, J.A., Soler, J.M.: Z. Phys. D**11**, 163 (1989)
4. Rubio, A., Balbás, L.C., Alonso, J.A.: Phys. Rev. B**45**, 13657 (1992); ibid **46**, 4891 (1992); Boll, A., Martín, G., López, J.M., Alonso, J.A.: Z. Phys. D**28**, 311 (993)
5. Selby, K., Kresin, V., Masui, J., Vollmer, M., de Heer, W.A., Scheidemann, A., Knight, W.D.: Phys. Rev. B**43**, 4565 (1991); Wang, C.R.C., Pollack, S., Cameron, D., Kappes, M.M.: J. Chem. Phys. **93**, 3738 (1990)
6. Blanc, J., Bonacic-Koutecky, V., Broyer, M., Chevaleyre, J., Dugourd, Ph., Koutecky, J., Sheuch, C., Wolf, J.P., Woste, L.: J. Chem. Phys. **96**, 1793 (1992)
7. Dahlseid, T.A., Koretsky, G.M., Pollak, S., Wang, C.R.C., Hunter, J., Alamedin, G., Cameron, D., Liu, S., Kappes, M.: Nuclear concepts in cluster phyiscs. Schmidt, R., Lutz, H.O., Dreizler, R. (eds.), p. 263. Berlin, Heidelberg, New York: Springer 1992
8. Kohn, W., Shan, L.J.: Phys. Rev. A**140**, 1133 (1965)
9. Ashcroft, N.W.: Phys. Lett. **23**, 48 (1966)
10. Lushnikov, A.A., Maksimenko, V.V., Simonov, A.J.: Electromagnetic surface modes. Boardman, A.D. (ed.). New York: Wiley 1982
11. Pollack, S., Wang, C.R.C., Dahlseid, T.A., Kappes, M.M.: J. Chem. Phys. **96**, 4918 (1992)
12. Bréchignac, C., Cahuzac, Ph., Leygnier, J., Sarfati, A.: Phys. Rev. Lett. **70**, 2036 (1993)
13. Serra, Ll., Bachelet, G.B., Van Giai, N., Lipparini, E.: Phys. Rev. **48**, 14708 (1993)
14. López, J.M., Ayuela, A., Alonso, J.A.: Density functionals methods in chemistry. Labanowski, J., Andzelm, J. (eds.), p. 373. Berlin, Heidelberg, New York: Springer 1991
15. Yannouleas, C., Jena, P., Khanna, S.N.: Phys. Rev. B**46**, 9751 (1992)

Z. Phys. A 349, 223–225 (1994)

Clustering states in the ^{62}Ni(^{58}Ni,^{58}Ni)^{62}Ni reaction

L. Vannucci [1], **U. Abbondanno** [3], **M. Bettiolo** [1], **M. Bruno** [4], **N. Cindro** [5], **M. D'Agostino** [4], **P.M. Milazzo** [4], **R.A. Ricci** [1,2], **G. Vannini** [3]

[1] Istituto Nazionale di Fisica Nucleare, Laboratori Nazionali di Legnaro, I-35020 Legnaro, Italy
[2] Dipartimento di Fisica dell'Università, via Marzolo 8, I-35131 Padova, Italy
[3] Dipartimento di Fisica dell'Università, Trieste, and Istituto Nazionale di Fisica Nucleare, Sezione di Trieste, via Valerio 2, I-34127 Trieste, Italy
[4] Dipartimento di Fisica dell'Università, Bologna, and Istituto Nazionale di Fisica Nucleare, Sezione di Bologna, via Irnerio 46, I-40126 Bologna, Italy
[5] Ruder Bošković Institute, 41001 Zagreb, Croatia

Received: 27 October 1993

Abstract. Excitation function and angular distributions of the ^{62}Ni(^{58}Ni,^{58}Ni)^{62}Ni elastic scattering have been measured at incident ^{58}Ni energies from 220.0 to 230.0 MeV in steps of 0.5 MeV. Evidence of two structures was found in the excitation function; a statistical analysis suggests a possible nuclear cluster quasi-molecular nature for these structures.

PACS: 01.30.Cc; 25.70.Ef

Within the still open questions in the field of the resonant behaviour of heavy ion systems, two problems are of particular interest. The first one is connected with the fact that resonances can be interpreted both as structural effects in the composite system (group-theoretical and anharmonic rotator-vibrator models [1, 2]) and as phenomena related to the dynamics of the interacting nuclei (Orbiting Cluster Model [3, 4]). The second one concerns the possible presence of resonances in composite systems with $A \cong 100$, that could demonstrate that this behaviour is a general feature of heavy-ion interaction [5]. The first question can be solved, for example, by comparing resonant states observed both in two and in three-body exit channels, whereas an appropriate beam-target combination can contribute to the understanding of the second problem.

Preliminary results concerning the ^{62}Ni(^{58}Ni, ^{58}Ni)^{62}Ni reaction, to be compared with the data of the $^{58}\text{Ni} + {}^{66}\text{Zn} \rightarrow \alpha + {}^{58}\text{Ni} + {}^{62}\text{Ni}$ process, are presented in this contribution. The measurements were performed at the XTU Tandem accelerator of the LNL (Laboratori Nazionali di Legnaro, Padua, Italy). Excitation function and angular distributions for the ^{62}Ni(^{58}Ni, ^{58}Ni)^{62}Ni reaction were measured from $E_{\text{lab}} = 220$ MeV to $E_{\text{lab}} = 230$ MeV, in steps of 0.5 MeV. The excitation function relative to selected events in the center-of-mass angular range from 70° to 94° is reported in Fig. 1; the overall error was evaluated to be smaller than 7%. The mass identification was obtained, with a resolution of $\pm 1u$, by means of a kinematic coincidence using two position sensitive Si detectors. In these measurements, the good angular resolution (1.5×10^{-3} rad) allowed us to discriminate between elastic and inelastic events by the difference of the relative angle, because the energy resolution (1.5%) was not sufficient to do it.

Two structures were observed in the elastic scattering excitation function at $E_{\text{cm}} = 115.7$ and $E_{\text{cm}} = 116.5$ MeV (Fig. 1). For what concerns the angular distributions, due to the large Coulomb contribution to the elastic cross section, the characteristic narrow oscillations usually related to the presence of a single-wave state were not observed (see, for example, Fig. 2). So, statistical methods using the autocorrelation and correlation functions were used to overcome this difficulty [6, 7].

We started by examining the autocorrelation function calculated for the differential cross-section $\sigma(\theta)$

$$C(\theta, 0, \Delta) = \overline{\left(\frac{\sigma(\theta)}{\langle\sigma(\theta)\rangle_\Delta} - 1\right)^2}$$

as a function of the median angular interval Δ to obtain the most appropriate value for averaging out the statistical fluctuations. This value was then used to calculate the correlation function

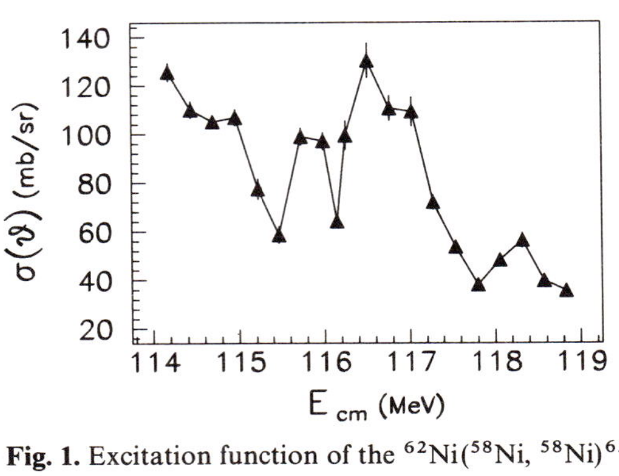

Fig. 1. Excitation function of the ^{62}Ni(^{58}Ni, ^{58}Ni)^{62}Ni reaction

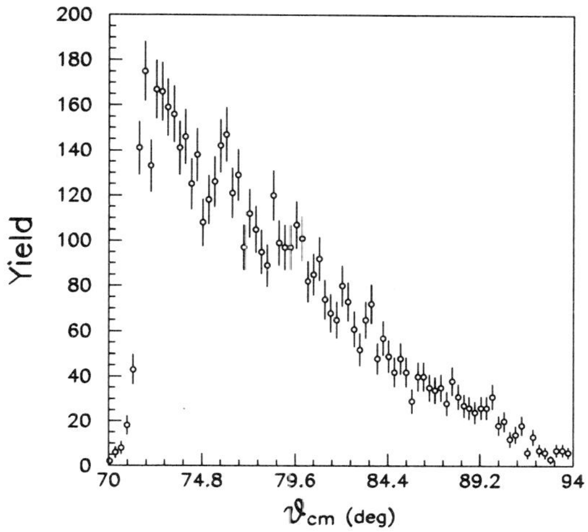

Fig. 2. Angular distribution of $^{62}Ni(^{58}Ni,\ ^{58}Ni)^{62}Ni$, measured at $E_{cm}=115.7$ MeV

$$C(\theta, \Delta\theta, \Delta)=\overline{\left(\frac{\sigma(\theta)}{\langle\sigma(\theta)\rangle_{\Delta}}-1\right)\left(\frac{\sigma(\theta+\Delta\theta)}{\langle\sigma(\theta+\Delta\theta)\rangle_{\Delta}}-1\right)}$$

varying the width of the angular interval $\Delta\theta$. It is in fact well known that, if the function $\sigma(\theta)$ has a period $\Delta\theta_{per}$, the correlation function too, plotted as a function of $\Delta\theta$, must have a period $\Delta\theta_{per}$.

An oscillatory behaviour of the correlation function is present (Fig. 3) for energies corresponding to the structures observed in the elastic excitation function ($E_{cm}=115.7$ and $E_{cm}=116.5$ MeV); at energies far from the region of the structures, such as, for example, $E_{cm}=118.6$ MeV, this behaviour is not observed. In order to reveal the periodicity of these oscillations and to extract its numerical value, a function χ^2 was calculated within the limits of the values that the function $C(\theta, \Delta\theta, \Delta)$ assumes at angles $\theta=\theta_n$ and $\theta=\theta_n+\theta_0$. This function χ^2, depending on the step θ_0, is given by

$$\chi^2(\theta_0)=\frac{1}{(N-1)}\sum_{n=1}^{N}\left(\frac{C(\theta_n, \Delta\theta, \Delta)-C((\theta_n+\theta_0), \Delta\theta, \Delta)}{\sigma}\right)^2$$

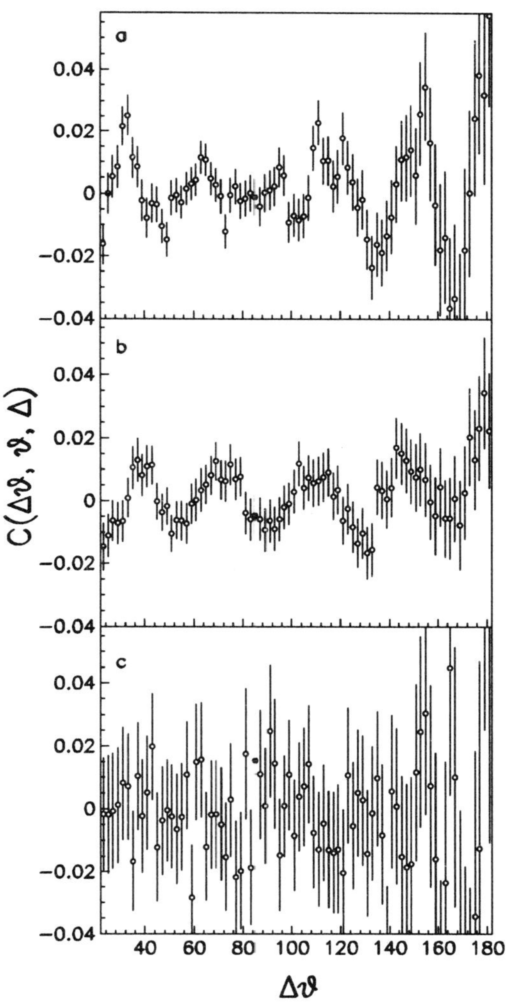

Fig. 3. Correlation functions $C(\theta, \Delta\theta, \Delta)$ vs. the angular interval $\Delta\theta$ for $E_{cm}=$ **a** 115.7, **b** 116.2, and **c** 118.6 MeV

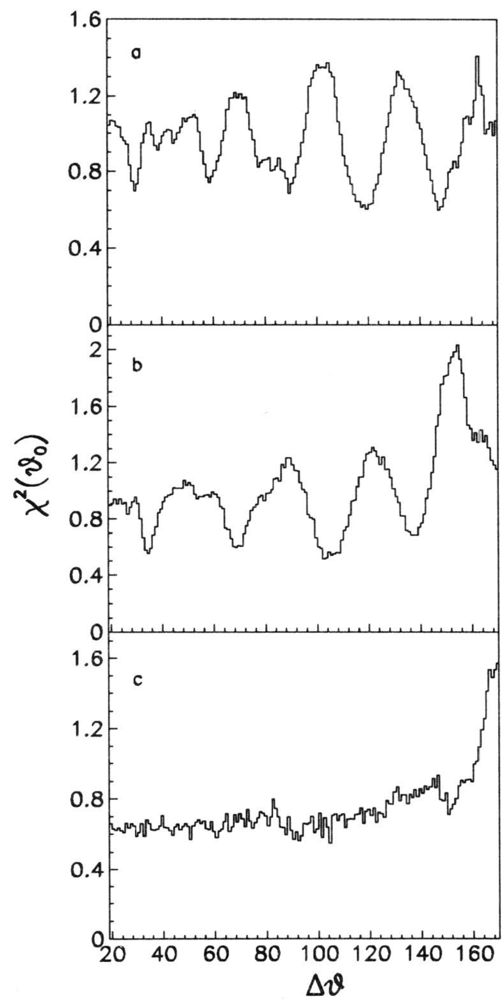

Fig. 4. The values of χ^2 plotted as a function of θ_0 for $E_{cm}=$ **a** 115.7, **b** 116.5, and **c** 118.6 MeV

Table 1. Analysis of the structures observed in the $^{58}Ni+^{62}Ni$ reaction

E_{lab}(MeV)	E_{cm}(MeV)	$\Delta\theta_{per}$(deg)	$J(\hbar)_{exp}$	$J(\hbar)_{calc}$
224.0	115.7	2.95	61 ± 1	57
225.5	116.5	3.42	53 ± 1	59

with

$$\sigma=\sqrt{\sigma^2(\theta_n)+\sigma^2(\theta_n+\theta_0)}.$$

The values of χ^2 plotted as a function of θ_0 at the same energies as the correlation function reported in Fig. 3 are presented in Fig. 4. The value of χ^2 reaches its minimum when the value of θ_0 matches the value of $\Delta\theta_{per}$, and is minimum again for each value of θ_0 multiple of $\Delta\theta_{per}$. The numerical value of the periodicity in the χ^2 function, when such a periodicity was observed, was calculated as the mean value of the ratios of the angular position at each minimum to the corresponding number of observed periods.

The systematic application of this analysis method to all the data of the present work gives a clear indication of periodicity for the angular distributions measured at energies corresponding to the two peaks observed in the excitation function. As the periodicity of the correlation function is the same as that of the angular distribution $\sigma(\theta)$, we could, then, temptatively assign value of $J=\pi/\Delta\theta_{per}$ to the predominant single waves associated to the angular distributions corresponding to the peaks in the excitation function.

The deduced J-values, together with the corresponding values calculated be means of the Orbiting Cluster Model [3], are reported in Table 1: within the crudity of the preliminary analysis and of the adopted models, the general agreement can be considered satisfactory, suggesting a possible quasi-molecular nature of the observed structures.

References

1. Iachello, F.: Phys. Rev. C **23**, 2778 (1981)
2. Cindro, N., Greiner, W.: J. Phys. G **9**, L175 (1983)
3. Cindro, N., Počsanić, D.: J. Phys. G **6**, 359 (1980)
4. Počanić, D., Cindro, N.: Nucl. Phys. A **433**, 531 (1985)
5. Abbondanno, U., Cindro, N.: Int. J. Mod. Phys. E **2**, 1 (1993)
6. Pappalardo, G.: Phys. Lett. **13**, 320 (1964)
7. Gadioli, E., Iori, I., Mangialaio, M., Pappalardo, G.: Nuovo Cimento **38**, 1105 (1965)

Z. Phys. A 349, 227–232 (1994)

Nuclear clusters and nuclear molecules

B.R. Fulton

School of Physics and Space Research, University of Birmingham, Edgbaston, Birmingham B15 2TT, UK

Received: 27 October 1993

Abstract. The idea that clusters of nucleons might exist in the nucleus has a long history and dates back to the earliest days of nuclear physics. Recently there have been significant advances in our theoretical treatment of clustering and in the experimental methods available to identify cluster structure in nuclei. These developments are reviewed with particular reference to the ^{24}Mg nucleus where a rich variety of cluster configurations is observed.

PACS: 24.30.-v,25.70.-z,27.30.+t

1. Introduction

The idea of clustering in the nucleus has a long and distinguished history. Speculations about the existence of clusters such as alpha particles have been around since the earliest days of nuclear physics, stimulated initially by the observation of alpha particle decay. Of course the observation of a complex fragment (or cluster) emerging from a nucleus need not necessarily imply the fragment existed as a pre-formed entity in the nucleus, and much discussion on this point can be found in the literature and conference precedings. Later, when reaction studies using heavy ion projectiles became possible, further evidence for clustering was obtained with the observation of selective population of particular nuclear states in nuclear reactions where clusters of nucleons were transferred between the two colliding nuclei. More recently the observation of resonances in the collision of even heavier nuclei (for example ^{12}C +^{12} C) suggested that even larger clusters might occur.

The earliest theoretical models were developed for alpha clusters in the nucleus and presented a geometrical (almost crystalline) picture of light nuclei built from alpha particles. More recently these models have been developed in a more rigorous quantum mechanical description and have predicted that cluster nuclei can be excited to many highly deformed shapes. At the same time new experimental approaches have been developed which can probe cluster structures by observing the breakup of the nucleus into the cluster components. These developments have lead to a renewed interest in clustering in the nucleus.

In this paper I will briefly review the early ideas on clustering before describing the results of the more recent theoretical and experimental work. The wide variety of highly deformed cluster structures which have been revealed will be discussed, ranging from the simple alpha cluster states to the extremes of orbiting nuclei (nuclear molecules) and the even more spectacular linear chains of 6 or 7 alpha particles (alpha chain states).

2. Clustering in nuclei

During the 1960's a number of theoretical groups began to work on the problem of describing the nucleus in terms of clusters of alpha particles and considerable advances were made. Two results from this period are quite illustrative. Figure 1 shows what has become known as the "Ikeda Diagram", and illustrates the possible cluster structures in a number of light nuclei (Ikeda et al. 1969).

Ikeda believed that each of the different cluster structures would be most noticeable in the energy region just above the relevant separation energy for the fragments. In the light of our recent discoveries, the predictions of this model seem remarkably far sighted. Figure 2 shows some predictions by Brink (Brink 1967) from around the same period in which he presented the light alpha conjugate nuclei as almost crystalline structures with specific arrangements of the alpha clusters.

In subsequent years the theoretical models of alpha clustering were developed on a more formal basis. In particular Brink (Brink 1966) continued to develop his model in which the nucleons in the nucleus are constrained to move as quartets in a 0s state of relative motion in a harmonic oscillator potential. The positions and width of the wells are allowed to vary until the energy of the system is minimised and this is taken to represent the ground state configuration of the nucleus. This is in essence a multi-centre shell model description and it was shown formally that in the limit of zero separation between the wells the wavefunction reduces to that of the shell model. Figure 3 is a density projection

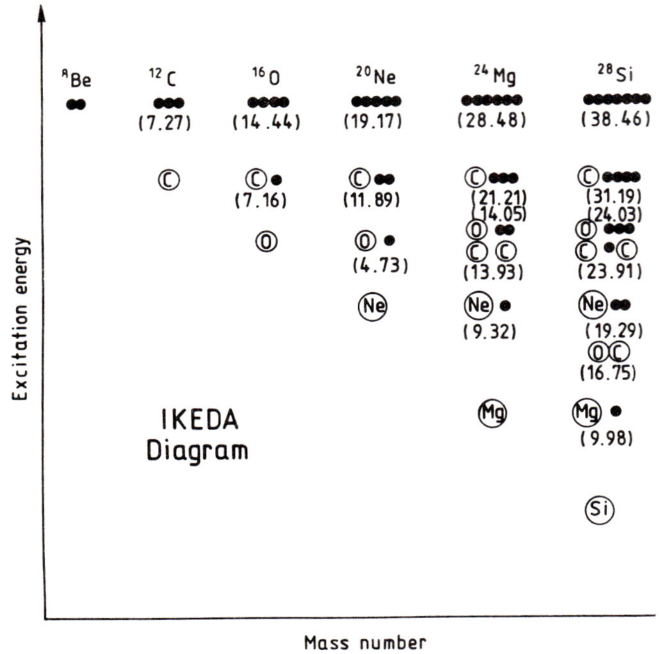

Fig. 1. The Ikeda diagram

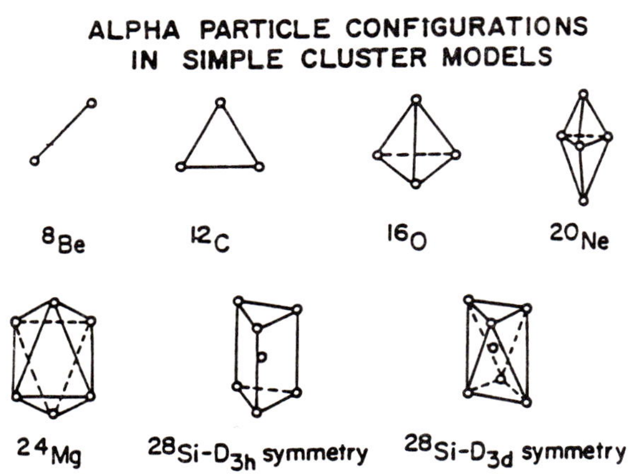

Fig. 3. Density projections from a cluster model calculation for ^{12}C

from a cluster calculation for ^{12}C. The density is shown before parity projection and the triangular arrangement is in accord with the known oblate shape of the ^{12}C ground state. Interestingly, the calculations also reveal another stable shape, although at a much higher excitation. This chain like arrangements of the well centres has been associated with the 0_2^+ state at 7.6 MeV in ^{12}C. But more on these chain states later.

One of the main difficulties at this time was that there was an absence of any experimental data with which to test these theoretical ideas. Although the models could reproduce the energies of low lying states in nuclei, and predict the spectrum of states seen in alpha transfer reactions, these

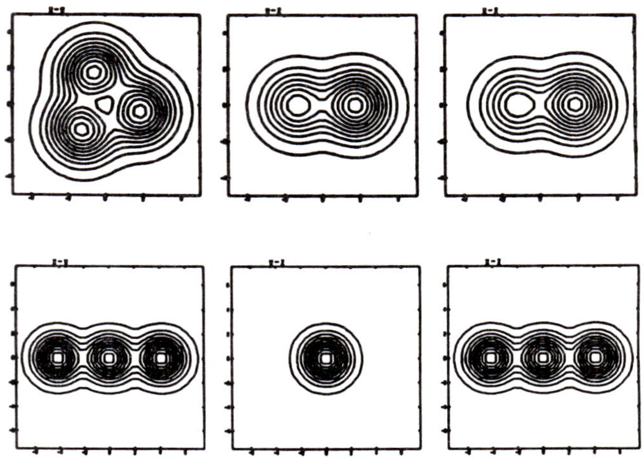

Fig. 2. Brink's early picture of alpha clustering in light nuclei

features could equally well be described by more straightforward models which did not require any cluster structure to be assumed. Possibly for this reason interest in the subject waned. It is only in recent years that new experimental techniques to probe alpha cluster structure have been developed and these results have rekindled interest in the topic (Fulton and Rae 1990).

At about the same time that Ikeda and Brink were developing their early models of alpha clustering, the first experiments which suggested that larger clusters might also exist were just beginning. The evidence came from some of the earliest experiments involving the collision of two heavy ions. When the yield of reaction products from the collision of two ^{12}C nuclei was measured, striking, narrow resonances appeared in the excitation function. These were correlated in a number of reaction channels, leading to the speculation that the two ^{12}C nuclei were forming an intermediate "nuclear molecule" in the reaction. This, then, would be an even more extreme form of cluster structure in the nucleus.

3. Experimental methods

The early experimental investigations of clustering usually involved measurements of scattering resonances. By colliding two nuclei and observing how the reaction yield varies as the bombarding energy is changed, enhancements in the yield are observed. The origin of these "resonances" is interpreted as the formation of a cluster state comprising the two colliding nuclei. The main problem with this approach is that many other reaction mechanisms contribute to the observed yield and it is very difficult to be certain the observations unambiguously point to cluster states in the combined system.

More recently a different approach has been used in which the nucleus is excited in a collision to cause it to break up into its component clusters. These experiments use a technique called "Resonant Particle Spectroscopy" in which the nucleus of interest is used as a projectile and excited to the high lying cluster states by scattering it off

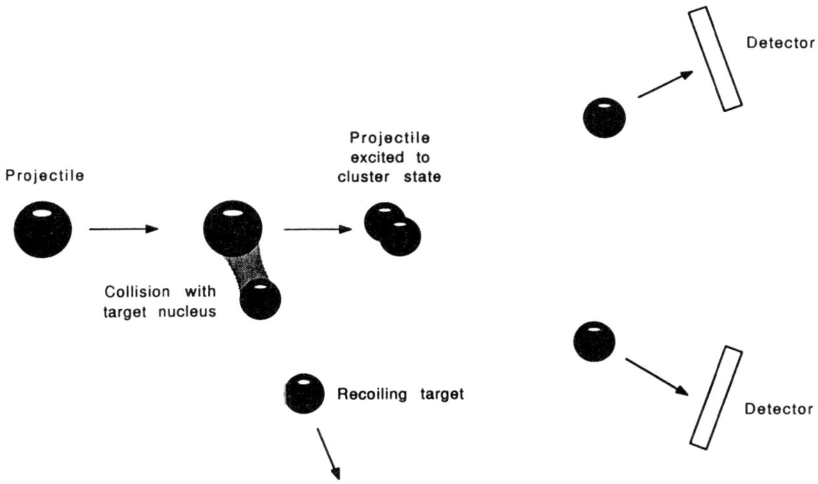

Fig. 4. Schematic outline of the breakup technique used to probe for cluster structure in nuclei

a light target nucleus (the light target serves to enhance the experimental efficiency as the reaction products are concentrated in the forward cone in the laboratory). The decay (or breakup) of the excited nucleus is then observed by detecting in coincidence the alpha particle and the residual daughter nucleus (see Fig. 4). If the energy and emission angle of the two fragments are recorded with sufficient accuracy the excitation energy of the alpha cluster state can be determined. In addition, an analysis of the angular correlation of the fragments can provide a measure of the angular momentum (spin) of the breakup state.

The main experimental requirements are to be able to identify the breakup fragments of interest in the high flux of reaction products and to register their energy and angle of emission with sufficient accuracy for the kinematic reconstruction of the event. For a reaction leading to three particles in the exit channel (the two breakup fragments and the recoiling target nucleus) it is only necessary to record two of the outgoing fragments, as the information on the unobserved fragment is provided by the requirements of conservation of energy and momentum.

Identification of the fragments is achieved by the normal energy loss process using a two (or three) detector telescope. The detectors register the energy of the particle and by using position sensitive detectors the angle of emission of each particle can be determined. For high efficiency it is preferable to use as many detectors as possible to enhance the coincidence count rate and this also aids the determination of the spin of the breakup states by providing a wider coverage of the angular correlation pattern.

4. Clustering in the ^{24}Mg nucleus

As an illustration of clustering in the nucleus we will concentrate on results for ^{24}Mg. A look back at Fig. 1 shows that Ikeda speculated a range of different cluster structures might occur in this nucleus: $\alpha + ^{20}$Ne, ^{8}Be + ^{16}O, ^{12}C + ^{12}C, ^{12}C + ^{12}C$_{\text{chain}}$ and a 6α linear chain state. This speculation

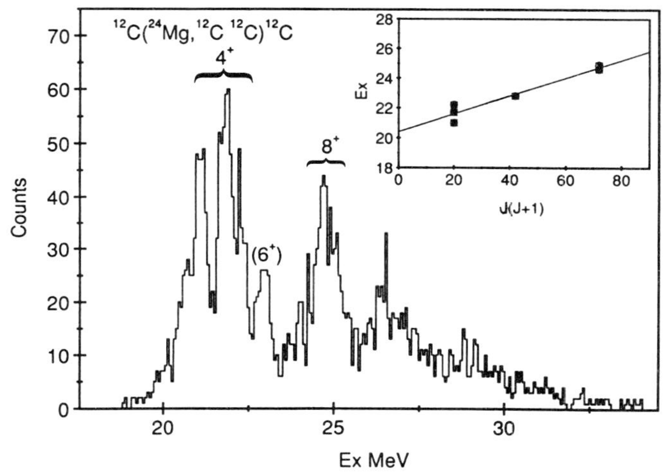

Fig. 5. ^{12}C – ^{12}C cluster states as revealed in the breakup of ^{24}Mg. The inset shows the energy-spin systematics of the states

has proved to be remarkably far sighted and there is now evidence for all these different structures.

Figure 5 shows an excitation energy spectrum for an excited ^{24}Mg nucleus which breaks up into two ^{12}C fragment nuclei (Fulton et al. 1991). The observation of discrete states in this spectrum is of interest since at the excitation energy region in question the level density is between 10^5 and 10^6 states per MeV. The suggestion is that some structural property makes them stand out in the continuum. The insert on this figure shows that the energy-spin systematics of these states are consistent with that expected from a rotating structure and the slope indicates a very high moment of inertia consistent with that expected from a structure of two touching ^{12}C nuclei.

These results provide strong evidence for the cluster configuration ^{12}C + ^{12}C in the ^{24}Mg nucleus predicted by Ikeda. Evidence for two of the other configurations is shown in figure 6 where preliminary data on the breakup of ^{24}Mg to α + ^{20}Ne and ^{8}Be + ^{16}O is shown. The remaining two

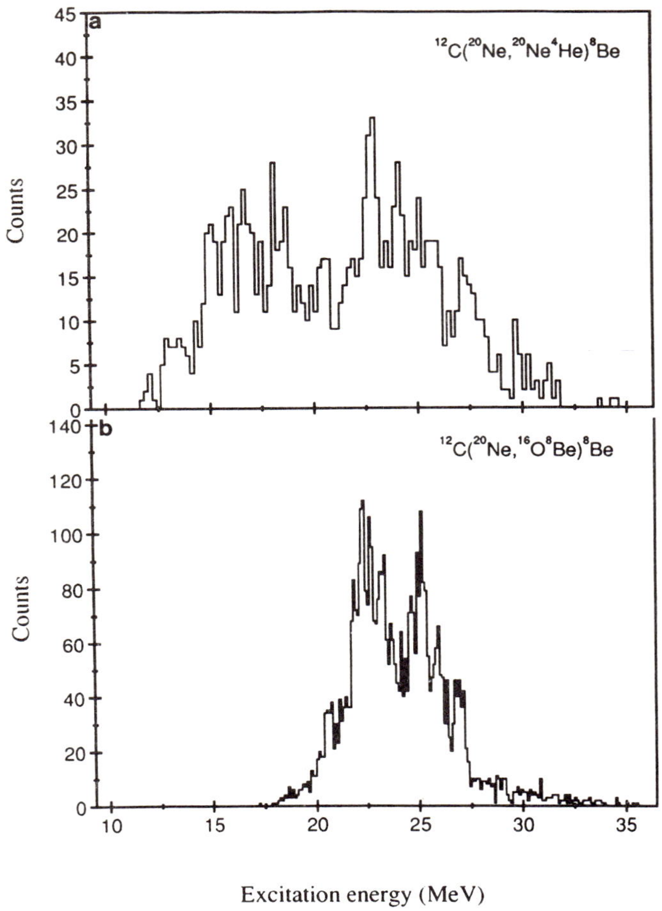

Fig. 6. **a** $\alpha - {}^{20}Ne$ and **b** ${}^{8}Be - {}^{16}O$ cluster states as revealed in the breakup of ${}^{24}Mg$

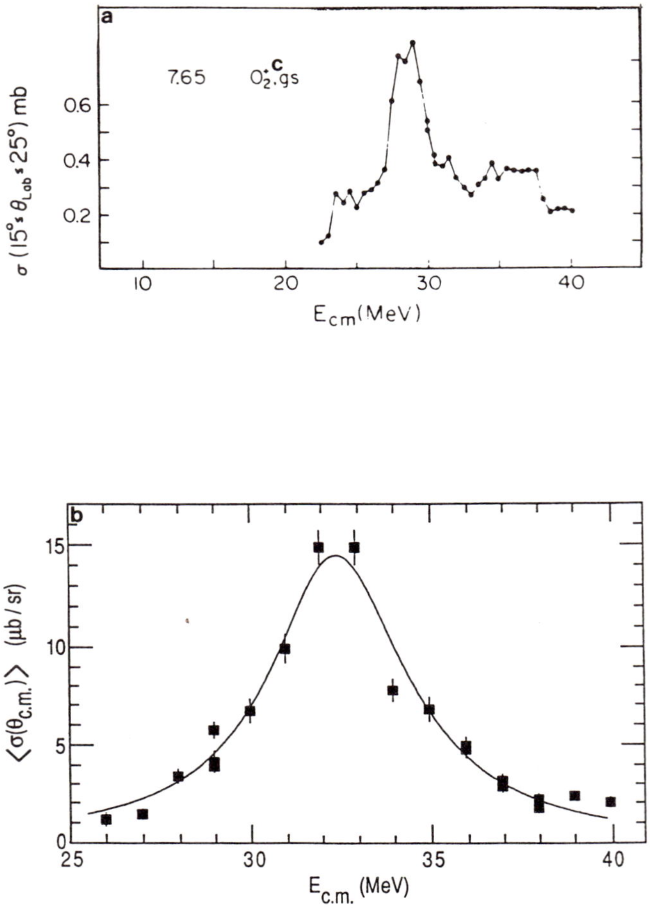

Fig. 7. Evidence for **a** the ${}^{12}C - {}^{12}C_{chain}$ state and **b** the 6α chain state in ${}^{24}Mg$ from ${}^{12}C + {}^{12}C$ scattering

configurations predicted by Ikeda involve chain-like clusters, the ${}^{12}C + {}^{12}C_{chain}$ configuration and the 6α chain state. Two difficulties have to be faced when looking for evidence of these structures - finding the best way to excite the state and deciding the best way to identify it when it breaks up. Since it is difficult to envisage a ${}^{24}Mg$ nucleus undergoing such a dramatic shape re-arrangement in a simple scattering, a more promising approach is to scatter two ${}^{12}C$ nuclei off each other. Since the chain state in ${}^{12}C$ is thought to be associated with the first excited 0^+ state, and since theoretical calculations suggest a few percent admixture of this in the ${}^{12}C$ ground state wavefunction, the argument is that some of the reactions may involve the ${}^{12}C$ colliding as alpha chains.

Figure 7 shows an excitation function for ${}^{12}C + {}^{12}C$ scattering, looking at the ${}^{12}C + {}^{12}C_{chain}$ yield (Fulton et al. 1980). The striking resonance apparent in the yield may well reflect the formation of the ${}^{12}C + {}^{12}C_{chain}$ structure. A similar approach may be adopted to search for the 6α chain state. A complication in this case is that we have no clue as to how such a state might decay, but it is a reasonable guess that it may split symmetrically into two ${}^{12}C$ nuclei in the chain state, which, since they are unstable, will immediately decay to three alphas. The experiment thus involves colliding two ${}^{12}C$ nuclei and looking for two jets of three alpha particles emerging on either side of the beam. The signature of the formation of the chain state would be a rapid variation in this yield as the energy of the beam is increased through the excitation energy of the state. This experiment was recently performed by a group from the Argonne Laboratory (Wuossma et al. 1992). In a beautiful experiment they not only confirmed the six alpha decay, but showed a near perfect Lorentzian resonance in the excitation function at an energy only a few MeV away from that predicted by the cluster calculations (Fig. 7).

In parallel with the experimental work, developments on the theoretical models have begun to provide us with further insight into these results. Figure 8 shows the result of cluster model calculations for the ${}^{24}Mg$ nucleus performed with an extended version of the Bloch-Brink alpha cluster model (Marsh and Rae 1986). In these calculations an unconstrained search is made which locates the positions of local energy minima in the potential energy surface. The calculations reveal a number of minima, the density projections of which are shown in the figure. Most of these configurations correspond to a small prolate deformation, which is consistent with the known shape of the ${}^{24}Mg$ nucleus in its ground state, but a number correspond to highly deformed shapes. Indeed the most extreme is the 6:1 axis ratio state

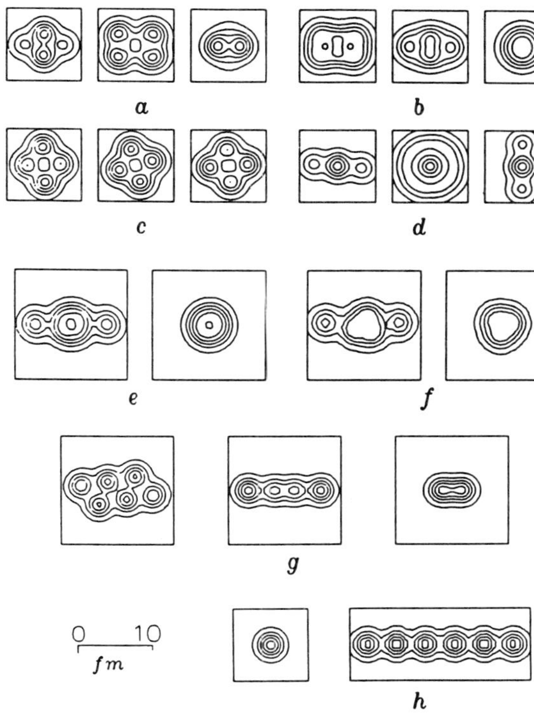

Fig. 8. Stable configurations in ^{24}Mg predicted by the cluster model calculations

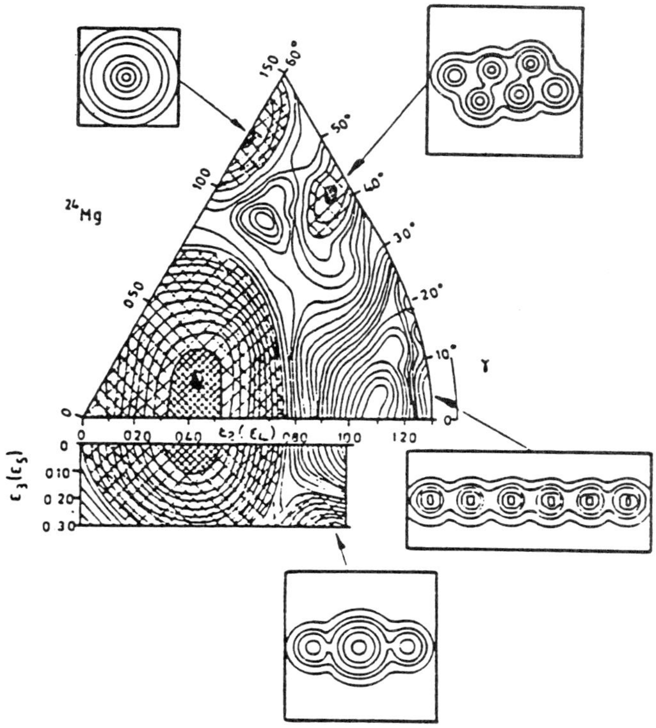

Fig. 9. The potential energy surface for ^{24}Mg showing the association between the secondary minima and the stable cluster model configurations

corresponding to all six alpha particles in a row - the alpha chain state (or "nuclear sausage") as it has been called. The calculations also include a cranking term which enables the stability of these structures to be explored as a function of angular momentum. The structures indeed turn out to be stable to quite high spins before the rotational stresses disrupt the structure. Here, then, are predictions of a range of highly deformed stable configurations in the ^{24}Mg nucleus on which rotational sequences are built. Moreover these fully quantum mechanical calculations confirm the cluster configurations predicted by Ikeda.

Interestingly the possibility that highly deformed shape isomers might exist in the ^{24}Mg nucleus had already been seen in other theoretical calculations. Figure 9 shows a potential energy surface for the ^{24}Mg nucleus based on a Nilsson-Strutinsky calculation (Leander and Larsson 1975). This shows a number of secondary minima in the potential energy surface, all at highly deformed shapes. The continuation of the potential energy surface into the lower panel below the normal ϵ-γ plane reflects the effect of including an octupole ($\epsilon 3$) degree of freedom in the calculation. One effect of this is to re-stabilise the minimum at a prolate 3:1 deformation which is otherwise disrupted because of the spin-orbit interaction. When the density distributions for the stable states found in the alpha cluster model calculations are compared with those corresponding to these secondary minima, there is a one to one correspondence (see figure 9). This agreement between two quite different theoretical descriptions is very gratifying, and adds support to the idea that highly deformed shape isomers exist in this nucleus which have a well established cluster structure.

Although the results I have discussed so far all refer to cluster structure in ^{24}Mg, this does not imply that the phenomenon is unique to that nucleus. Rather, it reflects the present concentration of work in this area. We do know from initial measurements on the breakup of ^{28}Si that it also shows states with a $^{12}C - ^{16}O$ cluster structure (Bennett et al. 1991), and in recent measurements we believe we may have located the seven alpha chain structure in this nucleus (Simmons 1993).

In fact both the cluster model and the Nilsson-Strutinsky calculations suggest the same features will be evident in a wide range of light nuclei. For example, Fig. 10 shows calculations for the potential energy surface in a wide range of alpha conjugate nuclei up to ^{40}Ca (Leander and Larsson 1975). All display a number of highly deformed secondary minima analogous to the ^{24}Mg nucleus. Cluster calculations also support the existence of these shape isomeric states and the likelihood is that there is a whole spectrum of highly deformed nuclear states awaiting experimental confirmation.

5. The future

And so to the future. The focus of interest is presently on the rather spectacular example of clustering evident in the chain states. The cluster model calculations predict that they will exist in a range of heavier nuclei and their identification and classification will be an exciting challenge. At the same time there is still much to be done on the highly deformed prolate cluster states. Can we, for instance, locate the same states in other nuclei? Calculations suggest they can occur

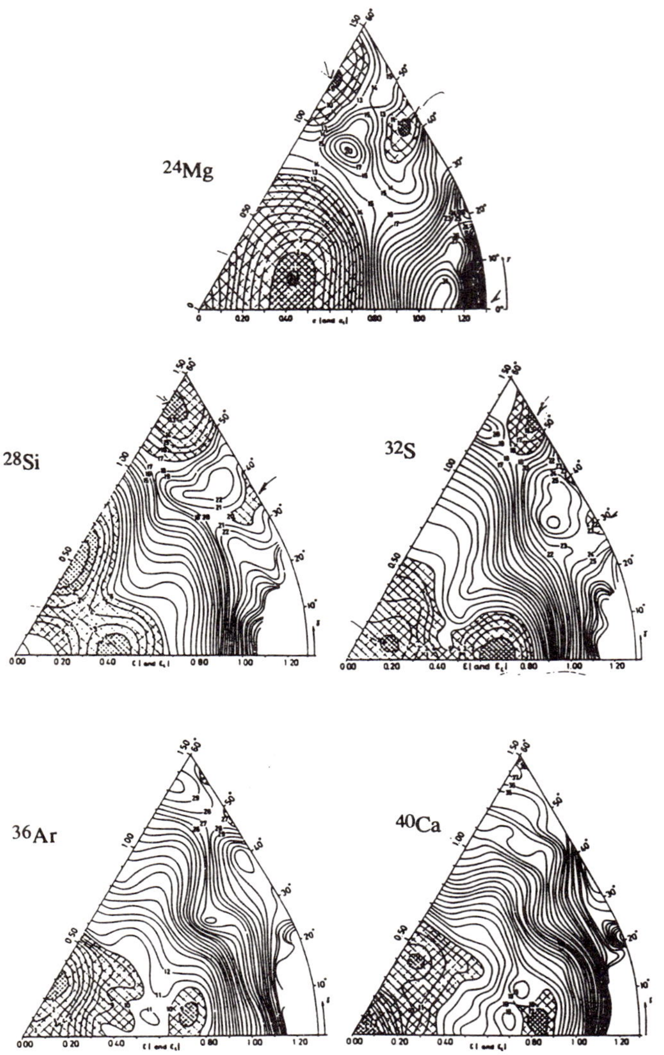

Fig. 10. The potential energy surface calculations for alpha conjugate nuclei from ^{24}Mg to ^{40}Ca

up to at least ^{56}Ni. Can we observe even longer chain structures involving clusters heavier than alphas? Cluster model calculations have indicated a likely candidate in ^{48}Cr which may have a $^{16}\mathrm{O} - {}^{16}\mathrm{O} - {}^{16}\mathrm{O}$ structure.

What about vibrations of the clusters? The spectra of atomic molecules are characterised by rotational and vibrational excitations - could the parallel exist in the case of nuclear molecules? We already have evidence for rotations of the clusters and it is interesting to speculate whether the splitting of the 4^+ states observed in Fig. 5 is evidence for a vibrational excitation. However much work needs to be done to confirm this speculation.

It is also possible that these cluster states will reveal novel decay modes. For example, it may be possible for alpha decay to occur between the highly deformed structures in neighbouring nuclei, such as the alpha conjugate nuclei in Fig. 9. Such decays would be a rather beautiful example of the physics of these secondary minima, and would be the particle decay analogue to the gamma decays observed within the secondary minima of heavier nuclei. In parallel with the experimental programme, there is much development needed on the theoretical side. We need to be able to explain why these states can have such narrow widths when they are so far above the particle decay threshold. We also need to be able to determine the branching ratios for decay to different cluster configurations. In addition we are also almost completely ignorant of the reaction mechanism through which these extreme nuclear states can be excited and a lot of effort is required in this area. Not only does this lack of knowledge hinder our experimental detection of the states, but also limits the structural information about them which reaction selectivities can reveal. Much work is needed in this area.

References

Ikeda K., Takigawa N., Horiuchi H.: Suppl. Prog. Phys. (Japan) Extra (1969) 464

Brink D.M.: Nucl. Phys.**A91**, 1(1967)

Brink D.M.: 1966 Int. school. of Physics Enrico Fermi XXXVII p247

Fulton B.R., Rae W.D.M.: J. Phys.**G16**, 333(1990)

Fulton B.R.,Bennett S.J.,Freer M.,Murgatroyd J.T., Gyapong G.J., Jarvis N.S., Jones C.D., Watson D.L., Brown J.D., Rae W.D.M., Smith A.E., Lilley J.S.: Phys. Lett.**267B**, 325(1991)

Fulton B.R., Cormier T.M., Harman B.J.: Phys. Rev.**C21**, 198(1980)

Wuossma A., Betts R.R., Back B.B., Freer M., Glagola B.G., Happ Th., Henderson D.J., Wilt P., Bearden I.G.: Phys. Rev. Lett. **68**, 1295(1992)

Marsh S., Rae W.D.M.: Phys Lett. **180B**, 185(1986)

Leander G., Larsson S.E.: Nucl. Phys. **A239**, 93(1975)

Bennett S.J., Freer M., Fulton B.R., Murgatroyd J.T., Woods P.J., Allcock S.C., Rae W.D.M., Smith A.E., Lilley J.S., Betts R.R.: Nucl. Phys. **A534**, 445(1991)

Simmons P.M.: Contribution to the Conference on Nuclear and Particle Physics, Glasgow 1993 (unpublished)

Z. Phys. D 31, 187–189 (1994)

Collisional process involving atomic cluster ions

J. Hirokawa, M. Ichihashi, S. Nonose, T. Kondow

Department of Chemistry, School of Science, The University of Tokyo, Bunkyo-ku, Tokyo 113, Japan

Received: 28 June 1993 / Final version: 16 March 1994

Abstract. Collision of Ar cluster ions, Ar_n^+ ($n=3$–16), with He and Ne atoms was investigated by use of mass spectroscopic techniques. The cross sections for the production of $Ar_{n'}^+$ ($n'<n$) were measured as functions of the size of the parent cluster ion and the collision energy (0.1–10 eV in the center-of-mass frame). These results were analyzed in the scheme of hard-sphere spectator collision with RRK theory. It was concluded that the reaction proceeds via collisional excitation of the parent cluster ion and following sequential loss of the constituent Ar atoms.

PACS: 36.40.+d

1. Introduction

Cluster collisions attract much attention in close analogy with nuclear heavy-ion collisions and are possibly explained in terms of concepts similar to those applied for understanding dynamical features of nuclear collisions. In this connection, various collision systems involving clusters have been explored [1-5]. These studies have revealed that the cluster collisions proceed via collisional excitation and following dissociation. The dynamics is characterized by the vibrational modes of the clusters associated with the collisions, and therefore varies depending on the collision energy and the cluster size. In order to clarify the essential features of the collision dynamics, the collisional reactions of Ar_n^+ with rare gas atoms were chosen as a prototype reaction since the structure of Ar_n^+ consisting of a trimeric core ion and solvent neutral Ar atoms is well studied by many theoretical and experimental studies such as photodissociation [6]. Previously, we have examined the collisional reaction of Ar_n^+ with Kr, in which charge transfer and fusion processes are observed as well as dissociation [7]. The present study deals with the dissociation process of Ar_n^+ in collision with He and Ne atoms. In slow collisions of 0.1–10 eV in the center-of-mass frame, it is expected that low frequency vibrational modes of the cluster ion play an important role in its collisional excitation, and the dissociation of the collisionally excited cluster ion is statistical. The cross section for the collisional dissociation and its dependences on the collision energy and the cluster size were interpreted on this basis.

This paper was originally submitted in connection with the 2nd. Int. Conference on Atomic and Nuclear Clusters held in Santorini from 28. June–2. July 1993 and is published here as a regular article after an independent refereeing procedure according to the standards of Z. Phys. D

2. Experimental

Figure 1 shows a schematic drawing of the apparatus. Cluster ions were produced by electron impact on a supersonic expansion of argon gas at a stagnation pressure of 5 atm, and then were size-selected by a quadrupole mass spectrometer with a mass resolution, $m/\Delta m \geqq 250$, at $m/z=1000$. The cluster ion, Ar_n^+, was transported into octopole ion guides and was allowed to collide with a

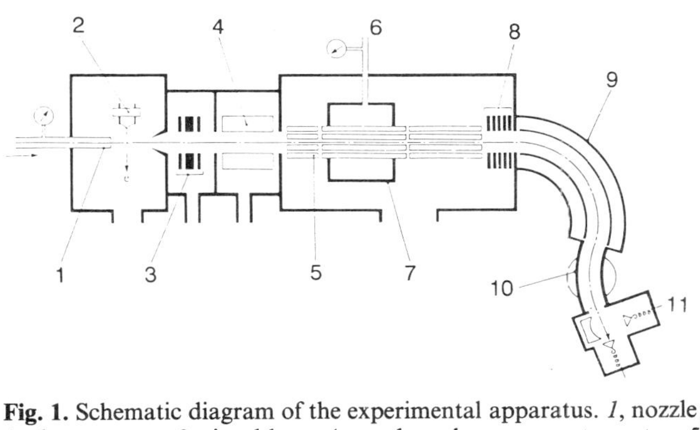

Fig. 1. Schematic diagram of the experimental apparatus. *1*, nozzle; *2*, electron gun; *3*, einzel lens; *4*, quadrupole mass spectrometer; *5*, octopole ion guides; *6*, inlet of the target gas; *7*, collision cell; *8*, acceleration lens; *9*, electrostatic sector; *10*, magnetic sector; *11*, detector

target gas He or Ne of $\sim 1\times10^{-5}$ Torr. The pressure of the target gas was measured by an MKS SRG-2 spinning rotor gauge. The collision energy was varied by changing the voltage between the ion source and the second octopole ion guide. The product ions were mass-selected and detected by a double-focussing mass spectrometer. The signals were processed by a multichannel scaler based on an NEC PC-9801 computer.

The temperature of the parent cluster ion was estimated from the unimolecular decay of the parent ion on the assumption of RRK theory as argued in Sect. 3. The temperature thus obtained was found to decrease with the cluster size from 150 to 50 K in the entire size range studied.

3. Results

In the collision of Ar_n^+ with Rg (= He or Ne), the product ions were Ar_p^+ ($p < n$) and hence the dominant process is considered to be evaporation of Ar atoms from Ar_n^+. The total cross section, σ_t, for the evaporation was calculated from

$$\sigma_t = -\frac{k_B T}{PL}\ln\frac{I_r}{I_0}, \quad (1)$$

where I_0 and I_r represent the intensities of the parent ion at the entrance and the exit of the collision cell, respectively, P is the pressure of the target gas, Rg, L is the effective path length, T is the temperature of the collision cell (300 K), and k_B is Boltzmann's constant. The ion intensities due to the unimolecular decay of the parent ion were subtracted from the observed ones, and I_0 and I_r were obtained. The partial cross section, σ_p, for the production of Ar_p^+ from the parent ion, Ar_n^+, was obtained from

$$\sigma_p = \sigma_t \frac{I_p}{\sum I_p}, \quad (2)$$

where I_p represents the intensity of the product ion, Ar_p^+.

The total cross section for the collisional dissociation of Ar_{10}^+ was measured by changing the collision energy in the range from 0.1 to 2.4 eV in the center-of-mass frame. Unless otherwise noted, the collision energy is given in the center-of-mass frame. As shown in Fig. 2, the total cross section for the He collision with Ar_{10}^+ increases with the collision energy, whereas the cross section for the Ne collision does not change appreciably with the collision energy. Figure 3 shows the total cross sections for the He and the Ne collisions plotted against the size of the parent cluster ion, n, in the logarithmic scale at the collision energy of 0.4 eV. In the $n > 7$ region, both the cross sections appear to depend on $n^{2/3}$.

The branching fractions of the product ions were also measured for the parent cluster ions. The branching fractions were found to be well reproduced by giving the most adequate value to the parameter, $\overline{\Delta E}$, on the assumption of sequential loss of the constituent Ar atoms from the collisionally excited parent cluster ion which was treated

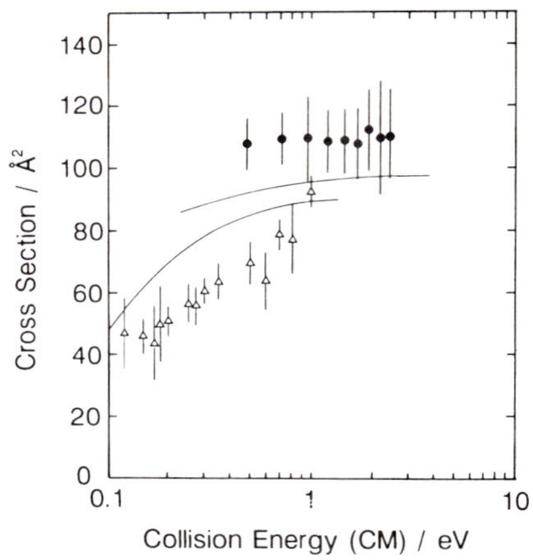

Fig. 2. Total cross sections for the collision of Ar_{10}^+ with He (Δ) and Ne (●) atoms are plotted as a function of the collision energy in the center-of-mass frame. The cross section for the Ne collision does not appreciably depend on the collision energy, while that for the He collision increases with the collision energy. The *upper* and the *lower solid curves* show the calculated cross sections for the Ne and the He collisions, respectively

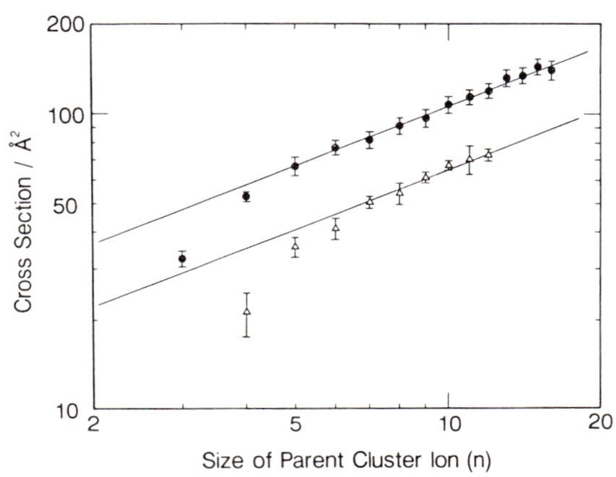

Fig. 3. Total cross sections are plotted against the size of the parent cluster ion, Ar_n^+, for the He (Δ) and the Ne (●) collisions at the collision energy of 0.4 eV. The *solid lines* show the $n^{2/3}$ dependence

with RRK theory [8]. Here $\overline{\Delta E}$ represents average transfer energy, an amount of energy acquired by the parent cluster ion in the collision. The assumption of the sequential loss is actually supported by the molecular dynamics simulation for the $Ar_n^+ + Ar$ collision system [9]. The most adequate values of $\overline{\Delta E}$ for the He and the Ne collisions thus obtained are plotted as a function of n at the collision energy of 0.4 eV in Fig. 4.

4. Discussion

The results indicate that the reaction proceeds via (a) collisional excitation and (b) subsequent dissociation of the parent cluster ion, Ar_n^+. In the process (a), an amount of energy acquired by the parent cluster ion in the collision, $\Delta E(b)$, should be given as a function of the impact

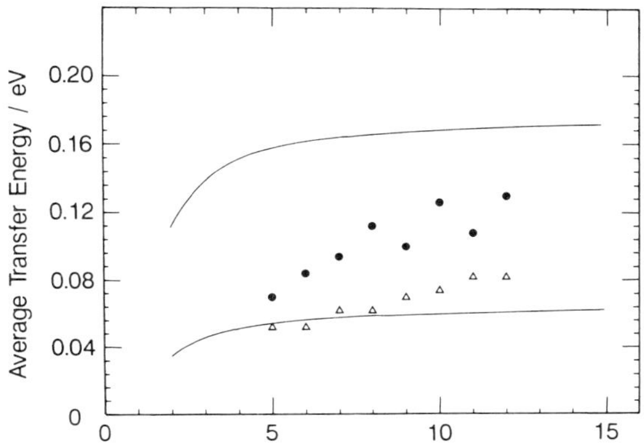

Fig. 4. Average transfer energies for the He (Δ) and the Ne (●) collisions are plotted as a function of the size of the parent cluster ion, Ar_n^+, at the collision energy of 0.4 eV. The *upper* and the *lower solid curves* represent size dependences of the average transfer energies calculated for the Ne and the He collisions, respectively, at the collision energy of 0.4 eV

parameter, b. In a hard-sphere collision model, the $\Delta E(b)$ for the Rg (= He or Ne) collision may be given as follows by taking into consideration of the impact parameter dependence of $\Delta E(b)$ for the ^{36}Ar collision obtained from the molecular dynamics calculation [9] as follows:

$$\Delta E(b) = \eta^{\mathrm{Rg}} E_{\mathrm{col}} \left[1 - \left(\frac{b}{b_{\mathrm{max}}^{\mathrm{Rg}}}\right)^2\right] \tag{3}$$

$$b_{\mathrm{max}}^{\mathrm{Rg}} = b_{\mathrm{max}}^{\mathrm{Ar}} - (r_{\mathrm{Ar}} - r_{\mathrm{Rg}}), \tag{4}$$

where b_{max}^X (X = Rg, Ar) represents the impact parameter for the X collision above which no collisional energy transmission occurs, and r_X represents the radius of X. The value, $b_{\mathrm{max}}^{\mathrm{Ar}}$, is given by the molecular dynamics simulation [9]. The parameter, η^{Rg}, is the conversion efficiency, which is the fraction of the collision energy, E_{col}, converted to the internal energy of the parent cluster ion, Ar_n^+, in the head-on collision ($b=0$). The conversion efficiency, η^{Rg}, depends on the masses of the projectile and the target, and is calculated by assuming that the target atom interacts mainly with a single Ar atom in Ar_n^+ since the Ar-Ar interaction is weak (spectator model). In the hard-sphere collision, the reaction always occurs in the $b \leqq b_c$ range; in the RRK scheme, b_c can be given in such a manner that almost all (> 99%) the parent cluster ions are dissociated to release at least one Ar atom in the time window of the reaction in the present experimental setup (~100 μs) when $\Delta E(b_c)$ is converted to them. By taking this into account, the total cross sections were calculated. In the Ne collision, the total cross sections were calculated to be $\sim 89 \times 10^{-16}\,\mathrm{cm}^2$ and $\sim 161 \times 10^{-16}\,\mathrm{cm}^2$ for $n=6$ and 13, respectively at the collision energy of 0.4 eV, and agree with the experimental ones with the deviation of ~15%. In the He collision, the agreement of the calculated and the experimental cross sections is not satisfactory, probably because of the errors of the internal energy estimation. The collision energy dependences of the total cross sections are also calculated for the $He - Ar_{10}^+$ and the $Ne - Ar_{10}^+$ systems as shown in Fig. 2. The collision energy dependences are qualitatively reproduced by the calculation based on the present scheme. The scaling law of the total cross section, $\sigma_t \sim n^{2/3}$, is explained as well by this collision scheme. Since the binding energies of argon cluster ions are almost same for $n \geqq 4$ [10], b_c increases almost linearly with the increase of b_{max}. As b_{max} depends approximately on $n^{1/3}$, the total cross section, πb_c^2, is approximately proportional to $n^{2/3}$, as expected from the scaling law.

The average transfer energy, $\overline{\Delta E}_{\mathrm{calc}}$, is given as,

$$\overline{\Delta E}_{\mathrm{calc}} = \frac{\int_0^{b_{\mathrm{max}}^{\mathrm{Rg}}} 2\pi \Delta E(b)\, b\, \mathrm{d}b}{\int_0^{b_{\mathrm{max}}^{\mathrm{Rg}}} 2\pi b\, \mathrm{d}b} = \tfrac{1}{2} \eta^{\mathrm{Rg}} E_{\mathrm{col}}, \tag{5}$$

where η^{Rg} is estimated on the basis of the spectator model. The $\overline{\Delta E}_{\mathrm{calc}}$ values obtained from (5) are shown in Fig. 4 (solid curves) as a function of the parent cluster size, n, at the collision energy of 0.4 eV; the upper and the lower curves correspond to those calculated for the Ne and the He collisions, respectively. This scheme predicts well the $\overline{\Delta E}$ vs. n dependence for the He collision, while does not for the Ne collision, probably because the Ne atom interacts with more than one Ar atom in Ar_n^+ in the collision.

In conclusion, the collisional dissociation of Ar_n^+ by He and Ne atoms proceeds via collisional excitation of the parent cluster ion, Ar_n^+, and sequential loss of the constituent Ar atoms. Here, the collisional excitation and the following dissociation processes are analyzed by a hard-sphere spectator model together with RRK theory.

References

1. Woodward, C.A., Stace, A.J.: J. Chem. Phys. **94**, 4324 (1991)
2. Campbell, E.E.B., Schneider, R.R., Hielscher, A., Tittes, A., Ehlich, R., Hertel, I.V.: Z. Phys. D**22**, 521 (1992)
3. Campbell, E.E.B., Ehlich, R., Hielscher, A., Frazao, J.M.A., Hertel, I.V.: Z. Phys. D**23**, 1 (1992)
4. Campbell, E.E.B., Schyja, V., Ehlich, R., Hertel, I.V.: Phys. Rev. Lett. **70**, 263 (1993)
5. Buck, U., Meyer, H.: J. Chem. Phys. **84**, 4854 (1986)
6. Nagata, T., Kondow, T.: J. Chem. Phys. **98**, 290 (1993)
7. Ichihashi, M., Hirokawa, J., Nonose, S., Nagata, T., Kondow, T.: Chem. Phys. Lett. **204**, 219 (1993)
8. Draves, J.A., Luthey-Schulten, Z., Liu, W.-L., Lisy, J.M.: J. Chem. Phys. **93**, 4589 (1990)
9. Ichihashi, M., Nonose, S., Nagata, T., Kondow, T.: J. Chem. Phys. (in press)
10. Hiraoka, K., Mori, T.: J. Chem. Phys. **90**, 7143 (1989)

Fragmentation of gold clusters stored in a Penning trap

St. Becker[1], G. Dietrich[2], H.-U. Hasse[1], N. Klisch[1], H.-J. Kluge[1], D. Kreisle[3], St. Krückeberg[1], M. Lindinger[1], K. Lützenkirchen[2], L. Schweikhard[1], H. Weidele[3] and J. Ziegler[1]

[1]Institut für Physik, Univerität Mainz, D-55099 Mainz, Germany
[2]Institut für Kernchemie, Univerität Mainz, D-55099 Mainz, Germany
[3]Institut für Physik, Univerität Konstanz, D-78434 Mainz, Germany

Received 28 June 1993

Abstract. The collision-induced dissociation of positively charged gold clusters (2 to 23 atoms) stored in a Penning trap has been studied. After collisions with rare gases, excited clusters predominantly decay by emission of one or two atoms. The loss of two atoms occurs most likely through the emission of a dimer rather than a sequential evaporation. The minimum kinetic energies of clusters required to induce dissociation exhibit a pronounced odd-even effect. Clusters with an even number of delocalized electrons are more stable than the odd ones.

PACS: 36.40.+d, 34.50.-s, 35.20.Gs

1. Introduction

The binding energy of atoms is one of the properties of clusters which are peculiar to the finite size of the system. On the average, for example, the binding energy is found to be lower than the energy needed to separate an atom from a bulk solid, which can be understood by a surface-tension effect. In addiotion, size-dependent alternations of the binding energy have been observed for small clusters [1]. Pronounced variations of the binding energy are also expected for clusters with closed shells of valence electrons.

Collision-induced dissociation is one way to study binding energies. Such investigations have been carried out for a number of metal clusters, e.g. aluminum [2,3] and nickel [4]. In the following we shall report on measurements of the collision-induced dissociation of singly charged gold clusters Au_N^+ with 2 to 23 atoms, stored in a Penning trap.

2. Experimental

Positively charged gold clusters are formed in a pulsed laser-vaporization source [5]. The clusters with positive charge are accelerated to an energy of 100 eV and guided towards a Penning trap, located in the bore of a 5-Tesla superconducting magnet.

The Penning trap consists of a segmented ring electrode and two end caps [6]. Each end cap has a hole in the center in order to inject/eject clusters into/out-of the trap. The clusters are captured in flight in the Penning trap [7] by suitably pulsing the potential of the entrance end cap. Typically, 10 to 100 cluster ions can be injected for each pulse of the desorption laser (Fig. 1a).

A cluster ensemble of defined size Au_N^+ is prepared by radial rf-excitation of all other clusters, which eventually collide with the walls of the ion trap (Fig 1b). Subsequently, the kinetic energy of the remaining Au_N^+ is adjusted by rf-excitation with a (variable) voltage V_{exc} at its cyclotron frequency, $\omega_c \simeq qB/M$.

The kinetic energy of the clusters is dissipated through collisions with krypton or xenon atoms (pressure $\simeq 6 \cdot 10^{-6}$mbar). Apart from the recoil energy imparted on the rare gas atoms, the kinetic energy is converted into excitation energy of the clusters' vibrational degrees of freedom. About 500ms after rf-excitation, the reaction products are identified in a time-of-flight mass-spectrometer (Fig 1c).

3. Measurements and Results

After collisions with xenon (Fig. 1c) Au_7^+, a cluster with an even number of (six) valence electrons, mainly decays into Au_5^+, with a minute amount of Au_6^+. Smaller clusters result from further decay of the reaction products [8].

The intensity of the reaction products has been observed as a function of the excitation voltage. The threshold excitation voltage needed to form Au_5^+ is found to be lower than the one for Au_6^+. Thus, Au_5^+ results from Au_7^+ by the evaporation of a dimer, Au_2, rather than by sequential evaporation of two monomers. Clusters with an odd number of delocalized electrons first evaporate an atom resulting in an even-electron cluster which further

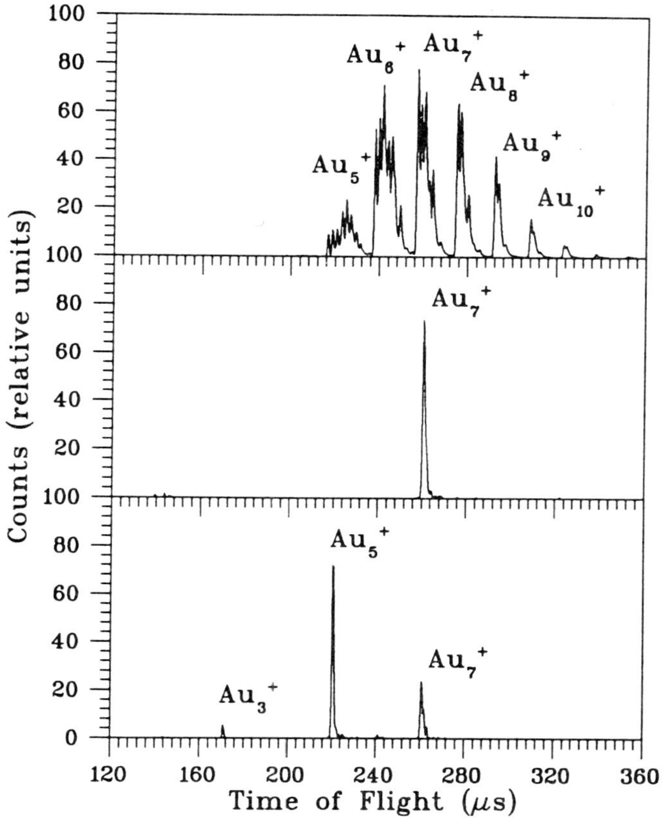

Fig. 1. Time-of-flight spectra of clusters ejected from the Penning trap at various stages of a collision experiment. (a) Trapping of Au_5^+ to Au_{12}^+. (b) Preparation of Au_7^+ by radial ejection of the other clusters. (c) Fragmentation of Au_7^+ after rf-excitation and collisions with xenon.

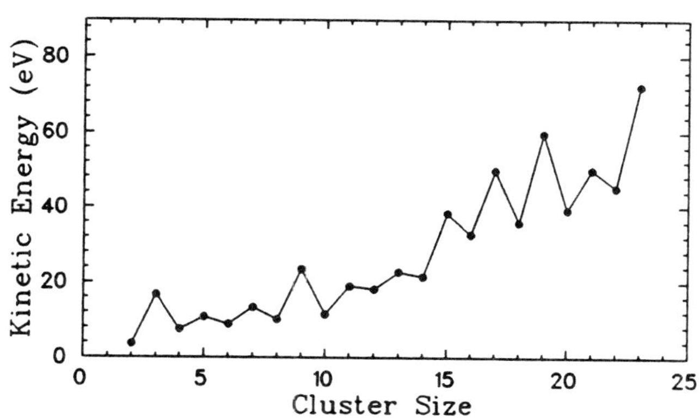

Fig. 2. Threshold kinetic energies for dissociation of gold clusters Au_N^+ induced by collisions with xenon.

decays by evaporation of dimers. For clusters larger than Au_{15}^+ only monomer evaporation is observed.

From the threshold excitation voltages the respective kinetic energies (laboratory system) are calculated, $E_{kin} \propto V_{exc}^2/N$, and depicted in Fig. 2. Since each cluster is likely to undergo several collisions prior to the first evaporation, it is not straightforward to determine the absolute excitation energies and, in turn, the dissociation energies from the threshold values of E_{kin}; a detailed knowledge of the inelastic energy loss mechanism would be required. Some progress may be expected from a study of the collision process based on quantum molecular dynamics calculations [9], which give information on the kinetic energy loss for a given impact parameter. The multiple collision process between clusters and rare gas atoms can then be simulated by the Monte Carlo technique and compared with experiments. Alternatively, absolute dissociation energies may also be obtained by inducing fragmentation with photons from a tunable laser.

The threshold kinetic energies show a clear odd-even alternation which, apart from some smooth function of the cluster size, reflects the alternations in the dissociation energy. Clusters with an even number of delocalized electrons have a higher dissociation energy than the ones with an odd number. This can be understood by a dependence of the highest occupied electronic states on the deformation of the clusters [10]. Au_3^+ and Au_9^+ are found to be particularly stable which is explained by closed shells of valence electrons. An additional discontinuity is observed for Au_{19}^+. These findings are in line with mass spectra of Au_N^+ [11], where pronounced discontinuities are reported for Au_3^+, Au_9^+, Au_{19}^+, and Au_{21}^+, which correspond to closed s-, p-, and d-valence electron shells.

References

1 Bréchignac, C., Cahuzac, Ph., Leygnier, J., and Weiner, J.: J. Chem. Phys. **90**, 1492 (1989)
2 Jarrold, M.F., Bower, J.E., and Kraus, J.S.: J. Chem. Phys. **86**, 3876 (1987)
3 Hanley, L., Ruatta, S.A., and Anderson, S.: J. Chem. Phys. **87**, 260 (1987)
4 Lian, L., Su, C.X., and Armentrout, P.B.: J. Chem. Phys. **96**, 7542 (1992)
5 Weidele, H., Frenzel, U., Leisner, T., and Kreisle, D.: Z. Phys. D **20**, 411 (1991)
6 Schweikhard, L., Lindinger, M., and Kluge, H.-J.: Int. J. Mass Spectrom. Ion Processes **98**, 25 (1990)
7 Schnatz, H., Bollen, G., Dabkiewicz, P., Egelhof, P., Kern, F., Kalinowsky, H., Schweikhard, L., Stolzenberg, H., and Kluge, H.-J.: Nucl. Instr. Meth. A **251**, 17 (1986)
8 Becker, St., Dietrich, G., Hasse, H.-U., Klisch, N., Kluge, H.-J., Kreisle, D., Krückeberg, St., Lindinger, M., Lützenkirchen, K., Schweikhard, L., Weidele H., and Ziegler, J.: to be published
9 Schmidt, R.: private communication
10 Penzar, Z., and Ekardt, W.: Z. Phys. D **19**, 109 (1991)
11 Katakuse, I., Ichihara, T., Fujita, Y., Matsuo, T., Sakurai, T., and Matsuda, H.: Int. J. Mass Spectrom. Ion Processes **74**, 33 (1986)

This article was processed using Springer-Verlag TEX Z.Physik D macro package 1.0 and the AMS fonts, developed by the American Mathematical Society.

Z. Phys. A 349, 233–236 (1994)

Binary nuclear molecules of an α-particle and a heavier cluster

M. Brenner

Åbo Akademi University, Department of Physics, FIN-20500 Turku, Finland

Received: 27 October 1993

Abstract. A simple nuclear molecule is used as a model to interpret the band structure observed in elastic α-scattering and α-transfer reactions in the sd-shell. As in spectra of halogen acid molecules the bands have even and odd spins and parities. Due to the broadening and fine structure of the rotational levels they are characterized as giant resonances. The big moment of inertia and the line broadening is discussed in terms of performed measurements of scattering and the (α, p) reaction. A fine structure of prominent peaks in the inelastic scattering cross-sections hints at the existence of rotational states with an α-particle orbiting an excited core.

PACS: 21.60.Gx; 24.30.Cz; 25.55.Ci

1. Introduction

The rotatioal nuclear molecules of one α-particle orbiting a heavier nucleus extend the Kepler-Bohr concept of a mass point orbiting another one several orders of magnitude heavier. As the orbital radius is only slightly larger than the radius of the nucleus encircled by the α-particle and the centripetal force is strong, the interaction between the two bodies is huge. In comparison, the tides of the moon-earth interaction and the interaction related to the atomic fine structure are small. The rotational nuclear molecules can be compared to atomic molecules with respect to the short distance between their constituents. As to the strong interaction, they are like exotic atoms consisting of a sigma hyperone orbiting a nucleus. Molecules consisting of two atomic clusters may have properties similar to the binary nuclear molecules, with respect to the multiparticle structure of their two members.

2. Rotational states

In the rotational spectra of halogene acids, the energy depends linearly on the spin squared. The energy E plotted versus $J(J+1)$ fits a line. The members of the rotational band have even and odd spins and parities, as the system is assymmetric. Hence we do not expect parity splitting of bands formed by the rotational α-cluster states to be discussed. The bands of the α-cluster states differ to this end from rotational bands of deformed nuclei, which usually have even spins and even parities. Separate even and odd parity bands of α-cluster states, however, occur. They have been observed in agreement with calculations based on local potentials, as reviewed by Hodgson [1]. The parity splitting results from the influence of the Pauli principle on the fundamental fermion interaction, which makes up the potential.

3. Experiments on elastic α-scattering and transfer reactions

For a review of the rotational like bands of even or odd parity in ^{16}O and ^{20}Ne an article by Buck et al. is referred to [2]. Parity splitting of bands has been well established experimentally and theoretically in ^{40}Ca and ^{44}Ti [3]. Bands of states with mixed parity were first deduced from resonances in α-particle scattering from ^{40}Ca targets [4]. The scattering yield or cross-section as a function of the bombarding energy, called excitation function, was in these pioneering works and still is a powerful tool to reveal the existence and the properties of the α-cluster states.

By another method Artemov et al. have brought about a systematic study of states in several nuclei between oxygen and sulphur. They applied a transfer reaction in which an α-particle is captured from ^{6}Li by the target nucleus. Results for targets from carbon to calcium were summarized in a scheme by this group [6]. As pointed out by Rudakov [5], mixed parity rotational bands have been seen in nuclei from ^{28}Si to ^{44}Ti, but not in ^{16}O or ^{20}Ne. Spin determinations of resonances in scattering excitation functions are in agreement with the transfer reaction measurements by Artemov et al. [7]. In the α-transfer to ^{28}Si the molecular states are observed as peaks

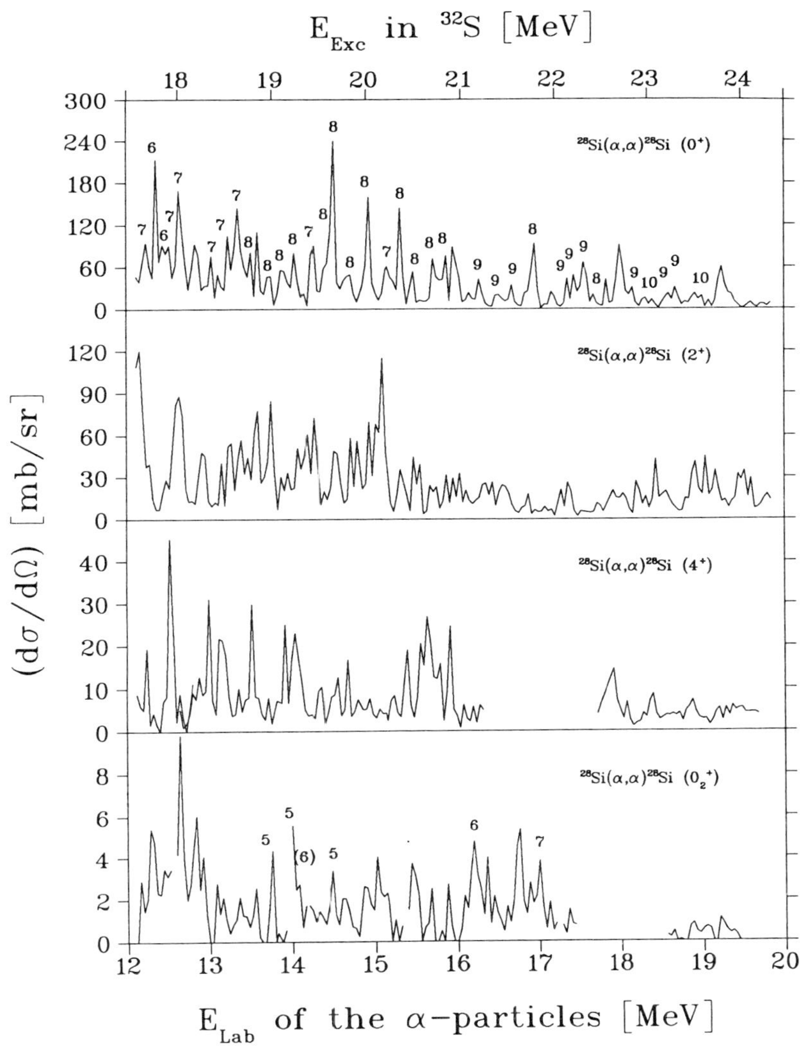

Fig. 1. Excitation functions at $\Theta = 173°$ with resonances in: elastic scattering of α-particles by ^{28}Si (uppermost curve), and inelastic resulting in the excitation of the first 2^+ and 4^+ and the second 0^+ state. In each of the two lowest curves is an empty region because of an oxygen impurity by which particles of the same energy as from the silicon are scattered

in the energy spectrum of coincident α-particles and deuterons.

There are indications that these peaks have a structure, which the experimental resolution is not able to resolve. The structure is, however, displayed in the α elastic scattering [8, 9], which exhibits a splitting of the α-cluster states into a fine structure of narrow peaks as seen in Fig. 1. Angular distributions measured at these peaks in the backward direction have been fitted by squared Legendre polynomials, to give the spin of the rotational state they belong to. Due to a large broadening of the cluster level, there are regions where two gross resonances overlap resulting in a mixing of peaks of different spins. A strong argument for the single α-particle nature of the broadened level is the sum of the reduced α-width of the fine structure levels, because it is of the order of one Wigner limit for single particle states [8]. The total width of this giant resonance is of the order of 1 MeV.

A line is obtained when the weighted means of the energies of peaks with the same spin J are plotted versus $J(J+1)$. This is shown in Fig. 2 for ^{32}S, which α-cluster states will be discussed in more detail in this presentation. The data are from recent thick target measurements of

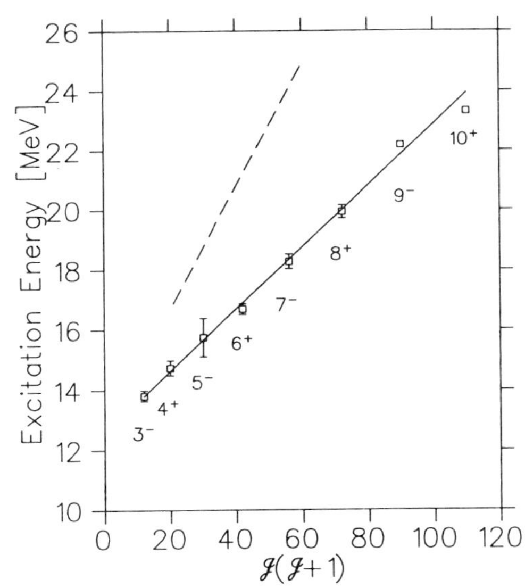

Fig. 2. Mean values of excitation energies of resonance states in ^{32}S with the same spin J versus $J(J+1)$. A *straight line* is fitted to the experimental points. The *broken line* is for an alpha particle orbiting the Si nucleus with a center to center radius $r = 1.25\,A^{1/3} + 1.6$ fm

α-particle elastic scattering [10]. The equation of the solid line is $E_{exc} = 0.103\,J(J+1) + 12.6$ MeV. From the coefficient 0.103 MeV the moment of inertia $4.85\,\hbar^2/\mathrm{MeV}$ is obtained.

The moment of inertia is a sum of the moment of inertia of the α-particle in orbit and 56% of the moment of the rigid ^{28}Si core. For comparison the dashed line in the same figure has been constructed for an α-particle only in orbit with the α-particle touching the encircled nuclear cluster, as necessary to allow for a centripetal attraction by the short ranged nuclear force. The radius $r = 1.25\,A^{1/3} + 1.6$ fm has been used.

4. The broadening of α-cluster resonances

An extensive presentation of line-broadening has been given by Lane [11]. It is a general phenomenon resulting from the interaction between an underlying special state and a background of ordinary states. The gross level of the special state is a giant resonance and the constituent small levels are its fine structure. In nuclear physics the first observed example of such a resonance is the dipole giant resonance. Others are the broadenings of single particle shell model states and doorway states of two-nucleon, one-hole; three-nucleon, two-hole etc., fission doorway states and isobaric analogue states. The background states, which give the fine structure can be bound or unbound. The rotational α-cluster molecule offers a new example of line-broadening comparable to the above mentioned ones. To find the cause of the broadening of the α-cluster states in ^{32}S, we first look for the compound states.

Fermi gas calculations [12] for the excitation energy region of Fig. 1 give a typical level density of $30\,\mathrm{keV}^{-1}$ for $J = 8$ levels. As seen in Fig. 1 the density of the fine structure resonance is of the order of $10\,\mathrm{MeV}^{-1}$. This is in disagreement with an assumption that the compound states cause the fine structure.

A strong interaction between the α-cluster states and the compound states would enhance the emission of neutrons and protons at the resonances. To find out if this is the case, spectra of the emitted protons were studied with varying energy in a region where the strong 14.6 MeV (E_{lab}) scattering resonance peak appears. As seen in Fig. 3 no resonance appears in the excitation function for the emission of protons summed over all proton energies. The proton emission resulting in the population of the lowest states in ^{31}P displays fluctuations but there is no correlation between the maxima in the proton intensity and the resonances in the elastic scattering.

Considering the facts above, interference between the α-cluster state and the compound states is not expected to be the cause of the line broadening. We are thus led to consider the existence of other states to find the background states related to the fine structure of the scattering excitation function in Fig. 1.

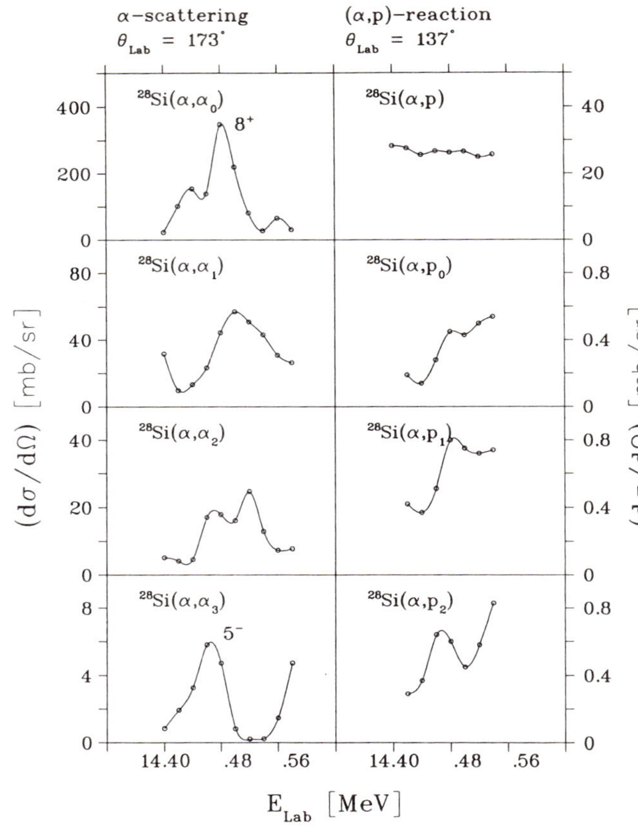

Fig. 3. *Left:* Differential cross-section at $\Theta = 173°$ for the emission of elastically scattered (uppermost) and the three lowest groups α_1, α_2 and α_3 of inelastically scattered α-particles. *Right:* Differential cross-section at $\Theta = 137°$ for the α-particle reaction resulting in the emission of protons of any energy (uppermost) and three groups p_1, p_2 and p_3 resulting in the population of the lowest $1/2^+$, $3/2^+$ and $5/2^+$ states in ^{31}P

5. The interaction with other α-cluster structures

When searching for another source of the line broadening the next choice is some other kind of cluster states in ^{32}S. This approach is encouraged by the many α-particle structures found lately e.g. in ^{24}Mg [13].

If we consider a single α-particle and a ^{28}Si core as one extreme of the α-cluster structure of ^{32}S, the other one would be a Bose-gas of 8 α-particles. For this system the Fermi gas calculation can be done the same way as for ^{8}Be, but with the sulphur radius replacing the beryllium radius. At high energy, the difference between Fermi-Dirac and Bose-Einstein statistics is negligible. The resulting density of $J = 8$ states at $E_{exc} = \sim 20$ MeV is $10\,\mathrm{MeV}^{-1}$ in agreement with the experiment. We conclude, accordingly, that the α-particle of the rotational state is influenced by the bosonic α-particle structure of ^{32}S instead of its fermionic compound nucleus. Thus the Pauli principle should not be applied in calculating the energies of these states. As the splitting of the bands into separate even and odd parity bands is a consequence of the Pauli principle, the conclusion explains why we see mixed parity bands.

6. Experiments on inelastic α-scattering

In the excitation functions for inelastic α-scattering, a fine structure of prominent peaks is observed (Fig. 1). The inelastic scattering resulting in the excitation of the second 0^+ state (0_2^+) allows for the assignment of the spins of the fine structure peaks from the backward angular distribution, in the same way as does the elastic scattering data. The spins so far determined are indicated in the figure. In comparison to the elastic scattering data they are lower by 2–3 units of $\hbar$.

No correlation is found in energy or spin between the peaks in the different excitation functions. This fact calls for a search for states formed by an α-particle orbiting cores of different state of excitation.

7. Conclusions and comments

Elastic α-scattering and transfer reactions have in earlier experiments given evidence for the existence of α-cluster states comprizing an α-particle orbiting a nuclear core eg. the *sd*-nuclei ^{24}Mg and ^{28}Si. The states occur over a broad energy region of excitation e.g. in ^{32}S ($\alpha + {}^{28}$Si) from 13 to 25 MeV at least. The states are observed in elastic α-scattering as broadened lines or giant resonances. The width is 1 MeV approx. The level spectra constitute rotational bands of even and odd spins i.e. mixed parities up to at least $J = 10\,\hbar$.

The strong interaction between the α-particle and the core has a classical analogue in a drag like the moon tidals. The core contributes to the moment of inertia of the two body system by half of its rigid moment. From the density of the fine structure it is concluded that the line broadening is due to an interaction with some more complicated alpha cluster states rather than with the compound states. Parity splitting of the bands does not occur, as the effect of the Pauli principle is reduced in the system dominated by bosons.

Excitation functions of inelastic scattering resulting in the excitation of the next 2^+, 4^+ and 0^+ excited states have a fine structure of mutually uncorrelated resonance peaks. They hint at the existence of single α-particle states with excited cores.

The rotational states comprizing an α-particle orbiting a core fit in the general picture of rotating two body systems. They have their closest resemblance in the exotic atoms with strong interaction between the orbital hadron and the nucleus, which results in the broadening of the X-ray lines.

The author is indepted to his collaborators V. Goldberg, K.-M. Källman, T. Lönnroth, P. Manngård, A. Pakhomov and V. Pankratov for the opportunity to use some results before their publication and to T. Ericson for his comments on some general features of rotational atomic two body systems. The spins in Fig. 1 and Fig. 2 were evaluated from thick target data by K.-M. Källman.

References

1. Hodgson, P.E.: Clustering Phenomena in Atoms and Nuclei. Int. Conf. on Nuclear and Atomic Clusters 1991, Turku 1991. Brenner, M., Lönnroth, T., Malik, F.B. (eds.). Berlin, Heidelberg, New York: Springer 1992
2. Buck, B., Dover, C.B., Vary, J.P.: Phys. Rev. **11**, 1803 (1975)
3. Yamaya, T., Saito, M., Fujiwara, M., Itahashi, T., Katori, K., Suehiro, T., Kato, S., Hatori, S., Ohkubo, S.: Phys. Lett. B**306**, 1 (1993)
4. Frekers, D., Santo, R., Langanke, K.: Nucl. Phys. A**394**, 189 (1983)
5. Rudakov, V.P.: Private communication
6. Brenner, M.: Clustering Phenomena in Atoms and Nuclei. Int. Conf. on Nuclear and Atomic Clusters 1991, Turku 1991. Brenner, M., Lönnroth, T., Malik, F.B. (eds.). Berlin, Heidelberg, New York: Springer 1992
7. Manngård, P.: Clustering Phenomena in Atoms and Nuclei. Int. Conf. on Nuclear and Atomic Clusters 1991, Turku 1991. Brenner, M., Lönnroth, T., Malik, F.B. (eds.). Berlin, Heidelberg, New York: Springer 1992
8. Artemov, K.P., Goldberg, V.Z., Petrov, I.P., Rudakov, V.P., Serikov, I.N., Timofeev, V.A.: Nucl. Phys. A**320**, 479 (1979)
9. Artemov, K.P., Brenner, M., Golovkov, M.S., Goldberg, V.Z., Källman, K.-M., Lönnroth, T., Manngård, P., Pankratov, V.V., Rudakov, V.P.: Sov. J. Nucl. Phys. **55**, 492 (1992)
10. Manngård, P.: Proc. 2nd Int. Conf. on Atomic and Nuclear Clusters '93, Santorini 1993. Z. Phys. A (in press)
11. Lane, A.M.: Isospin in Nuclear Physics. Wilkinson, D.H. (ed.). Amsterdam: North-Holland 1969
12. Bethe, H.A.: Rev. Mod. Phys. **9**, 69 (1937)
13. Fulton, B.R., Bennett, S.J., Freer, M., Murgatroyd, J.T., Gyapong, G.J., Jarvis, N.S., Jones, C.D., Watson, D.L., Brown, J.D., Rae, W.D.M., Smith, A.E., Lilley, J.S.: Phys. Lett. B**267**, 325 (1991)

Electronic structure and properties of designer clusters and cluster-assemblies

S.N. Khanna and P. Jena

Physics Department, Virginia Commonwealth University Richmond, VA 23284-2000, U.S.A.

Received 28 June 1993

Abstract. Using self-consistent calculations based on density functional theory, we demonstrate that electronic shell filling and close atomic packing criteria can be used to design ultra-stable clusters. Interaction of these clusters with each other and with gas atoms is found to be weak confirming their chemical inertness. A crystal composed of these inert clusters is expected to have electronic properties that are markedly different from crystals where atoms are the building blocks. The recent observation of ferromagnetism in potassium clusters assembled in zeolite cages is discussed.

PACS: 36.40.+d

Small clusters containing two to few dozen atoms have emerged as a new state of matter. Extensive experiments and theoretical investigations on isolated clusters are showing that the physical, electronic, magnetic or chemical properties of clusters [1] are often very different from individual atoms or bulk. The isolated clusters, however, have limited applications. For developing materials, the clusters have to be assembled together. When brought together, the clusters coalesce to form larger units and the properties characteristic of the smaller size evolve to bulk. The development of new materials using clusters, therefore, rests on the premise of designing clusters which will not coalesce when brought together. The properties of these materials will be dominated by those of individual clusters and are expected to be very different from those of the existing materials.

The purpose of this paper is two fold. First we propose how by controlling size and composition, it should be possible to design clusters which will be very stable [2] as well as chemically inert [3]. These may provide ideal systems for generating cluster materials where each site would be occupied by individual clusters. Secondly we offer an explanation to the recent discovery of ferromagnetism in Potassium clusters generated in zeolite cages. This part of our work relates to how the cluster assemblies can differ from bulk matter.

We begin by discussing how to design clusters for forming cluster materials. That it is indeed possible to obtain clusterswhich will not merge when brought together is shown by two recent examples. The first relates to the discovery of bucky balls and the fullerene - solids [4]. Here, the isolated C_{60} clusters weakly interacting via Van der Waal's forces assemble together to form body centered cubic (BCC) or face centered cubic (FCC) arrangements with C_{60} occupying each lattice site. The properties of the pure or doped fullerides [5] are very different from other carbon solids and now constitute a major area of research. The second example is the twenty atom clusters M_8C_{12} containing 8 transition metal (M) atoms and 12 carbon atoms discovered recently by Castleman and co-workers [6]. These clusters called the metallo-carbohederenes or met-cars are proposed to have distorted dodecahedral shapes and join together by sharing common pentagonal faces maintaining individual identity. In both these examples, the stable clusters were discovered experimentally, by accident.

We have been working to develop a general recipe for designing such clusters. Our arguments derive from two sets of experiments generating size selected clusters in beams. Firstly, experiments on smaller clusters of simple metals [7] show that clusters containing 2, 8, 18, 20, 40, ... atoms appear as dominant peaks in the mass spectra. These sizes called the magic numbers can be accounted for on the basis of a simple Jellium picture [7] where the cluster is modelled by a spherical Jellium. The magic numbers correspond to the sizes where the electronic shells are completely filled by the valence electrons. The added stability therefore reflects the electronic shell closure much in the same way as inert gas atoms in the periodic table. Secondly, experiments on larger simple metal clusters [8] showed that for clusters containing more than arounf 1000 atoms, the dominant cluster sizes correspond to clusters with complete geometric shells in an icosahedric or octahedric growth sequence. This indicates that for larger sizes the geometric effects dominates the cluster stability. Taken together, these experiments show that the cluster stability is governed by electronic and geometric shell closure with electronic effects dominating at small sizes while the geometric effects dominate at large sizes. It is this observation that lead us to propose [2] that a cluster would be most stable if one could simultaneously satisfy electronic and geometric shell closure. The electronic shell closure would further decrease the reactivity and render the clusters to be chemically inert. This would guarantee that when these clusters are brought together, they will retain shape and form weakly interacting solids much in the same way as buckyballs do.

In order to demonstrate that such a proposition indeed works we carried out ab-initio calculations on small prototype systems. One such possibility is a Al_{13} cluster. Al is a trivalent atom and Al_{13} has a icosahedric structure. However, it has 39 electrons, one shy of filling an electronic shell. This can be overcome by replacing an Al atom by a tetravalent atom like C or Si.

To examine the effect of shell closure we studied the ground state geometry and binding energy of Al_{13}, CAl_{12} and $SiAl_{12}$ clusters using a linear combination of atomic orbitals molecular orbital approach within the density functional framework. The details of the calculation are given elsewhere but in Table 1 we give the results of our investigations. It is interesting to note that replacement of the central Al by C in an icosahedral Al_{13} cluster increases the binding energy by as much as 4.4 eV. The effect is less dramatic for Si but the trend persists. Both these examples establish that compact clusters with filled shells do have enhanced stability. For these clusters to be able to form materials, it is futher necessary to establish that they will interact weakly. Ideally this requires a study of the interaction energy of two CAl_{12} clusters as a function of intercluster distance with geometry optimization at each separation. This is a monumental task with available computational resources. We therefore took a different approach. We calculated the binding energy of a single H-atom to an Al_{13} and a CAl_{12} cluster by bringing the H-atom along different directions. For Al_{13}, the H-atom was found to bind to on-top and bridge sites with a binding energies of 2.67 and 3.24 eV respectively. For CAl_{12} cluster, the H binding energies for the on-top, bridge and hollow sites were only 2.17, 1.5 and 1.46 eV respectively. Same calculations yielded a binding energy of 4.85 eV for an H_2 molecule. It takes 2.43 eV/atom to break an H_2 molecule. Since the H-binding energies to Al_{13} are greater than 2.43 eV, an H_2 molecule reacts dissociatively with Al_{13}. For CAl_{12}, the corresponding H-binding are less than 2.43 eV and therefore a CAl_{12} will not break an H_2 molecule. CAl_{12} therefore does appear to be less reactive than Al_{13}.

To summarize, we have demonstrated that by controlling size and composition, it should be possible to design cluster assembled material. Although we have only presented results on Al_{13} and $Al_{12}C$, one can use the recipe of filled electronic shells and closed atomic packing to design a large number of clusters. For example, a cluster A_nB_m with n atoms of type A and m atoms of type B will fulfill the above criteria if the two conditions are satisfied. Firstly, if x is the valence of A anf y the valence of B, then nx + my must equal one of the magic numbers N (N = 8, 20, 40, ...). Secondly, n + m should correspond to a closed packed system such as icosahedron or cuboctahedron. By assembling these clusters, it should be possible to form weakly interacting solids as in case of C_{60}.

Table 1. Our calculated bond length and binding energy (BE) in Al_{13}, $Al_{12}Si$, and $Al_{12}C$

Cluster	Bond Lenght (Å)	BE (eV)	HOMO-LUMO gap
Al_{13}	2.68	36.7	2.02
$Al_{12}Si$	2.67	39.2	2.01
$Al_{12}C$	2.53	41.1	1.93

We now focus on the second part of our work which relates to the properties of cluster materials. In the above we have outlined how to design weakly interacting clusters. When collected, these free clusters will assemble in Van der Waal solids. For other clusters which are not compact or have open electronic shells, the interactions can be fairly strong. For such clusters, the cluster assemblies have to proceed via alternate methods. One way is to generate clusters in cages of matrices. The cages prevent the clusters to agglomerate by maintaining large separations and permit only weak interaction between clusters. The properties specific to small clusters are therefore maintained and the assemblies can have very different properties than the bulk. One such example is the ferromagnetism in potassium clusters.

In a recent letter Nozue et al. [10] have reported the observation of ferromagnetism coupling below 4K between potassium clusters incorporated into the supercages of a LTA zeolite matrix. The supercages have a diameter of 11 Å and are connected to each other in a simple cubic lattrice by a window of 5 Å in diameter. Through optical reflection experiment, they identify each of these clusters to include about four 4s electrons. The effective magnetic moment per cluster is estimated to be about 0.13 μ_B. This discovery raises a fundamental question concerning the origin of magnetism. It is known from both theory [11] and experiment [12] that the ground state of alkali metal tetrameres is spin singlet. Thus, if the supercage consists of K_4 clusters as appears to be the case from optical measurements, it is unclear how these non magnetic clusters could couple ferromagnetically at any temperature.

Nozue et al. [10] have pointed out that the supercage in the K-LTA zeolite can accomodate up to 10 Potassium atoms and thus there is probably a distribution of cluster sizes in the zeolite. Of these clusters, the ones with odd number of electrons would possess a magnetic moment of at least 1 μ_B. The authors argue that the coupling between these clusters either through a direct exchange interaction or through an indirect Ruderman-Kittel-Kasuya-Yoshida (RKKY) mechanism could lead to the onset of ferromagnetic ordering. These assertions raise further questions. If there is a statistical distribution of magnetic potassium clusters in the zeolite, their densities must be quite low in order to produce a moment of 0.13 μ_B per supercage. One then wonders how the magnetic clusters well separated from each other would couple ferromagnetically. A second and perhaps equally important question is how such a distribution could yield four electrons per supercage. Thus it is difficult to reconcile the optical measurement that yields 4 electrons per supercage with the susceptibility measurements that yields 0.13 μ_B per supercage.

In this paper we propose an alternate explanation that is not only consistent with the susceptibility and optical measurements of Nozue et al. [10], but also provides an insight into magnetic ordering between the clusters. We show that the true ground state of a K_4 cluster is nonmagnetic and has a rhombus shape. In this configuration the K_4 cluster is too large to fit inside the K-LTA supercage and needs to be compressed. As the structure of K_4 is compressed from its ground state rhombus shape to that of a more compact squarish shape, the electronic ground state changes from a spin singlet to a spin triplet state. A study of the coupling between the two magnetic K_4 clusters as a function of inter-cluster distance illustrates the existence of a nonmagnetic ground state when the distance is less than $\sim$ 7.5Å but ferromagnetic at larger distances. It is fortunate that the supercages in the K-LTA zeolite are separated by a distance of 11Å. The ferromagnetic coupling is shown to partly originate from the overlap of electron densities between the two K_4 clusters. We are further able to show that the discrete energy levels of the magnetic K_4 cluster not only yield transition energies at 0.7 eV, 1.2 eV and 2.0 eV that are in excellent agreement with the optical data but also explain the broad band nature of the transition. These results are based on fully self-consistent ab-initio calculations [9] performed using the spin density functional theory. The total energies of the clusters with a given spin configuration and geometry were calculated self-consistently using the density functional theory and the local spin density approximation. In Fig. 1 we plot the total energy of a K_4 cluster for spin singlet and triplet configurations as a function of the bond angle θ. For each angle, θ the bond distance, d was optimized. This distance is a monotonic function of θ and varies from 3.93Å at $\theta = 90°$ to 4.18Å at $\theta = 130°$. It is clear that the ground state of the K_4 cluster is a spin singlet rhombus. This is consistent with the ground state structure of other alkali tetramers [11,13] such as Li_4 and Na_4. In this configuration, the optimized bond angle θ and the optimized bond distance between two K atoms are respectively 127° and 4.13Å. The size of the spin singlet K_4 cluster calculated by adding the diameter of the K atom (5Å) and the farthest distance between the two K atoms (12.5Å) is larger than the space available in the supercage of the K-LTA zeolite. Thus the cluster has to undergo structural compression in order to fit inside

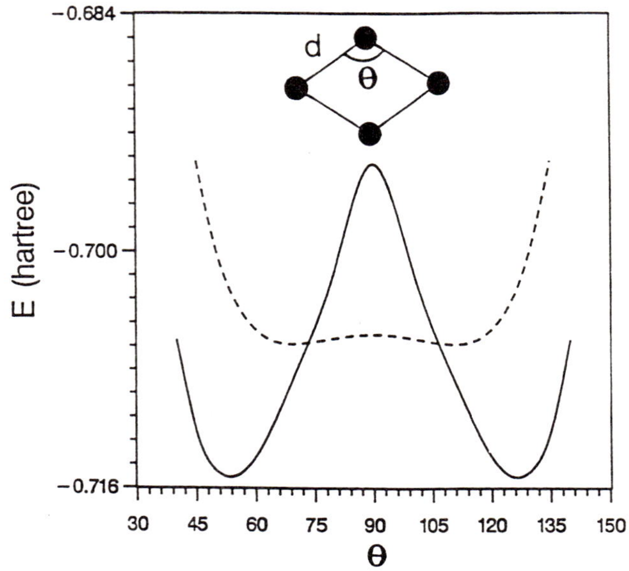

Fig. 1. Energy of the K_4 cluster in rhombus geometry as a function of angle θ for singlet (solid line) and triplet (dashed line) configurations. The bond length was optimised for each θ and spin multiplicity.

the supercage. As expected, the energy of the system increases as the structure is compressed as long as the spin zero state is maintained. however, the situation is different for the triplet structure. As the K_4 cluster is compressed to form a square, the p_x, p_y-like states become electronically degenerate and the Hund's rule coupling that favors higher spin multiplicity causes the energy to decrease. Thus at $\Theta \sim 106°$ the spin singlet and triplet states are isoenergetic and for $\Theta < 106°$ the spin triplet becomes the energetically preferred state. Note that the total energy of the spin triplet state for $70° < \Theta < 106°$ is fairly constant and for $\Theta = 90°$, the farthest distance between the K-atoms is 5.5Å. Thus, the size of the K_4 cluster is about 10.5Å and can easily fit inside the supercage. It is also possible that the K_4 cluster could assume a tetrahedral shape which has a preferred spin multiplicity of 3 and would require even less space. We have calculated the equilibrium geometry of K_4 tetrahedral cluster whose bond distance is 4.04Å. However, this structure is 0.06eV higher than the "squareship" shape discussed earlier. It should be pointed out that the relationship between spin multipicity and geometry of Li_4 and Na_4 clusters was demostrated quite sometime ago [13,14].

To provide additional evidence that the K_4 cluster inside the K-LTA zeolite has the "squareship" geometry with spin 1, we plot in Fig. 2 the molecular energy levels of the cluster for both spin up and down electrons. The possible transitions are indicated along with the energy differences of 0.69 eV, 10.6 eV and 1.86 eV for various transitions. Furthermore, there is a bunching of unoccupied energy levels to which transition could take place. Consequently, the structures in the optical spectrum instead of being sharp would be more "band" like. This is, indeed, the case as seen experimentally by

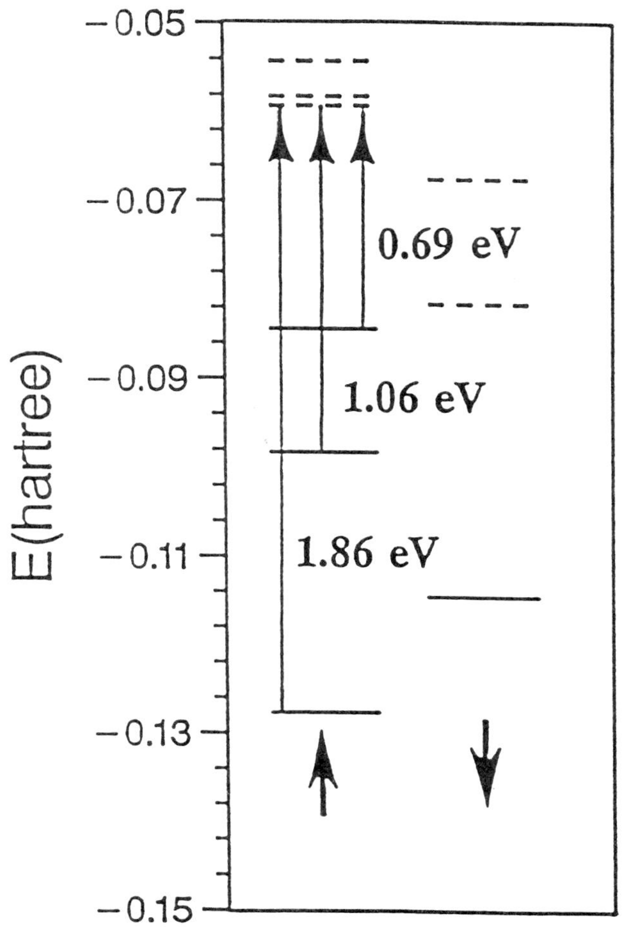

Fig. 2. One electron level in the metastable spin triplet configuration.

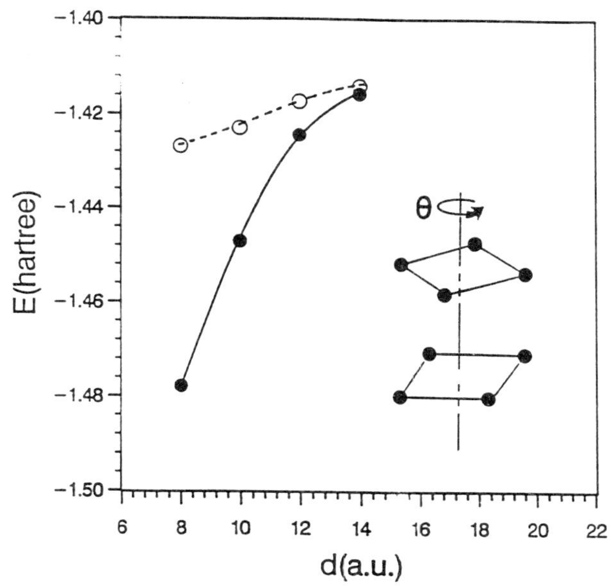

Fig. 3. Energy of two K_4 clusters in square geometry separated by a distance d. The solid and dashed lines represent S=0 and S=2 states respectively. For each d, the horizontal squares are allowed to rotate about a vertical axis to minimize the energy.

Nozue et al. [10]. In addition, their observed structures at 0.7, 1.15, and 2.0 eV are in agreement with the values in Fig. 2.

The next problem that has to be tackled is that experimentally the moment estimated per supercage is small (0.13 μ_B) while the net spin of the K_4 cluster is 2 μ_B. This probably arises from several factors. (i) The spin of the K_4 cluster is oriented in random directions. As the magnetic field is applied, it is only the z-component that survives, thus reducing the net spin. (ii) There are likely to be statistical distribution of K-clusters ranging from 0 to 4 atom clusters inside the zeolite supercages. Clusters bigger than 4 atoms are too large to fit in! Since the K_2 clusters have zero spin and K and K_3 have spin 1/2, the effective magnetic moment per supercage will be further reduced. Without a detailled knowledge of this statistical distribution, it will be difficult to calculate the effective magnetic moment per supercage. That the moment is much less than 2 μ_B is certainly understandable.

We now discuss the origin of ferromagnetic ordering. It is true that if two K_4 clusters are held apart from each other inside the matrix, their spins would couple antiferromagnetically. if the two K_4 clusters are close to each other so that they resemble a K_8 cluster, the net spin would be zero. In order to understand if there is a critical distance between the two K_4 clusters that would favor ferromagnetic ordering, we have calculated the total energies of two K_4 clusters as a function of intercluster distance for total spin = 0 and 2. The former corresponds to a nonmagnetic state while the latter corresponds to ferromagnetic coupling. We have restricted the clusters to have a square geometry since that is what could fit inside the supercage. We have, however, allowed the clusters to rotate with respect to each other at each intercluster distance. The resulting energies are plotted in Fig. 3. Note that the combined cluster prefers to have nonmagnetic configuration at small distances, but this preference becomes less obvious as distance between the K_4 clusters increases. There is clear indication that for distance larger than 7.5Å, the preferred coupling is ferromagnetic. It is gratifying to note that the distance between two supercages in the K-LTA zeolite is 11Å. Thus the observation of ferromagnetic coupling between K clusters in a zeolite is due to two factors: The limited space inside the supercage forces the K_4 cluster to assume the metastable spin triplet configuration and the adequate separation between the two supercages favors ferromagnetic coupling.

To shed light on whether the coupling is due to an indirect RKKY type interaction or through a direct exchange mechanism, we have plotted the charge density contours of the K_4 magnetic cluster in Fig. 4. Note that there is significant charge density even at a distance of 5.3Å from the cluster center. Thus, there is certainly room for charge overlap between K_4 clusters isolated at neighboring cages. It is, therefore, cur belief that the

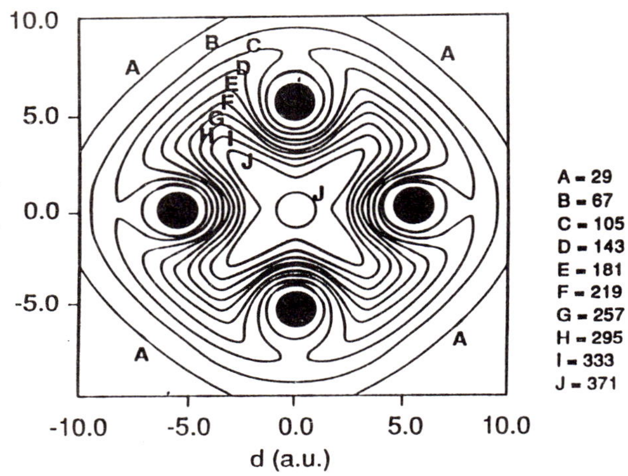

Fig. 4. Charge density contours (in arbitrary units) of a square K_4 cluster in a spin triplet configuration.

leading cause of ferromagnetic coupling is through a direct exchange mechanism.

In summary we have illustrated that structural transformation can induce magnetic transition in a K_4 cluster. This arises from a competition between the Jahn-Teller effect and Hund's rule coupling, two mechanisms by which a cluster can lower its energy. Clusters can have high structural symmetry that lead to electronic degeneracies and consequently to high multiplicity as required by Hund's rule coupling. Clusters can also lower energy by structural distortion that breaks the electronic degeneracy and splits the energy levels as required by Jahn-Teller effect. In a free K_4 cluster the Jahn-Teller effect wins and the K_4 cluster assumes a rhombus geometry. The energy levels are split to the extent that it is not favorable to promote an electron to a higher level in order to gain energy due to exchange coupling. As the cluster is forced to change its geometry to a square form, the splitting diminishes substantially. thus, promoting the electron to a higher level to gain exchange energy becomes profitable. This is what happens to K_4 cluster inside the K-LTA zeolite.

That the origin of ferromagnetism of K_4 clusters is due to geometrical confinement can be conclusively demonstrated by repeating the experiments where K_4 clusters are embedded into zeolites with supercage diameter of 13Å or larger or incorporating Li_4 and Na_4 clusters into the K-LTA zeolite. In either case, the room available inside the supercage can accomodate the cluster in its nonmagnetic rhombus structure and no ferromagnetic coupling should exist.

This work was supported in part by a grant from the Army Research Office (DAAL-03-89-K-0015).

References

1 Physics and Chemistry of Finite Systems: From Clusters to Crystals, Eds. Jena, P., Khanna, S.N., and Rao, B.K., (Kluwer, Dortrecht, 1992)

2 Khanna, S.N., and Jena, P.: Phys. Rev. Lett. **69**, 1664 (1992)

3 Khanna, S.N., and Jena, P.: (Submitted for publication)

4 Kroto, H.W., Heath, J.R., O'Brien, S.C., Curl, R.F., and Smalley, R.E.: Nature **318**, 162 (1985); Krätschmer, W., Lamb, L.D., Fostiropoulos, K., and Huffman, D.R.: Nature **347**, 354 (1990)

5 Hebard, A.F., Rosseinsky, M.J., Haddon, R.C., Murphy, D.W., Glarum, S.H., Palstra, T.T.M., Ramirez, A.P., and Kortan, A.R.: Nature **350**, 600 (1991)

6 Guo, B.C., Kerns, K.P., and Castleman, A.W.Jr.: Science **255**, 1411 (1992)

7 Knight, W.D., Clemenger, K., de Heer, W.A., Saunders, W.A., Chou, M.Y., and Cohen, M.L.: Phys. Rev. Lett. **52**, 2141 (1984)

8 Martin, T.P., Bergmann, T., Gohlich, H., and Lange, T.: Chem. Phys. Lett. **172**, 209 (1989)

9 Reuse, F., Khanna, S.N., de Coulon, V., and Buttet, J.: Phys. Rev. B **41**, 11743 (1990)

10 Nozue, Y., Kodaira, T., and Goto, T.: Phys. Rev. Lett. **68**, 3789 (1992)

11 Martins, J.L., Buttet. J., and Car, R.: Phys. Rev. B **31**, 1804 (1985); Bonacic-Koutecky, V., Fantucci, P., and Koutecky, J.: Phys. Rev. B **37**, 4369 (1988); Rao, B.K., and Jena, P.: Phys. Rev. B **32**, 2058 (1985)

12 Garland, D.A., and Lindsay, D.M.: J. Chem. Phys. **78**, 2813 (1983)

13 Rao, B.K., Khanna, S.N., and Jena, P.: Chem. Phys. Lett. **121**, 202 (1985); Khanna, S.N., Rao, B.K., Jena, P., and Martins, J.L.: in Physics and Chemistry of Small Clusters, Ed. Jena, P., Rao, B.K., and Khanna, S.N. (Plenum Press, 1987) p.435

14 Ghosh, D.K., Majumdar, C.K., and Rajagopal, A.K.: Ind. J. Phys. **43**, 282 (1969)

This article was processed using Springer-Verlag TEX Z.Physik D macro package 1.0 and the AMS fonts, developed by the American Mathematical Society.

Z. Phys. A 349, 237–240 (1994)

Nuclear clustering aspects in astrophysics

Shigeru Kubono

Institute for Nuclear Study (INS), University of Tokyo, 3-2-1 Midori-cho, Tanashi, Tokyo, 188 Japan

Received: 27 October 1993

Abstract. Atomic nuclear clusters play a crucial role in nucleosynthesis in the universe, especially in the main sequence of heavy element synthesis. Cluster aspects in nucleosynthesis are briefly discussed based on a Cluster-Nucleosynthesis Diagram proposed here. Two recent topics on critical α-induced thermonuclear reactions are reviewed; the first one is the $^{12}C(\alpha, \gamma)^{16}O$ reaction for the He burning stage and the other one is the $^{8}Li(\alpha, n)^{11}B$ reaction for the big bang nucleosynthesis. A new field of nuclear astrophysics using radioactive nuclear beams is also discussed.

PACS: 21.60.Gx

1. Nucleosynthesis in astrophysics

Nuclear reactions play a crucial role in the evolution of the universe, and are the main source of energies there. Nevertheless, many nuclear reactions of the universe are not well understood yet [1–4], specifically those of the explosive processes, where short-lived radioactive nuclei are involved. Thus, many important nucleosynthesis processes are still a "scenario".

There are three major sites for nucleosynthesis in the universe. The first one is believed to be in the early universe just after the big bang. There should have been a quark-hadron phase transition just after the big bang, and then the formation of nuclei began, which led to synthesis predominantly of light elements such as He and Li isotopes. These elemental abundance ratios also should have been modified by nuclear reactions in interstellar sites, mostly by high energy spallation reactions. This is the second site of nucleosynthesis in the universe.

The third and the major part for the nucleosynthesis is in the stars and the vicinities. In the early stage of the stars, there is nucleosynthesis which involves mostly stable nuclei and lasts for a long duration. A typical example known is the nuclear reactions of the *pp*-chain in the sun. As the stars evolve, the nucleosynthesis processes go faster, and finally they come to the point of a gravitational collapse if the mass of the star is large, which leads to a supernova explosion. Or, it may stop by becoming a white dwarf if the mass is small.

Each step of the stellar evolution is characterized by a specific nucleosynthesis. Thus, one can test critically the standard model of the stellar evolution by studying the nuclear reaction processes which are not well known yet. These works eventually give enough nuclear physics information to deduce the information on the stellar environment of interest. These are the main subjects of nuclear astrophysics. Of course, the nucleosynthesis scenarios have been checked by the astronomical observations, meteorite studies, cosmic rays, etc.

2. Clustering aspects in nucleosynthesis

In the universe, $4n$ nuclei, e.g., ^{4}He, ^{12}C, and ^{16}O, are the most abundant nuclides except for hydrogen, and often participate in the nucleosynthesis, where cluster states above the cluster threshold play a crucial role. Thus, the nuclear clustering of nuclei exists in nature, and it is favoured to minimize the energy of the nuclear systems because of the large binding energies of these nuclei. The nucleosynthesis in nature of course goes through a synthesis flow-path which obeys the same principle under the site condition of interest. Therefore, nuclear clustering feature is quite important in nucleosynthesis in the universe. There are a variety of nucleosynthesis scenarios proposed in different stages of evolution of the universe.

Let me first explain a typical scheme of nucleosynthesis by an α-capture reaction ($A+\alpha \rightarrow B+\gamma$). To understand better the nucleosynthesis which involves cluster states, we propose here a "Cluster-Nucleosynthesis (CN) diagram" in Fig. 1. This is similar to the so called cluster diagram in nuclei [5], but includes the dynamics of nucleosynthesis and the typical scenarios. This CN diagram explains very well the nature of nucleosynthesis which involves a variety of clusters in the universe. The ordinate

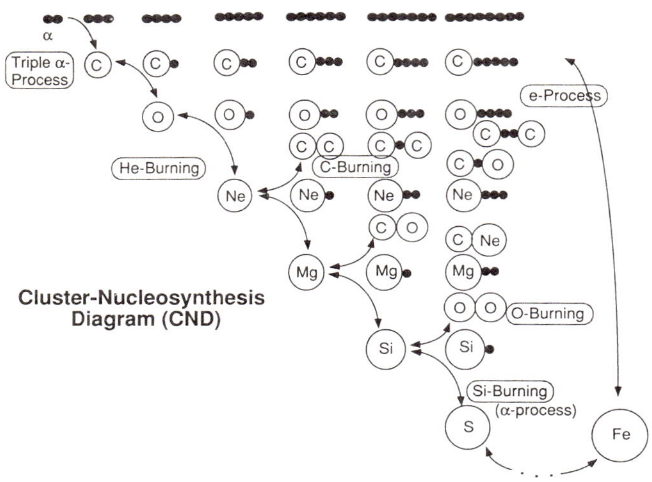

Fig. 1. Cluster-Nucleosynthesis (CN) diagram in the universe

measures the energy released or binding energy per nucleon, and so more energy will be released as the nucleosynthesis developes by going to the lower level until Ni or Fe. One should see the diagram from the α cluster threshold to the ground states of B, where the reaction goes through a resonant cluster state of $\alpha + A$ to the non-cluster ground state B. Similarly by going to the right and lower part in the CN diagram, the nucleosynthesis is favoured and generates energies in the astrophysical site. Of course, these flow-rates are determined by the cluster property as well as non-cluster property of the resonant and the ground states in nucleus B. In nuclear physics on the contrary, one would discuss the development of clustering in nucleus B by putting excitation energies to the nuclear system, B, although the basic physics is the same.

After the hydrogen burning in the main sequence, He, the ash to the *pp*-chain, can be the main fuel for the next nucleosynthesis, which is called He-burning. The first step is the synthesis of ^{12}C by the triple α-process, and then the $^{12}\mathrm{C}(\alpha,\gamma)$ reaction to produce ^{16}O, but it would not go much beyond because the next possible step, $^{16}\mathrm{O}(\alpha,\gamma)$, has a small reaction rate due to the nuclear structure just above the α threshold in ^{20}Ne. After the He burning, there would be nuclear burning processes which use ^{12}C and ^{16}O for the fuel, i.e., C-burning and O-burning. In the C-burning considerable ^{20}Ne will be produced by emitting α particles, as can be seen in the CN diagram. Massive stars evolve further and produce Si-core inside at some point, then there begins Si-burning. Here, a part of Si-nuclei photo-disintegrate to Mg$+\alpha$ and Ne$+2\alpha$, and simultaneously other Si-nuclei will capture these α particles successively, producing nuclei up to ^{56}Fe. Just before the gravitational collapse, the central core will be an equilibrium state of ^{56}Fe, αs', protons and neutrons, which is called *e*-process. This stage could include also a variety of cluster configurations in Fig. 1. Since all these nuclear reactions mentioned above take place at the temperature region of $1\times10^8 \sim$ a few $\times 10^9$ K, corresponding to a few hundred keV to a few MeV, the nuclear levels of such cluster nature and the continuum states just in these energy region above the threshold should be crucial for the reaction rates. As was mentioned above, one needs to know the details of the nuclear states of cluster as well as non-cluster nature.

3. Recent topics of critical α-induced thermonuclear reactions

We will discuss two α-induced reactions which involve stable and unstable nuclei. There still are many reactions of stable nuclei, which are very important, but not well determined primarily because of small cross sections. One of the most critical such reactions should be $^{12}\mathrm{C}(\alpha,\gamma)^{16}\mathrm{O}$.

Although this reaction is very important in the He-burning stage and the rate has a critical consequence for the scenario of the standard stellar model, the reaction rate is not determined yet at the temperature region of interest [6–15], i.e., $\approx 2\times10^8$ K or 300 keV above the threshold. The problem here is that the cross sections seem to be quite small ($\approx 10^{-8}$ nb/sr estimated at 300 keV), and there will be an interference of the tails of the subthreshold 1^- state at 7.117 MeV and the broad 1^- resonance at 9.585 MeV. The gamma decay is considered to be $E1$ together with some $E2$ strength.

Table 1 summarizes the astrophysical S-factors estimated from the experimental reaction rate of $^{12}\mathrm{C}(\alpha,\gamma)^{16}\mathrm{O}$ determined at Ec.m. $\geqq 0.94$ MeV by measuring the singles gamma rays [7–9] or the gamma rays and the recoil product ^{16}O in coincidence [10]. Here, the $E1$ and $E2$ strengths were separated by measuring angular distributions. The extrapolation procedures have a quite large ambiguity (more than 50%) depending on the theories used, as can be seen in Table 1.

Recently, a very small $S_{E1}(300)$ value was reported by Ouellet et al. [12], as shown in Table 1. This is much smaller than those previously reported. Extensive measurements were also performed for the β-delayed α decay of ^{16}N. This measurement has an advantage [6] for the β-decay to the lower excitation energies and the decay is also sensitive to the $E1$ strength. Buchmann et al. [13, 15] and Zhao et al. [14] reported S-factors obtained by analyzing the α spectrum from ^{16}N with K-matrix and R-matrix theories. The results are consistent each other and much larger than that in [12]. These are, however, roughly a factor of two smaller than those currently used for the stellar models [16]. It should be concluded that the es-

Table 1. S-factors of $^{12}\mathrm{C}(\alpha,\gamma)^{16}\mathrm{O}$ in unit of keV$\cdot b$

S_{E1}(300 keV)	S_{E2}(300)	S_{tot}(300)	Reference
140 (three-level R.)			Dyer [74]
80 (R.+optical M.)			
250	180	430	Kettner [82]
200	96	296	Redder [87]
$\geqq$140			Kremer [88]
1	40		Ouellet [92]
57 (K-matrix)			Buchmann [93]
79 (Global R-matrix)			
95 (R-matrix)			Zhao [93]
69 (Global R-matrix)			

timated values are scattered in a wide range, and thus it is poorly determined in both experiment and theory.

There are many experimental plans on this subject. One possibility is to use the Coulomb dissociation method [17], which has been successfully applied for determination of the reaction rate of $^{13}N(p,\gamma)^{14}O$ at low energies by using an ^{14}O beam at intermediate energies (87.5 MeV/u) [18]. Low energy capture cross sections were obtained by measuring the breakup cross sections at small relative momenta and converting them by using the detailed balance theorem. This method gives a quite large yield enhancement because of a large phase volume from the theorem as well as of a thicker target usable. One might go down to the region of Ec.m. = 0.5 MeV, although this method should be sensitive to the $E2$ part for this reaction. However, one has to deal carefully with final state interaction effects and the nuclear force contribution in a continuum region. These require both theoretical and experimental developments.

Of course one could go down in energy with the direct measurement or β-delayed α decay measurement with a higher-intensity beam with an improved detector system, such as the one in development at INS, University of Tokyo. This subject still deserves much intense work in both experiment and theory.

Now, let me move on the next subject. One of the key issues in the primordial nucleosynthesis is if the metals (^{12}C and heavier elements) were produced in the very early universe. The Inhomogeneous Big Bang (IBB) models [19–21] have succeeded in predicting nucleosynthesis of finite metallicity through the dominant flow path [20, 21],

$$^{4}He(t,\gamma)^{7}Li(n,\gamma)^{8}Li(\alpha,n)^{11}B(n,\gamma)^{12}B(\beta^{-}\nu)^{12}C.$$

In this flow, the $^{8}Li(\alpha,n)^{11}B$ reaction was completely unknown when this scenario was proposed since the target nucleus ^{8}Li is unstable.

Several experiments [22–25] were made for this process. One was performed by using the reverse reaction $^{11}B(n,\alpha)\,^{8}Li$ [22]. Some clear resonances were observed including the largest peak of $S = 8400\ \mathrm{MeV}\cdot b$ at about 580 keV, which is near the Gamow peak of $T = 10^9$ K. This is a typical temperature for heavy element synthesis in the primordial site. Although this S-factor was close to the one assumed in the IBB model calculation [20], it is clearly not a constant value as was assumed in the models. Further, this measurement includes only the neutron decay to the ground state in ^{11}B.

The decay property of this state was next studied [23, 24] by using the transfer reaction $^{9}Be(\alpha,p)^{12}B^{*}(n)^{11}B$, which was found to excite these resonant states with reasonable cross sections. The spin-parity of $J^{\pi} = 2^{+}$ was assigned tentatively for the state from the angular distribution and angular correlation measurements, which also gave the branching ratios, approximately $\Gamma_{n0}/\Gamma_{\mathrm{tot}} = 1/3$, $\Gamma_{\mathrm{nex}}/\Gamma_{\mathrm{tot}} = 1/3$, and $\Gamma_{\alpha}/\Gamma_{\mathrm{tot}} = 1/3$. Since the sum of the intensities of the neutron decays to the excited states are almost the same as the yield for the ground state, the reaction rate of the $^{8}Li(\alpha,n)^{11}B$ process should be larger than the previous estimate [22] roughly by a factor of two.

For this reaction, a direct measurement of $^{8}Li(\alpha,n)^{11}B$ was also tried [25] at RIKEN by using a low-energy radioactive beam of ^{8}Li. The beam was introduced to an ionization chamber, MUSIC, which had an admixture of ^{4}He as a target. The cross sections obtained were larger than the previous values [22] by a factor of 5 at Ec.m. $\geqq$ 1.7 MeV, suggesting that the cross sections at Ec.m. $\approx$ 0.58 MeV could be also larger than the value of [22]. This is consistent with the decay measurement mentioned above. It should be still of great importance to determine this reaction rate directly at around Ec.m. = 0.5 MeV.

4. Perspectives

The nuclear reactions which involve short-lived unstable nuclei should be quite important in explosive processes in the universe, e.g., supernova explosions and the primordial nucleosynthesis just after the big bang because radioactive nuclei are inevitably involved in such fast nucleosynthesis processes. One of the most clear evidences of unstable nuclei in the universe would be the light curve of the supernova 1987A, which is quite well explained by the nuclear decay of ^{56}Co [26].

A new field in nuclear astrophysics has been opened up by using unstable nuclear beams. Recent development of this field in nuclear astrophysics is summarized in [1–4]. The unstable nuclear beams can be provided now from very low energy to high energy from heavy-ion induced reactions (mostly by fragmentation processes), or by an ISOL (Isotope separator on-line) + post accelerator scheme. Although the quality of the secondary beams from the former method is limited, there are many interesting experiments to be made for nuclear astrophysics.

A more promising method could be the second one, i.e., an ISOL + post accelerator scheme. The beam quality to be provided is as good as for ordinary stable nuclear beams. The first facility has been established at Louvain-la-Neuve with a limited capability, where they have succeeded in the $^{13}N(p,\gamma)^{14}O$ experiment [27] by using a radioactive nuclear beam of ^{13}N. This method should provide more intense beams than the projectile fragment method, since here one can use much thicker target with high energy proton beams for unstable-nuclei production. The major concern here may be the efficiency of ion sources, which is under development. This type of unstable nuclear beam factory is under construction at INS, University of Tokyo, and the Oak Ridge National Laboratory in U.S.A. There are similar plans in Europe and Russia. Nuclear astrophysical experiments are definitely one of the major research subjects in these factories.

References

1. Kubono, S.: Proc. Fourth Int. Conf. Nucleus-Nucleus Collisions (Kanazawa, Japan, 1991); Nucl. Phys. **A538**, 505c (1992)
2. Kubono, S.: Comm. Astrophys. **16**, 287 (1993)
3. Proc. of the International Workshop on Unstable Nuclei in Astrophysics (Tokyo, 1991), Kubono, S., Kajino, T. (eds.). Singapore: World Scientific 1992

4. Proc. of the International Symposium on Origin and Evolution of the Elements (Tokyo, 1992), Kubono, S., Kajino, T. (eds.). Singapore: World Scientific 1993
5. Morinaga, H.: Private communication, Ikeda, K. et al.: Prog. Teor. Phys. Suppl. Extra Number (1968) 464
6. Barker, F.: Aust. J. Phys. **24**, 777 (1971)
7. Dyer, P., Barnes, C.A.: Nucl. Phys. A**233**, 495 (1974)
8. Kettner, K.U. et al.: Z. Phys. A**308**, 73 (1982)
9. Redder, A. et al.: Nucl. Phys. A**462**, 385 (1987)
10. Kremer, R.M. et al.: Phys. Rev. Lett. **60**, 1475 (1988)
11. Ji, X., Filippone, B.M., Humblet, J., Koonin, S.E.: Phys. Rev. C**41**, 1736 (1990)
12. Ouellet, J.M.L. et al.: Phys. Rev. Lett. **69**, 1896 (1992)
13. Buchmann, L. et al.: Phys. Rev. Lett. **70**, 726 (1993)
14. Zhao, Z. et al.: Phys. Rev. Lett. **70**, 2066 (1993)
15. Buchmann, L., Azuma, R.E.: Phys. Rev. Lett. (submitted)
16. Thielemann, F.-K., Rauscher, T.: Ref. [4]
17. Baur, G., Weber, M.: Nucl. Phys. A**504**, 352 (1989)
18. Motobayashi, T. et al.: Phys. Lett. **264**B, 259 (1991)
19. Applegate, J.H., Hogan, C.J.: Phys. Rev. D**31**, 3037 (1985)
20. Malaney, R.A., Fowler, W.A.: Origin and Distribution of the Elements, p. 76; Mathews, G.J. (ed.). Singapore: World Scientific 1988; Astrophys. J. **333**, 14 (1988)
21, Kajino, T., Mathews, G.J., Fuller, G.M.: Heavy Ion Physics and Nuclear Astrophyiscal Problems, p. 51. Kubono, S. et al. (eds.); Singapore: World Scientific 1988; Astrophys. J. **364**, 226 (1990)
22. Paradellis, T. et al.: Z. Phys. A**337**, 211 (1990)
23. Kubono, S. et al.: Z. Phys. A**338**, 459 (1991)
24. Kubono, S. et al.: Z. Phys. A**341**, 121 (1991)
25. Boyd, R.N. et al.: Phys. Rev. Lett. **68**, 1283 (1992)
26. Kumagai, S., Shigeyama, T., Nomoto, K.: Astron. Astrophys. **243**, L13 (1991)
27. Decrock, P. et al.: Phys. Rev. Lett. **67**, 808 (1991)

Z. Phys. A 349, 241–242 (1994)

Alpha clustering and the stellar nucleosynthesis of carbon

H. Oberhummer[1], H. Krauss[1], K. Grün[1], T. Rauscher[1], H. Abele[2], P. Mohr[2], G. Staudt[2]

[1] Institut für Kernphysik, TU Wien, A-1040 Vienna, Austria
[2] Physikalisches Institut, Universität Tübingen, D–72076 Tübingen, Germany

Received: 27 October 1993

Abstract. The astrophysical S–factor and reaction rates for the triple–alpha process are calculated in the direct–capture model. It is shown that the stellar carbon production is extremely sensitive to small variations in the N–N interaction.

PACS: 24.50.+g; 25.20.Lj; 97.10.Cv

1. Introduction

The most important cluster reaction in nuclear astrophysics certainly is the triple–alpha process occuring in helium burning of stars. Almost all of the carbon found in our universe is synthesized by this reaction. According to most biochemists belief, carbon is the only element that can form the basis of spontaneously generated life. Therefore, the stellar production of carbon is essential for the existence of life in our universe.

In the triple–alpha reaction three alpha particles merge in a two–step process to form the nucleus ^{12}C. In the first step the ground state of ^{8}Be is formed. Although the lifetime of this nucleus is extremely short ($\approx 10^{-16}$ s), it is anomalously long compared with the α+α collision time ($\approx 10^{-21}$ s). Therefore, the ^{8}Be is almost in equilibrium with the α–particles. During the life time of ^{8}Be a third α–particle is captured by the reaction ^{8}Be(α,γ)^{12}C. The first step depends sensitively on the properties of the ^{8}Be ground state, whereas the reaction rate of the second step is mainly a consequence of the properties of the 0_2^+ state in ^{12}C, which lies just above the threshold of the ^{8}Be(α,γ)^{12}C reaction.

2. The triple–alpha process

A detailed account of the calculation and results for the astrophysical S–factor and the reaction rate of the triple–alpha process obtained in the direct–capture (DC) model is given in [1]. The α–α scattering and the ground–state properties of ^{8}Be as well as the bound and resonant states in ^{12}C, which are relevant for the triple–alpha process are discussed in [2].

Direct capture is based on the description of the dynamics of nuclear processes by a Schrödinger equation with local potentials in the exit channel. The most important ingredients of DC are the wave functions for the scattering states and the bound and quasi–bound (resonant) states in the entrance and exit channels. The renormalized folding potentials used in the calculation of the triple–alpha process can be found in [2].

The astrophysical S–factor for the reaction ^{8}Be(α,γ)^{12}C is calculated using the DC model [3–5] with the computer code TEDCA [6]. One observes a resonant contribution of the dominating 0_2^+ state at 0.2875 MeV and a nonresonant background. These two contributions and the interference term are obtained simultaneously in our model. Approximating the cross section in the vicinity of the resonance energy of $E_R = 0.2875$ MeV by a Breit–Wigner formula we obtain the following widths: (i) $\Gamma(0_2^+) \approx \Gamma_\alpha(0_2^+) = 7.5$ eV in agreement with the experimental data $\Gamma(0_2^+)^{exp} = (8.3 \pm 1.0)$ eV and (ii) $\Gamma_\gamma(0_2^+) = 4.1$ meV which is comparable with the experimental value $\Gamma_\gamma(0_2^+)^{exp} = (3.7 \pm 0.5)$ meV [7].

We can now determine the reaction rates for the triple–alpha process using the results for the astrophysical S–factor calculated with the DC model. The low–temperature rates (10^7 K–10^8 K) including the non–resonant parts can be found in [1].

3. Variation of the nucleon–nucleon (N–N) interaction

The following advantages in applying the DC model together with the folding procedure for the calculation of the astrophysical S–factor and the reaction rates for the triple–alpha process are evident. The first and second step of the process is carried out consistently using the same model, no parameter has to be adjusted to the reaction data and the resonant and non–resonant part of the astrophysical S–factor is obtained simultaneously. These features also allow to demonstrate the extreme sensitivity of the triple–alpha reaction to variations in the strength of the underlying effective N–N interaction. A preliminary account of this calculation was given in [8].

Table 1. Effects of variations in the strength of the effective nucleon–nucleon interaction on the triple–alpha reaction rate

	E^{R} [keV] $^{8}Be_{g.s.}$	E^{R} [keV] $0_2^+(^{12}C)$	Γ [eV] $0_2^+(^{12}C)$	F_1 1. step	F_2 2. step	F_T total
0.2 %	66.8	215.5	0.2	18	$3.7 \cdot 10^3$	$6.7 \cdot 10^4$
0.1 %	79.2	251.7	1.5	4.2	60	250
0 %	91.5	287.5	7.5	1	1	1
−0.1 %	103.7	323.0	29	0.24	0.016	0.0038
−0.2 %	115.7	357.9	90	0.06	$2.5 \cdot 10^{-4}$	$1.5 \cdot 10^{-5}$

We now introduce small variations of the strength of the effective N–N interaction in the calculation of the ^{8}Be-ground state and the 0_2^+–state in ^{12}C without changing the renormalization factors λ in the folding potentials given in [2]. These factors, introduced originally to describe higher–order corrections to the simple folding procedure, are insensitive to small variations of the effective N–N interaction for states with a marked cluster structure. This result can be justified in the single-nucleon exchange approximation [9, 10]. In our calculations we also assume that the small variations of the effective N–N interaction will only modify the α–density distribution needed in the computation of the folding potentials in higher order. Furthermore, the changes in the wave function of the final bound states in ^{12}C should also be negligible. The reason for this assumption is that changes of energies and wave functions are of the order of $|\Delta E/E| \lesssim |(\epsilon V_0)/E|$, where ΔE is the change in the energy, E is the energy with respect to the threshold, ϵ is the change of the effective N–N interaction and V_0 is the depth of the corresponding folding potential. Obviously, the above change is much smaller for states which have a large energy $|E|$ with respect to the threshold.

The changes of the resonance energy of the ground state of ^{8}Be as well as the resonance energy and total width Γ of the 0_2^+–state in ^{12}C induced by variations of the effective N–N interaction are listed in Table 1. It results that the γ–width of the 0_2^+–state is almost unaffected by this variation. For the calculation of the reaction rates we will consider in the following only the resonant contributions. For a temperature of $1 \cdot 10^8$ K the changes of the reaction rate of the first, second and both steps in the triple–alpha process are given by F_1 and F_2 together with $F_T = F_1 \cdot F_2$, respectively and are listed in the last three columns of Table 1.

As can be seen from this table the reaction rates of the triple–alpha process are extremely sensitive to small variations in the strength of the effective N–N interaction. A variation of this interaction and therefore also of the depth of the folding potentials ($V_0 \approx 100$ MeV) by about 0.1% leads to a shift of the ^{8}Be ground–state and the 0_2^+–state in ^{12}C of the order of 10–30 keV. These energie shifts introduced in the Boltzmann–factors $\exp(-E_R/kT)$ for the first and second step of the triple–alpha proces result then in the drastic changes of the reaction rates shown in Table 1.

4. The anthropic principle and the triple–alpha process

The "anthropic principle" [11] states that of all possible universes the one we actually inhabit is tailor–made for the creation of life. The triple–alpha process plays a key role for the anthropic principle. This can be seen from the remarkable prediction made by F. Hoyle that the creation of carbon would only be possible, if this reaction would proceed resonantly. To our knowledge this is the only case in which the anthropic principle was used to predict the outcome of a laboratory experiment. Later on, the prediction of the 0_2^+–state in ^{12}C was indeed confirmed by experiment [12–14].

Investigations of the changes in nucleosynthesis of very light elements by variations of the strength of the N–N interaction have already been performed for the unbinding of the deuteron and binding of the di–proton [15]. In this case the limit was about 16 % for the deuteron and about 12 % for the di–proton in order that the anthropic principle is of significance for such variations. In the case of the triple–alpha process we show that a reduction of the N–N strength by only 0.1 % would already be of significance for the anthropic principle, because it reduces the stellar carbon production by about a factor of 250 at $1 \cdot 10^8$ K.

Acknowledgement. We want to thank the Austrian Science Foundation (FWF) (project P8806-PHY), the Österreichische Nationalbank (project 3924) and the DFG–project Sta290/2.

References

1 Krauss, H., Grün, K., Herndl, H., Oberhummer, H., Abele, H., Mohr, P., Staudt, G.: (to be published)
2 Mohr, P., Abele, H., Kölle V., Staudt, G., Oberhummer, H., Krauss, H.: Z. Phys. A. (in press)
3 Oberhummer, H., Staudt G.: In: Nuclei in the Cosmos. H. Oberhummer (ed.) p. 29. Berlin Heidelberg New York: Springer 1991
4 Mohr, P., Abele, H., Zwiebel, R., Staudt, G., Krauss, H., Oberhummer, H. Denker A., Hammer J. W., Wolf, G.: Phys. Rev. **C48**, 1420 (1993)
5 Kim, K. H., Park, M. H., Kim, B. T.: Phys. Rev. **C23**, 363 (1987)
6 Krauss, H.: computer code TEDCA, TU Wien (unpublished)
7 Ajzenberg–Selove, F.: Nucl. Phys. **A506**, 1 (1990)
8 Oberhummer, H., Grün, K., Hinterberger, R., Krauss, H., Abele, H., Staudt, G., Belyaev, V.: in Proceedings of the Seventh Workshop in Nuclear Astrophysics 22.–27. March 1993, Ringberg Castle, Bad Wiessee. E. Müller, W. Hillebrandt (eds.) p. 119. Garching: Max-Planck-Institut fuer Physik und Astrophysik
9 Satchler, G. R., Love, W. G.: Phys. Rep. **55**, 183 (1979)
10 Love, W. G., Owen, L. W.: Nucl. Phys. **A239**, 74 (1975)
11 Carter, B.: In: Confrontation of Cosmological Theories with Observation, M. S. Longmaier (ed.) Dordrecht: Reidel 1974
12 Dunbar, D. N. F., Pixley, R. E., Wenzel, W. A., Whaling, W.: Phys. Rev. **92**, 649 (1953)
13 Hoyle, F., Dunbar, D. N. F., Wenzel, W. A., Whaling, W.: Phys. Rev. **92**, 1095 (1953)
14 Hoyle, F.: Astrophys. J. **1**, 121 (1954)
15 Pochet, T., Pearson, J. M., Beaudet, G., Reeves H.: Astron. Astrophys. **243**, 1 (1991)

This article was processed using Springer-Verlag TEX Z.Physik A macro package 1.0 and the AMS fonts, developed by the American Mathematical Society.

The optical response of anionic alkali metal clusters

J.M. Pacheco[1], and W. Ekardt[2]

[1]Departamento de Fisica da Universidade 3000 Coimbra, Portugal
[2]Fritz-Haber Institut der Max-Planck Gesellschaft, Faradayweg 4-6, D-14195 Berlin, Germany

Received 28 June 1993

Abstract. The dynamic response of atomic clusters made up of alkali atoms of sodium and potassium, negatively ionized, is studied microscopically and compared to the corresponding response in neutral and positively ionized species, making use of the FULL-SIC-TDLDA. An excellent agreement is found with recent experimental findings. Furthermore, a careful analysis of the line width is carried out, in order to disentangle the most important contributions to the lifetime of these systems.

PACS: 36.40.+d, 31.50.+w, 33.20.Kf

The first measurements of the photo-response of free metallic clusters [1] were performed on neutral species. More recently, it has been more feasible to study the response in ionized clusters, mostly of positively charged ones (cations). For all these systems, the surface plasmon is excited below the ionization threshold. This well known feature is used at profit experimentally to heat up clusters, since by irradiating them with laser light of a carefully tuned frequency, the surface plasmon is first excited, decaying then by transferring its energy to the normal modes of the vibrating ions. This decay mechanism, which we denote by *resistivity damping* [2], seems to be the preferred decay channel for a neutral or cationic alkali cluster, seemingly providing the most important contribution to the lifetime of the plasmon excitation. Under the assumptions of ref. [2], this mechanism will contribute to the photo-absorption cross section with a width which increases linearly with the vibrational temperature of the clusters.

In small, negatively charged clusters (anions), the plasmon is excited above the ionization threshold [3], and therefore it can decay via electron detachment. This decay channel, the branching ratio of which has been recently measured [4], is entirely analogous to the mechanism responsible for the so-called escape width of the Giant Dipole Resonance (GDR) excited in atomic nuclei, that is, it results from the coupling of the collective excitation (GDR or surface plasmon) to a continuum of unbound particle-hole ($p-h$) excitations with an energy which is degenerate with the excitation energy of the collective mode. In infinite systems, this coupling leads to the well known phenomenon of ***Landau damping***, a name which is usually kept even in finite systems such as clusters and atomic nuclei. Because Landau damping implies a coupling to unbound excited states of the system, it works as to reduce the lifetime of the collective excitation, leading to enlarged line-shapes of the photo-absorption cross sections.

Similar to the Giant Dipole Resonance (GDR) in atomic nuclei, the surface plasmon excitation in metal clusters is well described, at a microscopic level, as a correlated superposition of $p-h$ excitations originating from transitions between occupied and unoccupied one-electron states moving in a single-particle, mean-field potential. In a more realistic, quasi-particle-like description [5], this mean-field potential is substituted by an orbital dependent potential. However, this refinement does not destroy the underlying physics of the optical response. In particular, the validity of the quasi-particle picture embodies the long lifetime of these one-electron states, as observed experimentally [6]. Therefore, linear response theory has proved an adequate tool to compute, at a fully microscopic level, the excitation of collective modes in both metal clusters [5] and atomic nuclei [7]. From the hole set of $p-h$ excitations out of which the collective mode is generated, one should distinguish those with energy below and those with energy above the ionization threshold for particle emission. Because the quasi-particle states are long lived, the coupling of that collective mode to $p-h$ transitions below the threshold will not contribute significantly to reduce the lifetime of the collective mode. However, and as is well known [8,9], this coupling leads to a *fragmentation* of the plasmon peak into a more complicated structure, every time such a bound $p-h$ transition occurs at an energy degenerate with that of the collective mode, therefore contributing to the linewidth of the photo-absorption line. Due to

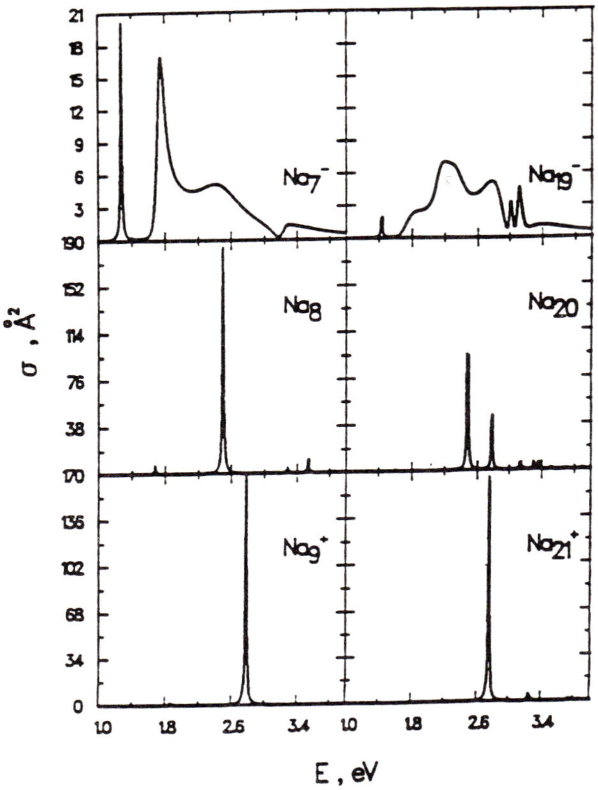

Fig. 1. Photoabsorption cross sections fot Na_7^-, Na_8 and Na_9^+ and for Na_{19}^-, Na_{20} and Na_{21}^+, in $\AA^2$, calculated making use of the FULL-SIC-TDLDA. In all cases an intrinsic numerical width of 10 meV was used (for details cf. ref. [5]).

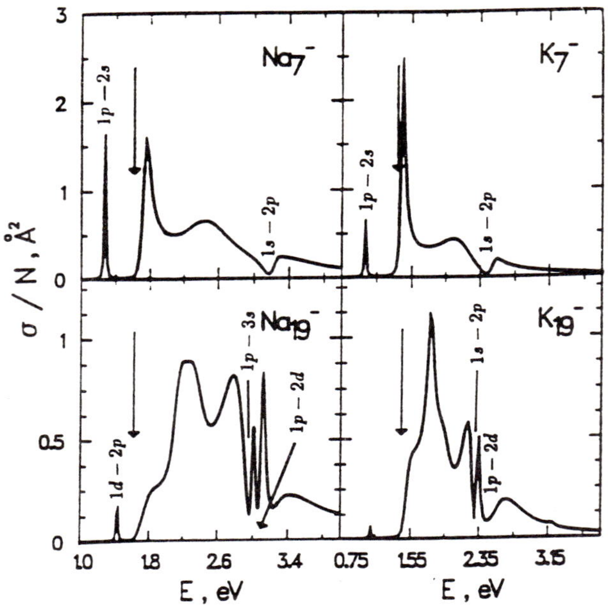

Fig. 2. Photoabsorption cross sections per valence electron of Na_7^-, K_7^-, Na_{19}^- and K_{19}^-, in $\AA^2$, calculated making use of the FULL-SIC-TDLDA. The vertical lines indicate the (size-dependent) ionization threshold. Note that the ionization threshold is growing in energy faster than the corresponding blue shift of the overall plasmon response as a function of size, indicating that for larger clusters ($N > 100$), the plasmon resonance in anionic clusters will be excited below the ionization threshold. The bound $p - h$ transitions are indicated at the associated energies.

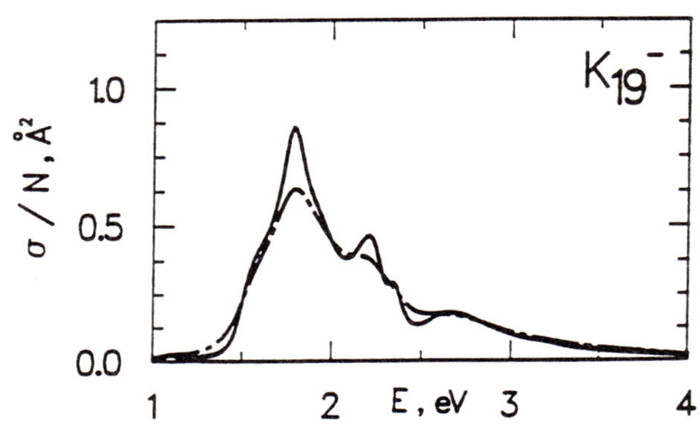

Fig. 3. Photoabsorption croos section per valence electron, in $\AA^2$, for K_{19}^-. The FULL-SIC-TDLDA line shape from fig. 1 has been broadened by the 2 extra mechanisms discussed in main text (for details, cf. ref. [3]). The broadening mechanisms were included at two different temperatures, namely 100K (solid curve) and 500K (chain dashed curve), being very effective in smoothing the resulting line shape.

the analogy with Landau damping, this phenomenon is called *Landau fragmentation*. Both Landau damping and fragmentation take place via the same coupling that produces a plasmon out of a set of $p - h$ excitations, being therefore present at the level of linear response theory, such as the *FULL - SIC - TDLDA* [5]. This is in contrast with the *resistivity damping* as well as with the coupling of the collective mode to the quantal [2] (at $T = 0$) and thermal [11,12] (at $T \neq 0$) fluctuations of the surface, which are not accounted for in linear response.

Fig. 1 shows the photo-absorption cross section of sodium clusters as a function of size and ionization charge, for the magic configurations with 8 and 20 valence electrons and for ionization charges from -1 to $+1$ (for details of calculation, cf. refs. [5,3]). A sizeable difference in the plasmon lineshape is observed between the anions and their neutral and cationic partners, mostly due to the Landau damped peaks in the anionic cross sections. Multi-peaked structures are observed in all ionization states, corresponding to Landau fragmentation. The agreement between the calculated cross sections and the experimentally determined peak structures is, where available, metter than 0.01eV. In Fig. 2 the cross-sections for sodium and potassium anions are shown. All the peaks above the ionization threshold (indicated by the vertical arrows) are Landau damped, whereas the multipeaked structure resulting from Landau fragmentation is identified by indicating the bound $p - h$ transitions responsible for the observed fragmented multi-peak structure.

To compare the computed results with experimental results, one should include the other two mechanisms which are not accounted for in linear response: Resistivity damping and coupling to surface fluctuations. This is done in fig. 3 for two different temperatures, showing that these temperature-dependent mechanisms can mask the multi-peak structure emerging from the linear response calculations, indicating that experimentally one should expect smooth line-shapes at temperatures of the order of room temperature, a feature which has been confirmed in recent experiments [4].

References

1 de Heer, W., et al.: Phys. Rev. Lett. **59**, 1805 (1987)

2 Yannouleas, C., Pacheco, J.M., and Broglia, R.A.: Phys. Rev. B **41**, 6068 (1990)

3 Pacheco, J.M., and Ekardt, W.: Phys. Rev. B **47**, 6667 (1993)

4 Haberland, H.: contribution to this conference

5 Pacheco, J.M., and Ekardt, W.: Annalen der Physik **1**, 2254 (1992)

6 Eaton, J.G., et al.: in "Nuclear Physics concepts in the study of Atomic Cluster Physics", (Bad Honnef, 1991), Lecture Notes in Physics, vol. 404, Springer Verlag, 1992

7 Bertcsh, G.F., and Tsai, S.F.: Phys. Rep. **18**, 125 (1975)

8 Ekardt, W.: Phys. Rev. B **31**, 6360 (1985)

9 Yannouleas, C., et al.: Phys. Rev. Lett. **63**, 255 (1989)

10 Pacheco, J.M., Broglia, R.A., and Mottelson, B.R.: Z. Phys. D **21**, 289 (1991)

11 Pacheco, J.M., and Broglia, R.A.: Phys. Rev. Lett. **62**, 1400 (1989)

12 In the adiabatic approximation, the coupling of the plasmon to thermal quadrupole surface fluctuations leads to a contribution to the line-width with a $\sqrt{T}$ dependence

This article was processed using Springer-Verlag TEX Z.Physik D macro package 1.0 and the AMS fonts, developed by the American Mathematical Society.

Z. Phys. A 349, 243–248 (1994)

Alpha chain states in 4N-nuclei

A.C. Merchant, W.D.M. Rae

Department of Physics, University of Oxford, Nuclear Physics Laboratory, Keble Road, Oxford, OX1 3RH, UK

Received: 27 October 1993

Abstract. We review the experimental evidence for alpha chain states in 4N-nuclei. Bloch-Brink cranked cluster model calculations suggest that such states should exist in all light A=4N nuclei. Recently a strong resonance has been observed in the $^{12}C(^{12}C,^{12}C(0_2^+))^{12}C(0_2^+)$ reaction at 33 MeV which has been taken as evidence of a 6-α chain state in ^{24}Mg. The angular distributions show strong peaks at 90° due to the simultaneous excitation of several nearly degenerate resonances with a range of L-values. We review our analysis of these data and present the results of calculations for the widths of resonances in hyperdeformed nuclei. We demonstrate that such resonances may have relatively narrow widths despite their high excitation energies. We surmise that the 33 MeV resonance in $^{12}C+^{12}C$ scattering has a chain state structure.

PACS: 24.10.Eq; 24.30.Gd; 21.60.Gx

The idea that linear alpha chain states might exist in the lightest $4N$-nuclei is very old. In 1965 Brink [1] considered the possibility of such configurations in ^8Be, ^{12}C and ^{16}O. A little later Horiuchi et al. [2] also suggested that linear alpha chain states should exist as stable structures in all $4N$-nuclei up to at least ^{44}Ti, with excitation energies slightly above the threshold for breakup into N constituent alpha particles. Later still in 1986 Marsh and Rae [3] used the Bloch-Brink cranked cluster model to predict that a rotational band of 6-alpha chain states should be present in ^{24}Mg, at excitation energies in excess of 17 MeV above the 6-alpha breakup threshold.

Until recently however, the experimental evidence in support of these theoretical expectations was limited to the very lightest $4N$-nuclei. The list of good N-alpha chain state experimental candidates in this region comprises; the ground state rotational band of ^8Be [0^+ (g.s.), 2^+(3.04 MeV) and 4^+(11.4 MeV)] (i.e. a 2-alpha chain state) the 0_2^+(7.65 MeV) state of ^{12}C, which has long been considered to contain a significant 3-alpha chain state component [1, 4, 5], and several states in ^{16}O which decay to $^8Be+^8Be$ and/or $^{12}C(0_2^+)+\alpha$ [which do not include any 0^+ states, but do contain several higher spin members, principal amongst which are the 2^+ (16.93 MeV), 4^+(18.02 MeV) and 6^+(19.32 MeV)]. The status quo was changed dramatically last year with the discovery of a strong resonance in $^{12}C(^{12}C, ^{12}C(0_2^+))^{12}C(0_2^+)$ by Wuosmaa et al. [6] at Argonne National Laboratory. This has been taken as evidence for a 6-α chain state [7]. More recently, the U.K. Charissa collaboration has studied the $^{16}O+^{12}C$ reaction [8, 9], and has made a preliminary search for evidence of a 7-alpha chain state at an excitation energy of about 59 MeV in ^{28}Si. The details of this experiment and the results will be presented at this Conference [10]. Perhaps surprisingly, there are as yet no reports of a 5-α chain state in ^{20}Ne. These experiments have revitalised interest in alpha chain state structures.

We would like to prove that the resonance seen in $^{12}C(^{12}C, ^{12}C(0_2^+))^{12}C(0_2^+)$ does indeed correspond to a 6-α chain state, but at the present time we can only surmise that this is true. However, we support our conjecture with the results of three different sets of calculations. First we review our Bloch-Brink calculations for chain states in $4N$-nuclei which predict the excitation energies. Secondly we review our analysis of the angular distribution for $^{12}C(^{12}C, ^{12}C(0_2^+))^{12}C(0_2^+)$. Finally we present the results of calculations of particle decay widths from hyperdeformed nuclei. We show that such resonances can have narrow widths despite their high excitation energies.

We have recently reported [11] completely unconstrained variational calculations for alpha chain states with a cranked version of the Bloch-Brink cluster model [1]. A detailed description of the nature of such calculations has already been published [3], as has an explanation of the application of the cranking approximation within the context of such a cluster model [12]. We calculated the energies of the chain band head within the cranked cluster model for all $4N$-nuclei from ^8Be to ^{60}Zn

Table 1. Calculated properties of alpha chain states in ^{8}Be$-^{60}$Zn. E_{max}=maximum energy above chain head and L_{max}=maximum angular momentum before instability sets in

Nucleus	E_x (MeV) Chain Head	E_{max} (MeV) relative to chain head	L_{max} $\hbar$
^{8}Be	0.0	16.7	4
^{12}C	7.9	48.8	12
^{16}O	18.9	71.4	21
^{20}Ne	31.7	107.3	35
^{24}Mg	45.5	126.0	50
^{28}Si	60.2	92.0	58
^{32}S	75.5	152.1	84
^{36}Ar	91.3	107.0	94
^{40}Ca	107.7	129.6	118
^{44}Ti	124.4	95.0	128
^{48}Cr	141.5	154.3	170
^{52}Fe	158.9	125.4	184
^{56}Ni	176.6	131.5	212
^{60}Zn	194.6	105.7	222

[11], and found that both the energy and the length of the chain before cranking are directly proportional to $N=A/4$. Table 1 lists some properties of the α-chain states predicted by the Bloch-Brink model. The excitation energy predicted for a 6-α chain state in ^{24}Mg is 45.5 MeV, close to the experimental energy of the ^{12}C(^{12}C, ^{12}C(0_2^+))^{12}C(0_2^+) resonance. Similarly, the energy predicted for the 7-α chain state in ^{28}Si is 60.2 MeV. Although these states are predicted to be well above the thresholds for N-α breakup, the discovery of the resonance in ^{12}C(^{12}C, ^{12}C(0_2^+))^{12}C(0_2^+) with a width of only 3–4 MeV (despite its excitation energy, some 17 MeV above threshold for breakup into 6 α-particles) suggests that there may be some hope of finding similar resonances in other nuclei.

As discussed above data have recently been published for the ^{12}C(^{12}C, ^{12}C(0_2^+))^{12}C(0_2^+) reaction [6]. We have recently published our analysis of these data [7]. To fit the data on resonance we needed a coherent contribution from many even partial waves. This suggests that the reaction may be exciting a highly deformed structure with a large moment of inertia. When the widths of the excited-resonant states are large compared to the rotational spacing the resonances can be considered essentially degenerate. We obtained our best fits by including $l=2$ (or possibly 4) up to $l=18$. We needed a very specific phase relationship of $(-1)^{l/2}$ between the different partial waves. This was found to be consistent not only with the cluster model but also with a partial-wave truncated delta function angle eigenstate. Having stated this much, and assuming that a highly deformed nucleus has been formed, we can already get an idea of the shape of this nucleus from the angular momenta involved. Classically there are two grazing angular momenta corresponding to the long and short axis of the deformed nucleus and all partial waves between these limits will be involved. We obtain from $k_f R_{max}=l_{max}$ and $k_f R_{min}=l_{min}$ a length of 18 fm and a width of around 2–4 fm (corresponding to $l_{min}\sim 2$–4). This length is somewhat smaller than the Bloch-Brink predictions for an α-chain state in ^{24}Mg. The low value of l_{min} is not inconsistent with the absorption of low angular momenta in the entrance channel.

In [7] we simplified further to obtain a simple closed form expression for the differential cross-section,

$$\frac{d\sigma}{d\Omega}=A\left[\frac{(l_{max}+1)!}{2^{l_{max}}[(l_{max}/2)!]^2}\,\frac{P_{l_{max}+1}(\cos\theta)}{\cos\theta}\right]^2 + B|(2L+1)P_L(\cos\theta)|^2 \tag{1}$$

and an excellent fit to the data, as shown in Fig. 1. In a later publication [13] we used the simple closed form expression to predict the energy dependence of the cross section. Interference between the background and resonant term is predicted in good agreement with the data [6, 14].

We now turn to the particle decay widths of resonances in hyperdeformed nuclei. Here we will show that a highly deformed state with an energy well above the threshold for particle decay is likely to have a relatively narrow width. Particle decay from a highly deformed nucleus is generally a very complex problem. A coupled channels model is required and this makes it difficult to obtain physical insight. We shall use the adiabatic approximation and consider only a single decay channel. The problem then simplifies to one of solving for the resonances in a deformed well with a fixed orientation. The adiabatic approximation requires one to neglect the excitation energies of the states in the rotational band of the deformed core. However, from the discussion above, the adiabatic approximation would appear to be applicable since the resonances are effectively degenerate. We shall ignore Coulomb potentials, and employ a highly simplified nuclear potential which has a certain constant value out to the boundary of the deformed nucleus, where it falls sharply to zero, i.e. a highly deformed "square well". We introduce the prolate spheroidal coordinate system which is particularly appropriate for carrying out this programme.

As we have stated we will use a potential fixed in space to represent the deformed nucleus. We need to describe the shape of this potential. This is not easy in spherical polar coordinates but there is another orthogonal coordinate system – prolate spheroidal coordinates – where the coordinate surfaces are not concentric spheres, right circular cones and half planes (as in spherical polar coordinates) but rather prolate spheroids, hyperboloids of two sheets and half planes [15]. This coordinate system is shown in Fig. 2. This coordinate system has the important property that it tends to the spherical polar system asymptotically and thus can be used for scattering calculations. The transformation equations from Cartesian coordinates are $x=a\sinh u\ \sin v\ \cos\phi=a\sqrt{(\xi^2-1)(1-\eta^2)}\cos\phi$, $y=a\sinh u\sin v\sin\phi=a\sqrt{(\xi^2-1)(1-\eta^2)}\sin\phi$ and $z=a\cosh u\cos v=a\xi\eta$.

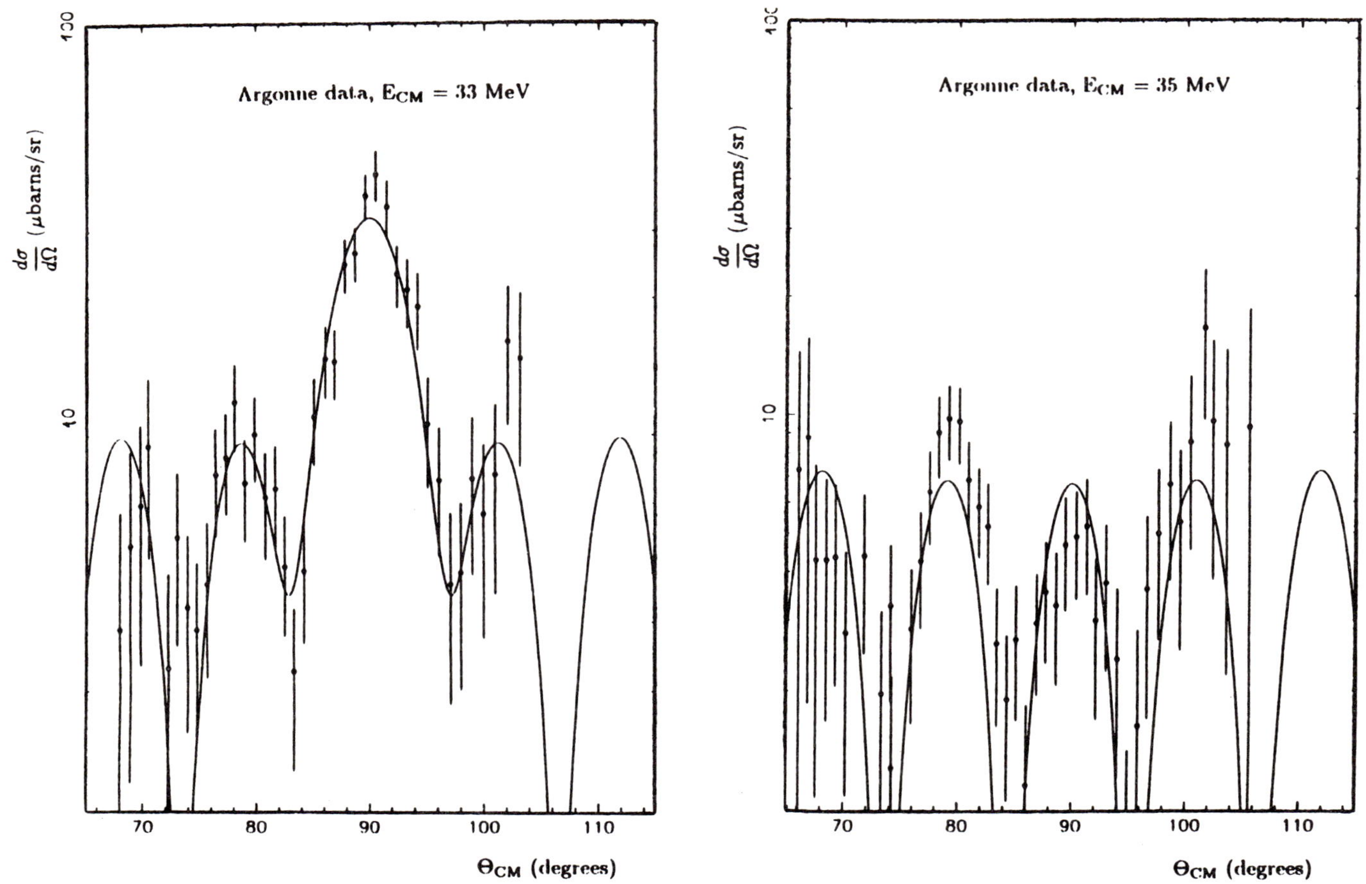

Fig. 1. Measured angular distributions for the $^{12}C(^{12}C, {}^{12}C)0_2^+))^{12}C(0_2^+)$ reaction [6] at 33 MeV (on resonance) and 35 MeV (off resonance). Equation (1) is used to fit the lower energy data [7], while a single squared Legendre polynomial is plotted with the higher energy data to guide the eye and to emphasize by contrast the unusual nature of the 33 MeV peak at 90°

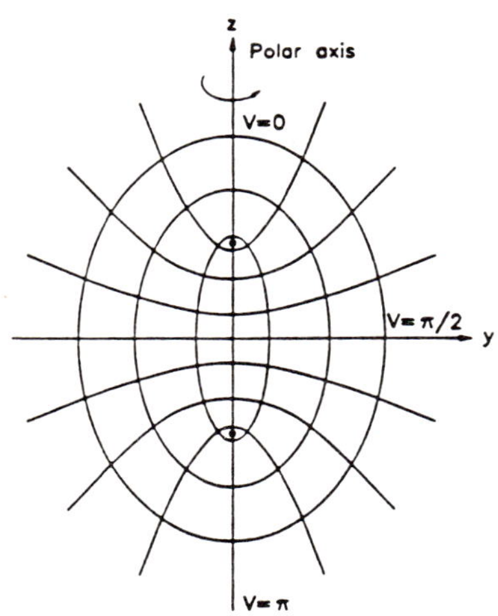

Fig. 2. The prolate spheroidal coordinate system

As ξ becomes large $\xi = \cosh u \to r$ and $v \to \theta$, yielding the familiar spherical polar coordinate radius and polar angle respectively. Here the distance between the foci is 2a and $\xi = 1/e$ where e is the eccentricity of the ellipse.

The dynamics of particles using a prolate spheroidal coordinate system requires the use of the formal Lagrangian and Hamiltonian methods of classical mechanics and the techniques of separation of the wave equation using curvilinear coordinates in quantum mechanics [15, 16]. However, if we use a square well approximation, the particles' trajectories are straight lines both inside and outside the potential (neglecting Coulomb effects). This leads to a much simpler semi-classical method of interpretation in familiar Cartesian coordinates. Indeed the classical problem of billiards in a spheroidal cavity has been well studied [17–20]. There are also many analogies with the more familiar spherical system. We shall discuss only $K=0$ orbits (i.e. orbits with $m=0$ along the symmetry axis) which reduces the problem to two dimensions classically, e.g. in the y-z plane.

Consider first the very simple case of a particle in a diamond shaped orbit in an ellipsoidal well V having the form, $V(\xi) = -V_0$ for $\xi \leqq \xi_0$ and $V(\xi) = 0$ for $\xi > \xi_0$ as in Fig. 3b. It is clear that for this orbit both p_y^2 and p_z^2 are conserved. If $(p_y^2 + p_z^2)/2\mu < V_0$, where μ is the reduced mass, the particle is trapped classically. But if both $p_y^2/2\mu$ and $p_z^2/2\mu$ are less than V_0 then the particle is still trapped classically although $(p_y^2 + p_z^2)/2\mu$ can be greater than V_0. This is actually quite familiar in a spheria

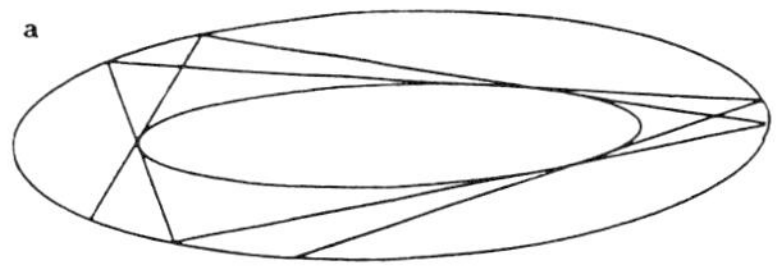

b

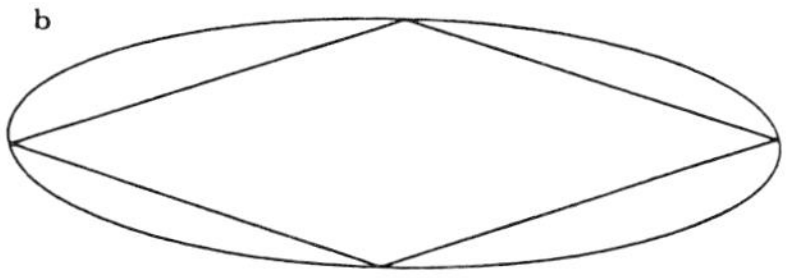

Fig. 3. **a** Orbits in an ellipsoidal square well which are specularly reflected from an ellipsoidal surface are tangential to a smaller confocal ellipsoid called a caustic. **b** Diamond shaped orbit in an ellipsoidal "square well"

cal well where we usually deal with the angular kinetic energy and the radial kinetic energy. For a classical particle inside the centrifugal barrier, the total kinetic energy can be greater than V_0 but if the radial kinetic energy is less than V_0 the particle is trapped.

Returning to spheroidal coordinates the analogy holds exactly. First it is convenient to work with dimensionless variables. We define $c^2 = 2\mu(E+V_0)a^2/\hbar^2$, and $c'^2 = 2\mu E a^2/\hbar^2$, where E is the kinetic energy of the particle for $\xi > \xi_0$. In analogy with the spherical system there is a conserved quantity λ which is quantised as is $\hat{l}^2$ in spherical systems. The eigenvalue λ is the equivalent of $l(l+1)$ and classically $\lambda_{12} = (\lambda - c^2)$ is the scalar product of the angular momentum around each focus, viz $\lambda_{12} = (\lambda - c^2) = \hat{l}_1 \cdot \hat{l}_2$. However, λ depends on c^2 and thus is not conserved when the potential depth changes – i.e. when the kinetic energy changes. In spheroidal coordinates, for $\lambda_{12} > 0$, classically the orbits pass around the outside of the two foci and if the spheroidal "radial" kinetic energy $p_\xi^2/2\mu < V_0$ then the particle is trapped by the spheroidal angular momentum barrier potential $\lambda_{12}/(\xi^2 - 1)$. The straight line orbits are tangential to an inner ellipsoid called a caustic [17] (see Fig. 3a). In a highly deformed well a stronger force is required to contain the particle at either end of the spheroid (near the foci, where $\eta = \pm 1$) since here the particle is reflected almost normally to the surface of the spheroid, while the particle makes only glancing collisions with the sides of the spheroid (near $\eta \sim 0$). This can be shown using the form of the classical angular momentum barrier potential given above and the metric scale factors for spheroidal coordinates [15].

For $l=0$ in spherical coordinates it is impossible to trap a particle with total kinetic energy $(p_y^2/2\mu + p_z^2/2\mu) > V_0$ (in the absence of a Coulomb barrier) since there is no centrifugal barrier. In spheroidal coordinates, classically there are orbits with $\lambda_{12} = \hat{l}_1 \cdot \hat{l}_2 < 0$ which pass between the foci. A particle on such an orbit is bounded by two hyperbolic caustics [17] and will escape after only a few collisions when the total kinetic energy is greater than the potential depth V_0. This can be seen from the spheroidal angular momentum barrier potential quoted above, which is now negative – essentially there is no barrier!

In spheroidal coordinates λ depends on the kinetic energy as noted above. So λ is not conserved if the particle actually crosses over the edge of the square well. This implies channel coupling at the surface in quantum mechanics. So as the particle crosses the potential edge it will make a transition from the state labelled by λ (which depends on c^2) into another angular state, λ', determined by the value of c'^2.

We use the above results to give a simple semiclassical interpretation of decay widths using methods similar to those of Buck et al. [22, 23] for alpha decay from heavy nuclei [9]. We assume that we have a particle in a *single* internal state λ with a specific n. This can not be achieved exactly in the proper quantum calculations (as below) where a spread in λ-values (or equivalently n values) is found, but usually a single value is dominant. We further assume that tunnelling through the barrier is only important at either end (the Gamov factors for tunnelling out the sides being about an order of magnitude smaller). Finally we use the classical spheroidal angular momentum barrier given above.

The major and minor axes of the potential for these calculations were determined from the predictions of the Bloch-Brink model for α-chain states, being equal to the length and width of the chain state, respectively [11]. The depth of the potential was chosen to fit the number of nodes required by the Pauli principle ($G=n$) between the two foci [21]. This yields a depth of $V_0 \sim 50$ MeV for alpha particles. The experimental Q-values were used where known, and the others taken from the Bloch-Brink model [11].

We have applied this model to α-chain states in ^{24}Mg and ^{28}Si. The results are shown in Table 2. Although the model is not applicable to lighter nuclei, where the energies of chain states are not degenerate, the widths predicted by this model agree to within a factor of two with the data for these lighter nuclei if the correct Q value for each state is used. The result for ^{24}Mg is encouraging, predicting a total width (assuming a simple incoherent sum) to within a factor 3–4 of that observed experimentally.

We have also performed exact quantum mechanical calculations for resonances and hence widths for particle emission from hyperdeformed nuclei. We will present de-

Table 2. Semi-classical calculation of widths for emission of α, ^{8}Be and $^{12}C(0_2^+)$ from α-chain states in ^{24}Mg and ^{28}Si obtained using prolate spheroidal coordinates

Decay	Calculated width (keV)
$^{24}Mg \to {}^{20}Ne + \alpha$	317
$^{28}Si \to {}^{24}Mg + \alpha$	173
$^{24}Mg \to {}^{16}O + {}^8Be$	444
$^{28}Si \to {}^{20}Ne + {}^8Be$	270
$^{24}Mg \to {}^{12}C + {}^{12}C$	461
$^{28}Si \to {}^{16}O + {}^{12}C$	399

Table 3. Calculated quantum widths for alpha particle emission from chain states in some $4N$-nuclei. Asterisks indicate that energies have been calculated using the Bloch-Brink prediction for the excitation energy of the 5-α chain state in ^{20}Ne

Chain decay	Length (fm)	E_{res} (MeV)	Γ_α (MeV)
$^{16}O \rightarrow ^{12}C + \alpha$	15.4	4.3	~ 1.2
$^{20}Ne \rightarrow ^{16}O + \alpha$	18.6	8.2*	~ 1.7
$^{24}Mg \rightarrow ^{20}Ne + \alpha$	23.8	5.4*	~ 1.2
$^{28}Si \rightarrow ^{24}Mg + \alpha$	27.8	3.0	~ 1.1

tails of these calculations elsewhere [24]. We have taken a nuclear potential depth of 54 MeV, and an inter-focal distance of 24 fm to represent a chain state in ^{24}Mg as suggested by the Bloch-Brink calculations. For other nuclei we have also applied the length predicted by the Bloch-Brink model, but fine-tuned its value by ~ 0.5 fm to obtain correct Q-values. We have chosen ξ_0 so that the volume of the nuclear ellipsoid should be equal to $16\pi(1.2)^3 N/3$, the spherical value for the same mass number $4N$. We find that the internal amplitudes and the scattering matrix show very strong peaks at certain energies. We identify these peaks as resonances for α-scattering off a hyperdeformed nucleus. We find that the resonances are to be found predominantly in the first λ-channel consistent with the Pauli restrictions discussed above, although neighbouring internal channels are weakly excited. We also find that the phase of the scattering matrix rises through π in a considerable number of exit channels simultaneously, because of the strong channel coupling, consistent with out expectations. The values of Γ_α, (the full widths at half maximum for alpha particle emission) are shown in Table 3. These are compatible with and systematically somewhat larger than the previous semi-classical results, but the widths are still not large, thus backing up our claim that resonances in hyperdeformed nuclei can have relatively narrow widths.

The α-widths we have calculated using this exact quantum mechanical model for hyperdeformed resonances are still smaller than the width observed experimentally for the $^{12}C(^{12}C, ^{12}C(0_2^+))^{12}C(0_2^+)$ resonance. However, if the quantum widths for the other decay channels (^8Be and $^{12}C(0_2^+)$) scale proportionally (viz. they are a factor of about 4 larger than the semi-classical results) the total width predicted quantum mechanically would become much closer to the experimental value. We must await the results of more complete calculations for ^8Be and ^{12}C emission before reaching any firm conclusions [24]. However, the definitive widths will come from calculations where all decay channels are coupled together.

To summarise, theoretical calculations have predicted the existence of linear alpha chain states in $4N$-nuclei for over 25 years. Until recently the only experimental evidence for such structures was restricted to the very lightest nuclei, ^8Be, ^{12}C and ^{16}O. However, recent experimental work at Argonne National Laboratory has found a candidate for a 6-alpha chain state in ^{24}Mg. The excitation energies of these resonances are in agreement with a series of band head energies calculated with the Bloch-Brink cluster model, suggesting that these resonances correspond to α-chain states. In addition, the Argonne result suggests that the very unusual angular distribution indicates that a highly deformed system has been excited.

At first sight it is surprising that states so high above the threshold for breakup into their constituent alpha particles (over 15 MeV for ^{24}Mg) should be narrow enough ($\Gamma = 3$–5 MeV) to be experimentally detectable. However, extensive rearrangement of the nuclear constituents is necessary to enable such states to decay to most of the lower-lying, considerably less deformed daughter states, whilst the number of lower-lying, hyperdeformed states is very small.

The calculation of the particle decay widths of these highly deformed states is a problem which we have begun to address in this paper. We have presented here some preliminary calculations of a highly simplified nature which, we believe, demonstrate the plausibility that such states may have particle emission widths of only a few MeV. Much more work is needed in this area to make these calculations more realistic, in particular the inclusion of Coulomb interactions, the employment of a diffuse nuclear potential, the coupling together of the various different types of particle decay channels, and ideally a greater flexibility in the treatment of the actual deformed nuclear shape. In conclusion we surmise that the 33 MeV resonance in $^{12}C + ^{12}C$ scattering has an α-chain state structure.

We would like to thank Dr. R.R. Betts and Dr. A.H. Wuosmaa for invaluable discussions and communications concerning their experimental data. We thank Peter Simmons and the Charissa group who set up and analysed the ^{28}Si experiments at Daresbury. We are indebted to Dr. Brian Buck who helped interpret the angular distributions and who suggested prolate spheroidal coordinates. We also thank Dr. Buck for many useful discussions as we came to grips with coupled channels calculations in this new coordinate system and for a critical reading of the manuscript. We acknowledge useful discussions with Neil Rowley and we would like to thank Jie Zhang for his enthusiasm and continued interest. ACM is supported by the SERC.

References

1. Brink, D.M.: In: Proceedings of the International School of Physics, "Enrico Fermi", Course XXXVI, Varenna 1965, Bloch, C. (ed) p. 247. New York: Academic Press 1966
2. Horiuchi, H., Ikeda, K., Suzuki, Y.: Prog. Theor. Phys. Suppl. **52**, 89 (1972)
3. Marsh, S., Rae, W.D.M.: Phys. Lett. **180 B**, 185 (1986)
4. Brink, D.M., Friedrich, H., Weiguny, A., Wong, C.W.: Phys. Lett. **33 B**, 143 (1970)
5. Takigawa, N., Arima, A.: Nucl. Phys. **A 168**, 593 (1971)
6. Wuosmaa, A.H., Betts, R.R., Back, B.B., Freer, M., Glagola, B.G., Happ, Th., Henderson, D.J., Wilt, P.: Bearden, I.G.: Phys. Rev. Lett. **68**, 1295 (1992)
7. Rae, W.D.M., Merchant, A.C., Buck, B.: Phys. Rev. Lett. **69**, 3709 (1992)
8. Simmons, B., et al.: In: IOP Converence on nuclear and particle physics, Glasgow, UK 1993, Programme and Abstract, p. 34, and this conference

9. Rae, W.D.M., Merchant, A.C.: "Nuclear cluster states" invited presentation in IOP Conference on nuclear and particle physics, Glasgow, UK 1993, IOP Conference Series Number 133, p. 35. MacGregor, I.J.D., Doyle, A.T. (eds.)
10. Simmons, P.M., Rae, W.D.M.: Contribution to this Conference
11. Merchant, A.C., Rae, W.D.M.: Nucl. Phys. **A549**, 431 (1992)
12. Caurier, E., Grammaticos, B.: J. Phys. **G9**, L 125 (1983)
13. Merchant, A.C., Rae, W.D.M.: J. Phys. **G19**, L 89 (1993)
14. Wuosmaa, A.H., Betts, R.R., Back, B.B., Freer, M., Glagola, B.G., Happ, Th., Henderson, D.J., Wilt, P., Bearden, I.G.: In: Argonne National Laboratory Report PHY-6908-HI-91
15. Arfken, G.: Mathematical methods for physicists, (2nd edn.) p. 103. New York: Academic Press 1970
16. Landau, L.D., Lifschitz, E.M.: In: Course of theoretical physics, Vol. 1 Mechanics (3rd edn.) p. 153. Oxford: Pergamon Press 1976
17. Keller, J.B., Rubinow, S.I.: Ann. Phys. **9**, 24 (1960)
18. Berry, M.V.: Eur. J. Phys. **2**, 91 (1981)
19. Arvieu, R., Brut, F., Carbonell, J., Touchard, J.: Phys. Rev. **A35**, 2389 (1987)
20. Ayant, Y., Arvieu, R.: J. Phys. **A20**, 397 (1987)
21. Arvieu, R., Ayant, Y.: J. Phys. **A20**, 1115 (1987)
22. Buck, B., Merchant, A.C., Perez, S.M.: Phys. Rev. Lett. **65**, 2975 (1990)
23. Buck, B., Merchant, A.C., Perez, S.M.: Phys. Rev. **C45**, 2247 (1992)
24. Rae, W.D.M., Merchant, A.C. (in preparation)

Z. Phys. A 349, 249–253 (1994)

Many-particle decays of α-chain structures in ^{24}Mg

A. H. Wuosmaa

Physics Division, Argonne National Laboratory*, Argonne, IL 60439, USA

Received: 27 October 1993

Abstract. We have searched for evidence of exotic cluster configurations in ^{24}Mg resembling a linear chain of α particles in various many-particle final states of the ^{12}C+^{12}C system, including ^{12}C(0_2^+)+^{12}C(0_2^+) and ^{8}Be+^{16}O*(4α). Such configurations are predicted to occur by a number of different theoretical models of the structure of ^{24}Mg. An array of highly segmented double-sided silicon strip detectors permits detailed, high resolution reconstruction of these many-charged-particle final states.

PACS: 25.70.Ef; 21.60.Gx; 27.30.+t

---

1. Introduction

Alpha-cluster structures are well established in light nuclei with even $N = Z$. Nuclei which show strong α-cluster character typically display a number of different coexisting shapes based upon the α-particle sub-unit. One such nucleus is ^{24}Mg. A number of theoretical calculations for the structure of ^{24}Mg which display the underlying cluster degrees of freedom have been performed. The theoretical frameworks employed include the Cranked Cluster Model (CCM)[1], and the Nilsson-Strutinsky[2] and Hartree-Fock[3] formalisms.

One very interesting prediction of many of these theoretical results is that there should occur in these light α-particle nuclei configurations that resemble linear chains of α particles[4, 5]. These configurations were suggested to exist as early as 1956, by Morinaga[6]. Experimental evidence for such chain structures in the ^{16}O has been reported by Chevallier, et al.[7] who identified a number of strong resonances in the ^{12}C(α,^{8}Be)^{8}Be reaction. The spin sequence of these resonances, as well as the fact that they possessed large decay branches to the ^{8}Be+^{8}Be final state, suggested this assignment. In addition, Ames[8] noted that these resonances possessed α-decay widths to the first excited 0$^+$ level in ^{12}C at 7.65 of MeV excitation energy.

In the nucleus ^{24}Mg, the CCM predicts that chain-like configurations should occur at excitation energies between 40 and 50 MeV, well above the threshold for decay of ^{24}Mg into six α particles, at E_X=28.48 MeV. Thus, any attempt to study such phenomena must contend with the rather difficult experimental problem of reconstructing a complicated six-α-particle final state. In ^{24}Mg, one possible simplification of this situation could be found by considering the excited 0$^+$ level in ^{12}C at E_X=7.65 MeV. This level has previously been associated with a linear three-α-particle chain structure[4, 9]. If populated, one likely decay mode for a highly elongated cluster structure in ^{24}Mg would be by symmetric breakup into two prolate-deformed ^{12}C clusters, corresponding to an exit channel containing two ^{12}C nuclei in their excited 0_2^+ states. A possible signature for the existence of such chain-like configurations in ^{24}Mg could then be the observation of resonance-like behavior in the excitation function for the inelastic scattering channel ^{12}C+^{12}C$\rightarrow$^{12}C(0_2^+)+^{12}C(0_2^+).

Further information about these chain structures in ^{24}Mg could be obtained by considering other reaction channels in the ^{12}C+^{12}C system. One reaction of interest is that leading to ^{8}Be+^{16}O, where the ^{16}O subsequently decays to α-unbound final states. The α+^{12}C scattering measurements of Chevallier et al., and Ames identified levels in ^{16}O which might be structurally related to those we seek to study in ^{24}Mg. If we are able to populate α-chain configurations in ^{24}Mg through ^{12}C+^{12}C scattering, these structures should also possess non-symmetric decay branches to the ^{8}Be+^{16}O*(4α) final state, leaving ^{16}O in one of these levels having a four-α chain character. Of course, since these states in ^{16}O are also unbound with respect to decay into four α particles, these reactions also present the experimental challenge of reconstructing two-body reaction kinematics from a six-charged-particle final state.

2. Experiment

In order to study reactions of this type, we have constructed an array of Double-Sided Silicon Strip Detectors (DSSDs)[10 – 12]. These detectors are fabricated from 5×5 cm^2 silicon wafers effectively segmented into 256 quasi-

* Work supported by the U. S. Department of Energy, Nuclear Physics Division, under contract W-31-109-Eng-38

pixel regions. We have performed measurements using a number of configurations of this array. For our studies of ^{12}C+^{12}C inelastic scattering, two DSSDs were used, placed at distances of 14 to 17 cm from the target, centered at angles from 12° to 35° in the laboratory. In order to search for four-α decaying levels in ^{16}O populated by α transfer, we used two additional DSSDs, yielding an array with a total solid angle of 430 msr, divided into 1024 quasi-pixels. For the case of ^{12}C+^{12}C scattering to six-α-particle final states, this array permits us to detect and identify as many as all six α particles, and measure each of their angles and energies with quite high precision ($\Delta E \approx$ 30 keV FWHM, $\Delta\theta_{lab} \approx 1°$). A time-of-flight measurement relative to the pulsed beam was made for each detected particle for particle identification purposes, although in practice, for events with more than four detected particles, energy and multiplicity were sufficient to identify six-α-particle final states.

For all of our experiments, ^{12}C beams from the ATLAS accelerator at Argonne National Laboratory bombarded 50μg/cm^2 ^{12}C targets at energies between E_{lab}= 52 to 80 MeV. These energies correspond to excitation energies in the composite nucleus ^{24}Mg ranging from ≈40 to 54 MeV, excitation energies at which the CCM predicts the occurrence of elongated α-particle structures in ^{24}Mg. Our excitation-function measurements were made in steps of ΔE_{lab}= 2 MeV, and angular distribution data were obtained in long runs at several selected energies between E_{lab}= 52 and 80 MeV. Data for the ^{8}Be+^{16}O α-transfer reaction were taken at five energies in the same range.

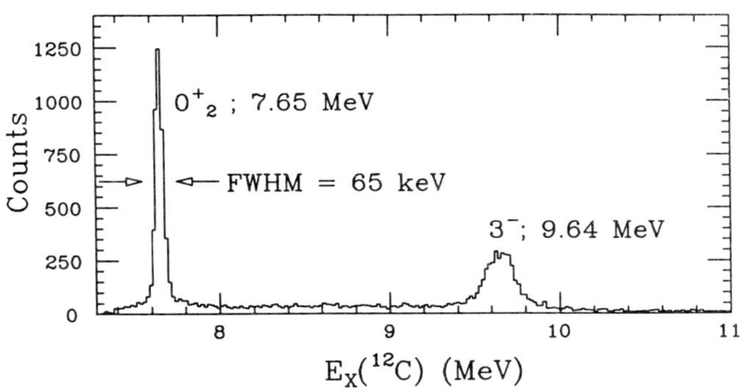

Fig. 1. Reconstructed ^{12}C excitation-energy spectrum at E_{lab}=65 MeV

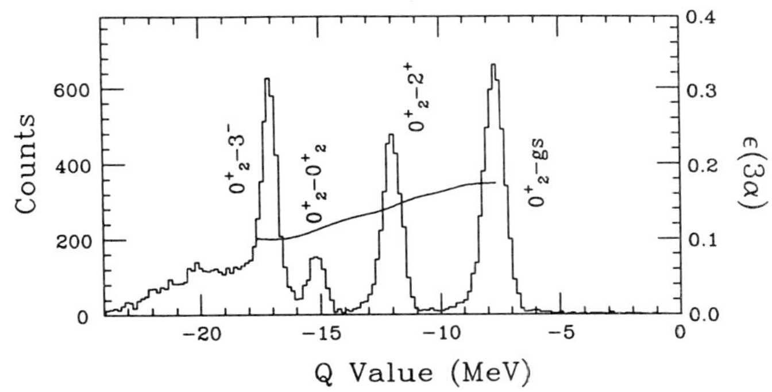

Fig. 2. Q-value spectrum for ^{12}C+^{12}C inelastic scattering at E_{lab}=65 MeV, where one ^{12}C nucleus is left in its 0^+_2 state. The ^{12}C(0^+_2)+^{12}C(0^+_2) mutual excitation is at Q=-15.30 MeV. The *solid curve* represents the three-α detection efficiency as determined from a Monte Carlo simulation of the detector setup. The efficiency scale is given on the right-hand axis

3. ^{12}C+^{12}C inelastic scattering results

To analyse the ^{12}C+^{12}C inelastic scattering data, we considered events in which at least three α particles were detected in one DSSD. By measuring the angles and energies of these three α particles, assuming that they result from the decay of an excited ^{12}C nucleus, we can reconstruct the kinetic energy, scattering angle and excitation energy of the decaying ^{12}C fragment. The Q value of the two-body inelastic scattering reaction is then determined from the kinematic relationship between kinetic energy and scattering angle of the decaying recoil nucleus[12].

An example of this reconstruction appears in Fig. 1, which shows a spectrum of the reconstructed excitation energy for a decaying ^{12}C , calculated for events in which three α particles hit one DSSD. The ^{12}C(0^+_2) level at 7.65 MeV is clearly identified, and is reconstructed with a resolution of approximately 65 to 70 keV FWHM. Figure 2 shows a Q-value spectrum for ^{12}C+^{12}C inelastic scattering, determined from the scattering angle and kinetic energy of a ^{12}C fragment which was identified as having been in its 0^+_2 excited state. Several peaks are apparent, and the reaction channel of interest, ^{12}C(0^+_2)+^{12}C(0^+_2), is clearly seen at Q=-15.30 MeV.

The measured excitation function for the mutual ^{12}C(0^+_2) inelastic scattering channel appears in Fig. 3. We observe a strong peak in the excitation function centered at $E_{c.m.}$= 32.5 MeV. This energy corresponds to an excitation energy of E_X=46.4 MeV in ^{24}Mg, near the energy at which α-chain structures are predicted to occur by the CCM[1]. This

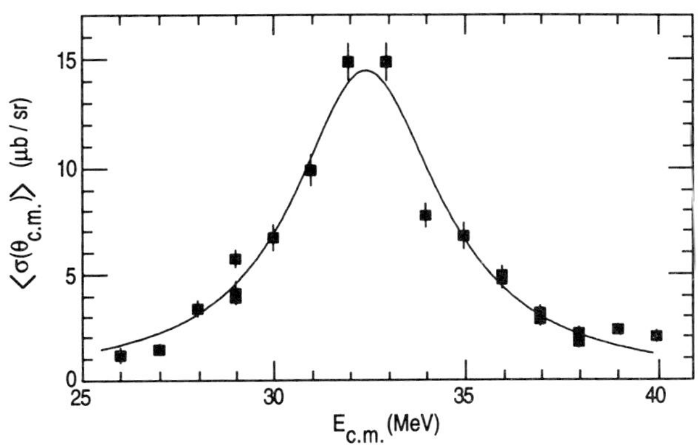

Fig. 3. ^{12}C+^{12}C→^{12}C(0^+_2)+^{12}C(0^+_2) inelastic scattering excitation function, reproduced from [12]

structure is quite wide, with $\Gamma_{c.m.}$ ≈4.5 MeV, and as such it is unlikely to arise from the population of a single level in the composite system, or to reflect the influence of a single resonating partial wave.

In order to determine which angular momenta contribute to this inelastic scattering channel, we have obtained detailed angular-distribution data at several energies across this ex-

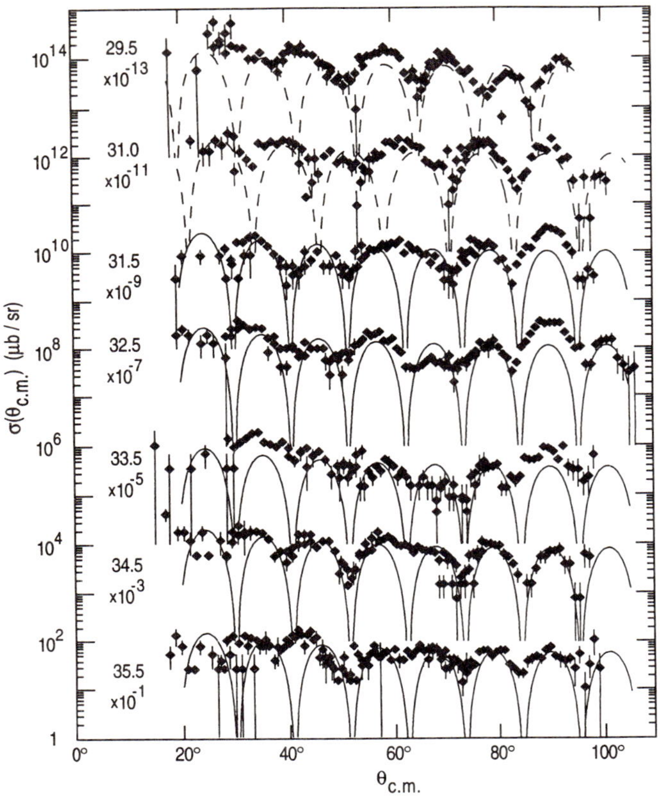

Fig. 4. $^{12}C+^{12}C \rightarrow ^{12}C(0_2^+)+^{12}C(0_2^+)$ inelastic scattering angular-distribution data. The numbers to the left indicate the center-of-mass energy in MeV, and the multiplicative factor to be applied to the cross section scale for each energy. The solid(dashed) curves represent pure squared Legendre polynomials of order 16(14)

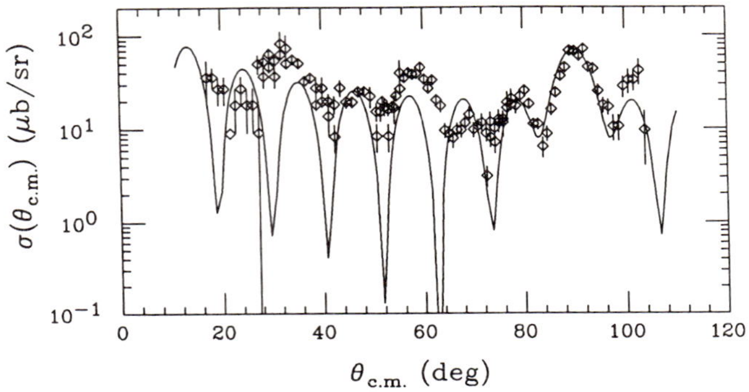

Fig. 5. $^{12}C+^{12}C \rightarrow$ $^{12}C(0_2^+)+^{12}C(0_2^+)$ angular distribution at $E_{c.m.}$=32.5 MeV, with fit from [13].

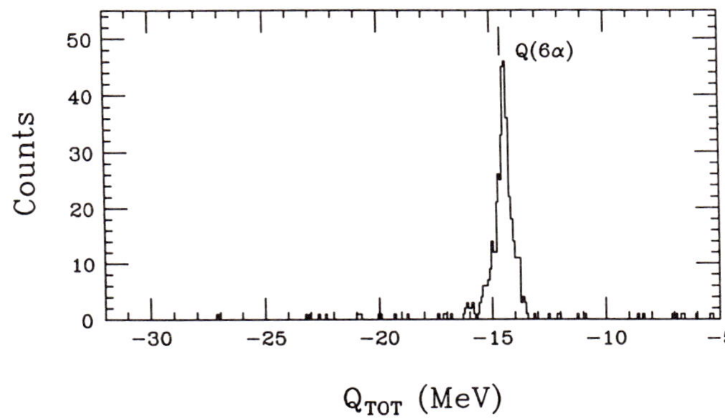

Fig. 6. Total reaction Q-value spectrum for $^{12}C+^{12}C \rightarrow 6\alpha$ at E_{lab}= 65 MeV

citation function peak. Figure 4 shows angular distributions obtained at seven energies between $E_{c.m.}$= 28.5 and 35.5 MeV. A simple comparison to pure Legendre polynomials squared suggests that the dominant angular momenta in this energy region are near 14-16 $\hbar$, placing the structure near the location of a crossing between a rotational band build upon a cluster configuration with a strong $^{12}C+^{12}C$ character, and one built upon a six-α chain structure.

The angular distribution data we obtain are, however, considerably more complex than those expected for a single dominant angular momentum. In particular, at the maximum of the excitation function $E_{c.m.}$=32.5 MeV, the angular distribution shows a strong peak at center-of-mass angles near 90°. This enhancement requires a coherent superposition of several partial waves at this energy. Rae,et al., have recently shown that such a superposition of angular momenta can come about through the population of nearly degenerate members of a rotational band built upon a highly elongated cluster configuration[13]. A fit to the data obtained at $E_{c.m.}$=32.5 MeV using the formalism described in reference [13] appears in Fig. 5. The agreement between theory and experiment is surprisingly good considering the simplicity of the model.

4. $^{12}C(^{12}C,^{8}Be)^{16}O^*(4\alpha)$ results

To examine the possibility of non-symmetric decays of chain structures in ^{24}Mg, we now turn to our results for the reaction leading to $^{8}Be+^{16}O^*$. We wish to identify levels at high excitation energy in ^{16}O that correspond to the possible four-α chain structures identified in this nucleus as resonances in $\alpha+^{12}C\rightarrow$ $^{8}Be+^{8}Be$ and $\alpha+^{12}C(0_2^+)$. Such reactions are significantly more difficult to study than the final states in $^{12}C+^{12}C$ inelastic scattering, for their identification requires the detection of at least five α particles, rather than only three for the $^{12}C+^{12}C$ case, and a more careful and detailed analysis.

We first consider events in which either five or six α particles have been detected in our array of four DSSDs. If one α particle is missed, its momentum and energy are reconstructed from the momenta of the other detected particles. Figure 6 shows a total reaction Q-value spectrum for six-α coincidence events. The peak at Q_{tot}=-14.55 MeV represents the energy needed to completely dissociate two ^{12}C nuclei into α particles.

Candidate $^{8}Be+^{16}O^*(4\alpha)$ events were those in which two α particles from the decay of a ^{8}Be in its ground state were observed on one side of the beam, and the four additional α particles on the other side. For these events, the two-body Q value of the transfer reaction was calculated from the reconstructed kinetic energy and scattering angle of the identified ^{8}Be. In addition, the excitation energy of the

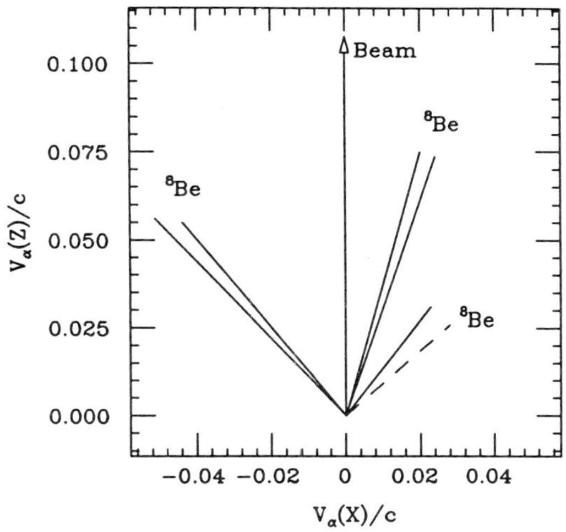

Fig. 7. Velocity diagram for an event identified as ^{12}C+^{12}C $\rightarrow$^{8}Be+^{16}O*(^{8}Be). The *solid lines* represent the magnitude and direction of the velocity of identified α particles, projected onto the X-Z plane. The *dashed line* represents the velocity of an unobserved alpha particle, and the beam velocity is indicated by the central arrow

decaying ^{16}O can be calculated from the measured angles and energies of the four remaining α particles. For each candidate event the excitation energy of the ^{16}O nucleus determined from the ^{8}Be kinetic energy and angle was required to match the excitation energy determined from the other four α particles to within one MeV.

Finally, all possible combinations of two and three α particles from the decaying ^{16}O were analysed to identify the ^{16}O decay branch, in order to determine whether the ^{16}O decayed to either two ^{8}Be nuclei, or to α+^{12}C(0_2^+). As an example of this analysis, Fig. 7 contains a diagram which shows the velocities of all α particles in an event identified as ^{12}C+^{12}C$\rightarrow$^{8}Be+^{16}O*(^{8}Be). The primary ^{8}Be goes to the left of the beam at a scattering angle of 42.9°, with a kinetic energy of 20.6 MeV. The recoil ^{16}O on the right side of the beam then decays to two additional ^{8}Be nuclei, which finally decay into four α particles. The high segmentation and excellent energy resolution of the DSSD makes possible a complete and detailed analysis of the six-body final state for this event.

Figure 8 shows the excitation-energy spectra obtained for various final states in ^{12}C(^{12}C,^{8}Be)^{16}O*(4α). Figure 8a contains data for all events obtained at E_{lab}= 62,65, and 67 MeV which could be identified as ^{8}Be+^{16}O*(4α). A prominent peak is observed at E_X= 19.4 MeV, and a smaller one near 17.5 MeV. Both structures persist if the ^{16}O decays to α+^{12}C(0_2^+). For the ^{8}Be+^{8}Be final state, the statistics are quite poor, however there does appear to be some enhancement near E_X= 17.5 MeV. These excitation energies are quite close to those of levels observed as resonances in α+^{12}C$\rightarrow$^{8}Be+^{8}Be and α+^{12}C(0_2^+). While angular correlation measurements are still required in order to determine the spins of the levels populated in this reaction, and identify them with the resonances observed in α+^{12}C scattering, it seems unlikely that we would be populating different states with similar structural characteristics in the same region of excitation energy in ^{16}O. Should further measurements

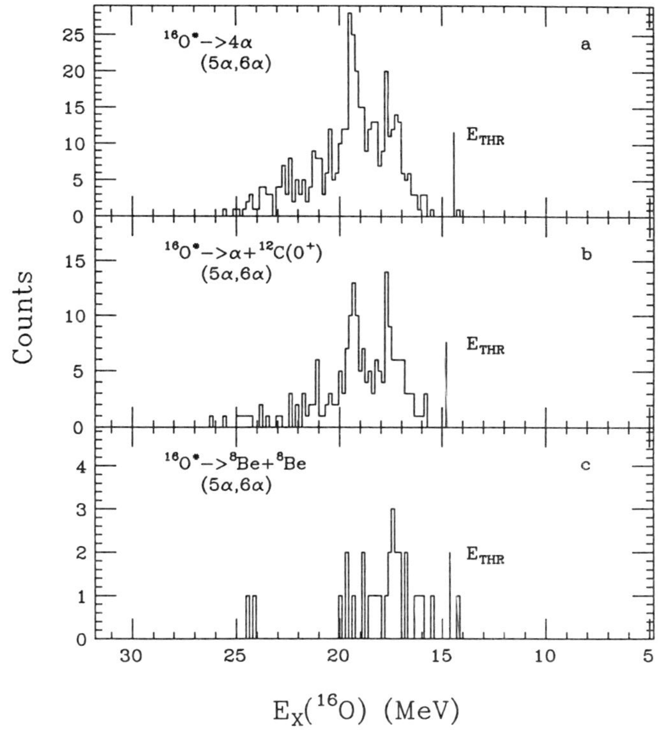

Fig. 8a,b,c. ^{16}O excitation-energy spectra for events where the ^{16}O decays into (**a**) any combination of four α particles, (**b**) α+^{12}C(0_2^+), and (**c**) ^{8}Be+^{8}Be. The decay thresholds for each channel are indicated by the solid lines

bear out the association of these peaks with four-α chain states in ^{16}O, this identification would provide considerable additional evidence for the presence of exotic elongated configurations in ^{24}Mg.

5. Summary and conclusions

We have used an array of DSSDs to make a detailed study of many-particle final states in the ^{12}C+^{12}C system. A strong peak is observed in the excitation function for ^{12}C(^{12}C,^{12}C(0_2^+))^{12}C(0_2^+) inelastic scattering. The energy of the excitation-function peak, as well as the shapes of the angular distribution data are consistent with the possibility that extended α-particle cluster configurations in ^{24}Mg are formed in this system. Furthermore, we observe that through the reaction ^{12}C(^{12}C,^{8}Be)^{16}O, levels in ^{16}O at high excitation energy which decay to ^{8}Be+^{8}Be and α+^{12}C(0_2^+) are populated at the same center-of-mass energies at which we observe the excitation function peak in the ^{12}C(0_2^+)+^{12}C(0_2^+) final state. Observation of these states in ^{16}O , which have been suggested to possess a four-α chain structure, could represent non-symmetric decays of similar configurations in ^{24}Mg.

The possibility that we are populating such unusual configurations in ^{24}Mg is an intriguing and compelling one, and suggests a number of avenues for future investigation. An angular correlation measurement for the four-α unbound levels in ^{16}O would make the association of these levels with resonances in α scattering more likely. Also important is a determination the beam-energy dependence of the yield for these final states in ^{16}O. Finally, we hope to pursue reaction calculations for the angular-momentum mismatched $^{12}C(0_2^+)+^{12}C(0_2^+)$ final state. This additional information should help to further clarify the situation for these potentially exotic and interesting reactions.

Acknowledgement. I wish to thank my collaborators, in particular M. Freer and R. R. Betts, for many useful discussions, and invaluable assistance in obtaining the data presented here.

References

1 S. Marsh, W.D.M. Rae: Phys. Lett. **180B**, 185 (1986)
2 G. Leander, S.E. Larsson: Nucl. Phys. **A239**, 93 (1975)
3 H. Flocard et al.: Phys. Lett. **95B**, 5 (1980)
4 D. M. Brink et al.: Phys. Lett. **33B**, 143 (1970)
5 W. Bauhoff et al.: Phys. Rev. **C29**, 1046 (1984)
6 H. Morinaga: Phys. Rev. **101**, 254 (1956)
7 P. Chevallier et al.: Phys. Rev. **160**, 827 (1967)
8 L. L. Ames, Ph. D. Thesis, Univ. of Wisconsin - Madison, 1979, (unpublished)
9 Strip detectors fabricated by Micron Semiconductor, Lancing, UK
10 N. Takigawa, A. Arima: Nucl. Phys. **A168**, 594 (1971)
11 A. H. Wuosmaa et al.: Proceedings of the Workshop on Large Gamma-Ray Detector Arrays, Chalk River Laboratories, 418 (1992)
12 A. H. Wuosmaa et al.: Phys. Rev. Lett. **68**, 1295 (1992)
13 W. D. M. Rae et al.: Phys. Rev. Lett. **69**, 3709 (1992)

This article was processed using Springer-Verlag TEX Z.Physik A macro package 1.0 and the AMS fonts, developed by the American Mathematical Society.

Z. Phys. A 349, 255–257 (1994)

Linear α-chain states in nuclei

L. Zamick [1], **D.C. Zheng** [2]

[1] Department of Physics and Astronomy, Rutgers University, Piscataway, NJ 08855, USA
[2] Department of Physics, University of Arizona, Tucson, AZ 85721, USA

Received: 27 October 1993

Abstract. We present results for the energies of the linear α-chain states in several light nuclei, ^{12}C, ^{16}O, ^{20}Ne and ^{24}Mg, using the Skyrme Hartree-Fock method. We discuss mainly the SKM* results but other interactions are also considered for comparison.

PACS: 21.60.Gx; 21.60.Jz; 21.60.Ev.

I. Introduction

It is easiest to picture the linear α-chain states in the context of the deformed oscillator (DO) model. The Hamiltonian is

$$H_{\rm DO}=H_x+H_y+H_z$$
$$=\left(\frac{p_x^2}{2m}+\frac{1}{2}m\omega_x x^2\right)+\left(\frac{p_y^2}{2m}+\frac{1}{2}m\omega_x y^2\right)$$
$$+\left(\frac{p_z^2}{2m}+\frac{1}{2}m\omega_z z^2\right). \tag{1}$$

We assume axial symmetry: $\omega_x=\omega_y\equiv\omega_\perp$. The energy is

$$E=\hbar\omega_x\Sigma_x+\hbar\omega_y\Sigma_y+\hbar\omega_z\Sigma_z, \tag{2}$$

where $\Sigma_x=\sum(N_x+\frac{1}{2})$ with N_x the number of quanta in the x direction, etc..

The ground state, say, for ^{16}O, is obtained by filling the lowest four levels: $(N_x, N_y, N_z)=(0,0,0)$, $(1,0,0)$, $(0,1,0)$, $(0,0,1)$. The linear chain state is obtained by Brink's prescription of "putting all the quanta in the z direction" [1], so for ^{16}O, the occupying states are $(0,0,0)$, $(0,0,1)$, $(0,0,2)$ and $(0,0,3)$. To get the deformation we must supplement the above with the Mottelson condition:

$$\frac{\omega_x}{\omega_z}=\frac{2\Sigma_z}{\Sigma_x+\Sigma_y}. \tag{3}$$

This condition leads to a minimization of the energy (2) provided that the volume is conserved, i.e., $\omega_0^3=\omega_x\omega_y\omega_z$ is fixed.

For the chain states, it is easy to show that [2]

$$E_{\rm chain}=\hbar\omega_\perp A+\hbar\omega_z\frac{A^2}{8}=\frac{2A^{4/3}}{2^{5/3}}\hbar\omega_0, \tag{4}$$

and that $\omega_x/\omega_z=A/4$, where A is the mass number. The above energy is interpreted as an intrinsic state energy from which we must project out the states of good angular momentum. If we assume that $\hbar\omega_0=45A^{-1/3}-25A^{-2/3}$ MeV, the above procedure yields intrinsic state energies of 9.8 MeV, 29.2 MeV and 102.2 MeV for ^{12}C, ^{16}O and ^{40}Ca, respectively.

II. The Skyrme Hartree-Fock methods

A. General Skyrme interactions

The Skyrme interaction can be written as:

$$V_{\rm Skyrme}=\sum_{i<j}V_{ij}+\sum_{i<j<k}V_{ijk}, \tag{5}$$

where

$$V_{12}=-t_0(1+x_0P^\sigma)\delta(\mathbf{r}_1-\mathbf{r}_2)$$
$$+\tfrac{1}{2}t_1[\mathbf{k}'^2\delta(\mathbf{r}_1-\mathbf{r}_2)+\delta(\mathbf{r}_1-\mathbf{r}_2)\mathbf{k}^2] \tag{6}$$
$$+t_2\mathbf{k}'\delta(\mathbf{r}_1-\mathbf{r}_2)\mathbf{k}+iW\mathbf{k}'\cdot\delta(\mathbf{r}_1-\mathbf{r}_2)\sigma\times\mathbf{k},$$

and

$$V_{123}=t_3\delta(\mathbf{r}_1-\mathbf{r}_2)\delta(\mathbf{r}_1-\mathbf{r}_3). \tag{7}$$

In the above, the t_0 term is attractive to hold the nucleons and the t_3 term is repulsive to prevent nuclear collapse. The W term is the spin-orbit interaction which causes a $j=l+\frac{1}{2}$ orbit to be lower in energy than the $j=l-\frac{1}{2}$ orbit. The t_1 and t_2 terms are finite-range terms. In Hartree-Fock calculations, V_{123} is equivalent to a density-dependent interaction $\frac{t_3}{6}\rho(R)\delta(\mathbf{r}_1-\mathbf{r}_2)$.

B. Zero-range Skyrme interactions and oscillator trial wave functions

One can achieve saturation with just the zero-range t_0 and t_3 terms of the Skyrme interaction. So we have a simplified Hamiltonian:

$$H=\sum_i \frac{\mathbf{p}_i^2}{2m}-\sum_{i<j} t_0\delta(\mathbf{r}_i-\mathbf{r}_j)$$
$$+\sum_{i<j<k} t_3\delta(\mathbf{r}_i-\mathbf{r}_j)\delta(\mathbf{r}_i-\mathbf{r}_k). \tag{8}$$

We use a single Slater determinant involving harmonic oscillator trial single-particle wave functions expressed in terms of a harmonic oscillator length parameter $b=\sqrt{\frac{\hbar}{m\omega}}$. The value of b for mass number A is roughly $1.0\,A^{1/6}$ fm.

The trial energy has the structure [8]

$$E_{\text{trial}}=\frac{\hbar^2}{2mb^2}\Sigma-t_0\frac{|A|}{b^3}+t_3\frac{|B|}{b^6}, \tag{9}$$

where $|A|$ and $|B|$ do not depend on b, and Σ is the oscillator sum $\Sigma=\sum_i^A(2n+l+\frac{3}{2})$. The minimum condition is $\frac{dE}{db}=0$. The nuclear incompressibility is $b^2\frac{d^2E}{db^2}$.

For the deformed oscillator, the trial wave function, assuming axial symmetry, is a Slater determinant involving states $|N_x,N_y,N_z\rangle$. The general structure of the b_x, b_z dependence of the trial energy can be obtained by a dimensional analysis as:

$$E_{\text{trial}}=\frac{\hbar^2}{2mb_x^2}(\Sigma_x+\Sigma_y)$$
$$+\frac{h^2}{2mb_z^2}\Sigma_z-t_0\frac{|A|}{b_0^3}+t_3\frac{|B|}{b_0^6}, \tag{10}$$

where $b_0^3=b_xb_yb_z$ and where $|A|$ and $|B|$ are independent of b_x and b_z.

Note that the potential energy depends only on b_0; there is no explicit dependence on b_z. We therefore have to minimize only the kinetic energy. The minimum condition $\left[\frac{\partial E_{\text{trial}}}{\partial b_z}\right]_{b_0}=0$ leads again to the Mottelson conditions, the same ones of the deformed oscillator model:

$$\frac{b_z^2}{b_0^2}=\frac{\omega_\perp}{\omega_z}=\frac{2\Sigma_z}{\Sigma_x+\Sigma_y}. \tag{11}$$

We thus have a close connection between the Skyrme model and the previously considered Deformed Oscillator or Nilsson model.

III. Results

We go beyond the simple deformed oscillator model for the linear chain states by expanding the single-particle orbitals:

$$\Psi=\sum_{n_z,n_\perp} C_{n_z,n_\perp}\phi_{n_z}(z/b_z)\phi_{n_\perp}(\rho/b_\perp). \tag{12}$$

We include 14 major shells.

The results are presented in Table 1. We give the intrinsic energies and the $J^\pi=0^+$ bandhead energies. The latter are obtained by the Villars result [9]

$$E_{0^+}=E_{\text{int}}-\frac{\hbar^2}{2\mathscr{T}}\langle J_\perp^2\rangle, \tag{13}$$

where we take for $\mathscr{T}$ (and this point deserves further attention) the *rigid-body* moment of inertia.

The calculated $J^\pi=0^+$ bandhead excitation energies for ^{12}C, ^{16}O, ^{20}Ne and ^{24}Mg are respectively 12.3 MeV, 25.8 MeV, 37.2 MeV and 58.6 MeV. The experimental values are 7.65 MeV for ^{12}C and 47.8 for ^{24}Mg. There was a report of a possible $J^\pi=0^+$ bandhead at 16.7 MeV in ^{16}O by W.D.M. Rae [11] in a previous cluster conference but according to P. Hodgeson at this reference one cannot get claim to have found the linear chain state in this nucleus.

Table 1. The ground and linear chain states in HF calculations with the SKM* interaction. "Diff" means the difference in the results for the Chain state and for the ground state

Nucl.	State	β	E_K	E_V	E_S	E_{so}	E_C	E_{int}	$\langle J_\perp^2\rangle$	$\mathscr{T}_{\text{rigid}}$	$-\frac{\langle J_\perp^2\rangle}{\mathscr{T}_{\text{rigid}}}$	E_{0^+}
^{12}C	G	0.00	183.4	−312.3	51.0	−23.5	8.1	−93.21	–	–	0.00	−93.21
	C	1.03	150.6	−269.2	47.6	−6.1	6.5	−70.71	58.7	2.88	−10.19	−80.90
	Diff		−32.8	43.1	−3.6	17.4	−1.6	22.50			−10.19	12.31
^{16}O	G	0.00	220.3	−413.3	53.1	−1.2	13.5	−127.59	–	–	0.00	−127.59
	C	1.25	204.0	−358.9	61.4	−7.7	10.2	−90.93	143.9	6.65	−10.82	−101.75
	Diff		−16.3	54.5	8.3	−6.5	−3.3	36.66			−10.82	25.84
^{20}Ne	G	0.37	298.2	−530.4	65.5	−11.0	20.3	−157.44	17.22	3.04	−2.83	−160.26
	C	1.36	258.3	−449.3	75.3	−9.5	14.4	−110.88	311.3	12.76	−12.20	−123.08
	Diff		−39.9	81.1	9.7	1.6	−5.9	46.56			−9.37	37.18
^{24}Mg	G	0.39	383.3	−663.7	80.1	−25.3	28.6	−196.95	19.32	3.88	−2.49	−199.44
	C	1.42	320.7	−549.5	93.6	−12.3	19.1	−128.34	520.6	20.89	−12.46	−140.80
	Diff		−62.6	114.2	13.5	13.0	−9.5	68.6			−9.97	58.64

Note that the very large deformation $\beta = \sqrt{\frac{\pi}{5}} \frac{Q_0}{A\langle r^2 \rangle}$, ranging from 1.03 for ^{12}C to 1.42 for ^{24}Mg. The relation of β to $q=\omega_x/\omega_z$ is [7]

$$q^2 = \frac{2\sqrt{\pi/5}+2\beta}{2\sqrt{\pi/5}-\beta} \tag{14}$$

Also of interest in the breakdown of the difference in the energy of the linear chain states and the ground state into various parts. For example, in ^{16}O, the Coulomb energy difference is -3.3 MeV, i.e., there is less Coulomb repulsion in the chain state than in the ground state because the nucleons are further apart. The difference in the surface energy is 8.3 MeV because the deformed linear chain state has a larger surface area than the ground state. The spin-orbit interaction helps lower the energy of the chain state by 6.5 MeV.

It may be that the α-chain energies come out too high with the SKM* interaction, especially for light nuclei. The results with other Skyrme interactions are similar. For example, the *intrinsic* excitation energy for the linear chain state in ^{16}O using the Skyrme 3 interaction is 38.4 MeV, even higher than the result of 36.7 MeV using the SKM* interaction.

We have also noted that one can bring the linear chain states down to a lower energy by reducing the finite-range terms, t_1 and t_2, which affect the surface energy and effective nucleon mass. For example, in the extreme case, $t_1=t_2=0$ (zero-range Skyrme) and with no spin-orbit interaction, the intrinsic state energy in ^{16}O is 20.4 MeV [8]. This is 16.3 MeV lower than the intrinsic state energy for SKM*.

We wish to thank our collaborators in some of this work: E. Moya de Guerra, J.A. Caballero, who have done a lot of work along these lines using the Nilsson model, S.J. Lee and D.W.L. Sprung. This work was supported in part by the U.S. Department of Energy Grant No. DE-FG05-86ER 40299 and by the U.S. National Science Foundation, Grant No. PHY-9103011.

References

1. Brink, D.M.: Many-Body Description of Nuclear Structure and Reactions, International School of Physics, Enrico Fermi, Course XXXVI, p. 247. Bloch, C. (ed.). New York: Academic Press 1966
2. Zamick, L., Zheng, D.C., Lee, S.J., Caballero, J.A., Moya de Guerra, E.: Ann. Phys. **212**, 402 (1991)
3. Nilsson, S.G.: K. Dan. Vidensk. Selsk., Mat. Fys. Medd. **29**, 16 (1955)
4. Skyrme, T.H.R.: Philos. Mag. **1**, 1043 (1956); Nucl. Phys. **9**, 615 (1959)
5. Vautherin, D., Brink, D.M.: Phys. Rev. C**5**, 626 (1972)
6. Beiner, M., Flocard, H., van Giai, N., Quentin, P.: Nucl. Phys. A**238**, 29 (1975)
7. Zamick, L., Moya de Guerra, E., Caballero, J.A., Berdichevsky, D., Zheng, D.C.: Phys. Lett. B**242**, 7 (1990)
8. Zamick, L., Zheng, D.C., Sprung, D.W.L.: Phys. Lett. B**293**, 1 (1992)
9. Villlars, F.: In: Nuclear Physics, Proceedings International School of Physics, Enrico Fermi, Course XXXVI, Varenna 1966. Block, C. (ed.). New York: Academic Press 1966
10. Mekjian, A., Zamick, L.: Phys. Rep. C**95**, 322 (1982)
11. Rae, W.D.M.: Proc. Fifth. Conf. Clustering Aspects in Nucl. and Subnucl. Systems, Kyoto, 1988; J. Phys. Soc. Jpn. S**58**, 77 (1989)

Z. Phys. A 349, 259–264 (1994)

Energetics of the fission process

Friedrich Gönnenwein

Physikalisches Institut, Universität Tübingen, Auf der Morgenstelle 14, D-72076 Tübingen, Germany

Received: 27 October 1993

Abstract. The mass asymmetry of fragments from nuclear fission of heavy nuclei is reviewed. While mass asymmetry is a common and well-known phenomenon for low-energy fission of the lighter actinides, more recent experiments have demonstrated that, for the heaviest actinides, the mass distribution switches to a symmetric one. On the other hand, it has been discovered that, though for fissioning nuclei with mass numbers $A \lesssim 225$ the mass distribution is basically symmetric, an asymmetric component is clearly to be identified for nuclei down to the Pb-region. In the absence of a generally accepted dynamical theory of fission, the above experimental findings are discussed in terms of static energy considerations. Triggered from the outset by the structure of the potential energy surface at the saddlepoint, the energy balance at the scission point between the available energy (Q-value) of the reaction and the Coulomb and deformation energy of the nascent fragments is shown to steer the characteristics of the fragment mass distributions.

PACS: 25.85. Ec

1. Introduction

The asymmetry of the mass split in binary disintegrations of the lighter actinides is a conspicuous feature of nuclear fission. At lower excitation energies of these nuclei, the asymmetry in the mass distribution of fission fragments is a universal phenomenon. In the framework of the liquid-drop model of fission, mass asymmetry was for many years a puzzle. First hints for an explanation came with the development of the shell model of nuclei. In fact, among the first applications of the cluster model of nuclear reactions was a study of mass asymmetry in fission [1]. Corroborated by various theoretical approaches, it is nowadays an accepted view that shell effects in deforming and fission prone nuclei are responsible for the asymmetry in mass split. Recent experiments on "cold fission" have once more underlined this fact.

Considerable progress has been achieved these last years in assessing the range of fissioning nuclei where asymmetric mass distributions prevail. In the heavier actinides, Fm to No, it was observed that shell effects give rise to a symmetric fission component with very special features, viz. an unusually narrow symmetric mass distribution coupled to extremely large kinetic energies of the fragments. In parallel, there exists a second, again symmetric, component with more standard characteristics. This phenomenon has been coined "bimodal fission". In the preactinides, on the other hand, it has been demonstrated that an asymmetric mass component may be traced down to fissioning nuclei as light as about Pb. This component is easily missed in the broad and dominant symmetric mass distribution being typical for lighter nuclei. It reveals itself as a fine structure in the mass distributions and, above all, as pecularities in the kinetic energy distributions of fragments. The asymmetric component appears to vanish rather abruptly for nuclei lighter than Pb.

A comprehensive theory of nuclear fission being not available, it is the purpose of the present review to point out that at least the gross features of mass distributions may be understood in physical terms by inspecting the static properties of potential energy surfaces of deforming nuclei and, especially, the energy balance between available energy and the potential energies of scission configurations. Not unexpectedly, shell-stabilized cluster fragments play a key role for the case in point.

2. Cold fission in the lighter actinides

A sample of fragment mass distributions for fission induced by thermal neutrons in some standard isotopes of the lighter actinides, U and Pu, is presented in Fig. 1. The masses have been obtained by a time-of-flight method and represent primary masses prior to neutron evaporation [2]. It is seen from the figure that the center of gravity of the heavy mass peak stays pretty constant, while the lighter mass peak shifts on average to larger mass numbers for increasing mass of the fissioning compound. The data exhibit considerably fine structure being correlated to even-odd effects in the fragment charge yields. This detail will not be discussed further here. Upon raising the excitation energy of the fissioning compound nuclei, the yield of the symmetric relative to

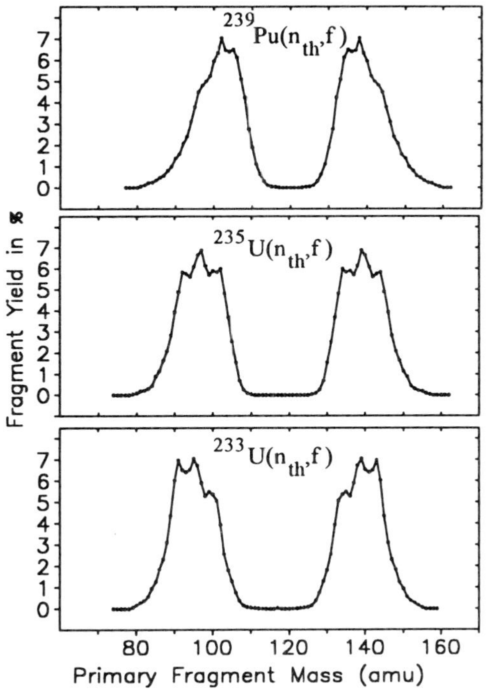

Fig. 1. Mass distributions of primary fragments for fission induced in ^{233}U, ^{235}U and ^{239}Pu by thermal neutrons [2]

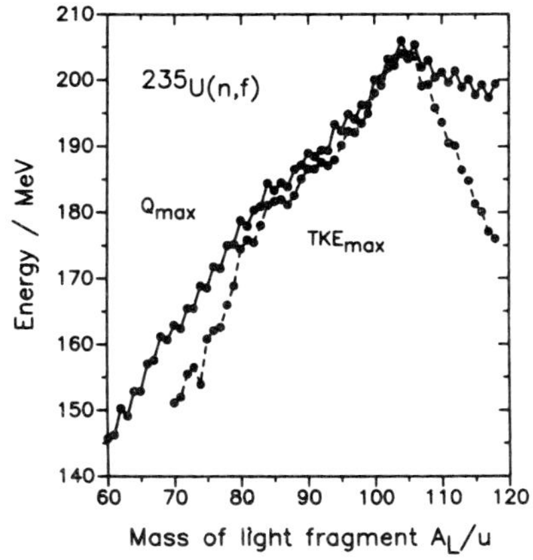

Fig. 2. Maximum Q-value Q_{max} and maximum total kinetic energy TKE_{max} as a function of fragment mass for ^{235}U(n$_{th}$,f) [7]

the asymmetric component grows and at excitation energies in excess of some 50 MeV eventually leads to an overall symmetric distribution [3].

In the following we limit the discussion to the ^{235}U(n$_{th}$,f) reaction with thermal neutrons as a typical case. With the energetics of the fission process in mind, the total energy made available, i.e. the Q-value of the reaction, will be of prime importance. For a given mass split the total energy release will still depend on the nuclear charge ratio of the fragments. The maximum Q-value, calculated from mass tables while taking the neutron binding energy into account, is shown in Fig. 2 as a function of fragment mass for the ^{235}U(n$_{th}$,f) reaction. It should be noted that in Fig. 2 the complementary data for the heavy fragments being strictly equal to the light ones have been omitted. The largest Q-value is attained for the fragment pair $^{132}_{50}$Sn$_{82}$ - $^{104}_{42}$Mo$_{62}$. The doubly magic $^{132}_{50}$Sn$_{82}$ nucleus catches the eye. It marks the light wing of the heavy mass peak (s. Fig. 1). But otherwise there is no clear correlation between the Q-value and the mass yield.

Further insight into the mechanism at work may be gained by analysing experimental results on "cold fission". By definition, cold fission is a limiting case of nuclear fission where virtually all of the available energy shows up in the total kinetic energy of the fragments. Hence, no excitation energy is left to the fragments, i.e. they have to be born in their ground states carrying neither intrinsic excitation nor deformation energy. The study of cold fission has been initiated by Signarbieux et al. [4] and since then been carried on by several groups. Beyond the fact that it is observable at all, it is surprising that only a limited range of mass splits is involved in cold fission. This is manifest in Fig. 2 where, as a function of the light fragment mass A_L, the maximum Q-value Q_{max} is compared to the maximum total kinetic energy TKE_{max} measured in experiment for the ^{235}U(n$_{th}$,f) reaction. From the figure it is observed that only for light masses ranging from about A_L = 80 up to about A_L = 106 the energies TKE_{max} come close to the respective Q-values within a few MeV. This is brought more clearly into evidence in Fig. 3, where the difference $(Q_{max} - TKE_{max})$ has been plotted as a function of fragment mass A_L. It should be noticed that $(Q_{max} - TKE_{max})$ is equal to the minimum total excitation energy TXE_{min} left to the system. Obviously, both for symmetric and very asymmetric mass divisions, cold fission is not attained. Most remarkably, however, the mass range of cold fission in Fig. 3 practically coincides with the mass range of sizeable yields in Fig. 1.

Cold fission may thus offer a clue to understanding mass distributions. An appealing feature of such an approach should be that cold fission is singled out by very simple scission configurations. In fact, the corresponding scission configurations may be visualized as a dimolecular state with both fragments being in their respective ground states and, hence, with no uncertain input data on deformabilities intervening.

In the model having been explored, the scission configurations are parametrized by two separate fragments facing each other and being aligned with their axes of elongations on a common axis. Crucial input parameters are the ground state deformations of the fragments involved. The shapes of nuclei were taken to be those calculated by Möller and Nix [5] in a macroscopic-microscopic approach. The interaction energy of the two adjacent fragments was determined following closely the prescriptions having been worked out by Shi and Swiatecki [6]. There are two contributions to the interaction energy V_{int}: the repulsive Coulomb potential V_C and the attractive proximity potential V_N due to nuclear forces. It should be emphasized that the proximity potential is defined for nuclei with a leptodermous surface where the nuclear density drops from 90% to 10% of its value at the center of the nucleus on a scale of about 2 fm near the surface. The only free parameter of the model, then, is the

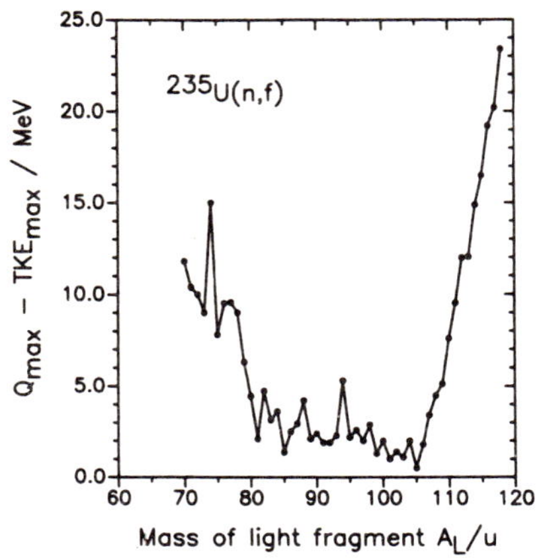

Fig. 3. Difference ($Q_{max} - TKE_{max}$) as a function of fragment mass for ^{235}U(n$_{th}$,f) [7]

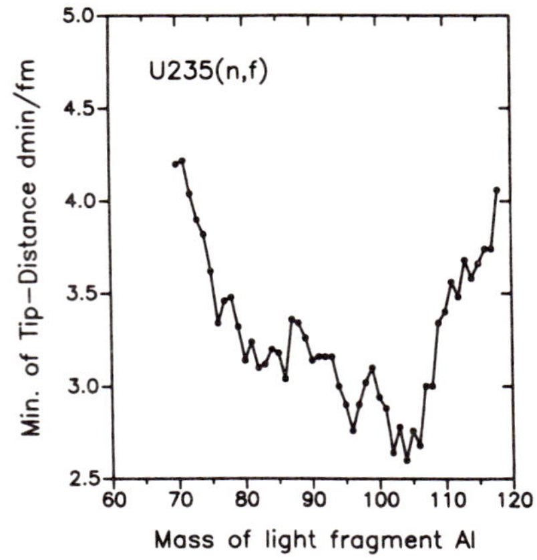

Fig. 4. Minimum tip distance as a function of fragment mass for ^{235}U(n$_{th}$,f) [7]

distance d between the tips of the two adjacent fragments with equivalent sharp surfaces. The distances of closest approach which correspond to the maximum kinetic energy release TKE_{max} are finally determined from the condition that the interaction energy V_{int} should be equal to the total available energy Q. For still smaller distances the large Coulomb repulsion will further raise the interaction energy V_{int} to reach values where it exceeds the available energy Q. These situations are excluded by energy conservation, since the scission configurations considered are conceived to be those of the exit point of fission, where intermediary barriers have been left behind in the elongation process of the nucleus.

Various fission reactions have been analyzed according to the scheme outlined [7]. The minimum tip distances d_{min} obtained for the ^{235}U(n$_{th}$,f) reaction are presented in Fig. 4 as a function of light fragment mass number A_L. The general structure of the dependence of d_{min} on A_L in Fig. 4 is strikingly similar to the one of ($Q_{max} - TKE_{max}$) or the minimum total excitation energy TXE_{min} in Fig. 3. In particular, the mass range of A_L for which cold fission is reached corresponds to minimum tip distances of about 3 fm. It is noteworthy that near both boundaries of the mass range in question, one of the fragments of the pair is a spherical magic nucleus, while the complementary fragment is strongly deformed in its ground state. At the lower mass edge it is the fragment pair $^{82}_{32}Ge_{50}$ - $^{154}_{60}Nd_{94}$, whereas at the higher mass edge it is the pair $^{104}_{42}Mo_{62}$ - $^{132}_{50}Sn_{82}$. In view of the diffuse skin of nuclei, tip distances $d_{min} \lesssim 3$ fm appear to be physically quite reasonable for a scission configuration to be valid. For ground state tip distances in excess of about 3 fm, the fragments will in contrast have to be deformed beyond their ground state shapes in order to reach a valid scission configuration. This deformation (elongation) will bring down the Coulomb interaction and, hence, decrease the kinetic energy release TKE to values below the total available energy Q. With excitation energies ($Q - TKE$) coming into play, cold fission should be outruled. This is precisely the correlation to be read off from Figs. 3 and 4 where, both for symmetric and very asymmetric fission, large ground state tip distances d_{min} parallel large minimum excitation energies.

Besides helping to understand the occurrence of cold fission, the model sketched in the foregoing proves to be valuable for learning something on mass distributions, i.e. the mass yields of fragments integrated over kinetic energy. It was shown that for the mass range pertaining to cold fission, the most compact scission configurations correspond to adjacent fragments in their ground states, while outside of this mass range the fragments have to be deformed. This tells that, in deformation space, the accessible phase space is fully open for fragments with masses exhibiting cold fission, while for fragments with other mass splits the phase space is narrowed down. Models exploiting the phase space properties at scission have become known as scission point models [8]. They have met with considerable success in explaining mass distributions. Intuitively, it is evident that large accessible phase spaces should entail large mass yields. Therefore, the coincidence in mass range where sizeable mass yields, on the one hand, and cold fission, on the other hand, are observed, is not surprising (s. Figs. 1 and 3). On the contrary, the study of cold fission has sharpened the attention to be given to fragment properties for the outcome of fission in general. Let us recall in this context the cluster properties of one partner of the fragment pairs marking the range in mass for both, substantial overall yield and cold fission.

Though fragment properties coming into view at scission appear to be crucial for explaining distinct features of fission, in a full-fledged theory the process should be tracked from the outset. It is a remarkable fact that in the lighter actinides the eventual mass asymmetry of fragments is steered already at the saddlepoint. Analyzing the properties of fission barriers it has been found that pear-shaped saddlepoints are lower in energy compared to reflection symmetric saddlepoints [8]. The foundations for the final mass asymmetry are thus seen to be laid at an early stage of fission.

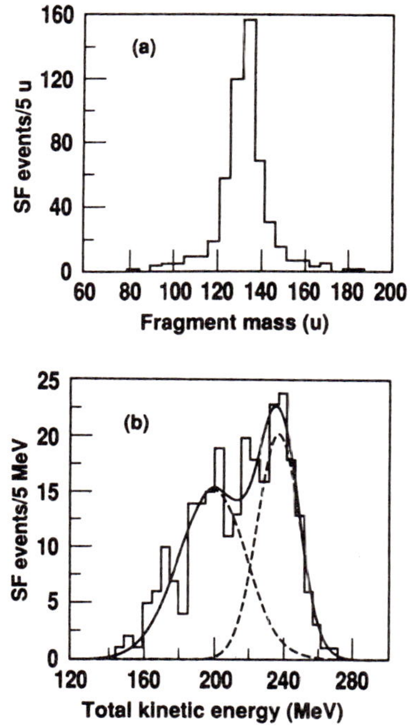

Fig. 5. Mass distribution (top) and kinetic energy distribution (bottom) for fragments from ^{262}No(sf) [10]

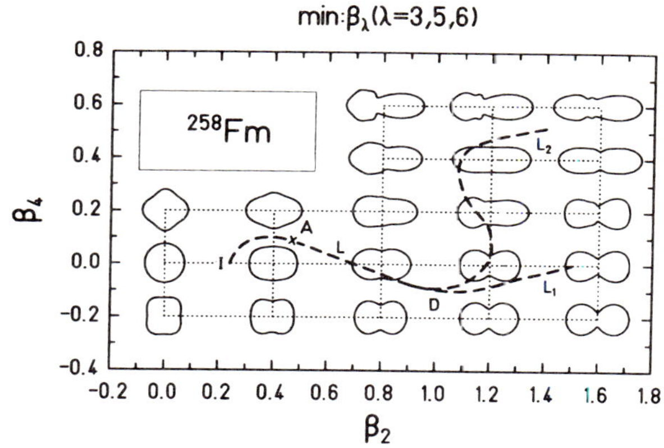

Fig. 6. Fission paths L_1 and L_2 and shapes in deformation space (β_2, β_4) for ^{258}Fm(sf) [11]

3. Bimodal fission in the heavy actinides

If, as intimated in the previous section, it is true that the properties of the final fragments (and especially those of cluster nuclei) are relevant in fission, then a special situation should obtain in the heavier actinides for an island of nuclei centered around $^{264}_{100}\mathrm{Fm}_{164}$. For this, so far evasive, isotope it is tempting to conjecture that it should preferentially disintegrate into two doubly magic fragments $^{132}_{50}\mathrm{Sn}_{82}$ giving rise to a symmetric mass distribution. Symmetric mass distributions have indeed been observed for spontaneous fission of some heavy isotopes of Fm, Md and No [9]. An example is provided in Fig. 5 for the ^{262}No(sf) reaction [10]. At first sight, the shape of the mass distribution in Fig. 5 is well in agreement with the above naïve expectation. However, the distribution of the total kinetic energies of the fragments, which is also shown in Fig. 5, reveals an unexpected feature. It is not a single Gaussian-like distribution as commonly observed in fission but, instead, appears to be composed of two distinct Gaussian distributions indicated in the figure. The high-energy component, or mode, is nearly exhausting the Q-value of the reaction, which means that cold fission is no longer, as in the lighter actinides, a limiting case, but has become a major contribution to the disintegration process. The low-energy mode, on the other hand, fits smoothly into the systematics of kinetic energy release all over the chart of nuclides. Mass distributions conditioned by kinetic energy are symmetric for both modes but differ in width, being narrow and wide for the high-energy and the low-energy mode, respectively [10].

The discovery of two separate modes, or bimodal fission, in the heavy actinides by E.K. Hulet et al. was not anticipated by theory. Evidently, also from the viewpoint put forward for the lighter actinides, it is difficult to imagine for the same mass splits two distinct modes by invoking only fragment properties at scission. The puzzle has been resolved by several groups of theoreticians, reaching all the same conclusion that the potential energy surface close to the fission barrier exhibits a peculiar structure for the nuclei in question. The result of such an investigation is presented in Fig. 6 [11]. In the potential energy surface two separate fission paths appear to be feasible which are depicted in the figure as dashed lines L_1 and L_2 in the deformation space (β_2, β_4). The corresponding shapes of the fissioning compound are visualized. The fission path L_1 leads to a symmetric and very compact scission configuration which should be identified as the high-energy mode. This mode is, by the way, the one expected from the cluster properties of fragments close to ^{132}Sn. The competing path L_2, on the other hand, is characterized by a nearly symmetric and strongly elongated scission shape. These features comply with the properties of the low-energy mode observed in experiment. In summary, theory is in a good shape to foresee the existence of two competing modes of fission in the Fm-No nuclei, though the question concerning the ratio of mode yields has so far not been solved satisfactorily. In any case, a message to be retained from the study of bimodal fission is that fragment properties by themselves may not always give a full picture of the complex fission characteristics.

4. Mass asymmetric modes in the preactinides

From experiment it is known that, as to fragment mass distributions, Ra-Ac nuclei play a pivotal role in fission. For near-barrier fission of neighbouring heavier nuclei the mass distributions are asymmetric, whereas for lighter nuclei the mass distributions are generally symmetric. Due to the lack of suitable target nuclei, the precise location of the turnover point is difficult to assess but, very probably, lies close to mass numbers $A \approx 220$ [3]. From the point of view of the energetics of the fission process in its simplest approach, i.e. the Q-values, these results are somewhat surprising. Reaction Q-values for spontaneous fission in a wide range of nuclei have been put together by Itkis et al. [12]. They are

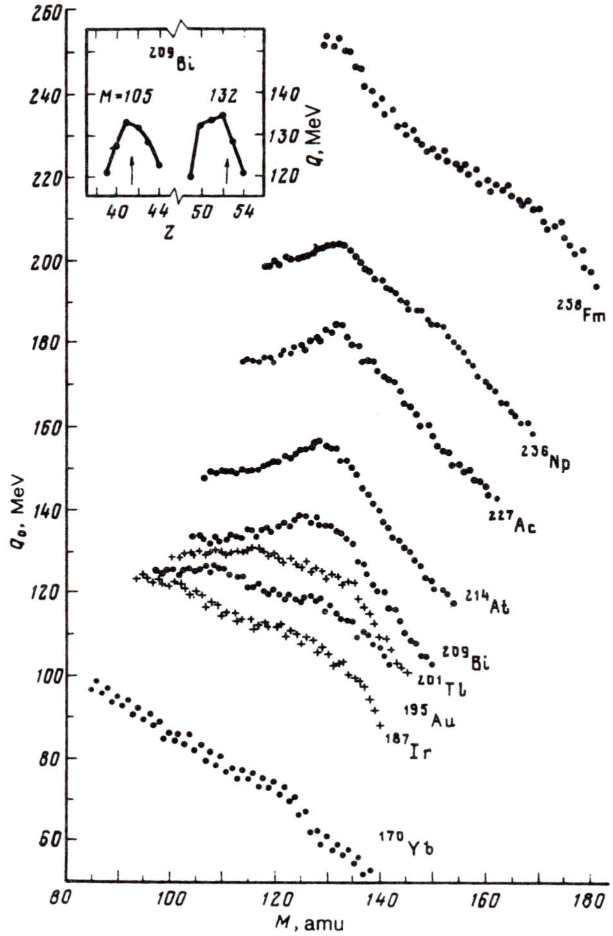

Fig. 7. Maximum Q-values as a function of fragment mass for spontaneous fission of nuclei indicated by labels [12]

portrayed as a function of the heavy fragment mass in Fig. 7 (note that, in actual practice, spontaneous fission has not been detected for most of the nuclei shown). It is apparent from this survey that the influence of the doubly-magic ^{132}Sn fragment, while dominant for Fm, is still perceptible down to about Bi.

Though, contrary to naïve expectation, the mass distributions for these lighter nuclei are basically symmetric, it is comforting that at least an asymmetric mass component with distinct features has been disclosed down to fissioning nuclei with masses around $A \approx 200$. In systematic studies performed by Itkis et al. [13] it has been realized that, for preactinide nuclei with $Z \gtrsim 82$ and $A \gtrsim 200$ at excitation energies of about 10 MeV above the barrier, the mass distributions deviate in shape from a simple Gaussian and, in particular, exhibit a shoulder for heavy fragment masses around $A = 132$. Even more revealing are structures in the distribution of kinetic energy, both in the average value and the variance, as a function of fragment mass. This is demonstrated in Fig. 8 for fissioning nuclei ranging from ^{213}At down to ^{201}Tl. The increase in total kinetic energy for fragments with $A \gtrsim 132$ recalls the compact scission configurations invoked for cluster fragments, whereas the pronounced peaks in the variance point to a superposition of at least two distinct modes, a symmetric one with lower energy and an asymmetric one with larger kinetic energy (a more careful analysis even uncovers two separate asymmetric modes). Hence, the phenomenon of multimodal fission reported for the Fm-No nuclei (s. Sect. 3) shows up again in the preactinides. Similar to the heavy actinides, also here the occurrence of multimodal fission is linked to structures in the potential energy surface near the saddlepoint. Distinct valleys in the energy surface for the nucleus ^{208}Pb have indeed been predicted many years ago by Pashkevich [14], with the barrier for symmetric fission being lower in energy than for asymmetric fission by 2-3 MeV. This prediction from theory is likewise corroborated by experiment: for the fissioning nucleus ^{213}At, the asymmetric modes disappear when fission is induced close to the (lower symmetric) barrier.

A final remark addresses the sudden disappearance of any asymmetric mode for nuclei lighter than $Z \approx 82$ and $A \approx 200$. This observational fact is not to be explained by the properties of the energy surface near the saddlepoint. Instead, arguments have been put forward [12] being very much in parallel to the reasoning presented in section 2 for understanding cold fission. From a systematic analysis it is concluded that precisely for nuclei with the above charge and mass numbers (and a fortiori for still lighter nuclei), the available reaction energy Q is not large enough to allow for the compact scission configurations being characteristic for cluster fragments and, hence, asymmetric fission. In other words, asymmetric fission is running out of phase space in nuclei lighter than Pb.

5. Conclusions

A dynamical theory of nuclear fission being still missing, the present survey of fragment mass and kinetic energy distributions for a wide range of fissioning nuclei aims at an interpretation in terms of the energetics of the process, i.e. the static properties of potential energy surfaces for deformed nuclei and the energy balance at scission. It is shown that basic characteristics of mass distributions are triggered at an early stage of the process, viz. by the structural properties of the potential energy surfaces close to the saddle point. In recent systematic studies of energy surfaces rather complex structures have emerged, with new valleys appearing even once the saddlepoint has been passed [15]. All the implications of this multi-faceted landscape on the outcome of fission could not be discussed here. Let us stress, however, that the peculiar structures of the potential energy as a function of deformation are essentially due to shell effects.

Distinct valleys in the energy surface will give rise to multimodal fission. The superposition of different modes is most conspicuous in the kinetic energy distributions of fragments. Since the Coulomb repulsion between nascent fragments contributes the main part to kinetic energy, this points to the importance of scission configurations with a broad choice of fragment shapes. Largely different scission shapes may be reached by different fission paths. An example is provided by bimodal fission in the Fm-No nuclei. At scission, fragment properties will play a role, with espe-

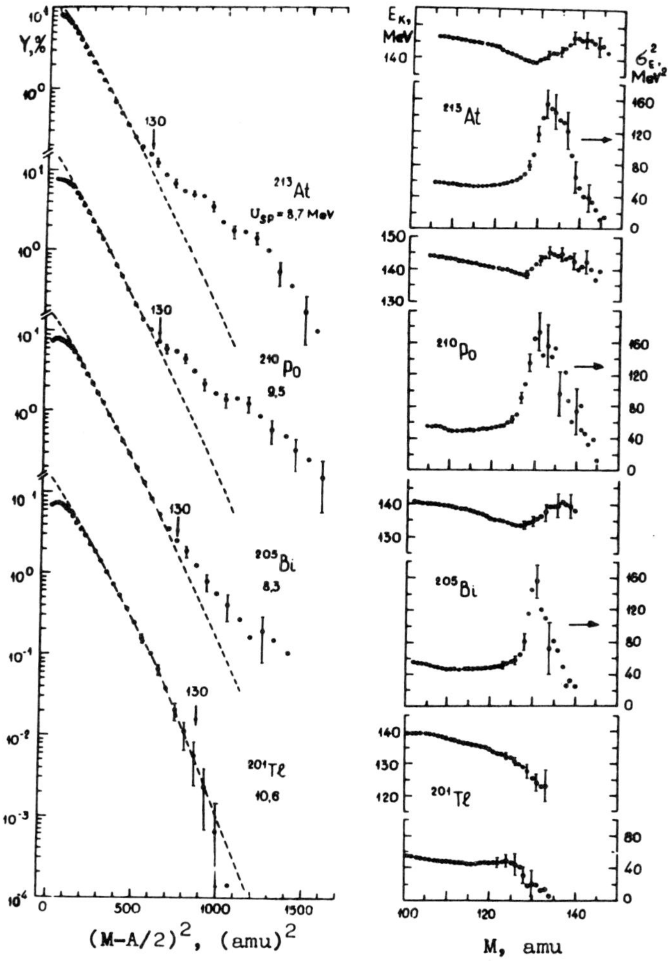

Fig. 8. Mass distributions (*left*) as a function of $(M - A/2)^2$ with M the fragment and A the compound mass. Average total kinetic energy E_K and variance σ^2_{EK} of energy distributions (*right*) as a function of fragment mass for compound nuclei indicated by labels [13]

cially cluster fragments coming into focus. It is shown that specific phenomena, like cold fission in the lighter actinides or the sudden disappearance of asymmetric modes for nuclei lighter than about Pb, are well understood from the energy balance between available energy Q and the potential energy of interaction for physically feasible scission configurations.

After all, it is astonishing that static energy considerations already give an at least qualitative account of some fundamental fission properties. The dynamics of the process does not come into the foreground and it may, therefore, be conjectured that fission is a slow process. However, dynamical calculations should certainly not be discouraged since quantitative predictions of mass or energy distributions, including weights to be given to competing fission paths and/or widths of distributions, are far beyond the scope of static models.

References

1 Faissner, H., Wildermuth, K.: Phys. Lett. **2**, 212 (1962)

2 Geltenbort, P., Gönnenwein, F., Oed, A.: Radiat. Effects **93**, 57 (1986)

3 Gönnenwein, F.: The Nuclear Fission Process. C. Wagemans (ed.) Boca Raton: CRC Press 1991

4 Signarbieux, C., Montoya, M., Ribrag, M., Mazur, C., Guet, C., Perrin, P., Maurel, M.: J. de Phys. Lett. **42**, L437 (1981)

5 Möller, P., Nix, J.R.: At. Data Nucl. Data Tables **26**, 165 (1981)

6 Shi, Y.-J., Swiatecki, W.J.: Nucl. Phys. **A464**, 205 (1987)

7 Gönnenwein, F., Börsig, B.: Nucl. Phys. **A530**, 27 (1991)

8 Moreau, J., Heyde, K.: The Nuclear Fission Process. C. Wagemans (ed.) Boca Raton: CRC Press 1991

9 Hoffman, D.C.: Nucl. Phys. **A502**, 21c (1989)

10 Hulet, E.K.: Proc. of Welch Foundation Conf. on Chem. Research XXXIV, "Fifty Years with Transuranium Elements", Houston USA, 1990, p 279

11 Cwiok, S., Rozmej, P., Sobiczewski, A., Patyk, Z.: Nucl. Phys. **A491**, 281 (1989)

12 Itkis, M.G., Mulgin, S.I., Okolovich, V.N., Rusanov, A.Ya., Smirenkin, G.N.: Sov. J. Nucl. Phys. **53**, 764 (1991)

13 Itkis, M.G., Okolovich, V.N., Smirenkin, G.N.: Nucl. Phys. **A502**, 243c (1989)

14 Pashkevich, V.V.: Nucl. Phys. **A169**, 275 (1971)

15 Brosa, U., Grossmann, S., Müller, A.: Phys. Rep **197**, 167 (1990)

This article was processed using Springer-Verlag TEX Z.Physik A macro package 1.0 and the AMS fonts, developed by the American Mathematical Society.

Z. Phys. D 31, 191–197 (1994)

Fission of highly charged alkali metal clusters

U. Näher, S. Frank, N. Malinowski, U. Zimmermann, T.P. Martin

Max-Planck-Institut für Festkörperforschung, Heisenbergstrasse 1, D-70569 Stuttgart, Germany

Received: 28 June 1993

Abstract. Highly charged metal clusters can be easily produced using the technique of multi-step photoionization. The critical sizes at which fission occurs are reported here for alkali metal clusters having charges as high as +7. An interpretation within the framework of the liquid droplet model reveals that the fission process is strongly asymmetric and thermally activated.

PACS: 36.40.+d

Introduction

Should highly charged metal clusters break apart into two more or less equally sized fragments as atomic nuclei? Or is the fission process in clusters more closely related to 'α-decay', i.e. the loss of a small charged fragment? Of course the energy and distance scales are quite different, but aside from this, many properties of clusters and nuclei are amazingly similar. The similarity between these two forms of condensed matter is not accidental. Both system consist of fermions moving, nearly freely, in a confined space. So, the shell model developed by nuclear physicists describes very nicely the electronic properties of alkali metal clusters. The giant dipole resonances in the excitation spectra of nuclei have their analogue in the plasmon resonances of metal clusters. Therefore, it is tempting to apply the models describing the fission of unstable nuclei to the fragmentation of highly charged metal clusters.

The history of fission of charged metal clusters has its roots in 1882 when Rayleigh studied the stability of charged metal droplets [1]. He was the first to notice that small metal droplets are held together by their surface tension and might break apart when they are charged. Under the assumption of a prolate ellipsoidal deformation of the spherical droplets he found a universal connection between stability and charge. Applied to metal clusters, the Rayleigh charging limit predicts that the minimum number of atoms in a stable cluster should be proportional to the square of the number of charges.

In the rapidly developing field of cluster research, fission of charged clusters was one of the first subjects to be investigated [2]. Critical sizes for many kinds of cluster material have been investigated. In the last few years several studies concerning the symmetry of the fission process have been published. Most of these studies have delt with the fission of doubly or triply charged clusters [3]. In the present article we are going to discuss fission of highly charged alkali metal clusters. The critical number of particles, n_c, at which fission occurs for M_n^{z+} clusters is investigated experimentally up to $z=7$. We try to determine whether Rayleigh was right when he postulated a symmetrical fission process for charged metallic droplets.

Experimental

Although we have studied the fission process for various alkali metals (Na, K, Rb, and Cs), exemplarily, we will describe the experimental setup used for cesium clusters.

The technique we have used is photoionization time-of-flight mass spectrometry. The cluster source is a low pressure, inert gas, condensation cell. Cesium vapor was quenched in cold He gas having a pressure of about 1 mbar. Clusters condensed out of the quenched vapor and were transported by the gas stream through a nozzle and through two chambers of intermediate pressure into a high vacuum chamber. The size distribution of the clusters could be controlled by varying the oven-to-nozzle distance, the He gas pressure, and the oven temperature. The clusters were photoionized with a 50 mJ, 15 ns, 157 nm (7.9 eV) F_2-laser pulse focused onto the neutral cluster beam with a 200 mm focal length MgF_2-lens. The main wavelength of an F_2-laser at 157 nm is far from the plasmon resonance of cesium clusters. However, apart from this main frequency, the beam of an F_2-laser contains other wavelengths at 635, 745 and 755 nm which contribute 5% to the total energy of the laser. Although we tried to suppress these wavelengths with a dielectric mirror, there was enough energy left to warm the clusters during the ionization process.

To produce multiply charged clusters we used multi-step photoionization. It has been shown that clusters with up to 14 positive charges can be produced with this technique [4]. Each electron removed requires a separate photon with a minimum energy. The energy necessary to remove an additional electron from a cluster that already has charge $+z$ can be written

$$I(z,n) = W + (\alpha + z)\, e^2/r_s n^{1/3}, \tag{1}$$

where W is the bulk work function, e is the electronic charge, n is the number of atoms and r_s is the Wigner-Seitz radius of the cluster material. Various values of α have been used in the literature. We will assume $\alpha = 0.5$ (classical limit) and point out that for large values of z the value of α used becomes unimportant.

Equation (1) gives an upper limit for the maximum charge of a cluster as a function of the photon energy. This means that a two-dimensional cluster space (n, z) can be divided by a line into clusters that can be formed with an F_2-laser and those that cannot, Fig. 1. Indicated in this figure is a second line dividing the space into stable and unstable clusters. Obviously the unstable region is accessible with the F_2-laser. The fission process can be investigated directly by multi-step ionization. This is not the case for Na clusters where an additional warming laser must be used as described in [5].

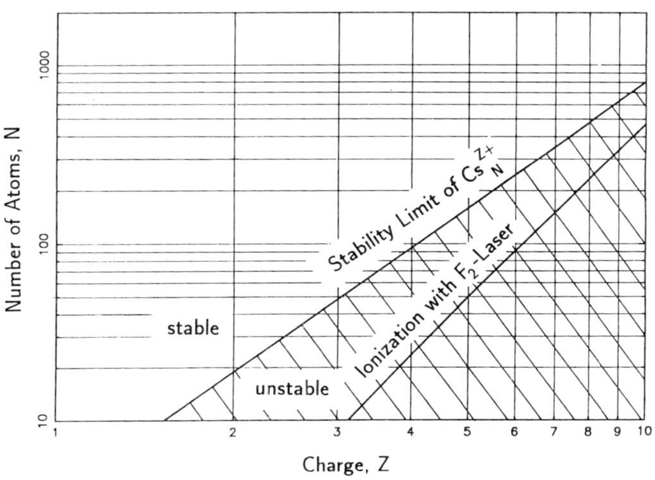

Fig. 1. The two-dimensional cluster space (n, z) can be divided by a straight line ('Ionization with F_2') into Cs_n^{z+} clusters which can (cannot) be ionized with the laser (7.9 eV). All clusters to the left of this line can be ionized. A second line divides the (n, z)-space into stable and unstable clusters. Notice, that all clusters that are stable can be ionized with 7.9 eV photons

Results

Figure 2 shows a typical spectrum obtained with the setup described above. The highest set of mass peaks belongs to clusters with integral values of n/z, for example singly charged clusters or doubly charged clusters with an even number of atoms. In the spectrum several groups of differently charged clusters can be identified. This can be seen more clearly in Fig. 3 where the spectrum of Fig. 2 is shown as a function of n instead of n/z. On this scale the groups of differently charged clusters can be distinguished much better. In fact, they separate into almost non overlapping domains, each domain containing clusters in a differently charged state. There is a simple explanation for this behaviour. To the left every domain is limited by the stability of the z-fold charged clusters. To the right, it is limited by the stability of clusters with charge $z+1$. This means that every cluster is charged as highly as possible.

To explain the observed structure we have assumed the following process as illustrated schematically in Fig. 4. A neutral cluster (a) is ionized and warmed by the laser

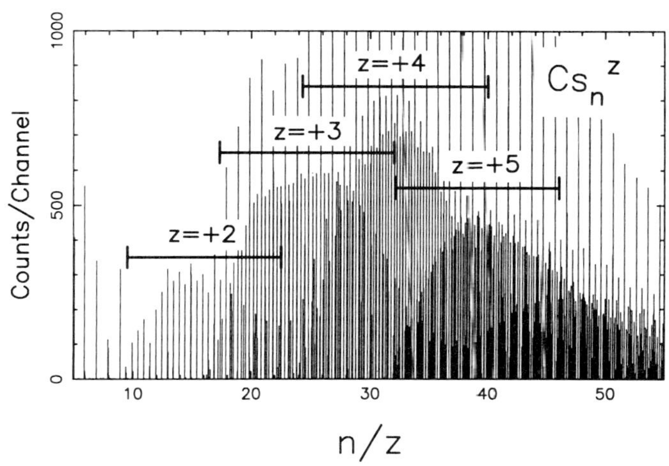

Fig. 2. A n/z spectrum of Cs_n^{z+}. Groups of multiply charged clusters can be identified

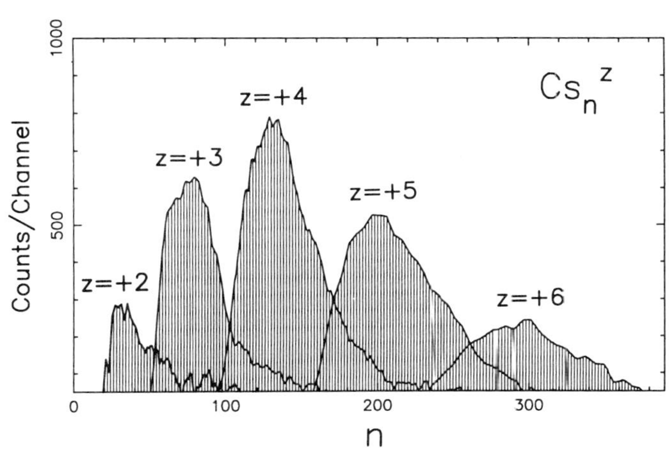

Fig. 3. The same spectrum as Fig. 2, now as a function of n (instead of n/z)

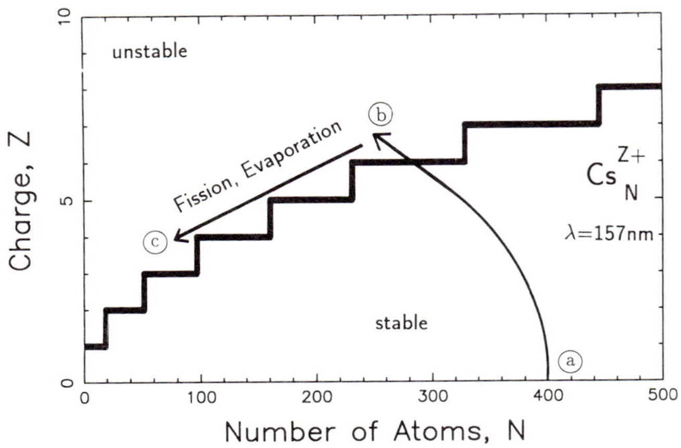

Fig. 4. Schematic view of the ionization process. A neutral cluster *(a)* is ionized and warmed by the laser pulse until it reaches the stability limit *(b)*. Further ionization steps are followed by the instantaneous loss of a small singly charged fragment. When the interaction of the cluster with the laser stops *(c)* the cluster is observed in a maximally charged state. The solid step-like line separates stable from unstable clusters

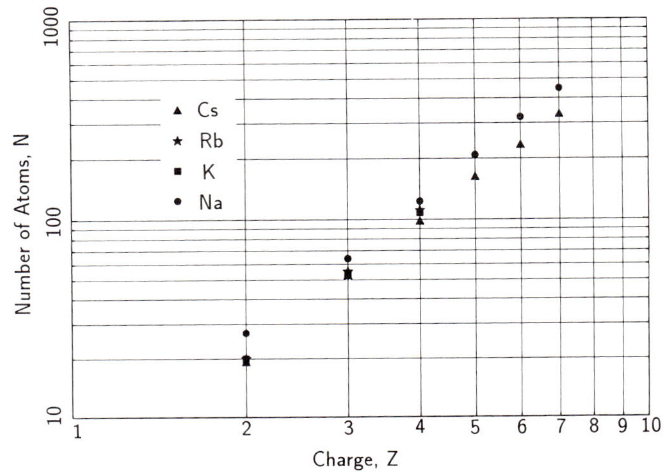

Fig. 5. The number of atoms in the smallest experimentally observed M_n^{z+} clusters (M = Cs, Rb, K, Na)

pulse. During this charging process the warmed cluster loses neutral atoms by evaporation. At the stability limit (solid line, (b)) the cluster cannot be charged further without becoming unstable. If we assume, as supposed by various authors [6–11], that fission of metal clusters corresponds to the evaporation of small, singly charged fragments, the total charge of the cluster will not change by an additional charging-step followed by the instantaneous loss of a singly charged fragment. In the $n-z$ diagram of Fig. 4 this process corresponds to a movement of the cluster to the left. When the interaction of the cluster with the laser stops (c), the cluster is observed in a maximally charged state.

Although this process explains the measured spectrum very nicely, it is no direct proof that the stability of highly charged clusters is limited by the loss of small singly charged fragments and not by 'Coulomb explosion', the symmetrical fission of the cluster into two equally charged fragments.

In similar experiments the critical sizes of all alkali metals have been investigated. The results are shown in Table 1. For cesium and sodium n_c could be observed to $z=7$. For the other elements, two or more isotopes complicated the spectra, so that critical sizes could be identified no further than $z=4$.

Discussion

Figure 5 shows the critical cluster sizes as a function of charge for alkali metal clusters. The data points for each material lie on a more or less straight line in the $(\ln z, \ln n)$-space. A linear fit of the form

$$\ln n_c = c \cdot \ln z + \text{const}. \quad (2)$$

gives a slope, c, of approximately 2.3 in contrast to the Rayleigh charging limit predicting $c=2$. This deviation is even more obvious in the relation

$$n_c(2):n_c(3):n_c(4) = 1:2.6:5.2 \quad (3)$$

instead of $n_c(2):n_c(3):n_c(4)=1:2.25:4$ (Rayleigh). For van-der-Waals clusters Echt finds $n_c(2):n_c(3):n_c(4) =1:2.2:4.1$ [3].

Figure 6 compares the critical radii, R_c, of the charged clusters which have been calculated accoridng to the relation

$$R_c = n_c^{1/3} \cdot r_s. \quad (4)$$

The values for r_s, the Wigner-Seitz radius, have been taken from [16]. Obviously, cesium clusters are larger than sodium, potassium, or rubidium clusters when they undergo fission. Qualitatively this is easy to understand. The surface tension which keeps the cluster together is lower for cesium than for the other alkali metals (Table 2). Additionally, the barrier for monomer evaporation which is in competition to the fission process is lower for cesium (see next paragraph).

Table 2. Surface tension for alkali metals [17]

	Li	Na	K	Rb	Cs
σ [eV/Å²]	0.025	0.0125	0.0069	0.0056	0.0044

Table 1. Critical sizes for fission

	$Z=2$	$Z=3$	$Z=4$	$Z=5$	$Z=6$	$Z=7$
Lithium	25 ± 1					
Sodium	27 ± 1	63 ± 1	123 ± 2	206 ± 4	310 ± 10	445 ± 10
Potassium	20 ± 1	55 ± 1	110 ± 5			
Rubidium	19 ± 1	54 ± 1	108 ± 3			
Cesium	19 ± 1	49 ± 1	94 ± 1	155 ± 2	230 ± 5	325 ± 10

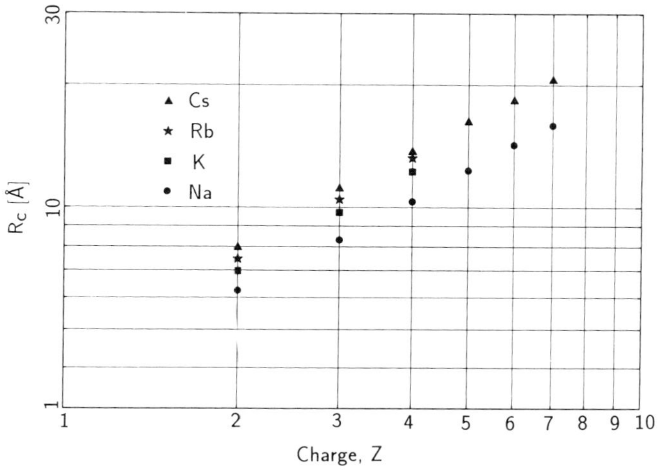

Fig. 6. The critical radii of the smallest experimentally observed clusters. For a given charge the critical size of the clusters scales with the inverse value of the surface tension of the bulk material

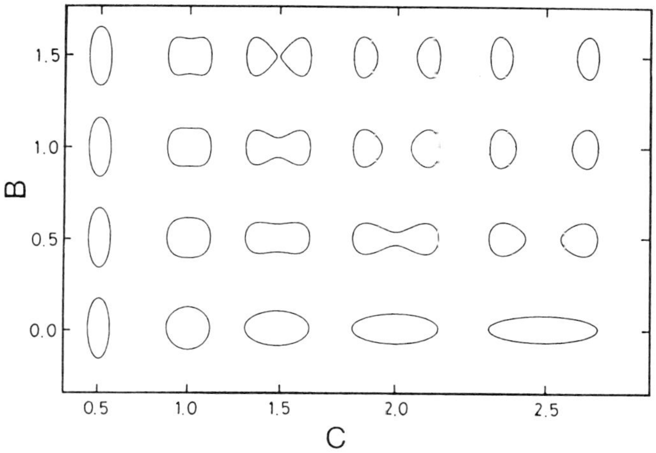

Fig. 7. The deformation of the clusters is described by an elongation parameter (C) and a neck parameter (B) [14]. A third parameter which is not shown in this figure describes the symmetry of the fission process, i.e. the relative position of the neck formation

Model

We have tried to interpret the data in terms of the classical liquid droplet model. Much better, self-consistent quantum-mechanical models have been used successfully to describe the fission process of clusters [11]. But, since great computational effort is required to investigate a multi-dimensional set of deformation parameters, a priori assumptions concerning the symmetry of the fission process had to be made. The great advantage of the classical droplet model is that no restrictions on the symmetry are necessary.

In the classical liquid droplet model the clusters are described as charged metallic droplets. Their energy is calculated as the sum of the Coulomb energy, E_C, and the surface energy, E_S. The surface energy is obtained by the integral

$$E_S = \int \mathrm{d}S\, \rho(r) \tag{5}$$

$\rho(r)$ is the surface tension. To keep the model as free of parameters as possible, we choose for ρ the value of the surface tension of the bulk solid without any corrections for the curvature.

The calculation of the Coulomb energy of a charged conducting droplet is a minimization problem. The equilibrium surface-charge distribution $\sigma(r)$ is obtained by minimizing the energy

$$E_C = \tfrac{1}{2} \int \mathrm{d}S' \int \mathrm{d}S \frac{\sigma(r)\cdot\sigma(r')}{|r-r'|} \tag{6}$$

under the constraint

$$Q = \int \sigma(r)\,\mathrm{d}S \tag{7}$$

Q is the surplus charge of the cluster ($Q = z\cdot e$). A detailed analysis of this problem has been published recently [13]. We chose an approximative method similar to that described in [14].

To describe the deformations of the droplet, a 3-dimensional parameter space has been introduced by nuclear physicists [15]. For a symmetric fission process two of these deformation parameters are shown in Fig. 7, a neck parameter, B, and an elongation parameter, C. The third parameter, γ, describes the symmetry of the fission process, i.e. the relative position of the neck formation.

We started with a calculation of the energy barrier for symmetric fission. The results for Na_{11}^{2+} and Na_{30}^{2+} are shown in Fig. 8. The fission trajectory starts at ($B=0, C=1$), which is the parametrization that corresponds to a sphere and has to cross the line of scission (thick line in Fig. 8). For Na_{11}^{2+} the energy surface is nearly flat. The fission has approximatively zero barrier. Na_{10}^{2+} should undergo fission spontaneously [14]. The calculated barrier for Na_{30}^{2+} assuming symmetric fission is 1.2 eV. As the critical size of doubly charged sodium clusters is near 30, the fission of warmed alkali clusters obviously is a thermally activated process. This idea has been proposed by various authors, especially Bréchignac et al. who interpreted the fission channel as competitive with the evaporation of monomers [8]. The fission barrier at the critical cluster size should have approximately the same height as the activation energy for evaporation (1.1 eV for sodium). As long as the fission barrier is higher than the activation energy for the evaporation of neutral atoms the latter process will be much more probable. The cluster will lose neutral atoms but no charged fragments. By this process the cluster will shrink. As the fission barrier decreases with decreasing cluster size, there is a cross-over of the barrier heights at $n=n_c$. From this cluster size on, fission will occur.

A cluster of 30 atoms has an internal energy of about $90\,k_B T$. At a temperature of 450 K (a value obtained from the evaporative ensemble theory for sodium clusters on our experimentally determined time scale) this corresponds to an energy of 3 eV. If one third of this amount of energy is localized in the fission channel, the fission barrier can be overcome.

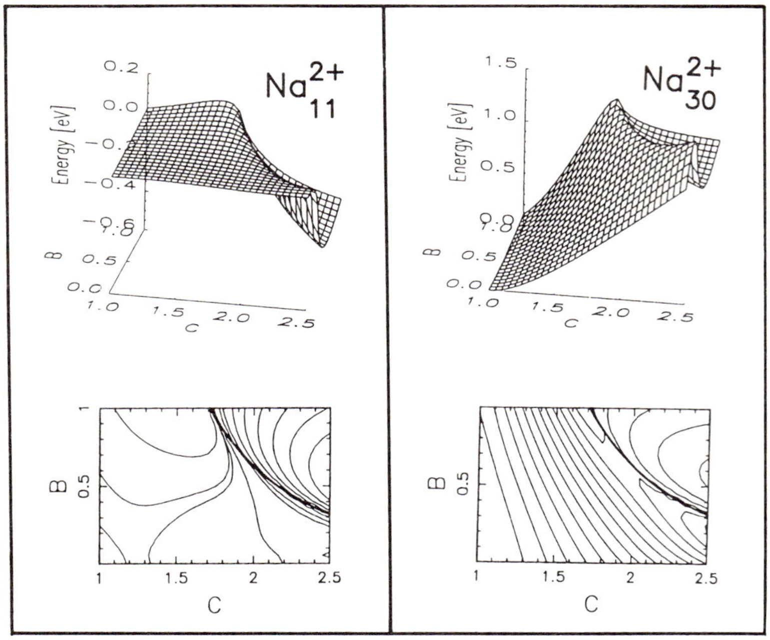

Fig. 8. Symmetrical fission of Na_{11}^{2+} and Na_{30}^{2+}. The deformation energy of the clusters is shown as a function of the parameters B and C. The fission of Na_{30}^{2+} has a barrier of about 1.2 eV. The energy surface for the deformation of Na_{11}^{2+} is nearly flat. The thick line in the contour plots *(lower pictures)* corresponds to the line of scission which has to be crossed by the fission trajectory

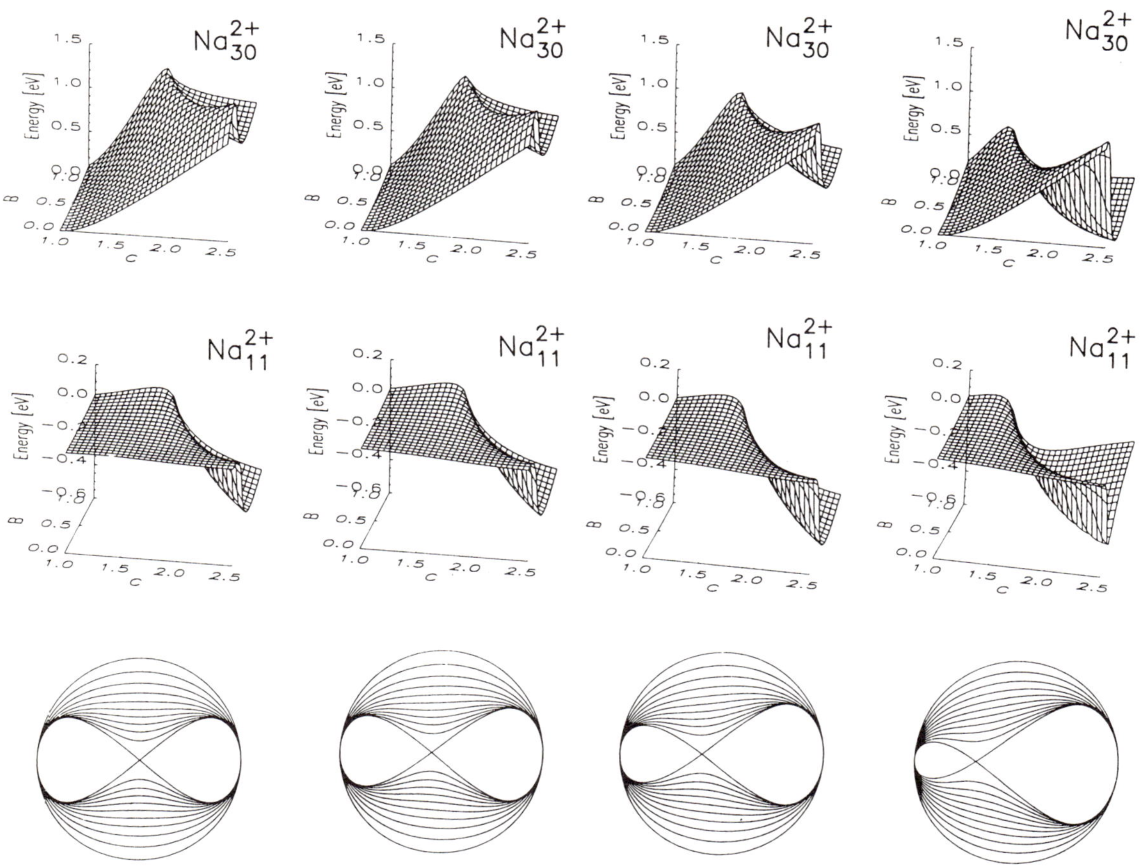

Fig. 9. The fission of Na_{11}^{2+} and Na_{30}^{2+} for various values of the symmetry parameter. The fission barrier of Na_{30}^{2+} decreases with increasing asymmetry. The fission energy of Na_{11}^{2+} is unaffected by the symmetry parameter

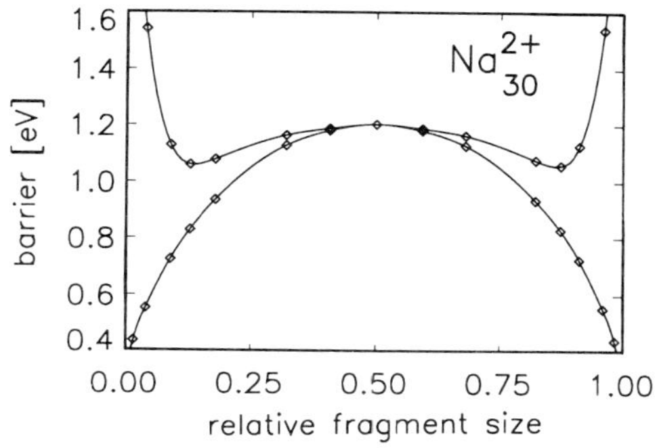

Fig. 10. The fission barrier of Na_{30}^{2+} as a function of the size of the fragment. The lower line corresponds to fission without any constraints concerning the charge of the fragment. In the upper line charge quantization has been considered. In this case the minimum fission barrier is obtained for a relative fragment size of about 0.1 (Na_3^+)

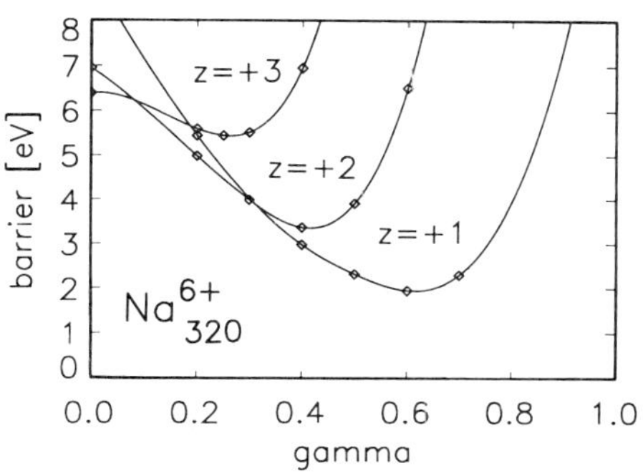

Fig. 11. The fission barrier of Na_{320}^{6+} as a function of the symmetry parameter, γ. $\gamma=0$ corresponds to symmetric fission, $\gamma=1.0$ to asymmetric fission. The three lines correspond to differently charged fragments. The singly charged fragment has the lowest fission barrier (1.9 eV) and a relative mass of about 0.01 ($\gamma=0.6$)

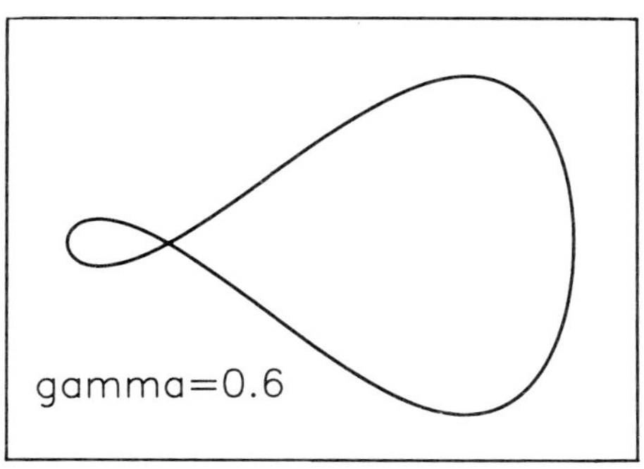

Fig. 12. The deformation that corresponds to $\gamma=0.6$. The parametrization used in the model is not able to describe strongly asymmetric deformations correctly. Since the deviations of the fragments from a sphere are too large, the fission barrier is overestimated

In a second calculation we investigated the symmetry of the fission process. The results are shown in Fig. 9. The energy barrier has been calculated for various symmetry parameters. For Na_{11}^{2+} the barrier is nearly independent of the symmetry. With the crude model used in this calculation no statement about the preferred fission channel is possible. Shell corrections should decisively influence the symmetry of the fission process for this case. For Na_{30}^{2+} asymmetric fission has a lower barrier than symmetric fission. Thermally activated fission is strongly asymmetric!

In classical models charge quantization is a problem. If the Coulomb energy is minimized according to (6) and (7) the small fragment in the asymmetric fission process of doubly charged droplets will not bear an integral charge, but only a small part of it. An additional constraint has to be introduced to force an integral charge onto this fragment. Figure 10 shows the barrier height of Na_{30}^{2+} under this constraint for various symmetries of the fission process. The upper line corresponds to fission with charge quantization, i.e. each fragment has one electronic charge. The energy minimum lies at a fragment size of 10%. For Na_{30}^{2+} we obtain Na_3^+ as a fragment. Even without shell corrections (favouring closed shell clusters like Na_3^+) the emission of Na_3^+ is the fission channel with the lowest energy barrier. This result has been found earlier by much more sophisticated models [10–12] and is experimentally confirmed [8].

For Na_{320}^{6+} the barrier height was calculated with the same model. There are three possibilities to split the six charges between the two fragments: 1:5, 2:4, and 3:3. As shown in Fig. 11 asymmetric fission is favoured here, too. The fragment is again in the size range of Na_3^+. However, the fission barrier (1.9 eV) is higher than the evaporation barrier. This effect might be caused by the selected parametrization which is unable to describe strongly asymmetric fission correctly. The standard 3-parameter model for the deformation of droplets was originally developed to describe symmetric fission or, at most, slight deviations from symmetric fission. For the fragmentation of very small particles the surface deformation is too large for this parametrization (Fig. 12) and the fission barrier is overestimated.

Conclusion

The critical size of charged alkali clusters (M_n^{z+}, $M=$ Li, Na, K, Rb, Cs) was measured for $z=2$ to 7 (Na and Cs only). Calculations in the framework of the classical liquid droplet model reveal that the fission of clusters is a thermally activated and a strongly asymmetric process. The asymmetry is a direct consequence of the high fission barrier.

References

1. Lord Rayleigh: Phil. Mag. **14**, 185 (1982)
2. Sattler, K., Mühlbach, J., Echt, O., Pfau, P., Recknagel, E.: Phys. Rev. Lett. **47**, 160 (1981); Sattler, K.: Surf. Sci. **156**, 292 (1985)
3. For a recent review see Echt, O.: In: Proc. of Conf. on Nuclear and Atomic Clusters, Turku. Brenner, Bergström (eds.). Berlin, Heidelberg, New York: Springer 1991
4. Näher, U., Göhlich, H., Lange, T., Martin, T.P.: Phys. Rev. Lett. **68**, 3416 (1992)
5. Martin, T.P., Näher, U., Göhlich, H., Lange, T.: Chem. Phys. Lett. **196**, 113 (1992)

6. Echt, O., Kreisle, D., Recknagel, E., Saenz, J.J., Casero, R., Soler, J.M.: Phys. Rev. A**38**, 3236 (1988)
7. Saunders, W.A.: Phys. Rev. Lett. **64**, 3046 (1990)
8. Bréchignac, C., Cahuzac, Ph., Carlier, C., de Frutos, M.: Phys. Rev. Lett. **64**, 2893 (1990); Bréchignac, C., Cahuzac, Ph., Leygnier, J., Sarfati, A.: Phys. Rev. B**44**, 11386 (1991); Bréchignac, C., Cahuzac, Ph., Carlier, C., de Frutos, M.: Phys. Rev. B**49**, 2825 (1994)
9. Katakuse, I., Ito, H., Ichihara, T.: Z. Phys. D**20**, 101 (1991)
10. Barnet, R.N., Landman, U., Rajagopal, G.: Phys. Rev. Lett. **67**, 3058 (1991)
11. Alonso, J.A., Barranco, M., Garcias, F., López, J.M.: Comments At. Mol. Phys. (to be published); Garcias, F., Alonso, J.A., López, J.M., Barranco, M.: Phys. Rev. B**43**, 9459 (1991)
12. Rao, B.K., Jena, P., Manninen, M., Nieminen, R.M.: Phys. Rev. Lett. **58**, 1188 (1987)
13. Krappe, H.J.: Z. Phys. D**23**, 269 (1992)
14. Nakamura, M., Ishii, Y., Tamura, A., Sugano, S.: Phys. Rev. A**42**, 22267 (1990); Koizumi, H., Sugano, S., Ishii, Y.: Z. Phys. D**26**, 264 (1993); Koizumi, H., Sugano, S., Ishii, Y.: Shell correction study of fission of doubly charged silver clusters (unpublished)
15. Brack, M., Damgaard, J., Jensen, A.S., Puli, H.C., Strutinsky, V.M., Wong, C.Y.: Rev. Mod. Phys. **44**, 320 (1972)
16. Ashcroft, N.W., Mermin, N.D.: Solid State Physics. Holt, Rinehart and Winston 1976
17. Handbook of chemistry and physics, 64th edn. Boca Raton: CRC Press 1984

Z. Phys. A 349, 265–268 (1994)

Multi-dimensional tunneling and nuclear fission process

Akira Iwamoto

Advanced Science Research Center, Japan Atomic Energy Research Institute, Tokai, Ibaraki, 319-11 Japan

Received: 27 October 1993

Abstract. Semiclassical treatment of the tunneling through a multi-dimensional potential barrier is given. We discuss first about methods to calculate the minimum action trajectory. Next, we discuss the adiabatic approximation to the minimum action trajectory. We apply these two methods to the calculation of spontaneous fission. Finally we discuss some problems which should be solved to calculate more realistic penetration problems.

PACS: 24.75.+i; 24.90.+d

1. Introduction

Calculation of the tunneling process through a multi-dimensional potential barrier is a necessity to understand several quantum problems. Typical example in the field of nuclear physics is nuclear fission. The fission process involves large distortions of nuclear matter and thus many shape parameters are required to calculate the potential surface. Thus fission calculation turns out to be a multi-dimensional penetration problem. We do not, however, have an established method to treat the multi-dimensional penetration problem. Choosing spontaneous fission process, which seems relatively simpler, we will attack this problem with semiclassical approach. The methods used so far for fission decay calculations [1,2,3] are based on numerical searching of all possible paths. These methods require considerable amounts of numerical calculation to get good results. Furthermore, it is hard to get a clear physical insight into the problem.

We consider semiclassical methods of fission half-life calculation, from the path-integral point of view and from the wave function [4]. The former method was developed by Coleman and Callan [5,6] and latter by Schmid [7]. Both were directed to solve the decay rate of metastable states in multi-dimensional space. The most probable fission path which minimizes the WKB action is determined in these methods.

We discuss next a new adiabatic model in which the path and the WKB action are obtained much easier than the preceding models.The model is convenient to get simply an approximate decay path when the mass tensors are functions of the coordinate space. The fission path is obtained by solving simple first order differential equations with suitable initial conditions.

2. Methods to obtain the least action trajectory

A standard model for the energy eigenvalue calculation proceeds as follow : We start from the imaginary time version of the partition function defined as

$$Z(T) \equiv tr(e^{-T\mathcal{H}/\hbar}) = \sum e^{-TE_n/\hbar}, \qquad (1)$$

where system Hamiltonian is given by $\mathcal{H}$ and its eigenenergy is E_n. We can easily find the relation

$$-\lim_{T\to\infty} \ln(Z(T))/T = E_0/\hbar. \qquad (2)$$

The partition function $Z(T)$ can be written in the path-integral form

$$Z(T) = \int \cdots \int \mathcal{D}q(\tau) \exp[-\frac{1}{\hbar}\int_{-T/2}^{T/2} \{\dot{q}^2/2 + V(q)\}d\tau], \qquad (3)$$

where from the character of trace, the boundary condition for $q(\tau)$ is written as

$$q(-T/2) = q(T/2). \qquad (4)$$

The stationary path satisfies the equation

$$-\ddot{q} + \frac{\partial V}{\partial q} = 0, \qquad (5)$$

which is a motion in an inverted potential. To obtain the final expression for the decay width, we need the following procedure, which the reader should refer tc Coleman and

Callan. [5,6] We have first to solve this equation and find the "bounce" solution. Next, we expand the path around this stationary path and then integrate with respect to the expansion coefficients. The problem of zero eigenvalue appears at this point, which should be solved when we integrate over the location of the bounce. Another technique necessary to get the decay width is to do analytic continuation of the path integral to the complex plain, where we make use of the saddle point approximation. Finally, after we sum up all the contribution of n-bounce solution, we get the decay width expression.

Essentially the same form of the expression is obtained from the approach of Schmid [7]. In the following part, we generalize the formalism to the case where mass tensors are functions of position, which is necessary in fission life time calculation [4]. We start from the Schrödinger equation,

$$\mathcal{H}\psi(\mathbf{q}) = E\psi(\mathbf{q}), \tag{6}$$

where $\mathbf{q}$ represents shape parameters and E is the energy eigenvalue. The Hamiltonian $\mathcal{H}$ is assumed to have the form

$$\mathcal{H} = -\frac{\hbar^2}{2}\frac{\partial}{\partial q^i}(M^{-1})^{ij}\frac{\partial}{\partial q^j} + V(\mathbf{q}), \tag{7}$$

where q^i represents a component of shape parameter $\mathbf{q}$ and M is the mass tensor. We will study a decay problem and thus the energy E is put

$$E = \hbar(\omega_t - i\Gamma)/2 \tag{8}$$

where ω_t represents zero-point energy in the case of decay from the ground state and Γ is the decay width. We put the wave function in the form,

$$\psi(\mathbf{q}) = \exp\{-W(\mathbf{q}/\hbar\}. \tag{9}$$

The exponent W is expanded in power of $\hbar$ in the form,

$$W(\mathbf{q}) = W_0(\mathbf{q}) + \hbar W_1(\mathbf{q}) + \hbar^2 W_2(\mathbf{q}) + \cdots. \tag{10}$$

This expansion is thought to be valid under quasi-classical conditions which justifies WKB approximation.

Under the quasi-stationary condition,

$$\Gamma << \omega_t, \tag{11}$$

we insert (10) into (6) and equate separately the terms of equal power of $\hbar$. In leading order, we obtain the eikonal equation,

$$H_0 = p_i(M^{-1})^{ij}\, p_j - V(\mathbf{q}) = 0, \tag{12}$$

where

$$p_i = \frac{\partial W_0}{\partial q^i}, \tag{13}$$

plays the role of momentum. In obtaining (12), the energy E in (6) was assumed to be much smaller than barrier height V_b. This E is of first order in $\hbar$ and is treated in the next higher order equation called transport equation,

$$\begin{aligned} H_1 = {} & \frac{1}{2}(\frac{\partial}{\partial q^i}M^{-1}_{ij})p_j + \frac{1}{2}M^{-1}_{ij}\frac{\partial^2 W_0}{\partial q^i \partial q^j} \\ & - M^{-1}_{ij}p_i p^{(1)}_j - \frac{1}{2}\omega_t = 0, \end{aligned} \tag{14}$$

where $p^{(1)}_i$ is given by

$$p^{(1)}_i = \frac{\partial W_1}{\partial q_i}. \tag{15}$$

The eikonal equation (12) is identical to the energy conservation condition for inverted potential, that is, $V \to -V$, and can be solved by the method of characteristics. The characteristics can be represented in parametric form as trajectories in the 2N-dimensional $\{\mathbf{q},\mathbf{p}\}$- space

$$q^i = q^i(\tau), \quad p_i = p_i(\tau), \tag{16}$$

such that they obey the canonical equations of motion

$$\frac{dq^i}{d\tau} = \frac{\partial H_0}{\partial p_i}, \tag{17}$$

and

$$\frac{dp_i}{d\tau} = -\frac{\partial H_0}{\partial q^i}, \tag{18}$$

with the initial condition $H_0 = 0$ at $\tau = 0$. We can introduce the concept of path which is defined as projections of the trajectories on N-dimensional $\{\mathbf{q}\}$-space. The path obeys the equations of motion

$$M_{ij}\ddot{q}^j = \frac{\partial V}{\partial q^i} - \dot{q}^k\frac{\partial M_{ij}}{\partial q^k}\dot{q}^j + \frac{1}{2}\dot{q}^j\frac{\partial M_{jk}}{\partial q^i}\dot{q}^k, \tag{19}$$

where the dot means a differentiation with respect to τ. Equation (19) is an extension of (4.7) of the paper of Schmid [7] to the coordinate-dependent mass tensors. It represents a classical equation of motion in the inverted potential, which is the same as (5) as it should be.

The boundary condition used to solve the eikonal equation is $\mathbf{p} = 0$ at the origin [7]. The origin corresponds to the ground state deformation of the fissioning compound nucleus which is taken as $\mathbf{q} = 0$. We show in Fig.1 an illustration showing the situation for the two-dimensional problem. Because the origin is a maximum point in our inverted potential, we need to boost the system infinitesimally to get meaningful solutions to (19). If we choose a suitable initial direction of $\mathbf{p}$ for the boosting, we are able to get a special path which can touch the outer equipotential surface given by $V(\mathbf{q}) = 0$, which is a line denoted by O in Fig.1., with zero velocity. This special path is called the escape path and the point at which this path touches the outer equipotential surface is the escape point [7], which is denoted by E in Fig.1. All other paths can not touch this outer equipotential surface. By trying all 2π directions for the initial boost, we can find the escape path by solving (19). This procedure is much simpler that the ones which have been used before, where all possible paths should be examined. The envelope of trajectory denoted by C is called the caustic. Therefore, beyond the caustic, there is no solution at this level of approximation.

Inside the caustic, we get solutions of (19) from which we can construct the wave function (9) in the following way. Using (13) and (17), we get the relation,

$$\frac{dW_0}{d\tau} = \frac{dq^i}{d\tau}\cdot\frac{\partial W_0}{\partial q^i} = \dot{q}^i\cdot M_{ij}\cdot\dot{q}^j. \tag{20}$$

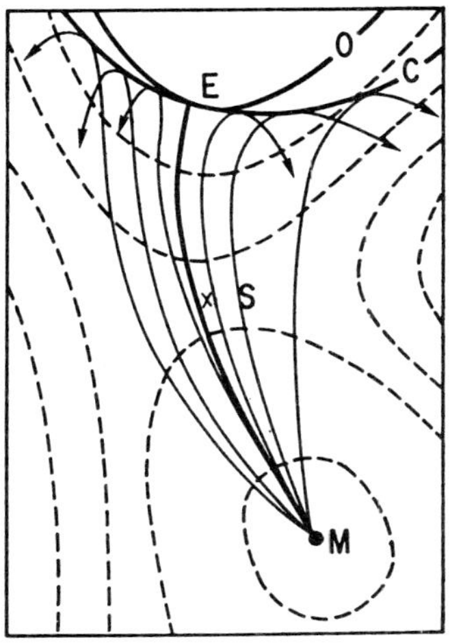

Fig. 1. An illustration which shows how to solve the eikonal equation and to obtain the minimum action trajectory. Arbitrary two-dimensional space is shown here. *Dotted lines* represent equipotential surfaces. Point M represents the maximum point in the inverted potential and point at $\times(S)$ represents the saddle point. The *lines* O and C touch each other at point E which is the escape point. The *bold line* which connects M and E is called escape path which gives the minimum action trajectory

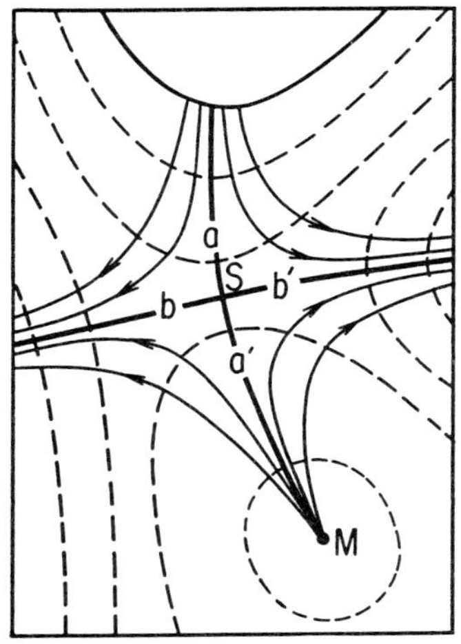

Fig. 2. Definition of the invariant adiabatic path. An arbitrary two- dimensional space is shown and *dotted lines* represent equipotential surfaces. Point M represents the maximum point in the inverted potential and S represents the saddle point of the potential. A *bold line* in the upper part is the equipotential surface whose energy is the same as point M. *Thin solid lines* with arrows are solutions of (24). The lines $a - a'$ and $b - b'$ are two asymptotes for the solution of (24), and the line a-a' defines the invariant adiabatic path

The most important trajectory is the escape path and we denote by S_{es} the value of W_0 which is obtained when we integrate from the origin to the escape point along the escape path,

$$S_{es} = \int_{es} \dot{q}^i \cdot M_{ij} \cdot \dot{q}^j \, d\tau. \tag{21}$$

The quantity S_{es} is the action integral and is equivalent to the popular form of the action integral S given as

$$S = \int_{\tau_i}^{\tau_f} [2|E - V| \cdot M_{ij} \cdot \dot{q}^i \cdot \dot{q}^j]^{1/2} d\tau. \tag{22}$$

In the lowest order, the fission half life $T_{1/2}$ is expressed as

$$T_{1/2} = (\frac{2\pi}{\omega}) \exp(-2S_{es} \,/\hbar), \tag{23}$$

where ω is the typical frequrency of the nuclear surface vibration and which is taken as $\hbar\omega/2 = 0.5\text{MeV}$ in our calculation.

3. Adiabatic approximation

In the previous section, we gave a method [4,8] to calculate the minimum action trajectory in a much easier and more transparent way than the method used befire [1,2,3]. We propose here an alternative model which is much easier in its structure and for numerical calculation but, at the same time, is invariant with respect to coordinate transformation. This model has the feature of the adiabatic model as will be shown below.

To define it, we start from solving the equation,

$$M\frac{d\mathbf{q}}{ds} = \text{const} \cdot \nabla V, \tag{24}$$

in the region where the following inequality holds,

$$(\frac{\partial V}{\partial q^i}) \cdot (\frac{\partial V}{\partial q^i}) \neq 0. \tag{25}$$

Physical meaning of (24) is discussed elsewhere [8]. We can generate a group of trajectories by solving (24). We show in Fig.2, the solution of this equation for a simple potential which has one minimum and one saddle point. The dashed lines in this figure are the equipotential energy surfaces and the solid lines with arrows are the solution of (24). Paths obeying this equation have two asymptotes, as shown in this figure by $a - a'$ and $b - b'$. We define the former asymptote $a - a'$ as the new adiabatic path.

The important feature of this adiabatic path is that it is invariant with respect to the coordinate transformation,

$$\mathbf{x} \to \mathbf{q} \tag{26}$$

where old and new coordinates are related to each other by

$$\mathbf{q} = \mathbf{q}(\mathbf{x}), \tag{27}$$

or

$$\mathbf{x} = \mathbf{x}(\mathbf{q}). \tag{28}$$

The invariance is easily checked by remembering the transformation property of mass tensor,

$$M_{ij} \to \mathcal{M}_{kl} = \frac{\partial x^i}{\partial q^k} \cdot M_{ij} \cdot \frac{\partial x^j}{\partial q^\ell}. \tag{29}$$

Thus our new adiabatic action is invariant with respect to the coordinate transformation, which is a necessity and is satisfied by the minimum action obtained in the previous section. We call this path the invariant adiabatic path, because in some cases, the path that is always orthogonal to the equipotential surface is erroneously called "adiabatic" path.

4. Application to spontaneous fission half life

In this section, we will apply the methods given in Sects.2 and 3 to the spontaneous fission calculation. Our concern is to check the applicability of the methods to nuclear fission; accordingly, we will not seek the quantitative solutions. We neglect shell and paring effects in the potential and mass tensor calculation, that is, we adopt the classical liquid drop model. Our aim is to calculate multi-dimensional tunneling, and we choose the 2-dimensional barrier penetration in a spontaneous fission process.

For the two-dimensional fission calculation, we choose two shape parameters: potential center separation z, spheroidal deformation parameter of fragments δ . We assume symmetric fission and the neck parameter ϵ is fixed at 0.5 [9]. The two center model is defined for a single-particle potential. From it, we calculate the shape of the equipotential surface corresponding to the Fermi energy and identify it with the shape of the fissioning nucleus. The potential energy is calculated by the Yukawa-plus-exponential model [10] and the mass tensors by the Werner-Wheeler method [11].

Calculating spontaneous fission of ^{236}U , we found the followings: the "traditional" adiabatic path, which is defined as the curve which is always orthogonal to the equipotential surface, deviates much from the minimum action trajectory. The origin of this difference is due to the neglect of kinetic energy in the adiabatic treatment. On the other hand, the invariant adiabatic path which was determined in the method given in Sect.3 almost coincides with the minimum action trajectory. It is a surprise that these two almost coincide with each other irrespective of the fact that they are defined quite independently.

The value of invariant adiabatic action calculated by the method of Sect.3 doesn't deviate much from the minimum action. For ^{236}U and ^{212}Po we examined, the adiabatic action is within 102% of the minimum action trajectory. On the other hand, the action determined from the traditional adiabatic path is more than 120 % of the minimum action. We found that invariant adiabatic path is very useful to get an estimate of the minimum action path and the value of the action.

5. Summary and remaining problems

We presented semiclassical methods to calculate decay width in a multi-dimensional parameter space. These methods worked well for the spontaneous fission of heavy nuclei. We found that the minimum action trajectory is obtained with our method much easier than the methods used before. We also found that the invariant adiabatic path is a good approximation to the minimum action trajectory.

It is of importance to extend the method we used here to obtain the minimum action trajectory to higher order. The lowest-order one treated here is eikonal equation (12), while the next order is transport equation (14). To get the pre-factor for the decay width expression, it is necessary to solve the transport equation. This is also necessary to join the wave function for the positive and negative energy region. Another interesting feature is to extend model to three or more dimensional problems. Formally, it is clear that the method can be applied to any dimensions, but the degree of difficulty to do the calculation should be examined.

Restricting ourselves to nuclear fission problem, we still have the following open problems. In this work, fission potential and mass tensors were calculated by the classical model. Shell and pairing effects play an important role in the nuclear fission problem and, if we include them, potential and mass tensors become more complicated. The methods proposed here should be checked beforehand if we wish to use them in such complicated geometry. Our methods are of interest not only for the double-humped barrier, which occurs quite generally in actinide nuclei, but also for bimodal fission. This follows because it is expected [12] that there are two or more trajectories which give minimum actions in bimodal fission.

References

1 Brack, M. et al.: Rev. Mod. Phys. **44**,320 (1972)
2 Ledergerber, T.: Pauli, H.C. Nucl. Phys. **A207**,1 (1973)
3 Baran, A. et al.: Nucl. Phys. **A361**,83 (1981)
4 Kindo, T. Iwamoto, A.: Phys. Lett. **B225**,203 (1989)
5 Coleman, S.: Phy. Rev. D**13**,2929 (1977)
6 Callan Jr., C.D., Coleman, S.: Phys. Rev. D**16**,1762 (1977)
7 Schmid, A.: Ann. Phys. **170**,333 (1986)
8 Iwamoto, A., Tomita, Y.: Prog. Theor. Phys. **87**,1171 (1992)
9 Maruhn, J., Greiner, W.: Z. Phys. **251**,431 (1972)
10 Krappe, H.J., Nix, J.R., Sierk, A.J.: Phys. Rev. C**20**,992 (1979)
11 Davies, K.T.R., Sierk, A.J., Nix, J.R.: Phys. Rev. C**13**,2385 (1976)
12 Möller, P., Nix, J.R., Swiatecki, W.J.: Nucl. Phys. **A469**,1 (1987)

This article was processed using Springer-Verlag TEX Z.Physik A macro package 1.0 and the AMS fonts, developed by the American Mathematical Society.

Z. Phys. D 31, 279–282 (1994)

The Cl^-NH_3, Cl^-H_2O, F^-NH_3 and F^-H_2O clusters and their photoelectron spectra

Uzi Kaldor

School of Chemistry, Tel Aviv University, 69978 Tel Aviv, Israel

Received: 28 June 1993 / Final version: 24 April 1994

Abstract. The geometries of the Cl^-NH_3, Cl^-H_2O, F^-NH_3 and F^-H_2O clusters are optimized using the coupled cluster method. The four lowest ionization potentials are then calculated, leading to the ground and low excited states of the neutral species. The first three IPs describe ionization from the external p state of the halogen atom, whereas the fourth corresponds to ionization from the NH_3 or H_2O moiety, leading to charge transfer complexes. These complexes were recently observed in the photoelectron spectrum of Cl^-NH_3, in full accord with our calculations.

PACS: 36.40.+d; 31.20.Tz; 33.60.Cv

I. Introduction

Clusters of halogen anions with water, $X^-(H_2O)_n$, have been investigated extensively, both experimentally [1] and theoretically [2–7]. Few studies of the ammonia analogs $X^-(NH_3)_n$ have been reported [8]. Most of these studies concern the energetics and geometrical structure of the clusters. Recently, Cheshnovsky and coworkers have measured vertical ionization potentials in mass-resolved clusters of the water [9] and ammonia [10] halides. The observation of a second band in the photoelectron spectrum of Cl^-NH_3, leading to a charge-transfer (CT) complex $Cl^-NH_3^+$, is of particular interest. The identification of the CT complex was confirmed by ab initio calculations [10].

Here we investigate the geometry and photoelectron spectrum of the X^-S clusters, where X is F or Cl and S is H_2O or NH_3. The coupled cluster method with single and double excitations (CCSD) [1] is used for geometry optimization, and the ionization potentials of the X^-S species, observed in the photoelectron spectrum, are obtained by the Fock-space CCSD [12, 13]. The geometry of the CT species X^-S^+ is also investigated.

II. Method

The Fock-space coupled cluster method has been described before [13], and only a brief outline is given here. The basic method follows Lindgren's [12] choice of a normal-ordered wave operator,

$$\Omega = 1 + S + \tfrac{1}{2}\{S^2\} + \cdots. \tag{1}$$

S is the excitation operator describing *connected* single, double,... excitations

$$S = S_1 + S_2 + \cdots$$
$$= \sum_{ij} \{a_i^\dagger a_j\} s_j^i + \tfrac{1}{2} \sum_{ijkl} \{a_i^\dagger a_j^\dagger a_l a_k\} s_{kl}^{ij} + \cdots \tag{2}$$

where $a_i^\dagger$ and a_j are creation and annihilation operators, $s_j^i, s_{kl}^{ij}, \ldots$, are excitation amplitudes, and the curly brackets denote normal order with respect to a reference (core) determinant. The summation is carried out over connected terms only. The equations determining the excitation amplitudes in a complete model space may be derived from the generalized Bloch equation [12]

$$[S, H_0] = (QV\Omega - \Omega V_{\text{eff}})_{\text{conn}} \tag{3}$$

$$V_{\text{eff}} = PV\Omega P \tag{4}$$

where P and Q are the usual projection operators, and H_0 and V result from the partitioning of the Hamiltonian,

$$H = H_0 + V. \tag{5}$$

The Fock-space method is designed to handle open-shell states. A closed-shell reference determinant is selected, and the excitation operator for states with n valence holes

Supported in part by the U.S.-Israel Binational Science Foundation

This paper was originally submitted in connection with the 2nd. Int. Conference on Atomic and Nuclear Clusters held in Santorini from 28. June –2. July 1993 and is published here as a regular article after an independent refereeing procedure according to the standards of Z. Phys. D

(electrons removed from the reference) and m valence particles (electrons added) is denoted by $S^{(n,m)}$. The total S operator may be written as

$$S=\sum_n \sum_m S^{(n,m)}. \quad (6)$$

Haque and Mukherjee [14] have shown that this partitioning allows for partial decoupling of the Fock-space CC equations. The equations for $S^{(i,j)}$ involve only $S^{(k,l)}$ elements with $k \leq i$ and $l \leq j$, and the very large system of non-linear coupled equations separates therefore into smaller subsystems which are solved consecutively. The equations for $S^{(0,0)}$ are first iterated to convergence, then $S^{(0,1)}$ and $S^{(1,0)}$ are solved using the known $S^{(0,0)}$, and so on. This separation is exact, and reduces the computational effort significantly. It should not be confused with the expansion (2) of S in terms of the number of virtual excitations, i.e. the *total* number of electrons excited from the P-space determinant. Each term of (6) has an expansion similar to (2), e.g.

$$S^{(0,2)}=S_2^{(0,2)}+S_3^{(0,2)}+\cdots \quad (7)$$

where $S_j^{(n,m)}$ describes a j-body excitation from a state obtained by removing n electrons from the reference determinant and adding to it m electrons.

The coupled cluster method with single and double excitations (CCSD) is employed here. It truncates the expansion (2) at

$$S=S_1+S_2. \quad (8)$$

The ground state of the X^-S anion is used as the closed-shell reference state. Its Hartree-Fock function is obtained, and it is then correlated by the CCSD method. The geometry of the anion is optimized at the CCSD level, using analytic first derivatives [15]. The ionization potentials (IP) of the anion are then calculated by Fock-space CCSD, using the four highest occupied molecular orbitals of the parent anion as valence holes in the (1,0) sector. The geometries of the neutral species XS and their CT excited states X^+S^- were optimized separately. This was done by the quasirestricted coupled cluster method [16]. All geometry optimizations were carried out by the ACES II program package [17].

In selecting the basis functions for the computations, a compromise was sought between accuracy and computational resources. The atomic natural orbital (ANO) sets of Widmark et al. [18] were used. They were truncated to $5s4p2d$ on the halogen atoms, as these bases give good CCSD electron affinities (3.668 eV for Cl and 3.22 eV for F, compared with the experimental 3.62 and 3.40 eV, respectively [19]). Since it is important to have a balanced basis on the different atoms in the molecule, the O and N bases included $4s3p2d$ ANOs for the Cl clusters and $5s4p2d$ ANOs for the F compounds. The hydrogen basis was composed of $3s1p$ ANOs. The quality of the NH_3 and H_2O bases was assessed by comparing the geometry and ionization potentials obtained for these molecules with experiment [20, 21]. The results are satisfactory (see Table 1). The geometry optimization of

Table 1. CCSD properties of NH_3 and H_2O

	R_e (Å)	θ_e	IP_{vert} (eV)
NH_3			
$4s3p2d/3s1p$	1.015	106.5°	10.68
$5s4p2d/3s1p$	1.015	106.4°	10.71
Exper. [20, 21]	1.012	106.7°	10.8
H_2O			
$4s3p2d/3s1p$	0.961	104.4°	12.40
$5s4p2d/3s1p$	0.960	104.6°	12.42
Exper.	0.956	105.2°	12.62

the open-shell X^+S^- species of the Cl clusters was done with a somewhat inferior basis, using double-zeta-plus-polarization functions [22] on the O, N and H atoms.

III. Results and discussion

The calculated properties of the four anions are summarized in Table 2. The optimized structures exhibit the result of two conflicting effects. The charge-dipole interaction would tend to put the halide ion on the symmetry axis of the ligand molecule, whereas the hydrogen bond requires an X-H-L angle of 180° for minimal energy, where X is the halogen and L is O or N. These angles are within a few degrees of 180° for the fluorine clusters, whereas the chlorine analogs give 166–168°. The four anions have C_s symmetry. The optimized geometries may be compared with previous correlated calculations for F^-H_2O [5] and Cl^-H_2O [7]. The F^-H_2O geometry obtained here is very close to the CASSCF and CISD results of Yates et al. [5], with differences below 0.02 Å and 1°. The results of Sapse and Jain [3], with a considerably smaller basis, are also quite close, except for the F-H-O angle which is 4° smaller than ours. Combariza et al. [7] optimized the geometry of Cl^-H_2O at the second-order perturbation theory level. Their H-Cl distance is 0.2 Å longer than ours, and their H-O-H angle is 1.5° larger. We are not aware of comparable computations for the ammonia clusters. Evans et al. [8] carried out a Hartree-Fock optimization of the Cl^-NH_3 cluster in a small basis. Their reported bond angles are within a few degrees of ours.

The binding energies shown in Table 2 are in good agreement with the measurements of Arshadi et al. [1] for F^-H_2O (1.01 eV) and Cl^-H_2O (0.57 eV). The experimental binding energy of Cl^-NH_3, 0.35 eV [8], is also very close to the calculated value. It should be noted that calculated dissociation energies do not include differences in zero-point vibrational energy. These should, however, be small, since the molecular structure of H_2O and NH_3 does not change significantly in the cluster, and the X^--S potential is rather shallow. The net effect should be to decrease slightly the calculated dissociation energies, bringing them somewhat closer to experiment.

Two ionization potentials are given in Table 2 for each of the four anions. The lowest IP belongs to ionization from the $2p$ shell of fluorine or the $3p$ shell of chlorine,

Table 2. Calculated properties of the X^-S clusters

	Cl^-NH_3	F^-NH_3	Cl^-H_2O	F^-H_2O
Dissociation energy (eV)	0.36	0.67	0.63	1.14
Vertical ionization energies (eV)				
to XS	3.88	3.97	4.17	4.74
to X^-S^+	6.36	5.42	7.19	5.75
Distances[a] (Å)				
X-L	3.372	2.681	3.137	2.452
X-H_1	2.363	1.626	2.174	1.413
H_1-L	1.026	1.056	0.982	1.041
L-H_2	1.017	1.018	0.961	0.958
Angles				
∠ (X-H_1-L)	167.8°	175.6°	166 4°	176.8°
∠ (H_1-L-H_2)	104.4°	105.0°	100.8°	102.4°
∠ (H_2-L-H_3)	104.9°	104.5°	-	-

[a] X is Cl or F, L is N or O. H_1 is closest to X

Table 3. Calculated geometries and dipoles of the X^-S^+ species. $Cl^-NH_3^+$ and $Cl^-H_2O^+$ were optimized at the CCSD level, the two others by 2nd-order perturbation theory

	$Cl^-NH_3^+$	$F^-NH_3^+$	$Cl^-H_2O^+$	$F^-H_2O^+$
Dipole moment (D)	7.18	4.87	4.03	2.88
Distances[a] (Å)				
X-L	3.062	2.770	2.823	2.475
X-H_1	1.993	1.719	2.674	1.351
H_1-L	1.069	1.051	1.151	1.125
L-H_2	1.019	1.003	0.979	0.969
Angles				
∠ (X-H_1-L)	180.0°	180.0°	176.7°	179.9°
∠ (H_1-L-H_2)	121.0°	120.9°	103.7°	105.3°
∠ (H_2-L-H_3)	118.0°	118.2°	-	-

[a] X is Cl or F, L is N or O. H_1 is closest to X

leading to the XS neutral species. These shells are split in the clusters, and three distinct IPs are calculated. They fall within a range of ~0.03 eV, and cannot be separated experimentally. The known IPs of this type are 4.3 eV for Cl^-H_2O [9] and 4.0 eV for Cl^-NH_3 [10]. The second IP in the table involves an electron residing mainly on the ligand molecule, leading to a charge transfer complex X^-S^+. Such a complex was recently observed in the Cl^-NH_3 system [10]. Its existence was deduced from the appearance of a broad band in the photoelectron spectrum of the parent anion, peaked at 6.45±0.15 eV, in excellent agreement with the calculated 6.36 eV. Similar bands may be expected to appear in the spectra of other clusters as well.

The CT complex appears as a broad band in the photoelectron spectrum of the anion because its equilibrium geometry differs greatly from that of the parent. The structures of the four CT complexes were optimized at the CCSD (for the Cl clusters) or second order perturbation theory (F clusters) level. The calculated geometries and dipole moments of the X^-S^+ species are given in Table 3. The CT complexes are all planar, with an X-H-L angle of 180° or very close to it. Insepction of Tables 2 and 3 shows that structure changes of the Cl clusters are much more pronounced than those of F clusters. In particular, the Cl-L distance goes down by 0.4 Å while the F-L separation is less affected. This reflects the fact that the F^-S structure is dominated by the hydrogen bond, with an F-H-L angle close to 180° and not changing much upon going to the CT complex, whereas the Cl-containing anions have a much smaller angle which increases considerably in the CT compound. Geometry changes in the H_2O clusters are smaller than in the NH_3 compounds. These observations should be qualified by noting that some of the X^-S^+ geometries were optimized with smaller bases and a lower theoretical level than X^-S structures, because open-shell systems require greater computational resources than closed-shell species.

IV. Summary and conclusion

The geometries and ionization potentials of the four anions Cl^-NH_3, Cl^-H_2O, F^-NH_3 and F^-H_2O were investigated by the coupled-cluster method. The first IP of each cluster corresponds to ionization from the halide anion, leading to the ground state of the neutral XS. The second IP describes ionization from an orbital centered mainly on the ligand molecule, resulting in the charge transfer complex X^-S^+. This complex has been observed recently for Cl^-NH_3. The calculated IPs are in very good agreement with experimental data, when the latter are available.

The author thanks Ori Cheshnovsky and Gil Markovich for helpful discussions.

References

1. See e.g. Arshadi, M., Yamdagni, R., Kebarle, P.: J. Phys. Chem. **74**, 1475 (1970); Keesee, R.G., Castleman, A.W.: Chem. Phys. Lett. **74**, 139 (1980); Hiraoka, K., Mizuse, S.: Chem. Phys. **118**, 457 (1987); Hiraoka, K., Mizuse, S., Yamabe, S.: J. Phys. Chem. **92**, 3943 (1988)
2. Diercksen, G.H.F., Kraemer, W.P., Roos, B.O.: Theor. Chim. Acta **36**, 249 (1975)
3. Sapse, A.M., Jain, D.C.: Int. J. Quantum Chem. **27**, 281 (1985)
4. Gao, J., Garner, D.S., Jorgensen, W.L.: J. Am. Chem. Soc. **108**, 4784 (1986)
5. Yates, B.F., Schaefer, H.F., Lee, T.J., Rice, J.E.: J. Am. Chem. Soc. **110**, 6327 (1988)
6. Zhao, X.G., Gonzales-Lafont, A., Truhlar, D.G.: J. Chem. Phys. **94**, 5544 (1991)
7. Combariza, J.E., Kestner, N.R., Jortner, J.: Chem. Phys. Lett. **203**, 423 (1993); (to be published)
8. Evans, D.H., Keesee, R.G., Castleman, A.W.: J. Chem. Phys. **86**, 2927 (1987)
9. Markovich, G., Giniger, R., Levine, M., Cheshnovsky, O.: J. Chem. Phys. **95**, 9416 (1991); Z. Phys. D**20**, 69 (1991)
10. Markovich, G., Cheshnovsky, O., Kaldor, U.: J. Chem. Phys. **99**, 6201 (1993)
11. For a review see Bartlett, R.J.: J. Phys. Chem. **93**, 1697 (1989)
12. Lindgren, I.: Int. J. Quantum Chem. S**12**, 33 (1978); Salomonsson, S., Lindgren, I., Mårtensson, A.-M.: Phys. Scr. **21**, 351 (1980); Lindgren, I., Morrison, J.: Atomic many body theory. Springer Series in Atoms and Plasmas 3. Berlin, Heidelberg, New York: Springer 1982
13. Mukherjee, D., Pal, S.: Adv. Quantum Chem. **20**, 291 (1989); Kaldor, U.: Theor. Chim. Acta **80**, 427 (1991); Rittby, C.M.L., Bartlett, R.J.: ibid **80**, 469 (1991)
14. Haque, A., Mukherjee, D.: J. Chem. Phys. **80**, 5058 (1984); Pramana **23**, 651 (1984)
15. Salter, E.A., Trucks, G.W., Bartlett, R.J.: J. Chem. Phys. **90**, 1752 (1989); Salter, E.A., Bartlett, R.J.: ibid **90**, 1767 (1989)
16. Gauss, J., Stanton, J.F., Bartlett, R.J.: J. Chem. Phys. **95**, 2639 (1991)
17. ACES II, Stanton, J.F., Gauss, J., Watts, J.D., Lauderdale, W.J., Bartlett, R.J.: Quantum Theory Project. University of Florida, Gainesville 1992
18. Widmark, P.-O., Malmquist, P.-Å, Roos, B.O.: Theor. Chim. Acta **77**, 291 (1990); Widmark, P.-O., Persson, B.J., Roos, B.O.: ibid **79**, 419 (1991)
19. Hotop, H., Lineberger, W.C.: J. Phys. Chem. Ref. Data **4**, 539 (1975)
20. Herzberg, G.: Electronic spectra and electronic structure of polyatomic molecules. Vol. III, p. 585, 609. New York: Van Nostrand 1966
21. Bowers, M.T. (ed.): Gas phase ion chemistry. Vol. II, p. 16. New York: Academic Press 1979
22. Dunning, T.H.: J. Chem. Phys. **53**, 2823 (1970); Redmon, L.T., Purvis, G.D., Bartlett, R.J.: J. Am. Chem. Soc. **101**, 2856 (1979)

Z. Phys. A 349, 269–271 (1994)

Nonlinear approach in the α-cluster model

K.A. Gridnev

Saint Petersburg State University, 198904 Saint Petersburg, Russia

Received: 27 October 1993

Abstract. Nonlinear Schroedinger Equation (NoSE) and Korteveg de Vries Equation (KdVE) are applied to the solving of problems of the nucleation and clustering.

PACS: 21.65.+f

In last years more and more attention is being paid to problems connected with nonlinear equations and their solutions. The aim of the present contribution is to apply Nonlinear Schroedinger Equation (NoSE) and Korteveg de Vries Equation (KdVE) to an interpretation of some effects of the clustering in nuclei. Earlier we applied the NoSE for the explanation such phenomena as the ALAS (Anomalous Large Angular Scattering) in interactions of α-particles with light nuclei [1]. Also we used the NoSE for the description of α-particle spectra in^{20}Ne [2]. In both cases we exploitated the NoSE in such form:

$$-(\hbar^2/2m)\nabla^2 X + VX + C \mid X \mid^2 X = \text{EX} \ , \tag{1}$$

where $C = K/9$; here K- the compressibility constant. It turned out that (1) is a particular solution of a general equation suggested by Bialynsky-Birula and Mysielsky [3]. Using instead of cubic term $\epsilon \log(a \mid \psi \mid^2)\psi$ they formulated the nonlinear quantum mechanics without any contradiction in classical quantum mechanics. By the way, Gaehler, Klein and Zeilinger [4] attempted to find the constant ϵ according to the predictions made in the work [3]. But it was made without any success. Our approach supposes the constant $C = K/9$ or ϵ (here both constants are the same) to be the compressibility constant. It was confirmed from the values of the constant K, obtained from different experiments with the help of this approach ($K \approx 270$ MeV) and also by the recent work of J.Braecher [5]. J.Braecher considered the nonlinear term in the NoSE from the point of view of the information theory. According to the work of Shannon and Weaver [6] the information I acquired upon measurements of the state$\mid \psi >$ is proportional to the logarithm of the probability $I = -\log_2(a \mid \kappa \mid^2)$ bits, where $a = \mid \psi_0 \mid^{-2}$.

The certain amount ϵ of energy per bit is expended, transferred, stored, associated with the information encoded. That's why it is necessary to add the term ϵI to the standard Hamiltonian:

$$H = H_0 + \epsilon I = \sum_{j=1}^{N} \frac{1}{2m} \nabla_j^2 + U - \epsilon \log_2(a \mid \psi \mid^2) \tag{2}$$

Now it is possible to show that this constant ϵ is very close to our compressibility constant. The total energy as the expectation value of the Hamiltonian (2) is:

$$E =< \psi \mid H > =< \psi \mid H_0 \mid \psi > + \epsilon < \psi \mid H_0 \mid \psi > \\ = E_0 \epsilon < \psi \mid \log_2(a \mid \psi \mid^2) \mid \psi > \tag{3}$$

The entropy for the subsystem is:

$$S = K < \psi \mid \ln(a \mid \psi \mid^2) \mid \psi > , \tag{4}$$

where K - the Boltzmann constant. Now we obtain the free energy E_0, the temperature T and constant ϵ :

$$E_0 = E - TS, \quad T = \left(\frac{\partial E}{\partial S}\right)_V, \quad \epsilon = KTln(2) \tag{5}$$

The value $\epsilon = KTln(2)$ is very close to the constant C. If we suppose the value KT for the average kinetic energy ≈ 30 MeV we obtain the reasonable value of compressibility constant. For two level systems with the difference of energies between two levels one can determine the temperature in the following way:

$$T = \epsilon(Kln(N_{unex}/N_{ex})^{-1} \tag{6}$$

where N_{unex} - the number of particles on the low level, N_{ex} - the number of particles on the upper level.

The higher temperature of the excited nucleus the larger the compressibility constant. The more particles are involved into the process of excitation the larger the compressibility constant. Of course, the temperature will depend upon the clustering.

Let's remember the Boltzmann's determination of the entropy: $S = KlnW$. Now it is easy to understand the quasi-elastic nucleon transfer reaction between heavy nuclei and the nuclear Josephson effect [7].

W. von Ortzen used the following formula for the X-nucleons transfer [7]:

$$W_{xn} = (W_{1n})^x \tag{7}$$

For the two simultaneous nucleon transfer we can write:

$$W_{12} = w_1 * w_2 = exp(s_1/k) * exp(s_2/k) = exp((s_1 + s_2)/k) = exp(2s/k) \tag{8}$$

Keeping in mind theorems of the theory of probability, it is easy to see that the cross section of the two step transfer of two nucleons is less than for simultaneous transfer. Solving the NoSE it is possible to calculate the probability of two or four nucleon transfer reactions $P = exp(< \psi \mid a \mid \psi \mid^2 \psi)$ of course, without the factor of enhancement.

One of the factors of the enhancement is the barrier, arised due to the overlapping of the dencities of the two interacting nuclei. The Josephson current might pass flow across this barrier. Differential equations, describing this process could be written in such form:

$$i\hbar\frac{\partial\psi_1}{\partial t} = H_0\psi_1 + C(\psi_1^2 + \psi_2^2)\psi_2$$
$$i\hbar\frac{\partial\psi_2}{\partial t} = H_0\psi_2 + C(\psi_1^2 + \psi_2^2)\psi_1 \tag{9}$$

where the indexes 1,2 correspond to the two nuclei.

Summing of these two equations yields the standard form of the NoSE.

$$i\hbar\frac{\partial\psi}{\partial t} = H_0\psi + C \mid \psi \mid^2 \psi \tag{10}$$

where $\psi = \psi_1 + \psi_2$. To get the Josephson current it is necessary to match the solutions inside and outside the barrier. Such procedure has been done for the superfluity in the work [8].

The cross section of the cluster transfer reaction was given in the work [7]:

$$\frac{d\sigma(\theta)}{d\Omega} =< \sigma_{el}(\theta) > P_t(\theta)F(Q,L) \tag{11}$$

The factorization was done in the quantities $< \sigma_{el}(\theta) >$ - elastic scattering, P_t - reaction (transfer) probability and $F(Q,L)$ - a matching factor normalized to the unit for well matched transitions. Here we want to suggest another way of getting of the cross section of multi nucleon transfer reaction. We shall follow Mitropolsky and Shuvaev [10]. In their approach the wave function $\psi(X,t)$ is a solution of the following Schroedinger equation:

$$i\hbar\frac{\partial}{\partial t}\psi(x,t) = \left(-\frac{\hbar^2}{2m}\frac{\partial^2}{\partial x^2} + U(x,t)\right)\psi(x,t) \tag{12}$$

$$\text{where} \quad U(x,t) = qL\int_{-\infty}^{\infty} dx\frac{\partial}{\partial t}j(x,t) \tag{13}$$

$$\text{with} j(x,t) = \frac{i\hbar}{2m}q\left(\psi\frac{\partial}{\partial x}\psi^*(x,t) - \psi^*\frac{\partial}{\partial x}\psi(x,t)\right). \tag{14}$$

Here arises an analogy between the passing a current through the coil with some induction. Let's consider the solution in the form of the running wave

$$\psi(x,t) = \exp\left(\frac{i}{\hbar}(mvx - (mv^2/2 - \omega)t)\right)\psi(x,t) \tag{15}$$

where the function solution of the NoSE is:

$$-\frac{\hbar^2}{2m}\phi'' - g^2Lv^2\phi^3 = -\omega\phi \tag{16}$$

The normalization of the wave function determines the parameter ω:

$$\omega = \hbar^2/2m\Delta^2, \qquad \Delta = 2\hbar^2/mv^2q^2L.$$

The soliton, as a solution of (16) is:

$$\phi = \frac{1}{\sqrt{2\Delta}}ch^{-1}((x - vt)/\Delta) \tag{17}$$

To get the cross section for the cluster transfer reaction we can use the same procedure as in Coulomb excitation theory in the first order of perturbation

$$d\sigma = Pd\sigma_R, \qquad P = \frac{1}{u_{i+1}} \mid b_{\text{if}} \mid^2 \tag{18}$$

where $b_{if} = \frac{1}{\hbar}\int_{-\infty}^{\infty} dt < f \mid u(x,t) \mid i > \exp i\omega t$.

Another problem which could be solved with the help of the NoSE it is a problem of the nucleation of clusters in the surface region of nucleus. Patashinskii and Shumilo [10] developed the theory of the nucleation based on the parameter of the order $\phi(x,t)$ with well known Hamiltonian:

$$H(\phi) = \frac{1}{2}\int_{-\infty}^{\infty}(C(\nabla\phi)^2 + \mu\phi^2 + \frac{g}{2}\phi^4 - 2\hbar\phi)dx \tag{19}$$

They got the NoSE for the function $\psi(x,t) = \phi(x,t)/\phi_s$:

$$\frac{\partial\psi}{\partial t} = \frac{\partial^2\psi}{\partial x^2} + \frac{n-1}{x}\frac{\partial\psi}{\partial x} + 2(\psi - \psi_s)^3 \tag{20}$$

with the solution

$$\psi(x,t) = -\text{ th}(x - x_0), x_0 = \text{ const}, \phi_s = (M/q)^{1/2} \tag{21}$$

Such solution was obtained by us [11] at the consideration of the problem of the hard core, connected with the Pauli principle. Besides NOSE there is another interesting equation, which was applied many times for the prediction of new forms of the motion in nuclei [12]. This is the KdV equation. KdV equation, was applied many times in nuclear physics [12-17].

In the work [12] the connection was established between the solution of the KdVE and the nuclear potential in Schroedinger equation. At the certain circumstances we can see the evolution of the nuclear potential with time.

G.N.Fowler, S.Raha and R.M.Weiner [15] attempted to predict the localized stable surface excitations in peripheral heavy ion collisions like in Rossby waves.

In the work [16] A.Sandulesky and W.Greiner obtained a new coexistence model consisting of the usual shell model and a cluster model, describing a soliton moving on the nuclear surface. They got the excellent description of experimental spectroscopic factors for cluster decays.

In its turn we considered the perturbation of the nuclear surface as a quadrupole deformation of the nuclear potential:

$$U = U_0 \mathrm{sech}^2((r-R)/a) = 2aU_0 \frac{df(2(r-R)/a)}{dr} \quad (22)$$

The evolution of this perturbation having the soliton form with time is given by the KdV equation [18]:

$$U_t - 6UU_x + \beta U_{XXX} = 0, \quad (23)$$

where the function U is the potential of Schroedinger equation

$$\beta\psi_{xx} - U\psi = 0, \qquad \beta = \hbar^2/2m \quad (24)$$

On the other hand this perturbation in the form of soliton can decay on the certain quantity of solitons with the small amplitude [29]. If we fix the KdVE in the form

$$U_t + UU_x + \beta U_{XXX} = 0, \quad (25)$$

its solution will be in the form:

$$U(x) = U_0 \mathrm{sch}^2(U_0/12\beta)^{1/2}X) \quad (26)$$

Now we apply the similarity principle. The initial condition we take in the form:

$$U(x, 0) = U_0\phi(x, l). \quad (27)$$

With new variables

$$\eta = U/U_0, \quad x = X/l, \quad \tau = U_0 t/l, \quad (28)$$

we obtain the new KdVE:

$$\eta_t + \eta\eta_x + \sigma^{-2}\eta_{XXX} = 0, \quad (29)$$

where $\sigma = l(U_0/\beta)^{1/2}$.

In the work [19] the critical value $\sigma = \sqrt{12}$ was obtained. With increasing number σ the corresponding perturbation breaks up into a larger number of solitons. The number of this solitons was determined by Karpman [20]

$$N = 2a\sqrt{U_0 6\beta}. \quad (30)$$

For the scattering of α -particles with kinetic energies in the range 4 and 12 MeV we can get for the nucleus ^{12}C the value $N \sim 5$. It corresponds to the number of existing states generated by the surface potential well: $K^+ = 0, 2, 4, 6, 8$. Gardner et al. [21] showed, that the amplitude of solitons determined by eigen values: $A = 2E_n$.

The potential having the soliton shape gives the following spectrum [22]:

$$E_n = -\frac{\hbar^2}{8m\Delta}\left(-(1-2n) + (1 + 8m\Delta U_0\hbar^{-2})^{1/2}\right). \quad (31)$$

For our example ($n = 0, 2, 4, 6, 8$) we can get the following values (MeV): $-15.57, -8.65, -1.75, 5.19, 12.11$, respectively. The smaller eigen value the lager soliton amplitude. Knowing the eigen values one can construct the total potential [23].

Thus we interpreted two equations NoSE and KdVE respectively to the problem of clustering. This work is only the first approximation to this problem because we used only one dimension. To our mind the next step should be the application of Kadomtsev-Petviashvily equation to this problem.

The most promising thing is the application of the Similarity principle to the linear alpha-chain states in nuclei [24].

References

1. Gridnev, K.A., Hefter, E.F., Mikulas, K., Semjonov, V.M., Subbotin, V.B.: Aust. J. Phys. **36**, 155 (1983); Nuovo Cimento **93**, 135 (1986)
2. Gridnev, K.A., Darwish, N., Ivanov, A.G., Subbotin, V.B., Fadeev, S.N., Heffer, E.F.: The notes of Ac.Sci of USSR (physics) **50**, 1980(1986)
3. Bialynicki-Birula, I., Mycielski, J.: Ann. Phys. **100**, 62 (1976)
4. Gaehler, R., Klein, A.G., Zeilinger, A.: Phys. Rev. **A23** 1611 (1981)
5. Bracher, J.D.: Int. J. Theor. Phys. **30**, 979 (1991)
6. Shannon, C.E., Weaver, W.: The Mathematical Theory of Communication, University of Illinois Press, Urbana, Illinois 1979
7. Oertzen, W.: Workshop on Nuclear Structure and Heavy-ion Reaction Dynamics, Preprint HMI-P **6/87**, 71 (1990)
8. Jacobson, D.A.: Phys. Rev. **A138**, 1066 (1965)
9. Mitropolsky, I.A., Shuvaev, A.G.: J. Theor. Phys. **61**, 1 (1991)
10. Patashinskii, A. Z., Shumilo, B.I.: The Theory of Nucleation Near the First Order Phase Transition Line, Preprint INP **N79-101**, Novosibirsk, 1 (1979)
11. Gridnev, K.A.: Clustering Phenomena in Atoms and Nuclei Inter Cout on Nuclear and Atomic Clusters, Turku, Finland, June 3-7 Berlin Heidelberg New York: Springer 1992
12. Gridnev, K.A., Hefter, E.F.: Phys. Lett. **A77**, 490 (1980)
13. Hefter, E.F., Mitropolsky, I.A.: J. Moscow Phys. Soe, **99** (1991)
14. Hefter, E.F., Gridnev, K.A.: Prog. Theor. Phys. **77**, 549 (1984)
15. Fowler, G.N., Raha, S., Weiner, R.M.: Phys. Rev. **C31**, 1515 (1985)
16. Sandulesku, A.,Greiner: Rep. Prog. Phys. **55**, 1423 (1992)
17. Kartavenko, V.G.: Sov. J. Nucl. Phys. **40**, 337 (1984)
18. Miura, R.M.: SIAM Review **18**, 412 (1976)
19. Berezin, Ju.A., Karpman, V.I.: Soviet Physics JETP **24**, 1045 (1967)
20. Karpman, V.I.: Phys. Lett. **A25** 708 (1967)
21. Gardner, C.S., Green, J.M., Kruskal, M.D., Miura, R.: Phys. Rev. Lett. **19**, 1095 (1967)
22. Landau, L.D., Lifshiz, I.M.: Quantum Mechanics, 97 (1974)
23. Gridnev, K.A., Semtchenkov, A.G.: The Theory of the Quantum Systems with the Strong Interaction, Tver 45 (1992)
24. Ikeda, R., Takigawa, N., Hariuchi, H: Suppl. Prog. Theor. Phys., Extra Number 464 (1968)

This article was processed using Springer-Verlag TEX Z.Physik A macro package 1.0 and the AMS fonts, developed by the American Mathematical Society.

Spectroscopy of microclusters in superfluid helium

Jaap H. M. Beijersbergen, Qin Hui, and Michio Takami

The Institute of Physical and Chemical Research (RIKEN) Wako, Saitama 351-01, Japan

Received 28 June 1993

Abstract. The optical spectra of microclusters in superfluid helium (HeII) are studied by a dissociation/ emission spectroscopy. The microclusters are produced by ablating solid samples immersed in HeII with a pulsed YAG laser. The absorption spectra are measured by dissociating the clusters with a pulsed dye laser and observing the recombination lines of atomic ions. Narrow excitation spectra observed in Ca and Sr are attributed to the absorption by the van der Waals clusters of these atoms.

PACS: 36.40.+d

---

1. Introduction

The spectroscopic study of clusters in superfluid helium (HeII) is of particular interest by the following reasons. First, all internal freedom of motions in the clusters are equilibrated with a well-defined liquid temperature around 1.5K. At this temperature clusters are distributed only in a small numbers of low lying vibrational levels. The slow mobility of impurity particles in HeII provides us with an opportunity to confine the clusters in a limited volume for a considerable lenght of time. Liquid He is transparent over a wide range of photon energy, from microwave to UV region, and has the weakest interaction with dissolved impurities which allows the study of clusters in the environment close to free space. Recently Yabuzaki and his collaborators reported a method to disperse atoms and molecules by laser ablation of solid samples immersed in HeII [1]. The importance of this technique is that the laser ablation produces small particles of almost any solid sample materials in HeII. We studied the optical absorption spectra of clusters by dissociating them with a pulsed dye laser and detecting atomic emission lines. The present paper reports our preliminary experimental results on the small Ca and Sr van der Waals clusters.

2. Experimental method

Superfluid helium was produced in a pyrex dewar by reducing the saturated vapor pressure of helium above the liquid. A pulsed YAG laser (10-20 Hz, 533 or 355 nm, 50-100 mJ/pulse) was focused onto a sample in HeII with a quartz lens of 1 m focal lenght. A second pulsed YAG laser dissociated the particles into atoms and small clusters. A pulsed dye laser excited atoms and molecules with 8 μs delay time to observe laser induced fluorescence. The emission along the laser beam was focused on the slit of a 50 cm monochromator and detected with a photomultiplier. An overview of the experimental arrangement is shown in Fig. 1. When necessary, the photomultiplier was gated to the dye laser pulse in order to avoid exposure of the photomultiplier to the strong scattered light from the YAG laser beam. The fluoresence signal was averaged with a Boxcar integrator.

3. Experimental results and discussion

We first observed the laser induced fluoresence excited from the lowest singlet and triplet states of Ca, Sr and Ba atoms in HeII. An unusual feature of the atomic transitions in HeII is that the absorption lines exist on the blue side of the emission lines. Thus the atomic lines were measured both by scanning monochromator while the laser wavelenght was fixed (emission spectrum) and by scanning dye laser while the monochromator was fixed (excitation spectrum). All the emission spectra were observed very close to the free atomic lines with approximately 1 nm line width. On the other hand, the excitation spectra were very broad and showed large blue shift from the free atomic lines. These features are similar to those reported previously [1,2]. According to the bubble model [3,4] which explains the general feature of atomic transitions in HeII semi-quantitatively, a neutral atom is trapped in a bubble-like cavity. The large blue shift and width are strongly correlated with the change in the bubble size induced by the increase

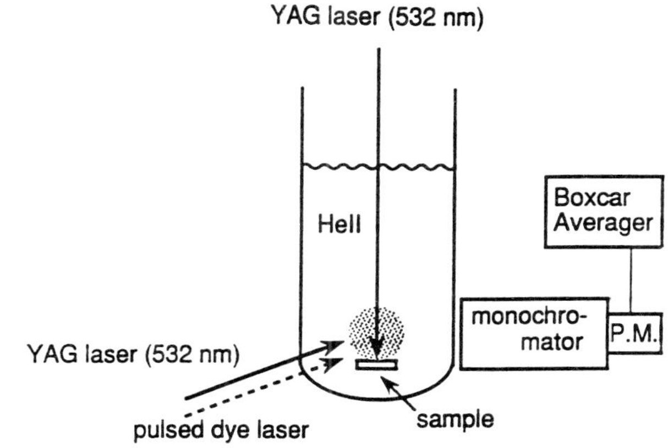

Fig. 1. The experimental arrangement for the spectroscopy of atoms, molecules, and clusters in HeII.

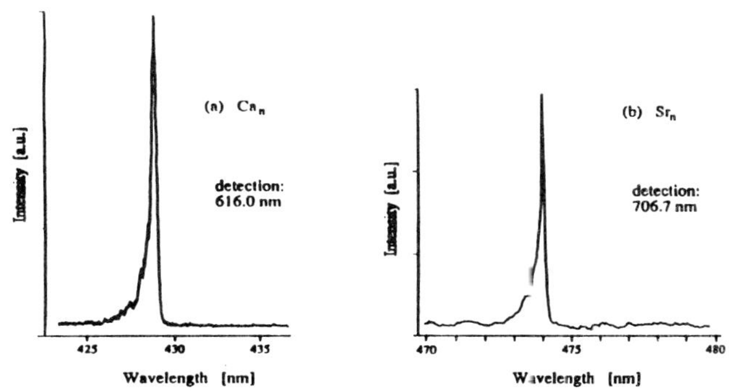

Fig. 2. A narrow excitation spectrum observed for Ca clusters (a) and for Sr clusters (b). The monochromator was set to the strongest component of the lowest of the lowest triplet transitions for Ca and to the singlet transition for Sr.

in the mean radius of excited electron orbit [5]. The observation of strong excitation spectra from the lowest metastable triplet state implies that the dissociation of small particles is associated with the production of atomic ions which recombine with electrons and relax to the lowest triplet as well as the singlet states by emitting photons.

The dissociation/emission spectra of clusters were first observed during an extensive survey of higher excited levels of Ca. The spectrum was found at 428.7 nm, slightly below the singlet transition $4s5p\ ^1P_1 \leftarrow 4s^2\ ^1S_0$ at 423.6 nm, while the monochromator was set to the strongest component of the lowest triplet transitions at 616.0 nm. This excitation spectrum (Fig. 2a) has characteristics entirely different from those of atomic spectra: much narrower width than those of atomic spectra and the emission both in the lowest singlet and triplet transitions. An extensive search for similar spectra in Sr and Ba revealed a very sharp line for Sr at 474.3 nm (Fig. 2b), again slightly below the singlet atomic transition, but not in Ba so far.

The simultaneous observation of the lowest singlet and triplet transitions suggests that the excitation is accompanied with an ionization process. The narrow width implies the existence of resonance in the excitation process. However, there is no corresponding atomic level in the singlet manifold. In the triplet manifold, there are some levels close to the observed transition energy, but the excitation of these levels is unlikely because they are already in the ionization continuum due to the reduction of ionization energy by about 1.8 eV in HeII [6]. Thus we tentatively attributed the spectra to the absorption by atomic clusters.

The small clusters of alkaline earth atoms are known to be van der Waals clusters [7]. Because the excitation spectra were observed only when the dissociation laser was present, we believe the spectra are originated from small vdW clusters. Suppose we excite an atom in Sr_n to the lowest singlet excited state (5s5p). This state has n-fold degeneracy corresponding to n different atoms though the degeneracy will be partially removed by a geometrical factor. The excited state will locate slightly below the atomic 5s5p state because the aggregation energy will increase by the excitation of the atom. This is consistent with the slight red shift observed in our measurement. The second excited state, in which another atom is excited, has the n(n-1)/2-fold degeneracy with about twice the energy of the first excited state. Thus there will exist a series of excited states with almost equal energy spacing. This allows a resonant multistep excitation of the cluster, providing the internal energy high enough to release ions into HeII. Although the above model seems consistent with the experimental observation, there still remains uncertainty due primarily to the lack of size selectivity in the present measurement. The study of these clusters in free space is highly desirable for confirmation of the model. Application of this technique to other systems as well as the extension of the study to the MW and IR region are in progress.

We are grateful to Professor T. Yabuzaki of Kyoto University for valuable discussion on the present work. One of us (JHMB) is grateful to Science and Technology Agency for STA Fellowship.

References

1 Yabuzaki, T., et al.: Atomic Physics **13**, ed. Walther, H., Hänsch, Th.H., and Neizert, B. (AIP New York, 1992) p.337; Fujisaki, A., et al.: Phys. Rev. Lett. **71**, (1993) 1039.
2 Bauer, H., et al.: Pys. Lett. A **146**, 134 (1990)
3 Cole, M.W., and Bachman, R.A.: Phys. Rev. B **15**, 1388 (1977)
4 Reycher, H.J. et al.: Phys. Lett. A **115**, 238 (1986)
5 Beijersbergen, J.H.M., Hui, Q., and Takami, M.: Phys. Lett. A, in press
6 Beau, M., et al.: XIV International Conference on Coherent and Nonlinear Optics, September 1991, St. Petersburg, Russia
7 See, for example, Pacchioni, G., and Koutecky, J.: Chem. Phys. **71**, 181 (1982)

This article was processed using Springer-Verlag TEX Z.Physik D macro package 1.0 and the AMS fonts, developed by the American Mathematical Society.

Z. Phys. D 31, 253–258 (1994)

Effects of the cluster surface on the electronic shell structure: faceting, roughness and softness

J. Mansikka-aho, M. Manninen, E. Hammarén

Department of Physics, University of Jyväskylä, P.O. Box 35, FIN-40351 Jyväskylä, Finland

Received: 28 June 1993 / Final version: 22 April 1994

Abstract. Several simple models have been used to study the effects of the surface on the electronic shell structure in metal clusters. The main results are as follows: The icosahedral clusters have the same electronic shell structure as the sphere up to about 1000 atoms. The surface roughness causes the distribution of the level spacings to be a Wigner distribution. By varying the softness of the potential we can obtain potentials where the simplest classical orbits are the 'five-point star' or even the 'three-point star'.

PACS: 36.40

1. Introduction

In the case of a spherical potential the separation of radial and angular coordinates leads to a shell structure, which is determined solely by the radial dependence of the potential. The characteristics of the shell structure are determined by the 'accidental degeneracies' of the eigenvalues with different radial and angular quantum numbers, and can be understood with a semiclassical analysis of the closed classical orbits [1, 2]. In spherical metal clusters the valence electrons are confined by the nearly spherical surface of the cluster. A rough estimate of the confining potential is a spherical square well with the depth $V_0 = \phi + \varepsilon_F$, where ϕ is the work function and ε_F the Fermi energy of the bulk metal. This model explains the observed shell structure, i.e. the smallest magic numbers, for alkali metal clusters. However, in a more detailed analysis of the electronic structure the effects of the discrete atoms has to be taken into account. This has two consequences: The potential is not homogeneous inside the metal and the surface is not smooth. It is known that for alkali metals the band structure calculations give a nearly spherical Fermi surface and free-electron-like bands [3]. Consequently the effect of the inhomogeneity of the potential inside the cluster has only a vanishingly small effect on the shell structure [4].

The surface, on the other hand, affects on the shell structure drastically. In nonmagic clusters the optimal shape of the surface is related to the shell filling, leading to spheroidal shapes according to the Nilsson model [5, 6], or to even more complicated shapes if a more general deformation is allowed [7, 8]. In large clusters the faceting of the cluster and the surface roughness disturb the shell structure. The shell structure is also sensitive to the r-dependence of the potential. The softness of the surface potential then plays a dominant role: It determines the closed classical orbits of the valence electrons. In the present paper we will study separately the effects of the surface faceting, roughness and softness on the electronic shell structure.

Martin et al. [9] has observed that the set of magic numbers for large simple metal clusters correspond to the geometrical packing of atoms in an icosahedral structure. We have studied in earlier papers [10, 11] the effect of the surface faceting on the electronic shell structure. We found that the icosahedral symmetry can keep up the shell structure even if the cluster has hundreds of atoms. This gives a possibility of simultaneous existence of the icosahedral structure and the shell structure of an essentially spherical potential. A similar conclusion has been obtained by Clemenger [12] and by Pavloff and Greagh [13]. Maiti and Falicov [14], Stampfli and Bennemann [15] and Glossman et al. [16] have also studied the grossover from the electronic shell structure to the geometric shell structure, but in these studies the effect of the nonspherical geometry on the shell structure is not taken into account.

A nearly spherical surface consisting of individual atoms is always rough in the sense that it consists of several steps on the surface. The exact surface structure

This paper was originally submitted in connection with the 2nd. Int. Conference on Atomic and Nuclear Clusters held in Santorini from 28. June–2. July 1993 and is published here as a regular article after an independent refereeing procedure according to the standards of Z. Phys. D

varies from cluster to cluster leading to different energy levels in different clusters. Since many average electronic properties depend on the density of states at the Fermi level, the statistics of the level spacing becomes important [17–20]. We have used a lattice gas Monte Carlo method to simulate metallic fcc clusters at finite temperatures [21].

The self-consistent jellium model describes well the observed supershell structure in alkali metal clusters [22]. However, it is more illustrative to use the semiclassical interpretation where the supershell structure is a result from the interference of the classical triangular and square orbits [2]. Recent experiments of Lerme et al. [23, 24] opened the question whether the star orbits could dictate the shell structure in some metal clusters. Although the results for aluminium can be explained better with geometrical packing [25] than with electronic shells, it is important to study more generally the possibility and the consequences of star-orbits.

2. Surface faceting

In the case of alkali metals the jellium model describes well the observed magic numbers. In this model the valence electrons move in a spherically symmetric effective potential. For large clusters the potential resembles a square well with a rounded edge [2, 26]. A reasonable approximation for the effective potential is then a finite potential well with a depth determined by a sum of the work function and the Fermi energy. For spherical clusters, the radius is determined by $r_0 = N^{1/3} r_s$, where N is the number of atoms in the cluster and r_s the Wigner-Seitz radius. For nonspherical clusters the well is assumed to have the same depth and volume as in the spherical cluster. This is the approach taken in [11]. In an earlier work we assumed the electrons to move in an infinite potential well and approximated the level structure using the Hückel model [10].

The finite potential well can be chosen to have the wanted geometry by adding a perturbation potential, which is defined as the difference between the potentials of spherical and nonspherical clusters. We can see from Fig. 1 that the difference between the spherical and nonspherical potentials is restricted within a narrow region close to the cluster surface. This allows us to use effectively the solutions of the spherical potential as basis functions in solving the nonspherical problem [11].

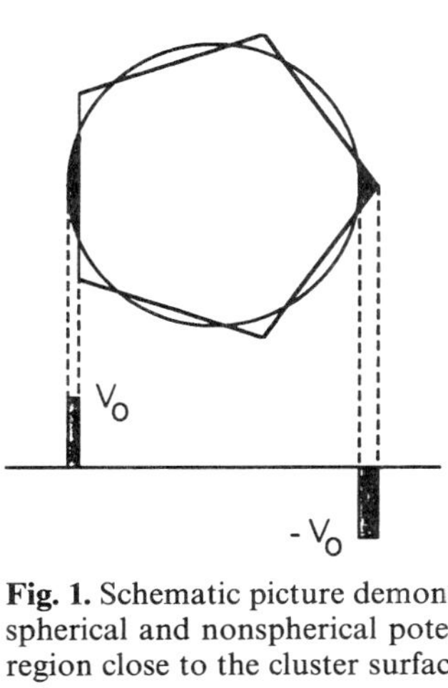

Fig. 1. Schematic picture demonstrating that the difference between spherical and nonspherical potentials is restricted within a narrow region close to the cluster surface

Figure 2 shows the densities of states for the 309-atom icosahedral and spherical sodium clusters. The discrete levels have been convoluted with a Lorentzian. The shell structure of the icosahedral cluster is similar to that of a sphere. However, we notice that in detals the subshell behaviour in the two cases is different. It should be noted that the grouping of the levels of different angular momentum values is sensitive to the details of the spherical potential. Therefore, it is not surprising that when the spherical potential is replaced with an icosahedral potential, the exact magic numbers can change even if the shell structure is still very similar to that of a sphere up to about 1000 atoms in the cluster [11]. This result is in agreement with the independent findings by Clemenger [12] and by Pavloff and Greagh [13].

3. Surface roughness

The effect of the surface roughness is different for each individual cluster (at a given time), but the experiments measure only average properties over a large cluster ensemble. This calls for a statistical analysis of the level spectrum [27, 28]. In the atomic clusters the distribution of the single particle levels in the immediate vicinity of the Fermi level is important in determining the thermodynamical properties. In order to be able to compute the level structure of a large number of clusters, a simple model of lattice gas with a tight binding electronic struc-

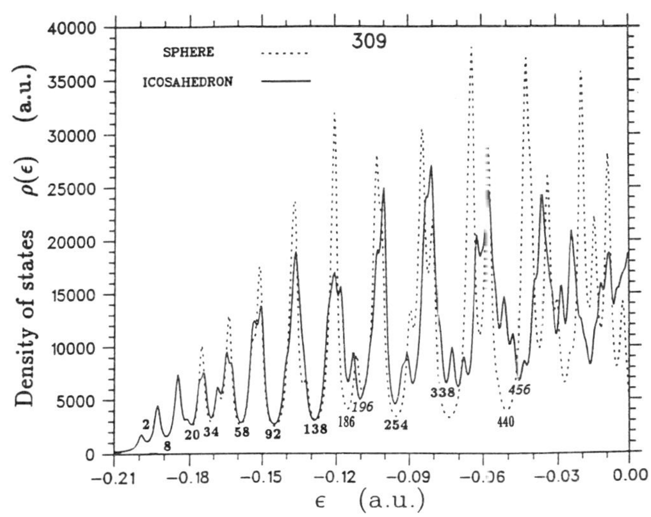

Fig. 2. Density of states of the 309-atom sodium cluster calculated for a spherical (*dashed line*) and icosahedral (*uniform line*) square well. The *numbers below the energy gaps* indicate the corresponding magic numbers. *Italic numbers* denote the magic numbers for the icosahedral potential. A *boldface number* is used when the spherical and icosahedral potential give the same magic number

ture is used. The lattice gas model has been successfully used to study the roughening of plane surfaces of solids [29].

The simplest model which takes into account the atomic positions is the Hückel model. It has been shown that close to the bottom of the band the energy levels calculated with the Hückel model are identical to those of the free electron model [10, 30]. However, it is useful to include p-electrons in addition to s-electron for each atom, because the simple Hückel model strongly overestimates the density of states close to the Fermi energy [10, 20, 21].

In our model [21] the overlap integrals are neglected and only the nearest neighbour resonance integrals are assumed to be nonzero. The six resonance integrals have been chosen by fitting the calculated energy bands to the free electron energy bands below the Fermi energy. In the simple Hückel model the matrix elements are 1 or 0 depending whether the corresponding atoms are nearest neighbours or not, respectively. The nearest neighbour matrix elements can have seven different values.

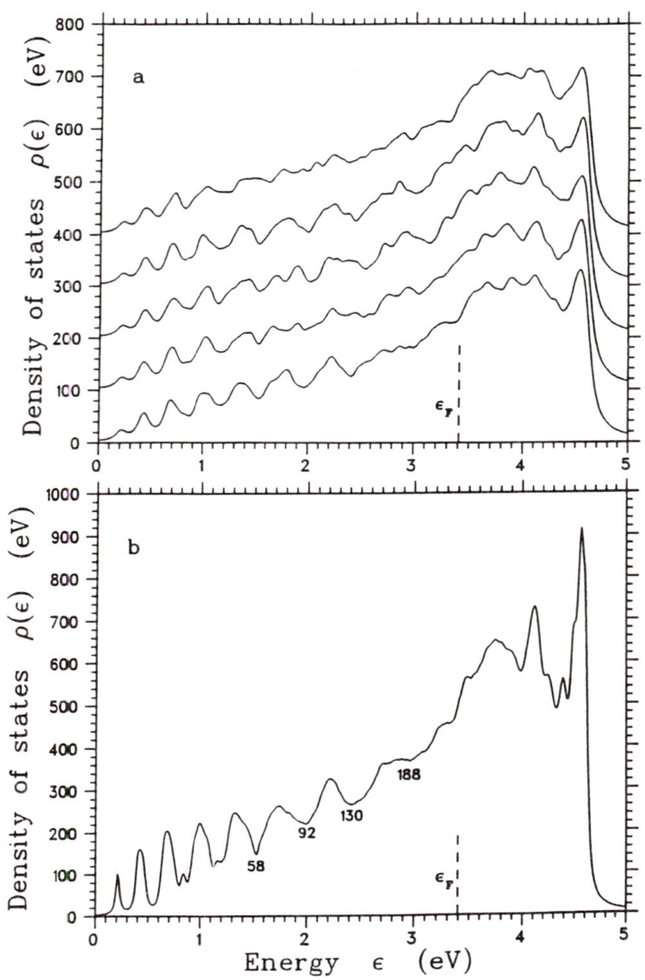

Fig. 3. a Density of states for five random clusters. **b** The average density of states for the ensemble of 970 clusters of 249 atoms

In large clusters the electronic shell structure is clearly seen close to the bottom of the band, but it is not visible all the way up to the Fermi energy. The reason for this is the roughness of the cluster surface. Even the most spherical fcc clusters seem to have so rough surfaces that the energy gaps related to the shell structure disappear at the Fermi level. This gives more confidence to the suggestions that the large clusters, where the shell structure has been observed, do not have the lattice structure of the bulk metal.

For the statistical analysis of the level structure we used the Monte Carlo method to create an ensemble of clusters with a fixed number of atoms and a given temperature. The starting geometry was chosen to be a sphere and the atoms were bound to be in the fcc lattice sites. At each Monte Carlo step one atom was moved from the cluster into an empty lattice site. The number of atoms was kept fixed by forbidding evaporation of atoms.

Figure 3a shows the density of states for 5 clusters chosen arbitrarily from the ensemble of 970 clusters at 800 K. Each cluster with a different surface shape has a different electronic structure, the main differences being at the Fermi level and at high energies. In the low energy part of the spectrum the density of states is very similar in each cluster. Figure 3b shows the average density of states for the whole ensemble of 970 clusters. The average shell structure shows clear minima at the free electron shell closings 92, 138 and 186 (the actual minima indicated at the figure are 92, 130 and 188). It is important to notice that the average density of states shows the shell structure even if all the individual clusters do not.

At a finite temperature the electronic properties of the clusters depend on the distribution of energy levels at the Fermi energy. To this end we determined the level spacing distribution by taking three level spacings below and three above the Fermi level [21]. From the 970 clusters we obtained about 6000 level spacings which was enough to determine the distribution. Figure 4

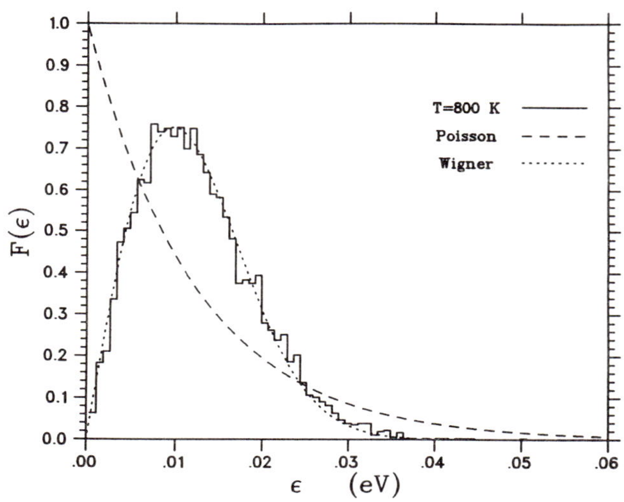

Fig. 4. Level spacing distribution for the 249 atom clusters generated using the Monte Carlo method at 800 K. The Poisson and Wigner distributions are shown for comparison

shows the level spacing distribution for an 249-atom cluster. The Poisson distribution (the dashed lines) and the Wigner distribution (the dotted lines) have been drawn for comparison. It is clear that the distribution fits to the Wigner distribution. This is also the result of earlier computations based on the Hückel model [18–20]. Ratcliff [17] has studied the effect of the surface roughness using a perturbation theory for a hard wall spherical potential box. Also his result is in accordance with a Wigner distribution at large energy eigenvalues. The increase of the temperature roughens the surface and makes the average level spacing narrower [21].

4. Potential softness

In the previous chapters the potential at the cluster surface has been approximated either by a step or by an infinite wall. A more realistic model is a softer potential. The effect of the potential softness have been studied recently by Nishioka et al. [2], Clemenger [12], and by Lerme et al. [31]. We have studied the effect of the potential softness by using two different potentials. First, we introduce a new potential

$$V(r) = -\frac{V_0}{1+(r/r_0)^{\beta}}, \qquad (1)$$

where V_0 is the depth of the potential well, r_0 the radius of the potential and β a parameter determining the potential softness. For making a connection to real metals we choose V_0 to be the sum of the Fermi energy and the work function. The radius of the potential is chosen to be $r_0 = r_s N^{1/3}$, r_s being the electron density parameter and N the number of electrons in the cluster. The calculations have been made for a potential which depth and the radius correspond to Al cluster with 1000 valence electrons.

Potential (1) has a long range for small values of β. For $\beta = 1$ it approaches the Coulomb potential at large distances. On the other hand, for small distances it is

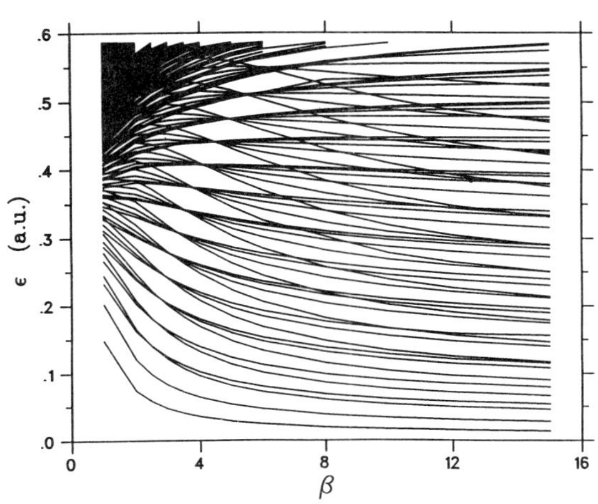

Fig. 5. Energy eigenvalues as a function of β for the potential in (1) corresponding to 1000 electron Al cluster

harmonic for $\beta = 2$. Figure 5 shows the energy levels as functions of the softness parameter β. For small β-values high-energy levels correspond to the long-range potential. When the β is increased the levels group together forming the harmonic-oscillator shells. With further in-

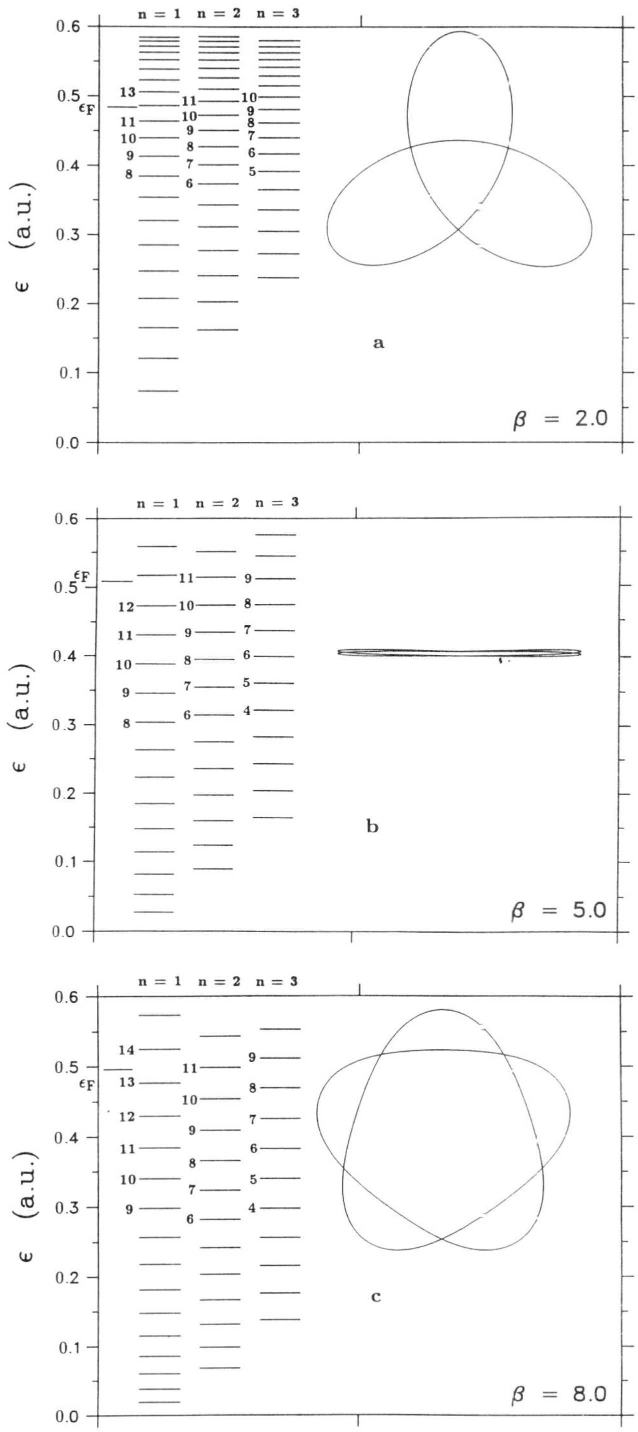

Fig. 6a–c. Electron energy levels for three different values of β. The radius and the depth of the potential correspond to a 1000 electron Al cluster. The shortest classical orbits at the Fermi energy are shown in each case

crease of the β-value the degenerate levels split and form first a shell structure typical of the soft potential well and eventually ($\beta > 15$) the hard-wall potential.

In order to investigate the level structures in detail, levels at three typical β-values are shown in Fig. 6 in separate columns for different n-values. We have derived numerically the classical closed orbits for different values of β. For $\beta = 2.0$ (Fig. 6a) the degenerate levels have the ratios $\Delta n : \Delta l = 2:3$ near the Fermi level. An increase of β moves the levels with $n=1$ down and those with $n=3$ up leaving the levels with $n=2$ nearly intact. The value $\beta = 2.0$ gives a propeller-type orbit ('three-point star'). For $\beta = 5.0$ (Fig. 6b) degenerate levels have the ratios $\Delta n : \Delta l = 1:2$, thus the harmonic-oscillator limit. In a harmonic potential all elliptic orbits are allowed. However, the value $\beta = 5.0$ gives only very narrow ellipse-type orbits, which are numerically very difficult to separate from "open" orbits where the ellipse rotates around the center. Figure 6b shows a borderline case where the ellipse has about the maximum width and has already started to rotate (two circles around the center are shown). At $\beta = 8.0$ (Fig. 6c) the levels of $n=3$ are nearly degenerate to those of $n=1$ with the angular-momentum difference $\Delta l = 5$, which is a signature of the classical five-point star orbit [1, 32]. The simplest nontrivial classical orbit is indeed the five-point star.

The second potential we have studied [32] is the Woods-Saxon potential

$$V_{WS}(r) = -\frac{V_0}{1+e^{\alpha(r-r_0)}} . \tag{2}$$

Figure 7 shows electronic level energies as functions of the softness parameter α. For small α-values levels group together forming the harmonic-oscillator shells. When the α is increased the degenerate levels split and levels with larger l values come down. With further increase of α levels from neighbouring harmonic-oscillator shells cross with each other, and eventually form a shell structure typical to the hard-wall potential. At an intermediate value of ($\alpha = 0.4$) the ratio of the quantum-number differences is $\Delta n : \Delta l = 2:5$, which corresponds to the classical five-point star orbit.

At the moment the density of states cannot be investigated experimentally expect for the Fermi energy region. The shell structure of the Al clusters have been observed in the mass spectrum after near-threshold ionization of the cluster beam [23, 24], thus only at the Fermi level. Therefore, the energy difference between the lowest unoccupied and the highest occupied levels is calculated for each N:

$$\Delta\varepsilon(N) = \varepsilon_{N+1} - \varepsilon_N . \tag{3}$$

To make the calculation simple the potential is fixed with the radius parameter at $N = 1000$, and electrons are put in up to each size of the cluster. The $\Delta\varepsilon(N)$ have been plotted as a function of $N^{1/3}$ in Fig. 8 for $\alpha = 0.4$. When the $\Delta\varepsilon(N)$ is large the ionization potential is expected to be large, and therefore the magic numbers should be observed. For $\alpha = 0.4$ each peak is divided into two peaks expect for the first three. It is important to note that the peaks are not equally spaced in the $\alpha = 0.4$ case. Consequently, if these two shells could be separated in experiments, the peaks (or minima, depending on the technique) would be not equally spaced as a function of $N^{1/3}$.

We have also studied the classical orbits in the case of the Woods-Saxon potential [32]. In large clusters the triangular orbit becomes dominating even at $\alpha = 0.4$. The signature of the five-point star orbit is seen, therefore, only in a limited size range.

We have used the self-consistent jellium model to study the shell structure in large Al clusters [32]. The degeneracy with the ratio 2:5 was observed in the level structure of 440 and 462 electron Al clusters only in a limited low-energy part of the spectrum and does not play any significant role in the shell structure. Related results have been provided recently ba a jellium-model calculation for Al cluster [22], which produced a similar shell structure as for alkali-metal clusters and no signature of the classical star-orbit was seen.

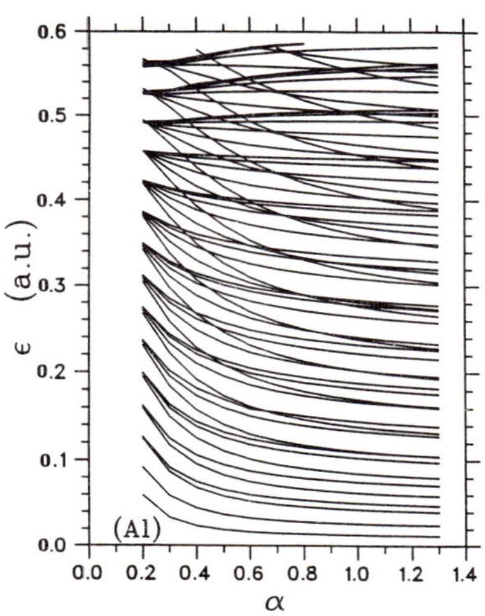

Fig. 7. Energy eigenvalues as a function of α for a Woods-Saxon potential corresponding to 1000 electron Al cluster

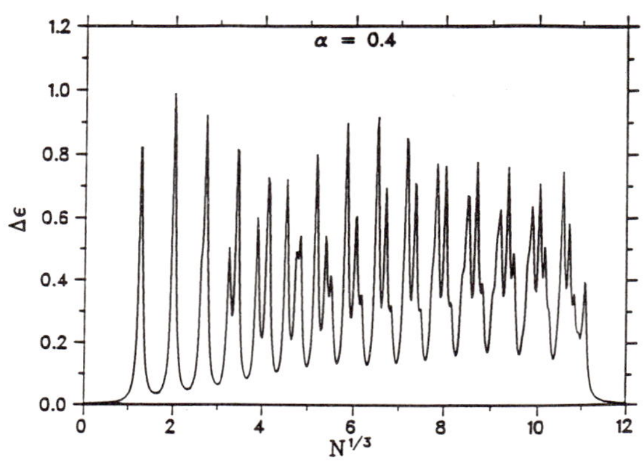

Fig. 8. The difference between the lowest unoccupied and the highest occupied energy level as a function of $N^{1/3}$, where N is the number of electrons in the cluster

5. Conclusions

We have studied the effects of the surface on the electronic shell structure of alkali clusters. The surface faceting of solid clusters destroys the shell structure when the number of atoms exceeds 1000. Up to now there is no experimental evidence of simultaneous icosahedral packing and electronic shell structure. The electronic shell structure is obtained in liquid clusters whereas solid clusters prefer icosahedral packing [33].

The Hückel model and a lattice gas was used to study the level spacing statistics in metallic clusters. At the Fermi level the level spacings have a Wigner distribution. It was observed that the fcc clusters have always so rough surface that the shell structure at the Fermi level is disturbed, even though just below the Fermi level it is clearly seen. An ensemble of clusters at a finite temperature can have an average density of states which shows clearly the shell structure although all of the individual clusters do not. The weakening of the shell structure due to the surface roughness might be an important effect in quantitative analysis of the experimental mass spectra.

The shell structure of large spherical clusters depends on the profile of the confining potential at the cluster surface. The soft potential gives rise to twice as many peaks as the hard potential. This can be explained semiclassically by inspecting the possible closed classical orbits in the potential wells. In the soft potential the most important classical orbit is the five-point star, which is about twice as long as the important orbits (triangle and square) of the hard potential. A soft potential with a long range can also have a 'three-point star' as a dominating classical orbit. Computations for real metals show that the potential is always too hard to make the star orbit important, and so far there is no experimental evidence of the importance of the star orbits.

The authors thank H. Nishioka and J. Suhonen for useful discussions. This work was supported by the Academy of Finland.

References

1. Bohr, A., Mottelson, B.R.: Nuclear structure, Vol. II. London: Benjamin 1975
2. Nishioka, H., Hansen, K., Mottelson, B.R.: Phys. Rev. B **42**, 9377 (1990)
3. Ashcroft, N.W., Mermin, N.D.: Solid state physics. New York: Holt, Reinhard and Winston 1976
4. Manninen, M., Jena, P.: Europhys. Lett. **14**, 643 (1991)
5. Clemenger, K.: Phys. Rev. B **32**, 1359 (1985)
6. Penzar, Z., Ekardt, W.: Z. Phys. D **19**, 109 (1990)
7. Hamamoto, I., Mottelson, B., Xie, H., Zhang, X.Z.: Z. Phys. D **21**, 163 (1991)
8. Frauendorf, S., Pashkevich, V.V.: Z. Phys. D **26**, S98 (1993)
9. Martin, T.P., Bergmann, T., Göhlich, H., Lange, T.: Chem. Phys. Lett. **176**, 343 (1991)
10. Mansikka-aho, J., Manninen, M., Hammarén, E.: Z. Phys. D **21**, 271 (1991)
11. Mansikka-aho, J., Hammarén, E., Manninen, M.: Phys. Rev. B **46**, 12649 (1992)
12. Clemenger, K.: Phys. Rev. B **44**, 12991 (1991)
13. Pavloff, N., Greagh, S.C.: Phys. Rev. B **48**, 18164 (1993)
14. Maiti, A., Falicov, L.M.: Phys. Rev. A **44**, 4442 (1991)
15. Stampfli, P., Bennemann, K.H.: Phys. Rev. Lett. **69**, 3471 (1992)
16. Glossman, M.D., Alonso, J.A., Iñigues, M.P.: Phys. Rev. B **47**, 4747 (1993)
17. Ratcliff, K.F.: Multivariable analysis V. Krishnaiah, P.R. (ed.) Amsterdam: North-Holland 1980
18. Tanaka, S., Sugano, S.: Phys. Rev. **34**, 740 (1986)
19. Bucher, J.P., Xia, P., Bloomfield, L.A.: Phys. Rev. B **42**, 10858 (1990)
20. Pavloff, N., Hansen, M.S.: Z. Phys. D **24**, 57 (1992)
21. Mansikka-aho, J., Manninen, M., Hammarén, E.: Phys. Rev. B **47**, 10675 (1993)
22. Genzken, O., Brack, M., Chabanat, E., Meyer, J.: (to be published)
23. Lerme, J., Pellarin, M., Vialle, J.L., Baguenard, B., Broyer, M.: Phys. Rev. Lett. **68**, 2818 (1992)
24. Pellarin, M., Lerme, J., Baguenard, B., Perez, A., Broyer, M., Vialle, J.L.: (submitted for publication)
25. Martin, T.P., Näher, U., Schaber, H.: Chem. Phys. Lett. **199**, 470 (1992)
26. Ekardt, W.: Phys. Rev. B **29**, 1558 (1984)
27. Kubo, R., Kawabata, A., Kobayashi, S.: Ann. Rev. Mater. Sci. **14**, 49 (1984)
28. Kimura, K.: Phase Transitions **24–26**, 493 (1990)
29. Häkkinen, H., Merikoski, J., Manninen, M., Timonen, J., Kaski, K.: Phys. Rev. Lett. **70**, 2451 (1993)
30. Manninen, M., Mansikka-aho, J., Hammarén, E.: Europhys. Lett. **15**, 423 (1991)
31. Lermé, J., Bordas, Ch., Pellarin, M., Baguenard, B., Vialle, J.L., Broyer, M.: Phys. Rev. B **48**, 12110 (1993)
32. Mansikka-aho, J., Manninen, M., Nishioka, H.: Phys. Rev. B **48**, 1837 (1993)
33. Martin, T.P., Näher, U., Schaber, H., Zimmermann, U.: J. Chem. Phys. **100**, 2322 (1994)

Free clusters of copper atoms : a study of structural size effects.

D. Reinhard, P. Berthoud, D. Ugarte, B.D. Hall[1] and R. Monot

Institut de Physique Expérimentale, EPFL, CH-1015 Lausanne, Switzerland
[1] Department of Physics & Biophysics, Massey University, Palmerston North, New Zealand

Received 28 June 1993

Abstract. The structures of unsupported ultrafine Cu particles produced by the inert-gas-aggregation technique, have been studied by 100 keV electron diffraction. Model diffraction patterns have been fitted to the data, using a simulated annealing scheme. The results show the coexistence of icosahedra and decahedra, below ~ 4 nm, and agree with recent molecular dynamics calculations. A slight expansion of the mean Cu-Cu distance is found.

PACS: 36.40.+d; 61.14.-x; 61.46.+w

The importance of structural studies in dependence of the sizes of atomic clusters has long been recognized by both theorists and experimentalists. Not only interesting in themselves, these topics are of relevance to several fields of condensed matter physics and chemistry : catalysis, nucleation and phase transitions, to name but a few examples. Structural size effects in small particles (e.g. < 10 nm in diameter) can be separated into two classes, concerned with changes in lattice parameter values on one hand and changes in symmetry on the other. Due mainly to their applications in catalysis, small metal clusters have been extensively studied, in regard of their structural properties. The most striking discoveries thus far are lattice contractions (several % in some cases [1, 2] and particles with fivefold symmetries, so called *Multiply-Twinned Particles*, or *MTP*. Both are found to be present in numerous metals, as well as other chemical species [3, 4]. However, many open questions remain, especially in the case of copper. While Cu lattice parameter changes have been investigated experimentally by several authors, with contradictory results [2, 5, 6], direct studies of Cu cluster symmetries [7] are very rare in the 1–10 nm range, due mainly to their high reactivity, compared to noble metals. High energy electron diffraction is the "classical" technique for both types of studies; however, the application of standard diffraction data interpretation schemes to small particle samples has been questioned by several authors [8]. The main problems arise because such methods (Scherrer's formula, indexed ring diameters) rely on an assumption of bulk-like crystallographic symmetries, which may very well be false at small sizes. EXAFS has recently proven to be an appealing alternative to diffraction techniques [1,6], because no structural hypothesis needs to be made in the interpretation. However, contradictory results for Cu have also been obtained with this method [6]. Furthermore, the standard EXAFS analysis has also been criticized lately, because of the possibility of strongly anharmonic atomic motions, which may mimic bond contractions [9]. Finally, interactions between sample particles and support (substrate or matrix) can never be ruled out completely, especially for very small particles, and constitute a fundamental problem common to all experiments to date. In this paper, we present results on structural properties of small Cu particles, obtained in a way which essentially avoids all the above-mentioned problems.

In our 100 keV electron diffractograph [10], we have generated molecular beams of Cu particles in He carrier gas, extracted from an inert-gas-aggregation source (IGAS) through two differential pumping stages. We have recorded the diffraction patterns (Debye-Scherrer rings, due to random orientations of the free particles) using a pair of linear CCD arrays, operated in a direct electron counting mode. For analysis by transmission electron microscopy, we have also obtained supported samples of Cu particles on thin amorphous carbon films, by direct exposure to the molecular beam. We thus have access to the distribution of particle sizes. The particle structures and size distributions have been influenced by varying nucleation and growth conditions in the evaporation chamber, via the crucible temperature T_c and the He pressure, P_{He}. The inert gas was injected at room temperature.

In Fig. 1. we present two experimental diffraction patterns, obtained at : a) $T_c = 1704 \pm 2$ °C / $P_{He} = 4.2 \pm 0.1$ mbar, and b) $T_c = 1643 \pm 2$ °C / $P_{He} = 1.4 \pm 0.1$ mbar, with contributions due to scattering from individual atoms (carrier gas, uncondensed Cu vapor) removed. The scattering parameter is defined as $s = 2 \sin \theta_B/\lambda$, with θ_B, the Bragg reflection angle and λ, the de Broglie wavelength of the electrons. Intensity units are arbitrary.

Our interpretations of these experiments are based on fits of model diffraction patterns, optimized by simulated annealing [4]. The building blocks for the models are kinematic diffraction patterns, calculated for geometrically perfect, complete-shelled particles. Three structural families are considered : cubo-octahedra (fcc single crystals), truncated decahedra (MTP) and icosahedra (MTP), with diameters between 0.7 nm (13 atoms) and 12.5 nm (~ 86000 atoms). Random size distributions are generated separately for each family and the resulting patterns compared with the experimental curves. The difference between experiment and model is squared and integrated over the whole pattern, yielding a quality measure Q, which is minimized.

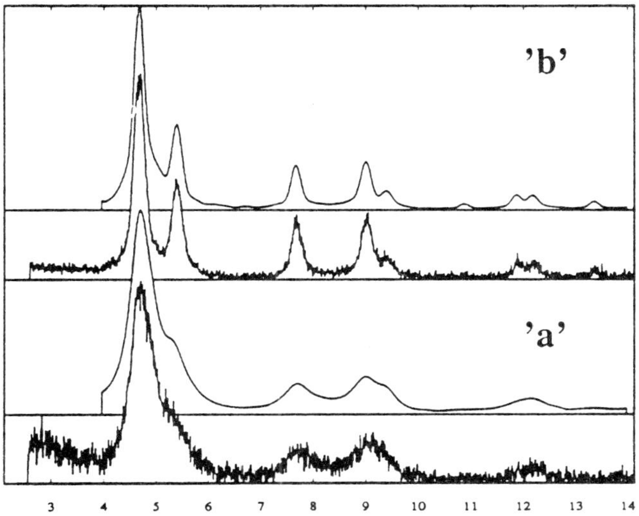

Fig.1. Experimental and model diffraction patterns. Axes: intensity (vert.) versus s [nm^{-1}] (horiz.)

Finally, we have simulated homogeneous expansions or contractions of our particles by scaling of the calculated curves, performing the whole fitting procedure each time. Fig. 2 shows that best fits are obtained unambiguously using slightly expanded particles, for both experiments (the continuous lines are guides for the eye).

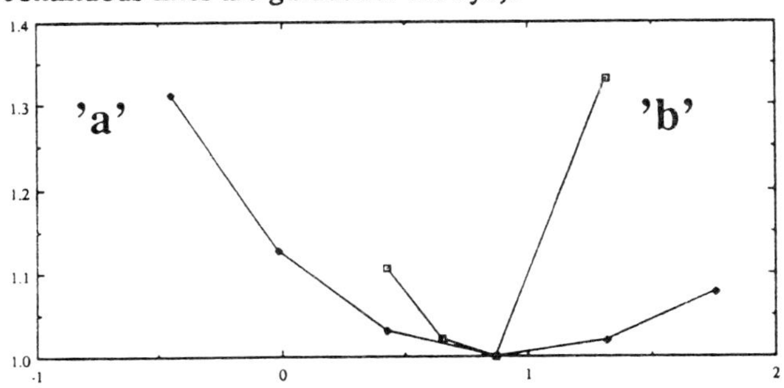

Fig. 2. Optimization of interatomic distance. Axes: normalized Q value (vert.) vs. relative Cu-Cu dist. change [%] (hor.)

The best fits to our experiments (lowest Q values) are visible as smooth curves in Fig. 1. Fig. 3 shows the corresponding structural compositions, in volume-weighted form, to reflect the actual relative contributions to the diffracted intensities. A volume-weighted experimental size distribution (supported sample), only available from experiment a), is also visible and is in good agreement with the model distributions.

The results presented here clearly show, for the first time, that under favorable nucleation and growth conditions, free Cu icosahedra and decahedra coexist in comparable size ranges. They are in full agreement with recent molecular dynamics calculations showing instability of Cu single crystals below ~ 3.6 nm, and of Cu icosahedra above ~ 3.8 nm [11]. The dependence on source conditions is surprising: an IGAS is usually expected to produce larger particles as temperature and pressure rise, and our previous experiments have generally confirmed this. Here, changing flow conditions could be responsible for the observed behavior.

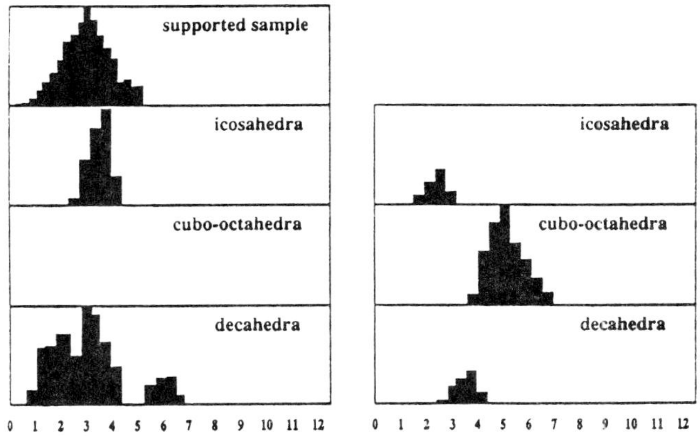

Fig.3. Volume-weighted size distributions. Left: a), right: b) Horizontal axis: particle diameters [nm]

Finally, our results do not support the idea of interatomic distance contraction in Cu particles of a few nm. They rather suggest that such particles could be in a state of weak dilatation, possibly due to temperature effects or changes in lattice dynamics, leading to an increase in mean-square atomic displacements [12]. However, it should be kept in mind that our model for expansion/contraction is rather crude, being homogeneous and affecting identically particles of all sizes and symmetries.

Support by Swiss National Research Found. is acknowledged

References

1. Montano, P.A.,Purdum, H., Shenoy, G.K.,Morrison, T.I., Schulze, W. : Surf. Sci. 156, 228 (1985); Solliard, C, Flüeli, M. : Surf. Sci. 156, 487 (1985)
2. De Crescenzi, M, Diociaiuti, M., Lozzi, L., Picozzi, P., Santucci, S. : Phys. Rev. B 35, 5997 (1987)
3. Gillet, M. : J. Micr. Spect. Electron. 11, 449 (1986); Farges J., de Feraudy, M.F., Raoult, B., Torchet, G. : J. Chem. Phys. 78, 5067 (1983); *ibid*, 84, 3491 (1986)
4. Hall, B.D., Flüeli, M., Monot, R., Borel, J.-P. : Phys. Rev. B 43, 3906 (1991)
5. Wasserman, H.J. Vermaak, J.S. : Surf. Sci. 32, 168 (1972)
6. Apai, G., Hamilton, J.F., Stohr, J., Thompson, A. : Phys. Rev. Lett. 43, 165 (1979); Montano, P.A., Shenoy, G.K., Alp, E.E., Schulze, W., Urban, J. : Phys. Rev. Lett. 56, 2076 (1986).
7. Reichelt, K., Schreiber, H. : Surf. Sci. 43, 644 (1974)
8. Briant, C.L., Burton, J.J. ; Surf. Sci. 51, 345 (1975); Solliard, C. : Surf. Sci. 106, 58 (1981); Paoletti, L., Diociaiuti, M., Picozzi, P., Santucci, S. : Thin Solid Films 145, 249 (1986)
9. Hansen, L.B., Stoltze, P., Nørskov, J.K., Clausen D.S., Niemann, W. : Phys. Rev. Lett. 64, 3155 (1990)
10. Hall, B.D., Flüeli, M., Reinhard, D., Borel, J.-P., Monot, R., : Rev. Sci. Instrum. 62, 1488 (1991)
11. Valkealahti, S., Manninen, M. : Phys. Rev. B 45, 9459 (1992)
12. Buffat, P.-A. : Solid State Commun. 23, 547 (1977)

Z. Phys. A 349, 273–278 (1994)

Very neutron-rich exotic nuclei

F. Pougheon

Institut de Physique Nucléaire, IN2P3-CNRS et Université Paris-Sud, F-91406 Orsay Cedex, France

Received: 27 October 1993

Abstract. A brief summary is done of the various types of experiment used in studies of the very neutron rich nuclei. Some highlights are given for the two-neutron halo and ^{11}Li nucleus and for the one-neutron halo and ^{11}Be nucleus.

PACS: 21.10.gr; 27.20.+n; 25.70.Np

1. Introduction

The limits of nuclear stability correspond roughly to zero binding energy for protons and neutrons and are commonly referred to as the drip-lines. During the last years one has been able to investigate nuclei close to the drip lines thanks to a large variety of techniques involving new accelerators and very good isotope separators. In this paper I shall focus on light nuclei located near or on the neutron drip-line where novel phenomena have appeared.

Experimentally, the neutron drip-line has been reached only for the lightest elements up to Neon [1]. In this part very striking effects appear. The essential new ingredient is that the low binding energy of the last neutron (or pair of neutrons) will permit quantum mechanical tunneling of the particle(s) away from the nuclear core and hence the formation of cloud called now commonly "halo". The consequence is that some principles of nuclear physics are violated:

– Mass and charge radii may differ by large amounts.
– Since the neutrons and protons are physically separated, electric dipole excitations of single particle strength appear at very low energy in strong contrast to the usual dominance of the Giant Dipole Resonance.
– There are theoretical indications that the three-body aspects play a role, leading to molecular or cluster features that are not fully accounted for in a mean field approximations. But theoretical aspects of the halo will not be treated in this paper.

I shall try to give a survey of the pehnomenology of the neutron halo. Indeed, it is impossible to present all experimental details and conclusions related to the phenomenon of the neutron halo. Thus I shall give only some examples of experiments probing the existence of such a halo via nuclear reactions.

I shall start with the two-neutron halo and the ^{11}Li nucleus (Sect. 2). The case of the one-neutron halo and ^{11}Be will be discussed in Sect. 3.

2. The two-neutron halo and ^{11}Li nucleus

The neutron halo was first predicted by Hansen and Jonson in 1987 [2], for ^{11}Li. Their model is a crude cluster model where ^{11}Li is treated as a ^{9}Li core plus two neutrons which form a point like dineutron. Since this date, more than two hundred theoretical papers have been published on this subject including very sophisticated models (for a review, look at the Proceedings of the Sixth International Conference on Nuclei Far from stability, Bernkastel-Kues, Germany, July 1992).

The neutron(s) halo happens for nuclei near or at the neutron drip-line where small binding energy of the last neutron(s) allows them to move away from the attractive nuclear potential and into free space by quantum mechanical tunneling. The surface of the nucleus is thus extended considerably and this dilute neutron cloud is what is termed the "neutron halo". Among nuclei at the drip-line, the ^{11}Li has the lowest two-neutron separation energy observed so far: $S_{2n} = 315 \pm 40$ keV (average of three measurements [3]).

Some words about the production, the identification and the acceleration of such exotic nuclei. The experimental facilities that enables production of exotic nuclei have undergone a rapid development during the last decade. This development has taken two main directions: target fragmentation and projectile fragmentation. By target fragmentation the exotic secondary nuclei are the reaction products of e.g. a proton beam incident on a target whereby the target nuclei are fragmented. The sec-

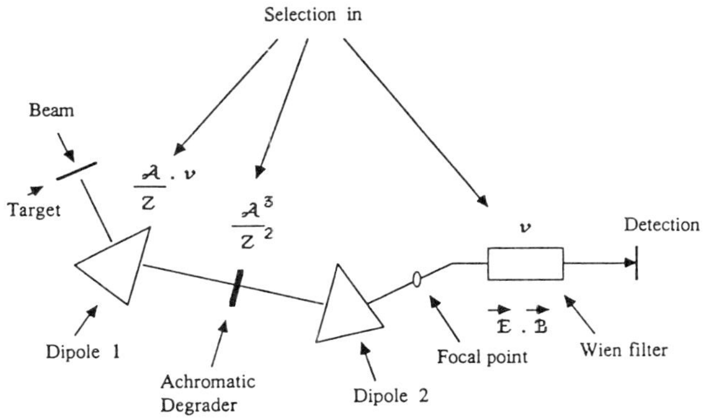

Fig. 1. Schematic view of the LISE 3 achromatic spectrometer installed at GANIL

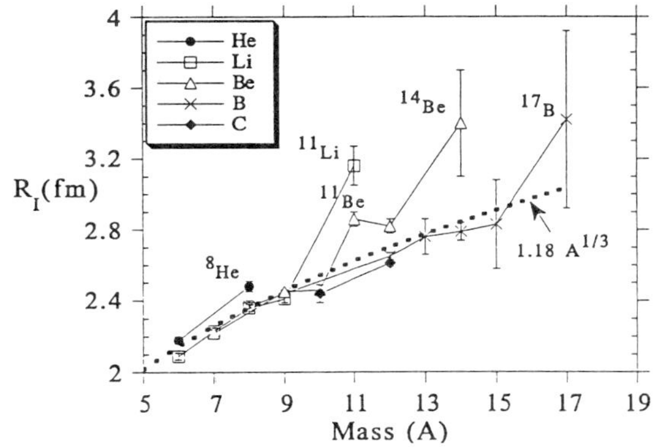

Fig. 2. Interaction radii for light nuclei of the p shell. *Dotted line* is the radius calculated from the liquid drop model [5b]

ondary nuclei emitted by the target are captured by an ion-source (secondary beam line) and separated and identified by means of magnetic spectrometry, electrostatic deflection, time of flight and (or) energy-loss. Target fragmentation is used at ISOLDE (Cern) and LAMPF (Los Alamos).

The method of projectile fragmentation is used at intermediate energy (MSU, RIKEN, GANIL) or high energy (BERKELEY, G.S.I., SATURNE). The secondary beams of exotic nuclei are here produced by the fragmentation of the projectile nuclei in a production target. The fragments that emerge with approximately beam velocity are separated and identified by means of two or more of the previously mentioned methods. A schematic view of the LISE 3 achromatic spectrometer installed at Ganil [4] is shown in Fig. 1.

2.1. *Cross-sections and radii*

The "easiest" type of experiments with a secondary radioactive beam is a measurement of its total interaction cross-section with a target. In a transmission experiment, the incident flux is compared to the one after the target. The method was pionnered at Berkeley [5] and over the years a wealth of data has been accumulated: interaction cross sections σ_1 are to day determined for many combinations of light projectiles and targets. At high energy the interaction cross-section is geometric and can be written as:

$$\sigma_1(p,t)=\pi[R_I(p)+R_I(t)]^2$$

The nuclear interaction radius of the target (t) and the projectile (p) can be separated and Fig. 2 summarises the obtained results for the He to B isotopes. This interaction radius is compared to the liquid-drop model parametrization. It clearly appears for the very neutron rich nuclei a strong increase of the interaction radius for ^{8}He, ^{11}Li, ^{14}Be and ^{17}B.

This increase may be due to a deformation of the nucleus or to a long tail in the matter distribution. For ^{11}Li three different measurements demonstrate that the second hypothesis is the good one and furthermore that the core of ^{9}Li is little affected by the presence of the halo neutrons:

– The magnetic dipole moments of ^{9}Li and ^{11}Li have been measured by a refined combination of optical and beta-decay techniques to be 3.439 and 3.668 nuclear magnetons, respectively, close to the Schmidt limit of 3.79 n.m. [6] for a proton in the $1p$ 3/2 state. This indicates that the proton system is little affected by the two additional neutrons in ^{11}Li and that this nucleus is not deformed.
– Recently it has been shown [7] that the electrical quadrupole moments also are identical for ^{9}Li and ^{11}Li within the experimental errors, indicating a same charge radius.
– Another experiment confirmed that the charge radius does not vary between ^{9}Li and ^{11}Li. Blank et al. [8] measured charge-changing reactions for ^{8,9,11}Li impinging on a Carbon target at 80 MeV/u and found that all have the same cross-section while the total cross-sections strongly increases with the Lithium mass (see Fig. 3).

It was predicted [2] that Coulomb stripping plays an important role for high-Z targets and at low energies since the cross-section varies as Z^2/SnEu where Eu is the beam energy per nucleon. At 29 MeV/u [9] Coulomb dissociation becomes dominant so that for a gold target the experimental cross section for the removal of two neutrons from ^{11}Li is: $5.0 \pm 0.8\,b$, two orders of magnitude above "normal" Coulomb cross-sections. Nuclear stripping is also strongly enhanced for a nucleus with a neutron halo. At 29 MeV/u, on a Beryllium target the neutron yield near zero degrees is more than one order of magnitude greater for ^{11}Li with 8 neutrons than it is for ^{9}Li with six neutrons [9].

2.2. *Momentum distributions*

The idea is that due to the "Heisenberg principle" $\Delta x \cdot \Delta p \geqq \hbar/2$, a large spatial extent of the halo must be related to a narrow momentum distribution width. So it was at first hoped that the momentum distributions might have a very simple interpretation.

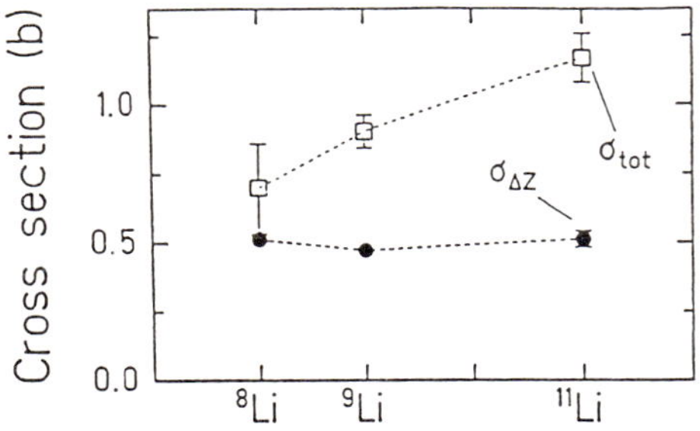

Fig. 3. Charge-changing (*full circles*) and total reaction (*open squares*) cross sections for ^{8,9,11}Li at 80 MeV/u from an experiment at SATURNE [8]

The main emphasis has been on measurements of the transverse momentum distributions of either the halo neutrons or the core from the "halo removal" channel. All the data show a very narrow spread in transverse momentum, an order of magnitude below typical nucleon momenta in a nucleus. But there has been a lot of discussion about a consistent interpretation of the data and their relation to the neutron momenta. It was recently pointed out that transverse momentum data are affected by Coulomb deflection and by multiple scattering [10]. Not affected by these effects, parallel momentum distributions are thought to reflect reliably the wave function of the projectile (through its Fourier transform of). Experimentally, it is difficult to measure parallel momenta but a group at MSU has overcome this problem. Figure 4 shows the obtained width for ^{11}Li at 66 MeV/u impinging on beryllium, Niobium and Tantalum targets. It is interesting to note that there is very little target dependence as it was the case for the neutron angular distribution obtained at GANIL [9]. This can suggests that the data reflect predominantly the momentum distribution of the ground state of the halo.

2.3. *Correlations of the two neutrons*

The nature of the correlation between the two neutrons is not known. ^{11}Li may be thought of as having a ^{9}Li core immersed in a halo of two neutrons, correlated as a di-neutron. Other models predict an anticorrelation between the halo neutrons, favouring emission with large opening angle between the directions of the two-neutrons.

The ultimate experimental answer implies a fully exclusive measurements in which all reaction partners in the entrance and exit channel, are identified and all the momenta measured. Such experiments are indeed difficult but recently such measurements have been done at MSU [11] and G.S.I. [12]. Kinematically complete measurements for Coulomb dissociation of ^{11}Li into ^{9}Li$+2n$ were made at 28 MeV/u [11]. The correlation function was deduced from the relative momentum spectrum between two neutrons in coincidence with a ^{9}Li fragment (Fig. 5). The fit results from a model calculation that assumes a spherical Gaussian source with a best fit value

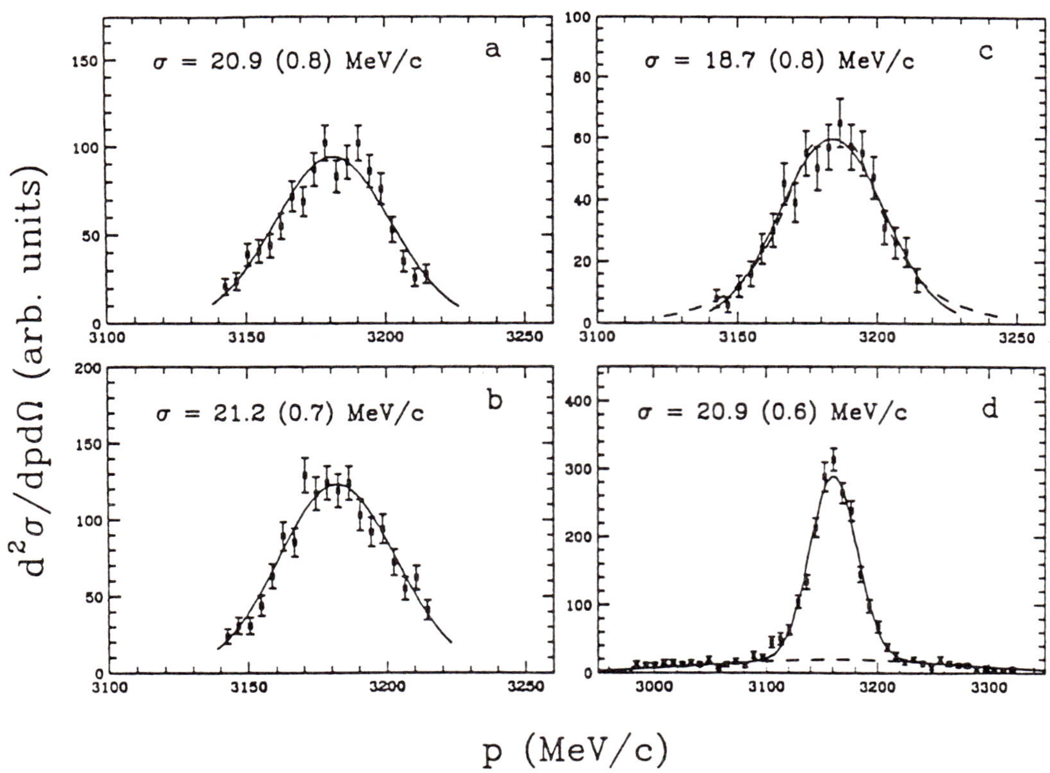

Fig. 4. Momentum distributions for ^{9}Li fragment (coming from ^{11}Li break-up) with single Gaussian fits using **a** ^{9}Be **b** ^{93}Nb and **c** ^{181}Ta targets **d** high acceptance mode distribution for a 10% range of ^{9}Li momenta using a ^{9}Be target [see 10]

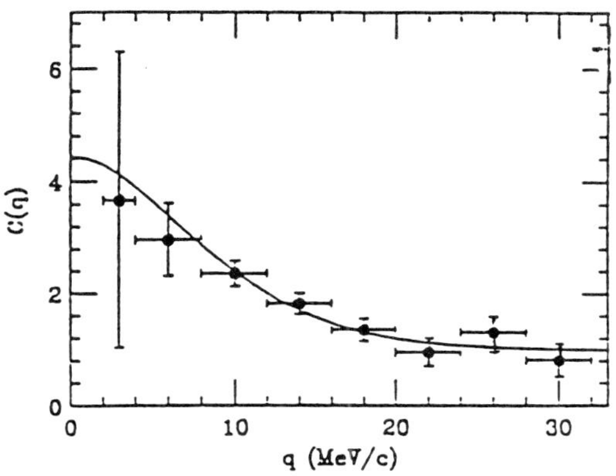

Fig. 5. Correlation function between the two neutrons. The line represents a calculation based on a spherical Gaussain source with radius 5.3 fm [11]

of 5.3+0.3 fm for its radius. Since the radius of ^{9}Li is only 2.3 fm these data constitute direct evidence of the halo structure of the valence neutrons in ^{11}Li, and are consistent with a non correlation between the neutrons.

In a recent paper [13] K. Riisager has done a compilation of all data on ^{11}Li concerning the momentum width of the neutrons and of the ^{9}Li core. When neutron and core widths are compared at the same energy one finds a ratio of approximately $\sqrt{2}$, which is what could be expected [14] if the two neutrons in the halo are uncorrelated.

2.4. Elastic scattering

It is of interest to study the neutron halo within the framework of the optical model, using data from the angular distribution of elastically scattered exotic beams. But due to the modest optical quality of the secondary beams made by projectile fragmentation, high precision measurements are not easy.

Three experiments have been performed up to now. The experiment at RIKEN [15] measures proton elastic scattering on ^{9,11}Li at 60 MeV/u under inverse kinematical conditions. At Ganil [16] beams of 30 MeV/u ^{7,11}Li at 60 MeV/u under inverse kinematical conditions. At Ganil [16] beams of 30 MeV ^{7,11}Li are scattered on a Silicium target. At MSU elastic scattering of ^{11}C, ^{11}Be and ^{11}Li on ^{12}C have been performed at 58 MeV/u [17]. All the data show a large difference in the angular distributions for the behaviour of ^{11}Li compared to the "reference" beams implying changes in the optical model parameters in link with the existence of the neutron halo.

3. The one neutron halo and ^{11}Be

The difficulty encountered in probing the halo in more detail by means of nuclear reactions lies in the interplay between nuclear structure and the reaction mechanism. In the ^{11}Be case the structure part is well understood and it is a two body problem with a core and one neutron.

The ground state of ^{11}Be is an example of what we today call an intruder state: the odd neutron is in the $2s\ 1/2$ shell. For a neutron the external wave function is simply a Yukawa function proportional to $\exp(-r/\rho)/r$ where the decay length $\rho = \dfrac{\hbar}{\sqrt{2\mu \mathrm{Sn}}}$ is given in terms of the reduced mass and the separation energy. The neutron separation energy of 504 keV leads to the external decay length $\rho = 6.8$ fm, well in excess of the root-mean-square radius of a p-shell nucleus of about 2.5 fm. The first and unique excited state $1p\ 1/2$ lies at 320 keV; its lifetime was measured [18] and was found to be the fastest E_1 transition between low-lying states known in nuclear physics. The reduced transition probability is 0.36 Weisskopf units.

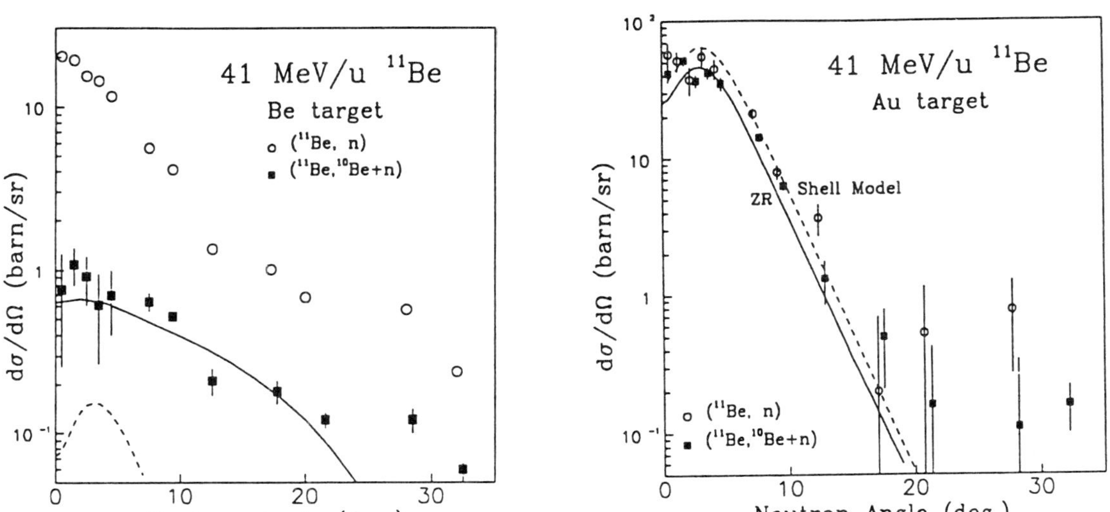

Fig. 6. Exclusive and inclusive neutron angular distributions from reactions with the gold and beryllium target

3.1. *Coulomb and nuclear dissociation of* ^{11}Be

In a recent experiment at Ganil [19] we have extended our investigations to the single-neutron halo nucleus ^{11}Be. The fast forward neutron in coincidence with a ^{10}Be recoil was detected by a neutron hodoscope.

For a heavy target such as gold the exclusive differential cross-sections are not very different from the single neutron inclusive distributions showing that Coulomb dissociation is a dominant process. For the beryllium target there is a strong difference and the exclusive distribution is very broad as is expected for diffraction dissociation on a small target nucleus (Fig. 6).

The exclusive angular distribution on three different targets are shown in Fig. 7, together with the results of two simple calculations (Coulomb dissociation for gold target and diffraction dissociation for Be target) based on a Yukawa wave function alone in the "zero range approximation" [19]. There are no free parameters in these calculations and the agreement between experimental and theoretical curves is quite good.

An important point to be underlined is the role of the binding energy Sn of the last neutron. In Fig. 8 one can see that the cross section near 0° is very sensitive to Sn suggesting that a neutron hodoscope coupled to a recoil spectrometer is a powerful tool for searching for extreme halo systems and to give a reasonable estimate of their mass.

3.2. *Coulomb excitation*

The aim of a very recent experiment performed at Ganil [20] was to search for bound excited levels in halo nuclei using Coulomb excitation. We took advantage of the very high polarisability of the loosely bound halo which is expected to give a very large cross-section for E_1 excitation.

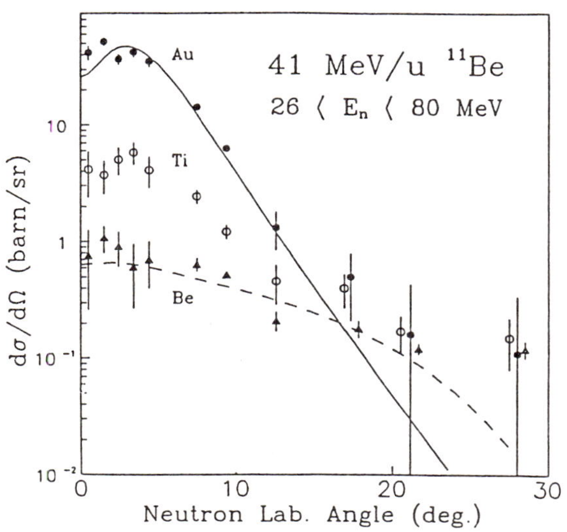

Fig. 7. Exclusive neutron angular distributions on three different targets (Au, Ti, Be)

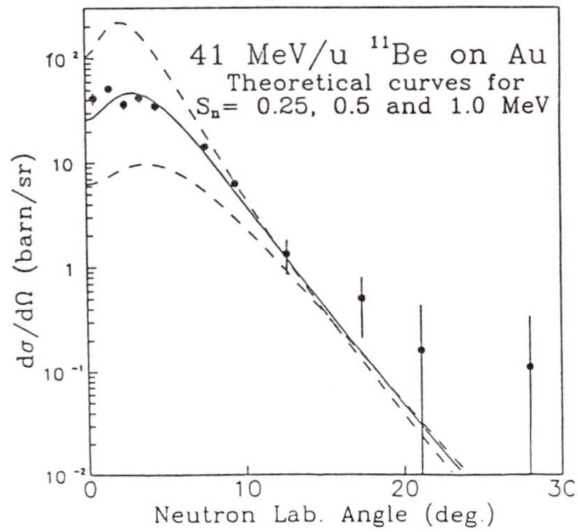

Fig. 8. Exclusive neutron angular distribution from the reaction of ^{11}Be at 41 MeV/u in a gold target. The theoretical curves are calculated for three different neutron binding energies

A good test case is the ^{11}Be since it has a known bound excited state at 320 keV with a large $B(E_1)$. This $1/2^-$ state has been populated and the γ transition back to the $1/2^+$ ground state observed. The preliminary results for the cross-sections are in agreement with theoretical predictions based on the measured half life and on the theory for high energy Coulomb excitation [21].

For the first time a Coulomb excitation with a high energy (45 MeV/u) radioactive beam has been performed. This positive result shows that Coulomb excitation experiments may constitute a powerful tool for nuclear structure studies of unstable nuclei.

4. Perspectives

For the near future one must perfect the existing techniques. Beam intensity, detector coverage and detector resolution are all elements that have been improved during the last few years and where one can hope for further progress.

Experiments with detection of all fragments should eventually yield a quite detailed picture of the halo structure but there still is plenty of work to be done in this domain.

All in all, we begin to understand the neutron halo and the interplay between structure and reaction mechanism.

All the light nuclei at the neutron drip line show a halo structure. Most of the work until now has concentrated on the best cases ^{11}Li and ^{11}Be but there must exist very interesting candidates beyond beryllium such as ^{19}B, ^{19}C (a candidate for one-neutron halo like ^{11}Be), ^{22}C and so on. They have been produced and identified but production rates of these species are small and the experiments will be hard. It would be valuable if it were possible to perform simple experiments that directly give an idea of the amount of halo in a given nucleus. The key is the separation energy of the last neutron(s) since the halo size scales with $\sqrt{S}$. The calculations using a Yukawa wave function [14] indicate that the Coulomb cross sections scales as $1/S^2$.

It will be quite important to know how much of the structural information is due to the special nature of shell closure at $N=8$. Halos in excited states or proton halos could also exists.

Post accelerated radioactive beams hold a great promise in this field.

References

1. Guillemaud-Mueller, D. et al.: Phys. Rev. **C41**, 937 (1990)
2. Hansen, P.G., Jonson, B.: Europhys. Lett. **4**, 409 (1987)
3. Thibault, C. et al.: Phys. Rev. **C12**, 644 (1975); Wouters, J.M. et al.: Z. Phys. **A331**, 229 (1988); Kobayashi, T.: Nucl. Phys. **A538**, 343C (1992)
4. Mueller, A.C., Anne, R.: Nucl. Instrum. Methods **B56**, 559 (1991)
5. Tanihata, I. et al.: Phys. Rev. Lett. **55**, 2676 (1985); Tanihata, I.: Nucl. Phys. **A520**, 411 (1990)
6. Arnold, E. et al.: Phys. Lett. **B197**, 311 (1987)
7. Arnold, E. et al.: Phys. Lett. **281B**, 16 (1992)
8. Blank, B. et al.: Z. Phys. **A343**, 375 (1992)
9. Anne, R. et al.: Phys. Lett. **250B**, 19 (1990)
10. Orr, N.A. et al.: Phys. Rev. Lett. **69**, 2050 (1992)
11. Ieki, K. et al.: Phys. Rev. Lett. **70**, 730 (1993)
12. Schrieder, G.: Proceedings of the Sixth International Conference on Nuclei far from Stability, Bernkastel-Kues, July 1992 (eds) Neugart, R., Wohr, A., p. 369
13. Riisager, K.: Proceedings of the Third International Conference on Radioactive nuclear beams, MSU, May 1993 (to be published)
14. Hansen, P.G.: Nucl. Phys. **A553**, 89c (1993)
15. Moon, C.B. et al.: Proceedings of the Sixth International Conference on Nuclei far from Stability, Bernkastel-Kues, July 1992 (eds) Neugart, R., Wohr, A., p. 277
16. Lewitowicz, M. et al.: Nucl. Phys. (in press)
17. Kolata, J.J. et al.: Phys. Rev. Lett. **69**, 2631 (1992)
18. Millener, D.J. et al.: Phys. Rev. **C28**, 497 (1983)
19. Anne, R. et al.: Phys. Lett. **B304**, 55 (1993); Dogny, S.: PhD thesis, Orsay, IPNOT 92.03
20. Corre, J.M. et al.: (to be published)
21. Winther, A., Alder, K.: Nucl. Phys. **A319**, 518 (1979)

Z. Phys. A 349, 285–290 (1994)

Clustering aspects of light exotic nuclei

J.S. Vaagen[1], I.J. Thompson[2], J.M. Bang[3], M.V. Zhukov[4], B.V. Danilin[5], D.V. Fedorov[6]

[1] Department of Physics, University of Bergen, N-5007 Bergen, Norway
[2] Centre for Nuclear Physics, University of Surrey, Guildford, Surrey GU2 5XH, UK
[3] The Niels Bohr Institute, University of Copenhagen, DK-2100 Copenhagen Ø, Denmark
[4] Department of Physics, Chalmers University of Technology, S-41296 Göteborg, Sweden
[5] The Kurchatov Institute of Atomic Energy, 123182 Moscow, Russia
[6] Institute of Physics and Astronomy, University of Aarhus, DK-8000 Aarhus C, Denmark

Received: 27 October 1993

Abstract. The discovery of light nuclei with neutron halos has opened new perspectives for nuclear structure and a possibility to study genuine few-body aspects of matter. A number of borromean systems where the system is bound but where no binary subsystems are bound, are realized by nature in the context of nuclei, ^{11}Li being the nucleus which has had most recent attention. Although a large number of calculations have addressed the detailed structure of the ^{11}Li neutron halo, no single model has yet explained all the data presently available. This is in part due to uncertainties in the n+^{9}Li interaction. For ^{6}He, also a borromean system, where the n + core interaction is known, the best three-body and reaction models succeed in reproducing the data. In the ^{11}Li case, a major uncertainty concerns the effect of possible $1s$ intruder states.

PACS: 25.60.+v; 21.45.+v

1. Introduction

This paper is about halo nuclei; nuclei with a dilute far-ranging matter halo which surrounds a more compact core object. Thus there is in terms of matter distributions pictorially a division into a core cluster and a 'halo cluster'. The calculations to be described below treat however the problem dynamically as one where the halo nucleons interact with each other and with the spectator core; the halo emerges from this mutual interaction. The calculated one-particle density of ^{11}Li [1] is shown in Fig. 1 which gives the individual core and valence halo contributions for different three-body calculations. The pronounced neutron-halo is clearly seen in this Figure.

Many experiments with neutron-rich radioactive nuclear beams have now confirmed that nuclei such as ^{6}He and ^{11}Li have large neutron halos, or dilute neutron skins, which extend to large radii. Such experiments have measured not only the total reaction cross sections for the radioactive projectiles, but also, following the breakup of ^{11}Li at high energies, the inclusive momentum distributions of the ^{9}Li-core and neutron fragments. Recent experiments now measure the kinematically-complete breakup into the valence neutrons and core fragments, and should give us the best insight into the halo structure.

These experiments are necessary. As yet there is no consensus as to the detailed angular momentum structure of the ^{11}Li halo. The neutron halo interpretation is supported by a number of recent theoretical calculations of the ^{11}Li ground state, but these calculations differ among themselves in important ways. We do not yet know enough about the $n-{}^{9}$Li interaction, in particular the energies of the $0p_{1/2}$ resonance(s) and any $1s_{1/2}$ virtual state. The ^{9}Li-core deformation and the relative strengths of pairing and spin-orbit forces in forming the two p-shell holes for the halo neutrons are also poorly determined. We think we understand the structure of the ^{6}He halo, largely because the $n-\alpha$ interaction is better known.

The existing calculations of the ^{11}Li g.s. have had to make assumptions on all the above points. Some calculations have treated the nucleus as a whole with mean-field theories[2,3,4] and shell models[5,6,7]. The best view of the halo, however, is obtained by treating ^{11}Li as an inert ^{9}Li core plus two valence neutrons[8,9,10,11,12]. The importance of the correlation energy of these neutrons means that mean-field approximations have restricted validity, and their weak binding means that shell models require many orders of $\hbar\omega$ excitations to be accurate. The best calculations will therefore treat the three-body dynamics explicitly, and will also include the effect of the Pauli Principle from the core neutrons. We will see, however, that the methods do so with varying precision, and the conclusions concerning the physical structure of the ^{11}Li halo have yet to be definitively drawn.

This conference takes place on what may be referred to as a fragmented 'halo island' the result of a violent non-man-made 'break-up' reaction 3600 years ago when the volcano erupted and 20-30 cubic kilometers of magma was blown into the air over a short period of time – tens or hundred of hours rather than years. The reconstruction of the previous island Strongyli (the round or circular one)

Correspondence to: J.S. Vaagen

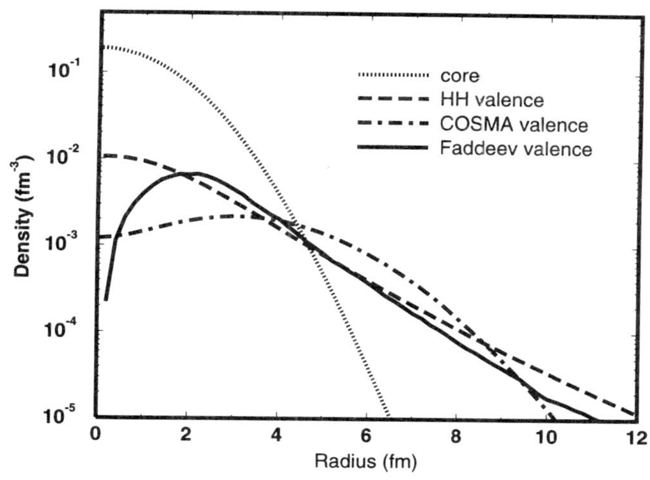

Fig. 1. Radial density distributions for ^{11}Li

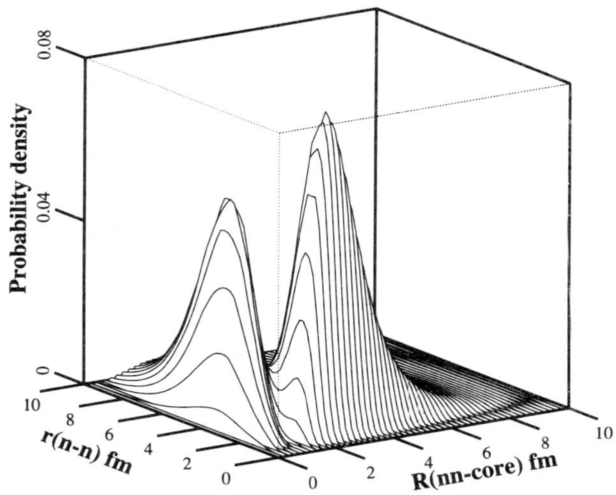

Fig. 2. Correlated density plot for ^{6}He. The nodal line follows from the fact that $p^2_{J=0}$ is the dominant state of the halo neutrons. With simple oscillator wave functions, this configuration is proportional to $\mathbf{r}_{1c}.\mathbf{r}_{1c}$, which, transformed to di-neutron coordinates $\mathbf{r}$ and $\mathbf{R}$, becomes $(\mathbf{R}+\mathbf{r}/2).(\mathbf{R}-\mathbf{r}/2)$, and gives a nodal line at $R = r/2$

with its nearly perfect halo (caldera) encircling a seafilled basin with a central volcano, has many similarities with the way we learn about halo nuclei. As alluded to above, our information has primarily come from (in our case repeatable) man made break-up reactions in the laboratory, the halo nuclei being produced as radioactive secondary beams on fixed target nuclei. The field owes its development to a fast experimental progress. We refer to the paper by F. Pougheon [13] at this conference for an overview.

2. ^{6}He: our benchmark halo nucleus

Although not considered for a long time as a halo nucleus, ^{6}He is a neutron-rich light nucleus which clearly exhibits a three-body structure. Numerous three-body calculations for A=6 nuclei[14,15,16,17] show that the ground and low-lying excited states have well developed three-body ^{4}He+N+N structures. These investigations have shown that most of the methods give similar descriptions of the binding energy and r.m.s. matter radii, as well as the main structure of the wave function (WF). They deviate from each other only in finer details of the WFs. The results of the hyperspherical harmonic (HH) and coordinate space Faddeev (CSF) methods with similar two-body potentials are[18,19] almost identical, and are similar to the predictions of other three-body calculations. This is not very surprising since all three-body approaches for ^{6}He use two-body potentials which correspond to rather complete sets of experimental data for the α-n and n-n channels. Thus minor differences in the three-body calculations have only a little influence on the WF structure.

The spatial correlation density plot for the bound state of ^{6}He is shown in Fig. 2. The two peaks are sometimes referred to as the dineutron and cigar-like components, the former having the two valence neutrons located together outside the core, the latter corresponds to neutrons positioned on opposite sides of the core.

The ^{6}He nucleus can be used to test models of the reactions of neutron-halo nuclei. The Serber (sudden break-up) model can be used to predict the momentum distributions of core and n fragments in inclusive reactions[20], and the fit to the transverse data[21] can be improved by taking final-state interactions into account[22]. Possible new experiments of $p_{||}$ and elastic scattering distributions should confirm the understanding of this three-body halo system.

Recently also the structure of the continuum of ^{6}He has been calculated within the HH formulation [18]. Results are discussed in B.V. Danilin's contributions to this conference [23] and in [24,25] which give the electric dipole response function and continuum cross sections for dipole excitation of ^{6}He at 800A MeV. The latter results, shown in Fig. 3, show that while there is strength accumulation at lower excitation energies, no narrow resonances that might have been referred to as a soft dipole mode are found even when final state interactions are taken into account.

3. ^{11}Li: the halo nucleus par excellence

Although ^{11}Li has a half life of 9 msec, many experiments have been performed in order to understand its structure. The large observed reaction cross section indicates a large matter radius[26], and measurements of ^{9}Li and ^{11}Li electric quadrupole moments[27] support the concept of a ^{9}Li core, showing that the halo involves only the neutrons. Narrow momentum distributions for the ^{9}Li core[28,29] and for neutrons[30] indicate that the wave functions of the halo neutrons are widely extended in space, and it is interest-

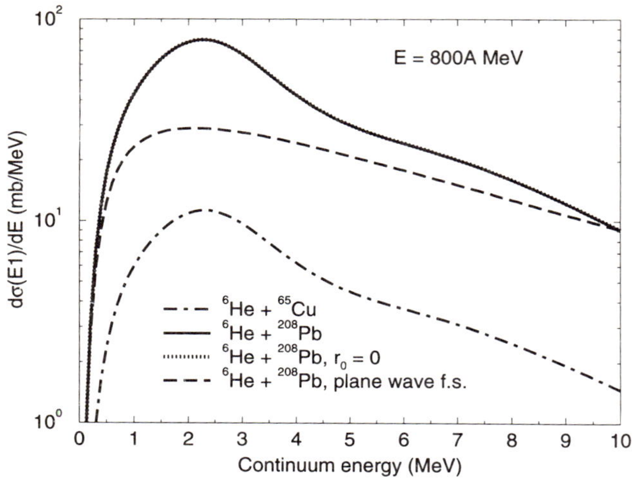

Fig. 3. Continuum cross sections for dipole excitation of ^{6}He at 800A MeV incident lab. energy

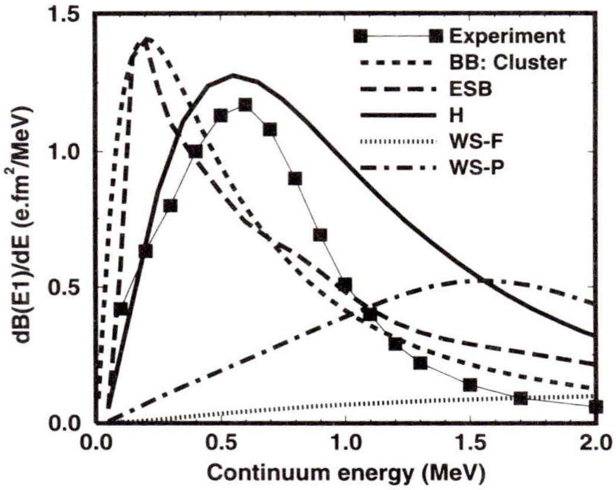

Fig. 4. Electric dipole response for ^{11}Li from the MSU experiment [31] (squares), with the Greens function [32] (ESB, long dashed) and cluster model [33] (BB, short dashed) calculations. The curves H (solid), WS-F (dotted), WS-P (dot-dashed) are the predictions from the current 3-body scenarios

ing that they suggest a size larger than that derived from the total reaction cross sections. Most recently, kinematically complete measurements of the breakup into core and two neutron fragments[31] show that there is a substantial concentration of B(E1) dipole strength in the continuum between 0.3 and 1 MeV.

The possible existence of a low-lying soft dipole mode has been a hot issue. Figures 4 compares various calculated electric dipole response functions with what has recently been extracted from experimental data from Michigan [31]. The results shown are those of a cluster calculation, the result of Esbensen and Bertsch (ESB) [32] using a Greens function method, and three HH calculations corresponding to the scenarios discussed below and three-body continuum calculations like those for ^{6}He. The theoretical situation definitely needs clarification; the calculation which comes closest to the data is that of the simple JJH scenario with relative s motion only. This somewhat embarrasing situation has a message that we return to below.

The many theoretical models of ^{11}Li will be discussed below, but all make n-n pairing correlations essential for the existence of the ^{11}Li bound state. They all show that the two neutrons, if they have a joint binding of $E \simeq -0.3$ MeV, must be in a state much larger than the core size. This supports the use of three-body models for ^{11}Li, which neglect the degrees of freedom internal to the ^{9}Li core, and can then treat the quantum dynamics of the core and halo neutrons in a fully correlated manner.

4. Is the core cluster a passive spectator nucleus?

The tightly bound alpha particle is not expected to be significantly distorted when embedded in a veil of valence neutrons forming the halo nuclei ^{6}He(0.975 MeV), ^{8}He(2.138 MeV) and ^{10}He(unbound), the numbers in parentheses are the measured two-neutron separation energies. In [19] a comparison is given between a number of calculated and measured geometric and electro-weak observables for ^{6}He, ^{6}Li and ^{6}Be, the results being quite satifactory. A slight expansion of the core is however required to obtain the precise binding energy.

Turning the problem around, the halo nuclei provide an interesting possibility for probing how a nucleus responds to a dilute environment of neutrons. This becomes particularly exciting in the case of ^{11}Li(0.340 MeV) which has one proton more than the unbound ^{10}He, but is a bound borromean system. The magnetic dipole moment has been measured both for ^{9}Li and ^{11}Li, respectively 3.4391 n.m and 3.6678 n.m and also the quadrupole moments [27] –27.4 mb and –31.2 mb, the latter being identical within the experimental errors Q(^{11}Li)/Q(^{9}Li) = 1.14(16). While the dipole moment for ^{11}Li is close to the Schmidt value of 3.71 n.m, that of ^{9}Li is significantly smaller. Both quantities do however only indirectly tell about the neutron distribution in the core. Although we feel that our three-body model has reasonable experimental support for quantitatively reasonable estimates of many observables, the core seems on the other hand not to be a completely passive ^{9}Li spectator. Its response is one of the objects of current studies.

5. ^{11}Li: what we have to assume

Any three-body model for ^{11}Li requires a number of theoretical inputs, many of which are presently only poorly constrained:

1. The potential between each halo neutron and the ^{9}Li core. Unlike the case of ^{6}He, insufficient experimental information is available on this two-body subsystem. The experimental evidence is conflicting, indicating n+^{9}Li

Table 1. Three-body calculations of ^{11}Li g.s. structure: physical two-body inputs and three-body results

Method	ref.	n+^{9}Li potential Physics input	Pauli Principle	$n+n$ potential	^{10}Li states (keV)	^{11}Li g.s. E (MeV)	rms (fm)	main WF	inclusive n, core widths	3-body breakup peak
Greens Ftn Method [10]	BE	WS Central, ls	$\perp$ to $0s$, $0p_{3/2}$	dens.-dep. contact	$0p_{1/2}$ at 800	–0.2		71% $(0p_{1/2})^2$	good	0.2 MeV
COSM [11]		Folded Central, ls	$\perp$ to $0s$, $0p_{3/2}$	HN no. 1	$0p_{1/2}$ at 800	+1	3.2	90% $(0p_{1/2})^2$		
Faddeev [12]	WS-F	WS Central, ls	$\perp$ to $0s$, $0p_{3/2}$	free	$0p_{1/2}$ at 200	–0.3	3.15	90% $(0p_{1/2})^2$	3× expt	4 MeV
Shell Model [6]	HG	Cohen-Kurath	full	Cohen-Kurath				93% $S=0$ 75% $(0p)^2$		
Faddeev [12]	WS-P	WS Central	$\perp$ to $0s$, $S=1$	free	$0p$ at 1400	–0.3	3.0	95% $[0p^2]_{S=0}$	2× expt	1.6 MeV
Cluster Model [40]	BB	Zero-range Central	none	dineutron assumed	none	–0.3 input	3.2	cluster $0s \times {}^1S_0$		0.2 MeV
Variational, HH [9]	JJH [8]	shallow –7.8 MeV	none	free	$0s_{1/2}$ at 100 [a]	–0.3	3.4	96% $(0s)^2$	good	0.6 MeV (good)
Faddeev [41]	WS-FI	WS Central, ls	$\perp$ to $0s$, $0p_{3/2}$	free	$1s_{1/2}$ at 10 [a]	–0.3	3.8	80% $(1s_{1/2})^2$ 17% $(0p_{1/2})^2$	good	

[a] The energies for s-wave virtual states are the energies at which $(\sin^2\delta)/E$ falls to half of its value at $E=0$

resonances at +0.8 MeV[34], at +0.4 MeV [35], at +0.2 MeV[36] and/or within 0.1 MeV of threshold[37]. Most theoretical calculations have assumed WS potentials of the usual form. Some[8,9], however, have used a shallow Gaussian potential ($V=-7.8$ MeV), but this has only an s-wave virtual state.

2. The set of occupied levels in the ^{9}Li core, against which the valence neutron state should be orthogonalised according to the Pauli Principle. Most models have assumed a neutron structure of $(0s)^2(0p_{3/2})^4$, which gives two p-shell holes as $(0p_{1/2})^{-2}$, having $S=1$ with $\frac{2}{3}$ probability. Some[6], however, indicate that the two p-shell holes should be almost entirely in an $S=0$ state.
3. The coupling to excited states of the ^{9}Li core. The g.s. electric quadrupole moment is known[27], but is largely determined by the unpaired $0p_{3/2}$ proton, and the neutron deformation could be different.

The overall matter deformation may well lead to positive-parity intruder states lying low in the ^{10}Li spectrum, as occurs in ^{11}Be. It may be that one or more of the resonances seen in ^{10}Li are $1s$ virtual states[38].

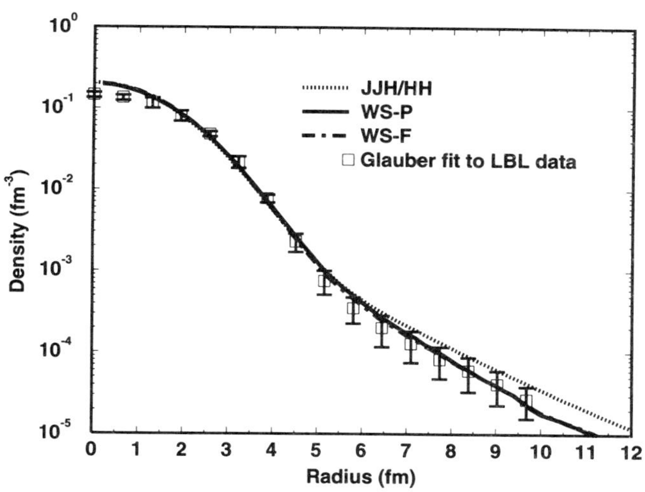

Fig. 5. Radial variation of the single-particle density of ^{11}Li for the JJH/HH, WS-F and WS-P calculations of Table 1. The error bars give the range of densities determined by Tanihata et al. [39] by a fit of Glauber reaction cross sections to experimental total cross sections

6. Models of halo structure

Once assumptions are made concerning the above, three-body equations can be solved to model the core + n + n structure. Table 1 summarises the different ways in which this has been done. The various cases will be referred to with the abbreviations in the second column.

What is noticeable here is the wide range of assumed physical inputs which can lead to a binding energy of 0.3 MeV and a r.m.s matter radius of 3.10 ± 0.17 fm[26]. The similarity of r.m.s. radii is illustrated in Fig. 5, where the radial part of the single-particle density $\rho(r)$ is plotted for the JJH/HH, WS-F and WS-P cases. The Glauber reaction cross section depends principally on this density. From the figure we see that the density distribution, and hence also reaction

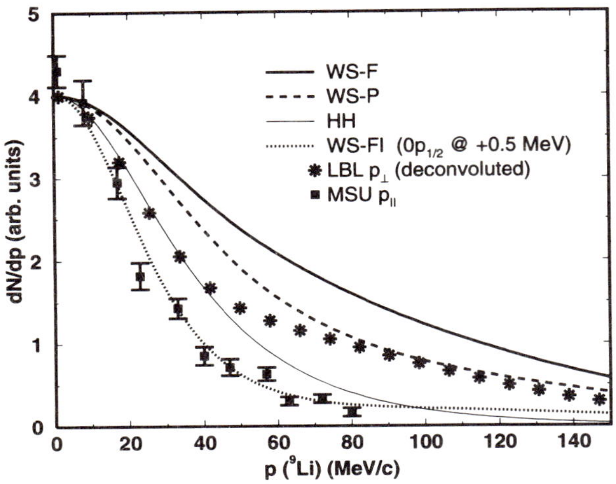

Fig. 6. Serber distributions for the rectilinear momentum of the ^{9}Li core following ^{11}Li fragmentation. The squares show $p_{||}$ data of [29], and the stars are the $p_\perp$ data of [28] corrected by them for multiple scattering in the target

cross sections, do not allow one to distinguish between the different cases.

The JJH/HH bound state has been replicated by solution of the Faddeev equations. The BE, COSM and WS-F methods should also produce similar results for similar inputs. The published BE result gives more binding because of its stronger (E = 0) binding of the singlet $n+n$ system, and because it neglects the repulsive potential in the $^3P_1$ channel. The COSM and WS-F cases, on the other hand, both give a pairing increment $\Delta \equiv 2E_{p_{1/2}} - E \sim 0.6$ MeV. We may conclude that a simple $0p_{1/2}$ resonance at +0.8 MeV is incompatible with the observed binding of ^{11}Li.

Although a number of the cases predict similar matter distributions, they have quite different weights for the internal angular-momentum quantum numbers. The BE, COSM, HG, WS-F and WS-P models have the valence neutrons mostly in the p-shell of the core mean field, whereas the JJH/HH, BB and WS-FI calculations have s-states dominating. These differences do not show up in measurements of total reaction cross section, but exhibit themselves in correlated two-particle densities and, to a lesser extent, in the distribution of fragment momenta seen in inclusive breakup experiments. See also the discussion in [22].

The inclusive momentum distributions are found to be largely independent of the choice of target nucleus[30,29,42], even though the reaction mechanisms will be very different. This can best be explained, if we ignore final-state interactions, by use of the Serber model. In this model, the inclusive distributions reflect the Fourier momentum distribution in the initial state in a manner largely independent of the precise reaction mechanism. This seems to be the case for the core fragment for ^{6}He (see Sect. 2) and for ^{11}Be ([43]).

The problem with ^{11}Li is that the Fourier distributions of the p-dominated wave functions are 2 or 3 times too wide, as seen in Fig. 6. The smallest widths are generated by models (JJH/HH, WSF-FI) which have the halo neutrons in s-wave states relative the core. As alluded to above, the JJH/HH scenario is the only case so far to have the calculated 3-body E1 strength function peak [31] at the correct energy (see Fig. 4). This scenario is unusual in that it has no explicit Pauli terms, but simply employs a shallow Gaussian potential which supports no occupied orbits. It may be that some kind of effective 'Pauli core' has thereby been included, but its success may also have derived from placing the valence neutrons in s states.

There is new evidence[37] that there is a nearly bound s-wave state in the ^{10}Li spectrum. This will not show up as an ordinary resonance, but will give rise to large cross sections at low $n-^9$Li relative energies. The effect of including this state is to completely change the nature of the ^{11}Li halo. If the state arises from a low-lying $1s_{1/2}$ mixture, as in ^{11}Be, then the ^{11}Li neutrons will now be in a cross-shell superposition of $(0p_{1/2})^2$ and $(1s_{1/2})^2$ configurations. This is the basis of the recent WS-FI calculations[41] listed in Table 1. The precise s and p-wave weights will depend on the position of the $0p_{1/2}$ resonance, but, if this is 0.4 – 0.8 MeV or more, the ^{11}Li halo will be dominated by s-wave configurations. These give adequately narrow inclusive momentum distributions for both n and core fragments, though final-state interactions have yet to be taken into account. The Serber model appears to be valid for the core momentum distribution, with a possible small correction from final-state interactions [22], and from sequential decay via sharp continuum resonances such as the 2^+ state in ^{6}He [44].

7. Conclusion

Although the theoretical activity related to radioactive beams and halo nuclei now is gaining momentum, we have only scratched the surface of the field. Calculations have yet to be performed which reproduce all the available data concerning the ^{11}Li halo neutrons, although predictions for the similar ^{6}He have been largely successful. Resolution of this problem will depend on clarification of the ^{10}Li states, and would also be facilitated by a direct measurement of the ^{11}Li binding energy. Only then might we hope to have an accurate understanding of this intriguing neutron-rich nucleus and of the next candidate in line ^{14}Be which is also a borromean nucleus and which also appears to have a halo.

Acknowledgement. We wish to acknowledge support from the Niels Bohr Institute and NORDITA, where much of this work was carried out and from ECT* in Trento where this work was completed. Comments from and discussions with Fangil Gareev, Jeff Tostevin and Jim Al-Khalili are appreciated. U.K. support from SERC grants GR/H 53648 and GR/H 53556 is acknowledged.

References

1 M.V. Zhukov, D.V. Fedorov, B.V. Danilin, J.S. Vaagen, J.M. Bang, I.J. Thompson: Nucl. Phys. **A552**, 353 (1993)
2 G.F. Bertsch, B.A. Brown, H. Sagawa: Phys. Rev. **C39**, 1154 (1989)
3 W. Koepf, Y.K. Gambhir, P. Ring, M.M. Sharma: CNRS-IN2P3 preprint ISN 91.24, Z. Phys. A (submitted)
4 C.R. Chinn, J. Dechargé, J.-F. Berger: Phys. Lett. B (submitted)
5 T. Hoshino, H. Sagawa, A. Arima: Nucl. Phys. **A506**, 271 (1990)
6 A.C. Hayes: Phys. Lett. **B254**, 15 (1991); A.C. Hayes, D. Strottman: Phys. Rev. **C42**, 2248 (1991); W.R. Gibbs, A.C. Hayes: Phys. Rev. Letts. **67**, 1395 (1991)
7 L.D. Skouras, H. Müther, M.A. Nagarajan: preprint nucl-th/9304011
8 L. Johannsen, A.S. Jensen, P.G. Hansen: Phys. Lett. **B244**, 357 (1990)
9 M.V. Zhukov, B.V. Danilin, D.V. Fedorov, J.S. Vaagen, F.A. Gareev, J.M. Bang: Phys. Lett. **B265**, 19 (1991)
10 G.F. Bertsch, H. Esbensen: Ann. Phys. **209**, 327 (1991)
11 Y. Tosaka, Y. Suzuki: Nucl. Phys. **A512**, 46 (1990)
12 J.M. Bang, I.J. Thompson: Phys. Lett. **B279**, 201 (1992); for errata see Surrey University preprint CNP93/4
13 F. Pougheon et al.: this conference
14 H.R. Weller, D.R. Lehman: Ann. Rev. Nucl. Part. Sci. **38**, 563 (1988)
15 J. Bang, C. Gignoux: Nucl. Phys. **A313**, 119 (1979)
16 V.I. Kukulin, V.M. Krasnopolsky et al.: Nucl. Phys. **A453**, 365 (1986)
17 B.V. Danilin, M.V. Zhukov, A.A. Korsheninnikov, L.V. Chulkov, V.D. Efros: Sov. Jour. Nucl. Phys. **49**, 351, 359 (1989)
18 B.V. Danilin, M.V. Zhukov, J.S. Vaagen, J.M. Bang, I.J. Thompson: Proc. 2nd Int. Conf. Radioactive Nuclear Beams, Louvain (1991), Adam Hilger, 185
19 M.V. Zhukov, B.V. Danilin, D.V. Fedorov, J.M. Bang, I.J. Thompson, J.S. Vaagen: Phys. Rep. **231**, 151 (1993)
20 M.V. Zhukov, L.V. Chulkov, B.V. Danilin, A.A. Korsheninnikov: Nucl. Phys. **A533**, 428 (1991)
21 T. Kobayashi: Nucl. Phys. **A538**, 343c (1992)
22 A.A. Korsheninnikov, T. Kobayashi: Riken preprint RIKEN AF-NP-148 (1993)
23 B.V. Danilin et al.: this conference
24 B.V. Danilin, M.V. Zhukov, J.S. Vaagen, J.M. Bang: Phys. Lett. **B302**, 129 (1993)
25 L.S Ferreira, E. Maglione, J.M. Bang, I.J.Thompson, B.V Danilin, M.V Zhukov, J.S Vaagen: Phys. Lett. **B316**, 23 (1993)
26 I. Tanihata et al.: Phys. Lett. **B206**, 592 (1988)
27 E. Arnold et al.: Phys. Lett. **B197**, 31 (1987); Lett. **B281**, 19 (1992)
28 T. Kobayashi et al.: Phys. Rev. Lett. **60**, 2599 (1988)
29 N.A. Orr et al.: Phys. Rev. Letts. **69**, 2050 (1992)
30 K. Riisager: 3rd Int. Conf. Radioactive Nuclear Beams, East Lansing, Michigan (May 1993)
31 K. Ieki et al.: Phys. Rev. Letts. **70**, (1993)
32 H. Esbensen, G.F. Bertsch: Nucl. Phys. **A542**, (1992) 310
33 C.A. Bertulani, A. Sustich: Phys. Rev. **C46**, 2340 (1992)
34 K.H. Wilcox et al.: Phys. Lett. **B59**, 142 (1975)
35 H.G. Bohlen et al.: Z. Phys. **A344**, 381 (1993)
36 A.I. Amelin et al.: Sov. J. Nucl. Phys. **52**, 782 (1990)
37 B.M. Young et al.: Phys. Rev. **C49**, 279 (1994)
38 F.C. Barker, G.T. Hickey: J. Phys. G **2**, L23 (1977)
39 I. Tanihata et al.: Proc. 2nd Int. Conf. Radioactive Nuclear Beams, Louvain (1991); Adam Hilger, 221; Phys. Lett. **B287**, 307 (1992)
40 C.A. Bertulani, G. Baur: Nucl. Phys. **A480**, 615 (1988)
41 I.J. Thompson, M.V. Zhukov: Phys. Rev. **C**, in press (1994)
42 R. Anne et al.: NFFS 6 Conference, Bernkastel-Kues, 19-24 July 1992
43 B. Cross: Nordita/Norfa Study Weekend on Halo Nuclei, Niels Bohr Institute, Copenhagen, March 1993
44 M.V. Zhukov, L.V. Chulkov, D.V. Fedorov, B.V. Danilin, J.M. Bang, J.S. Vaagen, I.J. Thompson: J. Phys. **G20**, 201 (1994)

This article was processed using Springer-Verlag TEX Z.Physik A macro package 1.0 and the AMS fonts, developed by the American Mathematical Society.

Z. Phys. A 349, 279–283 (1994)

Neutron-rich nuclei studied with AMD

H. Horiuchi, Y. Kanada-En'yo, A. Ono

Department of Physics, Kyoto University, Kyoto, 606-01 Japan

Received: 27 October 1993

Abstract. Antisymmetrized molecular dynamics (AMD) which is a newly constructed theory of quantum-mechanical microscopic simulation of heavy ion collisions is explained to present us with a very useful theoretical approach for the study of the structure-changes which occur when we go from ordinary nuclei to exotic neutron-rich nuclei. It is because of the flexible nature of the AMD wave function which can describe the formation and dissolution of clusters both in collision problems and in structure problems. After brief explanation of successful applications of AMD to heavy ion collisions and that of the frictional cooling technique for constructing minimum-energy wave functions, the results of the AMD analyses of Li, Be, and B isotopes are discussed. Very good reproduction of magnetic moments is obtained and is explained to be related with the structure-change between cluster structure and shell-model-like structure within each isotope. The superposition of AMD determinants is shown to reproduce neutron halo structure.

PACS: 21.10; 24.10; 27.20

1. Introduction

Our knowledge of nuclei far from stability line has been and is increasing by the use of unstable nuclear beams. For the study of unknown exotic nuclei, it is desirable to use a theoretical approach which is free from model assumptions and which can describe the systematic-changes of nuclear structure when we go from ordinary nuclei to exotic neutron-rich (or proton-rich) nuclei. Such theoretical framework is required to describe "formation and dissolution of clusters". This requirement is easily understood when we consider the Be isotopes. It is very well known that ^{8}Be has a well-developed α-α dumbbell structure while it is known that the ^{11}Be ground state is not deformed but has a spherical shell-model-like structure with an outermost neutron having a halo wave function.

In the field of intermediate and high energy heavy ion collisions, we know well that the formation of clusters is of fundamental importance and microscopic theories have been developed in order to describe the cluster production dynamics. Therefore such microscopic theory which treat the cluster production in heavy ion collisions can be a good candidate as a microscopic method to study the formation of clusters in nuclear structure problems. It was due to this reason that the antisymmetrized molecular dynamics (AMD) was used to study the formation and dissolution of clusters in nuclear structure problems [1, 2]. The AMD is a microscopic simulation theory of heavy ion collisions [1, 3, 4] and can be said to be an antisymmetrized version of the quantum molecular dynamics (QMD) [5]. The AMD is of quantum-mechanical nature since it describes the time-development of the system wave function. The AMD wave function is flexible enough to be able to describe variety of reaction processes including the cluster production process.

In this paper we report the results of the AMD analyses of the structure of Li, Be, and B isotopes. The construction of the wave functions is made by using the technique of frictional cooling. In calculations of heavy ion collisions by AMD, the initialization problem, namely the construction of the ground state wave functions of colliding nuclei, is treated by this frictional cooling technique. Due to the large flexibility of the AMD wave function, the AMD can describe variety of nuclear structure such as the double-closed-shell structure, prolate or oblate deformed structure, various clustering structure, neutron halo structure and so on. The AMD approach to the nuclear structure problem is a quite unique one since it is free from any model assumptions such as axial symmetry, existence of clusters and so on.

In the next section we briefly explain the framework of the AMD and in Sect. 3 we show some examples of the AMD calculation of heavy ion collisions. The frictional cooling technique is recapitulated in Sect. 4 and the discussion of the results of the AMD analyses of Li, Be, and B isotopes is made in Sect. 5.

2. Antisymmetrized molecular dynamics

In AMD, the wave function of the A-nucleon total system is described by a Slater determinant,

$$\Phi(\{\mathbf{Z}_j\}) = \frac{1}{\sqrt{A!}} \det[\varphi_j(\mathbf{r}_k)\chi_{\alpha_j}(\xi_k)],$$
$$\varphi_j(\mathbf{r}) = (2\nu/\pi)^{3/4} \exp\left[-\nu\left(\mathbf{r} - \frac{\mathbf{Z}_j}{\sqrt{\nu}}\right)^2 + \tfrac{1}{2}\mathbf{Z}_j^2\right]$$
$$\propto \exp\left[-\nu(\mathbf{r}-\mathbf{D}_j)^2 + i\frac{\mathbf{K}_j}{\hbar}\cdot\mathbf{r}\right], \tag{1}$$
$$\mathbf{Z}_j = \sqrt{\nu}\,\mathbf{D}_j + \frac{i}{2\hbar\sqrt{\nu}}\mathbf{K}_j,$$

where $\chi_{\alpha_j}(\xi)$ stands for the spin-isospin function and α_j represents the spin and isospin of j-th single particle state, $\alpha_j = p\uparrow,\ p\downarrow,\ n\uparrow$, or $n\downarrow$. The time developments of the coordinate parameters $\{\mathbf{Z}_j\}$ are determined by the time-dependent variational principle;

$$\delta \int_{t_1}^{t_2} \mathrm{d}t \frac{\langle\Phi(\{\mathbf{Z}_j\})|\left(i\hbar\frac{\mathrm{d}}{\mathrm{d}t} - H\right)|\Phi(\{\mathbf{Z}_j\})\rangle}{\langle\Phi(\{\mathbf{Z}_j\})|\Phi(\{\mathbf{Z}_j\})\rangle} = 0, \tag{2}$$

which leads to the equation of motion for $\{\mathbf{Z}_j\}$;

$$i\hbar \sum_{j\tau} C_{k\sigma,j\tau}\dot{Z}_{j\tau}$$
$$= \frac{\partial}{\partial Z^*_{k\sigma}} \frac{\langle\Phi(\{\mathbf{Z}_j\})|H|\Phi(\{\mathbf{Z}_j\})\rangle}{\langle\Phi(\{\mathbf{Z}_j\})|\Phi(\{\mathbf{Z}_j\})\rangle}, \tag{3}$$
$$C_{k\sigma,j\tau} \equiv \frac{\partial^2}{\partial Z^*_{k\sigma}\,\partial Z_{j\tau}} \log\langle\Phi(\{\mathbf{Z}_j\})|\Phi(\{\mathbf{Z}_j\})\rangle,$$

where $\sigma, \tau = x,\ y,\ z$. We can prove that the Hermitian matrix $\{C_{k\sigma,j\tau}\}$ is positive definite.

The above-described wave function and equation of motion of AMD are the same as those of the FMD (fermionic molecular dynamics) [6]. In AMD the following two problems are further treated in order to make AMD applicable to heavy ion collisions. One is the *initialization problem*, namely the construction of the ground state wave functions of colliding nuclei, and the other is the *incorporation of two-nucleon collisions.* The construction of the ground state wave function means to determine the values of all the coordinate parameters $\{\mathbf{Z}_j\}$ which minimize $\langle\Phi(\{\mathbf{Z}_j\})|H|\Phi(\{\mathbf{Z}_j\})\rangle/\langle\Phi(\{\mathbf{Z}_j\})|\Phi(\{\mathbf{Z}_j\})\rangle$. It is a multi-dimensional variational problem. The incorporation of two-nucleon collision is not a simple problem because the position parameters $\{\mathbf{D}_j\}$ do not represent nucleon positions and also the momentum parameters $\{\mathbf{K}_j\}$ do not represent nucleon momenta. This fact can be easily understood when we consider the following example. Let all $\mathbf{D}_j$ and $\mathbf{K}_j$ be vanishingly small in ^{16}O system. The resulting total wave function Φ is almost equal to the harmonic oscillator shell model wave function $\Phi_{\mathrm{c.s.}}$ of the double closed shell configuration of $0s$ and $0p$ shells, $(0s)^4(0p)^{12}$. Nucleons of the state $\Phi_{\mathrm{c.s.}}$ are neither all located at the space-coordinate origin nor all condensed to zero momentum state in spite of $D_j \approx 0$, $K_j \approx 0$ for all j.

The initialization problem is solved by the use of the frictional cooling method, which is discussed in Sect. 4. The incorporation of two-nucleon collisions is solved by the introduction of the physical nucleon coordinate parameters $\{\mathbf{W}_j\}$, which is discussed in Sect. 3.

The use of AMD in nuclear structure problems may be classified into two types. One is to use AMD in its original form, namely as a time-dependent theory. Various kinds of collective vibration have been studied in this way. The other is to use AMD in time-independent way. In the case of the latter type, the AMD wave function for the ground state constructed by the use of the frictional cooling is analysed in detail. The present paper reports the AMD study of the latter type.

3. AMD for heavy ion collisions

As mentioned in the previous section, the incorporation of two-nucleon collisions in AMD is made by introducing the physical nucleon coordinate parameters $\{\mathbf{W}_j\}$. The definition of $\{\mathbf{W}_j\}$ given in (4) is in mimicry of the definition of the canonical and physical coordinate in the time-dependent cluster model (TDCM) [7];

$$\mathbf{W}_j \equiv \sum_{k=1}^{A} (\sqrt{Q})_{jk}\mathbf{Z}_k,$$
$$Q_{jk} \equiv \frac{\partial}{\partial(\mathbf{Z}_j^*\cdot\mathbf{Z}_k)} \log\langle\Phi(\{\mathbf{Z}_l\})|\Phi(\{\mathbf{Z}_l\})\rangle. \tag{4}$$

The position coordinate $\mathbf{R}_j$ and the momentum coordinate $\mathbf{P}_j$ are the real and imaginary parts of $\mathbf{W}_j$; $\mathbf{W}_j = \sqrt{\nu}\,\mathbf{R}_j + \frac{i}{2\hbar\sqrt{\nu}}\mathbf{P}_j$. In the phase space of $\{\mathbf{W}_j\}$ there exists the Pauli-forbidden region just like in TDCM and the Pauli blocking of the two-nucleon collision is defined as the avoidance of entering into the Pauli-forbidden region. The Pauli-forbidden region is defined as the set of $\{\mathbf{W}_j\}$ for which there exists no $\{\mathbf{Z}_j\}$ satisfying (4). Since we can easily prove [3] the inequality,

$$\sqrt{\langle|\mathbf{W}_i - \mathbf{W}_j|^2\rangle_{\mathrm{average}}} = \sqrt{\frac{2(N_{\mathrm{osc}})_{\mathrm{min}}}{A-1}} \geqq \sqrt{2}, \tag{5}$$

the neighborhood region around the coordinate origin, $\mathbf{W}_j = 0$, $(j = 1 \sim A)$, is contained in the Pauli-forbidden region. Here $(N_{\mathrm{osc}})_{\mathrm{min}}$ stands for the minimum of the total number of oscillator quanta of A-nucleon system.

In trating heavy ion collisions the AMD theory is combined with the evaporation theory. Namely, AMD describes dynamical reation processes within several hundred fm/c but the decay processes of the excited fragments which are produced during the dynamical processes are described by the statistical multi-step decay theory. AMD was successfully applied to the studies of the nuclide distribution [3] and the energy spectra of fragments [4] in heavy ion reactions around the Fermi energy. One of the characteristic features of these AMD studies of heavy ion collisions is the ability of AMD to describe the shell effect in the fragment production during dynamical reaction processes. It is thanks to the quantum mechanical nature of AMD that treats the time development of the wave function of the system.

4. Frictional cooling

The frictional cooling method which we use for the construction of ground state wave functions in AMD is originally due to the cooling technique of Wilet et al. [8] in classical molecular dynamics approach which was used for example also in [9]. The AMD wave function to be constructed by the frictional cooling method can be extended into more general form as follows:

$$\Phi = \Phi(\{\mathbf{Z}_j\}) + C\Phi(\{\mathbf{Z}'_j\}) + C'\Phi(\{\mathbf{Z}''_j\}) + \dots . \tag{6}$$

Furthermore we can use parity projected state $\Phi^{\pm}$,

$$\begin{aligned}&\Phi^{\pm} = \Phi^{\pm}(\{\mathbf{Z}_j\}) + C\Phi^{\pm}(\{\mathbf{Z}'_j\}) + C'\Phi^{\pm}(\{\mathbf{Z}''_j\}) + \dots ,\\ &\Phi^{\pm}(\{\mathbf{Z}_j\} \equiv (1 \pm P)\,\Phi(\{\mathbf{Z}_j\}),\end{aligned} \tag{7}$$

where P is the parity operator.

In order to construct the wave function Φ for the ground state we must determine the value of all the parameters, $(\{\mathbf{Z}_j\}, (\{\mathbf{Z}'_j\}, (\{\mathbf{Z}''_j\},\dots, C, C',\dots)$, which minimize $\langle\Phi^{\pm}|H|\Phi^{\pm}\rangle/\langle\Phi^{\pm}|\Phi^{\pm}\rangle$. First we choose randomly parameters $\{u_k\} \equiv (\{\mathbf{Z}_j\}, (\{\mathbf{Z}'_j\}, (\{\mathbf{Z}''_j\},\dots, C, C',\dots)$. The wave function $\Phi^{\pm}$ (or Φ) with these parameters expresses a highly excited state with gaseous configuration of nucleons. This initial state is then cooled down to the energy minimum state by applying the equation of frictional cooling which is given as

$$i\hbar \frac{\mathrm{d}}{\mathrm{d}t} u_k = (\lambda + i\mu)\frac{\partial}{\partial u_k^*}\frac{\langle\Phi^{\pm}|H|\Phi^{\pm}\rangle}{\langle\Phi^{\pm}|\Phi^{\pm}\rangle}. \tag{8}$$

We can easily show the relation

$$\frac{\mathrm{d}}{\mathrm{d}t}\frac{\langle\Phi^{\pm}|H|\Phi^{\pm}\rangle}{\langle\Phi^{\pm}|\Phi^{\pm}\rangle} < 0, \quad \textit{for}$$
$$\mu < 0, \quad \lambda = \textit{arbitrary}. \tag{9}$$

Actual calculations have shown that the convergence to the minimum energy state is very fast (especially of course when a single AMD Slater determinant is adopted). The most important feature of the construction of the ground state with the frictional cooling method is that it is without prejudice i.e. free from model assumptions. It does not use any assumptions such as axial symmetry, existence of clusters, and so on.

Converged ground state configurations of the self-conjugate 4N nuclei have proved to be as follows [1]: They have negligibly small values for all K_j in all the cases of nuclei. In ^{8}Be, ^{12}C, and ^{16}O, ground state wave functions are almost identical to the Brink-type [10] two-alpha intrinsic wave function, the intrinsic state wave function of the Elliott SU(3) model [11], $(0s)^4(0p)^8(\lambda,\mu)=(0,4)$, and the double closed shell wave function, $(0s)^4(0p)^{12}$, respectively. These results are due to the calculations when a single AMD Slater determinant is adopted. But it was checked that when the linear combination of two AMD Slater determinants is adopted, the main AMD Slater determinant is almost the same as the AMD Slater determinant obtained by the calculation using a single AMD Slater determinant and the minor AMD Slater determinant mixes in with very small coefficient C. The above results show two things: First, the α-α structure of ^{8}Be has been proved theoretically without assuming the existence of any cluster structure. Second, the ground state wave functions expressed by the AMD theory are good and realistic.

The method of the frictional cooling can be extended so as to be used under given constraints [12]. An example of the appliction of the constrained frictional cooling method is the study of the structure-change of the ^{20}Ne nucleus along its yrast line for both positive and negative parities. The constraint in this case is that on the expectation value of the total angular momentum operator **J**. The structure change from $\alpha + {}^{16}$O clustering structure to oblately deformed shape and then to ^{12}C$+2\alpha$ was obtained.

5. Li, Be, and B isotopes

We studied the structure of Li, Be, and B isotopes with AMD combined with the frictional cooling method. Since the AMD wave function constructed with the frictional cooling method represents the intrinsic state, we make the angular momentum projection by performing the three dimensioal numerical integration. The two-nucleon force in the Hamiltonian H consists of the density-dependent nuclear force MV1-case 3 [13], the two-nucleon spin-orbit force G3RS [14], and the Coulomb force. The os-

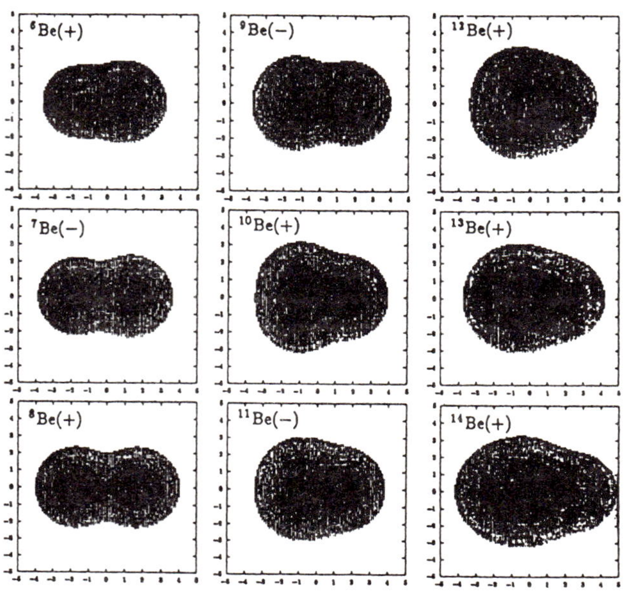

Fig. 1. Density distribution of Be isotopes

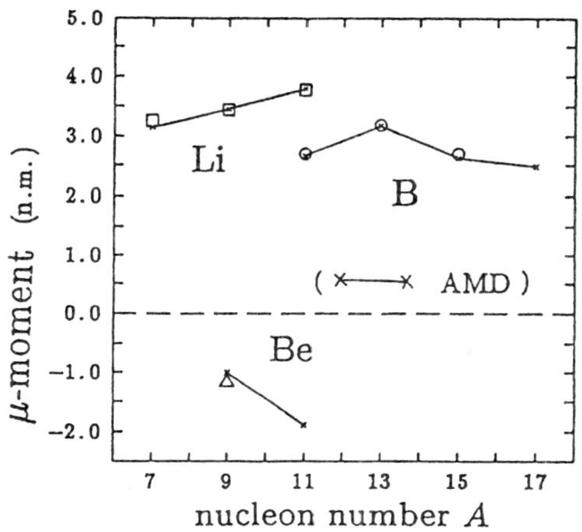

Fig. 2. Magnetic moments

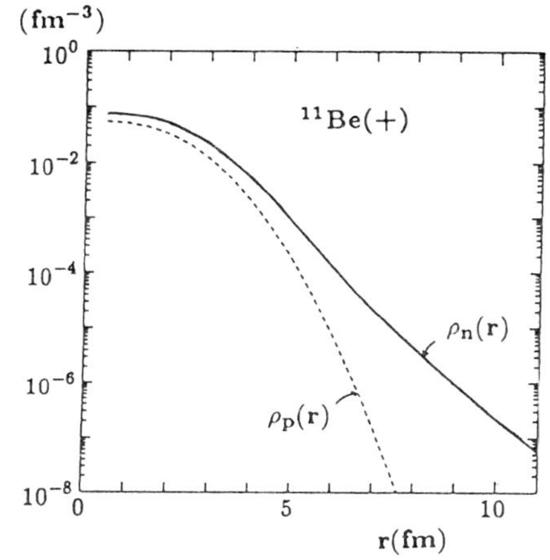

Fig. 4. Density distribution of ^{11}Be

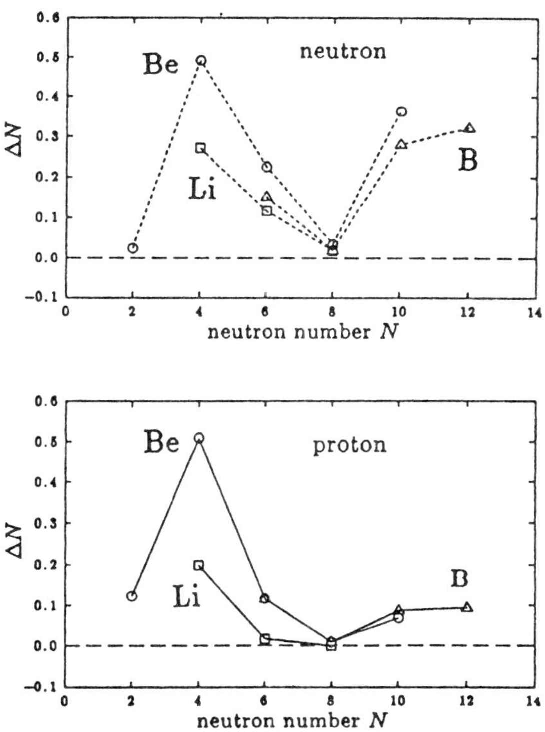

Fig. 3. Number of oscillator quanta

cillator constant ν was chosen so as to get the optimum binding energy for each intrinsic state with definite parity.

Calculations with linearly combined two AMD Slater determinants were compared to those with single AMD Slater determinant. In most cases it was found that the main AMD Slater determinant is almost the same as the AMD Slater determinant obtained by the calculation using a single AMD Slater determinant and the minor AMD Slater determinant mixes in with very small coefficient C. When we discuss the staste with neutron halo, however, the linear combination of only two AMD Slater determinants is not sufficient.

The energy level spectra are reproduced well like the results reported in [2]. In Fig. 1 the density distribution is shown for Be isotopes. We notice the deformation is small around ^{12}Be. Figure 2 shows good reproduction of magnetic moments which are close to the Schmidt value for nuclei with $N \approx 8$. This good reproduction is due to the change of the structure which has pronounced clustering for $N \approx Z$ but is shell-model-like for $N \approx 8$. This change is seen in Fig. 3 which shows the deviation of the number of total oscillator quanta from simple shell model values for protons and neutrons separately.

Since the single-particle wave function of AMD is a Gaussian wave packet with common oscillator constant, the long tail of neutron wave function of the neutron halo nucleus can not be sufficiently expressed by the linear combination of two AMD Slater determinants. Hence in the case of the positive parity state of ^{11}Be, we have superposed several AMD Slater determinants. The resulting density distributions of neutron and proton are shown in Fig. 4.

References

1. Horiuchi, H.: Nucl. Phys. A**522**, 257c (1991)
2. Horiuchi, H., Maruyama, T., Ohnishi, A., Yamaguchi, S.: Proc. Int. Conf. on Nuclear and Atomic Clusters, Turku (1991), Brenner, M., Lönroth, T., Malik, F.B. (eds.), p. 512. Berlin, Heidelberg, New York: Springer 1992; Proc. Int. Symp. on Structure and reactions of Unstable Nuclei, Niigata (1991), Ikeda, K., Suzuki, Y. (eds.), p. 108. Singapore: World Scientific 1992
3. Ono, A., Horiuchi, H., Maruyama, T., Ohnishi, A.: Phys. Rev. Lett. **68**, 2898 (1992); Prog. Theor. Phys. **87**, 1185 (1992)
4. Ono, A., Horiuchi, H., Maruyama, T., Ohnishi, A.: Phys. Rev. C**47**, 2652 (1993)
5. Aichelin, J., Stöcker, H.: Phys. Lett. **176**B, 14 (1986); Peilert, G., Stöcker, H., Greiner, W., Rosenhauer, A., Bohnet, A., Aichelin, J.: Phys. Rev. C**39**, 1402 (1989); Aichelin, J.: Phys. Rep. **202**, 233 (1991)

6. Feldmeier, H.: Nucl. Phys. A**515**, 147 (1990)
7. Saraceno, M., Kramer, F., Fernandez, F.: Nucl. Phys. A**405**, 88 (1983)
8. Wilet, L., Henley, E.M., Kraft, M., MacKellar, A.D.: Nucl. Phys. A**282**, 341 (1977)
9. Boal, D.H., Glosli, J.N.: Phys. Rev. C**38**, 1870 (1988)
10. Brink, D.M.: Proc. Int. School of Phys. "Enrico Fermi", p. 247, **course 36** (1965), Bloch, C. (ed.)
11. Elliot, J.P.: Proc. Roy. Soc. (London) A**245**, 128, 562 (1958)
12. Horiuchi, H., Ono, A., Kanada, Y., Maruyama, T., Ohnishi, A.: Proc. First Joint Italian-Japanese Meeting within the INFN-RIKEN Agreement on Perspectives in Heavy Ion Physics, Catania (1992), Di Toro, M., Migneco, E. (eds.), Italian Physical Society Conference Proceedings **38**, 223 (1993)
13. Furutani, H. et al.: Prog. Theor. Phys. Supple. No. **68**, Chap. III (1980)
14. Ando, T., Ikeda, K., Tohsaki, A.: Prog. Theor. Phys. **64**, 1608 (1980)

This article was processed using Springer-Verlag TEX Z. Physik D macro package 1.0 and the AMS fonts, developed by the American Mathematical Society.

Ionic hydrogen clusters : structure and fragmentation

B. Farizon, M. Farizon, M.J. Gaillard, E. Gerlic, and S. Ouaskit*

Institut de Physique Nucléaire de Lyon, IN2P3-CNRS et Université Claude Bernard, 43, boulevard du 11 Novembre 1918, 69622 Villeurbanne Cedex, France

Received 28 June 1993

Abstract. The structure of H_n^+ hydrogen clusters which are a nucleation of H_2 molecules around a H_3^+ core has been theoretically studied for n=5-13 (odd) using ab initio methods. For n larger than 9, the predicted structures can be thought as an addition of H_2 molecules around a weakly deformed H_9^+ core. Besides, mass-selected ionic hydrogen cluster beams of high energy are now available at the cluster acceleration facility at the Institut de Physique Nucléaire de Lyon. Fragmentation of 60-keV/amu - H_n^+ clusters induced by single collision on helium atoms is investigated for n = 9, 25 and 31. The deduced inclusive mass distributions of the charged fragments are measured from mass 3 to n-2. These distributions appear strongly different from those obtained at lower energy. An important production of intermediate-mass fragments is observed.

PACS: 36.40.+d

1. Introduction

Positive hydrogen cluster ions are of considerable experimental [1-10] and theoretical [11-14] interest since they are probably the simplest example of molecular clustering. The H_5^+ ion was first identified by Dawson and Tickner in 1962 [4] by mass spectrometry. Larger odd mass clusters were afterwards observed by Buckheit and Henkes [1] in TOF mass spectra and by Clampit and Gowland [2] using low-energy electron bombardement on solid hydrogen. Other workers [4] have observed odd mass hydrogen clusters in various nass distributions from low-energy collisional fragmentation studies. Kirchner and Bowers [4] have studied the dynamics of metastable H_5^+ fragmentation. Directly pertinent to the theoretical studies are experimental measurements [6-7] of the dissociation energies for the following process

$$H_3^+(H_2)_m \longrightarrow H_3^+(H_2)_{m-1} + H_2$$

with m $\leq$ 20. From a chemical point of vue, the most important experimental paper to date is the one written by Okumura, Yeh and Lee [9] on the infrared absorption frequencies measurements. The vibrational spectra of the H_n^+ cluster (n=3 to 13) have been observed near 4000 cm^{-1} by vibrational predissociation spectroscopy. In other respects, the interaction of fast ionic hydrogen clusters with solid target [10] has been studied recently, giving rise to theoretical development on cluster matter interaction.

Up to now, collisions of ionic hydrogen clusters with helium atoms have been mostly studied at low velocity (few eV/amu) [4] where molecular scattering is involved inducing mainly unimolecular dissociation of the incident cluster. However, the dissociation of the ionic hydrogen clusters by collisions with argon atoms has been studied at 20 keV/amu for cluster masses less than 25 [15]. At the present time, the availability of hydrogen clusters beams [16] up to 1.55 v_0 (v_0 being the Bohr velocity) opens a new field in the collision induced fragmentation of these clusters. Indeed, in this velocity range, the relative velocity between the projectile and the target atom is around or greater than the velocity of the velocity of the electrons in the cluster ($\approx$ the Bohr velocity). Then, the typical time for a collision with a target atom is short enough, compared to the typical time of the motion of the protons in the cluster, so that during the collision the protons can be considered to be stationary in the projectile frame [17]. Moreover, in this velocity range, the collision induces mainly electronic excitation up to ionization. Thus, processes such as the ionization of the incident cluster followed by the dissociation of the transient multicharged cluster can be involved.

After a brief description of these ionic hydrogen cluster from a theoretical point of view, results on the fragmentation [18] of fast ionic hydrogen clusters H_n^+ by single collision with helium atoms for various incident cluster masses will be presented.

* Permanent address : Université Hassan II, Faculté des Sciences II, B.P. 6621- sidi Othamane, Casablanca, Maroc

2. Structure

Among a substantial number of theoretical studies on positive hydrogen clusters, the investigation of Yamaguchi, Gaw, Remington and Schaefer [13] (YGRS) for the H_5^+ specie is of much interest. Larger H_{2n+1}^+ species have been studied at both levels of theory, the self-consistent field (SCF) and the configuration-interaction including all single and double excitations (CISD) [11.13]. Another study of Huber [12] has given near Hartree-Fock energies and structures of H_{2n+1}^+ (n=1-6) applying the floating orbital geometry optimization (FOGO).

The most recent ab initio SCF calculations [14] have been carried out for H_{2n+1}^+ (n=1-6) clusters using a triple-zeta plus polarization (TZP) basis set. This basis set was the Pople's one including five primitive *s* functions, contracted to 3 *s* functions, and *p* functions with the orbital exponent 0.75. Correlation effects were considered variationally by means of CISD calculation. This CISD method is well known to suffer from the absence of size consistency. To take into account this problem, size consistency corrections (SCC) have been estimated according to Davidson's scheme. Full geometry optimizations have been done, the optimization being stopped when the magnitude of the gradient is below 10^{-5} a.u.

Fully optimized structures and energies of H_{11}^+ and H_{13}^+ are presented in fig. 1. These structures can be thought as the addition of H_2 molecules to a weakly deformed H_9^+. Dissociation energies as a function of cluster size follow the pattern established experimentally by Hiraoka and Mori [7]. Nevertheless, our energy results on the biggest clusters suffer from the lack of size-consistency of CISD calculations. Moreover, since some of these potential hypersurfaces are extremely flat in the region of the stationary points, a harmonic vibrational analysis may be required to ascertain the character of each stationary point (local minimum or saddle point). Analytic gradient techniques have been used to locate stationary point geometries and to predict harmonic vibrational frequencies and infrared intensities at the two levels of theory, SCF (n=1 to 6) and CISD (n=1 to 4) both with TZP basis sets. Of special interest are the new vibrational modes of H_{11}^+ and H_{13}^+, which have no counterpart in the H_9^+ cluster. The predicted frequencies compare fairly well with the experimental results of Okumura, Yeh and Lee [9].

3. Fragmentation

Fragmentation phenomena are of experimental interest in various fields of physics and are the object of a fast developing field of statistical mechanics which concerns many systems of the microscopic world including polymers, gels, atomic clusters or atomic nuclei [19]. Until now, fragmentation of atomic clusters has been mostly studied by photon interaction [20] and low energy atomic collision (few eV/amu) [21,4]. Mainly atomic or molecular evaporation has been observed. Recently, fission process of a doubly charged cluster has been observed on Van der Waals [22] and mettalic [23] clusters. Mass-selected ionic hydrogen clusters of 60 keV/amu are now available up to mass 31 at the high-energy cluster-beam facility of the Institut de Physique Nucléaire de Lyon [16]. We report in the following on the experimental studies of the cluster dissociation and of the charged fragment production following the fragmentation of 60-keV/amu H_n^+ induced by single collision on helium.

3.1. Experimental set up

A new apparatus has been built in order to study hydrogen cluster fragmentation in single collision conditions. The experimental set-up is shown in fig. 2. Neutral clusters are formed in a cryogenic source and then ionized by electron impact. The beams from 0.3 to 1.86 MeV energies are delivered by a new radiofrequency (RF) quadrupole post accelerator associated to a Cockroft-Walton accelerator. The beam is pulsed with a cycle combining the cluster source cycle and the RF power one. After momentum analysis by a magnetic field, the incident beam is defined by two collimating apertures which ensure an angular dispersion of ± 0.8 mrad. Before reaching the gas target, the beam goes through two parallel plates. A voltage which depends on the energy of the beam can be applied or set to zero between these two plates with a fast voltage amplifier. Then, a cluster burst can either be deflected in the monitoring surface barrier (SB) detectror or reach the target. The intensity of the collimated incident cluster beam in a burst corresponds to about 1000 clusters per second. The target is a gas jet described in detail previously [24]. It is mounted on a micrometric goniometer that allows all possible translations (X,Y,Z) of the jet by steps of 50 μm. A differential pumping system is able to maintain a residual gas pressure of 5.10^{-7} Torr in the region close to the gas jet and along the beam line. The emergent beam is magnetically analyzed one meter after the target in a detection chamber which corresponds to a time of flight of 0.3 μs. The undissociated clusters and the charged fragments are then detected with movable SB detectors. Care was taken to ensure a complete collection of the cluster ions for a given q/m range. The detector is connected to a pulse height analyzer, the signal being proportional to the total energy of the detected fragment. The contamination of the incident cluster beam in the range of masses 15-19 can be rejected by the detection system. Indeed, as previously observed [25], the surface-barrier detector signal is different for a molecular ion including oxygen, carbon or nitrogen from the signal obtained with an hydrogen cluster of same mass and velocity.

3.2. Absolute dissociation cross section

Measurements of dissociation cross sections are necessary to determine the single-collision condition for

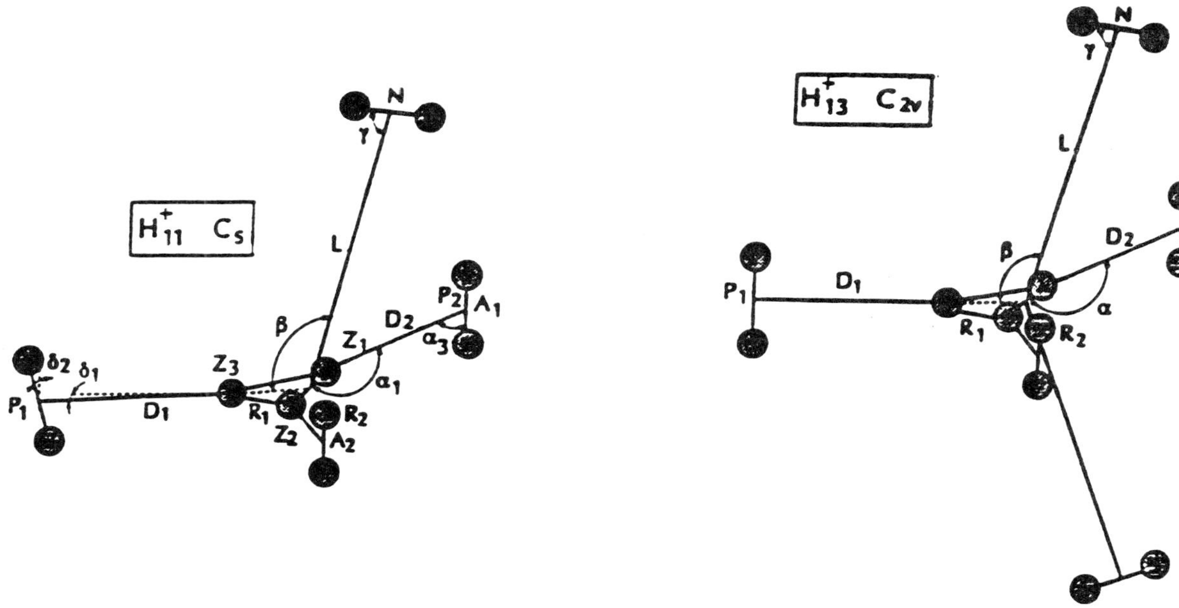

Level of theory	Energy (a.u.)	ZPE (a.u.)	SCC (a.u.)	R_1 (pm)	R_2 (pm)	D_1 (pm)	D_2 (pm)	P_1 (pm)	P_2 (pm)	L (pm)	β (deg)	γ (deg)	N (pm)	δ_1 (deg)	δ_2 (deg)	α_1 (deg)	α_2 (deg)	α_3 (deg)
SCF	-5.847 082	0.076 385		87.51	87.74	177.88	176.95	74.28	74.28	286.34	88.87	112.42	73.70	0.25	0.05	174.11	84.11	87.79
CISD	-6.024 303		-0.011 376	88.74	89.04	167.28	165.56	74.96	74.96	273.58	88.72	112.56	74.24	0.20	0.06	174.46	84.46	88.03

Level of theory	Energy (a.u.)	ZPE (a.u.)	SCC (a.u.)	R_1 (pm)	R_2 (pm)	D_1 (pm)	D_2 (pm)	P_1 (pm)	P_2 (pm)	L (pm)	β (deg)	γ (deg)	N (pm)	α (deg)
SCF	-6.980 708	0.088 384		87.38	87.84	178.75	176.74	74.26	74.26	289.42	109.30	89.02	73.70	153.02
CISD	-7.189 196		-0.016 283	88.46	89.20	168.52	165.17	74.88	74.90	275.08	109.68	88.84	74.18	152.24

Fig. 1. Fully optimized structures and energies of H_{11}^{+} and H_{13}^{+}

the fragment production. For a given mass-selected H_n^+ beam, the H_n^+ transmitted fraction F(x) is measured for various target thicknesses x. The thickness of the gas jet depends on two parameters, theinlet pressure which is measured with a precision better than 0.3% and the position of the jet capillary with respect to the incident beam which is known with a precision better than 4%. The position of the jet capillary with respect to the incident cluster beam is kept the same for the various incident cluster masses. The dissociation rate of the mass-selected incident H_n^+ in the beam line by interaction with the residual gas or by molecular evaporation has been measured for all incident clusters. It has been found to increase with the cluster mass from 1% up to 5% of the incident-beam intensity. In all cases, the transmitted fraction F(x) is found to decrease exponentially when the target thickness x increases. The dissociation cross section σ_n^{diss} is then deduced from the relation

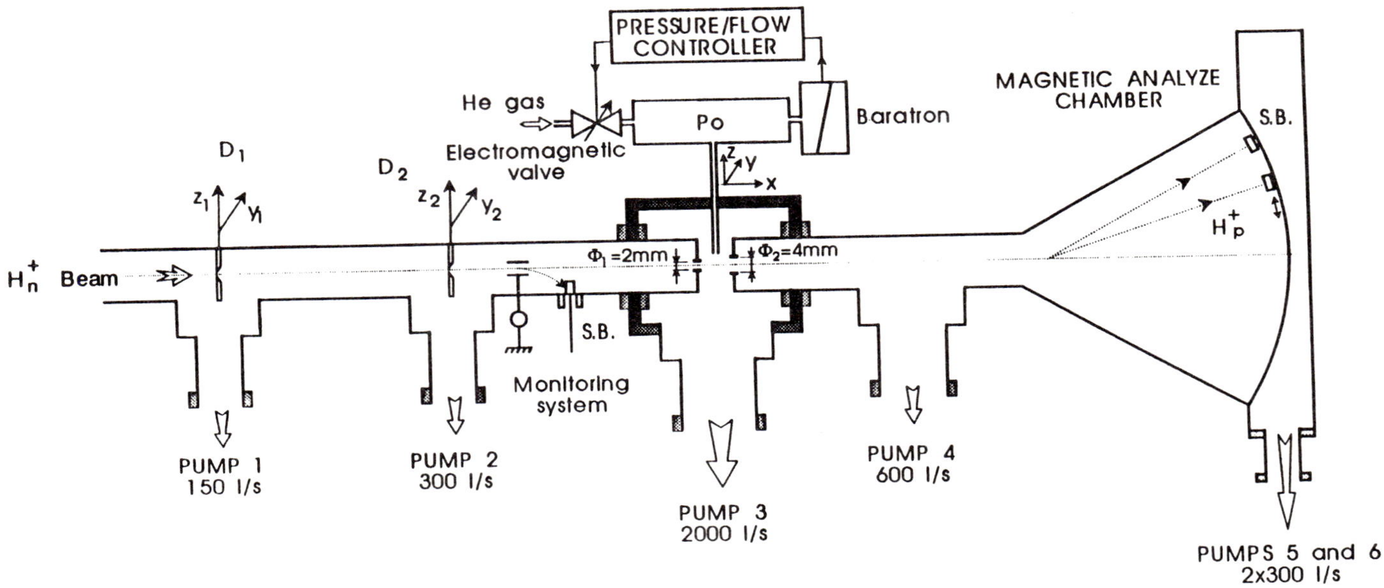

Fig. 2. The experimental set-up

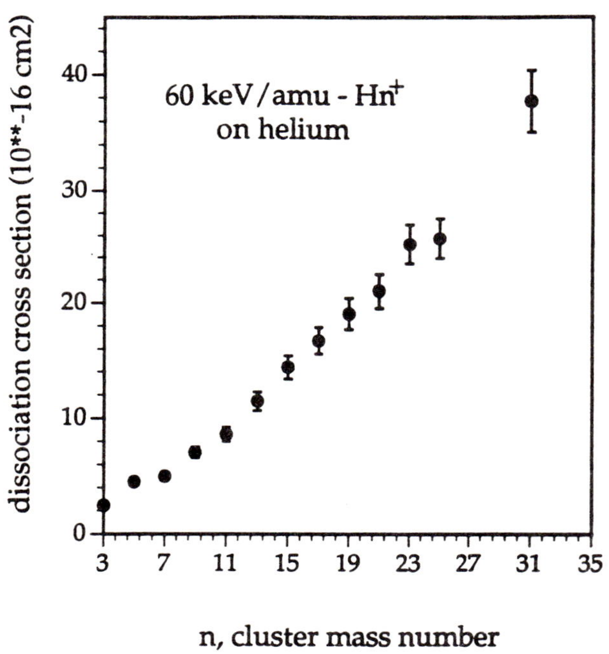

Fig. 3. H_n^+ dissociation cross sections versus n, at 60 keV/amu

$$F(x) = F(0)exp(-\sigma_n^{diss}x)$$

The determination of the target thickness x and tests of the absolute calibration are described in ref. [24]. The errors in the dissociation cross section measurements ($\pm 10\%$) arise from the uncertainty in the target thickness and from the beam intensity fluctuations.

The dissociation cross sections for the H_n^+ clusters, n ranging from 5 to 25, and 31, at 60 keV/amu. (1.55 Bohr velocity) are presented in fig. 3. The value obtained for H_3^+, $\sigma_3^{diss} = (2.5 \pm 0.2)10^{-16}cm^2$, is in agreement with the value (2 ± 0.3 $10^{-16}cm^2$) lineraly extrapolated from the only available data [26] obtained at higher energy (>120 keV/amu). As expected, the dissociation cross section increases with the mass of the cluster. While shell effects have been predicted in the theoretical description of the structure of the ionic hydrogen clusters [14] and have been observed in dissociation energy experiments [7], no shell effect is observed in the present results. Such fine effect connected to a small energy gap ($\approx$ 0.1 eV) between open and closed shells is probably hidden by the large range of impact parameters and energy transfers which can lead to the dissociation. For $n > 9$, the dissociation cross section increases proportionally to n. The $n^{2/3}$ dependence previously observed [15] in the same velocity range in collisions with argon atoms is not observed. The linear dependence of the dissociation cross section by collision with helium versus n seems to show that the H_2 molecules in the cluster contribute in an equivalent way to the dissociation, independently of their position in the incident cluster. This must be connected to the fact that these small clusters cannot be approximated by spherical droplets and is propably due, as already mentioned, to an effect of average over the various impact parameters.

3.3. Absolute fragment production cross section

The inclusive mass spectrum of charged fragments from mass p = 5 to 23 obtained by collision of H_{25}^+ incident clusters with helium atoms is displayed in fig. 4. This spectrum lets appear clearly 10 separated peaks where the total energies are ranging from 0.3 to 1.5 MeV. The highest energy peak corresponds to the detection of the transmitted H_{25}^+ clusters. Fragments of odd mass

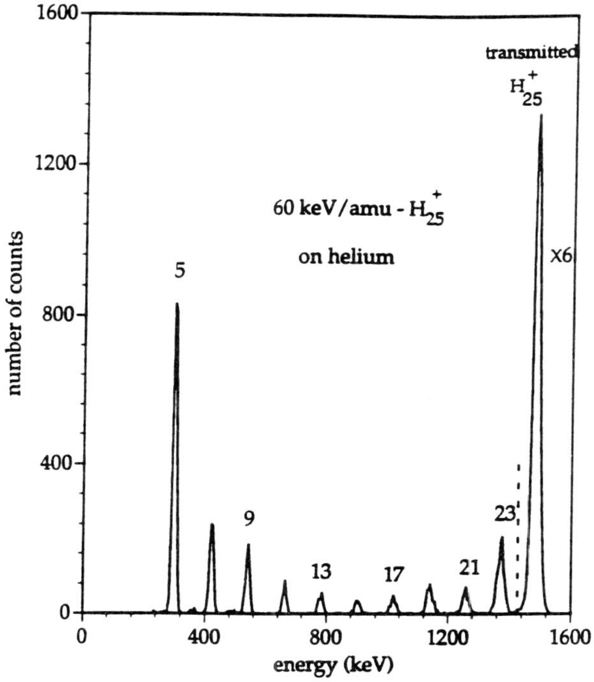

Fig. 4. The mass spectrum of charged fragments from n=5 to 23 measured by collision of H_{25}^+ incident clusters with helium atoms at 60keV/amu. The highest energy peak (to be multiplied by 6) corresponds to the detection of the transmitted H_{25}^+ clusters

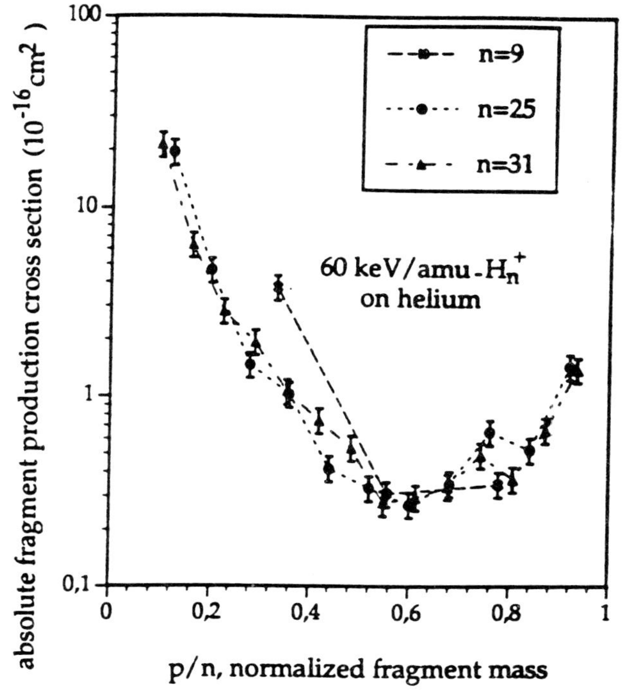

Fig. 5. The production cross section of fragment of mass p resulting from the fragmentation of H_n^+ clusters colliding with helium at 60 keV/amu versus p/n, the normalized fragment mass

number are mostly obtained, as already observed in the spectrum of the cluster source. The width of the peaks increases with the size of the detected fragment. This effect is connected to the response of the detector. The spectrum is obtained with a target thickness of $0.95 \pm 0.09\ 10^{14} atoms/cm^2$ and for a given number of incident clusters. We have also measured the spectrum obtained without gas target for the same number of incident clusters. The spurious events corresponding to about 3% of the incident clusters are likely due to collisions of incident clusters with the residual gas and to the spontaneous evaporation of a H_2 molecule after the last beam magnetic analysis. This last magnetic analysis takes place $10\mu s$ after the cluster ionization and $1\mu s$ before the detection.

From the dissociation cross section value, we can deduce that only 22% of H_{25}^+ incident clusters are dissociated in the gas jet. Therefore, since the cluster dissociation cross sections decrease with the cluster size, the double collision processes are negligible and the fragment yield is measured in single collision conditions. The absolute cross sections of the fragment production are deduced by dividing the number of fragments per incident cluster by the target thickness after substraction of spurious events. The errors in the absolute cross sections deduced from these measurements vary from ±10% to ±15%, taking into account both the target thickness and the statistical ubcertainties. Similar studies have been carried out for H_{31}^+ and H_9^+ incident clusters.

The production cross sections of mass-p fragments (p=3 to n-2) resulting from the fragmentation of H_n^+ clusters colliding with helium at 60 keV/amu are presented in fig. 5 versus p/n, the normalized fragment mass. The most surprising main feature of these spectra is the important production of intemediate-mass fragments (IMF) for p/n ranging from 0.15 to 0.6. The behavior of this fragment production is clearly different from the one obtained in low-energy collision [4] (few eV/amu) of H_{25}^+ clusters on helium (fig. 6) where mainly molecular evaporation results from the collision. This last process leads to a fragment distribution where the yield of each fragment increases with its size since a small energy transfer in the collision is more probable than a large one. In our case, the molecular evaporation process can be connected to the right side of the fragment mass distribution but cannot explain the entire distribution. In addition, for H_{25}^+ and H_{31}^+ incident clusters, the inclusive mass distributions of the fragment yield are found to be quite superimposed and, for p/n values smaller than 0.4, they can be parametrized with a power law $(p/n)^{-2.5}$. We point out that these distributions look similar to those obtained for high energy proton (a few GeV) [19] induced fragmentation of nuclei which have been connected to the existence of nuclear multifragmentation process. In the same way, the distribution corresponding to low-energy collision where mainly evaporation is involved could be compared to nuclear evaporation reaction resulting from collision with protons of about a few hundred MeV energy [19].

Besides this general behavior, one can remark some specific behavior specially in the p/n range from 0.7 to

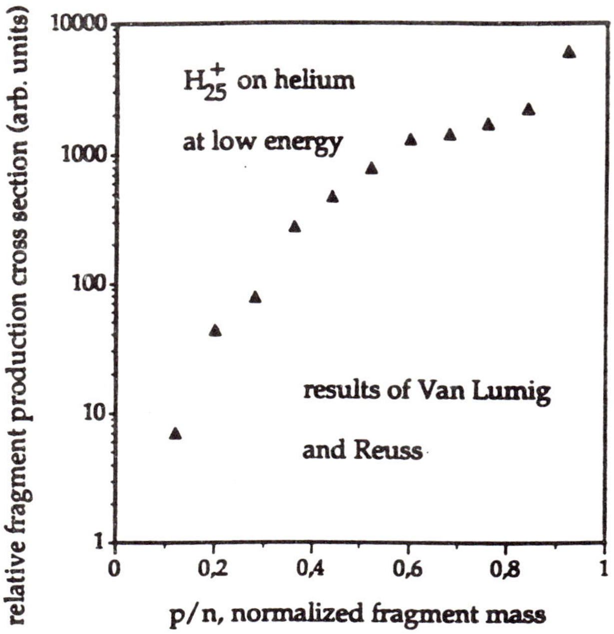

Fig. 6. The relative production cross section of fragment of mass p resulting from the fragmentation of H_{25}^+ clusters in low-energy collision [4] (few eV/amu) on helium versus p/n, the normalized fragment mass

0.8. In addition, one can notice that the H_9^+ fragment production is enhanced compared to H_7^+ and H_{211}^+ production as clearly seen on fig. 1 in the H_{25}^+ case. This can be related to the shell effects predicted in the structure of ionic hydrogen clusters [14] as discussed previously.

4. Conclusion

In conclusion, an important intermediate-mass fragment production resulting from a single collision of mass-selected ionic hydrogen clusters at 60 keV/amu on helium is observed. The only molecular evaporation process cannot explain the whole mass fragment distribution. Moreover, the behavior of the inclusive mass fragment distribution leads to expect a cluster multifragmentation. Exclusive measurements should uncover other cluster multifragmentation channels. Besides, some specific shell effects observed here have to be investigated with other incidents clusters and could be a helpful guide to theoretical work on structures of bigger size clusters.

The authors wish to thank R. Genre, J. Martin, R. Filliol, J.P. Lopez, H. Mathez for their expert technical assistance.

References

1 Buckheit, K., and Henkes, W.: Z. Angew. Physik **24**, 191 (1986)

2 Clampit, R., and Gowland, L.: Nature **223**, 815 (1968)

3 Kulazhenkova, N.A., Larin, I.K., and Talroze, V.L., Dokl. Phys. Chem. **207**, 915 (1972)

4 Thomson, J.J.: Phil. Mag. **24**, 209 (1912); Dawson, P.H., and Tickner, A.W.: J. Chem. Phys. **37**, 672 (1962); Van Deursen, A., and Reuss, J.: Intern. J. Mass Spectrom. Ion Phys. **11**, 483 (1973); Van Lumig, A., and Reuss, J.: Intern. J. Mass Spectrom. Ion Phys. **27**, 197 (1978)

5 Arifov, U.A., Pozhavov, S.L., Chernov, I.G., and Mukhamediev, Z.A.: High Energy Chem. (USSR) **5**, 69 (1971)

6 Bennet, S.L., and Field, F.H.: J. Am. Chem. Soc. **94**, 8669 (1972); Elford, M.T., and Milloy, H.B.: Australian J. Phys. **27**, 795 (1974); Elford, M.T.: j. Chem. Phys. **79**, 59 (1983); Johnsen, R., Huang, C.M., and Biondi, M.A.: J. Chem. Phys. **65**, 1539 (1976); Beulher, R.J., Ehrenson, S., and Friedman, L.: J. Chem. Phys. **79**, 5982 (1983)

7 Hiraoka, K., and Kebarle, P.: J. Chem. Phys. **62**, 2267 (1975); Hiraoka, K.: J. Chem. Phys. **87**, 4048 (1987); Hiraoka, K., and Mori, T.: Chem. Phys. Letters **157**, 467 (1989)

8 Oka, T.: Phys. Rev. letters **45**, 531 (1980)

9 Okumura, M., Yeh, L.I., and Lee, Y.T.: J. Chem. Phys. **83**, 3705 (1985); Okumura, M., Yeh, L.I., and Lee, Y.T.: J. Chem. Phys. **88**, 79 (1988)

10 Mazuy, B., Belkacem, A., Chevallier, M., Gaillard, M.J., Poizat, J.C., and Remillieux, J.: Nucl. Instr. Methods B **28**, 497 (1987); Farizon, M., Clouvas, A., de Castro Faria, N.V., Denis, A., Desesquelles, J., Farizon-Mazuy, B., Gaillard, M.J., Gerlic, E., and Ouerdane, Y.: Phys. Rev. A **43**, 121 (1991); Farizon, M., de Castro Faria, N.V., Farizon-Mazuy, B., and Gaillard, M.J.: Phys. Rev. A **45**, 179 (1992); de Castro Faria, N.V., Farizon-Mazuy, B., Farizon, M., Gaillard, M.J., Jalbert, G., Ouaskit, S., Clouvas, A., and Katsanos, A.: Phys. Rev. A **46**, 3594 (1992)

11 Yamabe, S., Hirao, K., and Kitaura, K.: Chem. Phys. Letters **56**, 546 (1978); Hirao, K., and Yamabe, S.: Chem. Phys. **80**, 237 (1983)

12 Huber. H.: Chem. Phys. Letters **70**, 353 (1980); Huber, H., and Szekely, D.: Theoret. Chim. Acta **62**, 499 (1983); Huber, H.: J. Molecular Structure **121**, 287 (1985)

13 Yamaguchi, Y., Gaw, J.F., and Schaefer, H.F. III: J. Chem. Phys. **78**, 4047 (1983); Yamaguchi, Y., Gaw, J.F., Remington, R.B., and Schaefer, H.F. III: J. Chem. Phys. **86**, 5072 (1987)

14 Farizon, M., Farizon-Mazuy, B., de Castro Faria, N.V., and Chermette, H., Chem. Phys. Letters **177**, 451 (1991); Farizon, M., Chermette, H., and Farizon-Mazuy, B.: J. Chem. Phys. **96**, 1325 (1992)

15 Chevallier, M., Clouvas, A., Frischkorn, H.J., Gaillard, M.J., Poizat, J.C., and Remillieux, J.: Z. Phys. D **2**, 87 (1986)

16 Gaillard, M.J., Schempp, A., Moser, H.O., Deitinghoff, H., Genre, R., Hadinger, G., Kipper, A., Madlung, J., and Martin, J.: Z. Phys. D **26**,S 347 (1993)

17 Kanter, E.P., Cooney, P.J., Gemmell, D.S., Groeneveld, K.O., Pietsch, W.J., Ratkowski, A.J., Vager, Z., and Zabransky, B.J.: Phys. Rev. A **20**, 834 (1979)

18 Ouaskit, S., Farizon, B., Farizon, M., Gaillard, M.J., and Gerlic, E.: to be published in Phys. Rev. A; Ouaskit, S., Farizon, B., Farizon, M., Gaillard, M.J., and Gerlic, E.: submitted

19 Campi, X.: Proceedings of the 107 course of the Int. School of Physics "Enrico Fermi" on the Chemical Physics of atomic and molecular clusters, (Varenna, 28 Juin - 7 Juillet, 1988, Italie); Campi., X.: Nucl. Phys. A **495**, 259c (1989)

20 Geusic, M.E., and Freedman, R.R.: J. Chem. Phys. **88**, 163 (1988) and ref. therein

21 Jarrold, M.F., Bower, J.E., and Kraus, J.S.: J. Chem. Phys. **86**, 3876 (1987)

22 Kreisle, D., Echt, O., Knapp, M., Recknagel, E., Leiter, K., Märk, T.D., Säenz, J.J., and Soler, J.M.: Phys. Rev. Lett. **56**, 1551 (1986); Gotts, N.G., and Stace, A.J.: Phys. Rev. Lett. **66**, 21 (1991)

23 Bréchignac, C., Cahuzac, Ph., Carlier, F., and de Frutos, M.: Phys. Rev. Lett. **64**, 2893 (1990); Katakuse, I., Ito, H., and Ichihara, T.: Int. J. Mass. Spectrom. Ion. Processes **97**, 47 (1990)
24 Ouaskit, S., Farizon-Mazuy, B., Farizon, M., Gaillard, M.J., Gerlic, E., and Stern, M.: in press in Phys. Rev. A (1993)
25 Ray, M.: Thesis, Université Lyon I, 1991, unpublished
26 Wolf, W., Coelho, L.F.S., Wolf, H.E., and de Castro Faria, N.V.: Phys. Rev. A **45**, 2978 (1992)

This article was processed using Springer-Verlag TEX Z.Physik D macro package 1.0 and the AMS fonts, developed by the American Mathematical Society.

Z. Phys. A 349, 291–296 (1994)

Statistical multifragmentation

D.H.E. Gross

Hahn-Meitner-Institut, Bereich Kern- und Strahlenphysik, Glienickerstrasse 100, D-14109 Berlin, and Fachbereich Physik der Freien Universität, Berlin, Germany

Received: 27 October 1993

Abstract. Two scenarios for the fragmentation of nuclei in collisions at intermediate energies are conceivable: A fast direct "shattering" or "abrasion – ablation" of a cluster has frequently been proposed in high-energy nuclear collisions. On the other hand there is an increasing number of data that suggest a nearly complete equilibration of the excited system before it breaks "simultaneously" into several pieces. The size-distribution, the size-correlations, the energy-spectra as well as the angular distributions are consistent with a loss of memory of the entrance channel. Because the latter scenario is ruled by the accessible phase-space and therefore by thermodynamics of a closed small system, which is a challenging problem of its own right, we treat only the *statistical* fragmentation of small equilibrized sources. A very similar situation is expected of the fragmentation of multiply charged metal clusters like Na_n^{z+}. With the help of *Microcanonical* Metropolis-Monte Carlo (MMMC) we study multifragmentation as a realistic example of thermodynamics of a finite many-body system subjected to long-range forces using the most basic prinicples of statistical physics. MMMC is an alternative to molecular dynamics but it has the advantage that one only samples the final states one is interested in and to get the relevant branching ratios.

PACS: 24.60; 24.75

1. Introduction

Meanwhile nuclear multifragmentation is experimentally well established as a dominant decay mode of a heavier nucleus opening after fission at higher excitation energies, e.g. [1, 2, 3]. Many details seem also to indicate that multifragmentation is a more "simultaneous" disassembly of a hot *equilibrized* source. To avoid unneccessary and fruitless discussions: these decays are fast, "simultaneous", on the time scale of disintegration of the system in the long range Coulomb field. On the time scale of traversing the short range of the nuclear field they may very well be sequentially binary. The essential point here is that subsequent decays remain coupled at least by the long-range Coulomb field in contrast to assumptions of conventional theories of binary decay.

In this simplified view nuclear fragmentation is an example of a small system in statistical equilibrium under the action of *long range forces.* Such systems are presently barely explored but very important. Multifragmentation of multiply charged molecular clusters and selfgravitating systems like the solar planetary system are important examples. In this respect the study of statistical nuclear fragmentation is highly relevant.

2. Nuclear thermodynamics

Under the condition of the decay of an equilibrized source at freeze-out one may formulate nuclear thermodynamics. However, ordinary thermodynamics distinguishes intensive variables like temperature, pressure and chemical potential from extensive ones like energy, entropy, volume and particle number. These definitions are useful in the thermodynamic limit for large systems with short range forces. As we are dealing with a finite system which cannot be extrapolated to the thermodynamic limit because of the long range Coulomb-force we need a more flexible definition of thermodynamics than usual.

We define a system to be at statistical equilibrium if it populates uniformly all accessable configurations in phase-space, i.e. all quantum states, which are consistent with the fundamental conservation laws of energy, momentum, angular-momentum, charge and mass. A key quantity of a statistical description of a many-body system at equilibrium is the partition-sum, Z, the number of all accessible quantum states of the system. Semiclassically, it is the sum of all different realisations of the system

Invited talk presented at the 2nd International Conference on Atomic and Nuclear Clusters 93, Santorini (Greece), June 28–July 2, 1993

in phase-space measured in units of $(2\pi\hbar)^d$. d is the number of degrees of freedom of the system.

In the case of fragmentation of nuclei at medium energies the total number of degrees of freedom is of the order of ~ 1000 and we have the chance to calculate Z by a clever Monte Carlo (Metropolis) routine directly. E.g., the entropy of the system is given by

$$S = \ln(Z) \tag{1}$$

and the temperature T by

$$\frac{1}{T} = \frac{\partial S}{\partial E} \tag{2}$$

which can be evaluated by sampling $\langle \partial/\partial E \rangle$ [4]. Integrating $1/T(E)$ over E gives then S or $Z = e^S$. Other thermodynamic quantities are obtained similarly.

2.1. *Microcanonical Metropolis Monte Carlo*

Of course, one cannot simply count the number Z of possible configurations, which is typically of the order of $\sim 10^{20}$ to $\sim 10^{30}$. However, expectation-values like

$$\langle F \rangle = \frac{\mathrm{tr}(F\rho)}{\mathrm{tr}(\rho)}, \tag{3}$$

are evaluated in a *Microcanonical* Metropolis Monte Carlo (MMMC) [5] routine as the average of F over a sufficiently large sample of *the most important configurations* of the system at given total energy, momentum, angular-momentum, mass, and charge *contributing to* $\langle F \rangle$ which occur in the sample with a frequency proportional to their *relative* phase-space volume. MMMC allows to find this *systematically* in an acceptable CPU-time. $\rho(E)$ is the microcanonical statistical operator with $\mathrm{tr}(\rho) = Z$. The mathematical formulation of a realistic though exact microcanonical Metropolis sampling was first given by [6]. An alternative possibility to simulate a microcanonical system is molecular dynamics (MD), or also quantum-molecular dynamics (QMD) [7]. At multifragmentation typically of the order of 10^{20} different exit channels are open. As one does not know the important *initial* configurations which contribute to the desired *final* observable one has to compute all transitions. This is of course not possible. Moreover, in contrast to MMMC, MD and likewise QMD do not provide any systematic way to obtain the branching ratios (the difference of entropy) of selected fragmentation channels. This can only be achieved by a diffusion through the phase-space guided by the relative contributions to the desired exit channels (importance sampling) as in MMMC.

2.2. *Physe-transition towards multifragmentation*

In thermodynamics phase-transitions are especially interesting. The reasons are twofold: first, phase-transitions are quite dramatic changes of the thermodynamic behaviour of the system, on the other hand phase-transitions

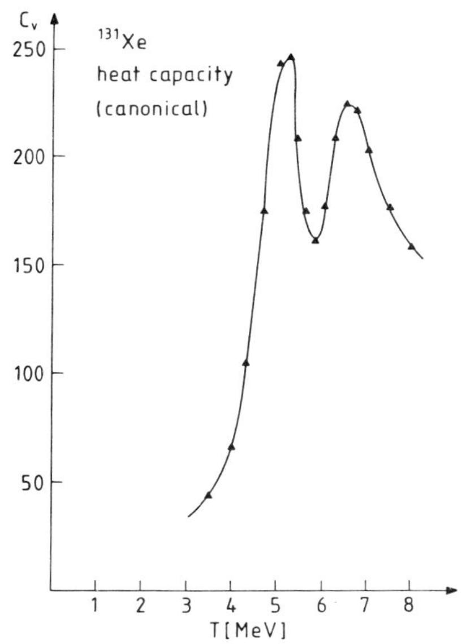

Fig. 1. Canonical Metropolis simulation of the heat-capacity c_v of ^{131}Xe at constant freeze-out volume as a function of the temperature

of second order are especially simple, as details of finite-range forces are unimportant compared to the long range of the critical correlations. Many quite different systems fall into a few universality classes near to the critical point of the transition depending on the geometry only. Here we are interested in phase-transitions of *microcanonical finite* systems. In finite systems phase-transitions of first and of second order are indistinguishable when the correlation length is of the order of the size of the system. As a function of the excitation energy the entropy of the system increases quickly at the transition.

In a canonical simulation of the heat-capacity $c_v(T)$ of ^{131}Xe at constant (freeze-out-) volume (Fig. 1) we see distinguishable maxima at $T \approx 5$ MeV and at $T \approx 6.5$ MeV which signal proper phase-transitions. The first one corresponds to the onset of fragmentation into two larger fragments (hot fission) and the higher one into three or more fragments of intermediate mass, IMF's, while several small fragments with $A < 10$ may be produced simultaneously.

2.3. *The effect of the Coulomb-force*

The ratio of the surface to the leading volume energy and of the ratio of the long-range self-energies to the volume energy depend oppositely on the size A of the fragments ($A^{-1/3}$, $A^{2/3}$ resp.) and on the temperature T: While at low temperature the droplet is held together by the surface-tension it fissions at higher temperature where the surface tension gets reduced due to surface-fluctuations. Now, and this is new, it will fragment into more than two pieces at even higher temperatures because then only smaller fragments are stabilized by the then weaker surface-tension.

In Fig. 2 we show the relative yields of decays of ^{197}Au* into final channels classified by the number of

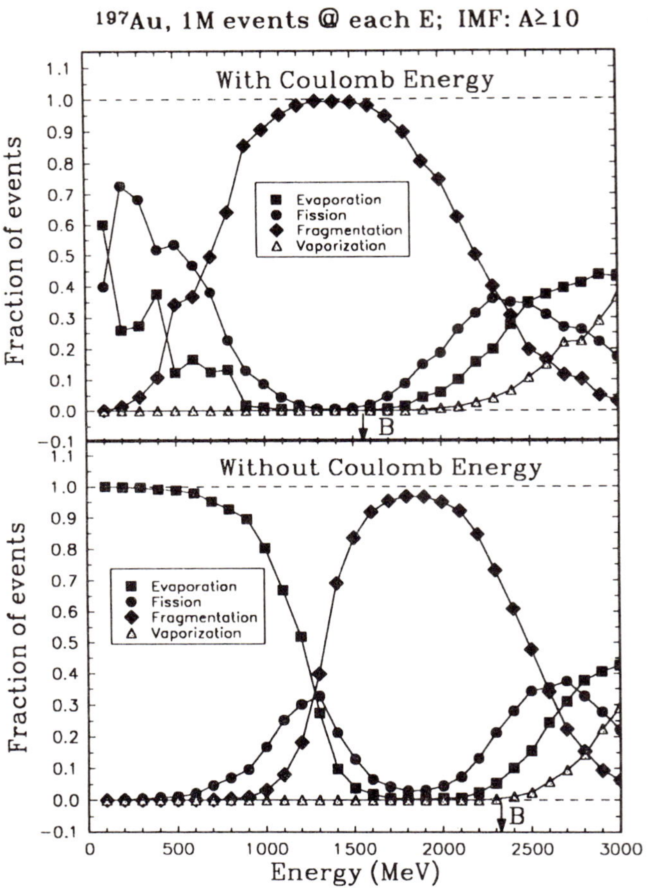

Fig. 2. Relative yield of decays of ^{197}Au into different final channels distinguished by the number of fragments with $A \geqq 10$, see text

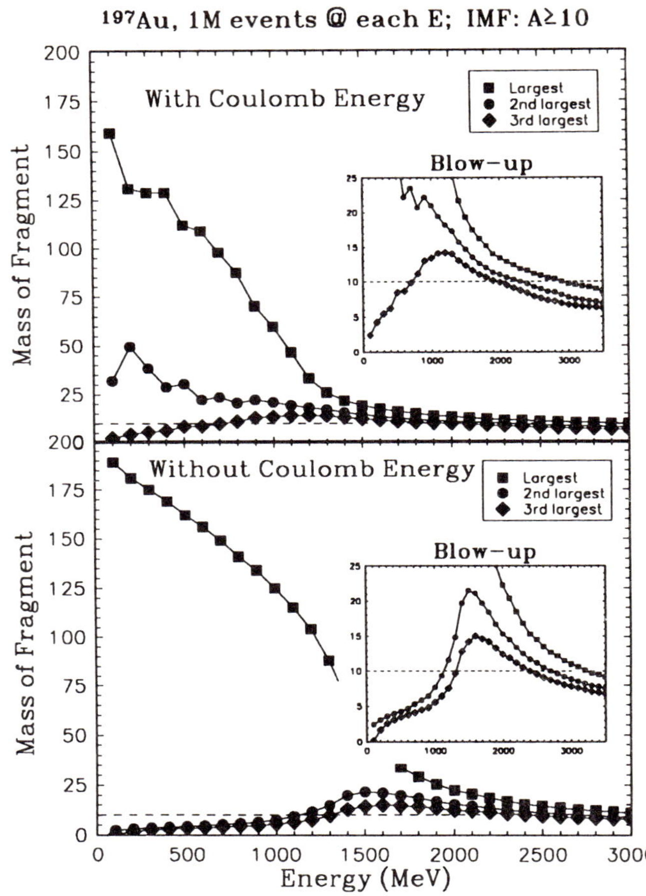

Fig. 3. Mean values of the mass of the targest, second largest and third largest fragment vs. excitation energy for ^{197}Au

large or intermediate-mass (IMF) fragments with mass $A \geqq 10$. We call these decays complete vaporisation (V) if there are no such fragments, evaporation (E), with one fragment (evap. residue) with $A \geqq 10$ out of many other small ones, fission events (F), with two fragments $A \geqq 10$, and finally multifragmentation (M), with three or more fragments $A \geqq 10$. Again, we see the rapid transition to multifragmentation at $E^* \approx 700$ MeV.

This is quite different without Coulomb-interaction, c.f. lower part of Fig. 2. *The transition is shifted considerably downwards in energy from* ≈ 1500 MeV *to* ≈ 700 MeV *by the long-range Coulomb-repulsion*, roughly by about the Coulomb self-energy. Also the production of the second or third large fragment changes considerably as can be seen in Fig. 3 where the average masses of the three largest fragments are plotted for various excitation energies. Without Coulomb, only in the more narrow energy-interval of $1600\,\text{MeV} \leqq E^* \leqq 2300\,\text{MeV}$ the second and third largest fragments are slightly larger than just α-particles. That is the reason why we think that the repulsive long-range Coulomb-interaction triggers fragmentation into several *larger* fragments. We call it the proper multifragmentation.

It is further interesting that on average over all decay channels up to about $E^* \approx 6 * A$ MeV the largest fragment is at least twice as large as the second largest one. I.e. *even the hot fission is far from being symmetric.* This was also noticed in experimental Dalitz plots of the Aladin data [2] where the ternary breaking seems to evolve from the corners and not from the middle of the sides (locus of symmetric fission). In context to recent findings of large cross-section for fragments $A \geqq 100$ [8] it is interesting that the mass of the heaviest fragment is on average $\geqq 100$ for $E^* \leqq 700$ MeV.

Moretto [9] gave recently another interesting argument why it is just the Coulomb instability that leads to multifragmentation. Following Lord Rayleigh [10] he shows that higher multipoles in uniformly charged droplet become unstable at high fissility *before* the lower multipoles.

3. Correlations and fluctuations

Many-body correlations and critical fluctuations are characteristic indicators of phase-transitions. These observables go far beyond any single particle description as, e.g., BUU. They are closely related to bifurcations when new substantial parts of the phase-space become "suddenly" accessible during the time evolution of the system or when the energy is raised. Moreover, as the relative contribution of fluctuations is $\propto 1/\sqrt{A}$ they also go quite beyond the normal thermodynamical description by canonical distributions of just the mean values. Very likely, in finite systems they are more of sensitive to the constraints imposed by conservation laws and long-range forces.

3.1. Fragment-size correlations

Correlations between fragments of different sizes within the events are strong and important. At lower energies one has fission events and the correlation between the fission fragments is evident. Also at higher excitation in the multifragmentation region it is interesting to know which fragments are produced together.

E.g., correlation between the masses of fragments, the so called Campi-scatter-plot [11, 12, 13] gives this information. We calculated this within our microcanonical Monte Carlo sampling in [14, 15]. The $\ln(\bar{Z})$ of the largest fragment in each event is plotted vs. $\ln(S_2) = \ln(M_2^*/M_1^*)$ where

$$M_k^* = \sum Z_i^k - \bar{Z}^k. \tag{4}$$

If one plots the same quantity for the droplet distribution in a macroscopic liquid to gas transition one distinguishes two different regions: One linear branch at large masses is the undercritical branch at the gas-liquid coexistence. The largest droplet corresponds to the condensate. This branch descends from the large masses down and to the right towards large fluctuations S_2. The second branch is also linear and goes downwards from medium masses and large fluctuations towards the origin of $\ln(\bar{Z})$ and $\ln(S_2)$. This is the supercritical branch and corresponds to events above the critical point of the liquid to gas transition. Both branches meet at the critical point. (In Fig. 4 this is near to $\ln(P) \approx 3$; $\ln(S_2) \approx 1.2$). Here the fluctuations (S_2) are maximal.

In the simulation of nuclear fragmentation this correlation is more complicated (Fig. 4): First, because of the finiteness of the system there is considerably more

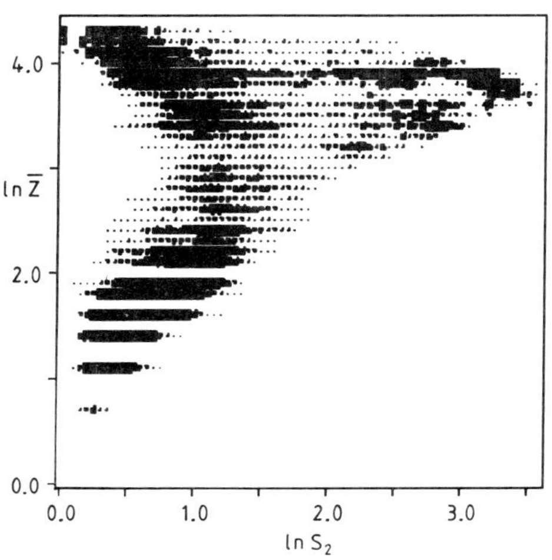

Fig. 4. Theoretical Campi-scatter plot for ^{197}Au of the charge of the largest fragment $\ln(Z)$ against to the second moment of the charge distribution of other fragments $\ln(S_2)$. The distribution of excitation energy corresponds to $1 * A$ GeV Au on emulsion [25]

fluctuation in the scatter plot from event to event. Moreover, there is a second group of critical events, the hot fission events at large S_2. These do not occur in a normal macroscopic liquid-gas transition due to the absence of the long-range Coulomb force which drives the fission in the nuclear case. This is a further example of how the long-range Coulomb interaction modifies the phase-transition.

3.2. IMF-IMF velocity-correlation functions

IMF-IMF velocity correlations at 180° have been used to infer the charge-distribution of the remaining part of the source in [16]. We concluded that in this experiment one just sees the onset of multifragmentation. Recently more work is being done to investigate IMF-IMF velocity correlations theoretically, e.g., [17] and experimentally, e.g. [18, 19].

The two-particle correlation function is defined by

$$C(a_1, a_2) = \frac{\Pi(a_1, a_2)}{\Pi(a_1)\Pi(a_2)}. \tag{5}$$

$a_i = (p_i, m_i, Z_i, \ldots)$ stand for momentum, mass, charge and other variables characterising fragment *i*. $\Pi(a_1, a_2)$ is the probability of coincidence of two particles and $\Pi(a_1)$ is the probability of one fragment. $C(a_1, a_2)$ depends on correlations existing already at the time of emission inside the source and of correlations between the final states of the two particles and the remainder of the source.

Due to the Coulomb interaction of the observed IMF with the rest of the system the IMF-IMF correlation function can be used to get information about the rest of the source. The rise of the correlation-function above one at small relative velocities signals whether the system is multifragmented or not.

3.3. Secondary evaporation and the effect of the fluctuations

At freeze-out the fragments are still hot. The treatment of the secondary decay of the hot fragments is straight forward in principle but there are still severe problems which reflect the great uncertainty about the level-density of the fragments at temperatures of $T \approx 5$ MeV or more. At low energies the level-densities are fitted by a more or less corrected Bethe formula,

$$\rho(E, A) = \frac{cc}{E^{1.25} a^{0.25}} e^{2\sqrt{aE}}, \tag{6}$$

Mostly, the level-density parameter is chosen simply as $a = A/8$. This formula is obtained from the grand-canonical partition function of a Fermi-gas $Z(\beta, \mu)$ via double Laplace back-transform [20] in saddle-point approximation. For *several* fragments in equilibrium at a given total excitation-energy we need the *folding* in E of several factors $\rho(E)$. It is evident that this cannot be done with the Bethe formula (6) because of the divergent prefactor. With some suitable regularisation of the diver-

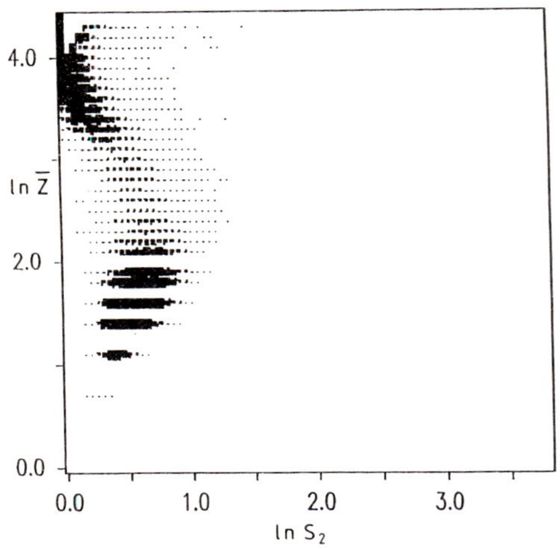

Fig. 5. Same as Fig. 4 but without Coulomb

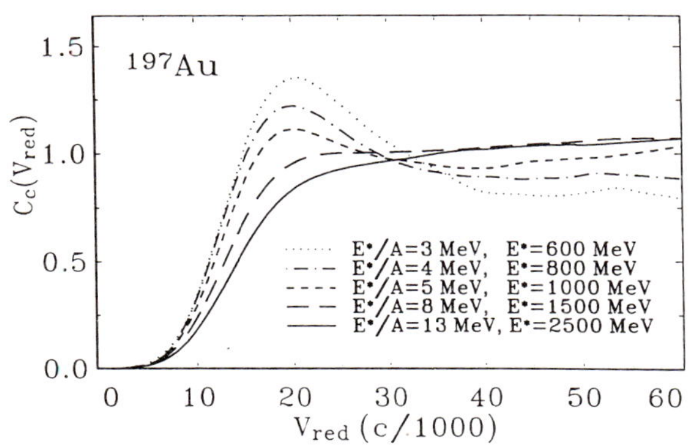

Fig. 6. IMF-IMF correlation integrated over all angles as function of the reduced relative velocity for $v_{red} = v_{rel}/\sqrt{Z_1 + Z_2}$ for ^{197}Au at various excitation energies

gence the entropy of the combined system does not depend on the total volume ($\sum A_i$) only but still depends on the individual partition into the A_i. This is in contradiction to the original form of the grandcanonical partition function one started with. Evidently, the use of the asymptotic form (the Bethe-formula) in the folding integral is not allowed. The integral runs also over intervals in E which are not in the asymptotic region of large E. This is a serious problem and the fluctuations in $\beta = 1/T$ and μ which are incorrectly treated in the Bethe-formula for finite nuclei are important for the folding of the densities of several nuclei in contact.

In view of this and of the fact that the nuclear level-density is experimentally mostly unknown at excitations $E/A > 2$ MeV – this is especially so for all unstable fragments – we use the following recipe [21]: The bound states of the fragments below the lower of the neutron threshold and the proton Coulomb-barrier B_{coul} are taken from the experimental data if available. If not, an empirical level-density formula (Truran-Cameron-Hilf [22] or Dilg [23]) with pairing-, shell- and deformation-corrections is used. The internal unbound phase space of the fragments is populated by a random number of evaporated nucleons. It is assumed that each nucleon has two possible spin orientations and moves with its classical energy $\varepsilon = p^2/2m + V + B_{coul} > 0$, B_{coul} resp. inside a potential pocket which is $V = -50$ MeV deep and has the radius $R = 1.224\,A^{1/3}$ fm. The total number of occupied phase space-cells is proportional to $[\Omega_f/(2\pi\hbar)]^n/n!$ where n is the total number of evaporated neutrons/protons and Ω_f is the total volume of all fragments together. These nucleons then leave the fragments after freeze out, as described below. I.e.: the higher excited states of the fragments are the product of the several unbound s.p. states times the bound states of the evaporation-residue and the entropy is the sum of the entropy of the unbound nucleons and the entropy of the bound residue. *In this approximation the internal entropy (level density) of the unbound single-particle states depends only on the total volume of all fragments and does not change when the fragments are split or combined.* This has the further advantage that one does not need to sample unstable fragments in the freeze-out configuration with mostly unknown binding energies and level densities but only well known stable nuclei as secondary residues. Secondly, *fluctuations* of the individual excitation energies, which are important for the production of neutron rich very light nuclei, are automatically incorporated. An interesting side-result is that at multifragmentation the fragments are considerably cooled by the coupling to the other degrees of freedom of the system like new surfaces etc.. We are working on improvements of this approach to take a surface correction of the entropy of the primary fragments into account and also to replace the Bethe formula for the level-density by a function which has the right properties under folding in A and E.

The critical fluctuations at the phase-transition and the intermittency of the size distribution of the fragments are discussed in detail in [24].

References

1. Bowman, D.R., Peaslee, G.F., Souza, R.T. de, Carlin, N., Gelbke, G.K., Gong, W.G., Kim, C.D., Lisa, M.A., Lynch, W.G., Phair, L., Tsang, M.B., Williams, C., Colonna, N., Hanold, K., McMahan, M.A., Wozniak, G.J., Moretto, L.G., Friedman, W.A.: Phys. Rev. Lett. **67**, 1527 (1991)
2. Ogilvie, C.A., Adloff, J.C., Begemann-Blaich, M., Boissou, P., Hubele, J., Imme, G., Iori, I., Kreutz, P., Kunde, G.J., Leray, S., Lindenstruth, V., Liu, Z., Lynen, U., Meijer, R.J., Milkau, U., Müller, W.F., Ngo, C., Pochodzalla, J., Raciti, G., Rudolf, G., Sann, H., Schüttauf, A., Seidel, W., Stuttge, L., Trautmann, W., Tucholski, A.: Phys. Rev. Lett. **67**, 1214 (1991)
3. Hagel, K., Gonin, M., Wada, R., Natowitz, J.B., Lou, Y., Ruiz, J., Drain, D., Chambon, B., Cheynis, B., Guinet, D., Hu, X.C., Demeyer, A., Pasteur, C., Giorni, A., Lieres, A., Stassi, P., Viano, B., Gonthier, P.: Progress in Research, Cyclotron Institute Texas A & M University, College Station, Texas (1991)
4. Gross, D.H.E., Zhang, X.Z.: Nucl. Phys. A**495**, 231c (1989)
5. Gross, D.H.E.: Rep. Progr. Phys. **53**, 605 (1990)
6. Zhang, X.Z., Gross, D.H.E., Xu, S.Y., Zheng, Y.M.: Nucl. Phys. A**461**, 668 (1987)
7. Aichelin, J., Peilert, G., Bohnet, A., Rosenbauer, A., Stöcker, H., Greiner, W.: Preprint (1987)
8. Aleklett, K., Liljenzin, J.O., Loveland, W., Sihver, L.: Nucl. Phys. at GANIL (1992)

9. Moretto, L.G., Wozniak, G.J.: In: Nuclear and Particle Science, Jackson, J.D. (ed.). Palo Alto: Annual Reviews, Inc. 1993
10. Rayleigh, F.R.S.: Philos. Mag. **14**, 184 (1882)
11. Campi, X.: J. Phys. A**19**, 917 (1986)
12. Campi, X.: Phys. Lett. B**208** B, 351 (1988)
13. Campi, X.: J. Phys. **50**, 183 (1989)
14. Jaqaman, H.R., Papp, G., Gross, D.H.E.: Nucl. Phys. A**514**, 327 (1990)
15. Jaqaman, H.R., Gross, D.H.E.: Nucl. Phys. A**524**, 321 (1991)
16. Gross, D.H.E., Klotz-Engmann, G., Oeschler, H.: Phys. Lett. B**224**, 29 (1989)
17. Bauer, W., Gelbke, C.K., Pratt, S.: Ann. Rev. Nucl. Part. Sci. **42**, 77 (1992)
18. Kim, Y.D. et al.: Phys. Rev. C**45**, 338 (1992)
19. Kim, Y.D. et al.: Phys. Rev. C**45**, 387 (1992)
20. Bohr, Aa., Mottelson, B.R.: Nuclear Structure. New York: Benjamin Press 1969
21. Zhang, X.Z., Gross, D.H.E., Xu, S.Y., Zheng, Y.M.: Nucl. Phys. A**461**, 641 (1987)
22. Truran, J.W., Cameron, A.G.W., Hilf, E.: Cern Report **70–30**, 215 (1970)
23. Dilg, W., Schantl, W., Vonach, H.: Nucl. Phys. A**217**, 269 (1973)
24. Gross, D.H.E.: In: Proceedings of the Second European Biennial Workshop on Nuclear Physics, Megeve (France), March 29–April 2 1993. Guinet, D. (ed.) Singapore: World Scientific 1993
25. Waddington, C.J., Freier, P.S.: Phys. Rev. C**31**, 888 (1985)

Probing the structure and dynamics of cage-like clusters: from water to Met-Cars

A.W. Castleman,Jr.

Department of Chemistry, The Pennsylvania State University, University Park, PA 16802, U.S.A.

Received 28 June 1993

Abstract.

PACS: 36.40.+d

1. Introduction

Elucidating the unique properties characteristic of systems of finite size provides a challenge of considerable current scientific interest [1]. In the case of hydrogen bonded clusters, studies of water clusters provide basic insight into the fundamentals of hydrogen bonding and solvation effects, as well as phenomena of importance in other fields such as atmospheric science [2]. The relationships between the physical and chemical properties and structures of water clusters have been studied under various experimental conditions using techniques ranging from cloud chambers to cooled flowtubes and supersonic molecule beams [3-11]. Other quantitative studies also have been made including ones devoted to measuring formation and dissociation rate constants, as well as rates of various intracluster reactions of hydrate clusters. A topic of considerable interest pertains to the origin of magic numbers, discontinuities at certain sizes in otherwise smoothly varying mass spectral distributions. A number of experiments and theoretical calculations provide evidence which shows that the magic numbers observed in the cluster distributions relate to stable structures, but until recently, other than the work of Kebarle [12] and Michl [13], there has been a paucity of experimental data on water cluster cations at large degrees of hydration, especially with regard to their thermochemical properties.

In the case of metals and metal compounds, studing materials comprised of clusters is particularly revealing in that one has the oppotrunity, for example, to investigate when discrete molecular properties begin to coalesce and display collective behavior characteristic of solids (e.g., band structure). Particularly interesting are systems whose composition can be selectively altered, thereby leading to varying electronic properties which can transcend semiconductor to metallic-like behavior. Such systems can display optical absorption features attributable to quantum confinement and ultimately the onset of giant dipole collective resonances [14,15].

Our recent work on metal compounds led to the discovery of a new class of metal-carbon clusters [16-19] which are of finite size and have specific geometry, but exhibit varying electronic character because of the different metals of which they can be comprised. We term these metallo-carbohedrenes or Met-Cars for short. This paper reviews the progress made in elucidating the structures if these two classes of clusters which seem to be quite different, but have some interesting commom features involving structural considerations.

2. Studies of the structure and bonding of water clusters

The study of water clusters has been a subject of long-standing interest, particularly in the context of providing a series of test cases for theoretical approaches to understanding nucleation phenomena and understanding hydrogen bonded structures of clusters. One cluster that has been the subject of numerous investigations, but for which there has been continuing debate rregarding its structure, is that of 21 water molecules bound to a proton [3,8,20]. It is often seen as a magic number in both laboratory studies and is observed in the atmosphere as well.

2.1. Titration techniques and magic numbers

An important finding fron our laboratory has been the development of a "titration" technique [21] for use in elucidating the structures of clusters from which hydrogen atoms or protons extend. This work provided the first definitive evidence concerning the structures of the magic number for protonated water clusters, namely $(H_2O)_{21}H^+$ and the adjacent interesting structure

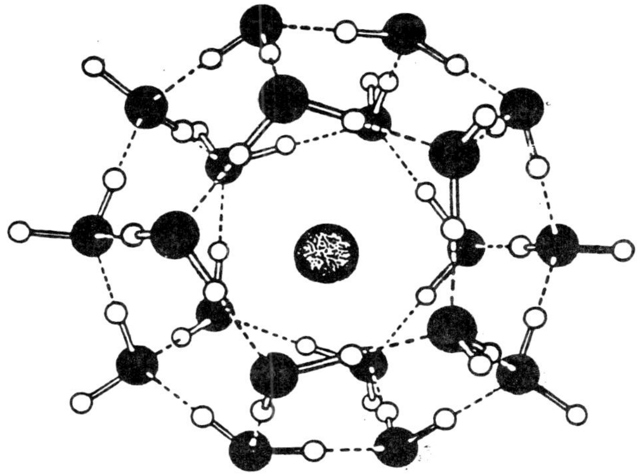

Fig. 1. Proposed dodecahedral structure for $M^+(H_2O)_{20}$ (M = H_3O^+). White spheres, hydrogen; blach spheres, oxygen; gray sphere denotes either H_3O^+ or a metal cation; dashed lines, hydrogen bonds.

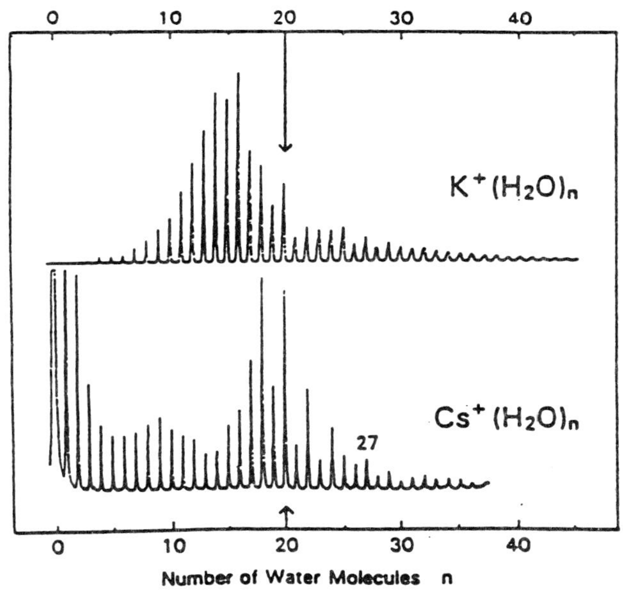

Fig. 2. Mass spectra at thermalized alkali metal cation-water: (a) $K^+(H_2O)_n$ at -110°C; (b) $Cs^+(H_2O)_n$ at -117°C.

$(H_2O)_{20}H^+$. (See references [21-23].) Many speculations existed in the literature [8,9,20,24,25] concerning the stability of this species and the reasons for its observation in studies ranging from those of expansions of neutral water molecules and subsequent electron impact ionization, to expansion of protons in molecular beams, and even following the sputtering of ions from ice-like surfaces. Our work using the titration method now has established that both the 20 and 21mer are clathrate-like cages. See Figure 1. Proton tunneling serves to further stabilize the lattice cage for the 20-mer, while it is established that H_3O^+ becomes trapped within a pentagonal dodecahedron clathrate of water and effects the stability of the 21-mer. (The latter finding nicely complements some theoretical work of some years ago [25].) Our new results bear on a number of issues concerning reasons for the appearance of magic numbers in mass spectral distributions of clusters.

As an extension of our investigatios of the structures of clusters, we have also undertaken studies of their interaction with common metal ions. (See reference 26.) A fast-flow reactor was used to produce and investigate ionic water clusters with the general composition $M^+(H_2O)_n$; where M is one of the alkali metals Li, Na, K or Cs and n ranges up to 45. A series of magic numbers was observed (see Figure 2), especially pronounced for $K^+(H_2O)_{20}$ and Cs^+ bound to 20 water molecules, and for some other cluster sizes, too. Weaker, but still completely reproducible features occur at n=16, 18, 22, 24, 27 and 29. Suprisingly, results indicate the existence of a magic number for $Li^+(H_2O)_{20}$, but none for sodium/water clusters. Evidence was obtained that these stable cluster ion gas phase hydrates have direct analogy to the protonated species discussed above, and the well-known clathrate hydrates which exist in the condensed phase [27].

In other related studies to determine cluster structures, mixed water-methanol clusters were also investigated utilizing a laser based reflection time-of-flight mass spectrometer. (See reference [28].) The intensity distributions of $(H_2O)_n(CH_3OH)_mH^+$ showed magic numbers at $n+m=21$, $0 \leq m \leq 8$ due to the enhanced stabilities of the dodecahedral cage structures in the mixed clusters. Studies of the metastable dissociation of $(H_2O)_n(CH_3OH)_mH^+$ $(n+m<40)$ provided evidence that the dissociation channels are governed by the competition of effects dominated by dipole-dipole and dipole-polarizability interactions compared to those arising due to the hydrogen bonded nature of the cluster, and the cluster structure. The mixed clusters $(H_2O)_n(CH_3OH)_mH^+$ were found to display a magic number of hydrogen bonds and its closed network structure. The titration experiments showed that $(H_2O)_{21}H^+$ has 10 nonhydrogen-bonded hydrogen atoms extending outward from the cage, which would be expected to be replaceable by CH_3 groups of the methanol molecules, CH_3OH. Therefore, based on our considerations, $(H_2O)_n(CH_3OH)_mH^+$ clusters would be expected to form stable mixed dodecahedral cages corresponding to $n+m=21$, as observed. Although as many as 10 water molecules can be replaced by alcohols while retaining the clathrate structure, the clusters with 8 replaced waters are more prominent, possibly due to the statistical probabilities associated with the switching of the waters by alcohol, but more likely due to steric hidrance being minimized for replacing only 8 positions on the surface of the lattice. Further details are given in reference [28].

2.2. The surface composition of clusters through studies of metastable dissociation

Another important aspect of studying the metastable dissociation of these clusters [28] is the measurements enable a determination of the surface composition of mixed systems which is important in studying the heterogeneous chemistry of aqueous surfaces. The loss channel of H_2O is found to be open to all $(H_2O)_n(CH_3OH)_mH^+$ except $(H_2O)(CH_3OH)_mH^+$ for which the water loss is relatively small. In the small cluster size region ($p < 10$), there is also another channel which is open to those water-rich ($n > m$) clusters; it corresponds to losing CH_3OH. By contrast, in the large cluster size ($p > 20$) the only dominant dissociation channels are loss of H_2O at all compositions. The dissociation patterns of medium size clusters ($10 < p < 20$) are very complicated. For this size regime, water-rich clusters ($n > m$) have three dissociation channels available, the dominant loss of H_2O, a CH_3OH loss channel, and a minor one corresponding to $(H_2O)(CH_3OH)$. However, for methanol-rich clusters ($m > n$) the loss channels are mainly dominated by H_2O loss.

The various findings are in agreement with considerations of the differences in the dipole moments and polarizabilities of water molecules compared to methanol. Polarizability is likely to be the key factor for governing the strength of the molecular interactions within these systems, because there is only a slight difference between the dipole moment of water and the dipole moment of methanol. Hence, a facile H_2O loss channel is expected to be nearly always available, as is found at all cluster sizes and composition regions. For the water-rich composition mixed clusters, it is suggested that water molecules have a tendency to build a cage structure in the cluster size region $p < 21$, with some methanols as well as waters being located on the outer periphery. The methanol loss channel becomes open for the water-rich mixed clusters ($p < 21$), and in the medium size region ($10 < p < 20$) there is also a weak channel available corresponding to the loss of $(H_2O)(CH_3OH)$. Since there are no metastable patterns involving the loss of $(H_2O)_2$ or $(CH_3OH)_2$, $(H_2O)(CH_3OH)$ might not be lost sequentially but as a bonded pair. In the large water/alcohol mixed clusters, water molecules have a tendency to form a flexible wall on the outside of the clusters. Therefore, only a metastable loss channel involving water can be observed in the large cluster region.

2.3. Thermochemical studies

In order to further contribute to understanding the origin of magic numbers, we undertook an investigation of their thermochemical properties using several different approaches. One of these was based on a time-of-flight reflector technique whereby the decay fractions of dissociating clusters could be measured. Employing the Klots evapotative ensemble model [29] and the measured decay fractions in a given time window, the average binding energies of water clusters were derived. Then, through an investigation of the steady-state distibution of clusters in a flow tube reactor under well-defined thermal conditions, we ascertained intensity ratios and

Table 1. Thermochemical data based on steady state distribution

Cluster Size (n)	$-\Delta G^*$ Steady State	Current Work	
		$-\Delta H^*$	$-\Delta S^\dagger$
6	5.34	16.09	82.69
7	5.23	12.03	52.31
8	5.42	9.77	33.46
9	5.28	8.78	26.92
10	5.53	8.88	25.77
11	5.45	9.30	26.92
12	5.46	9.83	33.62
13	5.37	10.30	37.92
14	5.42	10.60	39.85
15	5.47	10.72	40.38
16	5.34	10.82	42.15
17	5.38	10.76	41.38
18	5.47	10.75	40.62
19	5.33	10.66	41.00
20	5.39	10.54	39.62
21	5.56	10.43	37.46
[illegible]	[illegible]	[illegible]	[illegible]
23	5.39	10.45	38.92
24	5.41	10.28	37.46
25	5.36	10.06	36.15
26	5.45	9.72	32.85
27	5.39	9.41	30.92
28	5.45	9.06	27.77

* Units are kcal/mol
† Units are cal/mol·K (e.u.)

concentrations which enabled us to deduce values of their free energy of formation. By combining these measurements, we were able to obtain values of bond energies and entropies as a funcion of size for clusters ranging from 6 to 28 water molecules; see Table I. (Details are given in reference [30].)

The mass spectral data obtained during the course of the studies showed that the intensity of the $(H_2O)_{22}H^+$ parent ion is smaller than its neighbor, while that of its doughter ion is larger. This effect leads to an observed increase in the measured decay fraction of $n = 22$. Assuming that the activation energies of metastable dissociations are equivalent to the binding energies of clusters, and by fitting the evaporative dissociation value in terms of the known bond energy for the $n = 6$ cluster, binding energies of water clusters ($6 \leq n \leq 29$) are obtained.

Considering the various thermochemical data in Table I suggests that the "magic" number is likely to be due to an entropic effect. This hypothesis was born out through a series of calculations based on standard statistical mechanical equations, employing others made using the liquid drop (Thompson equation) [31] to deduce estimates of entropy and enthalpy trends with cluster size. At the 21-mer, where the hydronium ion enters the interior of the dodecahedral cage, the libations of the hydronium ion are of a lower frequency and the total entropy is less negative. A value of about $75 cm^{-1}$ for the torsional frequency of the encaged H_3O^+ is compatible with the experimentally deduced entropy change between the 20- and 21-mer. This value for the libation of an entrapped H_3O^+ in a cage is in good agreement with the expected frequencies.

The binding energies of the clusters represent the effects of a cluster's structure and bonding characteristics. At small cluster sizes $n < 8$, the charge in the cluster is the dominant bonding effect. For cluster sizes $8 \leq n \leq 12$, the cluster is more random in nature and therefore approaches the liquid drop calculations. For $n > 12$ rings, and cages start to form, hydrogen bonding is maximized and structural effects begin to govern the binding energy. This effect is most prominent for $13 \leq n \leq 25$. At higher cluster sizes with $n > 30$, there are both cages and random structures, and the binding energy once again approaches the liquid drop limit. The deviations in binding energy from the liquid drop limit are due to increased hydrogen bonding and structural effects.

3. Metallo-carbohedrenes and related cluster materials

During the course of undertaking detailed studies of dehydrogenation reactions of hydrocarbons induced by titanium ions, atoms, and clusters [32-38], we discovered [16-19] the formation of an unusually abundant and stable cationic species having a molecular weight of 528 amu, which we thereafter established contains 8 titanium atoms and 12 carbons. Subsequent work revealed the existence of the neutral molecular cluster, its anion analogue, the stability of other transition metal-carbon complexes of identical stoichiometry, and thereafter a general class of multicaged molecular clusters [16-19,39,40].

3.1. New class of molecular clusters

The experimental method employed in the original work leading to the discovery was based on a newly-developed MS/MS reaction system, described elsewhere [41]. The first Met-Car molecular cluster ion was generated through reactions of titanium with vapors of a wide range of small hydrocarbon molecules, utilizing a laser-based plasma reactor concept. Figure 3 shows a mass spectrum of clusters containing titanium metal atoms and carbons formed from reactions of titanium with two of the various different hydrocarbon molecules we have used thus far in our work; Figure 3a was obtained with methane, and Figure 3b with acetylene. As seen from Figure 3, a peak at 528 amu is completely dominant ("super magic") in both mass spectra. No other prominent peaks are observed in the mass range below 1200 amu. Reactions with ethylene, benzene and propylene, for example, generated similar cluster

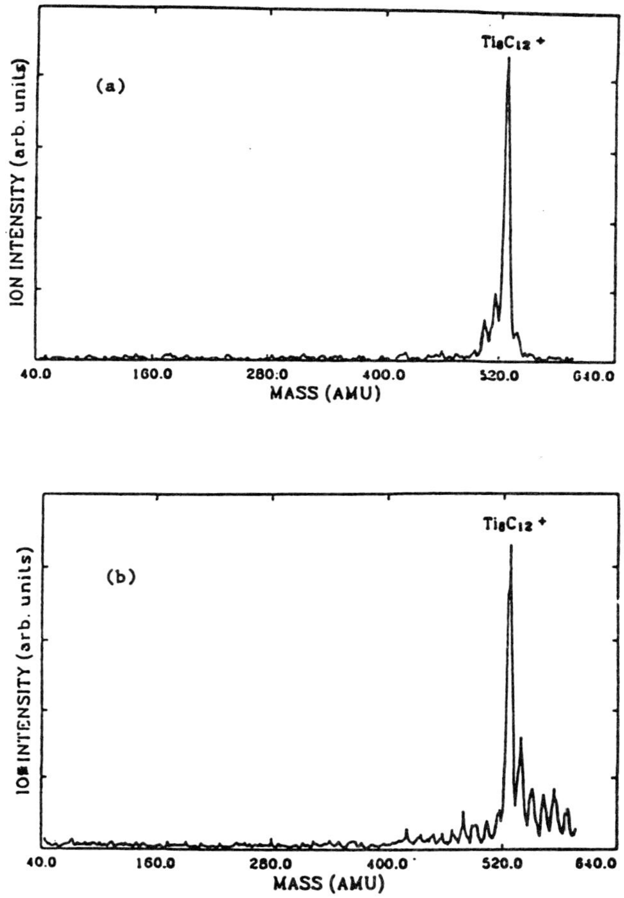

Fig. 3. (a) Mass distribution of $Ti_mC_n^+$ clusters generated from the reactions of titanium with CH_4. Note the "super magic" peak corresponding to $Ti_8C_{12}^+$. (b) Mass distributions of $Ti_mC_n^+$ clusters generated from the reactions of titanium with C_2H_2. Note the "super magic" peak corresponding to $Ti_8C_{12}^+$.

distributions, as did ones conducted later with rods of Ti and graphite powders [42].

Following these observations, we first undertook a series of studies with hydrocarbons of varying isotopic composition in order to definitively establish the identity of this unusually stable species. Isotope labelling experiments with deuterium showed that the clusters corresponding to the "super magic" peak did not contain any hydrogen atoms, while ^{13}C labelling established that the cluster accomodated exactly 12 carbon atoms. Subsequent high resolution isotope distribution pattern analyses supported the presence of 8 Ti atoms, leading to the assignment of $Ti_8C_{12}^+$. Based on our findings as well as others described below, we suggested that this species, as well as its corresponding neutral molecule discovered later, might have a pentagonal dodecahedron structure similar to that of $(H_2O)_{20}H_3O^+$. In this arrangement the Ti atoms would occupy eight unique positions that are identically coordinated which would account for all of the observations (see discussion below). In this structure, there are 12 pentagonal rings and each of the the rings contains two titaniums and three carbons. This suggestion implies that each of the titanium atoms can bond to three carbon atoms through Ti-C bonds, and each of the carbons may bond to its adjacent carbon through a C-C bond in addition to bonding to two titanium atoms. Substantial delocalized electronic character is expected, in keeping with our finding concerning the ease of ionizing Met-Cars. Alternatively bonded structures have been suggested [43-48] and experiments are required to determine the extra structure.

Thereafter, we conducted experiments [49] to "titrate" the metal atoms which are available for bonding to provide supporting evidence for a cage structure. This method is similar to that employed for resolving the structure of water clusters. The Met-Car molecular clusters were mass selected and injected into a thermal reaction cell containing individually, ND_3, H_2O and CH_3OH. Importantly, in each case the Met-Cars were found to accomodate exactly 8 polar molecules, showing that the titaniums are exposed at the cluster surface, and similarly coordinated. These findings are supportive of the proposed dodecahedral structure. Gas phase beam studies show that the Met-Cars are very unreactive with O_2, taking up only one molecule and without any degraded reaction products. Findings [17] of magic peaks corresponding to $M_8C_{12}^+$ (where M is V, Zr, and Hf) along with prior observations of $Ti_8C_{12}^+$, established metallo-carbohedrenes as a general class of stable molecular cluster ions, and now other members of the species have been found [50] including ones containing Mo, Cr and Fe. Subsequent studies revealed the existence of both neutral clusters of the above-mentioned compositions, as well as stable anions with the same metal to carbon ratios.

In order to gain some further insight into the stability of Met-Cars, we also undertook preliminary investigations of their metastable decay [51] using a reflectron time-of-flight mass spectrometer, and collision-induced dissociation [52] (CID) using a triple quadrupole apparatus. The metastable experiments reveal that Ti_8C_{12} loses Ti upon evapotarive dissociation; alternatively Ti_8C_{14} loses C_2. The CID experiments also display a dominant Ti loss channel from the Ti_8C_{12}, [in general agreement with photodissociation experiments of Pilgrim and Duncan [50(b)] made at 532nm]; a bond energy of $\leq 9eV$ for Ti loss is derived from our CID study. While not proving the dodecahedral structure, all of the results so far are in agreement with predictions based on this bonding arrangement of carbon and the transition metal atoms.

Following the foregoing work, we undertook investigations of larger cluster systems [18]. Interestingly, these revealed a unique structural growth pattern found in the system of Zr_mC_n, i.e., the formation of multicage structures. Careful examination of the experimental evidence showed that the first cage closes at Zr_8C_{12}. Suprisingly, subsequent cluster growth does not lead to the enlargement of the cage size as it usually does in the case of pure carbon clusters and water clusters, for example. Rather, multi-cage structures are developed, i.e., a double-cage at $Zr_{13}C_{22}$ and $Zr_{14}C_{21}$, a triple-cage at $Zr_{18}C_{29}$, and a quadrupole-cage at $Zr_{22}C_{35}$. This unique feature distinguishes this class of metallo-carbohedrenes from the regular doped fullerenes. See Figure 4. This growth pattern is also in keeping with the proposed dodecahedral structures.

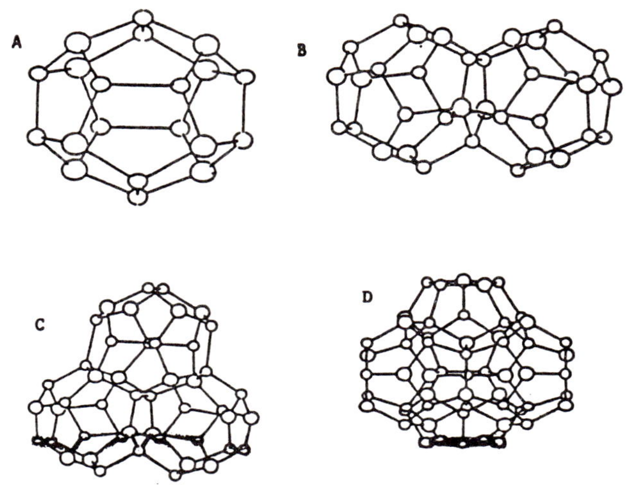

Fig. 4. (A) Idealized structure of pentagonal dodecahedron structure, T_h point group, proposed to account for the observed stability and ammonia uptake by $Ti_8C_{12}^+$. Note that the 8 titanium atoms (large spheres) which appear at the edge of a cube-like arrangement, are identically coordinated to 3-carbon atoms (small spheres) at all equivalent positions. (B) A proposed double-cage structure of $M_{14}C_{21}$. Note that when one of the 14 metal sites (large spheres) is occupied by one carbon atom, that is $M_{13}C_{22}$, the structure is also very stable. Two additional carbon atoms can form a pentagon through bridging between metal sites on two balls, leading to $M_{14}C_{23}$. (C) A proposed triple-cage structure of $M_{18}C_{29}$. (D) A proposed quadruple-cage structure. Note that other arrangements of the metal and carbon atoms are possible and that the structures are idealized, that is, no attempt is made to represent the sizes of the atoms or the bond lengths. However, the weaker metal-metal compared to the metal-carbon-bonded arrangements are excluded.

3.2. Other metal-carbon cluster systems

It is instructive to compare and contrast these findings with other metal-carbon systems. For example, those comprised of W, Co, and Ta do not assemble into the Met-Car structures. An interesting contrast is provided by considering the tantalum-carbon system which builds through the formation of cubic clusters. A typical time-of-flight mass spectrum of Ta_mC_n clusters displayed a series of magic numbers, all of which were consistent with cubic structures [18]. This observation differs from the magic number pattern seen for Met-Cars [19], further revealing their unusual structure. Similar findings have arisen from our studies of the titanium nitride system [33]. The failure to observe similar patterns in clusters Zr_mC_n implies that the cubic structures do not exist in that system, but rather, the cage-like structures (shown in Figure 4a-c) are responsible for the magic numbers of metallo-carbohedrenes. Hence, while it might be suggested that a cubic structure with eight metals at the corners and 12 carbons connecting in-between could also explain the magic peaks we find at [8,12], such an explanation is not compatible with the growth mechanism [18,19] of the metallo-carbohedrenes; it cannot account for the very definitive pattern found for the development of multi-cage structures.

3.3. The production of bulk quantities of Met-Cars

Pursuant to their discovery, we embarked on a study of methods to produce Met-Cars in the solid state [42,53]. The lack of any hydrogen component in the observed mass spectra of the $M_mC_n^+$ suggested that efficient dehydrogenation reactions take place in the plasma. This raised the likelihood that either the hydrocarbons are cracked and lose their hydrogen atoms in multiple collisions with energetic electrons or other plasma constituents, or they are efficiently dehydrogenated by reactions with the transition-metal atoms or ions present in the plasma. These first studies of the reaction mechanisms raised questions about the possible role of partially dehydrogenated hydrocarbons as intermediates in the formation of Met-Cars. Also, working out an efficient synthesis routine for producing bulk amounts of metallo-carbohedrenes depends on knowledge of the mechanisms and knowing what form of carbon should be introduced into the reactor.

The above considerations prompted us to attempt the synthesis of the Met-Cars using only metal and carbon as reactants [42]. In the first trial experiments, the frequency-doubled output of a Nd:YAG laser was focused onto the surface of a titanium and carbon-containing rod. The titanium/carbon rods were prepared from powder, along with graphite powder in an approximate 3:1 (Ti/C) molar ratio. The high-temperature plasma, which contains ionic and neutral Ti, C, and Ti-carbon clusters, was carried in a supersonic gas jet produced by pulsing high-pressure He (ca. 6 atm) over the rod. The clusters were cooled in a supersonic expansion effected into vacuum through a conical-shape noozle. After passing through a skimmer, the neutral clusters were photoionized by the frequency-tripled output (355 nm) of a second Nd:YAG laser utilizing a typical fluence of 10 mJ/cm^2. (We have previously ascertained that fragmentation of these strongly bound clusters is negligible under the conditions employed.)

The finding of copious quantities of Met-Cars in the mass spectra of the above experiments using only carbon and titanium prompted us to attempt synthesis employing more conventional arc discharge methods [53] using an apparatus similar to ones used to synthesize fullerenes. TOF-MS analyses of the resulting samples revealed the presence of Met-Cars of Ti and V, depending on the metal used to prepare the rods. The Met-Car-containing soot samples have been found to be stable in air. This is an important clue to the electronic nature of these clusters as well as an affirmation of their predicted stability. These soot samples are inhomogeneous and C_{60}, and to a lesser extent C_{70}, are also sometimes present in the soot.

Although laser desorption can sometimes generate a range of products depending on the sample, we established through a number of tests that we are actually desorbing Met-Cars from the soot and not producing them during analysis. Producing M_8C_{12} (M = Ti, V) molecules within the sample with one laser pulse and desorbing them with subsequent pulses would be unlikely, and furthermore the Met-Car signal is observed immediately, whereupon the signal intensity then decreases with time. If the met-Cars were generated during desorption via gas phase growth processes, we would also expect to see their characteristic building pattern as is observed in the molecular beam experiments [17-19] and in the mass distribution of the laser desorption experiments of pure graphite samples in which fullerenes are produced [54]. No such building patterns are observed in the analysis of the soot. Furthermore, if Met-Cars were generated during analysis, we would expect to see M_8C_{12} from a mixed sample of pure metal and graphite powders studied over a wide range of desorption laser powers, but we have not seen any masses corresponding to Met-Cars in these test experiments. Similar tests were performed with heated titamium carbide with no indication that Met-Cars were observed to be formed in the desorption of these materials.

4. Conclusions

Clusters often assemble into cage-like structures, even for such diverse substances as water and metal-carbon compounds. The $(H_2O)_{20}$ clathrate is stabilized by the presence of the H_3O^+ guest, while M_8C_{12} Met-Cars arrange into cages due to the strong metal-carbon and carbon-carbon bonding. Studying these unique cluster materials further reveals the unusual nature of systems of finite size, providing new information on their formation, properties and structures.

Financial support by the Atomospheric Sciences Division of the National Science Foundation, Grant No. ATM-9015855, and the U.S. Department of Energy, Grant Nos: DE - FGO2 - 92ER14258, DE - FGO2 - 93ER61602, and DE - FGO2 - 88ER60668, is gratefully acknowledged.

References

1 (a) Khanna, S.N., and Jena, P.: Phys. Rev. Lett. **69**, 1664 (1992); (b) Castleman, Jr. A.W.: J. Cluster Sci. **1**, 3 (1990); (c) Martin, T.P.: in Elemental and Molecular Clusters (Benedek, G., Martin, T.P., and Pacchioni, G., Eds.) Springer-Verlag, Berlin, Heidelberg, pp 2-11 (1988)

2 Yang, X., and Castleman, Jr. A.W.: J. Geophys. Res. **96**, 22573 (1991)

3 Searcy, J.Q., and Fenn, J.B.: J. Chem. Phys. **61**, 5282 (1974)

4 Udseth, H., Zmora, H., Beuhler, R.J., and Friedman, L.: J. Phys. Chem. **86**, 612 (1982)

5 Buehler, R.J., and Friedman, L.: J. Chem. Phys. **77**, 2549 (1982)

6 Haberland, H.: in Electronic and Atomic Collisions, (Eichler, J., Hertel, V., Stolterfoht, N., Eds.) Elsevier Science, pp 597-604 (1984)

7 Lin, S.S.: Rev. Sci. Instrum. **44**, 516 (1973)

8 Hermann, V., Kay, B.D., and Castleman, Jr. A.W.: J. Chem. Phys. **72**, 185 (1982)

9 Dreyfuss, D., and Wachman, H.Y.: J. Chem. Phys. **76**, 2031 (1982)

10 Stace, A.J., and Moore, C.: Chem. Phys. Lett. **96**, 80 (1983)

11 Echt, O., Kreisle, D., Knapp, M., and Recknagel, E.: Chem. Phys. Lett. **108**, 401 (1984)

12 Lau, Y.K., Ikuta, S., and Kebarle, P.: J. Am. Chem. Soc. **104**, 1462 (1982)

13 Magnera, T.F., David, D.E., and Michl, J.: Chem. Phys. Lett. **182**, 363 (1991)

14 (a) de Heer, W.A., and Knight, W.D.: in Elemental and Molecular Clusters (Benedek, G., Martin, T.P., and Pacchioni, G., Eds.) Springer-Verlag, Berlin, Heidelberg, pp 45-63 (1988); (b) Ekardt, W., and Pacheco, J.M.: in Clustering Phenomena in Atoms and Nuclei (Brenner, M., Lönnroth, T., and Malik, F.B., Eds.), Springer-Verlag, Berlin, Heidelberg, p 87 (1992); (c) Wang, C.R.C., Pollack, S., Hunter, J., Alameddin, G., Hoover, D., Cameron, D., Liu, S., and Kappes, M.M.: Z. Phys. D **19**, 13 (1991); (d) Bréchignac, C., Cahuzac, Ph., Leygnier, J., and Sarfati, A.: Phys. Rev. Lett. **70**, 2036 (1993); (e) Selby, K., Voiimer, M., Masui, J., Kresin, V., de Heer, W., and Knight, W.: Phys. Rev. B **40**, 5417 (1989)

15 Hertel, I.V., Steger, H., deVries, J., Weisser, B., Menzel, C., Kamke, B., and Kamke, W.: Phys. Rev. Lett. **68**, 784 (1992)

16 Guo, B.C., Kerns, K.P., and Castleman, Jr. A.W.: Science **255**, 1411 (1992)

17 Guo, B.C., Wei, S., Purnell, J., Buzza, S., and Castleman, Jr. A.W.: Science **256**, 515 (1992)

18 Wei, S., Guo, B.C., Purnell, J., Buzza, S., and Castleman, Jr. A.W.: Science **256**, 818 (1992)

19 Wei, S., Guo, B.C., Purnell, J., Buzza, S., and Castleman, Jr. A.W.: J. Phys. Chem. **96**, 4166 (1992)

20 Holland, P.M., and Castleman, Jr. A.W.: J. Chem. Phys. **72**, 5984 (1980)

21 Wei, S., Shi, Z., and Castleman, Jr. A.W.: J. Chem. Phys. **94**, 3268 (1991)

22 Yang, X., and Castleman, Jr. A.W.: J. Am. Chem. Soc. **111**, 1684 (1989)

23 Yang, X., and Castleman, Jr. A.W.: J. Phys. Chem. **94**, 8500 (1990); "Erratum", J. Phys. Chem. **94**, 8974 (1991)

24 Nagashima, U., Shinohara, H., Nishi, N., and Tanaka, H.: J. Chem. Phys. **84**, 209 (1986); Shinohara, H., Nagashima, U., Tanaka, H., and Nishi, N.: ibid. **83**, 4183 (1985)

25 Kassner, Jr. J.L., and Hagen, D.E.: J.Chem. Phys. **64**, 1860 (1976)

26 Selinger, A., and Castleman, Jr. A.W.: proc. International Symposium on the Physics and Chemistry of Finite Systems: From Clusters to Crystals, pp 1137-1140 (1992)

27 Davidson, D.W.: in Water, A Comprehensive Treatise (Franks, F., Ed.) Plenum, NY (1973)

28 Shi, Z., Wei, S., and Castleman, Jr. A.W.: Chem. Phys. Lett. **200**, 142 (1992)

29 Klots, C.E.: Z. Phys. D **21**, 335 (1991)

30 Shi, Z., Ford, J.V., Wei, S., and Castleman, Jr. A.W.: "Water Clusters: Contributions of Binding Energy and Entropy to Stability", J. Chem. Phys., submitted

31 Holland, P.M., and Castleman, Jr. A.W.: J. Phys. Chem. **86**, 4181 (1982)

32 Chen, Z.Y., Walder, G.J., and Castleman, Jr. A.W.: "Dehydrogenation of Gaseous Hydrogen Molecules: The Formation of Elemental and Molecular Clusters", Phys. Rev. B, submitted

33 Chen, Z.Y., and Castleman, Jr. A.W.: J. Chem. Phys. **98**, 231 (1993)

34 Guo, B.C., and Castleman, Jr. A.W.: J. Am. Soc. Mass Spectrom. **3**, 464 (1992)

35 Guo, B.C., and Castleman, Jr. A.W.: Int. J. Mass Spectrom. Ion. Proc. **113**, R1 (1992)

36 Guo, B.C., Kerns, K.P., and Castleman, Jr. A.W.: Int. J. Mass Spectrom. Ion. Proc. **117**, 129 (1992)

37 Guo, B.C., Kerns, K.P., and Castleman, Jr. A.W.: J. Phys. Chem. **96**, 4879 (1992)

38 Guo, B.C., and Castleman, Jr. A.W.: J. Am. Chem. Soc. **114**, 6152 (1992)

39 Guo, B.C., Wei, S., Chen, Z., Purnell, J., Buzza, S., Kerns, K.P., and Castleman, Jr. A.W.: J. Chem. Phys. **97**, 5243 (1992)

40 Chen, Z.Y., Guo, B.C., May, B., Cartier, S., and Castleman, Jr. A.W.: Chem. Phys. Lett. **198**, 118 (1992)

41 Guo, B.C., Kerns, K.P., and Castleman, Jr. A.W.: J. Chem. Phys. **96**, 8177 (1992)

42 Chen, Z., Walder, G.J., and Castleman, Jr. A.W.: J. Phys. Chem. **96**, 9581 (1992)

43 Dance, I.: "geometric and Electronic Structures of $[Ti_8C_{12}]$: Analogies with C_{60}", J. Chem. Soc., Chem. Commun., 1779 (1992)

44 Methfessel, M., van Schilfgaarde, M., and Scheffler, M.: Phys. Rev. Lett., **71**, 209 (1993)

45 Chen, H., Feyereisen, M., Long, X.P., and Fitzgerald, G., "Stability, Bonding, abd Geometric Structure of Ti_8C_{12}, Ti_8N_{12}, V_8C_{12}, and Zr_8C_{12}", Phys. Rev. Lett., submitted

46 Reddy, B.V., and Khanna, S.N.: Chem. Phys. Lett. **209**, 104 (1993)

47 Ceulemans, A., and Fowler, P.W.: J. Chem. Soc. Faraday Trans. **88**, 2797 (1992)

48 Pauling, L.: Proc. Natl. Acad. Sci. **89**, 8175 (1992)

49 Guo, B.C., Kerns, K.P., and Castleman, Jr. A.W.: J. Am. Chem. Soc. **115**, 7415 (1993)

50 (a) Pilgrim, J., and Duncan, M.: "Metallo-Carbohedrenes: Chromium, Molybdenum and Iron Analogs", J. Am. Chem. Soc., in press; (b) Pilgrim, J., and Duncan, M.: J. Am. Chem. Soc. **115**, 4395 (1993)

51 Wei, S., Guo, B.C., Purnell, J., Buzza, S., and Castleman, Jr. A.W.: J. Phys. Chem. **97**, 9559 (1993)

52 Guo, B.C., Kerns, K.P., and Castleman, Jr. A.W.: "Collision Induced Dissociation (CID) of the Met-Car $Ti_8C_{12}^+$", Chem. Phys. Lett., submitted

53 Cartier, S.F., Chen, Z.Y., Walder, G.J., Sleppy, C.R., and Castleman, Jr. A.W.: Science **260**, 195 (1993)

54 O'Keefe, A., Ross, M.M., Baronowski, A.P.: Chem. Phys. Lett. **130**, 17 (1986); Creasy, W.R., Brenna, J.T.: Chem. Phys. **126**, 453 (1989)

This article was processed using Springer-Verlag TEX Z.Physik D macro package 1.0 and the AMS fonts, developed by the American Mathematical Society.

Z. Phys. D 31, 213–217 (1994)

Water clusters: the $(H_2O)_{64}$ case

Stavros C. Farantos

Department of Chemistry, University of Crete and Institute of Electronic Structure and Laser, Research Center of Crete, FORTH, Iraklion, Crete 711 10, Greece

Received: 28 June 1993 / Final version: 9 May 1994

Abstract. Minimum energy structures and dynamics of $(H_2O)_{64}$ cluster are investigated by molecular dynamics techniques and an empirical potential energy surface. Using plots of the cluster, the oxygen-oxygen distance distribution function, and power spectra, we show the existence of minima with deformed cubic arrangement. These results together with previous calculations on smaller clusters support the conclusion that the cubic type geometries are common among water aggregates.

PACS: 82.20.Fd; 82.20.Rp; 82.20.Wt

1. Introduction

There is a considerable amount of work on small water clusters as is revealed from the literature of the last years [1–17]. This is not only due to our desire to approach macroscopic water from the clusters point of view, but also to the fact that small clusters are amenable to ab initio electronic structure calculations [2–4, 9–11, 14, 15] and to detail dynamical studies [5, 13, 16–19]. This work has revealed that small water clusters can have minima with geometries quite different from the structures of the macroscopic states of water [20].

Recently, we have carried out a systematic study of $(H_2O)_n$ clusters with $n=3$–18 [21, 22]. One of the main objectives was to find minimum energy geometries which dominate in the dynamics of the cluster. These studies showed that the tetramer with a square geometry for the oxygen atoms and the free hydrogens in *trans*-position, and the octamer with the oxygen and hydrogen atoms forming cubes of S_4 or D_{2d} symmetry, appear to be more stable compared with other clusters. This conclusion supported by plotting the energy per hydrogen bond, caloric curves and root mean square fluctuations of the oxygen-oxygen bonds, as well as from power spectra and Maximal Lyapunov Exponents [21].

Furthermore, the study of larger clusters has shown that the square tetramer provides the building up element to construct stable cubic type structures for the 12-mer, 16-mer, and 18-mer. For clusters which do not have the appropriate number of water molecules to form cubic geometries, minima were found with a cubic core surrounded by the extra molecules; this is the case of 9-mer, 13-mer, and 17-mer [22].

These results have been obtained with the empirical potential function of Cieplak et al. [23]. However, we have tested other potential functions [24–26] and have confirmed the structural stability of the cubic type geometries of water clusters. The cubes of S_4 and D_{2d} symmetry for the octamer have been verified by ab initio calculations [10, 11]. Our conclusion that the tetramer and octamer can be used to construct stable cubic type conformations for larger clusters has also been supported by the calculations of Wales and Ohmine [16, 17]. They produced stable minima for the 20-mer by lining up cubes of S_4 or D_{2d} symmetry.

The importance of the square tetramer and the cubic octamer has been increased after the proposition of a thermodynamic model by Benson and Siebert [27], who assumed an equilibrium between $(H_2O)_8$ and $(H_2O)_4$ in order to explain the peculiar behavior of the heat capacity of liquid water.

In our previous work [22], we have argued that the existence of larger clusters with cubic conformations may help to circumvent some deficiencies of the Benson-Siebert model with respect to other properties of liquid water, such as the small average hydrogen bond per water molecule and the small molar volume, which the assumption of single cubes implies.

It is reasonable to assume that if the cubic type clusters exist in liquid water, then cubic arrangements also appear

---

Supported from the European Community SCIENCE program (Grant No SC1*0331-C)

This paper was originally submitted in connection with the 2nd. Int. Conference on Atomic and Nuclear Clusters, Santorini from 28 June–2 July 1993 and is published here as a regular article after an independent refereeing procedure according to the standards of Z. Phys. D

in larger clusters. The purpose of this note is to investigate this assumption further by examining a number of quenched geometries of a medium size cluster such as the $(H_2O)_{64}$. Of course, extrapolation of the results of this study to liquid water can not be made from an aggregate of this size, thus, the conclusions should be considered as providing assistance in the analysis of the results of proper simulations of liquid water.

2. Methods

For the location of the stationary points of the potential energy surface we use the quenching technique [28]. That means, we integrate a trajectory at a specific energy and at regular time steps the kinetic energy is quenched, thus, the system approaches the underlying minimum. This method provides a systematic way to investigate those minima which are dynamically connected.

To solve the equations of motion, the molecules are treated as rigid bodies. We use quaternions [29, 30] to describe the orientations of the molecules and cartesian coordinates for their centers of masses. As an integrator we employ a variable time step, and variable order predictor-corrector algorithm given by Shampine and Gordon [31]. This is an efficient algorithm and was used in our previous studies [21, 22]. The trajectories are integrated for about 100 ps during which averaged quantities are calculated.

Here, we concentrate on the oxygen atom velocity autocorrelation function

$$C(t)=\sum_{i}^{N}\frac{\langle \mathbf{v}_i(t)\cdot\mathbf{v}_i(0)\rangle}{\langle \mathbf{v}_i(0)\cdot\mathbf{v}_i(0)\rangle}, \tag{1}$$

the Fourier transform of which reveals aspects of the dynamical behavior of the system

$$I(\omega)=\int_{-\infty}^{\infty} C(t)\exp(i\omega t)\,\mathrm{d}t. \tag{2}$$

N denotes the number of water molecules in the cluster. Geometrical characteristics are examined by plotting the number distribution function

$$n(R)=\frac{2}{N(N-1)}\sum_{i<j}\delta(R-R_{ij}). \tag{3}$$

R_{ij} is the oxygen-oxygen distance between the ith and jth molecule.

The potential function of Cieplak, Kollman, and Lybrand (CKL) [23] is used.

This is the sum of a pairwise potential, consisting of the electrostatic interactions, based on a model with charges placed on the two hydrogens and on the bisector

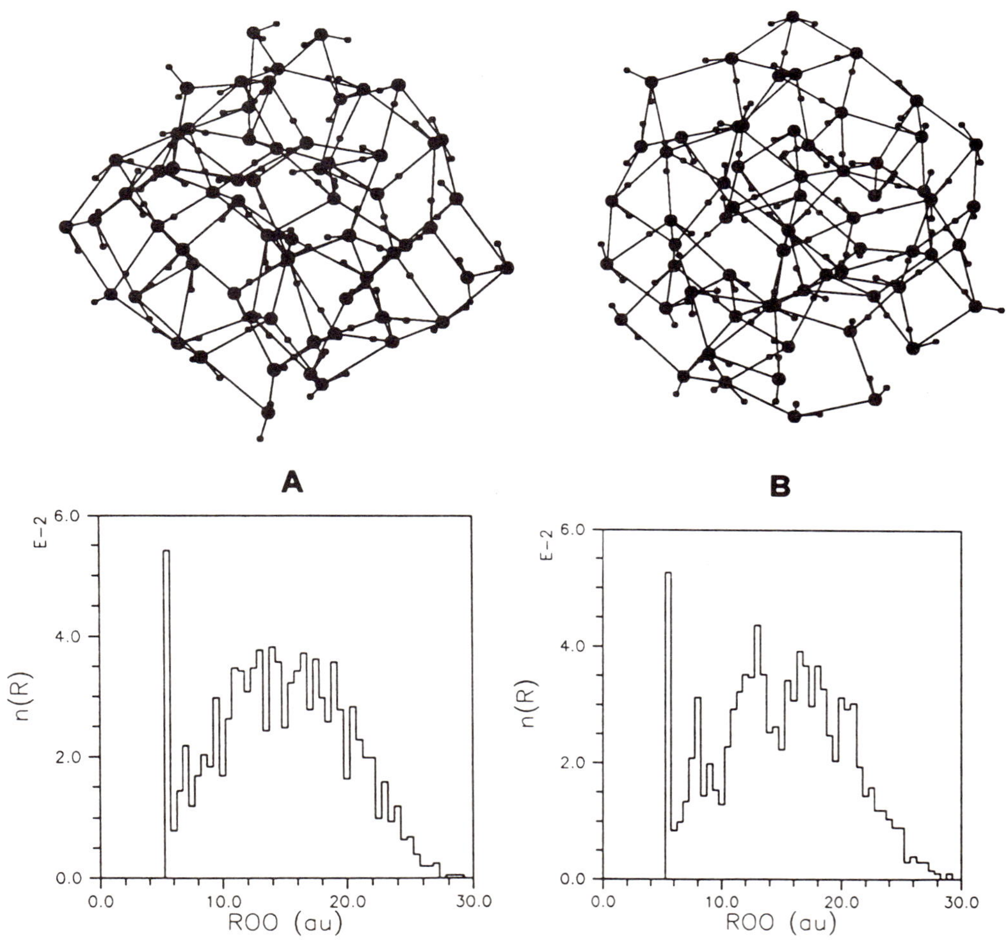

Fig. 1A, B. Minimum energy structures of $(H_2O)_{64}$ cluster, and the distribution of the O-O distances, $n(R)$, for these geometries. The energy of minimum **A** is -614.5 kcal/mol, and **B** -593.4 kcal/mol

of the HOH angle, oxygen-oxygen repulsion, Morse-type attractions among hydrogens and oxygens, and dispersion interactions, plus polarization terms. The latter are formed by dipole-dipole interactions. In the previous studies of small water clusters [21, 22] an extensive comparison was made of the results obtained with the CKL potential and those obtained with other potential functions. At least for the cubic conformations we do find agreement which reveals structural stability for this type of geometries.

In [21] we have examined the role of the many body polarization forces in the structure of minima and dynamics. It was found that the polarization terms make a contribution of about 14% in the total energy and do not affect the geometry of the minimum, although the variations of the absolute energies may cause changes in the order of some relative minima. The present calculations are performed with the additive part of the potential function.

3. Results and discussion

Initial conditions for the trajectories are selected either from a cubic lattice of the oxygen atoms and a random orientation of the hydrogen atoms through the assignment of three Euler angles per water molecule or by starting from a minimum and increasing the total energy of the cluster by scaling the momenta of the atoms. About one hundred initial arrangements were examined and all calculations were performed with zero total linear and angular momentum.

$(H_2O)_{64}$ has a large number of minima and it is out of our capability to make an exhaustive search. We are not even able to confirm that the lowest energy minimum found is the absolute minimum of the cluster. In the application of the quenching technique we have benefit a lot from the PVM (Parallel Virtual Machine) package of programs which allow us to use simultaneously several computers of different architecture [32]. The way we implement it is to run a trajectory on the fastest computer and pass the data for quenching (at every 5 ps run) to a number of workstations. Once the minimization procedure has been completed, then, the geometry of the quenched conformation is examined visually and by computing the distribution of the O-O distances.

The study of the quenched structures, spanning the energy range from -614.5 to -587.8 kcal/mol, shows that they can be broadly classified into two categories, for simplicity they will be denoted as (A) and (B) respectively. Representatives are shown in Fig. 1 together with the $n(R)$ functions. The energies of these minima are -614.5 (A) and -593.4 (B) kcal/mol respectively. We can see that $n(R)$ shows a sharp peak at 5.5 a_0 for

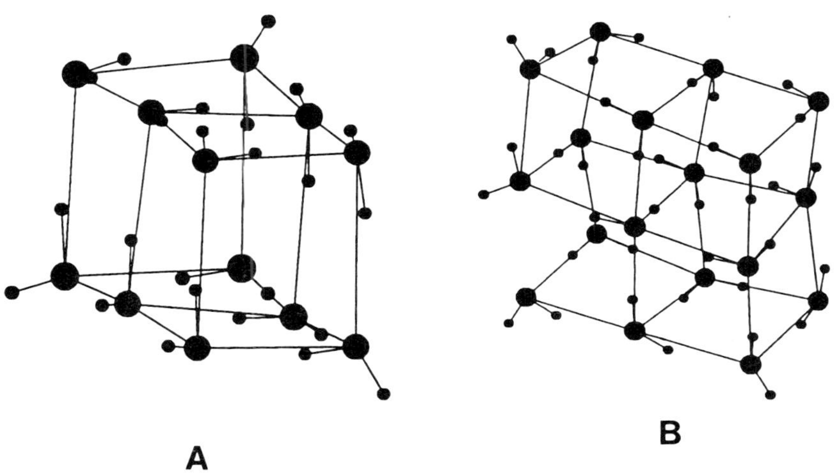

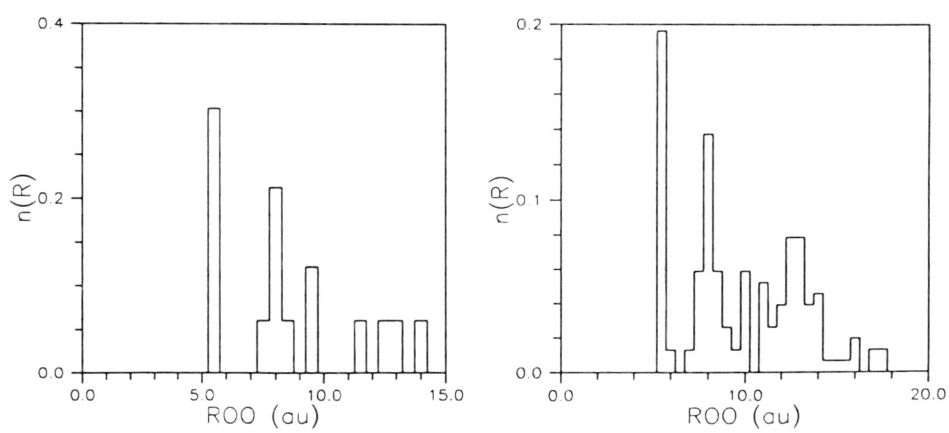

Fig. 2. The distributions of the O-O distances for **A** the dodecamer, and **B** decaoctamer

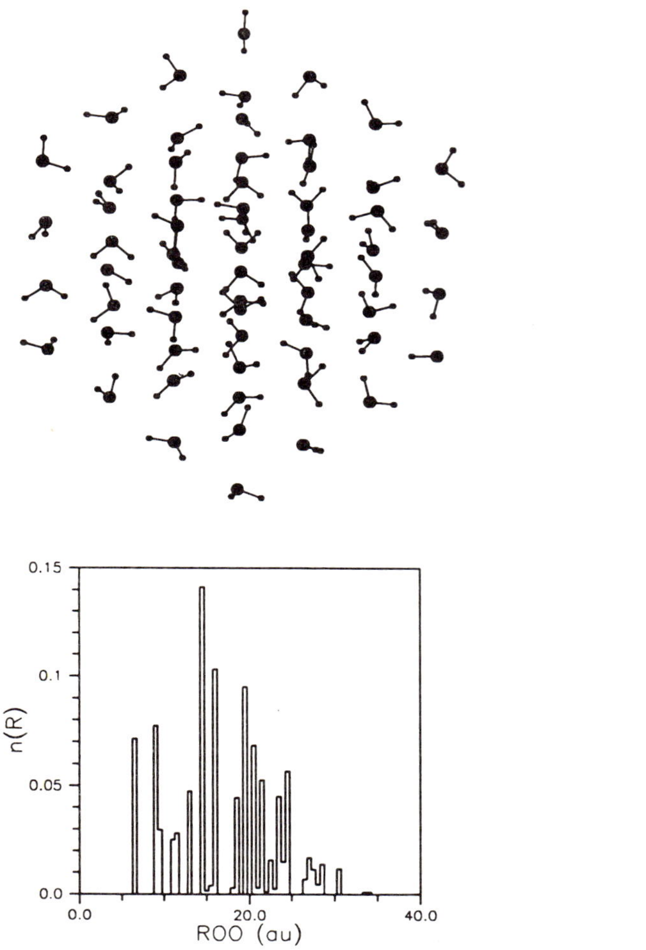

Fig. 3. An initial configuration of $(H_2O)_{64}$ cluster for the quenching process and with energy equal to -49.2 kcal/mol

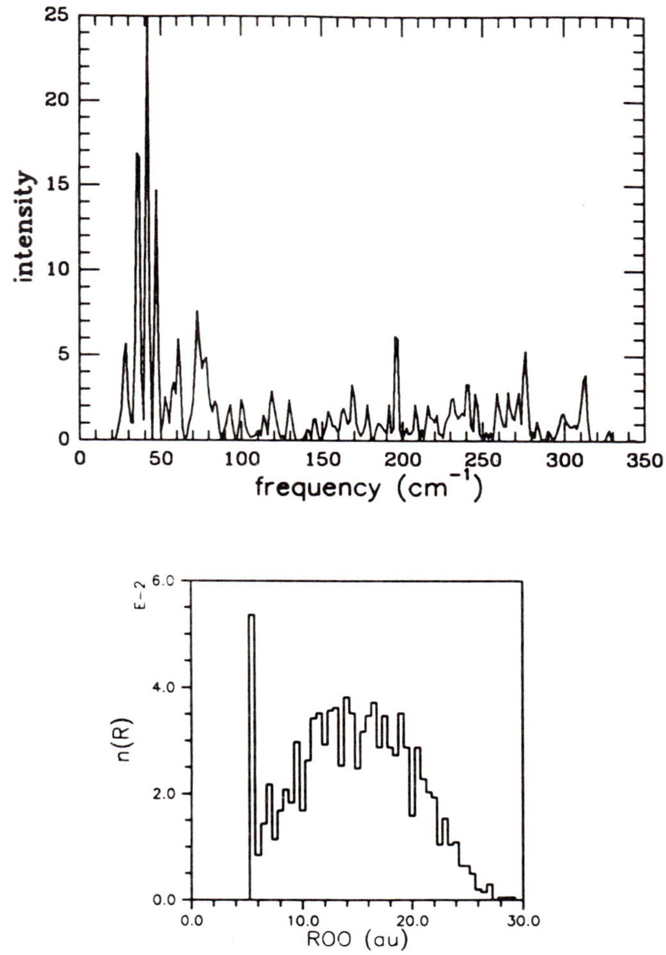

Fig. 4. The power spectrum and the time averaged O-O distances are shown for a trajectory initialized from the minimum of Fig. 1A and total energy -609.4 kcal/mol

both of them, which is the typical distance of two hydrogen bonded water molecules, and a broad peak spanning the range of distances from 6 to 29 a_0. For comparison in Fig. 2 we show the corresponding distributions of the cubic geometries of dodecamer and decaoctamer water clusters [22]. In Fig. 3 we show one typical initial geometery, from which the quenching procedure started, with the oxygen atoms initially forming a cubic lattice and then, randomly orienting the molecule choosing three Euler angles.

The higher energy relative minima (B) have an arrangement closer to a cubic structure, whereas the lower minima (A) tend to compress the molecular units, thus, increasing the number of tetrahedrally bonded molecules. This is reflected in the $n(R)$ function where for the -593.4 kcal/mol minimum (B) the second peak at 8 a_0 and a third peak at about 13 a_0 are in agreement with the structure of the number distribution functions of dodecamer and decaoctamer. This pattern remains for all conformations of 64-mer which are characterized as cubic like (B).

Minima of type (A) have a second peak at 7 a_0, but the whole pattern shows a rather broad band of values centered at 14 a_0. Trajectories initialized in the area of the two minima show that the system spends considerable time visiting minima of the same type.

Figures 4 and 5 show the power spectra obtained from trajectories initiated from the minima A and B respectively, and with energies -609.4 and -589.5 kcal/mol. Figure 4 resembles the power spectra found for smaller clusters, i.e., the largest intensities occur at small frequencies. In contrast, minima of type B seem to excite both the low frequency region as well as the region from 100–300 cm^{-1} which is usually associated with the stretching modes of the hydrogen bonds.

The low frequency band, here emerging from 20 cm^{-1}, has been a subject of discussion in the literature. Vij and Hufnagel [33] have observed in the power absorption coefficient of water a broad shoulder centered at 30 cm^{-1}, and assigned it to the librational modes of clustered molecules in condensed water. The calculations on water clusters indeed reveal vibrational modes with frequencies in this range. Even more importantly, it has been shown that the low frequency band appears from tetramers and

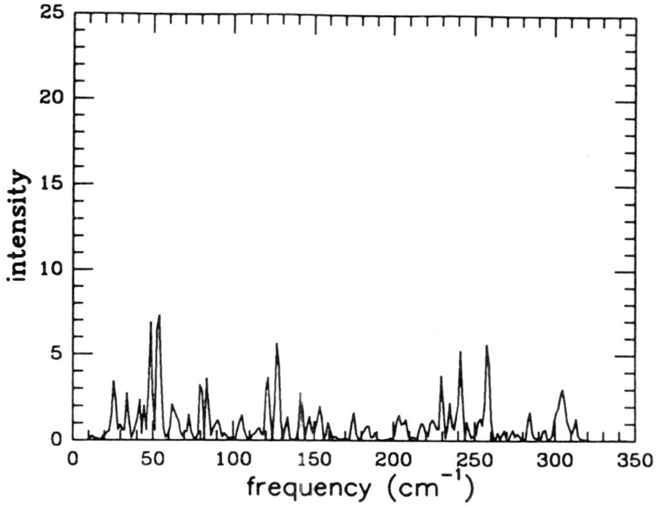

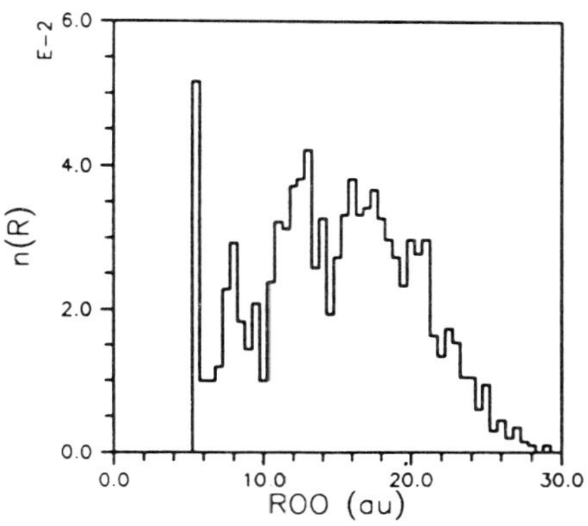

Fig. 5. As Fig. 4 but for a trajectory initialized from minimum of Fig. 1B and with total energy −589.5 kcal/mol

above. For the tetramer it has been found that the mode at 40 cm^{-1} is associated with the tortional motion of the two planes formed by the four oxygen atoms [21].

In summary, examination of a number of minimum energy structures of $(H_2O)_{64}$ cluster shows that there are conformations with deformed cubic geometries, and with a first and second peak in the O-O distance distribution function at 5.5 and 8 a_0. Lower energy conformations tend to squeeze the cluster giving more amorphous structures, with one peak at 5.5 a_0 and a broad peak extended from 6 to 29 a_0 in the $n(R)$ function.

These results demonstrate that although no regular cubic arrangements of the oxygen atoms have been obtained in this study, still subunits of the cluster may form deformed cubic structures. The connection of these findings with liquid water would require the examination of the behavior of these subunits by increasing the energy of the system and perhaps the study of larger clusters with periodic boundary conditions, questions which we shall address in future publications.

References

1. Belford, D., Campbell, E.S.: J. Chem. Phys. **86**, 7013 (1987)
2. Hermansson, K.: J. Chem. Phys. **89**, 2149 (1988)
3. Chalasinski, G., Szczesniak, M.M., Cieplak, P., Scheiner, S.: J. Chem. Phys. **94**, 2873 (1991)
4. Wojcik, M., Clementi, E.: J. Chem. Phys. **85**, 3544 (1986)
5. Plummer, P.L.M., Chen, T.S.: J. Chem. Phys. **86**, 7149 (1987)
6. Schröder, K.P.: Chem. Phys. **123**, 91 (1988)
7. Kim, K.S., Dupuis, M., Lie, G.C., Clementi, E.: Chem. Phys. Lett. **131**, 451 (1986)
8. Dykstra, C.E.: J. Chem. Phys. **91**, 6472 (1989)
9. Mhin, B.J., Kim, H.S., Kim, H.S., Yoon, C.W., Kim, K.S.: Chem. Phys. Lett. **176**, 41 (1991)
10. Tsai, C.J., Jordan, K.D.: J. Chem. Phys. **95**, 3850 (1991)
11. Tsai, C.J., Jordan, K.D.: J. Phys. Chem. **97**, 5208 (1993)
12. Brink, G., Glasser, L.: J. Phys. Chem. **88**, 3412 (1984)
13. Buffey, I.P., Brown, W.B., Gebbie, H.A.: J. Chem. Soc. Faraday Trans. **86**, 2357 (1990)
14. Burke, L.A., Jensen, J.O., Jensen, J.L., Krishnan, P.N.: Chem. Phys. Lett. **206**, 293 (1993)
15. Laasonen, K., Parrinello, M., Car, R., Lee, C., Vanderbilt, D.: Chem. Phys. Lett. **207**, 208 (1993)
16. Wales, D.J., Ohmine, I.: J. Chem. Phys. **98**, 7245 (1993)
17. Wales. D.J., Ohmine, I.: J. Chem. Phys. **98**, 7257 (1993)
18. Reimers, J.R., Watts, R.O.: Chem. Phys. **85**, 83 (1984)
19. Plummer, P.L.M., Chen, T.S.: J. Phys. Chem. **87**, 4190 (1983)
20. Franks, F.: Water, a comprehensive treatise. Vol. 1. New York: Plenum Press 1975
21. Vegiri, A., Farantos, S.C.: J. Chem. Phys. **98**, 4059 (1993)
22. Farantos, S.C., Kapetanakis, S., Vegiri, A.: J. Phys. Chem. **97**, 12158 (1993)
23. Cieplak, P., Kollman, P., Lybrand, T.: J. Chem. Phys. **92**, 6755 (1990)
24. Matsuoka, O., Clementi, E., Yoshimine, M.: J. Chem. Phys. **64**, 1351 (1976)
25. Jorgensen, W.L., Chandrasekhar, J., Madura, J.D., Impey, R.W., Klein, M.L.: J. Chem. Phys. **79**, 926 (1983)
26. Niesar, U., Corongiu, G., Clementi, E., Kneller, G.R., Bhattacharya, D.K.: J. Phys. Chem. **94**, 7949 (1990)
27. Benson, S.W., Siebert, E.D.: J. Am. Chem. Soc. **114**, 4269 (1992)
28. Stillinger, F.H., Weber, T.A.: Phys. Rev. A**25**, 978 (1982)
29. Evans, D.J., Murad, S.: Mol. Phys. **34**, 327 (1977)
30. Evans, D.J.: Mol. Phys. **34**, 317 (1977)
31. Shampine, L.F., Gordon, M.K.: Computer solution of ordinary differential equations. San Francisco: Freeman 1975
32. Beguelin, A., Dongarra, J., Geist, A., Jiang, W., Manchek, R., Sunderam, V.: PVM 3.0, Parallel Virtual Machine, ORNL/TM-12187 (1993)
33. Vij, J.K., Hufnagel, F.: Chem. Phys. Lett. **155**, 153 (1989)

Superconductivity in fullerenes: buckyballs

R. Friedberg[1], T.D. Lee[1], and H.C. Ren[1,2]

[1]Columbia University, New York, N.Y. 10027 and
[2]The Rockfeller University, New York, N.Y. 10021

Received 28 June 1993

Abstract. The molecular orbits on a buckyball, C_{60}, are studied analytically. It is found that the LUMO level and the second LUMO level form a closely spaced parity doublet. The implications of this parity doublet on the polarizability of C_{60} ions and the superconductivity pairing mechanism of the alkali doped C_{60} crystals are analyzed.

PACS: 31.20.-d, 74.60.-w

1. Introduction

The remarkable discoveries of the fullerenes [1], C_{60} (buckyballs), C_{240}, C_{540}, etc) and the superconductivity in the alkali doped buckyballs [1,2] present new challenges to both theoretical and experimental investigations. In this lecture, we shall review our theoretical works in this field for the past two years [3].

The molecular structure of a C_{60} is a trancated icosahedron consisting of sixty carbon atoms, shown in Fig. 1. There are two kinds of bonds, the bond shared by two hexagons and the bond shared by one hexagon and one pentagon. The difference in the bond lengths is tiny, which we shall neglect in the following discussions, and the average bond length is about 1.4 $\mathring{A}$. Among the four valence electrons of each carbon, three of them form σ-bonds with the neighbouring carbon atoms. The remaining one, refered to as π-electron is largely in $2p$ state with its wave function symmetric with respect to the radial vector at the carbon site and can hop through the entire molecule. The radius of a C_{60} molecule is about 3.5 $\mathring{A}$.

The crystal structure [4] of the alkali compound of buckyballs, say K_3C_{60} in cubic where C_{60}^{---} ions occupy the face centered sites, one-third of K^+ occupy the octahedral sites and the rest of K^+ occupy the tetrahedral sites. The lattice constant (the side length of a fundamental cube) is about 14.2 $\mathring{A}$. The crystal is an ionic conductor with three conducting electrons per unit cell.

Using the molecular orbit approximation, the one electron energy levels and wave functions of a C_{60} molecule can be obtained analytically based on the symmetry analysis. Among sixty molecular orbits, thirty of them are filled by sixty π-electrons (two electrons per orbit due to spin degeneracy) for a neutral C_{60}. The LUMO level and the second LUMO level are closely spaced two triplets with opposite parities and will be refered to as the parity doublet. The closeness of the parity doublet makes it natural to isolate their response to strong interactions, separate from other distant levels. It provides a convinient means to derive analytical expressions for many of the important parts of strong interaction effects. For example, we can calculate the polarization energy of C_{60}^{-} and the Coulomb energy of C_{60}^{--}. The latter suggests an interesting pairing mechanism.

In the next section, we shall describe the analytical treatment of the molecular orbits. The parity doublet and its applications will be discussed in section 3. The energy balance and the pairing mechanism in alkali doped C_{60} crystal will be analyzed in section 4 and some speculations on the superconducting phase will be jotted down in section 4.

2. An analytical treatment of the molecular orbits of C_{60}

The symmetry group of C_{60} molecule is the full icosahedral group, I_h, which is a direct product of the icosahedral group I and the inversion. The former consists of sixty discrete rotations which leave an icosahedron invariant (there are six 5-fold axes, ten 3-fold axes and fifteen 2-fold axes). The sixty proper rotations fall within five classes and therefore there are five inequivalent irreducible representations. Being a real subgroup of $SO(3)$ (the rotation group in continuum), the representations of I are closely related to those of $SO(3)$, labeled by the rank l of the standard spherical harmonics, $Y_{lm}(\theta, \phi)$. The representations of $l = 0, 1, 2$ remain irreducible when restricted within the icosahedral

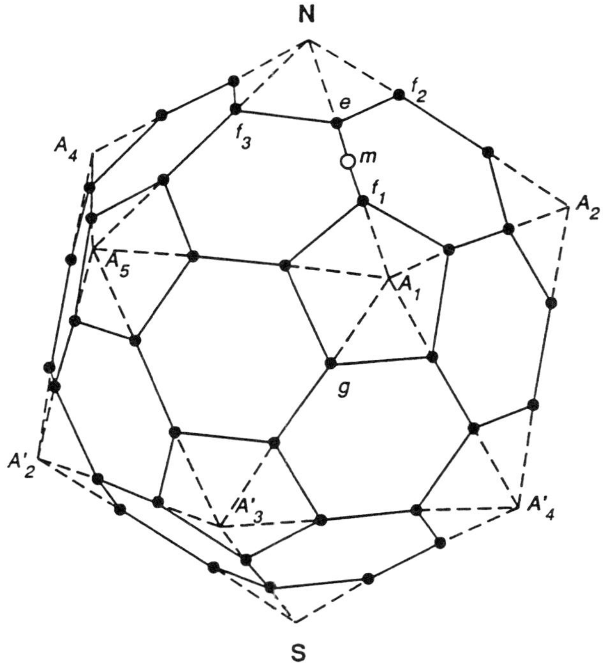

Fig. 1. The structure of a C_{60} molecule. The $A_1, A_2, \ldots, A'_2, \ldots$ are the icosahedral vertices and the dots denote the positions of carbon atoms. Each carbon atom corresponds to an element in the proper icosahedral group. For example, e is the unit element of the group and f_1, f_2 and f_3 are the three group elements which bring the radial vector $\hat{e}$ to its three nearest neighbors.

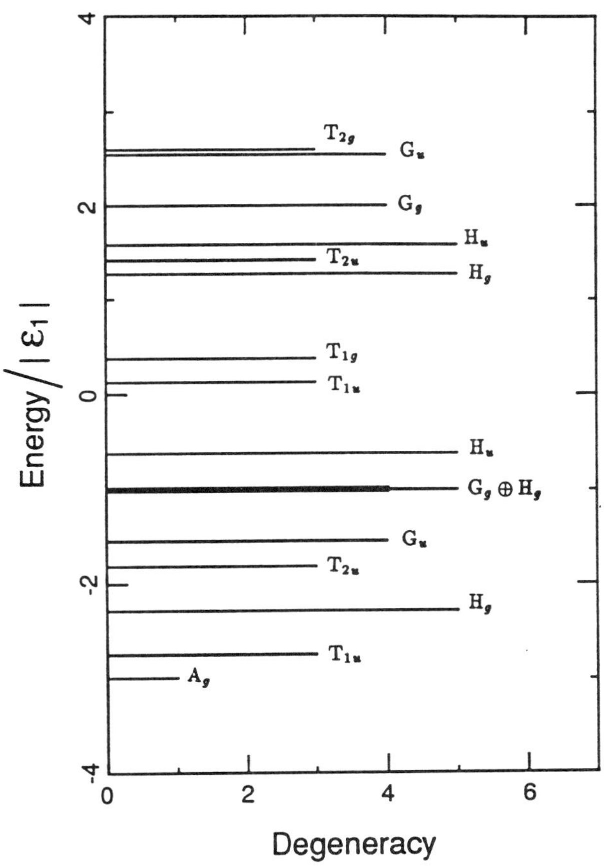

Fig. 2. The energy levels of the molecular orbits of C_{60}. The energy is measured in the unit of $|\epsilon_1|$ and the length of each bar represents the multiplicity of the level. The HOMO level, the LUMO level and the second LUMO level are marked explicity.

rotations. These representations are denoted by A, T_1 and H in the literature and their dimensions are 1, 3 and 5 respectively. The other two representations of I are decomposed from the $l = 3$ representation of $SO(3)$ and are denoted by T_2, H with dimensions 3, 4. The representations of I_h are then, $A_P, T_{1P}, T_{2P}, G_P, H_P$ with $P = u, g$ standing for odd and even parities.

Table 1. The eigenvalues of the matrix h

Representation	Eigenvalue λ
A	3
T_1	$\frac{1}{2}(\frac{3+\sqrt{5}}{2} \pm \sqrt{\frac{19-\sqrt{5}}{2}}$ $\frac{1}{2}(-3+\sqrt{5})$
T_2	$\frac{1}{2}(\frac{3-\sqrt{5}}{2} \pm \sqrt{\frac{19+\sqrt{5}}{2}}$ $\frac{1}{2}(-3-\sqrt{5})$
G	$\frac{1}{2}(-1 \pm \sqrt{17})$ 1 -2
H	$\frac{1}{2}(1 \pm \sqrt{13})$ $\frac{1}{2}(-1 \pm \sqrt{5})$ 1

The effective Hamiltonian for a π-electron reads:

$$H = -\frac{1}{2m}\nabla^2 + \sum_g v(\mathbf{r} - R_0\hat{g}) \tag{1}$$

where $v(\mathbf{r} - R_0\hat{g})$ is the mean field potential produced by the carbon atom residing at $R_0\hat{g}$ with R_0 the radius of the C_{60} molecule. Within the molecular orbit approximations [5], the one electron trial wave function can be written as

$$\psi(\mathbf{r}) = \sum_g C(g)\chi(\mathbf{r} - R_0\hat{g}) \tag{2}$$

where $\chi(\mathbf{r} - R_0\hat{g})$ is the atomic $2p$ wave function pointing in $\hat{g}$ direction.

The variational energy is

$$\epsilon = \frac{\int d^3\mathbf{r}\psi^*(\mathbf{r})H\psi(\mathbf{r})}{\int d^3\mathbf{r}|\psi(\mathbf{r})|^2} = \epsilon_0 + \epsilon_1 C^\dagger hC \tag{3}$$

where C is a 60×1 column matrix, h is a 60×60 nearest hopping matrix and ϵ_0, ϵ_1 are two energy integrals. Experimentally, $\epsilon_1 = -1.5 \sim -3eV$. The last step of (3) follows from Huckel approximation [5] in which all the cross terms in the denominator and all the cross terms beyond the nearest neighbors in the numerator are dropped.

A one to one correspondence between the sixty elements of the group I and the sixty carbon sites can be easily established. The unit element of I may be associate with any carbon atom whose radial vector is denoted by $R_0\hat{e}$. A proper rotation $g \in I$ brings $\hat{e}$ to $\hat{g}$. Since $\hat{e}$ does not coincide with any symmetry axes of the icosahedron, all $\hat{g}$'s generated this way are different and run over all carbon atoms. Therefore the Hilbert space of h supports the regular representation of I and the number of times each representation appears is equal to the dimensionality of the representation. The symmetry adapted basis are

$$|J, M, \lambda> = \sqrt{\frac{d_J}{60}} \sum_g |g> D^J_{M\lambda}(g)^* \tag{4}$$

where $D^J_{M\lambda}(g)$ is the matrix elements of the irreducible representation $J(= A, T_1, T_2, G, H)$ and d_J is the dimensionality of the representation J. The hopping matrix h is block-diagonalized, i.e.

$$< J, M, \lambda | h | J', M', \lambda' > = \delta_{JJ'} \delta_{MM'} f_{\lambda\lambda'} \tag{5}$$

The $d_J \times d_J$ matrix $f_{\lambda\lambda'}$ can be further block-diagonalized according to the parities and the eigenvalues and the eigenvectors of h are readily evaluated and are listed in the Table 1. The numerical values of ϵ are plotted in Fig. 2, which is consistent with the results of other groups [6].

As is shown in Fig. 2, the sixty π-electrons fill up the thirty orbits up to the HOMO-level H_u. The LUMO-level, T_{1u} and the second LUMO-level, T_{1g} are closely spaced and of opposite parities. The energy difference between LUMO and the second LUMO is

$$\Delta\epsilon = 0.2434|\epsilon_1| \tag{6}$$

This parity doublet is the main subject we will address in the subsequent sections.

3. The parity doublet and the apllications

Geometrically, the parity doublet can be understood in the following way: After diagonalizing the matrix (5) in the subspace of the triplet representation T_1, we obtain three eigenvalues $\lambda_0 = 2.7566$, $\lambda_- = -0.1386$, $\lambda_+ = -0.3820$ and three triplets eigenfunctions:

$$\psi_\alpha(\mathbf{r}) = \sum_g \mathbf{e}_\alpha(g) \chi(\mathbf{r} - R_0\hat{g}) \tag{7}$$

where each $\mathbf{e}_\alpha(g)$ is a triplet of eigenvectors of h with $\alpha = 0, \pm$. Regarded as a vector in physical space, $\mathbf{e}_0(g)$ lies almost parallel to (making only $1°$ angle with) the radial vector $\mathbf{g}$ at g), $\mathbf{e}_-(g)$ lies on the plane defined by $\mathbf{e}_0(g)$ and $\mathbf{g}$ and

$$\mathbf{e}_+(g) = \mathbf{e}_0(g) \times \mathbf{e}_-(g) \tag{8}$$

$\mathbf{e}_\alpha(g)$ and $\mathbf{e}_\alpha(g')$ are related by the icosahedral rotation which brings g to g'.

Under an inversion $\mathbf{e}_0(g)$ and $\mathbf{e}_-(g)$ switch their sign but $\mathbf{e}_+(g)$ does not. Furthermore, $\mathbf{e}_0$ changes slowly as one goes from g to its nearest neighbors. Both $\mathbf{e}_\pm$ change rapidly but in similar fashions (as can be seen in Fig. 1 for the sign switch from $g = e$ to $g = f_1$). Therefore the level corresponding to $\mathbf{e}_0$ should lie down below in the closed shell of sizty π-electrons and the levels corresponding to $\mathbf{e}_-$ and $\mathbf{e}_+$ should lie close to each other, forming a parity doublet. The near degeneracy of the T_{1u} and T_{1g} levels allows us to ignore all other levels except these two at first and to include strong interactions to all orders. This approximation also requires that the interaction energy should be smaller than the differences between these two levels and other levels which is not quite justified for some interactions we shall discuss below. In that case, we postulate that many physical properties of the C_{60} compound are given qualitatively by the parity doublet approximation. The inclusion of other levels brings about merely quantitative corrections.

One application of the parity doublet approximation is to calculate the polarizability of C_{60}^- in which the extra electron will reside on one of LUMO state. Under an external field, the mixing of the LUMO level and the second LUMO level implies a large polarizability for a weak field and a permanent dipole for a strong field. The standard perturbation formula yields a polarizability:

$$\alpha = \frac{1}{3} \frac{e^2 R_0^2}{\Delta\epsilon} \tag{9}$$

where $\Delta\epsilon$ is given by (6). We may compare this polarizability with that of a conducting sphere of the same radial, the latter being $\alpha_C = 4\pi R_0^3$. We find $\frac{\alpha}{\alpha_C} = 3.6$ for $\epsilon_1 = -1.5eV$ and $\frac{\alpha}{\alpha_C} = 1.8$ for $\epsilon_1 = -3eV$. The former value of α nearly saturate the Clausius-Mossotti limit of the C_{60} crystal ($\alpha_{C.M} \cong 48R_0^3$). In any case, C_{60}^- ion is expected to be highly polarizable.

The other application of the parity doublet is to calculate the Coulomb energy of C_{60}^{--}, which plays an instrumental role to our superconducting mechanism. The Coulomb interaction between the two electrons is given by

$$V(g, g') = \begin{cases} U_0 & \text{for} \quad g' = g; \\ \frac{e^2}{4\pi R_0 |\hat{g}' - \hat{g}|} & \text{for} \quad g' \neq g. \end{cases} \tag{10}$$

where the on-site Coulomb energy, U_0 is estimated to be $10 \sim 20$ eV. The two electron Hamiltonian can be diagonalized within the parity doublet approximation by forming appropriate Clebsch-Gordan combinations of T_{1u} and T_{1g}. The average Coulomb energy over all of these combinations is $\frac{U_0}{100} + 3.6eV$. The lowest states are a doublet of spin singlet scalar combinations (invariant under an icosahedral rotation) with opposite parities and of nearly degenerate energies 2.4eV. The orbital wave functions for these two states are

$$F_{\pm}(\mathbf{r},\mathbf{r}') = \sum_{g,g'} S_{\pm}(g,g')\chi(\mathbf{r} - R_0\hat{g})\chi(\mathbf{r}' - R_0\hat{g}') \tag{11}$$

where

$$S_{+}(g,g') = \hat{e}_{+}(g) \bullet \hat{e}_{+}(g') - \hat{e}_{-}(g) \bullet \hat{e}_{-}(g') \tag{12}$$

and

$$S_{-}(g,g') = \hat{e}_{-}(g) \bullet \hat{e}_{+}(g') + \hat{e}_{+}(g) \bullet \hat{e}_{-}(g') \tag{13}$$

The reduction of the Coulomb repulsion for the correlations given above can be understood geometrically. Both functions vanish when $g = g'$ and the on-site repulsion is thereby avoided. For the nearest neighboring g and g' the radial vectors $\hat{g}$ and $\hat{g}'$ can be approximated by a common vectror $\hat{z}$ and

$$\hat{e}_{+}(g) \cong \hat{z} \times \hat{e}_{-}(g) \tag{14}$$

$$\hat{e}_{+}(g') \cong \hat{z} \times \hat{e}_{-}(g') \tag{15}$$

Upon the substitution of (14) and (15) into (11) and (12), it follows that

$$\begin{aligned} S_{+}(g,g') \;&\cong\; (\hat{z} \times \hat{e}_{+}(g)) \bullet (\hat{z} \times \hat{e}_{+}(g')) \\ &\quad - \hat{e}_{-}(g) \bullet \hat{e}_{-}(g') \;=\; 0 \end{aligned} \tag{16}$$

and

$$\begin{aligned} S_{-}(g,g') \;&\cong\; \hat{e}_{-}(g) \bullet (\hat{z} \times \hat{e}_{+}(g')) \\ &\quad + (\hat{z} \times \hat{e}_{+}(g)) \bullet \hat{e}_{-}(g') \;=\; 0 \end{aligned} \tag{17}$$

and the short distance repulsion is thereby reduced. This mechanism provides a counter of Hund's rule in atomic physics.

The existence of the closely spaced triplets arround the HOMO level can also be demonstrated by the analog with the graphite structure and it works better for large fullerenes [7].

4. The pairing mechanism

The pairing mechanism is based on the following variational arguments on the hopping of conducting electrons. An electron can hop independently of each other and its wave function is of usual Bloch type within tight binding approximation, i.e.

$$\Psi_{\mathbf{k},\pm}(\mathbf{r}) = \frac{1}{\sqrt{N}} \sum_{\mathbf{R}} e^{i\mathbf{k}\mathbf{R}} \psi_{\pm}(\mathbf{r} - \mathbf{R}) \tag{18}$$

with

$$\psi_{\pm}(\mathbf{r} - \mathbf{R}) = \sum_{g,i} C_i(\mathbf{k})\hat{e}_{\pm}(g)_i \chi(\mathbf{r} - R_0\hat{g} - \mathbf{R}) \tag{19}$$

where $\mathbf{R}$ is the position of a C_{60} and N is the total number of C_{60}, which will be sent to ∞. The two bands thus formed will be refered to as T_{1u}-band and T_{1g}-band and they are both fermion bands. According to the band structure calculation [8], these two bands are nonoverlapping and closely spaced with the energy difference at Γ point $\mathbf{k} = 0$ about 0.4eV. The width of each band is about 0.5eV. The trial ground state built from these uncorrelated electrons would be the ordinary Fermi sea with the T_{1u}-band half filled. The advantage of this type of hopping is to reduce the kinetic energy to maximum extent (delocalization effect). The Coulomb energy on each C_{60}, however takes the mean-field value, $\frac{U_0}{100} + 3.6eV$ per pair, and is not minimized.

Alternatively, the electrons can hop through correlated pairs in order to minimize the Coulomb energy on each C_{60}. The two electron Bloch wave functions read

$$\begin{aligned} \Phi_{\pm}(\mathbf{r},\mathbf{r}') = &\frac{1}{\sqrt{N}} \sum_{\mathbf{R}} e^{i\mathbf{k}\mathbf{R}} \sum_{g,g'} S_{\pm}(g,g') \\ &\chi(\mathbf{r} - \hat{g}R_0 - \mathbf{R})\chi(\mathbf{r} - \hat{g}'R_0 - \mathbf{R}) \end{aligned} \tag{20}$$

and where $S_{\pm}(g,g')$ are given in (12) and (13). The corresponding bands will be refered to as scalar band (Φ_{+}) and the pseudo-scalar band (Φ_{-}). Since the correlations (12) or (13) are maintained during the hopping, the Coulomb energy per pair is expected to be lowered than that of two uncorrelated electrons by 1.2eV (for $U_0 = 10eV$). The reduction of kinetic energy, however, is not spectacular because of the narrow band width for the correlated hopping.

There are other energies to be considered: An excitation energy is required to promote an electron from T_{1u}-band to T_{1g}-band to form a pair. The van der Waals energy between two neighboring C_{60}'s, which is estimated to be $-4.5 \times 10^4\ n^2/d^6\ eV\ \mathring{A}^6$ per C_{60} with n the number of uncorrelated electrons per C_{60} and d the nearest neighboring distance favors the uncorrelated hopping.

Let $E(3e)$ represent the energy of the uncorrelated hopping with three T_{1u} electrons per C_{60} in average and $E(be)$ represent the energy of the correlated hopping with one boson and one T_{1u} electron per C_{60} in average. Based on the foregoing considerations, we end up with the estimate

$$E(3e) - E(be) \cong (\epsilon_F - 0.32eV)/C_{60} \tag{21}$$

where ϵ_F is the Fermi energy measured from the bottom of T_{1u} band. This estimate, though very crude, does suggests that there exist two boson bands, which may lie very close to, or overlapping with the T_{1u} fermion band. Because of the large polarizability of C_{60} ions, the charge fluctuations induced by hopping are not expected to offset qualitatively the energy balance.

The key factor for the pairing mechanism, the reduction of Coulomb repulsion through the correlated hopping may find its applications in other superconducting materials.

5. The superconductivity

Based on the discussions of previous sections, a simple model of superconductivity can be readily proposed. Let $\alpha_{\mathbf{k},s}, \alpha^{\dagger}_{\mathbf{k},s}$ stand for the annihilation and creation operators of the electrons in T_{1u}-band with momentum $\mathbf{k}$ and with the index s, labeling both the spin and the triplet components. Likewise, $\alpha'_{\mathbf{k},s}, {\alpha'_{\mathbf{k},s}}^{\dagger}$ are the corresponding operators in the T_{1g}-band. The energy of these two bands is described by the Hamiltonian:

$$H_0 = \sum_{\mathbf{k},s}(\epsilon_{\mathbf{k}}\alpha^{\dagger}_{\mathbf{k},s}\alpha_{\mathbf{k},s} + (\epsilon_{\mathbf{k}} + \delta\epsilon)\alpha'_{\mathbf{k},s}{\alpha'_{\mathbf{k},s}}^{\dagger}) \tag{22}$$

where we have assumed the same dispersion relation $\epsilon_{\mathbf{k}}$ for both bands. To simulate the low Coulomb energy preference of the pairing state on each C_{60}, we introduce

$$H_1 = -G_- \sum_{\mathbf{R}} \phi^{\dagger}_{\mathbf{R}}\phi_{\mathbf{R}} - G_+ \sum_{\mathbf{R}} {\phi'_{\mathbf{R}}}^{\dagger}\phi'_{\mathbf{R}} \tag{23}$$

where both $G_{\pm}$ are positive and nearly equal to each other, $\phi_{\mathbf{R}}$ and $\phi'_{\mathbf{R}}$ are pseudoscalar and scalar boson operators given by:

$$\phi_{\mathbf{R}} = \frac{1}{\sqrt{6}}\sum_{\mathbf{R},i}(\alpha'_{\mathbf{R},i,\downarrow}\alpha_{\mathbf{R},i,\uparrow} + \alpha'_{\mathbf{R},i,\uparrow}\alpha_{\mathbf{R},i,\downarrow}) \tag{24}$$

and

$$\phi'_{\mathbf{R}} = \frac{1}{\sqrt{6}}\sum_{\mathbf{R},i}(\alpha_{\mathbf{R},i,\downarrow}\alpha_{\mathbf{R},i,\uparrow} - \alpha'_{\mathbf{R},i,\uparrow}\alpha'_{\mathbf{R},i,\downarrow}) \tag{25}$$

with $\alpha_{\mathbf{R},s}, \alpha'_{\mathbf{R},s}$ the corresponding annihilation operators in the coordinate space. The total Hamiltonian is then

$$H = H_0 + H_1 - \mu\sum_{\mathbf{k},s}(\alpha^{\dagger}_{\mathbf{k},s}\alpha_{\mathbf{k},s} + \alpha'_{\mathbf{k},s}{\alpha'_{\mathbf{k},s}}^{\dagger}) \tag{26}$$

with μ the chemical potential.

The appearance of two kinds of bosons, pseudo-scalar and scalar, suggests two long range orders, coexisting at low temperatures. Denote these two long range orders by B_- and B_+, they are given by

$$B_- =< |\phi_{\mathbf{R}}| > \tag{27}$$

and

$$B_+ =< |\phi'_{\mathbf{R}}| > \tag{28}$$

where both B_- and B_+ are complex with a relative phase θ. Since the pseudo-scalar bosons do not mix with T_{1u}-electrons and the scalar bosons do, B_- is like Bose-Einstein condensate and B_+ is like BCS condensate.

To calculate the excitation spectra, we write

$$\phi_{\mathbf{R}} = B_- + (\phi_{\mathbf{R}} - B_-) \tag{29}$$

$$\phi'_{\mathbf{R}} = B_+ + (\phi'_{\mathbf{R}} - B_+) \tag{30}$$

where the parentheses on the right hand sides of (29) and (30) represent the fluctuations. Expanding the Hamiltonian (26) to the linear order of the fluctuations, the result Hamiltonian contains a constant part and a part quadratic in the fermion operators. The diagonalization of the quadratic part yields two branches of excitation energies:

$$E_{\pm} = [(\omega + \frac{1}{2}\delta\epsilon)^2 + \frac{1}{4}\delta\epsilon^2 + |g_+|^2 + |g_-|^2 \pm \sqrt{(\omega + \frac{1}{2}\delta\epsilon)^2 + |g_-|^2(\delta\epsilon^2 + 4|g_+|^2 sin^2\theta)}\ ]^{\frac{1}{2}} \tag{31}$$

where, $g_{\pm} = \frac{1}{\sqrt{6}}G_{\pm}B_{\pm}$, $\omega = \epsilon_{\mathbf{k}} - \mu$.

It is interesting to examine two limiting cases. The first one is that $g_- = 0$, $g_+ \neq 0$ and μ cuts in the lower band. The energies in (31) become:

$$E_- = \sqrt{\omega^2 + |g_+|^2} \tag{32}$$

and

$$E_+ = \sqrt{(\omega + \delta\epsilon)^2 + |g_+|^2} \tag{33}$$

The spectrum of the lower branch is of BCS-type with a gap energy $|g_+|$. The second case is that $g_+ = 0$, $g_- \neq 0$ and μ cuts in the lower band. The energies in (31) become then:

$$E_- = |-\frac{1}{2}\delta\epsilon + \sqrt{(\omega + \frac{1}{2}\delta\epsilon)^2 + |g_-|^2}\ | \tag{34}$$

and

$$E_+ = \frac{1}{2}\delta\epsilon + \sqrt{(\omega + \frac{1}{2}\delta\epsilon)^2 + |g_-|^2} \tag{35}$$

There is no gap in the lower branch. The Fermi level is lowered on account of the occurence of pseudo-scalar bosons which undergo the ordinary Bose-Einstein condensation. This is the same mechanism in the boson-fermion model of superconductivity [9].

The excitation spectra (31) is valid for weak, $G_\pm$. For strong $G_\pm$, the system can be described to the leading order in $G_\pm^{-1}$ by a lattice boson model with $\phi_{\mathbf{R}}, \phi'_{\mathbf{R}}$ replaced by local boson operators subject to certain on-site repulsions [10] and the excitation spectra will be of phonon type at low momentum. The situations at intermediate $G_\pm$, which is perhaps more relevant to the reality, are being investigated.

Finally we would like to argue that the observed dependence of the transition temperature on the lattice spacing [11] is also consistent with the picture of Bose-Einstein condensation. As the lattice spacing d is increased, the Van der Waals energy is weakend as d^{-6}. Therefore the uncorrelated hopping is less favored and more electrons will be evaporated to bosons, which in turn rases the transition temperature. Using the observed slope

$$\frac{\partial ln T_C}{\partial ln d} \cong 26 \tag{36}$$

we deduce the ratio between the boson density n_b and the total charge density n to be

$$\frac{n_b}{n} \cong \frac{1}{9} \tag{37}$$

for K_3C_{60}.

This work is supported in part by DE-FG02-92 ER40699 and DOE-91 ER40651, Task B.

References

1 Kroto, H.W., et al.: Nature **31**, 162 (1985); Kratschmer, W., et al.: Nature **347**, 354 (1990); Ajie, H., et al., J. Phys. Chem. **94**, 8630 (1990)

2 Hebard, A.F., et al.: Nature **350**, 600 (1991); Rosseinsky, M.J., et al.: Phys. Rev. Lett. **66**, 2830 (1991); Holczer, K., et al.: Science **252**, 1154 (1991)

3 Friedberg, R., Lee, T.D., and Ren, H.C.: Phys. Rev. **B46**, 14150 (1992); Friedberg, R., Lee, T.D., and Ren, H.C.: in preparation

4 Stephens, P.W., et al.: Nature **351**, 632 (1991)

5 Albert Cotton, F., Chemical Applications of Group Theory, 2nd edition (Wiley Interscience, 1963)

6 Bochvar, D.A., and Galpern, E.G.: Dokl. Akad. Nauk. SSSR Chem. Eng. **209**, 239 (1973); Haddon, R.C., Brus, L.E., and Raghavachari, K.: Chem. Phys. Lett., **125**, 459 (1986)

7 Gonzalez, J., Guinea, F., and Vozmediano, M.A.H.: Phys. Rev. Lett. **69**, 172 (1992)

8 Huang, M.Z., Xu, Y.N., and Ching, W.Y.: Phys. Rev. **B44**, 13171 (1991)

9 Friedberg, R., Lee, T.D.: Phys. Rev. **B40**, 6745 (1989)

10 Friedberg, R., Lee, T.D., and Ren, H.C.: Equivalence between Spin Waves and Lattice Bosons, Columbia University preprint CU-TP-587

11 Sparn, G., et al.: Science **252**, 1829 (1991); Zhou, O., et al.: Science **255**, 833 (1992); Fleming, R.M., et al.: Nature **352**, 787 (1991)

This article was processed using Springer-Verlag TEX Z.Physik D macro package 1.0 and the AMS fonts, developed by the American Mathematical Society.

Synthesis of C_{60} atomic clusters and magnetization studies of superconducting Rb_3C_{60}

P. Papanikolaou[1], S. Zolotas[1], M. Galatsida[1] and C. Politis[1,2]

[1]University of Patras, School of Engineering, 26110 Patras, Greece
[2] Kernforschungszentrum Karlsruhe, Institut für Nukleare Festkörperphysik, Karlsruhe, Germany

Received 28 June 1993

Abstract. Magnetization curves for superconducting Rb_3C_{60} with $T_c = 27.5K$ were obtained under various magnetic fields and temperatures. The temperature dependence of the upper critical field H_{c2} was determined. The zero temperature value of this field was calculated to be $H_{c2}(0) = 465 \pm 50kOe$.

The magnitudes of the coherence length and the penetration depth were calculated as $\xi(0) = 26.7 \pm 3\mathring{A}$; $\lambda_L(0) = 2150 \pm 100\mathring{A}$ at zero temperature. Good agreement between the measured temperature dependences of this parameter and Ginzburg-Landau prediction was found at $T \geq 23K$.

PACS: 74.30Ci; 74.70Jm

1. Introduction

The atomic cluster C_{60} (fullerene) is a new stable modification of carbon. C_{60} has a soccer-ball-like structure with 12 pentagonal and 20 hexagonal faces. Such stable clusters form a spherically close-packed solid with a face-centered cubic structure (a=1.418nm) which are held together by van der Waals forces.

The discovery of superconductivity in alkali-doped fullerenes resulted in wide interest in this new non-cooper and non-oxide high-T_c superconductors. C_{60} fullerene doped with the alkali metals K, Rb and Cs are superconducting materials with critical temperatures $T_c = 18K$ [1,2] $28K$ [3] and $33K$ [4,5] respectively.

The critical current density, lower and upper critical magnetic fields, but also penetration depth and coherence length are among the most interesting quantities characterizing this new superconductor.

Large quantities in the range of gramms of atomic clusters of $C_{60} - C_{70}$ mixtures were prepared by the Krätschmer-Fostiropoulos method [6]. After separation of the C_{60} from the C_{70} high purity clusters and crystalline materials were prepared. The Rb_3C_{60} samples were prepared by direct reaction of C_{60} powders with high purity liquid Rb. The prepared Rb_3C_{60} fullerite has a single, stoichiometric phase with a fcc structure (a=1.424nm). More details of the sample preparation and characterization may be found in Ref. [7].

2. Results and discussion

For the present investigations the method of measurement of dc magnetization was used, which is the most direct and facile method to study all the above properties with the help of a high sensitive SQUID magnetometer (Quantum Design). The experiments were performed at a magnetic field $-50kO_e \leq H \leq 50kO_e$ within the temperature region $4.5K \leq T \leq 300K$. Specimens for magnetic measurements with a weight of 20 to 30 mg was sealed in a thin wall perspex cylindrical container under 1 to 1.2 bar helium as exchange gas.

The temperature dependence of the magnetization $M(T)$ for superconducting Rb_3C_{60} material was measured in different applied magnetic fields, as well as hysteresis loops at different temperatures from that of liquid helium up to critical temperature T_c. Typical curves of the magnetic susceptibility χ are shown in Fig. 1 where both zero-field-cooling (ZFC) and field-cooling (FC) curves for $H_{ext} = 10Oe$ are plotted.

It is clear that the appearence of a diamagnetic signal is due to the transition from the normal to the superconducting state of this material, with a transition temperature of $T_c = 27.5K$ for this sample.

The data of the ZFC curve show the screening effect of the sample, while the data of the FC curve represent the Meissner effect. The big difference between ZFC and FC curves indicates the flux creep and, hence, type-II superconductivity. This is also confirmed by the magnetic field dependence of the magnetization.

In order to evaluate the upper critical magnetic field, we used the method described in Ref. [8], where H_{c2} was determined from the temperature dependence of the FC magnetization. The upper critical field H_{c2} here is equal to the applied external field $H_{c2} = H_{ext}$; the results

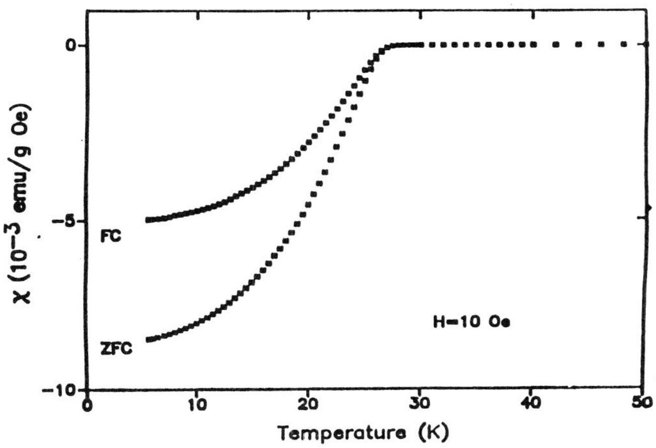

Fig. 1. Temperature dependence of the zero-field-cooled (ZFC) and field-cooled (ZC) d.c. magnetic susceptibility $H_{ext} = 10Oe$.

are shown in Fig. 2. The linear dependence, apparent in fields larger than $10kO_e$, leads to the results:

$$\frac{dH_{c2}}{dT} = -24.5kOe/K$$

By extrapolating $H_{c2}(T)$ to zero temperature, we can estimate $H_{c2}(0)$, using the Werthamer-Helfand-Hohenberg (WHH) formula [9]

$$H_{c2}(0) = 0.7(\frac{\partial H_{c2}}{\partial T})|_{T_C} \cdot T_c \quad (1)$$

which gives $465 \pm 50kOe$. The well known relation $H_{c2} = \Phi_0/2\pi\xi^2$ (where $\Phi_(0) = h/2e$ is the flux quantum) results in the coherence length $\xi = 26.7 \pm 3\AA$.

From these data and the value $H_{c1}(0) = 162Oe$, using the formula

$$H_{c1}(0) = (\frac{\Phi_0}{4\pi\lambda_L^2}) ln(\frac{\lambda_L}{\xi}) \quad (2)$$

we obtain $\lambda_L(0) = 2150 \pm 100\AA$.

This value for the penetration depth is close to $\lambda_L(0) = 2400\AA$, found in Rb_3C_{60} [10]. The results were also confirmed for the same sample, on the basis of the high-field data [11]. More details about coherence length and penetration depth may be found in Ref. [12]. The large upper critical field ($H_{c2} = 465kOe$) combined with the small lower critical field ($H_{c1} \leq 200Oe$) shows unambiguously that the Rb_3C_{60} fullerite compounds are extreme type-II superconductors.

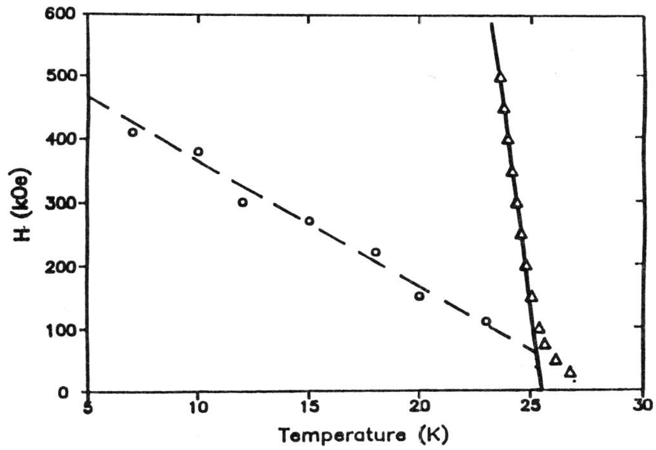

Fig. 2. The temperature dependence of the upper critical field H_{c2} (△) and the field H_{dev}, above which the critical current density became dramatically decreased (o). The dotted line is a guide for the eye.

References

1 Hebard, A.F., Rosseinsky, M.J., Haddon, R.C., Murphy, D.W., Glarum, S.H., Palstra, T.T.M., Ramirez, A.P., and Kortan, A.R.: Nature **350**, 600 (1991)
2 Holczer, K., Klein, O., Huang, S-M., Koner, R.B., Fu, K-J., Whetten, R.L., and Diederich, F.: Science **252**, 1154 (1991)
3 Rosseinsky, M.J., Ramirez, A.P., Glarum, S.H., Murphy, D.W., Haddon, R.C., Hebard, A.F., Palstra, T.T.M., Kortan, A.R., Zahurak, S.M., and Makhija, A.V.: Phys. Rev. Lett. **66**, 2830 (1991)
4 Kelty, S.P., Chen, C-C., and Lieber, C.M.: Nature **352**, 223 (1991)
5 Tanigaki, K., Ebbesen, T.W., Saito, S., Mizuki, J., Tsai, J.S., Kubo, Y., and Kuroshima, S.: nature **352**, 222 (1991)
6 Krätschmer, W., Lamb, L.D., Fostiropoulos, K., and Huffman, D.R.: Nature **347**, 354 (1990)
7 Politis, C., Buntar, V., Krauss, W., and Guverich, A.: Europhys. Lett. **17**, 175 (1992)
8 Holczer, K., Klein, O., Grüner, G., Thompson, J.D., Diederich, F., Whetten, R.L.: Phys. Rev. **67**, 271 (1991)
9 Werthamer, N.R., Helfand, E., and Hohenberg, P.C.: Phys. Rev. **147**, 295 (1966)
10 Sparn, G., Thompson, J.D., Whetten, R.L., Huang, S-M., Kaner, R.B., Diederich, F., Grümer, G., and Holczer, K.: Phys. Rev. Lett. **1228** (1992)
11 Politis, C., Sokolov, A.I., and Buntar, V.: Mod. Phys. Lett. **6**, 351 (1992)
12 Politis, C., Buntar, V., and Seminozhenko, V.P.: Intern. Journal of Modern Physics B **7**, 2163 (1993)

This article was processed using Springer-Verlag TEX Z.Physik D macro package 1.0 and the AMS fonts, developed by the American Mathematical Society.

Xe^+-ion capture by neutral Ar and Ne clusters

P. Scheier, B. Schiestl, M. Lezius, M. Foltin, M. Kolibiar and T.D. Märk

Institut für Ionenphysik, Universität Innsbruck, Austria

Received 28 June 1993

Abstract. A supersonic jet of cold neutral van der Waals clusters consisting of Ar and Ne atoms was sent through a Nier-type ion source and allowed to interact with thermal Xe ions produced therein via dissociative ion capture (DIC). A comparison of the resulting ions of the type Ar_nXe^+ and Ne_nXe^+ with pure Ar and Ne cluster ion distributions produced by electron impact (EI) of the same neutral vdW cluster precursor distribution shows characteristic differences in the overall shape of the distribution and in the ion stabilities (i.e. in the occurence of magic numbers in the mass spectra). To shed more light onto these differences we performed classical molecular dynamic calculations (MD) on the dynamics and Kinetics of these two ionization processes (DIC and EI) and found good agreement between the experimental results and the results deduced from these simulations.

PACS: 36.40.+d

1. Introduction

The mass spectrometer used is a double focussing sector field instrument with a mass range of 10000 Dalton at a nominal acceleration voltage of 3 kV and a maximum resolution of 25000. The cluster source consists of a temperature controlled stagnation chamber, a 20μm nozzle, a skimmer and a differentially pumped vacuum chamber. In the case of argon clusters the temperature was set to 110K using liquid nitrogen as a coolant. For the production of neon clusters the gas was cooled down to 60K using a closed He-cycle refrigerator. The stagnation pressure in the cluster source was typically in the case of argon clusters 1.5bar and for neon 3.5bar. The resulting pressure in the first differentially pumped chamber between the noozle and the skimmer is in the order of 10^{-3}Pa for Ar and 0.12Pa for Ne. Behind the skimmer another 500l/s turbo-molecular pump evacuates the ion source chamber (typically 10^{-5}Pa). The neutral supersonic jet then enters a Nier-type ion source and is crossed at right angles with an electron beam of variable energy.

The ion source has a standard gas inlet which is a glass capillary ending directly in the ionization chamber. Using this gas inlet the ionization region in the ion source can be filled with thermalized stagnant gas. The temperature of the ion source surfaces are stabilized typically at 550K. For ion capture investigations the total ion source pressure (pressure from the cluster beam and from the stagnant Xe) was set to 10^{-3}Pa. Under these conditions only two-body reactions of the type (R ... Ar or Ne)

$$Xe^+ + R_m \rightarrow R_{n-1}Xe^+ + (m-n+1)R \tag{1}$$

contribute to the $R_{n-1}Xe^+$ ion signal. We arrive at this conclusion from the experimental result that the ion signal $R_{n-1}Xe^+$ shows a linear dependence on the Xe pressure in the ion source in this pressure range. This observation excludes any three-body association reactions of the type

$$Xe^+ + R_m + R(orXe) \rightarrow R_mXe^+ + R(orXe) \tag{2}$$

because the pressure dependence of three-body association reactions is quadratically and limits the DIC reactions to true sticking collisions of the type

$$Xe^+ + R_m \rightarrow (R_mXe^+)^* \tag{3}$$

It is clear that subsequent randomization of the excess energy in the collision complex will heat up the cluster ion. This will lead eventually to the sequential evaporation (dissociation) of several argon atoms

$$\begin{aligned}(R_mXe^+)^* &\rightarrow (R_{m-1}Xe^+)^* + R \\ &\vdots \\ &\rightarrow R_{n-1}Xe^+ + (m-n+1)R\end{aligned} \tag{4}$$

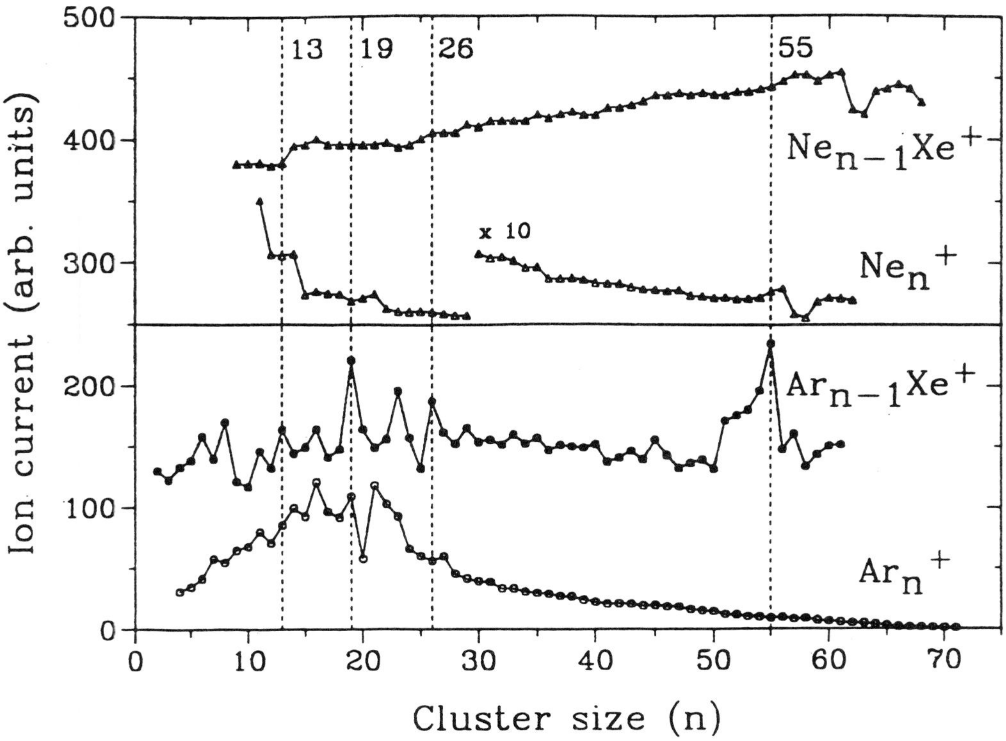

Fig. 1. Comparison of mass spectra produced by EI ionization of a neutral argon (∘) and neon (⋄) cluster beam and mixed $Ar_{n-1}Xe^+$ (•) and $Ne_{n-1}Xe^+$ (⋆) ion series produced by DIC of Xe^+ by the same cluster beam

as described in the evaporation ensemble model of Klots [1]. As the ion capture and the electron impact ionization are occurring in the same spatial region in the ion source it seems at first glance to be impossible to distinguish between $R_{n-1}Xe^+$ ions produced by capture of Xe^+ to R_m (reaction (1)) and $R_{n-1}Xe^+$ produced from other processes. However using electron energies below the appearance energy of R cluster ions (i.e. 14.6 eV for Ar [2]) only Xe will be ionized. By variation of the electron energy, the electron current and the Xe gas pressure it was possible to demonstrate that conditions exist for which $R_{n-1}Xe^+$ ions formed by ion capture can be isolated from all other production processes such as the two step reaction (neutral Xe capture and charge transfer)

$$R_m^+ + Xe \rightarrow R_{n-1}Xe^+ + (m-n+1)R \tag{5}$$

which may occur above the ionization potential of the respective rare gas cluster used. At the time it cannot totally be excluded that $R_{n-1}Xe^+$ ions may be produced via electron impact ionization of XeR_n (being formed by a kind of pick up reaction between the cluster beam and the neutral gas target), but the balance of arguments points towards reaction (1) as being the dominant production channel for $R_{n-1}Xe^+$.

2. Results and Discussion

Fig. 1 shows a comparison of mass spectra produced by EI ionization of a neutral argon (open circles) and neon (open diamonds) cluster beam and mixed $Ar_{n-1}Xe^+$ (closed circles) and $Ne_{n-1}Xe^+$ (star) ion series produced by DIC of Xe^+ by the same cluster beam, respectively. In the argon case the mixed cluster ion spectrum shows the icosahedral shell closure at n=55 and n=13 which cannot be observed in the pure Ar_n^+ ion series produced via EI ionization. Secondly, the overall shape of the cluster ion spectrum shows higher ion abundances for larger cluster sizes in the case of Xe ion capture. Especially this second observation shows that DIC is a "softer" ionization technique than EI.

A confirmation of this point has been obtained by molecular dynamic calculations whose results are summarized in Fig. 2. The picture shows the time evolution of the fragmentation pattern of Ar_{29} after EI ionization and DIC, respectively. As the energy release in case of EI is much larger the fragmentation pattern is very broad and shifted to lower sizes compared to the DIC result. In contrary to the observation that dissociative Xe ion capture to neutral Ar clusters is leading to well pronounced icosahedral shell closures (which are not visible in the pure Ar_n cluster distribution), the mixed $Ne_{n-1}Xe^+$ ion distribution does not show the magic numbers at n=14 and n=56 known from the pure Ne cluster distribution. An explanation for this fact is the large difference in size of Xe^+ and neutral Ne atoms. The central Xe ion cannot be surrounded by 12 Ne atoms in a geometrical

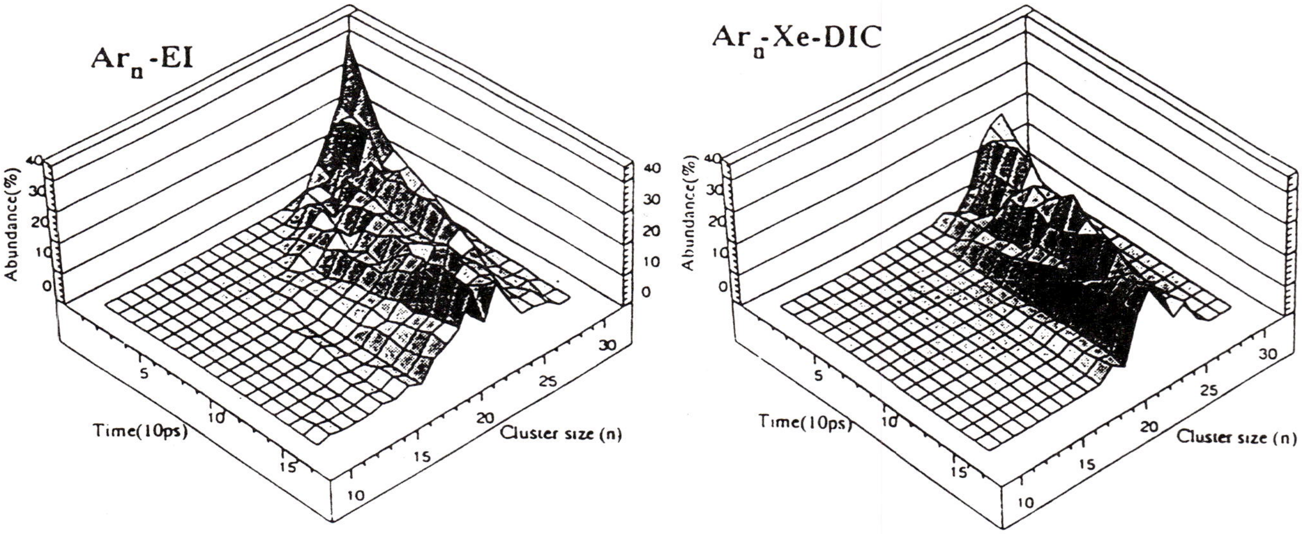

Fig. 2. Time evolution of the fragmentation pattern of Ar_{29} after EI-ionization and DIC of Xe^+

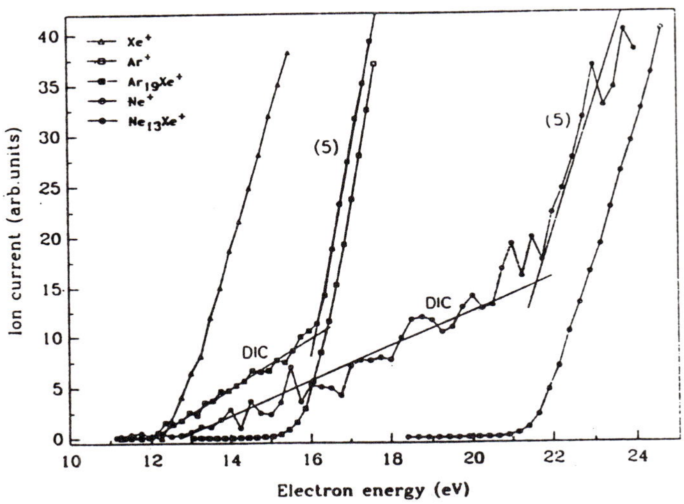

Fig. 3. Ion current close to the threshold for different ions as a function of electron energy

symmetric pattern (leading to a kind of shell closure and thus to a relatively stable structure), because the Ne-Ne bond lengths would exceed their normal value, but instead a kind of closed shell structure is observed at n=62. Fig 3 shows the ion current for different ions as a function of the electron energy. One can see for both $Ar_{n-1}Xe^+$ and $Ne_{n-1}Xe^+$ the onset of a second process at the appearance energy of the respective pure cluster. This indicates that indeed reaction (5) the capture of a neutral Xe by a rare gas cluster ion (Ar or Ne) and subsequent charge transfer in the collision complex occurs at high enough electron energies. Suprisingly the mass spectra do not change (same magic numbers and also a shift of the ion signal to larger cluster size) above this point.

Acknowledgement. Work Partially supported by the österreichischer Fonds zur Förderung der Wissenschaftlichen Forschung, the Bundesministerium für Wissenschaft und Forschung and the Österreichische Nationalbank, Wien. It is a plesure to thank U. Landman, Georgia Inst. Technology, for assistence in setting up the molecular dynamics program.

References

1 Klots, C.E.: Z. Phys. D **5**, 83 (1987)
2 Helm, H., Stephan, K., and Märk, T.D.: Phys. Rev. A **19**, 2154 (1979)

This article was processed using Springer-Verlag TEX Z.Physik D macro package 1.0 and the AMS fonts, developed by the American Mathematical Society.

Z. Phys. A 349, 357–358 (1994)

Higher nodal α-cluster band in ^{40}Ca

T. Yamaya [1], **S. Ohkubo** [2]

[1] Department of Physics, Tohoku University, Aoba-ku, Sendai, 980 Japan
[2] Department of Applied Science, Kochi Women's University, Kochi, 780 Japan

Received: 27 October 1993

Abstract. In the (^{6}Li,d) reaction on ^{36}Ar, the strongly excited seventeen levels were observed at excitation energies above 11 MeV in ^{40}Ca. On the basis of the result of the DWBA analysis, these levels may correspond to the members of the higher nodal α-cluster $K^\pi = 0^+$ band with prinicpal quantum number $N = 14$.

PACS: 21.10.J; 21.60.Gx; 24.50.+g

In the ^{36}Ar(^{6}Li, d)^{40}Ca reaction at $E = 50$ MeV [1], seventeen levels were strongly populated at excitation energies above 11 MeV. These levels consist of the strongly excited peaks in some blocks with fine structures as indicated by shaded peaks in Fig. 1. Such strongly excited states in the α-transfer reaction show that the α-particle wave function in ^{40}Ca may extend considerably further out than those of bound states. In the measurements of deuteron energy spectra from the reaction, no contaminant peaks were found at any angles because of the pure enriched ^{36}Ar gas target.

Angular distributions for these strongly excited levels were obtained at interval 2.5° in a region of $\theta_{\text{lab}} = 6°$–25° and 5° in a region of $\theta_{\text{lab}} = 25°$–35°. These angular distributions were analyzed within a framework of the distorted wave Born approximation (DWBA) to determine transferred angular momentum of an α-particle. The wave functions of relative motion between an α-particle and the ^{36}Ar core in ^{40}Ca were calculated as a resonance state for the local potential with about 180 MeV in depth.

Using these α-particle wave functions, DWBA curves were calculated to study the strongly excited levels in

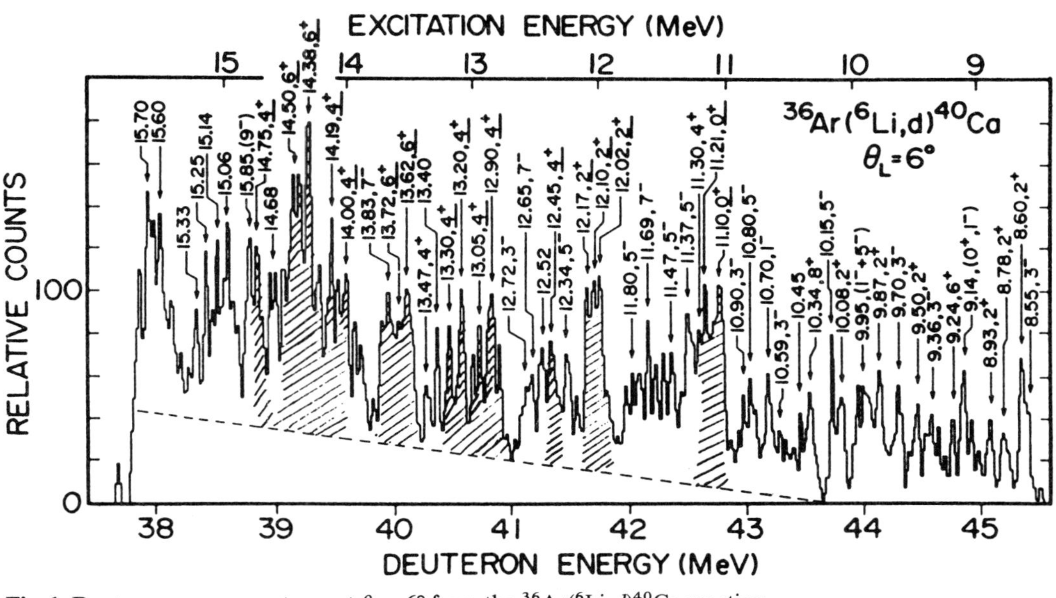

Fig. 1. Deuteron energy spectrum at $\theta_L = 6°$ from the ^{36}Ar(^{6}Li, d)^{40}Ca reaction

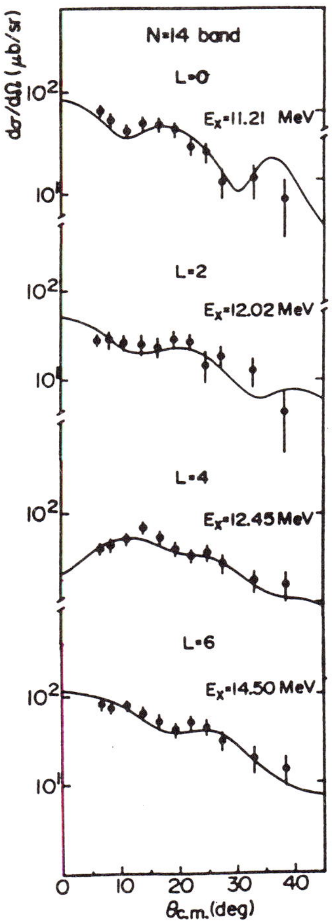

Fig. 2. Angular distributions for the states being typical of the members of the higher nodal $K^{\pi}=0^{+}$ band with $N=14$. Curves are results of DWBA calculations using the resonant α-particle wave functions with $N=14$

the reaction. For the α-particle wave functions with principal quantum numbers $N=12$ and 13, the DWBA curves calculated with any transferred angular momenta, L, did not reproduce the experimental angular distributions of these levels. The α-particle wave functions with $N=14$ were calculated at the resonance energies. These relative wave functions extend the tail of the wave function considerably further out than those with $N=12$ and 13. As the results of DWBA calculations using the wave functions with $N=14$ in ^{40}Ca, fits of the DWBA curves to the experimental angular distributions are shown in Fig. 2 for the transferred angular momenta from $L=0$ to 6.

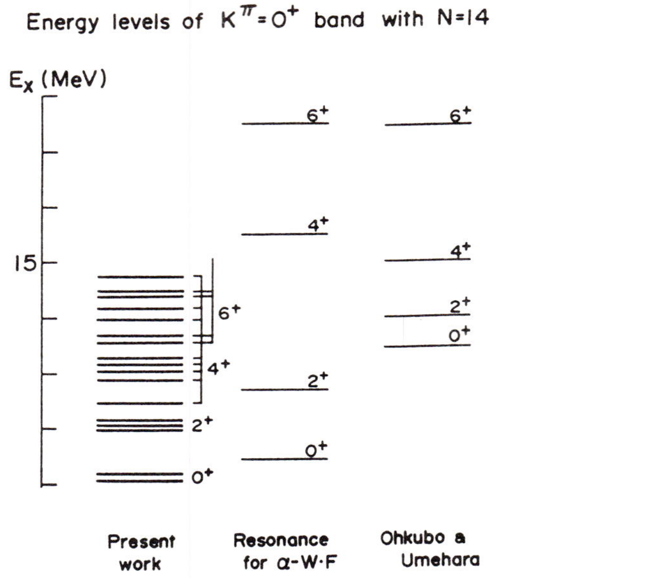

Fig. 3. The energy levels of the $K^{\pi}=0^{+}$ band with $N=14$. These levels are strongly populated in the (^{6}Li, d) reaction

These calculated curves reproduce well the experimental angular distributions. The observed energy levels and the resonance energies of the calculated α-particle wave functions are shown in Fig. 3, together with the results of the theoretical calculations by Ohkubo and Umehara [2].

The spectroscopic factor, S_{α}, for each spin state was deduced from the fit to the angular distributions. These summed S_{α} are 1.2, 0.47, 0.52 and 0.1 for $J^{\pi}=0^{+}$, 2^{+}, 4^{+} and 6^{+} states, respectively. On the basis of these results, a block of the 0^{+} states at excitation energy near 11 MeV can be considered as a strong candidate for the band head of the $N=14$ band with large α-cluster strength and the observed positive parity states with $N=14$ may constitute the members of the higher nodal $K^{\pi}=0^{+}$ band.

In the present work, the members of the higher nodal α-cluster band in ^{40}Ca were more clearly observed than in the established examples of *sd*-shell nuclei with α-clustering. This result suggest that the α-cluster model is applicable for *fp*-shell nuclei.

References

1. Yamaya, T., Saito, M., Fujiwara, M., Itahashi, T., Katori, K., Suehiro, T., Kato, S., Hatori, S., Ohkubo, S.: Phys. Lett. B**306**, 1 (1993)
2. Ohkubo, S., Umehara, K.: Prog. Theor. Phys. **80**, 598 (1988)

Z. Phys. A 349, 307–308 (1994)

Nuclear structure effects in cluster radioactivity

D. N. Poenaru[1,2], **W. Greiner**[1], **E. Hourani**[3], **M. Hussonnois**[3]

[1] Institut für Theoretische Physik der Universität, Robert-Mayer-Strasse 8-10, D - 60325 Frankfurt am Main, Germany
[2] Institute of Atomic Physics, Bucharest, Romania
[3] Institut de Physique Nucléaire, Orsay, France

Received: 27 October 1993

Abstract. Hindered or favoured transitions have been experimentally detected in cluster radioactivity of odd-mass nuclei, in a similar way to what has been previously observed in spontaneous fission or in the fine structure of α-decay. By combining the experimental study with a suitable theoretical description one can develop a powerful method to get direct spectroscopic informations on the deformed states of emitting nuclei. We are analyzing the possibility to explain within a superasymmetric fission model why the ^{14}C transition from ^{223}Ra toward the first excited state of ^{209}Pb is stronger than that toward its ground state.

PACS: 23.70.+j; 25.85.Ca; 27.90.+b

Cluster emission leading to excited states of the final fragments have been considered for the first time [1] in 1986. When the fine structure of ^{14}C radioactivity of ^{223}Ra has been discovered [2] it was shown that the transition toward the first excited state of the daughter nucleus is stronger than that to the ground state. In other words, like in spontaneous fission of odd-mass nuclei, or in fine structure of α-decay, one has a hindered and a favoured transition, respectively. Up to now theoretical considerations [3,4] have only been partly successful in explaining the experimental result.

Like in α-decay, or in spontaneous fission of odd-mass or odd-odd nuclei, one can define a hindrance factor [5] of a given transition

$$H = T^{exp}/T_{e-e} \qquad (1)$$

where T^{exp} is the measured half-life, and T_{e-e} is the corresponding quantity for a hypothetical even-even equivalent, estimated either from a systematics ($\log T$ versus $Q^{-1/2}$ for example [6]) or from a model. A transition is favoured if $H \simeq 1$, and it is hindered if $H \gg 1$.

Within our analytical superasymmetric fission model (ASAFM) we can easily simulate the even-even assumption. We also can take into account the angular momentum, l, determined from the spin and parity conservation:

$$|J_f - J_i| \leq l \leq J_f + J_1 \ ; \ \pi_i \pi_f (-1)^l = 1 \qquad (2)$$

The spin and parity of the (initial) ground state of ^{223}Ra is $J_i^{\pi_i} = 3/2^+$. In the final states of ^{209}Pb one has $J_f^{\pi_f} = 9/2^+$ (gs), $11/2^+$ (first excited state, at 0.779 MeV), and $15/2^-$ (second excited state, at 1.423 MeV). Consequently the allowed angular momenta are 4 and 6 units of $\hbar$ in the first two cases, and 7 and 9 in the last one.

The Q-values are 31.85, 31.07 and 30.43 MeV. If the intensities of the above mentioned ^{14}C transitions are 15 %, 81 % and 4 %, and the total half-life for cluster radioactivity is $10^{15.20}$ s, the corresponding partial life-times are $10^{16.02}$, $10^{15.29}$, and $10^{16.60}$ s. By ignoring the angular momentum, we get within ASAFM hindrance factors of 417; 2.4, and 2.5, respectively; they become 263; 1.5; 0.5 for l = 4; 4; 7, and 131.8; 0.8; 0.2 for l = 6; 6; 9, if the angular momentum is taken into account.

The favoured transitions are explained within the present version of ASAFM (one should allow an accuracy of $\pm$ 1 order of magnitude of any theoretical calculations), but the hindered one needs a larger action integral, K. Such an increase can be obtained with a larger potential barrier, due to the so-called *"specialization energy"* [7], in a similar way to what has been done for spontaneous fission of odd-mass nuclei [8,9]. If we introduce the correction, this would produce a larger area of the barrier $E(R) - E_{cor}$ integrated over the range of separation distances from the first to the second turning point, (R_a, R_b). We need a difference of about 15.76 MeV·fm, in order to get an hindrance factor of 417.

The specialization energy arises from the conservation of spin and parity of the odd particle during the fission process. With an increase of deformation (distance between fragment centers in the TCSM) the odd nucleon may not be transfered on a low energy at a level crossing, if in this way it cannot conserve the spin and parity. The corresponding barrier becomes higher and wider, compared to that of the even-even neighbour. The examples given in [7-9] are showing how the barrier height and width become larger due to this nuclear structure effect.

At every deformation ε, in a system with axial symmetry, in which the intrinsic angular momentum about the symmetry axis, Ω, is conserved, the specialization energy is defined as

$$E^s_\Omega(\varepsilon) = |E_\Omega(\varepsilon) - E^{lowest}_{\Omega^*}(\varepsilon)| \qquad (3)$$

where the last term corresponds to an angular momentum projection Ω^* changing at each level crossing, in a way that assures the lowest value of Fermi energy. Such a situation is met in even-even nuclei, in which $\Omega = 0$ state is the lowest at all deformations. In odd-mass and odd-odd nuclei, Ω is determined by the spin of the last impaired nucleon, hence to achieve its value at every level crossing, the system has to be (particle or hole)-excited.

$$V_{odd} = V_{e-e} + E^s_\Omega \qquad (4)$$

In this way the barrier increases and the fission half-life becomes longer.

Large hindrance factors have been measured not only in the fine structure of α-decay, but also in spontaneous fission [10,11]. Spontaneous fission rates of odd-A nuclei are slower than that of e-e nuclei, indicating a higher fission barrier, due to shell effects. In both kind of theories the physical explanation relies on the single-particle spectra of neutrons or/and protons. If the uncoupled nucleon is left in the same state both in parent and heavy fragment, the transition is favoured. Otherwise the difference in structure leads to a large hindrance.

Technically in a fission theory the calculations are performed as mentioned above, and in a many-body theory of α-decay the numerical value of the square of the overlap integral of the initial and final wave-functions is taken as a measure of the hindrance factor, because it reflects the difference in the cluster preformation amplitude.

Unlike in α-decay, where the initial and final states of the parent and daughter are not so far one from the other, in ^{14}C radioactivity of ^{223}Ra, one has a unique possibility to study a transition from a well deformed parent nucleus with complex parity-mixed odd-neutron orbitals to a spherical nucleus with pure shell model wave function.

In this way, one can get *direct spectroscopic information* on spherical components of deformed states. For example ^{223}Ra is known to be pear-shaped [12] with a possible set of deformation parameters [13] $\beta_2 = 0.129$, $\beta_3 = 0.10$, $\beta_4 = 0.075$, $\beta_5 = 0.01$, and $\beta_6 = 0.004$. According to the reflection-asymmetric rotor model the ground-state of ^{223}Ra is built from $3/2^+$ state emerging from $i_{11/2}$ orbital with small admixtures from $g_{7/2}$, $j_{15/2}$, and $g_{9/2}$.

References

1. Greiner, M., Scheid, W.: J. Phys. G. **12**, L229 (1986)
2. Brillard, L. et al.: C. R. Acad. Sci. (Paris) **309**, 1105 (1989)
3. Hussonnois, M., Le Du, J.F., Brillard, L., Ardisson, G.: Phys. Rev. C**44**, 2884 (1991)
4. Sheline, R.K., Ragnarsson, I.: Phys. Rev. C**44**, 2886 (1991)
5. Poenaru, D.N., Greiner, W. (eds.) Nuclear Decay Modes (to be published)
6. Hourani, E.: Chap.6 in [5]
7. Wheeler, J.A.: In: Niels Bohr and the Development of Physics, Pauli, W., Rosenfeld, L., Weisskopf, V. (eds) p. 163, London: Pergamon Press, 1955
8. Randrup, J. et al.: Phys. Rev. C**13**, 229 (1976)
9. Lojewski, Z., Baran, A.: Z. Phys.A**322**, 695 (1985)
10. Hoffman, D.C., Hamilton, T.M., Lane, M.R.: Chap.8 [5]
11. Wagemans, C.: The Nuclear Fission Process. Boca Raton: CRC Press 1991
12. Nazarewicz, W., Ragnarsson, I.: In: Handbook of Nuclear Properties, Poenaru, D., Greiner, W. (eds.). Oxford: Oxford University Press (to be published)
13. Sheline, R.K., Ragnarsson, I.: Phys. Rev. C**43**, 1476 (1991)

This article was processed using Springer-Verlag TEX Z.Physik A macro package 1.0 and the AMS fonts, developed by the American Mathematical Society.

Z. Phys. A 349, 309–310 (1994)

Spontaneous emission of ^{14}C clusters from $A = 221$ nuclei

R. Bonetti, C. Chiesa, A. Guglielmetti, C. Migliorino, P. Monti

Istituto di Fisica Generale Applicata dell'Universita di Milano, INFN, Sezione di Milano, Milano, Italy
(Fax +39 22 66 57 17, Tel.: +39 27 06 30 8 54)

Received: 27 October 1993

Abstract. We present results on ^{14}C radioactivity of ^{221}Ra and ^{221}Fr obtained by irradiating at ISOLDE a 4π assembly of solid state track detectors. The preliminary branching ratios relative to α decay come out to be $(1.9 \pm 1.3)\ 10^{-12}$ and $(1.0 \pm 0.2)\ 10^{-12}$, respectively.

PACS: 23.90.+w

1. Introduction

Since 1989, our group, partly in collaboration with P.B. Price, has been involved in measurements of half lives of cluster radioactivity of heavy nuclei by using the solid state nuclear track detector technique.

Recent experiments [1], such as those on ^{23}F emission from ^{231}Pa, on ^{24}Ne and ^{28}Mg from ^{233}U, on ^{14}C from ^{223}Ra and ^{225}Ac [2], have pointed out the importance of nuclear structure effects, in addition to barrier penetration, in determining the half lives of odd-A exotic emitters.

Briefly, it was found that, in close analogy with α decay, such nuclei exhibit a generally larger partial half life when compared with even-even emitters (having the same Q-value for heavy cluster emission), the amount of forbiddance (the so called hindrance factor) being related to the degree of overlap between the initial and the final odd nucleon single particle wave functions. The present paper reports the first results on two new cases of ^{14}C radioactivity from odd emitters, i.e. ^{221}Ra and ^{221}Fr.

2. Description of the experiment

Due to their short half lives, 28 s and 4.9 min, respectively, the only possibility was to perform two on line measurements at the electromagnetic isotope separator Isolde of Cern. A first irradiation was performed by implanting on a thin Al catcher the 60 keV $^{221}(Ra + Fr)$ beam produced by spallation reactions induced by 600 MeV protons on a thick ThC_2 target. The catcher was sorrounded by a sphere, $\phi = 13.8$ cm, whose inner surface was covered with BP-1 plates, a kind of phosphate glass acting as track detector [3]. The second irradiation was performed with the same detector set-up but in absence of the primary proton beam. ^{221}Fr was therefore indirectly obtained by the decay of the long-lived ($T_{1/2} = 10.0$ d) ^{225}Ac produced in the target by the previous irradiation, thus avoiding any ^{221}Ra contamination because of the short (8 min) half life of its progenitor ^{225}Th.

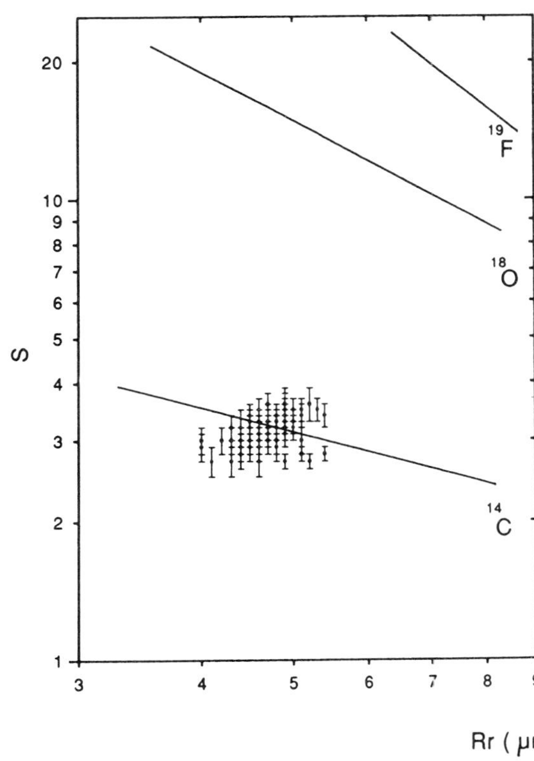

Fig. 1. Comparison of experimental data and accelarator calibration curves for the decay of ^{221}Fr; S is the adimensional sensitivity and Rr the particle residual range

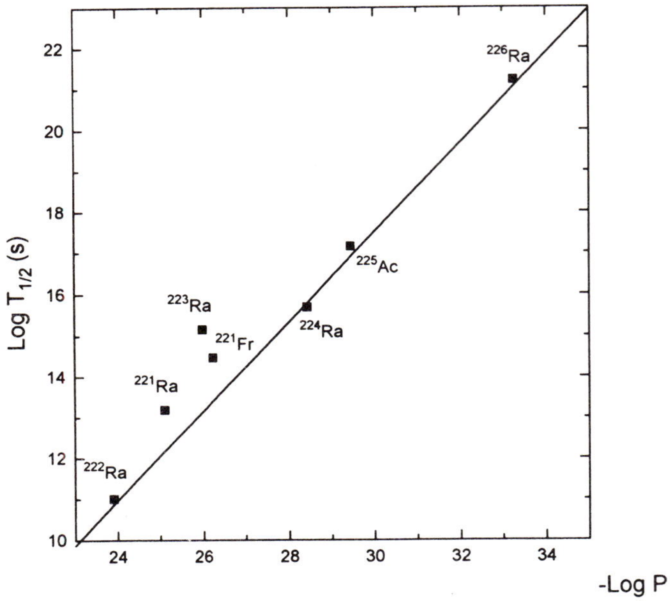

Fig. 2. Geiger Nuttal plot for all Carbon emitters discovered up to now; $T_{1/2}$ is the experimental half-life and P is the WKB penetrability

The irradiated detectors were etched in HBF_4, 50%, 65 °C and scanned with an optical microscope by using an automated analyzing system. In the case of ^{221}Fr we found 65 tracks due to cluster decay in 284 cm^2 of scanned surface. All the events can be unambiguosly attributed to *C* by comparison with accelerator calibration curves (see Fig. 1). The preliminary branching ratio relative to α decay came out to be $(1.0 \pm 0.2)\ 10^{-12}$.

The analysis of ^{221}(Ra + Fr) detectors is still on progress. A preliminary result, obtained by subtracting the ^{221}Fr contribution from the 23 carbon events measured up to now in $\cong$30% of the total surface, gives for ^{221}Ra $B_\alpha = (1.9 \pm 1.3)\ 10^{-12}$. The above numbers represent the smallest branching ratios measured up to now for ^{14}C radioactivity.

3. Discussion

A nice method to point out any "structure" effect in processes governed by barrier penetration, such as α or heavy cluster decay, is to plot data in the form of Geiger-Nuttal curves.

This is shown in Fig. 2 for all ^{14}C emitters including those reported in the present paper. The vertical deviation exhibited by odd-*A* emitters from the curve drawn for even-even emitters is a graphical evidence of the hindrance factor discussed in the Introduction, being further exemplified by ^{221}Ra and ^{221}Fr.

While a description of hindrance factor in microscopic terms has been suggested in [4], attempts towards an alternative description based on macroscopic variables are now on progress.

References

1. Bonetti, R., Guglielmetti, A.: Handbook of Nuclear Decay Modes. Boca Raton: CRC Press (in press)
2. Bonetti, R., Chiesa, C., Guglielmetti, A., Matheoud, R., Migliorino, C., Pasinetti, A.L., Ravn, H.L.: Nucl. Phys. A (in press)
3. Bonetti, R., Chiesa, C., Guglielmetti, A., Migliorino, C.: Nucl. Tracks Radiat. Meas. **18**, 325 (1991)
4. Hussonnois, M., Le Du, J.F., Brillard, L., Ardisson, G.: Phys. Rev. C**42**, R495 (1990)

Z. Phys. A 349, 311–314 (1994)

^{14}C cluster emission: a tool for the study of nuclear structures

M. Hussonnois[1], **G. Ardisson**[2]

[1] Institut de Physique Nucléaire, B.P. n° 1, F-91406 Orsay Cedex, France
[2] Laboratoire de Radiochimie, Université de Nice, F-06034 Nice Cedex, France

Received: 27 October 1993

Abstract. A brief survey of the ^{14}C cluster emission of different radium isotopes will remind the first evidence for an odd-even effect in ^{14}C radioactivity. The discovery at the I.P.N. Orsay, of a fine structure in the ^{14}C emission of ^{223}Ra enhanced the development of this new investigation field. The qualitative interpretation of the favoured transitions to the first two excited states of the ^{209}Pb daughter nucleus offers a new possibility to better understand nuclear structures of heavy deformed nuclei. For instance, an accurate determination of the energy of the ^{14}C clusters emitted by ^{225}Ac and ^{221}Fr must allow to improve our knowledge of ground state configurations of these nuclei.

PACS: 23.90.+w; 27.90.+b; 21.60.Gx

1. ^{14}C radioactivities of radium isotopes: odd-even effect

In 1984, Rose and Jones from the Oxford University published, in Nature, their discovery of the ^{14}C emission by ^{223}Ra [1]. They measured, with a $\Delta E \times E$ telescope, a ^{227}Ac source in which ^{223}Ra was in secular equilibrium. In 189 days, they detected 11 events with a total energy of about 30 MeV, attributed to a ^{14}C emission by ^{223}Ra. The branching ratio, relative to the α decay, was determined as $R = (8.5 \pm 2.5)\ 10^{-10}$.

At the Institute of Nuclear Physics of Orsay, we had a quite unique possibility to check on rapidly this result. The superconducting solenoidal spectrometer, SOLENO, permitted to select unambiguously and to focus, with a large solid angle ($\Omega = 115$ msr), the ^{14}C clusters on a telescope, rejecting the intense flux of α particles emitted by the source. So we could measure with a 300 times stronger ^{227}Ac source than the one used at Oxford. Moreover, SOLENO being installed at the Orsay tandem accelerator, beams of ^{12}C, ^{14}C and ^{16}O ions were used to calibrate precisely our telescope. In a run of only 5 days, 11 events were detected and fell inside the expected location of ^{14}C clusters emitted by ^{223}Ra [2]. The branching ratio, determined as $R = (5.5 \pm 2)\ 10^{-10}$, was in good agreement with the value of Rose and Jones.

Several other groups also confirmed this ^{14}C emission, using different techniques. Alexandrov et al. [3] repeated the measurement of a ^{227}Ac source with a simple $\Delta E \times E$ telescope. Price et al. [4] used solid state tracks detectors, radium isotopes being produced in spallation reactions at CERN and separated by the ISOLDE separator. Kutschera et al. [5] focussed the ^{14}C ions emitted by ^{223}Ra on a gas telescope with a spit-pole magnetic spectrometer. Unfortunately, the poor energy resolution – due to the preparation of the ^{227}Th source – in this experiment did not allow to determine whether excited states of the ^{209}Pb daughter nucleus were populated.

At Orsay, soon after our clear confirmation of this new mode of radioactivity, we planed to search other ^{14}C emitters. During the preparaton of this second experiment, the Berkeley group published their experimental data [4] which confirmed the ^{14}C radioactivity of ^{223}Ra and gave evidence for the ^{14}C emission by two other radium isotopes ^{222}Ra and ^{224}Ra. We confirmed the ^{14}C radioactivity of ^{222}Ra, by the measurement, with SOLENO, of a ^{230}U source in equilibrium with its daughters. But mainly in this experiment – about 100 years after the discovery of ^{226}Ra by Marie Curie – the Orsay group put in evidence the ^{14}C radioactivity of this isotope [6]. From the 4 events registered in 9 days with a 90 MBq source, we deduced a branching ratio of $(3.2 \pm 1.6)\ 10^{-11}$. This value was precised, with improved statistics, in a measurement of the ^{226}Ra source with polycarbonate detectors [7].

From these experimental results, *an odd-even effect – similar to the one well known for the α decay – was put in evidence in ^{14}C radioactivity* [7]. The ^{14}C emission rate of odd ^{223}Ra is hindered as compared to the ones measured for neighboring even-even isotopes. This effect is clearly evident when the logarithm of the measured half-lives are plotted versus $Q^{-1/2}$, Q being the desintegration energy. The even-even isotopes determine a straight line, whereas the ^{223}Ra value is located above this line.

2. Evidence for a fine structure in the emission of ^{14}C by ^{223}Ra

In all these experiments and all half-life calculations performed with different fission or cluster models, everybody considered that the ^{14}C decay left the daughter nucleus in its ground state. The decay energy was the main parameter and any detail of nuclear structure was taken into account.

So in 1989 we decided to look for a possible feeding of excited states, with a relative probability of few percents. The best test case was naturally the ^{14}C emission by ^{223}Ra, but it was necessary to improve both the energy resolution and the statistics, the separation between ground state and first excited state of ^{209}Pb being only 779 keV. In order to register some hundreds of ^{14}C in a reasonable amount of time, a strong ^{223}Ra source was prepared. Starting from an old (not treated since 25 years) supply of 150 mg of ^{231}Pa, 70 μg (185 MBq) of ^{227}Ac were first recovered. About 40 days after ^{227}Ac purification, the ingrown ^{227}Th (150 MBq) was radiochemically separated and electrodeposited on Pt as a thin source. Immediately after its preparation and at the end of the experiment, intensity and energy resolution (35 keV for α particles) of this source were measured. Used two weeks after its preparation, the ^{223}Ra activity was relatively constant (66 ± 2 MBq) during the experiment.

From our previous experiments we knew that the SOLENO magnetic spectrometer selected unambiguously ^{14}C clusters. So we decided to use, instead of a telescope, a large surface barrier detector (450 mm^2) selected for good energy resolution with heavy ions. Moreover this increase in size of the detector permitted to use SOLENO with a solid angle of $\Omega = 200$ msr at the maximum of the transmission curve.

In 11 days, 410 ^{14}C clusters were registered in several partial spectra, measured with various currents in SOLENO in order to change transmission efficiencies of the different predicted transitions [8]. All these spectra exhibit three peaks (Fig. 1): *the energy spectrum of* ^{14}C *clusters emitted by* ^{223}Ra *presents a* ***Fine Structure****, similar to the one of the α-emission*, discovered 60 years ago by Rosenblum. From the energy calibration of our detection system with a ^{14}C beam, these peaks were attributed to ^{14}C groups feeding the ground state and the first two excited states of ^{209}Pb daughter nucleus. After corrections of efficiencies, we obtained that the ^{209}Pb ground state is weakly fed (15%) whereas the first and second excited states are populated with respective 81% and 4% branching ratios. Surprisingly, as for α emission of odd nuclei, *the dominant* ^{14}C *decay of odd* ^{223}Ra *does not populate the ground state but the first excited state.* The ^{14}C/α branching ratio value was determined to be 6.4×10^{-10}, in agreement with all previous results. So the logarithm of ^{14}C total half life (in seconds) is 15.2, but the log of partial half-lives for the transitions to the ground state and the two excited states are respectively 16.0, 15.3 and 16.6 (Fig. 2). The ground state value disagrees with that admitted (15.2) to normalize all the theoretical predictions.

By analogy with the α-emission, to characterize the degree of hindrance of the different ^{14}C branches, we have calculated, in two different ways, the "hindrance factors":

i) Strickly experimentally, we plotted the logarithm of half-lives versus $Q^{-1/2}$ for all radium isotopes. Data for even-even isotopes ^{222}Ra, ^{224}Ra and ^{226}Ra are on a straight

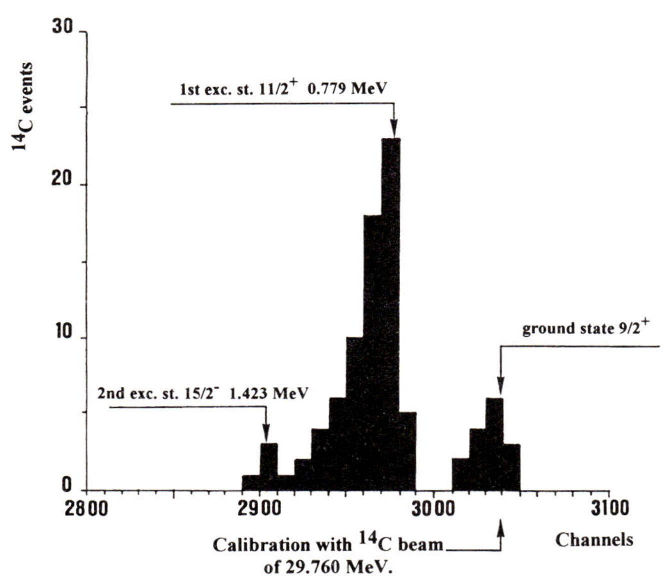

Fig. 1. Energy spectrum of ^{14}C particles emitted by ^{223}Ra. The arrows indicate the positions of the transitions to the ground state and the first two excited states of ^{209}Pb, obtained by the energy calibration with a ^{14}C beam

410 ^{14}C clusters recorded in 11 days

$\lambda_{^{14}C} / \lambda_{\alpha} = (6.4 \pm 0.4) \cdot 10^{-10}$

$\log T_{tot.} = 15.2$

Level	Percentage	$\log T_{1/2}$(s)	H.F.
Ground state	15 ± 3	16.0	~600
1st excited	81 ± 6	15.3	~3
2nd excited	4 ± 2	16.6	~3

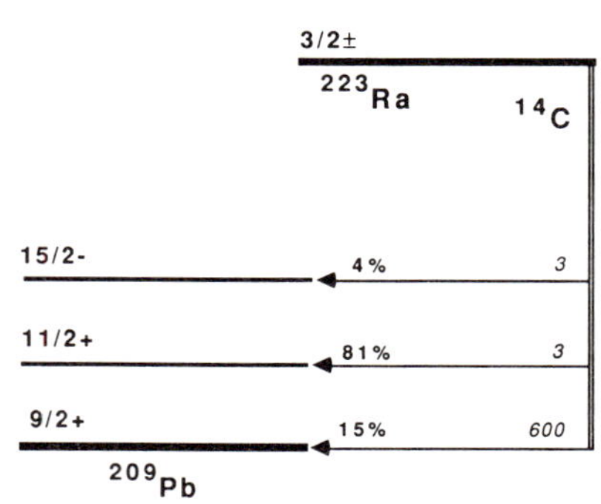

Fig. 2. Main results deduced from the precise energy measurement of the ^{14}C clusters emitted by ^{223}Ra and first level scheme from the ^{14}C decay of odd ^{223}Ra nucleus

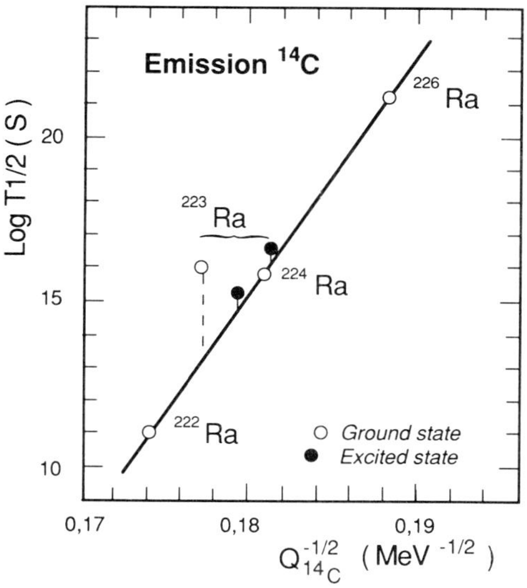

Fig. 3. Diagrams of decimal logarithm of the ^{14}C half-life versus $Q^{-1/2}$, for different Ra isotopes. Vertical distances between the ^{223}Ra points and the straight line, determined by even-even isotope values, represent the Hindrance Factors

line, whereas those for ^{223}Ra are apart. Hindrance factors are the vertical distances between the ^{223}Ra points and the straight line (Fig. 3). Low values (HF$\approx$3) for ^{14}C transitions feeding of the two excited states, and a high hindrance factor (HF$\approx$600) for the ground state were obtained [8].

ii) Like in the α decay theory, we defined the hindrance factors as the ratio of the average of ground state reduced widths of the neighboring even-even nuclei to the reduced width of the odd isotope, i.e., $F_{^{14}C} = 1/2[\gamma_0^2(A+2) + \gamma_0^2(A)]/\gamma_0^2(A+1)$, where the reduced widths γ_0^2 are given by $\gamma_0^2 = h \ln 2/2\,T_{1/2}P$, with $T_{1/2}$ in seconds and the barrier penatrability P expressed in function of the Gamow factor G. The more precise HF values are respectively 583, 3.9 and 4.6 [9], in good agreement with the previous ones. *The ^{14}C emission of odd ^{223}Ra is similar to the α emission of odd nuclei with a High Hindrance Factor for the transition to the ground state and Favoured Transitions to some excited states.*

3. Search for fine structure in the ^{14}C emission of even-even emitters

After the above discovery, it seemed important to check the possibility of existence of a fine structure for an even-even nuclei. ^{222}Ra was studied first because of its favorable branching ratio and of the possibility to obtain it easily as a member of the ^{230}U$\rightarrow$^{226}Th$\rightarrow$^{222}Ra radioactive series.

Pure ^{230}U can be radiochemically separated from its ^{230}Pa parent, itself synthesized by proton irradiation of thick targets of thorium. By addition of activities produced in two successive irradiations, a strong (85 MBq) but thin ^{230}U source was prepared.

Careful measurements of this source with SOLENO permitted to detect 210 ^{14}C events in 16 days, all except one being grouped in a single peak. Using the three energies of ^{14}C clusters emitted by ^{223}Ra to calibrate the detector, we deduced a 30.930$\pm$0.090 MeV energy for the ^{14}C emitted by ^{222}Ra, in good agreement with the energy calculated for the decay to the ^{208}Pb ground state (30.969$\pm$0.08 MeV). Accounting for the single event far from the peak, the relative population of the 3^- first excited state of ^{208}Pb, at 2614 keV, was estimated to be less than one percent. *The ^{14}C emission of even-even ^{222}Ra feeds essentially the ground state of ^{208}Pb daughter nucleus, like the α decays of even-even nuclei* [10].

Hourani et al. [11] have measured, with SOLENO, a very strong source of ^{224}Ra, prepared at the on-line separator ISOLDE at CERN. All the 149 ^{14}C events, recorded in 6 days, are in a single peak at the position expected for transition to the ^{210}Pb ground state.

4. Interpretation of the fine structure

Immediately after the discovery of the fine structure, by analogy with the interpretation of the α decay of odd nuclei, we suggested a qualitative explanation of the most significant result related to the ^{14}C decay of ^{223}Ra: very low ($\approx$3) hindrance factors for ^{14}C branches to the first two excited states in ^{209}Pb at 779 ($I^\pi = 11/2^+$) and 1423 keV ($I^\pi = 15/2^-$), but very high ($\approx$600) HF for the transition to the ground state ($I^\pi = 9/2^+$).

We thought that *the behavior of a odd nucleus ^{14}C decay should be mainly governed by the overlap between the initial and final configuration of the odd nucleon* [12, 13, 14, 9].

In our case, the configurations of the unpaired neutron in ^{209}Pb ground state and first excited states are well known:

ground state	0 keV	$9/2^+$	$\nu(2g_{9/2})^1$
1st exc. state	779 keV	$11/2^+$	$\nu(1i_{11/2})^1$
2nd. exc. state	1423 keV	$15/2^-$	$\nu(1j_{15/2})^1$.

Taking in account the recent measurement of ^{223}Ra ground state nuclear spin as $I = 3/2$, a reinvestigation of the structure of this nucleus pointed out that experimental data are better reproduced by assuming (in addition to a quadrupole deformation $\beta_2 \approx 0.13$) a static octupole deformation $\beta_3 \approx 0.1$. This stable octupole deformation gives rise to parity-mixed bands, so-called "parity doublet", $K^\pi = 3/2^\pm \{3/2[631] \otimes 3/2[761]\}$, describing both the ground state and the first excited band. In other words, the 143rd neutron of ^{223}Ra ground state can be considered as sharing its time between the two 3/2[631] and 3/2[761] Nilsson orbitals which arise respectively from the high spin $1i_{11/2}$ and $1j_{15/2}$ shell models orbits.

So, in the ^{14}C decay of ^{223}Ra, the transition favoured with regard to the odd neutron configuration should be those to the first and second excited states of ^{209}Pb, as it was observed experimentally.

Sheline and Ragnarsson [15], following later a parallel way, performed calculations, in the framework of the reflexion-asymmetric model, showing that the ^{223}Ra ground state has a very large content of the shell-model orbital $1i_{11/2}$ and a very low content of $2g_{9/2}$, in agreement with our qualitative interpretation of HF for the first excited and ground states. But they don't obtain a large content of $1 j_{15/2}$ to reproduce the favoured transition to the second excited state.

Another calculation, carried out by Briançon et al. [16], can account for the low hindrances factors observed in the experiment, but unfortunately cannot bear out the high hindrance observed for the ^{209}Pb ground state.

In our opinion, these calculations, performed in the framework of the reflexion asymmetric rotor model, allow a good qualitative interpretation of hindrance factors observed in α-decay, between two deformed nuclei, but partly fail to describe those obtained in ^{14}C emission, transition from a deformed nucleus to a quasi-spherical one. They have to be improved.

5. ^{14}C emission of ^{225}Ac and ^{221}Fr

In the paragraph above, we pointed out the possibility to extract informations on the ground state configuration of odd N actinides from the ^{14}C emission. The case of odd Z actinides is also interesting to examine since few data are known about the proton mean field in this mass region. However, up to 1991 searches of exotic decay of $Z<91$ odd nuclei, such as ^{225}Ac or ^{221}Fr gave negative results [7], although theoretical calculations led to branching ratio values of about 10^{-12}.

Recently Bonetti et al. discovered the ^{14}C emission by ^{225}Ac [17]. They measured a ^{225}Ra intense source produced by spallation of a Th target with 600 MeV proton beam and mass separated with the on-line separator ISOLDE at CERN. With an hemispherical phosphate glass detector they recorded 305 ^{14}C events, and determined a branching ratio of $(7.2 \pm 1.6) \times 10^{-12}$ underestimated by all theoretical models by practically one order of magnitude. However, solid state track detectors have not enough energy resolution to distinguish between ^{14}C clusters emitted by ^{225}Ac or by ^{221}Fr simultaneously present in the ^{225}Ra parent source.

This relatively high probability of the decay $^{225}\mathrm{Ac} \rightarrow {}^{211}\mathrm{Bi} + {}^{14}\mathrm{C}$ was interpreted by Bonetti et al. by the similarity between the initial and final states. Indeed, if one assumes, as in earlier work [18], that ^{225}Ac is a symmetric rotor, with quadrupole and/or hexadecapole deformations, its ground state is described by the 3/2[532] Nilsson configuration which originates from the $h_{9/2}$ spherical shell at zero deformation. As the 83rd proton of the near spheric ^{211}Bi nucleus is described by the $1 h_{9/2}$ shell of the spherical shell model, the ^{14}C emission is favoured by a transition between g.s.$\rightarrow$g.s., with the highest possible Q-value.

Conversely, the assumption of a ground state octupolar shape ($\beta_3=0.104$, $\beta_5=0.038$, $\beta_7=0.013$) lead to the proton state described by the $\Omega=3/2$ orbital ($n_\Omega=13$) [19] which originates from the $1 f_{7/2}$ shell at zero deformation. Therefore the ^{14}C transition to the ^{211}Bi g.s. should be hindered, while the decay to the first excited state ($E=0.405$ MeV, $I^\pi=7/2^-$) should be favoured.

The second nuclide involved in the Milano group experiments is ^{221}Fr, which decays by ^{14}C emission to ^{207}Tl. The systematic of odd thallium isotopes proves that the ^{207}Tl ground state is the hole state $(3 s_{1/2})^{-1}$. The spin of ^{221}Fr ground state has been measured by collinear laser spectroscopy as $I=5/2$. Following the prediction of the reflexion asymmetric rotor of Leander and Chen [20], the ^{221}Fr ground state is assumed to be the $I^\pi=5/2^-$ member of a $K=1/2$ parity-doublet band, which has a decoupled $h_{9/2}$-like character. Consequently the ^{14}C transition to the ^{207}Tl ground state should be hindered as well as those to the first excited state at 0.305 MeV, interpreted as the hole state $\pi\,(d_{3/2})^{-1}$. A different interpretation can be found in the paper of Cwiok and Nazarewicz [19] who used a Woods-Saxon potential and a monopole pairing force, with deformation parameters up to β_7. They found that the ^{221}Fr ground state is described by the $n_\Omega=18$ proton orbit, with $\Omega=1/2$ and having mainly a negative parity content, which originates from the $1 f_{7/2}$ subshell at zero deformation.

So we think that a precise measurement of the energy spectrum of the clusters emitted by $^{225}\mathrm{Ac}-{}^{221}\mathrm{Fr}$ source would add new data to our knowledge of the wave functions describing the ground state of these nuclei.

References

1. Rose, H.J., Jones, G.A.: Nature **307**, 245 (1984)
2. Gales, S., Hourani, E., Hussonnois, M., Schapira, J.P., Stab, L., Vergnes, M.: Phys. Rev. Lett. **53**, 759 (1984)
3. Alexandrov, D.V., Belyatski, A.F., Gluhov, A.Y., Nikolski, E.Y., Novatsky, B.V., Oglobin, A.A., Spepanov, D.N.: JETP Lett. **40**, 909 (1984)
4. Price, P.B., Stevenson, J.D., Barwick, S.W., Ravn, H.L.: Phys. Rev. Lett. **54**, 297 (1985)
5. Kutschera, W., Ahmad, I., Armato III, S.G., Friedman, A.M., Gindler, J.E., Henning, J.E., Issit, T., Paul, P., Rehm, K.E.: Phys. Rev. C**32**, 2036 (1985)
6. Hourani, E., Hussonnois, M., Stab, L., Brillard, L., Gales, S., Schapira, J.P.: Phys. Lett. **160B**, 375 (1985)
7. Barwick, S.W., Price, P.B., Ravn, H.L., Hourani, E., Hussonnois, M.: Phys. Rev. C**34**, 362 (1986)
8. Brillard, L., Elayi, A.G., Hourani, E., Hussonnois, M., Le Du, J.F., Rosier, L.H., Stab, L.: C.R. Acad. Sci. Paris **309**, Ser. II, 1105 (1989)
9. Hussonnois, M., Le Du, J.F., Brillard, L., Ardisson, G.: Phys. Rev. C**42**, 495 (1990)
10. Hussonois, M., Le Du, J.F., Brillard, L., Dalmasso, J., Ardisson, G.: Phys. Rev. C**43**, 2599 (1990)
11. Hourani, E., Rosier, L.H., Berrier-Ronsin, G., Elayi, A., Mueller, A.C., Rappenecker, G., Rotbard, G., Renou, G., Liebe, A., Ravn, H.L.: Phys. Rev. C**44**, 1424 (1991)
12. Ardisson, G., Hussonnois, M., Le Du, J.F., Brillard, L.: Proc. Int. School-Seminar on Heavy Ion Physics, Dubna 1989, 336 (1990)
13. Ardisson, G., Hussonnois, M.: C.R. Acad. Sci. Paris **310**, Ser. II, 367 (1990)
14. Hussonnois, M., Le Du, J.F., Brillard, L., Ardisson, G.: J. Phys. G**16**, 177 (1990)
15. Sheline, R.K., Ragnarsson, I.: Phys. Rev. C**43**, 1476 (1991)
16. Briançon, C., Cwiok, S., Eid, S.A., Green, V., Hamilton, W.D., Liang, C.F., Walen, R.J.: J. Phys. G**16**, 1735 (1990); Briançon, C.: Private communication
17. Chiesa, C.: Thesis
18. Lederer, C.M., Shirley, V.: Table of Isotopes. New York: Wiley 1978; Akovali, Y.A.: Nucl. Data Sheets **60**, 617 (1990)
19. Cwiok, S., Nazarewicz, W.: Nucl. Phys. A**529**, 95 (1991)
20. Leander, G.A., Chen, Y.S.: Phys. Rev. C**37**, 2744 (1988)

Z. Phys. A 349, 315–333 (1994)

Summary of the conference

Walter Greiner

Institut für Theoretische Physik, J.W. Goethe-Universität, Robert Mayer-Strasse 8–10, D-60054 Frankfurt am Main, Germany

Received: 27 October 1993

General remarks

Santorini is a wonderful place. Here, where a giant volcanic explosion 3500 years ago created this circular shaped harbour and made the Minoic culture, spread over various islands, dissapear, G.S. Anagnostatos and W. v. Oertzen organized this conference on "Atomic and Nuclear Clusters 93". We all agree, that this was a brilliant idea! Always when two major fields of physics are meeting (in this case atomic and nuclear physics) an intellectually alert atmosphere develops. We enjoyed it very much during this week. But not only scientific work characterized this conference: a welcome party, Barbecue-dinners, wine tasting, excursion to the volcanic reminders around here and a concluding gala dinner party additionally raised the excitement of the participants during these warm, sunny days.

Clusters are a very general phenomenon. They appear not only in the atomic and nuclear world, but also in the subnuclear and in the cosmic world. Baryons (e.g. nucleons and mesons are clusters of quarks. For the last 20 years it was believed that the elementary clusters, as I would like to call them, disappear if nuclear matter is highly compressed ($\rho \gg (5\text{–}8)\,\rho_0$) and heated ($T \gtrsim 200$ MeV). A phase transition to a quark-gluon-plasma is expected. Only free quarks and gluons should then be present. The heating and compression of nuclear matter is achieved via nuclear shock-waves in high energy collision of large nuclei (heavy ions). This unique mechanism had been proposed by the Frankfurt school of Theoretical Physics in the late 60'ies/early 70'ies and has been in detail confirmed experimentally in the early 80'ies. Well, about 10 years ago Theis, Maruhn, Stöcker et al. observed, that exactly such a phase transition also occures within the meson field theory, where only baryons and mesons exist, but no quarks and gluons. The puzzle was solved only last year when D. Rischke analysed lattice-gauge calculations and found, that the so-called quark-gluon plasma is not a gas of free quarks and gluons, but dominated by colour singulett clusters, i.e. by baryons, mesons and glueballs: One deals with a *clusterplasma!* A series of graphs assembled in Fig. 1 illustrates these statements.

In cosmological dimensions stars are assembling in galaxis, i.e. huge *clusters of stars* ($\sim 10^{14}$). Our milky way is such a cluster. But this is not all. These galaxies are again clustering up. So does our galaxy belong to the cluster called *local group.* Our nearest neighbours are the Magellanic clouds, the Andromeda nebula, the so-called M5-galaxis, etc. One has recently discovered tremendous clusters of galaxies ($\sim 10^5$ galaxies in one cluster), and many cases where the clustering seems to occur along lines *(cosmic strings).* It is speculated that they appear

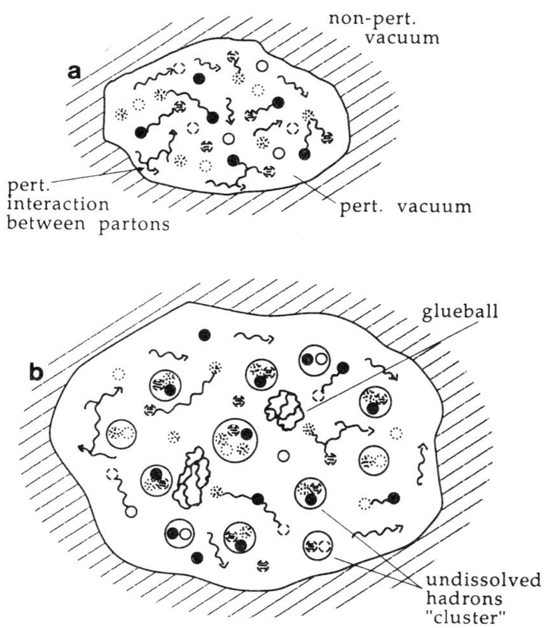

Fig. 1a, b. View of the so-called quark-gluon plasma. **a** Free quarks and gluons, **b** the quark-gluon-cluster plasma. The latter picture is the result of QCD-lattice calculations. Also clusters of strange matter are expected to appear. Such clusters are called strangeletts (Carsten Greiner, H. Stöcker) or Multistrange Exotic Mesonic Objects (MEMOS). Multi-Λ-hypernuclei do also belong into this class of clusters [1]

because of remnants of the old vacuum forming during the so called inflational phase, which ought to be seen in analogy of freezing water where also between various pieces of ice *strings* and *walls* of water (old phase) might exist. In cosmology these strings and walls of old vacuum have tremendously high energy density and therefore attract strongly the material around, i.e. they attract galaxies, which therefore clump together along cosmic strings.

We heard two talks by *S. Kubono* and *H. Oberhummer* concerning astrophysical questions, i.e. the building up of elements in the early stage of the development of our world. Two figures from Kubono's fine talk might recall some of the messages (Figs. 2, 3). Here are still very many questions open, both experimentally and theoretically. A detailed solution of the rate equations mimicing the production of heavy elements is still away.

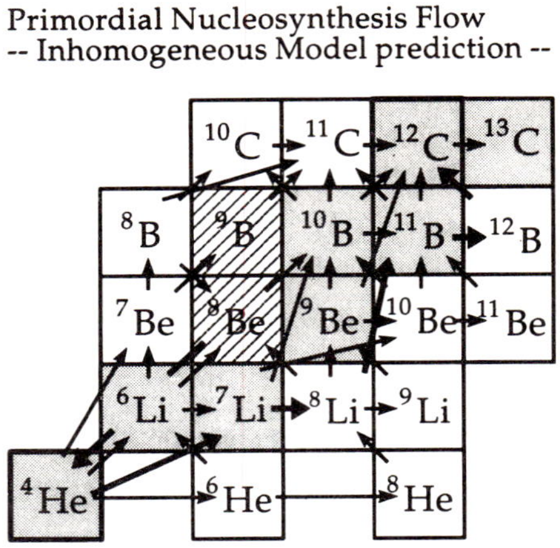

Fig. 3. The so-called inhomogeneous model predicts the synthesis of heavy elements because the primordial site has *n*-rich and *p*-rich regions due to quark-gluon phase transition. Many details are yet unknown. E.g. to study ^{8}Li$(\alpha, n)^{11}$B one needs the unstable nucleus ^{8}Li. This reaction is on the main sequence of heavy elements synthesis. ^{11}B$(n, 2\alpha)$ experiment$\rightarrow S = 8400$ MeV$\cdot b$ at $E_{cm} = 580$ keV; ^{9}Be$(\alpha, n)^{12}$B* (n) experiment$\rightarrow S = 16\,000$ MeV$\cdot b$ at $E_{cm} = 580$ keV; ^{8}Li$(\alpha, n)^{11}$B experiment$\rightarrow S = 40\,000$ MeV$\cdot b$ at $E_{cm} \geqq 1.7$ MeV. Hence one needs a good ^{8}Li beam for the problem. The radioactive beam facilities, which are planned or build at various places, help in this direction

Nuclear clusters – nuclear molecules

The α-clustering of light nuclei and the successive α-capture reactions play in the synthesis of light elements in the various burning processes an important role. It is therefore no wonder that α-type molecular clustering rouses the interest of nuclear physicists. Several talks, beginning with *P. Hodgson*, *B.R. Fulton*, *M. Brenner*, *R.W. Zurmühle*, *A.C. Merchant*, *A.H. Wuosmaa* and *L. Zamick* dealt with this aspect (α-chain molecules, -ring molecules, hollow α-nuclei and α-heavy cluster binary molecules). Also ^{12}C-^{12}C molecular configurations ob-

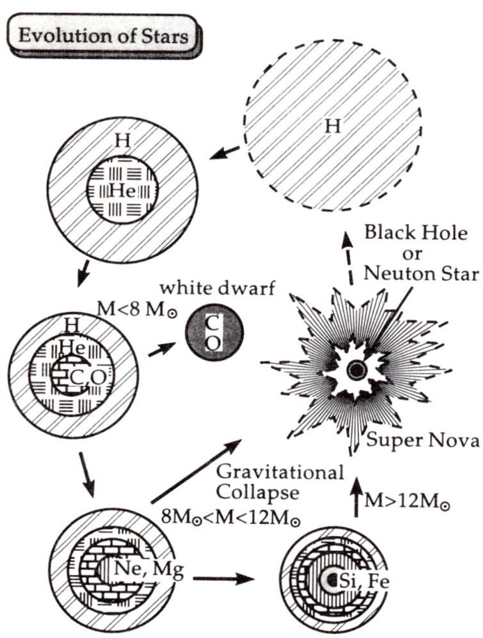

Fig. 2. Standard Model of Stellar Evolution. Each stage of evolution is well characterized. Nuclear reactions have a decisive role in the evolution of stars. Nevertheless, nuclear reactions in the temperature region of importance for the synthesis are not well determined experimentally. Thus, the study of such reactions give a stringent test for the standard stellar model

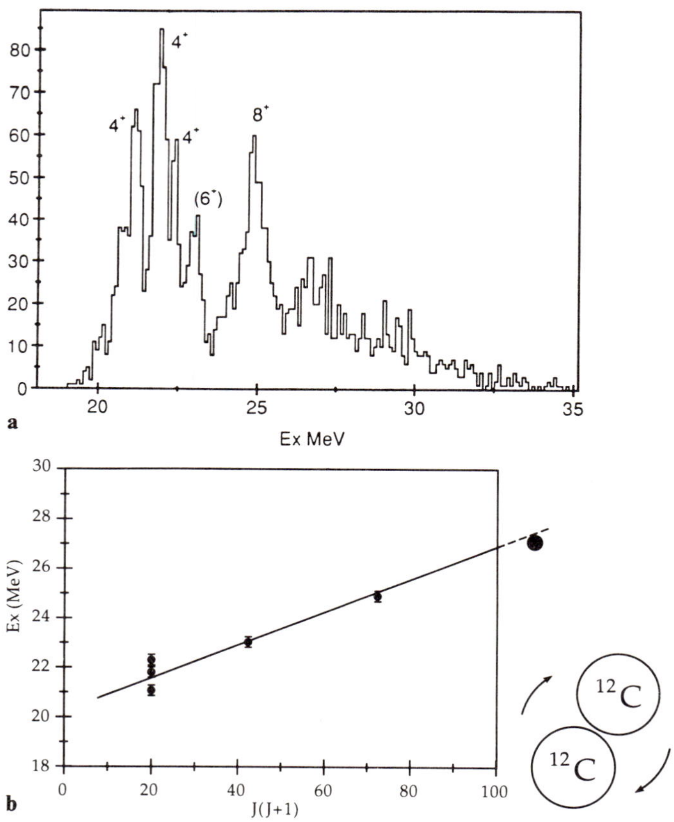

Fig. 4a, b. ^{12}C – ^{12}C cluster states observed in the breakup of ^{24}Mg. The spin systematics are consistent with that expected from a rotating structure and the slope indicates a high moment of inertia consistent with that from a structure of two touching ^{12}C nuclei [2]. From the rotational band one deduces the moment of inertia $J = 2.3 \times 10^{-42}$ MeV s^2, which is very close to the one of two touching nuclei rotating around each other: $J = 3 \times 10^{-42}$ MeV s^2. Hence one is convinced that one sees the ^{12}C – ^{12}C molecular band in ^{24}Mg

served in the breakeup of ^{24}Mg are important in this context. *Fulton* showed Fig. 4a, b.

Already in 1980 Fulton et al. looked at the reaction $^{12}C(^{12}C, ^{12}C)^{12}C(0_2^+)$ and found a sharp resonance at about $E_{cm} \approx 28$ MeV. It is suggestive to interprete this that one of the carbons has become α-chain-like (Fig. 5).

Wuosmaa in a very nice talk presented a new experimental set-up at Atlas (Argonne) and measured the excitation function $^{12}C(^{12}C, ^{12}C(0_2^+))^{12}C(0_2^+)$ – see Fig. 6. The broad structure in the $3\alpha + 3\alpha$-yield is probably fragmenting, but in the sense of what has just been said before, it suggests a α-chain-molecular-resonance in ^{24}Mg around $E_{cm} \approx 32.5$ MeV. Even though one is not absolutely sure (also a clover-type structure may break up into 3α) the picture is suggestive. In this connection the molecular dynamics calculation of Horiuchi is most interesting. He finds (Fig. 7) that the excited ^{12}C ion is *not* breaking up in stretched form, but first an α-cluster is strippen away from the carbon and the residual ^{8}Be is flying on, eventually decaying later.

Zurmühle worked further on his famous molecular resonance in the ^{24}Mg-^{24}Mg system, which has been theoretically interpreted in a very fine theoretical work by Maas and Scheid [5]. It is one of the most elongated

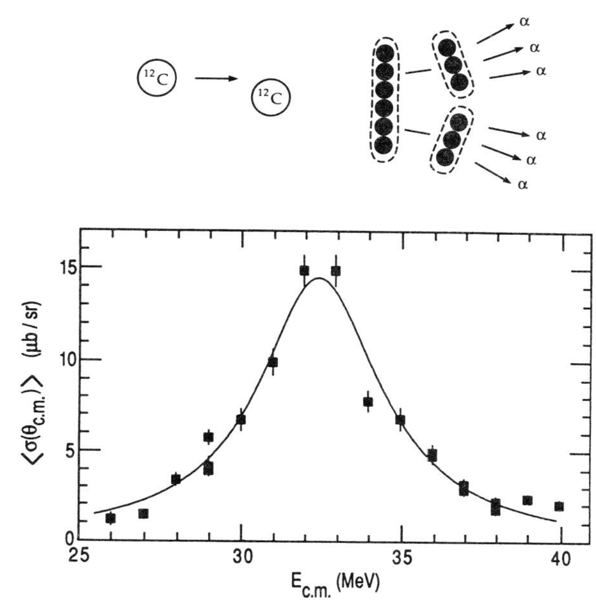

Fig. 6. Excitation function for $^{12}C - ^{12}C$ scattering looking at the $^{12}C_{0_2^+} + ^{12}C_{0_2^+}$ $(3\alpha + 3\alpha)$ yield. The striking resonance may reflect the formation of the 6α chain structure in ^{24}Mg. Namely, the excited 0^+ level in ^{12}C at 7.65 MeV has been suggested to have a linear 3α-chain-structure, and therefore a 6α-chain-structure in ^{24}Mg might appear as a resonance structure in the mentioned reaction. The resonance energy is within 1 MeV of the excitation energy for the chain structure in the cluster model calculations [4]

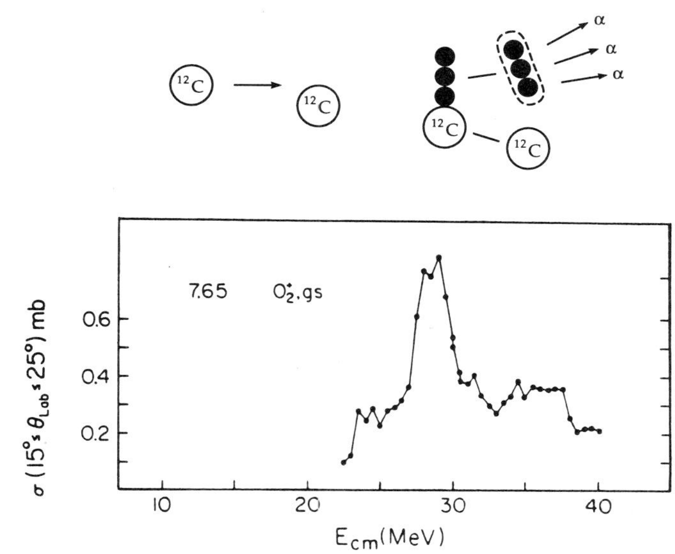

Fig. 5. Excitation function for $^{12}C - ^{12}C$ scattering looking at the $^{12}C + ^{12}C_{0_2^+}$ yield. The striking resonance may reflect the formation of the $^{12}C + ^{12}C_{chain}$ cluster structure in ^{24}Mg [3]

nuclear systems ever seen so far with an axis ratio of 3 : 1 (see Fig. 8). Since the last cluster conference Zurmühle identified the other channels (missing strength) – see Table 1. A whole set of molecular channels seems to couple in. Some of them he interpretes as containing hyperdeformed molecular states in ^{44}Ti.

M. Brenner measured the excitation function of $\alpha + ^{28}$Si, and identified fragmented (groups of) states with angular momenta 6, 7, 8 and 9 (Fig. 9a). In his analysis he made something new, he took the mean energy of these fragmented states and plotted them against $I(I+1)$ (Fig. 9b). This speaks for an α-^{28}Si-molecule!

The nuclear molecular model is alive as ever. I am convinced that nuclear molecules are a general phenomenon to be found in practically all systems. Let me close this subsection by drawing your attention to the question of giant nuclear molecules, like U + U, U + Th, etc. (Figs. 10, 11). It is most important to study the nuclear

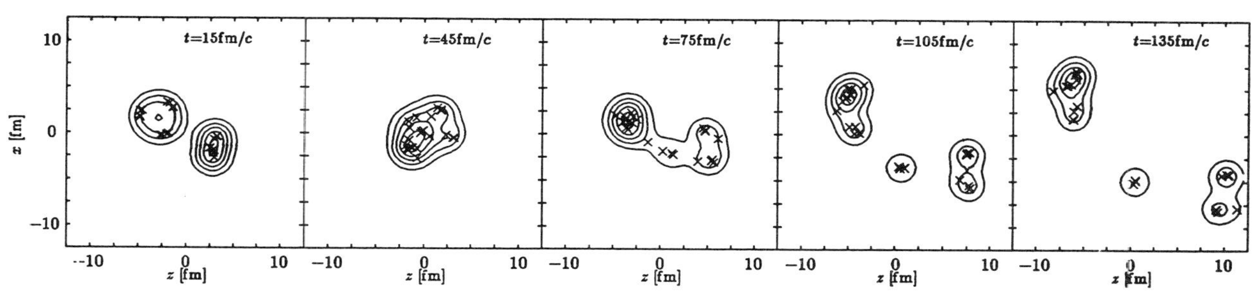

Fig. 7. Horiuchi's molecular dynamics calculations for $^{12}C - ^{12}C$ collisions at 28.7 MeV. It does not look like a 6α-chain

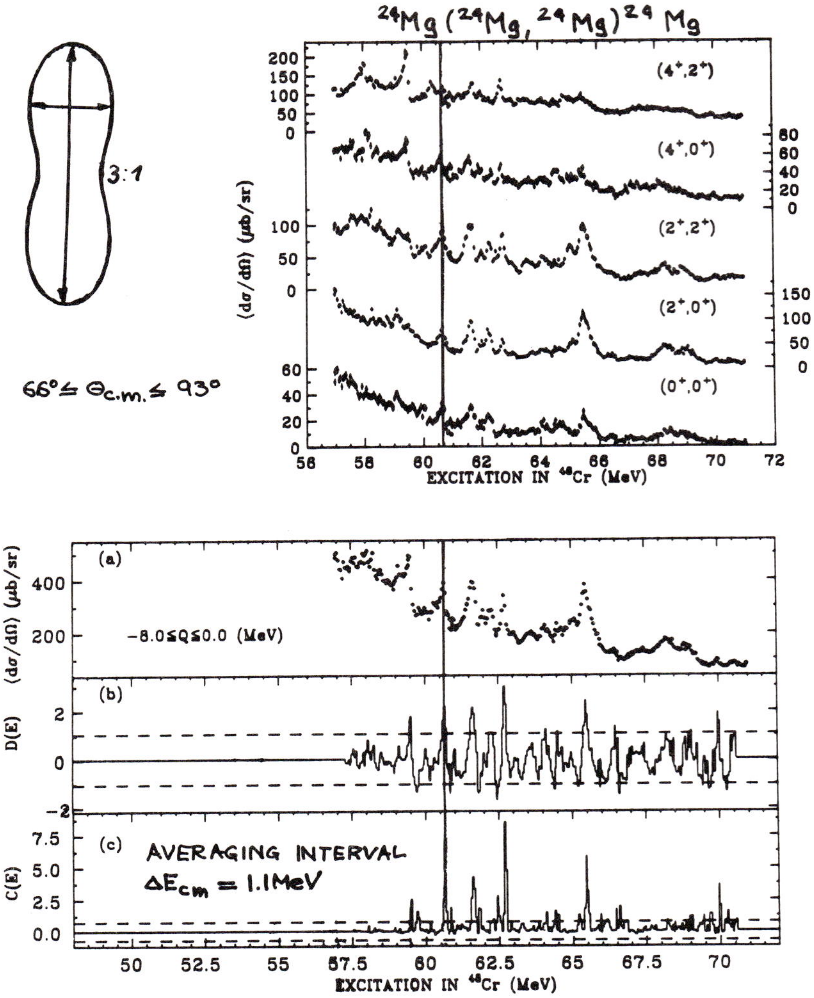

Fig. 8. Various excitation functions in $^{24}Mg-{}^{24}Mg$-scattering show at the excitation energy ≈ 60.06 MeV a molecular resonance, which corresponds to an exotically deformed molecular state (axis ratio 3:1)

Table 1. The various branching strengths for this resonance indicate that the molecular channel contributes only a small fraction (14%). Other strength is connected with a superdeformed state in ^{44}Ti. It must be superdeformed, because it decays into $^{24}Mg+{}^{20}Ne$, $^{28}Si+{}^{16}O$, and $^{32}S+{}^{12}C$

Decay mode	Branching strength
$^{24}Mg+{}^{24}Mg$	14%
$^{28}Si+{}^{20}Ne$	3%
$^{24}Mg+{}^{20}Ne+\alpha$	17%
$^{28}Si+{}^{16}O+\alpha$	25%
$^{32}Si+{}^{12}C+\alpha$	8%
Total observed strength	67%

structure of such giant systems at the Coulomb barrier, because it is most important to know wether there is a potential pocket or not in such giant systems. Only with a pocket the sticking time of two heavy nuclei can be long enough to observe the long predicted decay by positron emission of the supercritical vacuum in quantum electrodynamics. I urge our chairman W. v. Oertzen to engage in this problem! The colleagues at GSI and other places measuring e^+e^- spectra I urge to forget the particle idea which has captured their atention far too long. The peaks they see, if real at all, are most likely indicating nuclear conversion processes.

Binary nuclear molecules are theoretically best described within the two center shell model; chain molecules and other configurations by multi-center-shell-models. I mention in this context the report by *S. Åberg* and Fulton, but also refer to early work by *J. Hahn*, *Lustig* and *W. Greiner* [6].

Giant resonances in nuclei and metallic clusters

Many features of nuclear structure physics have their parallels in metallic cluster physics: Giant resonances, cluster-cluster collisions, fission, fusion, molecules, to mention a few.

I begin with the giant resonances. *R. Broglia* calculated it for the C_{60}-cluster utilizing Kohn-Sham equations and a $10^3\times 10^3$ $1p-1h$-basis for the 1^- states. It is a general fully triaxial RPA-calculation. Figure 12 compares the results for C_{60} with the measurements of Hertel et al., Fig. 13 compares the calculations for Na_{12}-clusters with experiments by Knight et al.

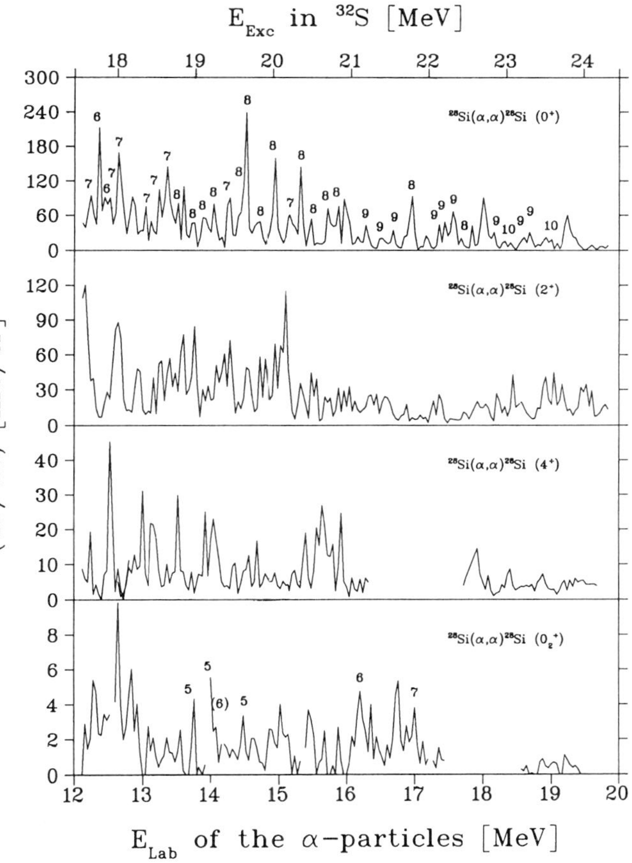

a

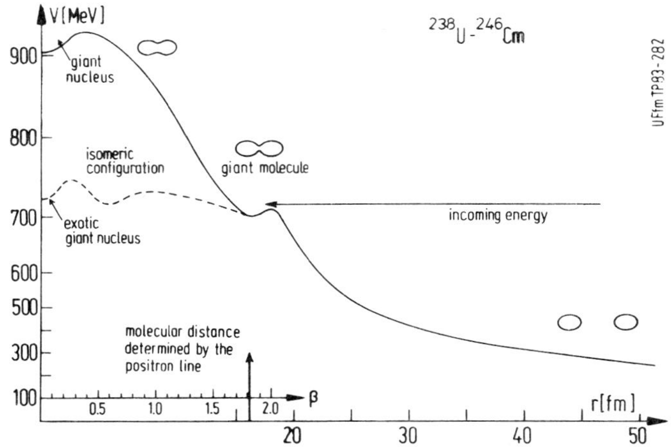

Fig. 10. The potential for an $^{238}U-^{246}Cm$ giant nuclear molecule shows a pocket at the touching configuration. This is supported by some theoretical models. Others predict just a plateau

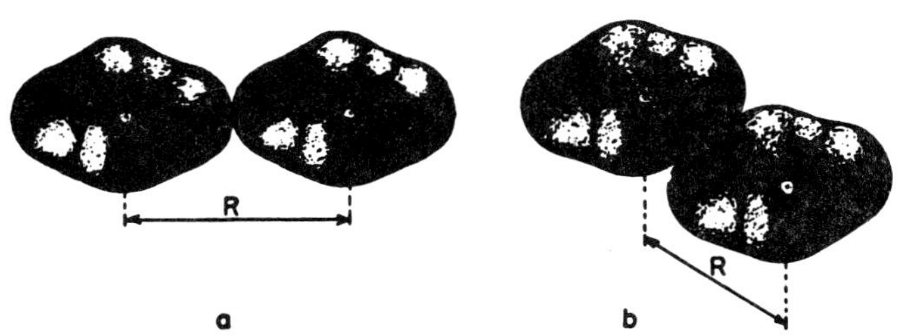

Fig. 11a, b. Typical positions for an U-U giant nuclear molecule, which are important in connection with the supercritical decay of the e^+e^--vacuum

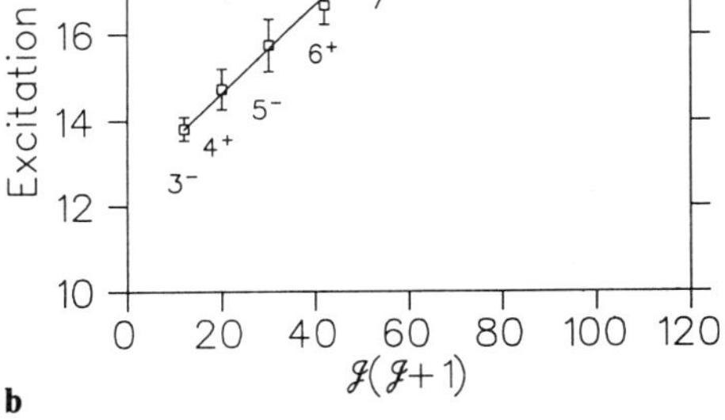

b

Fig. 9. a Fine structure of rotational $\alpha + {}^{28}Si$ states in ^{32}S with spins 6, 7, 8 and 9. Groups of states with the same angualr momentum are recognized. The molecular strength is obviously spreaded. **b** Mean energy of the spread rotational resonances in ^{32}S versus $J(J+1)$ *(solid line)*. The mixed parity band with definite moment of inertia can be recognized. The moment of inertia is larger then for α-particles in orbit with $M = 1.25\,A^{1/3} + 1.6$ fm *(dashed line)* but agrees with touching α – Si-molecule *(full line)*. The molecule is orbiting around its c.m., of course

I emphasize in this connection the Dynamic Collective Model in nuclear physics. It was worked out by *M. Danos* and *W. Greiner* in the early 60'ies and couples the giant dipole phonons $(\hat{d}^+)$ with the quadrupole phonons of the surface vibrations $(\hat{\beta}^+)$. There is strong interaction between dipole and surface motion. Therefore the states $\hat{d}^+|0\rangle$, $(d^+\beta^+)^{1-}$, $[d^+(\beta^+\beta^+)^{0,2}]^{-1}|0\rangle$ etc. do strongly mix and the naive Brink picture, that giant dipoles are build on lower energy states is not correct. The giant dipole resonance is split up into many states. I also point to elastic and inelastic photon scattering which tests the mixing of low lying states with the giant dipole state. *Nathan* from the University of Illinois has in beautiful work confirmed these ideas experimentally and in particular the Dynamic Collective Model (see [9], where the nuclear giant dipole physics is described in detail). Making use of a particle-hole-basis of about $10^3 \times 10^3$ 1^- configurations one performes the diagonalization of the Coulomb interaction in the Random Phase Approximation. Because C_{60} is not spherical one uses a general fully triaxial RPA. The results are shown in the figure in comparison with data by Hertel et al. In the experiments they would see only the dipole excitations above 7 eV.

Mrs. Bracco in her talk on *giant resonances in hot nuclei* came close by studying the effect of the shape and orientation fluctuations on the giant resonances in *Er-*

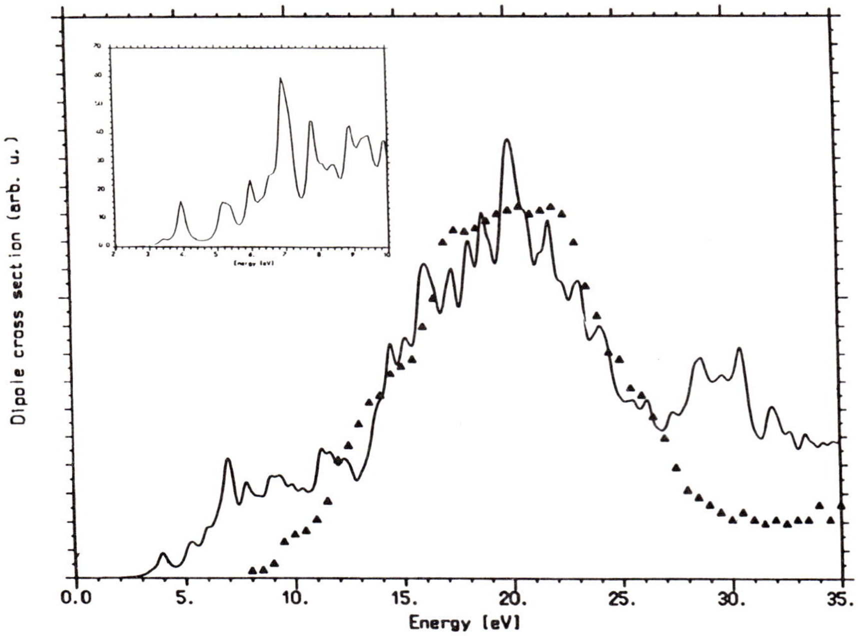

Fig. 12. The calculations shown were done by the Milano group (F. Alasia, L. Serra, E. Roman, J. Pacheco, G. Colo and R.A. Broglia). They correspond to a Kohn-Sham self-consistent calculation using the correct position of the 60 C-atoms in the icosahedron. One used all the so-called "pseudo-potentials" located on each ion. The pseudo-potentials describe the 4 valence electrons without describing explicitly the lower closed shell electrons. The resulting potential of this Kohn-Sham calculation is l-dependent quite complicated. It gives a closed shell for C_{60} with a shell closure of 1.8 eV as measured experimentally

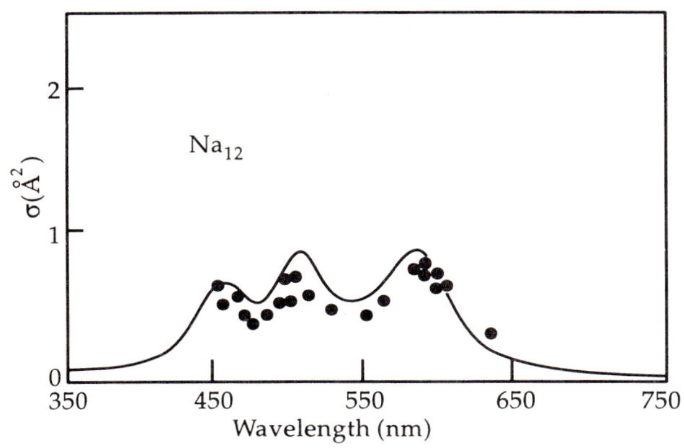

Fig. 13. Photoabsorption cross section for the cluster Na_{12} (i.e. 12 atoms of Na) measured by the groups of W. Knight [7] in comparison with calculations carried out by the Milano group (M. Bernath, C. Yannouleas and R.A. Broglia [8]). The cluster Na_{12} has a triaxial equilibrium deformation shape. This is the reason for the three peaks. The width associated with each peak is due to thermal fluctuations of the cluster surface. In other words: One can view Na_{12} as an ensemble of shapes. This is completely analogue to the so-called *Dynamic Collective Model* of Nuclei. One then calculates the response at each shape, and adds all of them in the final result weighting them with the Boltzmann factor $\exp(-F/T)$. It is worthwile to mention that no three peak photoabsorption cross section of giant dipole resonances in nuclei have been observed but only two peaked ones. This implies that nuclei are only dynamically triaxial and their triaxiality is rather small (see [9])

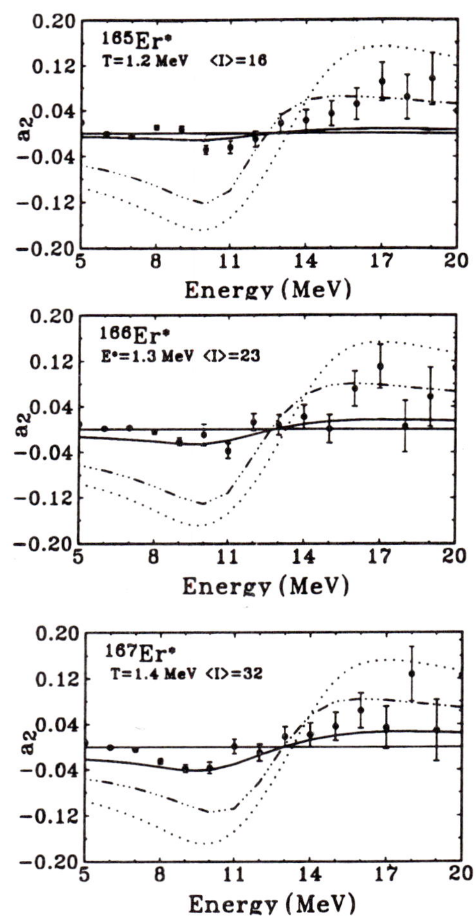

Fig. 14. $^{165-166}$Er are formed with heavy ion-fusion reactions. Also indicated are the temperature and the average spin of the compound nucleus. The a_2-coefficient of the angular distribution of photons relative to the beam axis is calculated under the influence of shape vibrations and rotations on the giant dipole resonances. Both are needed to explain the experiments

isotopes obtained in heavy-ion fusion reactions – see Fig. 14.

As *Danos* and *Okamoto* have predicted for nuclei and as has been manyfold confirmed experimentally, the giant dipole resonance should split in an axially deformed nucleus into two peaks. Depending whether the nucleus is prolate or oblate, the higher, respectively the lower of the peaks carries double strength (see Fig. 15). In clusters this is similar, and *J. Pedersen* has nicely assembled the data (Fig. 15). Obviously clusters are spherical at magic numbers but become immediately deformed if one goes away from magic numbers (prolate above, oblate below – this rule is violated above the magic number 40, yet unexplained). This is due to the lacking of the pairing force in clusters, which hinders in nuclei the immediate deformation of nuclei in the neighbourhood of magic nuclei.

Pedersen also reported about calculations of S. Frauendorf and V.V. Pashkevich who calculated the shapes of metallic droplets (Fig. 16). Some octupole deformed shapes can be seen. One should clarify the coupling of giant resonances to such shapes!

The giant resonances in hot and cold Na-clusters have also been reported by *Haberland.* Cold oblate nuclei "smear out" when becoming hot and dipole strength is pushed to higher energies (Fig. 17a). On the other hand cold prolate nuclei become oblate, structure disappears (Fig. 17b). Also comparisons with Eckhardt's SIC-improved calculations are shown (Table 2).

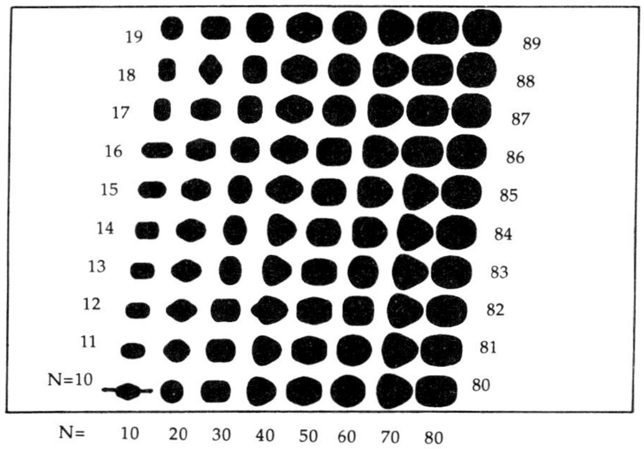

Fig. 16. Calculation of Sodium cluster shapes using the Strutinski formalism. Only axially symmetric distortions are included ($\alpha_2, \ldots, \alpha_6$). In their calculations, the magic size 40 is actually strongly octupole deformed. The symmetry axis is the horizontal line. The calculations were carried out by Frauendorf and Pashkevich

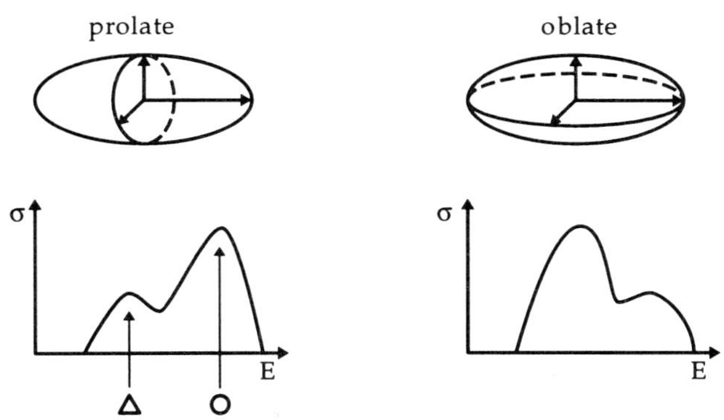

Table 2. The new improved calculations of W. Ekhardt eliminate a *mistake*, i.e. the selfinteraction (SIC = Self Interaction Correction) by subtracting the density δu^{ii} of the i-th electron from the response density of all electrons δu. The before last column shows the newly calculated energies as compared to the experiments (last column). It is desirable, however, to check the SIC improved calculations also for other states

N	TDLDA	SIC-TDLDA	FULL SCI	EXP
8	1.41	1.52	1.69	1.73 ± 0.03
20	1.34	1.46	1.61	1.58 ± 0.04
40	1.30	1.41	1.51	1.56 ± 0.04

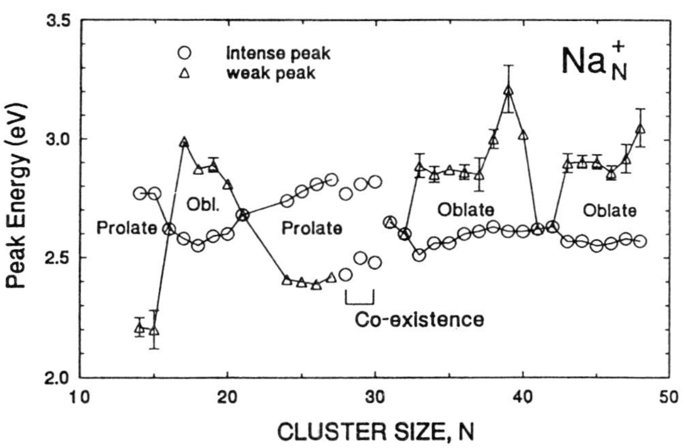

Fig. 15. The figure shows the experimentally measured resonance energies. In an axially symmetric cluster the area ratios of the two peaks will be ~2:1. In prolate systems the high energy component is two-fold degenerate. In the oblate system the low energy peak is two-fold degenerated. The shapes deduced from the figure show prolate deformation just above a shell closing (8, 20) and oblate shapes below a shell closing (20, 40). Above 40 this systematic behaviour is changed, yet unexplained

Shells and supershells in metallic clusters

The jellium shell structure was adressed in several talks. *M. Broyer* studied the supershells in Gallium clusters and found that a diffuse jellium potential resulted in a shift of the supershell closer to experiments (Fig. 18a, b). Among various explanations proposed for the diffuseness the surface oscillations of the positiv ionic background charge is quite plausible (Fig. 19).

Mansikka-aho adressed the question of how the roughness of the cluster surface influences the shell structure. Quite plausibly he finds that the shell structure near the Fermisurface is affected by the roughness (Fig. 20). I believe that also the levels above the Fermisurface, close to the continuum must be affected. *Mansikka-aho*'s answer in this respect has not been clear.

D. Reinhardt investigated the transition from cluster structure to crystal structure. When he had cut out a segment of a spherical cluster it became softer. This is – in my mind – quite plausible: If some surface segment, holding the nucleus together, is cut out, the residual cluster has to become softer towards bending it apart.

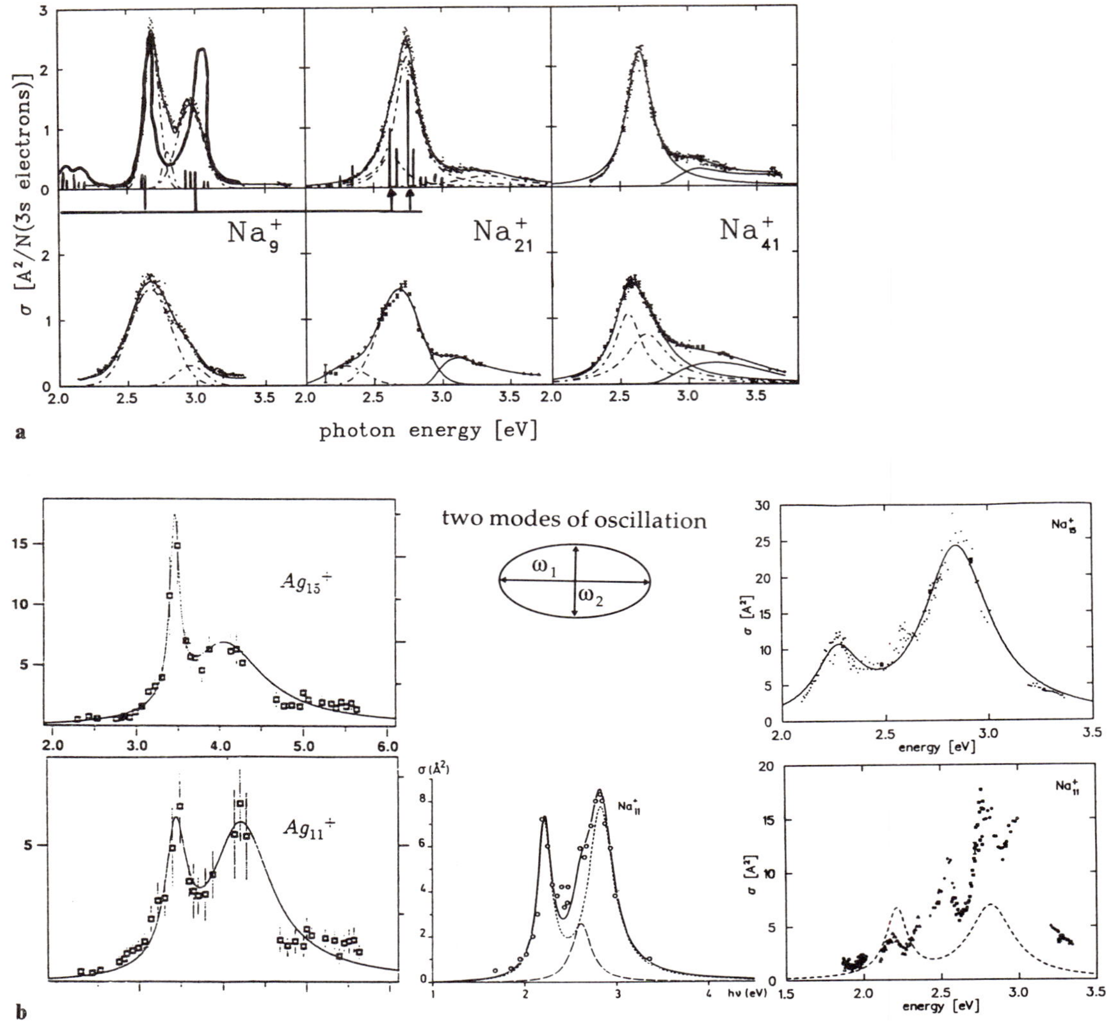

Fig. 17. **a** Optical Spectra of Spherical Na Clusters with 8, 20, 40 valence electrons (Haberland); *first row:* hot clusters: 1. one peak at ~80% of $h\omega$ (bulk-plasmon)/($\sqrt{3}$). 2. shoulder between $h\nu = 3$ to 3.5 eV; *second row:* cold clusters: 3. small blue shift due to thermal contraction $\omega^2 \sim \frac{n(e^-)}{V}$ 4. Na_9^+ ($8e^-$) splits into two peaks; the splitting is difficult to reconcile with pseudo-potential Jellium theories. 5. Na_{21}^+ ($20e^-$), good agreement with quantum chemistry TDLDA/SIC (Ekardt et al.) 6. Na_{41}^+ ($40e^-$) excellent agreement with TDLDA/SIC; **b** Non-Spherical Clusters (Haberland) $10e^-$, $14e^-$, *hot:* splitting of plasmon line; compatible with Jellium, *cold:* Na_{15}^+, compatible with jellium; Na_{11}^+ ($10e^-$), doubling of the number of peaks; *hot:* 2 peaks; *cold:* 4 peaks, possibly 5. The strength for cold Na_{11} is much smaller then that for cold ones. Where is it gone? To higher energies? The normalization of the cross section should be once more checked

Fission of nuclei and metallic clusters

About nuclear fission we heard an enthusiastic talk by *Goennenwein.* One can rather nicely have a general view of the various fission phenomena, including cluster radioactivity (news about that has been reported by *D. Poenaru*), the essential features of which I would like to present. Basic to all fission and fusion phenomena is the Asymmetric Two Center Shell Model (ATCSM) developed by *J. Maruhn* and collaborators (Fig. 21). Using the liquid drop model and adding the shell corrections the collective potential energy surface is calculated, utilizing cold valley which are important for fission and fusion processes. A representative example of such a collective potential energy surface is shown in Fig. 22.

Typical cuts along the two-center-separation R and the fragmentation coordinate $\eta = (A_1 - A_2)/(A_1 + A_2)$ show groundstate, shape isomer, superdeformed states, fission isomers and nuclear molecular configurations (Fig. 23). Since fission is a rather adiabatic process, one can put $\dot{R} \approx 0$ and thus obtain (near the scission point) the wave function for mass fragmentation from

$$[\tfrac{1}{2} B_{\eta\eta} \hat{p}_\eta^2 + V(\eta)]\, \Psi(\eta) = E\Psi(\eta)$$

after quantization in curvilinear coordinates. A typical wave function $\Psi(\eta)$ is qualitatively sketched and $|\Psi(\eta)|^2 \sqrt{B_{\eta\eta}}\, d\eta$ yields the probability for symmetric, asymmetric, superasymmetric fissioning of the nucleus. For extremely large asymmetry fission becames what one

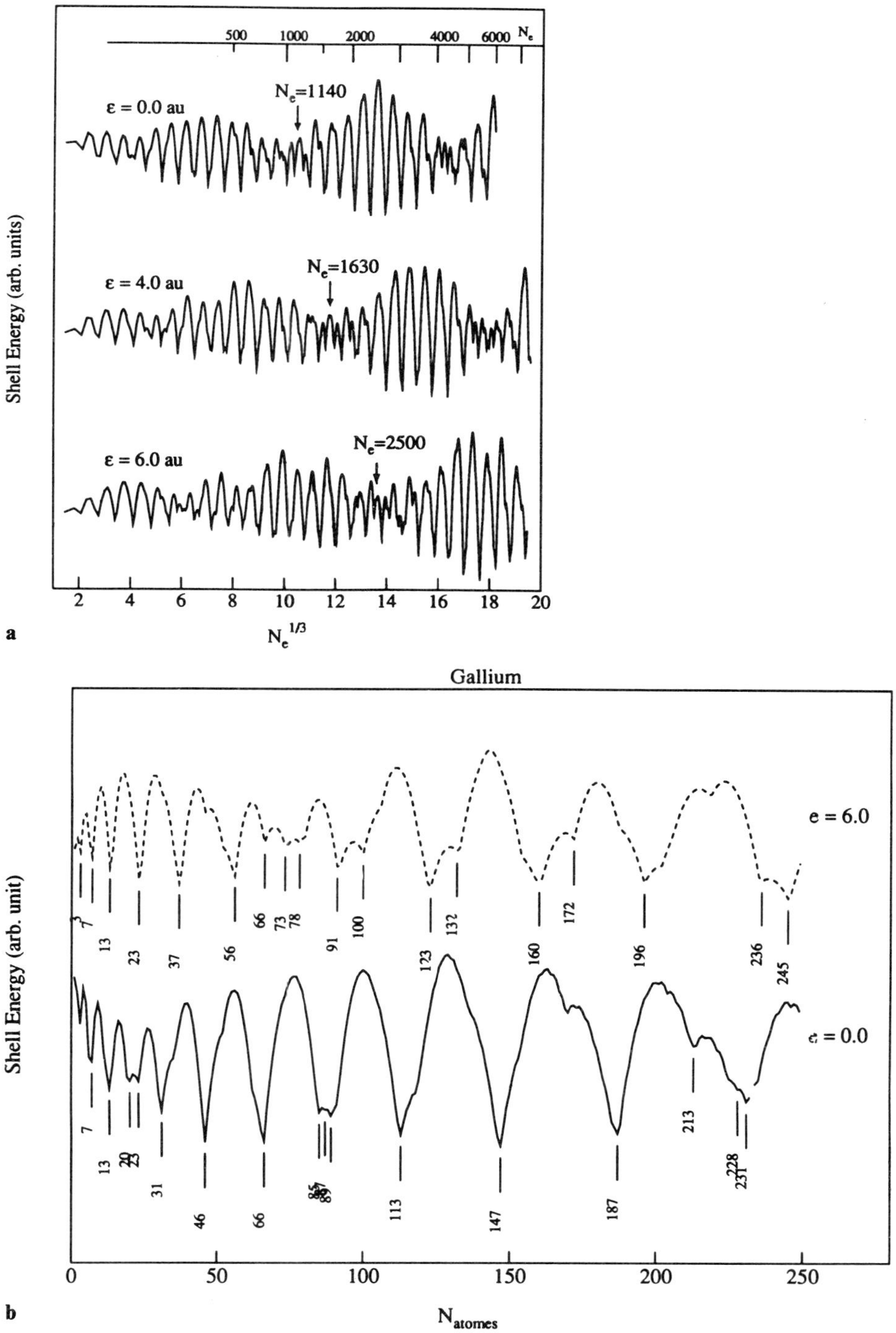

Fig. 18. a Dependence of supershell as a function of diffuseness ε: As diffusenes ε increases, supershell shift. **b** Supershells in jellium *(full line)* and diffuse jellium *(dashed line)* potentials

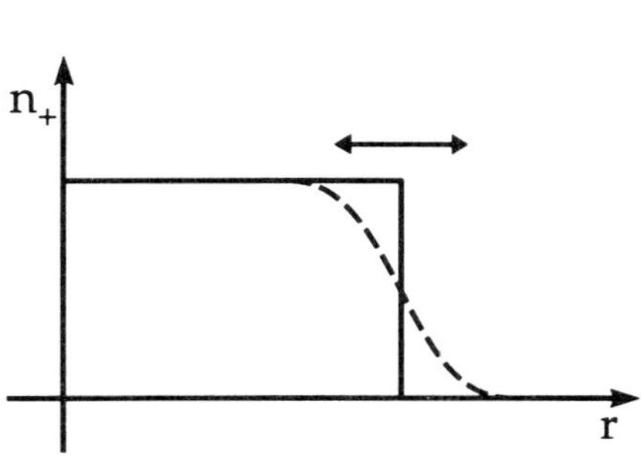

Fig. 19. The diffuse jellium potential could arise from surface oscillations of the ionic background

ordinarily identifies with radioactivity. This was the root of our prediction of *cluster radioactivity* [10] which was experimentally verified 3–4 years later by H. Rose and Jones. Within the fragmentation theory the mass distributions of various fissioning nuclei has been calculated 20 years ago by *J. Maruhn* and associates (Fig. 24), utilizing also the prediction of superasymmetric fission. As *Goennenwein* has shown, this has recently been nicely identified by M.G. Ithis et al. [11] – see Fig. 25. Earlier indications of such superasymmetric fission were contained in the measurements of an Indian group.

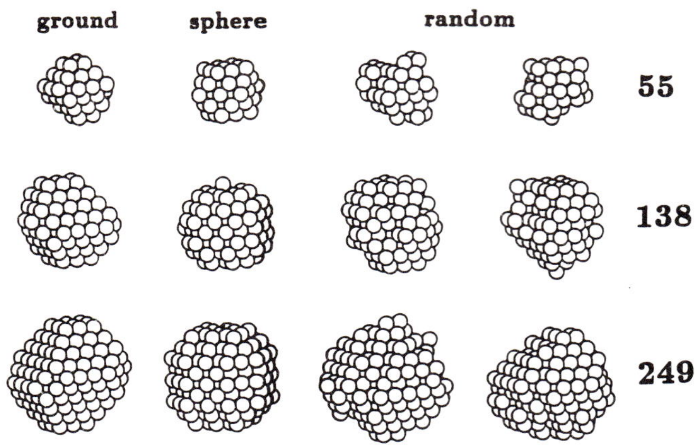

Fig. 20. Mansikka-aho's study of roughness of the cluster surfaces. Cluster shapes from ground state, and spherical towards random surface structure

Within fragmentation theory the performation probability is given by $(\int \Psi^*_{GS}(\mathbf{R},\eta)\, \Psi_{GS}(\mathbf{R},\eta) \sqrt{B_{RR}}\, d^3R) \times \sqrt{B_{\eta\eta}}\, d\eta$ where the integral goes only over d^3R; a cut through the potential surface along R at a fixed η yields the tunneling potential in the sense of Gamow. It is clear, that a definite final fragmentation can be reached from practically all η-values inside depending on the action integral amplitude. Valleys in the collective potential help in this direction. Extremely asymmetric fission near $\eta \sim 0.9$–1.0 is commonly called *cluster radioactivity.* Also *bimodal fission* (discovered by *Hulet* et al.) is naturally explained within fragmentation theory (see Fig. 26). In such calculations the mass tensor is essential. Its importance has been stressed by *Iwamoto* at this conference. He correctly stated that the fission path needs not always to be identical with the minimal path, because of the metric which is determined by the masses.

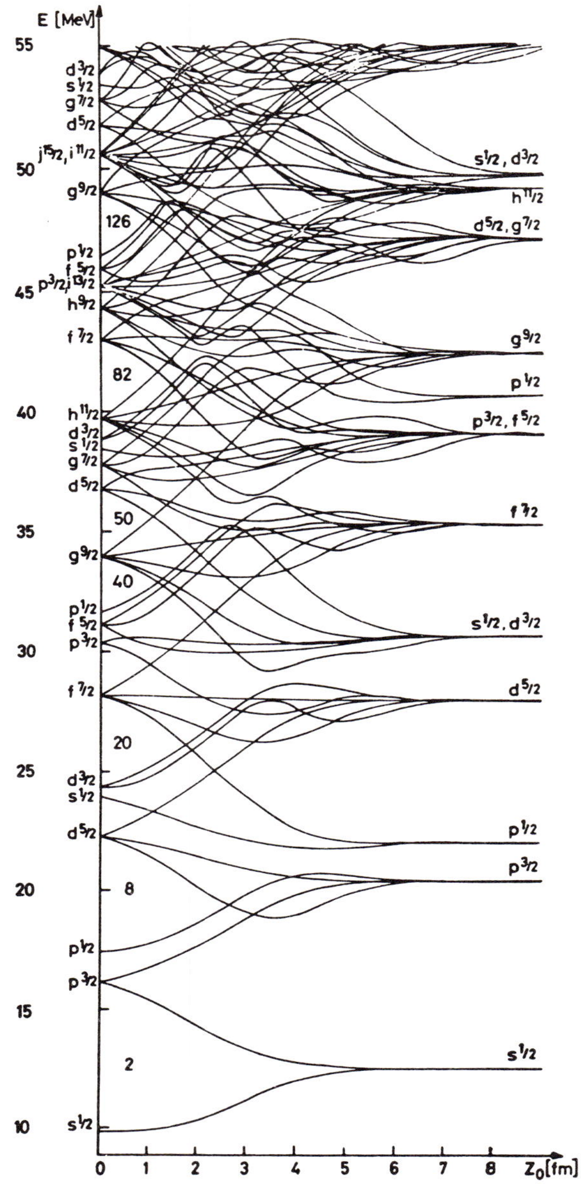

Fig. 21. Typical TCSM-level structure indicates that magic shells of the combined final fragments reach inwards up to the height of the fission barrier. This shell structure gives rise to the "cold valleys"

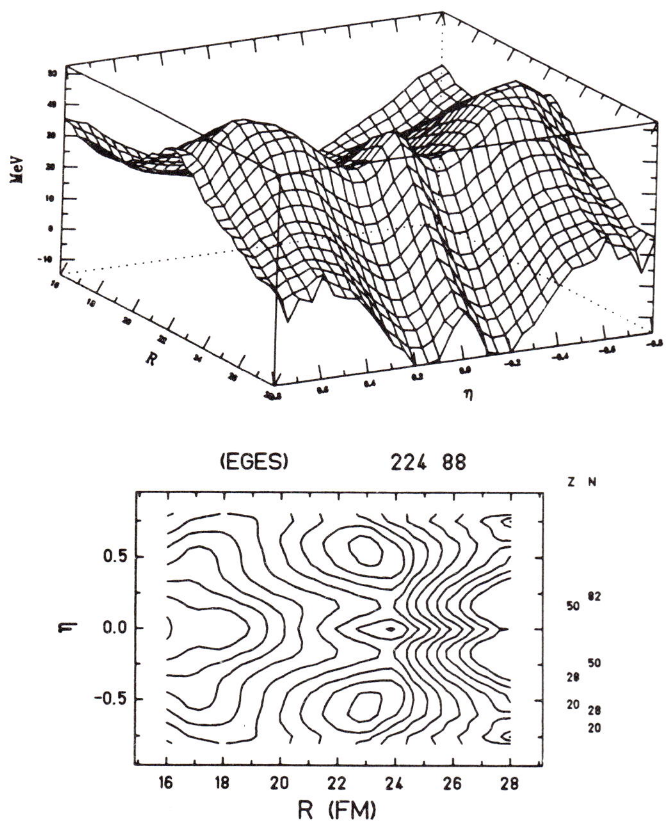

Fig. 22. Collective potential energy surface for ^{224}Ra. Two "cold valleys" can be seen; the fragmentation in these valleys is indicated (after D. Schnabel and J. Maruhn)

In our early papers on cluster radioactivity we had quantitatively predicted the various half lives (Fig. 27). The qualitatively and quantitatively new features of clus-

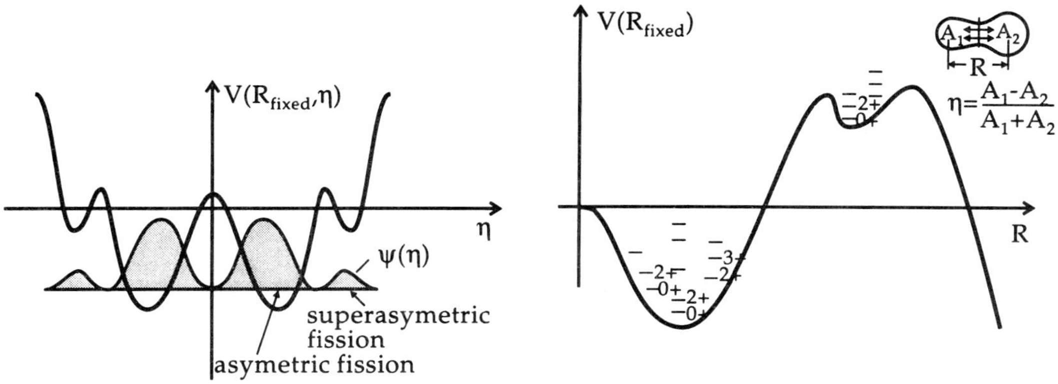

Fig. 23. Elements of mass distributions in fission: Structure of the Hamiltonian, typical potential energy curve as a function of the fragment separation R, and mass fragmentation potential $V(R_{\text{fixed}}, \eta)$

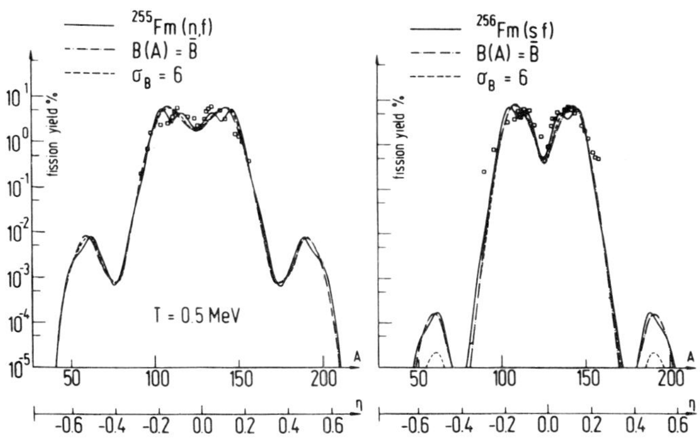

Fig. 24. Asymmetric and super-asymmetric fission of Fm-isotopes as calculated by J. Maruhn et al. [14]

ter radioactivity as compared to α- and β-radioactivity are depicted in Fig. 28.

The presently observed cluster decays, theoretical predictions for the half lives and their comparison with experiments are summarized in Fig. 29. The agreement is excellent!

Let me now speak about fission and decays of metallic clusters and assume that the pure liquid drop model (Coulomb plus surface energy) governs these processes. Figure 30 shows the liquid drop potentials at the touching configuration of two spheres as a function of the mass asymmetry. Depending on the charge of the drop one recognizes a maximum at $\eta=0$ (corresponding to symmetric fission) for low charges and a minimum for high charges. The turning point in between is known as the Bussinaro-Gallone point. Thus lowly charged drops fission very asymmetric; i.e. they evaporate small drops. This resembles cluster radioactivity. Highly charged drops fission symmetrically, but show also evaporation of small clusters.

Fission of charged metallic clusters has been one of the early subjects in the rapidly developing fields [14]. *T.P. Martin* and *U. Näher* reported about fission of highly charged large Na-clusters (Fig. 31a, b) with up to 448 atoms.

What they observe is what one calls in nuclear physics induced fission, because they heat up the cluster before the fission occurs. This is actually what *O. Hahn* and *F. Strassman* observed when they studied $U(n, f)$. Spontaneous fission was observed years later by *G. Flerov*.

Mrs. *C. Brechignac* gave a wonderful talk, studying the competition between evaporation and fission in doubly charged small clusters (Fig. 32a, b). She sees everything from spontaneous emission of small clusters (what she calls evaporation, but what is, actually, the analogue of nuclear cluster radioactivity) and also the transition from asymmetric to symmetric fission as the cluster size increases (the cluster charge is kept constant; hence the fissility parameter $x=Z^2/N$ decreases for N getting larger!). This is directly to be compared with Fig. 30!

For small sizes $C>\Sigma \Rightarrow$ symmetrical fission is expected. For large sizes $C<\Sigma \Rightarrow$ asymmetrical fission is expected. As cluster size increases the fission goes from symmetrical to asymmetrical fission.

Radii of light neutron rich nuclei

Nuclei obtain exotic features if they are far enough away from their normal relatives, i.e. if they are off the stability line. One such a feature is the fact that nuclei with loosely (i.e. weakly) bound neutrons have large radii, which are outside the normal $R=r_0 A^{1/3}$-law. *F. Pougheon* summarized the status of these investigations (Fig. 33). She stressed that the binding energy of the last neutrons plays an important role. For me this is obvious, because the binding energy determines the tail of the wavefunction $\sim\exp(-\sqrt{2mE_B/\hbar}r)$. No wonder that *Gregers-Hansen* finds great sensitivity of the cross sections to E_B and that one can turn the argument around and deduce from the measured cross sections the binding energy.

From a theoretical side the problem was adressed by *J. Vaagen* who performed Fadeev-type calculations. For the case of ${}^6_2He_4$ the two neutrons outside the α-core show a density distribution as depicted in Fig. 34. *Horiuchi* also

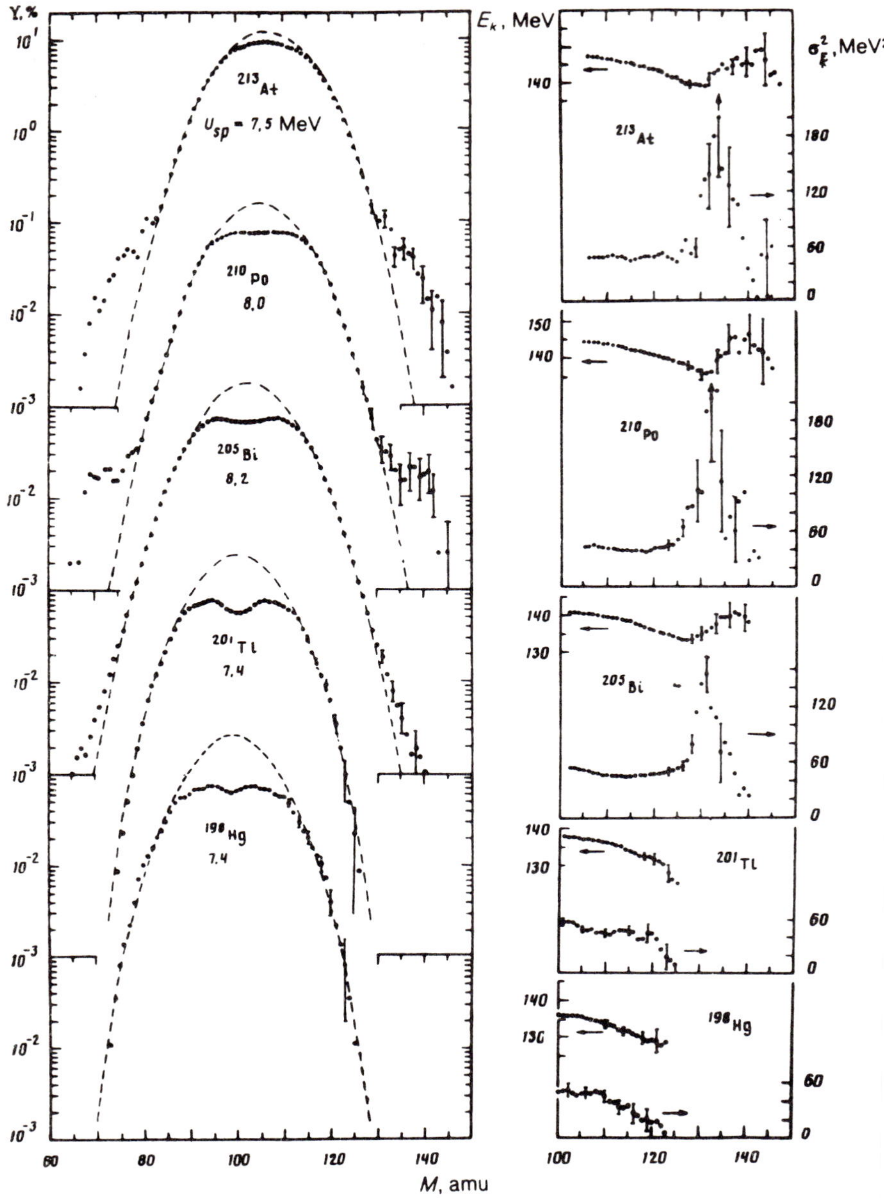

Fig. 25. Symmetric and super-asymmetric fission of various isotopes measured by Itkis et al. (see report of Goennenwein). Some isotopes show asymmetric and indications of super asymmetric fission. An early indication of superasymmetric fission had been found by Rao et al. [13]

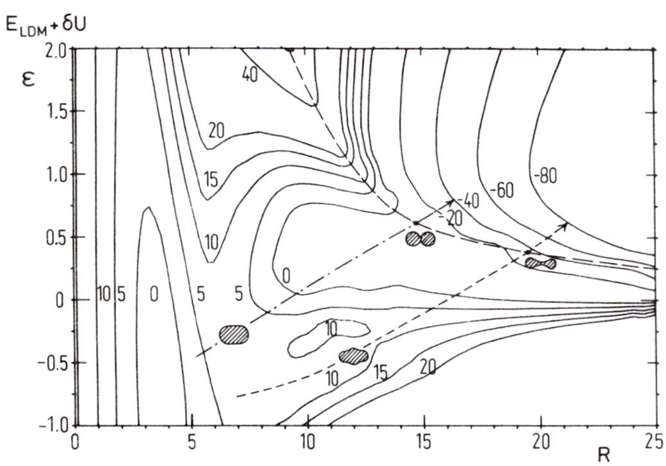

Fig. 26. Typical fission paths from bimodal fission (according to J. Maruhn et al. – similar calculations are known from A. Sobiczewski). The additional new fission channel is characterized by a vanishing neck, so that the kinetic energy of the fragments is high and their internal (excitation) energy small. This resembles the cold fission process which, in turn, is typical for cluster radioactivity

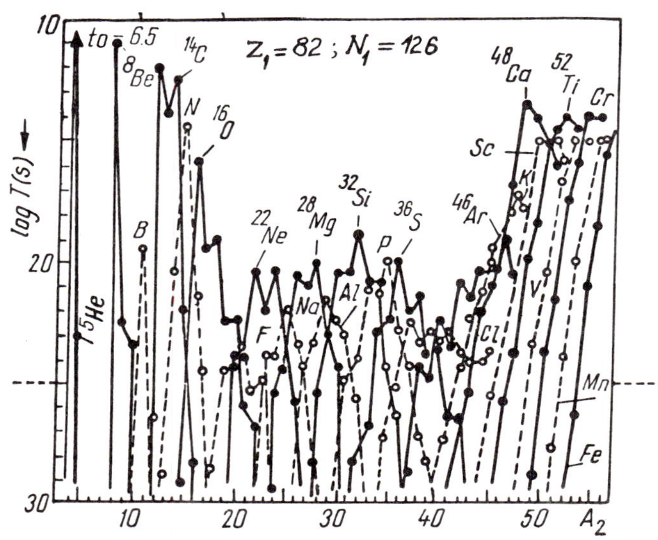

Fig. 27. Half life predictions for various cluster decays of mother nuclei leading to ^{208}Pb as one of the final fragments (Poenaru, Greiner, Sandulescu)

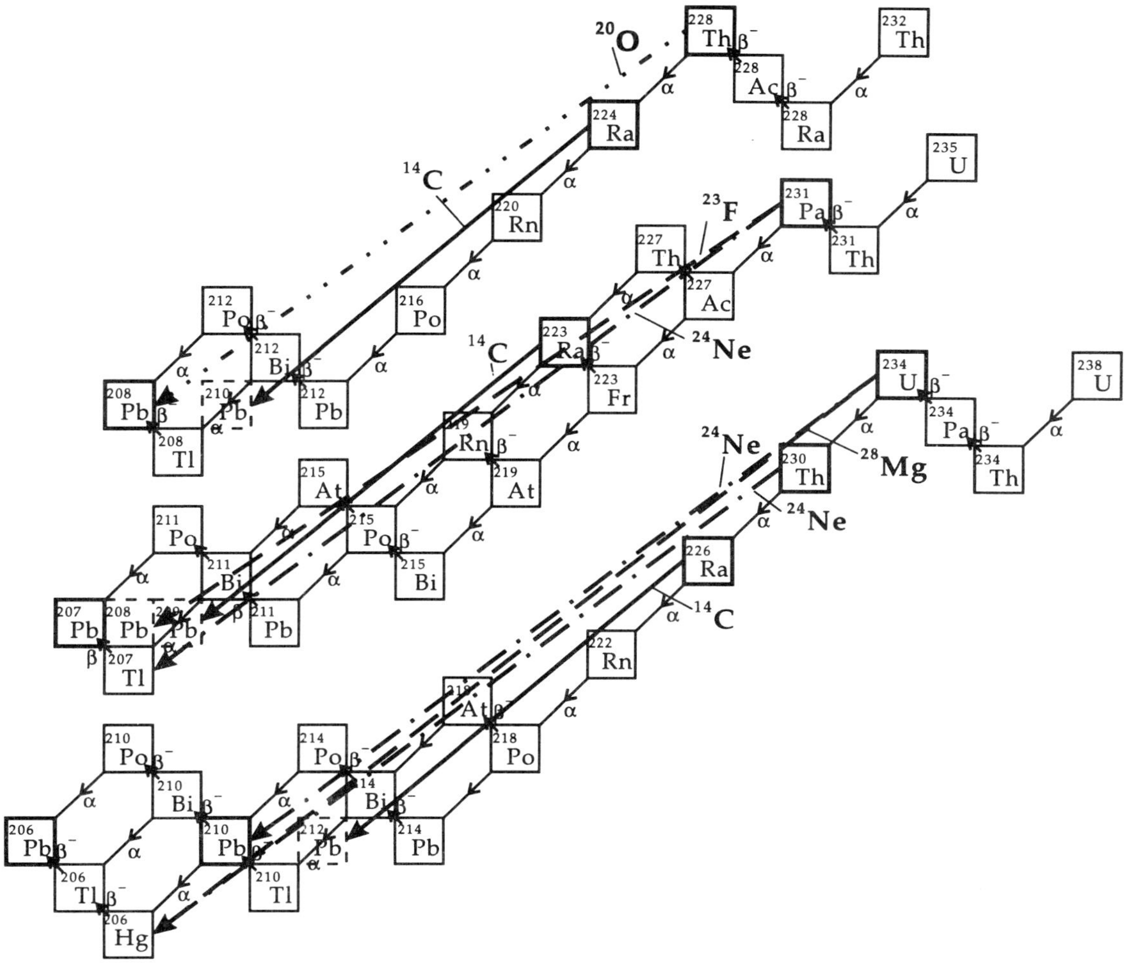

Fig. 28. Some of the cluster emitters, already experimentally confirmed, are members of the three natural radioactive families. While α- and β-radioactive decays go in relatively small steps, cluster radioactivity bridges large nucleon numbers

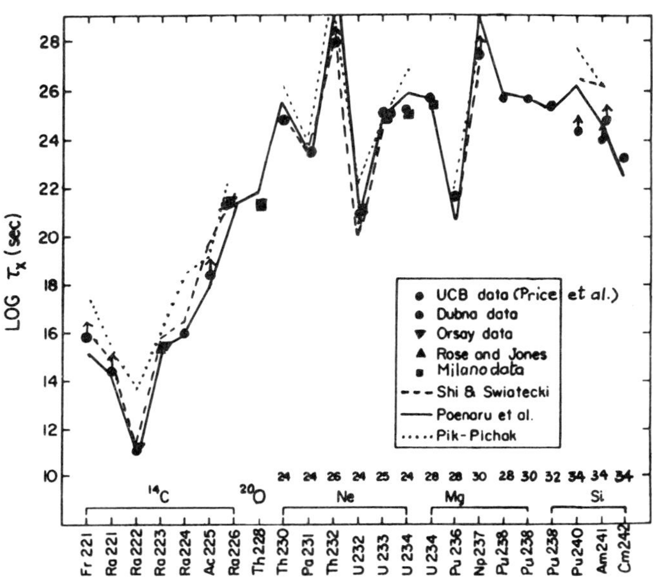

Fig. 29. Half lifes for various cluster decays of nuclei as compared to experiments

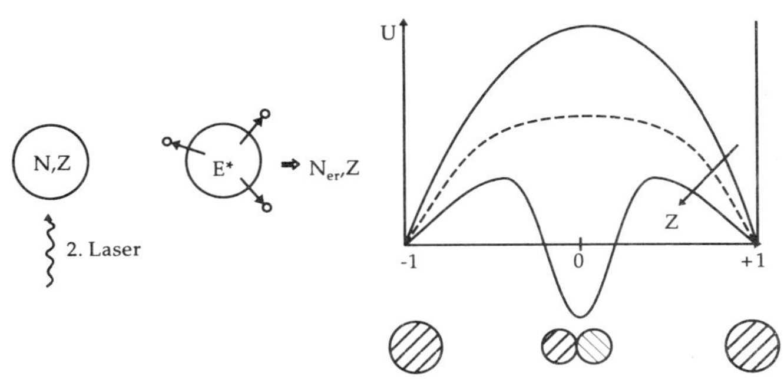

Fig. 30. The liquid drop potential of a charged drop breaking up into two spheres at their touching point as a function of the mass asymmetry $\eta = \dfrac{N_1 - N_2}{N_1 + N_2}$ where N_ν are the number of atoms in a cluster. The *upper curve* is for low charge Z, the *lower curve* for high charges. The turning point in between is called the Bussinaro-Gallone point

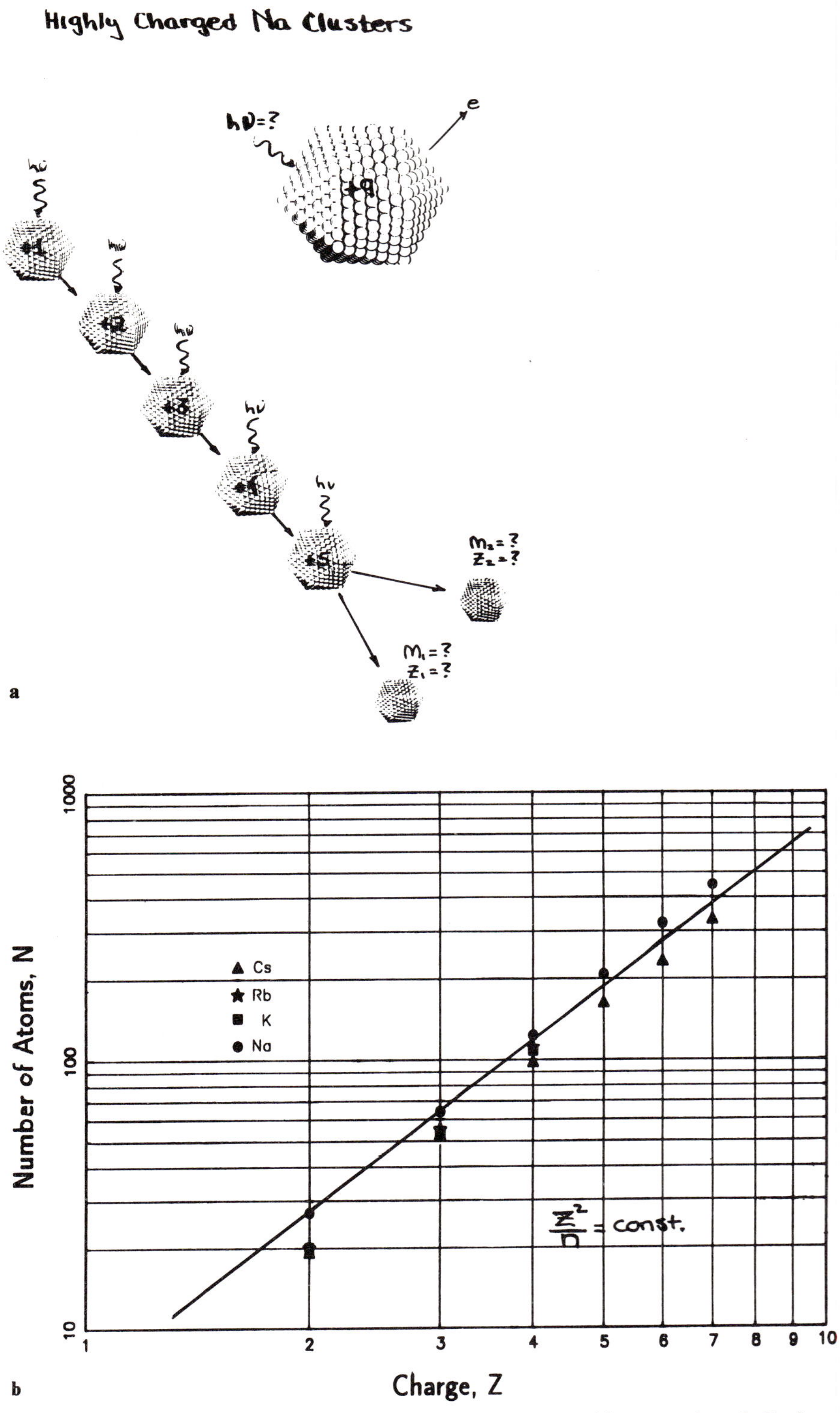

Fig. 31. **a** Electrons can be removed from metal clusters by multistep photoionization. In fact, the clusters can become so highly charged that they undergo fission. It is relevant to ask: "In what way does this fission resemble the same process in nuclei?" and "In what way do the two processes differ?" One difference to be kept in mind is that the clusters prepared by multistep photoionization are *hot*; so hot that they are evaporating atoms. The ORSAY group has clearly demonstrated that the fission process must compete with the evaporation process. The process with the lower barrier dominates. This means that in clusters, like nuclei, fission is 1) thermally activated, 2) characterized by a high barrier equal to evaporation energy. The consequence of these two characteristics is that the fission is strongly asymmetric, one fragment containing typically only three atoms; **b** In spite of these differences, many aspects of fission are similar for clusters and nuclei. If the critical size for fission of alkali metal clusters is plotted against cluster charge it becomes immediately clear that the data obey the rule $Z^2/N = \text{const.}$ Clearly, the fission process can be qualiatively described as the competition between the surface tension and electrostatic repulsion in a charged liquid drop

Fig. 32. a Competition between evaporation and fission, according to C. Brechignac. $n > n_c^{++}$: Dissociation energy is lower than the Coulombic energy ⇒ evaporation prevails. The kinetic energy of the atom is $\sim kT$; $n \simeq n > n_c^{++}$: Competition between evaporation and fission. In the case of fission the kinetic energy of each fragment is $\gg kT \simeq$ energy barrier; $n < n_c^{++}$: Fission dominates with barrier or without barrier for very small masses. $n < n_b^{++}$: For the Potassium case one has $n_c^{++} \simeq 20$ and $n_b^{++} = 7$. The cluster temperature $\simeq 500$ K. **b** Energetics of the fission process (after C. Brechignac).

$$\Delta = \underset{\substack{\Sigma \\ \text{surface term}}}{E_T(K_n^{++})} - \underset{\substack{C \\ \text{Coulomb term}}}{E_T(K_{n-p}^{+})}\, E_T(K_p^{+})$$

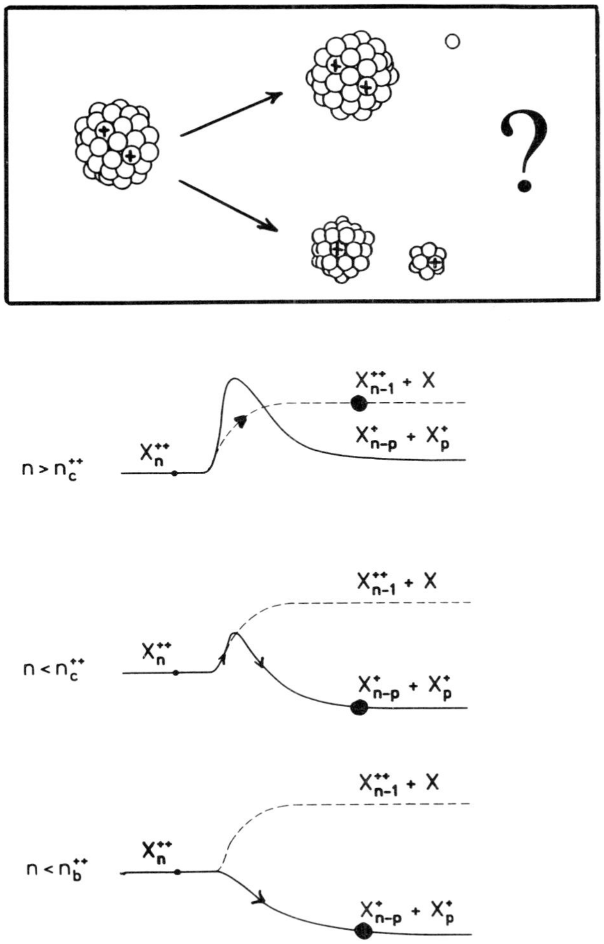

a

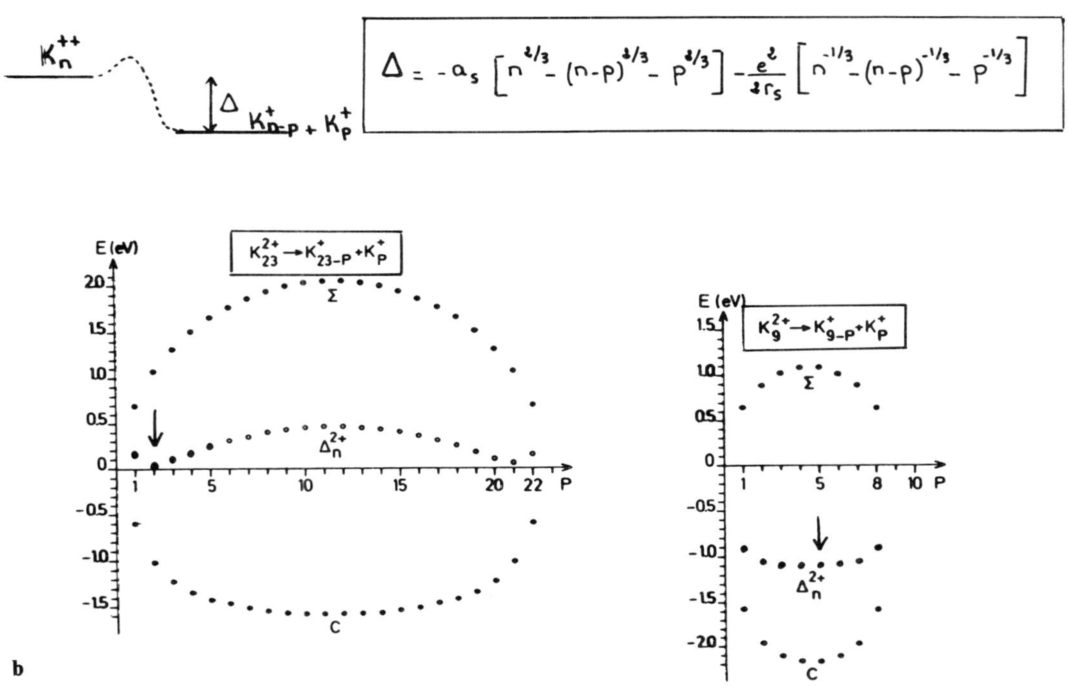

b

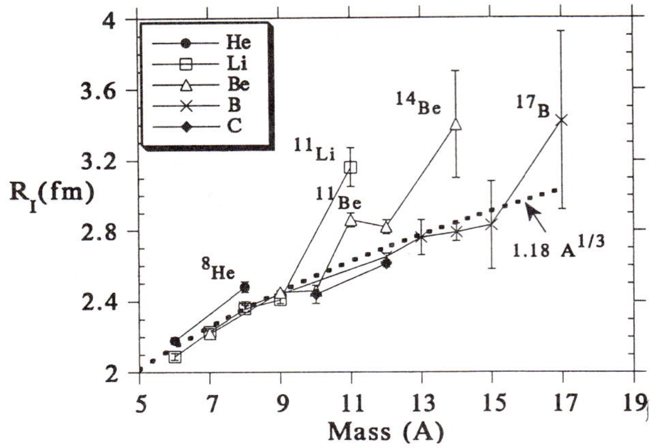

Fig. 33. The radii can be measured though the interaction cross section which is geometric at high energy. A lot of data has been accumulated and summarized here. For "normal nuclei" the interaction radii follow the simple $R = r_0 A^{1/3}$-law, expressing incompressibility of nuclear matter in its ground state. One can clearly see that for ^{14}Be, ^{11}Li, ^{17}B the radii are strongly increased. A deformation or not, that is the question

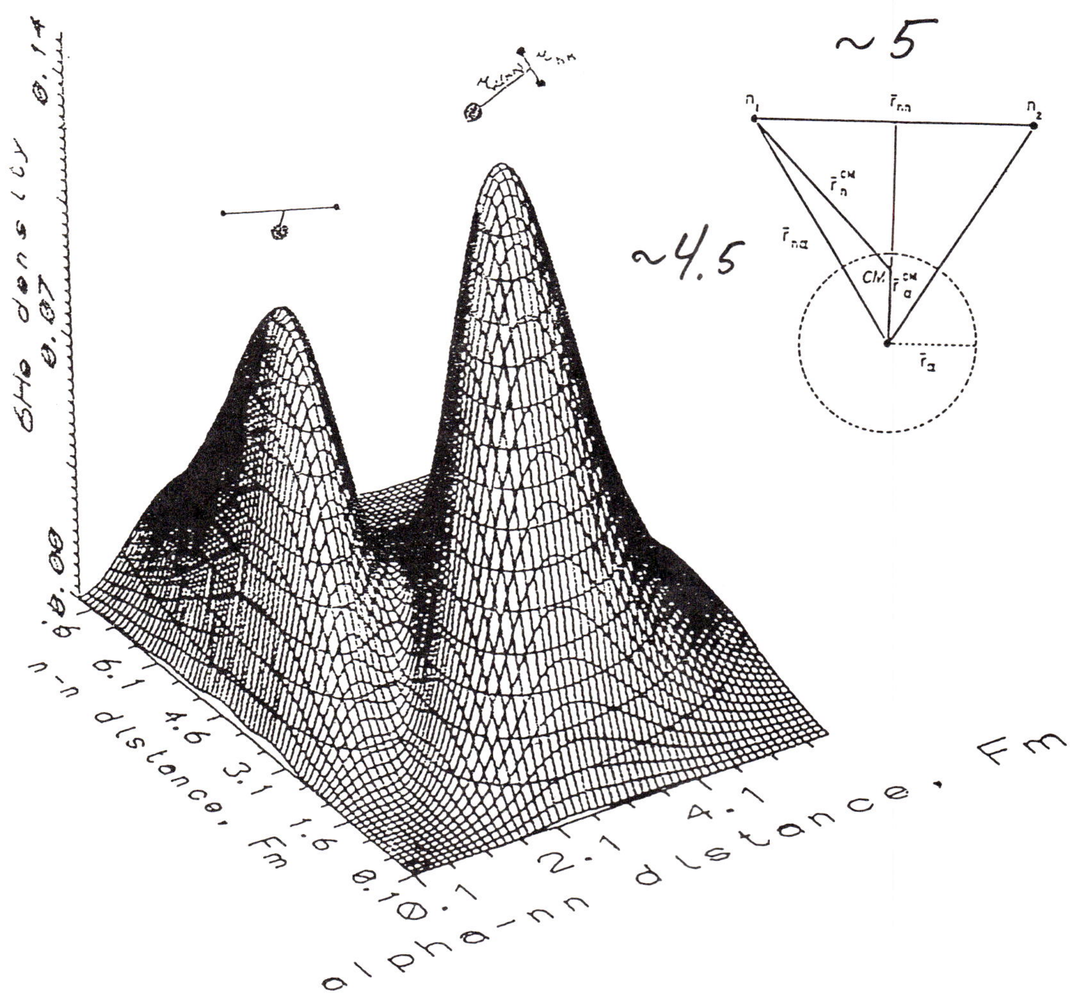

Coordinate	nn nα Method	GPT SBB HH	GPT WS HH	SSC SBB HH	SSC WS HH	SSC WS CSF	SSC WS CSF*
$\overline{r_{nn}}$		4.83	4.72	5.03	4.99	4.77	4.58
$\overline{r_{n\alpha}}$		4.43	4.36	4.50	4.50	4.52	4.19
$\overline{r_n^{CM}}$		3.46	3.40	3.54	3.53	3.67	3.43
$\overline{r_\alpha^{CM}}$		1.24	1.22	1.25	1.25	1.29	1.18
$\overline{R_{matter}}$		2.59	2.53	2.65	2.64	2.57	2.45

Fig. 34. J. Vaagen's cigar-shaped and di-neutron shaped density maxima for the two-neutron-density outside the α-core of $^{6}_{2}He_4$

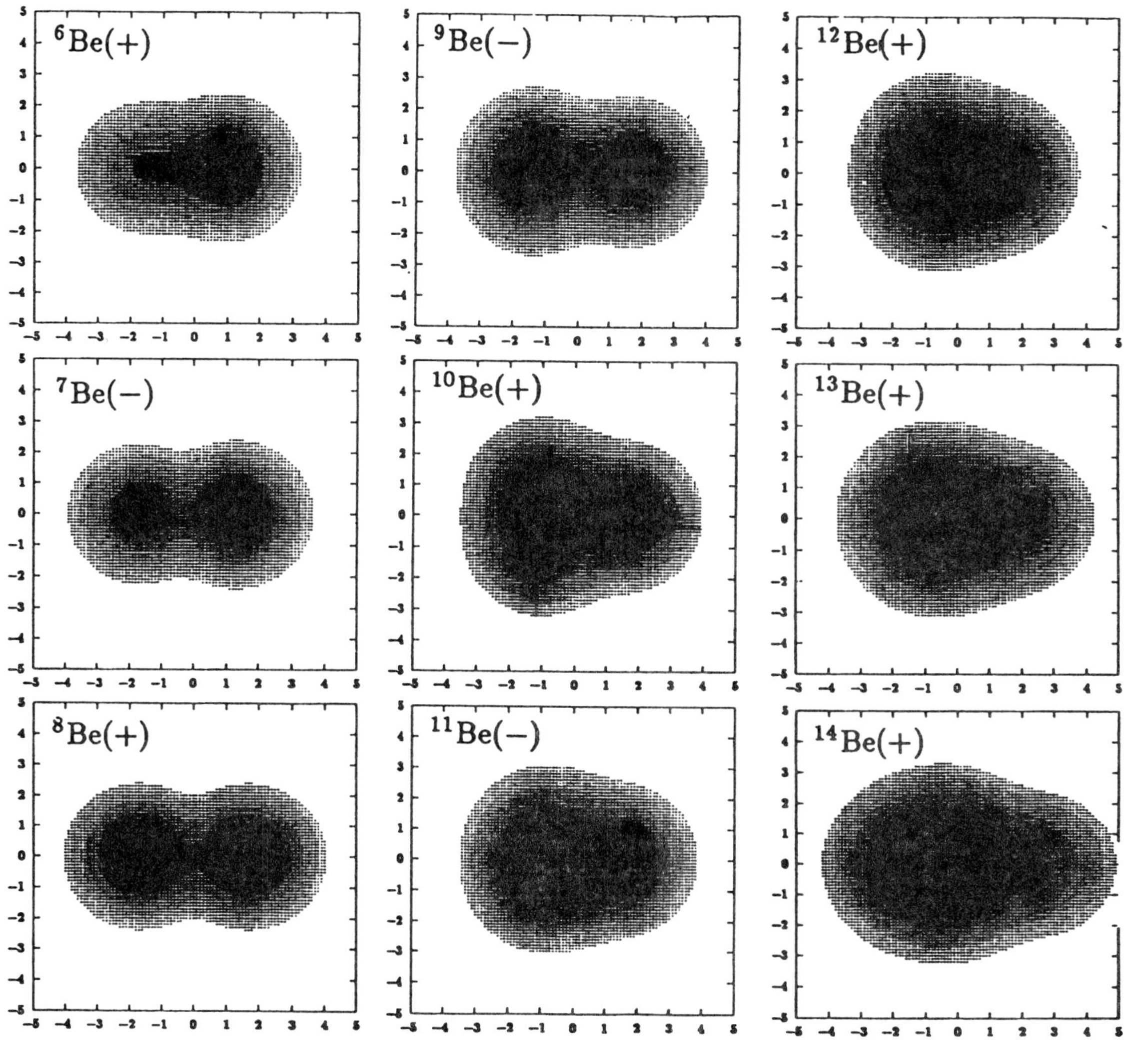

Fig. 35. Density distribution for Be-isotopes according to Horiuchi. Various degrees of deformation and shape are recognized

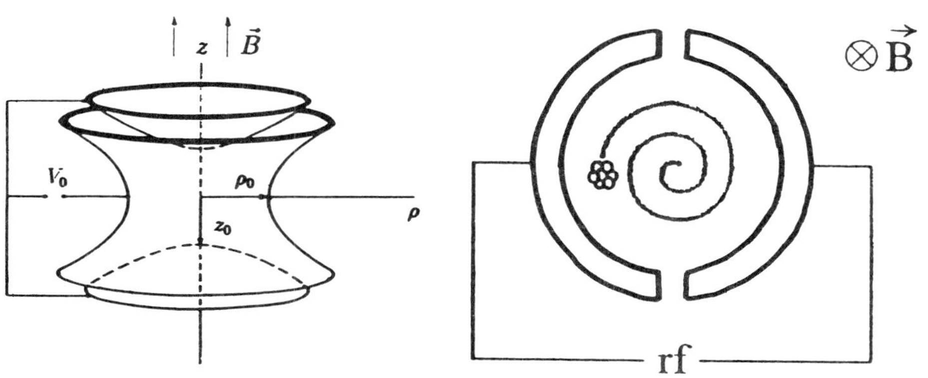

Fig. 36. Scheme of the Penning trap and of a cluster-cluster collision experiment according to Lützenkirchen. Traps are especially suited for the study of phenomena with long life times and to accumulate clusters for reactions with low cross sections. Penning Trap is an electromagnetic bottle for charged particles. It also can be used for Mass separation (q/M-filter, $q=1$) and by varying the kinetic energy to study collision processes

adressed this problem. His calculated density distributions for the Be-isotopes are shown in Fig. 35.

Aspects pointing into the future

Charged clusters trapped in Penning-traps. The principle of the method has been explained by *K. Lützenkirchen* (Fig. 36). A magnetic field **B**, around which the charged clusters are orbiting with the cyclotron frequency $\omega_c = qB/m$ and an electric quadrupole potential do the job. The former is responsible for the radial, the latter for the longitudinal confinement. With that device *Lützenkirchen* measured the TOF spectra of products ejected from the trap after exciting Au_7^+ clusters by collisions (Fig. 37).

The threshold values of the kinetic energy are mea-

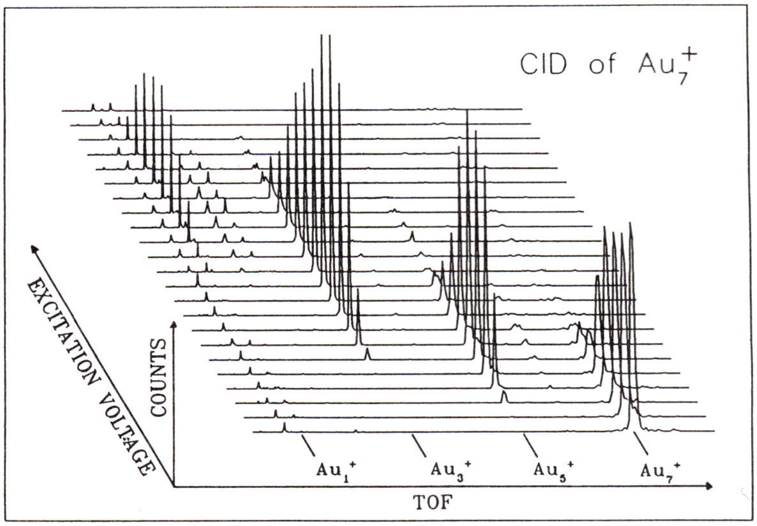

Fig. 37. Excitation of Au_7^+ by collisions: TOF-spectra of products ejected from the trap. The 2nd coordinate (excitation voltage) measures the kinetic energy of Au_7^+

sured and from them the binding energy of the various clusters can be deduced. One recognizes that dominantly Au_2 (not $2 \times Au_1$) is evaporated from the original Au_7. A strong odd-even effect, indicating tight binding for Au_2, is recognized. This device can be used to study radiative cooling of thermally excited clusters: $\hbar\omega \rightarrow$wait 1–100 ms$\rightarrow \hbar\omega \rightarrow$evaporate atom. Even more intrigueing is the possibility to study the fusion of clusters in cluster + cluster collisions, e.g. $Au_{10}^+ + A_{11}^+ \rightarrow A_{22}^+$. Such reactions have been theoretical investigated, using quantum molecular dynamics by *R. Schmidt*. A sample of his results is seen in Fig. 38a, b.

The study of cluster-cluster collisions are important for three reasons: First they constitute the inverse of the fission processes, but the details of the dynamics is different. Secondly they may show possibilities to produce even heavier clusters and, third, those reactions have to be seen in analogy to heavy ion reactions. Transfer of atoms, deep inelastic processes, and at high energies shock waves and compression of clusters can be expected (see for the nuclear analogue [9] and [15]). This may be of practical importance in fusion physics (pellet-pellet collisions).

Quite different in nature, but extremely interesting is

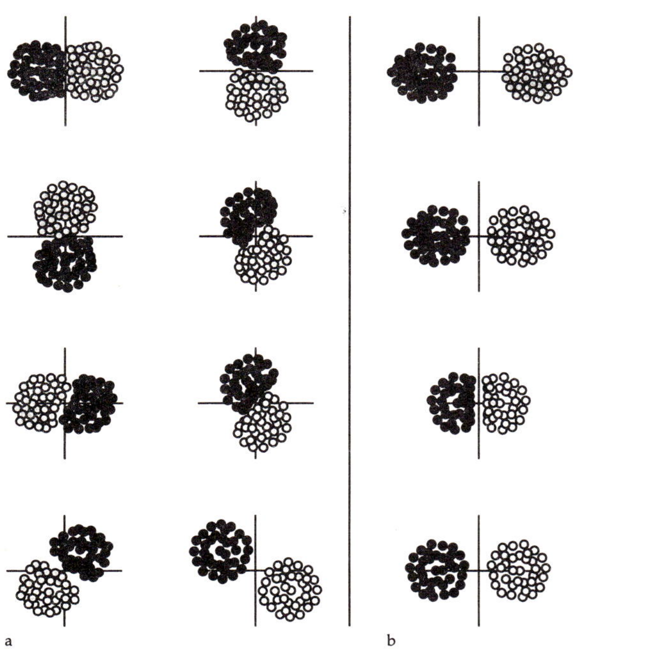

Fig. 38. a MD simulation of the fusion process between two C_{60} clusters at 100 eV c.m. incident energy and impact parameter $b = 1.2\,R(C_{60})$. **b** MD simulation of deep inelastic scattering between two C_{60} clusters at 20 eV c.m. incident energy and zero impact parameter

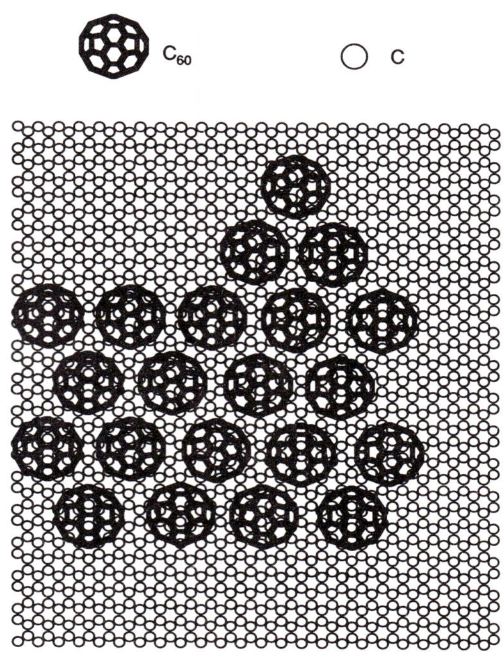

Fig. 39. Schematic of a well-ordered 2-D island of C_{60} clusters on graphite at room temperature (structural parameters from low energy electron diffraction scans). Palmer et al. can grow ordered "cluster assembled materials" with thickness from 1 to $\geqq 10$ monolayers

the application of low energy electron scattering to growth of clusters assembled in (several) layers on surfaces. *R.E. Palmer* reported about it. From elastic electron scattering he is able to observe the diffraction pattern and hence the structure of the material, while inelastic electron scattering yields the excitation spectrum. Figure 39 reminds us of the principle *Palmer* et al. are using.

It seems possible to grow cluster arrays using $(\text{Alkali})_N$ and $(\text{Noble Metal})_N$ and to create new electronic, magnetic or photonic materials. Of course, remembering *Jenas*'s futuristic talk, the growth of cluster-crystals (if they exist, i.e. if cluster do not dissolve if closely packed) may proceed along the lines R.E. Palmer has demonstrated.

The exciting talk of *A. W. Castleman* on cage-like clusters has given us an impression how chemists contribute to the fascinating field of cluster physics. From water to metcars: cage-like cluster molecules are very fascinating. We have seen model pictures of well established water clusters, e.g. a protonated 21-mer, which is more appropriatly written as $(H_2O)_{20} \cdot H_3O^+$ where H_3O^+ is located in the center of a pentagonal dodecahedron – simply fascinating!

Conclusion

We have had a wonderful conference. It was intellectually very challenging. Cluster physics is an exciting field. The close analogy of nuclear cluster phenomena and metallic cluster phenomena fulfills me with pleasure. Both fields are intimately related, even though quantitatively different. The same theoretical methods can be applied in both areas, even though differences do occur, of course. *P.G. Reinhardt* has quite beautifully pointed this out by comparing RPA-methods in both fields.

Before closing, I would like to thank the two chairmen Prof. *G.S. Anagnostatos* and Prof. *W. v. Oertzen* for arranging this wonderful conference with that excellent program. I have learned a lot during this week. Thank you all!

References

1. Schaffner, J. et al.: Prog. Part. Nucl. Phys. **30**, 327 (1993)
2. Fulton, B.R. et al.: Phys. Lett. B**267**, 325 (1991)
3. Fulton, B.R. et al.: Phys. Rev. C**21**, 198 (1980)
4. Wuosmaa, A. et al.: Phys. Rev. Lett. **68**, 1295 (1992)
5. Maas, R., Scheid, W.: J. Phys. G**18**, 707 (1992)
6. Hahn, J., Lustig, H.J., Greiner, W.: Z. Naturforsch. **32**a, 215 (1977)
7. Knight, W. et al.: Phys. Rev. B**31**, 2539 (1985)
8. Bernath, M., Yannouleas, C., Broglia, R.A.: Phys. Lett. A**156**, 307 (1991)
9. Eisenberg, J.M., Greiner, W.: Nuclear Theory, Vol. 1, Nuclear Models (3rd edn.), Amsterdam 1985
10. Sandulescu, A., Poenaru, D.N., Greiner, W.: Sov. Part. Nucl. **11**, 528 (1980)
11. Ithis, M.G. et al.: Sov. Nucl. Phys. **52**, 601 (1990)
12. Maruhn, J. et al.: J. Phys. G**6**, L25 (1980)
13. Rao, V.K. et al.: Phys. Rev. C**9**, 1506 (1974)
14. Sattler, K. et al.: Phys. Rev. Lett. **47**, 160 (1981); Echt, O. et al.: Phys. Rev. A**38**, 3236 (1988)
15. Stöcker, H., Greiner, W.: Phys. Rep. **137**, Nos. 5 and 6, 278 (1986)

Part II

Contributions/Posters

Possible water cluster formation by dilution and successions

G.S. Anagnostatos[1], P.Pissis[2], and K. Viras[3]

[1]Institute of Nuclear Physics, National Center for Scientific Research "Demokritos", Aghia Paraskevi Attikis, 153 10 GREECE
[2]Physics Department, National Technical Univerity of Athens, Polytechneiopolis, Zografou, 157 73 Athens, GREECE
[3]Physical Chemistry Department, Kapodistrian Univerity of Athens, Panepistimiopolis, Kouponia 157 71 Athens, GREECE

Received 28 June 1993

Abstract. While the formation of stable water clusters in no-water environment is well established, the present study, for the first time, proposes an approach of producing stable water clusters in water environment. The approach involves sequential dilutions and successions of water extracts, where here chamomile extract is taken as an exmple. The approach uses solutions far beyond the Avogardo number in order to secure that such formations are not those formed around the impurity (chamomile) molecules. By using dielectric relaxation spectroscopy and differential scanning calorimetry techniques experimental evidence was obtained for the existence of such water clusters.

PACS: 82.20. Fd, 82.20. Rp, 82.20. Wt

1. Introduction

As it is known water molecules can form structures (clusters) which exhibit stability [1]. Such clusters have been obtained by following different experimental techniques. However, all such techniques refer to free or isolated clusters, that is, to clusters formed in a no-water environment. It was tempting to us to examine possible stability of water formations in a water environment. That is, if such formations could be stable inside water under conditions of thermodynamic equilibrium between bulk water and water in these formations.

2. The model

As starting point for obtaining such water clusters was the knowledge that the presence of an impurity in water differentiate the water molecules in the vicinity of the impurity (mostly in the first layer) from the water molecules far from the impurity [2].Thus, the presence of an impurity leads to two classes of water molecules, those far from the impurity called free or bulk water, and those in the vicinity of the impurity called bound or modified water [2].

The remaining problem for creating water clusters in water was to examine if it was possible to separate the impurity from its surrounding water formation. For simplicity a single hydrophobic neutral (not ionized) impurity was assumed, where the impurity is situated at the center of the water formation made of several water layers forming an extended network via hydrogen bonds. It is apparent that there is a substantial difference in kinematic behaviour between the impurity and this water formation. Thus by applying a sufficiently forceful succussion on the container of the above solution we could obtain the desirable separation. It is apparent that the relocated impurity forms a new water formation round it. Hence, by applying repeated succusions on our solution one obtains more and more formations (clusters). Thus, starting from a single impurity in a large mass of water (that is from a situation effectively made of bulk water) one reaches to a situation where many such water formations are present in the water (that is, one reaches to a mixture of free water and bound water). After applying a large number of succussions the presence of the initial single impurity is negligible in comparison to the large number of water formations.

For a verification of such hypothetical water formations (clusters) by applying the above procedure, experimental justifications are necessary capable to distinguish between bulk water and a mixture of bulk and bound water.

3. Experiment

The model describes a hypothetical sample derived by a single molecule of impurity in a large quantity of water following a specific procedure. In practice a simulation of this hypothetical sample is obtained as follows. We start with an extract of chamomile in doubly distilled and de-ionized water, where the macromolecules dissolved in water play the role of the impurities. By repeated dilutions and succussions we reach solutions much beyond the Avogardo number, where effectively no chamomile molecules are present, i.e. we arrive to a situation very similar to the hypothetical one.

Two different well established experimental methods measuring different physical properties and both capable of distinguishing between bulk and bound water were used. That is, dielectric relaxation spectroscopy (DRS) [3] and differential scanning calorimetry (DSC) [4]. A standard Perkin-Elmer 4 equipment was used for DSC, while the details of the equipment for DRS are given in ref. [5].

4. Results and discussion

In DRS we make use of the well established fact that the dielectric relaxation time for the reorientation of water molecules is longer for bound than for bulk water [3]. We use the dielectric technique of thermally stimulated depolarization currents (TSDC) which is characterized by high sensitivity and high resolving losses as a function of temperature at a fixed frequency of about $10^{-3} Hz$. The low frequency, necessary to increase the resolving power of the method, implies that the measurements must be carried out in the solid phase. Figure 1 displays TSDC thermograms in the temperature region of interest measured on bulk water chamomile extract and a 10^{-60} dilution, all of them in the solid (frozen) phase. The bulk water thermogram exhibits a single peak at about 130K due to the reorientation of free water molecules [5]. The chamomile extract thermogram shows the free water peak and two additional peaks at higher temperatures, i.e. longer relaxation times, corresponding to the reorientation of bound water molecules [5]. The 10^{-60} dilution thermogram shows a complex band consisting of two overlapping peaks of almost equal magnitude which correspond to bulk and bound water, respectively. The conclusion is that a significant fraction of water molecules in the 10^{-60} dilution behaves dielectrically like bound water. Since effectively no chamomile impurity molecules are present in this dilution, the bound water is due to the presence of water formations obviously produced by the procedure followed.

DSC is the characteristic technique which offers the possibility of detecting phase transitions [4]. In our case this should correspond to the destruction of the water formations presumably existing in the solution, which here is taken to be the same as above, i.e. 10^{-60} dilution of chamomile water extract. Figure 2 displays a DSC plot of the 10^{-60} dilution with reference to the initial pure bulk water obtained during heating at a rate of 10K/min in the temperature region of interest. The plot exhibits an endothermic peak at 31°C, which we attribute to destruction of existing water formations, despite the fact that effectively there is not even a single molecule of chamomile impurity present. The measurements were repeated after twenty days anf showed the peak at the same temperature but with reduced magnitude. This shows the persistence (stability) of the water clusters formed possessing a decay rate. It is satisfying that both well established methods clearly show the presence of water formations (clusters) in a dilution far beyond the Avogardo number by independently measuring two different properties.

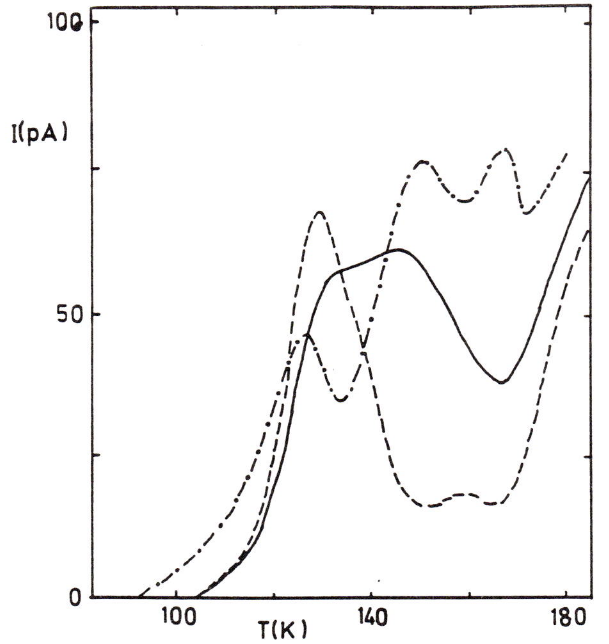

Fig. 1. Thermally stimulated depolarization current (TSDC) thermograms of (a) pure bulk water (broken line), (b) chamomile water extract (dash-dot line) and (c) 10^{-60} dilution of the extract (solid line).

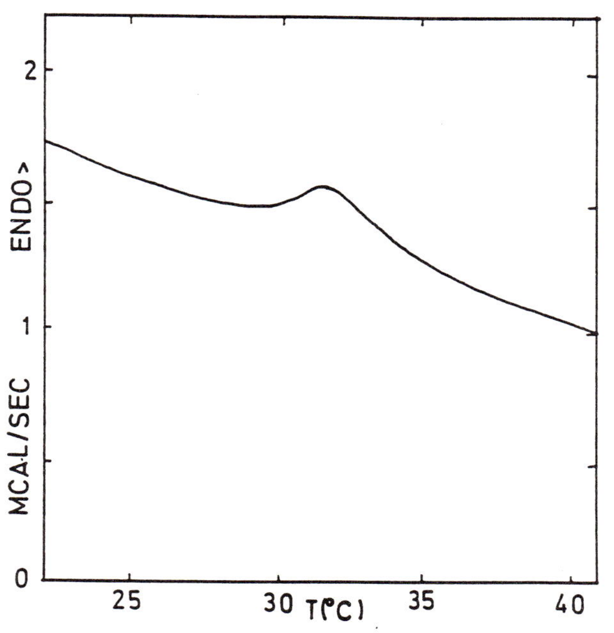

Fig. 2. Differential scanning calorimetry (DSC) thermogram restricted to the region of interest, of the 10^{-60} dilution of the chamomile water extract obtained during heating at a rate of 10K/min.

5. Conclusions

Despite the fact that more experimental techniques should be employed for a further verification, the present two methods [3-5] used suggest the formation of water clusters by applying the above approach, i.e. by sequential dilutions and succussions.

References

1 Wei, S., Shi, Z., and Castleman, Jr. A.W.: J. Chem. Phys. **94**, 3268 (1991)

2 Kuntz, I.D., and Kautzmann, W.: Advan. Protein Chem. **28**, 239 (1974)

3 Kaatze, U., and Pottel, R.: J. Mol. Lig. **49**, 225 (1991)

4 Hofer, K., Meyer, E., and Hodge, I.M.: J. Non-Cryst. Solids **139**, 78 (1992)

5 Pissis, P., Anagnostopoulou - Konsta, A., Apekis, L., Daoukaki - Diamanti, D., and Christodoulides, C.: J. Non-Cryst. Solids **131**, 1174 (1991)

This article was processed using Springer-Verlag TEX Z.Physik D macro package 1.0 and the AMS fonts, developed by the American Mathematical Society.

Z. Phys. A 349, 337–338 (1994)

The quadrupole moment of the neutron-halo nucleus ^{11}Li

E. Arnold [1], J. Bonn [1], A. Klein [1], P. Lievens [2,★], R. Neugart [1], M. Neuroth [1], E.W. Otten [1], H. Reich [3], W. Widdra [3]

[1] Institut für Physik, Universität Mainz, D-55099 Mainz, Germany
[2] PPE Division, CERN, CH-1211 Genève 23, Switzerland
[3] Fachbereich Physik, Universität Marburg, D-35037 Marburg, Germany

Received: 27 October 1993

Abstract. The quadrupole moment ratio of ^{9}Li and ^{11}Li was measured by a combination of in-beam laser induced nuclear polarization and β-NMR in $LiNbO_3$. The result $|Q(^{11}\text{Li})/Q(^9\text{Li})| = 1.14(16)$ is consistent with cluster models describing ^{11}Li as composed of a ^{9}Li core and a far extended halo of two loosely bound neutrons.

PACS: 21.10K; 21.60G; 27.20

Large nuclear matter radii for light nuclei at the neutron drip line were obtained from projectile fragmentation experiments [1]. This was interpreted as the development of a far extended neutron distribution for the weakly bound nuclei such as ^{6,8}He and ^{11}Li. A first theoretical interpretation was given by Hansen and Jonson, who considered the two-body interaction between the nuclear core and a neutron cluster at large distance from the core. In this way both the large matter radius and the small binding energy could be reproduced [2]. Since then nuclear reaction studies of the neutron rich nucleus ^{11}Li yielding large cross sections for electromagnetic dissociation [3, 4] and narrow momentum distributions of forward neutrons [4, 5], turned out to be consistent with models describing ^{11}Li as being clustered into a ^{9}Li core and a halo formed by the two extra neutrons [6].

Important structural aspects of nuclei are tested by the electromagnetic ground state properties like spin, magnetic dipole moment and electric quadrupole moment. Preceding experiments on radioactive Li-isotopes yielded the spin and magnetic moments of ^{11}Li and the quadrupole moment of ^{9}Li [7, 8]. In the present study we measured the quadrupole moment of ^{11}Li.

The experiment was performed at the ISOLDE isotope separator facility at CERN, by a technique implying collinear laser-induced nuclear orientation and β-NMR in a non-cubic $LiNbO_3$ single crystal. A ^{11}Li ion beam of about 10^3 atoms per second was produced by 600 MeV proton induced fragmentation of Ta, surface ionization and mass separation. After neutralization on a Na-vapour, the Li-atoms were polarized by in-beam optical Zeeman pumping in the presence of a small longitudinal magnetic field with circularly polarized laser light in the transition $2s\,^2S_{1/2} - 2p\,^2P_{1/2}$ at 670.7 nm. Due to the hyperfine interaction the atomic polarization is transferred to the nucleus. Then the atoms are implanted into a $LiNbO_3$ single crystal in a magnetic field H_0 of about 1 kG. The nuclear polarization is detected via the angular asymmetry in the β-decay in the directions parallel and antiparallel to H_0.

At the position of the crystal, a magnetic rf-field perpendicular to H_0 is generated by a pair of coils. The nuclear polarization will be reduced at resonance by transitions induced between the different magnetic sublevels. This leads to a reduction of the measured asymmetry, i.e. the normalized difference between the count rates in the two β-detectors.

In a non-cubic crystal there will be an additional quadrupole shift of the magnetic sublevels, resulting for nuclei with spin $I = 3/2$ (like the odd Li isotopes) in an NMR-spectrum with an unshifted central component at the Larmor frequency $\nu_L = \gamma H_0$ (γ the gyromagnetic ratio) and two satellites at $\nu_L - \Delta_Q$ and $\nu_L + \Delta_Q$. Since the quadrupole moment Q is proportional to Δ_Q, the ratio of $Q(^9\text{Li})$ and $Q(^{11}\text{Li})$ can be deduced from the measurement of Δ_Q for both isotopes. Due to the low production rate such an NMR measurement was not possible for ^{11}Li. However, the sensitivity could be significantly increased by a method involving simultanious irradiation with 3 different rf-fields of equal strength: one frequency is locked to the known Larmor frequency, while the other two are scanned up and down starting from ν_L. When these coincide with $\nu = \nu_L \pm \Delta_Q$, all three are in resonance, and the β-asymmetry is distroyed completely, yielding one strong resonance signal. Such spectra for ^{9}Li and ^{11}Li are shown in Fig. 1. This figure also includes the results of a χ^2 curve fitting procedure leading to values for the quadrupole splitting $\Delta_Q(^9\text{Li}) = 19.3(9)$ kHz and

★ *Present address:* Laboratorium voor Vaste Stof-Fysika en Magnetisme, K.U. Leuven, B-3001 Leuven, Belgium

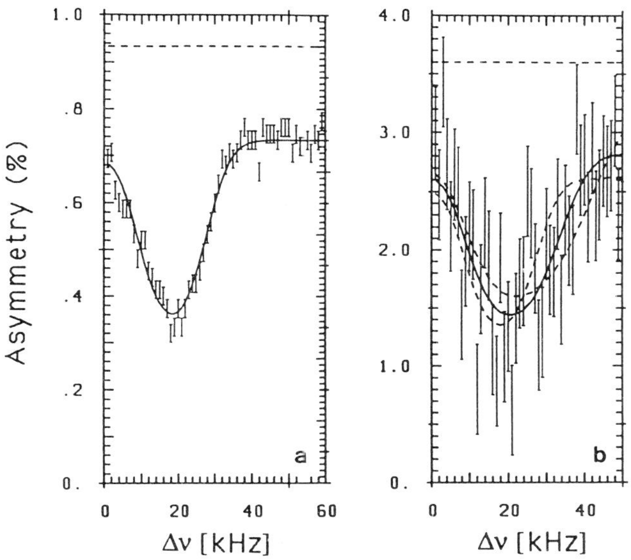

Fig. 1. Triple-resonance signals of ^{9}Li (**a**) and ^{11}Li (**b**) in $LiNbO_3$. The solid curves represent the best fits to the signals and the dashed curves correspond to one standard deviation in the determination of the quadrupole splitting. The dashed horizontal lines give the β-asymmetry without rf

Table 1. Electromagnetic moments of lithium isotopes

Isotope	I	μ_I(n.m.)	$\lvert Q\rvert$(mb)
^{7}Li	3/2	3.2564268(17)[a]	40.0(3)[b]
^{8}Li	2	1.653560(18)[a]	31.1(5)[c]
^{9}Li	3/2	3.4391(6)[a]	27.4(1.0)[c]
^{11}Li	3/2	3.6678(25)[d]	31.2(4.5)[e]

[a] Ref [11], [b] Ref. [10], [c] Ref. [8], [d] Ref. [7, 11], [e] This work and [9]

$\Delta_Q(^{11}\text{Li}) = 22(3)$ kHz, and to the ratio of the quadrupole moments $|Q(^{11}\text{Li})/Q(^9\text{Li})| = 1.14(16)$ [9]. The quadrupole moment of ^{11}Li can then be calculated using the moments of ^{7}Li and ^{9}Li for calibration [7, 8, 10, 11]. The electromagnetic moments of the lithium isotopes are summarized in Table 1.

Up to now only very few attempts have been made to calculate nuclear moments at the drip line. Already the different values of the quadrupole moments of ^{7}Li and ^{9}Li show that the extreme single particle model considering only the $p_{3/2}$ valence proton in the odd Li-isotopes is oversimplified and that shell model calculations should be performed in a larger configuration space. In such a shell model treatment, van Hees and Glaudemans could reproduce quite well the quadrupole moments of 9 isotopes in the $A = 4-16$ region, but no predictions were made for nuclei not previously measured [12]. Using a relativistic mean field approach including axial deformation, Koepf et al. recently published the only existing calculation – to our knowledge – of the ^{11}Li quadrupole moment [13]. Like in every mean field calculation in this region, it was not possible to reproduce the exceptionally large matter radius of ^{11}Li. The calculated binding energies are in quite good agreement with experiment. As expected the deformation is smallest for the closed shell isotope ^{11}Li, while a maximum deformation is predicted for ^{7}Li in the oblate case and for ^{9}Li in the prolate case. Both for oblate and prolate shapes the quadrupole moment ratio $|Q(^9\text{Li})/Q(^{11}\text{Li})| \approx 2$, in sharp contrast with our experimental result that the nuclear charge distribution is essentially unaffected by the addition of the 2 neutrons to ^{9}Li. This is incorporated in a hybrid model by Tosaka et al. [14] where the shell model is combined with three-body aspects. Valence neutrons form a dineutron far from the core, but stay in shell model orbits close to the core. This model should be able to predict properly the ratio of the quadrupole moments of ^{9}Li and ^{11}Li.

This work has been funded by the German Federal Minister for Research and Technology (BMFT) under the contract number 06 MZ 188 I.

References

1. Tanihata, I. et al.: Phys. Rev. Lett. **55**, 2676 (1985)
2. Hansen, P.G., Jonson, B.: Europhys. Lett. **4**, 409 (1987)
3. Kobayashi, T. et al.: Phys. Lett. B**232**, 51 (1989)
4. Anne, R. et al.: Phys. Lett. B**250**, 19 (1990)
5. Kobayashi, T. et al.: Phys. Rev. Lett. **60**, 2599 (1988)
6. Tanihata, I.: Nucl. Phys. A**522**, 275c (1991)
7. Arnold, E. et al.: Phys. Lett. B**197**, 311 (1987)
8. Arnold, E. et al.: Z. Phys. A**331**, 295 (1988)
9. Arnold, E. et al.: Phys. Lett. B**281**, 16 (1992)
10. Voelk, H.-G., Fick, D.: Nucl. Phys. A**530**, 475 (1991)
11. Raghavan, P.: At. Data Nucl. Data Tables **42**, 189 (1989)
12. Hees, A.G.M. van, Glaudemans, P.W.M.: Z. Phys. A**315**, 223 (1984)
13. Koepf, W. et al.: Z. Phys. A**340**, 119 (1991)
14. Tosaka, Y., Suzuki, Y., Ikeda, K.: Prog. Theor. Phys. **83**, 1140 (1990)

Z. Phys. A 349, 353–355 (1994)

Time-dependent Schrödinger approach to sub-barrier fission

N. Carjan, O. Serot, D. Strottman

Centre d'Etudes Nucléaires de Bordeaux, F-33175 Gradignan Cedex, France, and Theoretical Division, Los Alamos National Laboratory, Los Alamos, NM 87545, USA

Received: 27 October 1993

Abstract. The time-dependent Schrödinger equation for the fission of ^{213}At through a one-dimensional potential barrier was solved numerically within a finite spatial grid. Several initial quasi-stationary vibrational states at and below the top of the barrier were considered. The time dependence of the decay rate $\lambda(t)$ was calculated. The non-exponential evolution of the process in the very early stages ($\leqq 10^{-20}$ s) differs considerably from the asymptotic (stationary) one. Fission is therefore delayed by the sum of this preexponential time, of the tunneling time and of the time necessary for the descent from the outer turning point to the scission point. These non-dissipative quantum times play the same role as the classical transient times usually included in the analyses of experimental data on pre-scission emission of light particles and γ rays.

PACS: 25.85.−w; 25.70.−z

It has been recently pointed out that the presence of dissipation in large amplitude collective motions has important consequences for the de-excitation of hot compound nuclei. It hinders and delays the fission process, thereby allowing other channels to compete more efficiently than expected from standard statistical model considerations. Detailed measurements of pre-scission neutrons [1] light charged particles [2] and gamma rays [3] can therefore give insight into the time scale of the fission process at its very beginning. Up to now, this type of analysis has been done using classical approaches (namely Fokker-Planck [4] and Langevin [5–7]) and the observed high pre-scission multiplicities have been exclusively related to the presence of strong dissipation.

Quantum effects are however expected to be important since in all cases of interest the nuclear temperature T is smaller than the barrier height B. Besides the trivial increase of the total fission rate λ due to a non-zero tunneling probability, there is also an expected modification of λ due to the well-known deviations from the exponential decay law in the very early stages of a decay process [8]. Finally, the existence of a finite (non-zero) tunneling time [9] will lead to an additional delay of the fission process. To study these dynamical effects quantitatively, it is necessary to go beyond the stationary Hill-Wheeler approach [10] and treat the fission rate time-dependently. For this reason the time-dependent Schrödinger equation

$$i\hbar \frac{\partial}{\partial t} \psi(q,t) = -\frac{\hbar^2}{2m} \frac{d^2}{dq^2} + V(q)\,\psi(q,t) \tag{1}$$

for the collective coordinate $q = d_{c\cdot m}/R_0$ was solved numerically in this paper. The quantity $d_{c\cdot m}$ is the distance between the mass centers of the nascent fission fragments, $R_0 = r_0 \cdot A^{1/3}$ is the radius of the spherical nucleus and m is inertia in the fission direction that was taken in this preleminary calculation equal to the reduced mass $\left(\frac{A}{4}\right)$. $V(q)$ is the potential energy of deformation which was recently calculated [5] as function of two deformation coordinates for the symmetric fission of ^{213}At. In the present study we have used the values corresponding to the bottom of the potential valley (see Fig. 1) and have fitted them with 3 smoothly joint curves:

$$V(q) = \begin{cases} V_{GS} + \frac{1}{2} m\omega_1^2 R_0^2 (q - q_{GS})^2 & q \leqq a \\ V_{SP} - \frac{1}{2} m\omega_2^2 R_0^2 (q - q_{SP})^2 & a \leqq q \leqq b \\ \dfrac{Z^2 e^2}{4 d_{c\cdot m}} - Q & q \geqq b . \end{cases} \tag{2}$$

Knowing $V_{GS} = 0$, $V_{SP} = 9.8$ MeV, $q_{GS} = 0.75$, $q_{SP} = 1.61$, $Q = 135$ MeV [11] and imposing the continuity of the potential and of its derivative in a and b, one can obtain the other parameters:

$$\hbar\omega_1 = 3.78 \text{ MeV}, \qquad \hbar\omega_2 = 0.63 \text{ MeV},$$

$$a = 0.77 \quad \text{and} \quad b = 3.03 .$$

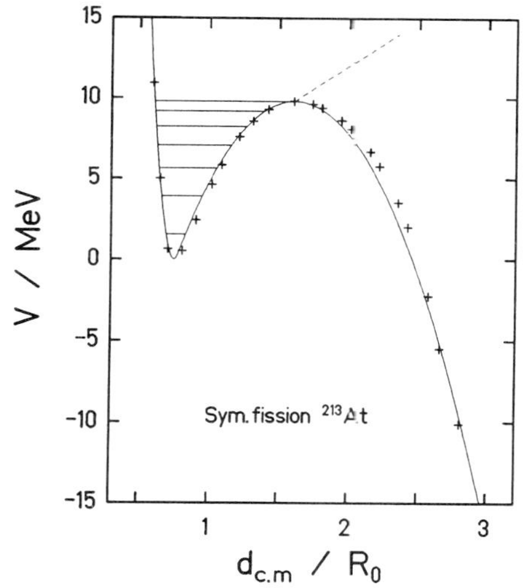

Fig. 1. Quasi-stationary vibrational states in the potential energy of deformation. The *dotted curve* represents the modification of the potential that was used to calculate these initial states

Equation (1) was numerically integrated within a spatial and temporal grid ($q_{min}=0.35$, $q_{max}=40.35$, $\Delta q=0.01$, $\Delta t=10^{-25}$ sec.) using the iterated leap-frog method [12], a multistep predictor-corrector method with excellent stability and accuracy. As initial wave functions, $\psi(q,0)$, four quasi-stationary vibrational states (i.e., eigenstates of a modified potential) ϕ_6, ϕ_5, ϕ_4 and ϕ_3, lying at and below the top of the barrier ($E_6=9.8$ MeV, $E_5=9.19$ MeV, $E_4=8.22$ MeV and $E_3=7.04$ MeV) have been chosen (see Fig. 1). It has been already shown in the case of alpha decay [13] that the results are essentially independent on the particular way of modifying the potential.

Figure 2 shows how fission occurs in time: defomations larger than the saddle-point values are increasingly populated at the expense of smaller deformations.

With the help of $\psi(q,t)$ one can indeed calculate the fission rate time dependently:

$$\lambda(t,q_B)=\frac{1}{1-\rho(t,q_B)}\,\frac{d\rho(t,q_B)}{dt}. \tag{3}$$

$\rho(t,q_B)$ is the fraction of the wave function that has tunneled through the barrier at time t:

$$\rho(t,q_B)=\int_{q_B}^{\infty}|\psi(q,t)|^2\,dq \tag{4}$$

and q_B is the arbitrary border between internal and external regions that will be treated as parameter. Usually $q_B=q_{SP}$ or $q_B=q_{SC}$ (the deformation at scission).

The result of applying (3) and (4) to the time evolution of $\psi(q,0)=\phi_6(q)$ with energy equal to the top of the fission barrier is shown in Fig. 3. A non-exponential behaviour of about 9×10^{-21} s and a shift of 3×10^{-21} s between saddle and scission are observed. Consequently, there is a delay ($t_{preexp}+t_{ssc}$) of 1.2×10^{-20} s before the asymptotic value of λ is reached at scission which is not due to dissipation.

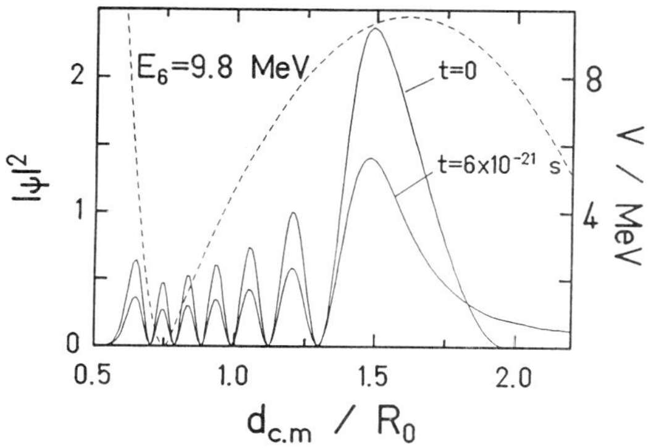

Fig. 2. Wave function with energy E_6 equal to the fission barrier height at $t=0$ (ϕ_6^2) and 6×10^{-21} s later. The *dashed curve* represents $V(q)$

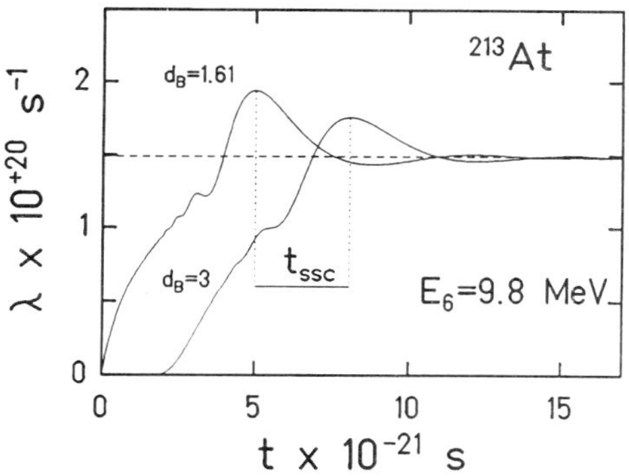

Fig. 3. Comparison of the fission rates calculated at saddle ($q_B=1.61$) and at scission ($q_B=3.0$) for the same initial conditions as in Fig. 2. t_{ssc} represents the saddle-to-scission time. The asymptotic value of λ is represented by *dashed line*

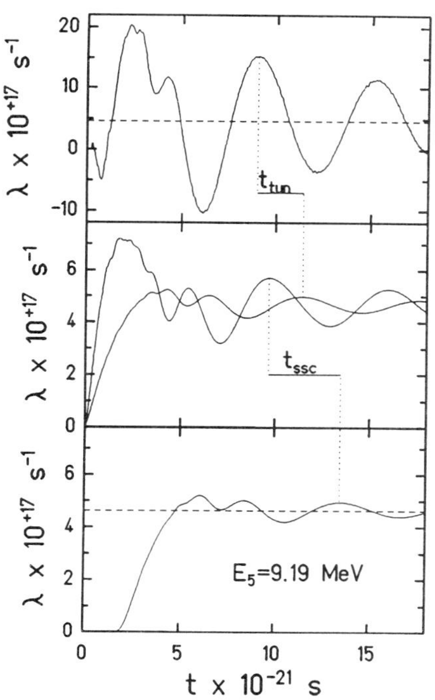

Fig. 4. Time dependence of the fission rate in the case of $\psi(q,0)=\phi_5(q)$. Four different positions of the borderline q_B have been considered, *from top to bottom:* internal turning point (1.4), saddle point (1.61), outer turning point (1.82) and scission point (3.0)

Table 1. Summary of results for 4 initial quasi-stationary vibrational states

E_n (MeV)	λ_{HW} (s^{-1})	$\lambda_{TD}(\infty)$ (s^{-1})	t_{preexp} (10^{-21} s)	t_{tun} (10^{-21} s)	t_{otsc} (10^{-21} s)
9.80	–	1.49×10^{20}	8.8	–	3.0
9.19	4.65×10^{17}	4.63×10^{17}	≈6	2.45	2.04
8.22	4.05×10^{13}	4.15×10^{13}	≈4	1.99	1.58
7.04	3.62×10^{8}	$\approx 4 \times10^{8}$	≈5	1.90	0.83

Figure 4 represents the fission rate for the initial state $\phi_5(q)$ with energy E_5 below the top of the barrier. For small values of $q_B(\leqq q_{SP})$, $\lambda(t)$ is strongly fluctuating around its asymptotic value for a long period of time. For larger values of q_B these fluctuations are however attenuated and the behaviour of λ becomes similar with that from the previous more energetic case. The tunneling time t_{tun} can be defined as the shift between the curves calculated at the internal (1.4) and outer (1.82) turning points. In the same way one can estimate quantum mechanically the time t_{otsc} of descent from the outer turning point to the scission point.

If one further decreases the energy of the vibrational state, the time evolution of its fission rate changes. Instead of a suppression during the non-exponential regime there is an enhancement as compared with the asymptotic value. The fluctuations have now larger relative amplitudes and frequencies and they are less attenuated.

Table 1 contains all calculated characteristic fission times. It also shows the remarkable agreement over 12 orders of magnitude between the Hill-Wheeler stationary fission rate λ_{HW} and the asymptotic value of the time-dependent approach $\lambda_{TD}(\infty)$.

In conclusion, the time dependent Schrödinger equation for the case of nuclear fission can be solved numerically. Starting with quasi-stationary vibrational states at and below the top of the barrier, the time-dependence of the fission rate has been calculated for the symmetric fission of ^{213}At. Even in the absence of dissipation, there is a considerable delay ($=t_{preexp}+t_{tun}+t_{otsc}$) of the order of 10^{-20} sec before the asymptotic (=stationary) value of the fission rate is attained. This delay affects the fission process in the same way as the transient time recently introduced to explain the observed high multiplicities of pre-scission light particles and γ rays. One should however emphasize that this explanation is based on the assumption that a certain initial state can be prepared and that the measurement starts when this state is ready. If this situation cannot be realized experimentally these kind of times will not be observable.

References

1. Hinde, D.J., Hilscher, D., Rossner, H.: Nucl. Phys. **A502**, 497c (1989)
2. Ikezoe, H., Shikazono, N., Nagame, Y., Sugiyama, Y., Tomita, Y., Ideno, K., Iwamoto, A., Ohtsuki, T.: Phys. Rev. **C42**, 342 (1990); Lestone, J.P., Leigh, J.R., Newton, J.O., Hinde, D.J., Wei, J.X., Chen, J.X., Elfstrom, S., Popescu, D.G.: Phys. Rev. Lett. **67**, 1078 (1991)
3. Butsch, R., Hofman, D.J., Montoya, C.P., Paul, P., Thoennesen, M.: Phys. Rev. **C44**, 1515 (1991)
4. Grange, P., Hassani, S., Weidenmüller, H.A., Gavron, A., Nix, J.R., Sierk, A.J.: Phys. Rev. **C34**, 209 (1986)
5. Carjan, N., Wada, T., Abe, Y.: AIP Conference Proceedings **250**, 230 (1992)
6. Tillack, G.R., Reif, R., Schulke, A., Fröbrich, P., Krappe, H.J., Reusch, H.G.: Phys. Lett. **B296**, 296 (1992)
7. Wada, T., Abe, Y., Carjan, N.: Phys. Rev. Lett. **23**, 3538 (1993)
8. Peres, A.: Ann. Phys. (N.Y.) **129**, 33 (1980)
9. Carjan, N., Serot, O., Strottman, D.: Jpn. J. Appl. Phys. Ser. **9**, 86 (1993)
10. Hill, D.L., Wheeler, J.A.: Phys. Rev. **89**, 1102 (1953)
11. Möller, P., Nix, J.R.: At. Data Nucl. Data Tables **39**, 213 (1988)
12. Gutschick, V.P., Nieto, M.M.: Phys. Rev. **D22**, 403 (1980)
13. Serot, O., Carjan, N., Strottman, D.: Nucl. Phys. **A569**, 562 (1994)

Z. Phys. A 349, 351–352 (1994)

The semimicroscopic algebraic cluster model of atomic nuclei

J. Cseh

Institute of Nuclear Research of the Hungarian Academy of Sciences, Debrecen, Pf. 51, Hungary-4001

Received: 27 October 1993

Abtract. The idea of the semimicroscopic algebraic cluster model of atomic nuclei was presented at the previous meeting of the same kind [1]. In this description phenomenological interactions are combined with the model space of the microscopic $SU(3)$ cluster model, which is free from the Pauli forbidden states and spurious excitations. Here we review the present status of this approach.

PACS: 21.60.Fw; 21.60.Gx

In the semimicroscopic algebraic cluster model [1,2] the internal structure of a cluster is described in terms of the $SU(3)$ shell model [3], therefore its wave function is characterised by the $U_C^{ST}(4) \otimes U_C(3)$ symmetry, where C refers to cluster, and $U^{ST}(4)$ is Wigner's spin-isospin group [4]. The relative motion of the clusters is accounted for by the vibron model with $U_R(4)$ group structure [5]. The representation labels of the group-chain

$$\begin{aligned} &U_{C_1}^{ST}(4) \otimes U_{C_1}(3) \otimes U_{C_2}^{ST}(4) \otimes U_{C_2}(3) \otimes U_R(4) \\ &\supset U_C^{ST}(4) \otimes U_C(3) \otimes U_R(3) \supset U_C^S(2) \otimes U(3) \\ &\supset U_C^S(2) \otimes O(3) \supset U(2) \supset O(2) \end{aligned} \tag{1}$$

provide us with the quantum numbers for the basis states of a two-cluster system. From this set we have to skip those states, which are Pauli forbidden, or which correspond to spurious excitations of the center of mass. To this end we can take into account the norm kernel eigenvalues of the microscopic $SU(3)$ cluster model [6], or one can apply a matching requirement between the quantum numbers of the shell model basis of the whole nucleus and its cluster model basis (within the harmonic oscillator approximation) [7,1,2]. For the low-lying major shells, where the influence of the Pauli blocking is the strongest, the latter calculation is usually simpler.

When the internal structures of the clusters are described by single $U_C^{ST}(4) \otimes U_C(3)$ representations, then the physical operators of the system can be obtained in terms of the generators of the $U_{C_1}^{ST}(4) \otimes U_{C_1}(3) \otimes U_{C_2}^{ST}(4) \otimes U_{C_2}(3) \otimes U_R(4)$ group. In such a case the description is algebraically closed, i.e. the matrix elements can be deduced with group theoretical techniques. In the limiting case when the Hamiltonian is given by the invariant operators of (1), then the eigenvalue problem has an analytical solution, and a $U(3)$ dynamical symmetry is said to hold.

If one or both of the clusters are $U_C^{ST}(4)$ and/or $U_C(3)$ scalars, then the problem becomes simpler. In [8] the formalism is presented in detail for the $U_{C_1}(3) \otimes U_R(4)$ and $U_{C_1}(3) \otimes U_{C_2}(3) \otimes U_R(4)$ models, as well as for the restricted $U_{C_1}^{ST}(4) \otimes U_{C_1}(3) \otimes U_R(4)$ model. In this latter case the restriction implies that only spin and isospin free interactions and a single $U_C^{ST}(4)$ representation are considered. If both of the clusters are $U_C^{ST}(4)$ and $U_C(3)$ scalars, the model reduces to that of the simple vibron model with a basis truncation corresponding to the Wildermuth condition [9,10].

Although the interactions applied in this approach are phenomenological ones, they can be related to the effective two-nucleon forces, due to the use of the microscipic $SU(3)$ cluster model basis. In [11] the vibron model Hamiltinian of the $U(3)$ dynamical symmetry (up to quadratic terms) was obtained from the most often used effective two-nucleon forces. It turns out that this simple phenommenological Hamiltonian can reproduce the matrix elements of the microscopic interactions (in a range of ca 100 MeV) reasonably well.

The applications of the semimicroscopic algebraic cluster model were carried out so far within the dynamical symmetry approximation.

The $T = 1$ states of the ^{18}O have been described in terms of core–plus–alpha–particle configuration [12]. 34 energy levels of the experimental spectrum were identified with model states, and assigned to 11 bands. This spectrum is more complete, than those of the previous calculations. Reduced $E2$ and $E1$ transition probabilities were also determined. Their agreement with the experimental data is similar to those of the best microscopic or semimicroscopic calculations.

The low–lying $T = 0$ states and the molecular resonances of ^{24}Mg are treated together in a ^{12}C + ^{12}C cluster description

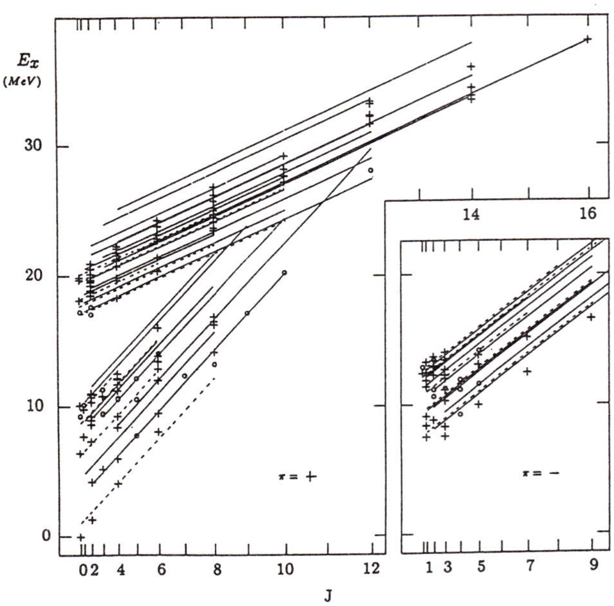

Fig. 1. Positive- and negative-parity $T = 0$ energy levels of the ^{24}Mg nucleus displayed separately in rotational diagram form [8]. *Circles* (∘) stand for states with uncertain J^π assignment. The *lines* denote the position of the calculated model bands. (*Dashed lines* indicate bands with $K = 0$, which contain only every second possible J value.)

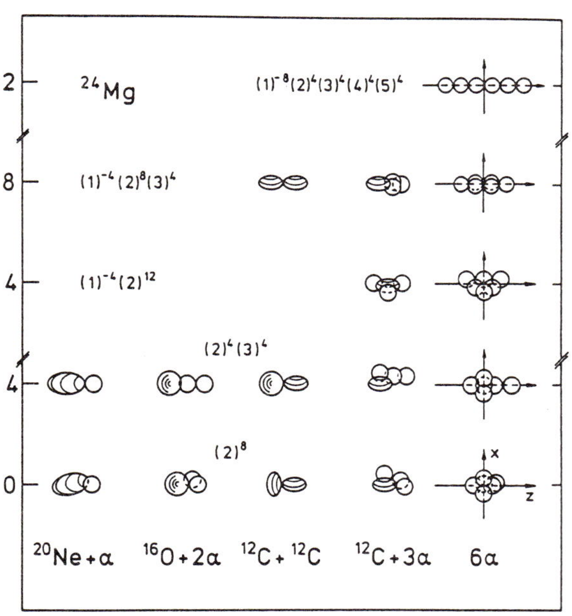

Fig. 2. Possible cluster configurations of some highly deformed states of ^{24}Mg [15]. Only the directions of the symmetry axes and the directions of amalgamations of the clusters carry physical content, while the relative distances do not. The y axis is perpendicular and coming out of the plain.

in [13]. About 150 experimental levels, in the energy range of 0 – 40 MeV (Fig. 1), and a large number of $E2$ transitions are described reasonably well.

The extension of the model to heavy nuclei seems to be possible based on the pseudo $SU(3)$ scheme [14]. The ^{210}Pb + ^{14}C clusterization was found to be present in the ground state wave function of the ^{224}Ra nucleus. In addition the low-energy spectrum could also be reproduced in terms of this cluster configuration.

The $U(3)$ dynamical symmetry of this approach has an interesting new feature: usually it is related to several cluster configurations. The consequences have been studied in two cases. i) The link between the superdeformed and cluster states of alpha-like nuclei ($N = Z = even$) was explored [15] (Fig. 2). ii) The allowed and forbidden binary fission modes of ground-state-like configurations in sd shell nuclei were determined [16].

The effects of a kinematic symmetry breaking have also been studied. In particular, the quantum deformed $U_q(3) \supset O_q(3) \supset O_q(2)$ dynamical symmetry has been applied for the description of the ^{16}O + α system [17]. Although the quantum algebraic treatment gave a better fit to the experimental data, than the Lie algebraic one, the improvement was not significant.

Extensions of this model to several directions are possible and desirable. The relative advantages and disadvantages can be deduced only after more systematic applications. Nevertheless, based on the first studies, mentioned here, this approach seems to be promising for the description of complex cluster systems.

Acknowledgement. This work was supported by the OTKA under Grant No 3008.

References

1. Cseh, J.: Proc. Int. Conf. on Nuclear and Atomic Clusters, Turku 1991, eds: Brenner, M., Lönnroth, T., Malik, F. B. Berlin Heidelberg New York: Springer, p. 184
2. Cseh, J.: Phys. Lett. B **281**, 173 (1992)
3. Elliott, J., P.: Proc. Roy. Soc. A**245**, 128 562 (1958)
4. Wigner, E. P.: Phys. Rev. **51**, 106 (1937)
5. Iachello, F., Levine, R. D.: J. Chem. Phys. **77**, 3046 (1982)
6. Horiuchi, H., Ikeda, K.: Cluster Models and Other Topics. Singapore: World Scientific 1986
7. Wildermuth, K., Kanellopoulos, Th.: Nucl. Phys. **9**, 596 (1958/59)
8. Cseh, J., Lévai, G.: Ann. Phys, (N.Y.) **230**, 165 (1994)
9. Cseh, J., Lévai, G.: Phys. Rev. C **38**, 972 (1988)
10. Cseh, J.: J. Phys. Soc. Jpn. Suppl. **58**, 604(1989)
11. Varga, K., Cseh, J.: Phys. Rev. C **48**, 602 (1993)
12. Lévai, G., Cseh, J., Scheid, W.: Phys. Rev. C **46**, 548 (1992)
13. Cseh, J., Lévai, G., Scheid, W.: Phys. Rev. C **48**, 1724 (1993)
14. Cseh, J., Gupta, R. K., Scheid, W.: Phys. Lett. B **299**, 205 (1993)
15. Cseh, J., Scheid, W.: J. Phys. G **18**, 1419 (1992)
16. Cseh, J.: J. Phys. G **19**, L97 (1993)
17. Cseh, J.: J. Phys. G **19**, L63 (1993)

This article was processed using Springer-Verlag TEX Z.Physik A macro package 1.0 and the AMS fonts, developed by the American Mathematical Society.

Z. Phys. A 349, 301–303 (1994)

Some predictions of new spontaneous cluster emitters and mechanism of cluster decay

W.Furman[1,*], S.Kadmensky[2,], Yu.Tchuvilsky[3,***]**

[1] Frank Laboratory of Neutron Physics, JINR, Dubna, Russia
[2] Voronezh State University, Voronezh, Russia
[3] Moscow State University, Moscow, Russia

Received: 27 October 1993

Abstract. Two basic mechanisms of cluster radioactivity are discussed. The theoretical predictions for the new region of cluster emitters with $A \sim 110$ are reported. The preliminary experimental results have indicated in favour of a nonadiabatic α-decay-like mechanism developed by the authors earlier.

PACS: 23.70.+j;27.90.+b

Different theoretical approaches [1-8] (see also review papers [9,10] and refs. therein) describe the experimental probabilities of spontaneous cluster emission from heavy nuclei more or less satisfactorily, on the whole, with the probable exception of the fine structure of cluster decay [11]. It is rather difficult to draw any definite conclusion about the nature of cluster radioactivity only on the basis of a formal comparison of available experimental data, especially with the results of the calculations in the framework of proposed theoretical approaches which exploit different basic assumptions as to the mechanism of cluster decay.

However, it is possible, in principle, to select the basic mechanism of cluster radioactivity by an analysis of the selfconsistency of various theoretical approaches or by a verification of the respective theoretical predictions outside the known mass region of the parent nuclei.

In general there are two alternative approaches to the description of cluster decay. The first one [1,2,4] uses a fission-like (adiabatic) mechanism to reproduce the main features of the process. The second [5-8] assumes that a cluster, preformed by a nonadiabatic mechanism inside the parent nucleus, penetrates the potential barrier created by the Coulomb and nuclear interactions between it and a daughter nucleus.

The main argument in favour of the latter approach is the discovery [11] of the fine structure of cluster decay (well-known in the case of α-transitions) revealing essential variations of cluster emission probability for different states of the daughter nucleus. In other words, there is a dependence of these probabilities on the shell structure of the initial and final states of nuclei involved in the process under investigation. An important point also consists of the existence [2,5] of the odd-even effect in the reduced probability of cluster emission from odd (even) parent isotopes. All of these facts can be naturally reproduced in the frame of the microscopic shell model approaches [5,9], but the inclusion of these effects in a fission-like scenario of cluster decay needs new adjustable parameters (specified for each final state in the case of a fine structure phenomenon).

One can bring very simple arguments against the adiabatic mechanism of cluster decay. Indeed, if we consistently include spontaneous cluster emission in the picture of nuclear fission we have to say that cluster decay is simply superasymmetric cold compact binary fission with a Q-value sufficiently less than the height of the respective Coulomb barrier between the cluster and the daughter nucleus. But it is known that even for cold compact fission with a Q-value close to the height of the Coulomb barrier, the yield of very asymmetric fragments is about $10^{-4} - 10^{-6}$ with respect to that from the peak of fragment mass distribution. If one compares, as it is shown in Fig. 1, the yields of α particles, Ne and Mg clusters and, finally, the fragments from the spontaneous fission of ^{234}U, it becomes very difficult to adopt a fission-like mechanism for the spontaneous emission of clusters lighter than ^{40}Ca (see discussion in [9] and below).

To realize another possibility to discover the mechanism of cluster radioactivity we calculated [12] the branching ratios of cluster decay for the mass region of parent nuclei $A \approx 110$, first indicated in [10]. The results are greatly different from those for the known mass region $A > 208$, where $T_{1/2}(\alpha)/T_{1/2}(cluster) \ll 1$. For example, in the case of $^{114}\mathrm{Ba} \rightarrow ^{12}\mathrm{C} + ^{102}\mathrm{Sn}$ decay, the branching ratio of respective widths is $log(\Gamma_C/\Gamma_\alpha) = -0.4$. The half-life of ^{12}C-decay,

* 141980 Dubna, Moscow Region, Russia, Tel.: (7-095)9243914, Fax: (7-09621)65085, E-mail: furman@lnp20.jinr.dubna.su
** 394030 Voronezh, Russia, Tel.: (7-0732)566571, E-mail: phys@vucnit.voronezh.su
*** Moscow, Russia, Tel.: (7-095)9390896, E-mail:tchuvl@compnet.msu.su

Table 1. ^{12}C and ^{16}O emission from proton-rich Ba isotopes (notation $1.0 \times 10^{\pm n}$=1.0(± n))

Parent nucl.	Q_α (MeV)	$T_{1/2}(\alpha)$ (sec)	$Q_{^{12}C}$ (MeV)	$T_{1/2}$(C) (sec)	T_c/T_α	$Q_{^{16}O}$ (MeV)	$T_{1/2}$(O) (sec)	T_c/T_α
	3.43	2.4(+3)	18.9	5.8(+3)	2.4	26.5	1.5(+5)	6.2(+1)
^{114}Ba	2.87	7.6(+7)	18.3	1.2(+5)	1.7(-3)	26.0	2.9(+6)	3.8(-2)
	4.65	4.4(-4)	20.1	1.7(+2)	3.8(+4)	27.7	1.0(+3)	2.3(+6)
	2.96	1.1(+7)	17.0	3.3(+8)	3.3(+1)	24.4	4.3(+9)	4.3(+2)
^{116}Ba	2.78	4.8(+8)	16.8	9.5(+8)	2.0	24.2	9.9(+9)	2.0(+1)
	3.39	4.1(+3)	17.4	2.3(+7)	5.6(+3)	24.8	4.6(+8)	1.1(+5)
^{118}Ba	2.4	5.5(+12)	15.1	1.7(+14)	3.1(+1)	21.1	1.3(+15)	2.4(+2)

Table 2. Most favourable examples of cluster radioactivity from some isotopes Ce, Nd and Sm (notation $1.0 \times 10^{\pm n}$=1.0(± n))

Parent nucl.	Cluster X	Q_X (MeV)	$T_{1/2X}$ (sec)	$T_{1/2\alpha}$ (sec)	Q_α (MeV)	$T_{1/2\alpha}/T_{1/2X}$
	^{12}C	17.8	8.0(+8)			0.49(-2)
^{118}Ce	^{16}O	29.3	8.8(+2)	3.8(+6)	3.18	0.45(+4)
	^{20}Ne	34.4	1.1(+8)			0.35(-1)
	^{16}O	27.5	5.1(+8)			0.36(+5)
^{122}Nd	^{20}Ne	36.6	3.4(+7)	1.8(+14)	2.55	0.52(+7)
	^{24}Mg	46.3	4.6(+5)			0.40(+9)
	^{24}Mg	46.2	8.1(+9)			0.82(+8)
^{126}Sm	^{28}Si	56.7	6.9(+7)	9.0(+16)	2.46	0.13(+10)
	^{32}S	59.6	6.6(+12)			0.14(+5)

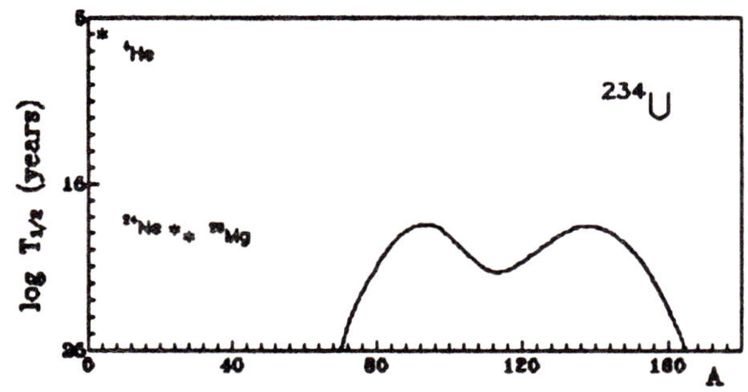

Fig. 1. Log. $T_{1/2}$ for α-decay, cluster emission and spontaneous fission of a ^{234}U nucleus

$T_{1/2} = 6 \times 10^3$ sec, and a ratio, $(\Gamma_C/\Gamma_{tot}) \approx 10^{-4}$, make this decay observable. The characteristics of the other example, $^{118}Ce \rightarrow {}^{16}O + {}^{102}Sn$, $\log(\Gamma_O/\Gamma_\alpha) = 3.7, T_{1/2} = 8.8 \times 10^2$sec, $(\Gamma_O/\Gamma_{tot}) \approx 10^{-3} - 10^{-4}$ are as good as the previous one.

The results of our calculations are summarized in Tables 1 and 2. It is clear that the branching ratios obtained strongly depend on the experimentally unknown values, Q_α and $Q_{cluster}$. We used the same mass systematics as in [13].

Preliminary experimental results [15,16] confirm our predictions [12] and disagree sharply (by many orders of magnitude) with the previous ones [13] and very recent calculations [14] based on the fission-like approach. What does this mean? In the spirit of the consideration of [17] one can say that the effective cluster preformation factors in the variants [1,13,14] of a fission-like theory have an incorrect dependence on the parent nucleus and cluster masses outside the known area of cluster emitters. It is possible in the framework of this approach to reproduce the experimental results [15,16] by an adjustment of the existing parameters(more than ten) or by the introduction of additional ones. It is more preferable, however, to conclude that the mechanism of cluster radioactivity has a nonadiabatic character for emitted clusters up to ^{32}Si. It is appropriate to note that in the present calculations (see also [12]) the same parameters of the theory were used as in the calculation of the old mass region of cluster radioactivity (see [5,9]). To understand the results on the new cluster emitters in more detail it is instructive to consider the dependence of theoretical cluster preformation (spectroscopic) factors on the masses of the nuclei involved. As follows from Table 3 the effective prefomation factors of the most selfconsistent fission-like model [4] are close to those calculated in the frame of the nonadiabatic mechanism [8,9,12]. We could not find the respective S_X numbers for numerous calculations performed on the basis of the so-called ASAFM and discussed, for example, in [17]. But in principle, the results of this comparison have to be rather similar (at least in the sense of relative mass dependence).

A very simple semiphenomenological formula of [8] gives S_X values close to ours, up to ^{28}Mg for the region of $A_{parent} > 215$, where its parameters were fitted. But outside the region of known masses of the clusters and parent nuclei, some difference appears. It is worth noting that in spite of a relatively small distinction between our spectroscopic factors and those calculated in accordance with [8], the difference in branching ratios predicted for the ^{12}C-decay of ^{114}Ba, for example, is over six orders of magnitude. This is due to the very low penetrability obtained with the optical potential [19] which [8] proposes. For clusters heavier than ^{28}Mg, the microscopic shell model's spectroscopic factors begin to fall off rather sharply in comparison with those from [8]. This is connected with the appearence of the so-called structural hindrance (see [9]) for these clusters which

Table 3. Mass dependence of calculated cluster preformation factors for even-even parent nuclei (notation $1.0 \times 10^{\pm n}$=1.0(± n))

Cluster X	Preformation (spectroscopic) factors S_X(A>215)				$S_X(A \leq 130)/S_X(A>215)$	
Reference	[8]	[9]	Extracted from [4]	[18]	[8]	[9,12]
^{8}Be	1.3(-6)	0.7(-6)			1	8.0
^{12}C	8.5(-9)	1.4(-9)			1	20.7
^{14}C	2.9(-10)	0.6(-10)	0.2(-10) 0.9(-10)	0.5(-18) 1.6(-18)	1	
^{20}Ne	1.2(-14)	1.3(-14)			1	376
^{24}Ne	1.3(-17)	7.0(-19)	1.0(-17) 5.6(-18)	6.8(-27)	1	1.8(+3)
^{28}Mg	3.2(-20)	0.2(-20)	2.9(-20) 4.0(-20)	2.3(-28) 4.3(-28)	1	6.5(+3)
^{32}Si	1.8(-23)	5.1(-25)		4.1(-28)	1	2.1(+4)
^{40}Ca	2.4(-29)	6.0(-33)			1	2.5(+5)

strongly inhibits the spectroscopic factors ensuring, possibly, the natural transition from nonadiabatic cluster decay to very asymmetric binary fission.

References

1. Poenaru, D.N., Ivascu, M., Sandulescu, A., Greiner, W.: J. Phys. G.**10**, L183 (1984)
2. Shi Yi-Jin, Swiatecki, W.J.: Phys. Rev. Lett. **54**, 300 (1985)
3. Rubchenya, V.A., Yavshits, S.G.: Yad. Fiz. **40**, 649 (1984)
4. Pik-Pichak, G.A.: Yad. Fiz. **44**, 1421 (1986)
5. Kadmensky, S.G., Furman, W.I., Tchuvilsky, Yu.M.: Izv. Akad. Nauk SSSR, Ser. Fiz. **50**, 1786 (1986)
6. Gupta, R.K., Gilaty, S., Malik, S.S. et al.: J. Phys. G., **13** L27, (1987)
7. Blendowske, R., Fliesbach, T., Walliser, H.: Nucl. Phys. A**464**, 75 (1987)
8. Blendowske, R., Walliser, H.: Phys. Rev. Lett. **61**, 1930 (1988)
9. Zamyatnin, Yu.S., Mikheev, V.L., Tretyakova, S.P. et al.: Sov. J. Part. Nucl. **21**(2), 231 (1990)
10. Price, P.B.: Nucl. Phys. A**502**, 41 (1989)
11. Brillard, L., Elayi, A.J., Hourani, E. et al.: C.R. Akad. Sci. **309**, 1105 (1989)
12. Kadmensky, S.G., Mikheev, W.L., Furman, W.I., Tchuvilsky, Yu.M.: In: Proc. Conf. on Nucl. Spectr. and Nucl. Structure, Alma-Ata 1992, Nauka, St. Petersburg, 327 (1992); Izv. Akad. Nauk., Ser. Fiz. **57**, 12 (1993)
13. Poenaru, D.N. et al.: At. Data Nucl. Data Tables, **48**, 231 (1991)
14. Poenaru, D.N., Greiner, W., Gherghescu, R.: Phys. Rev. C.**47**, 47 (1993)
15. Oganessian, Yu.Ts., Lazarev, Yu.A., Mikheev, V.L. et al.: In: Heavy Ion Phys., Sci. Reports 1991-1992, JINR, E7-93-57, Dubna, 49 (1993)
16. Oganessian, Yu.Ts.: Talk at the conference "Dynamical Aspects of Nucl. Fission", Smolenice, Bratislava (1993)
17. Poenaru, D.N., Greiner, W.: J.Phys. G.; Nucl. Part. Phys.**17**, 443 (1991)
18. Gupta, R.K., Singh, S., Puri, R.K. et al.: ibid. **18**, 1533 (1992)
19. Christensen, R.P., Winter, A.: Phys. Lett. B **65**, 19 (1976)

This article was processed using Springer-Verlag TEX Z.Physik A macro package 1.0 and the AMS fonts, developed by the American Mathematical Society.

Z. Phys. D 31, 205–207 (1994)

On the localization of electrons and holes in Hg_n and rare-gas clusters

M.E. Garcia, K.H. Bennemann

Institut für Theoretische Physik der Freien Universität Berlin, Arnimallee 14, D-14195 Berlin, Germany

Received: 28 June 1993 / Final version: 7 March 1994

Abstract. We have used a microscopic theory to study the size dependence of the degree of localization of the valence electrons and holes in neutral an ionized rare-gas- and Hg_n clusters. We discuss under which circumstances localization of the electrons and holes is favoured. We have calculated the ionization potential of Xe_n, Kr_n and small Hg_n clusters. Good agreement with experiments is obtained. We have also determined the dependence of the ionization potential on cluster structure.

PACS: 36.40.+d; 71.30.+h; 33.80.+Eh

The size evolution of the electronic structure of Hg_n clusters is one of the most interesting problems in cluster physics. From the point of view of the electronic structure in the atom (closed shell), small Hg_n clusters should be expected to be van der Waals bonded. However, since the bulk band widths, and consequently the hopping elements, are large, one cannot exclude the possibility of covalent bonding even in the very small clusters. It has been shown [1, 2], that a correct theoretical description of the electronic properties of small Hg_n clusters is only possible if one takes into account both the van der Waals and covalent bonding, and there interplay, on the same electronic level. Thus, a microscopic theory has been developed [1] which takes into account both bonding mechanisms. By using this theory one obtains that small Hg_n clusters with $n \lesssim 10$ are mainly van der Waals bonded, whereas in Hg_n with $n \gtrsim 20$ the bond character is mainly covalent [1]. Furthermore, there is a clear interdependence between the degree of localization of the valence electrons and the bond character. In small Hg_n clusters ($n \lesssim 10$), and in rare-gas clusters, electrons are almost completely localized around the atoms, whereas in larger Hg_n clusters with $n \gtrsim 20$ a delocalization of the electrons takes place together with the transition to covalent bonding. Concerning the changes in the electronic structure of van der Waals clusters like rare-gas clusters or small Hg_n clusters upon ionization, many important questions need be answered. For instance, the dependence of the ionization potential on cluster size and cluster structure, and its dependence on the degree of delocalization of the hole produced in the ionization process. In this paper we use a microscopic theory to shed light on this problem.

The essential changes in the ionized van der Waals clusters with respect to the neutral ones can briefly illustrated as follows. Neutral rare-gas and small Hg_n clusters are characterized by having local charge neutrality and no net dipole moments. Only dipole fluctuations are present, which give rise to the van der Waals interactions. In the ionized clusters, there is a hole which can polarize its environment, inducing dipole moments on the neighbouring neutral atoms. The physics mentioned below is contained in the following many-body model Hamiltonian [2, 3]

$$H = H_{vdW} + H_{cov} + H_{Q-P} + H_{P-P}. \quad (1)$$

In (1), H_{vdW} describes the van der Waals interactions, as arising from the coupling between local *sp* electronic excitations [1]. H_{cov} refers to the covalent interactions, represented by a two-band Hubbard Hamiltonian containing *s* and *p* on-site energies, hopping elements and Coulomb repulsion. H_{Q-P} and H_{P-P} refer to the charge-dipole and dipole-dipole interactions, respectively. The first two terms are the most important ones in the neutral clusters, whereas for the ionized clusters all four terms have important contributions. The ionization potential of a cluster with *n* atoms is calculated as [3]

$$I_p(n) = I_p(1) + |D_n^0| - |D_n^+|, \quad (2)$$

where $|D_n^0|$ and $|D_n^+|$ are the binding energies of the neutral and ionized clusters, respectively, and $I_p(1)$ is the atomic ionization potential.

This paper was originally submitted in connection with the 2nd. Int. Conference on Atomic and Nuclear Clusters held in Santorini from 28. June–2. July 1993 and is published here as a regular article after an independent refereeing procedure according to the standards of Z. Phys. D

We calculate the binding energy of the neutral clusters by using the slave boson method in the saddle-point approximation, as described in [1]. The calculation of D_n^+ is more complicated, due to the presence of many interactions at the same time. However, the electronic structure is basically determined by the competition between the kinetic energy of the hole and the polarization energy of the neutral atoms surrounding it. Delocalization of the hole lowers the kinetic energy, while polarization energy is maximal when the hole is completely localized. In order to take into account this competition, we use the *ionic core model*, proposed by Haberland [4] and already used by many authors to calculate properties of charged rare-gas clusters [5]. This model is based on the assumption that the hole delocalizes only within a subcluster of m atoms (ionic core), while the rest of the cluster remains neutral. This physical picture can be applied to the Hamiltonian of (1). We write the dipole and charge operators as [3]

$$\mathbf{P}_k = \langle \mathbf{P}_k \rangle + \delta \mathbf{P}_k ,$$
$$Q_l = \langle Q_l \rangle + \delta Q_l , \qquad (3)$$

i.e., as the expectation values plus fluctuations around them. The charge-dipole fluctuations can be neglected in van der Waals clusters. Moreover, the dipole-dipole fluctuations contribute to the van der Waals interactions in the ionized clusters. Assuming that the strength of the van der Waals interactions is approximately the same in the neutral and in the ionized cluster, the ionization potential is then roughly independent of the van der Waals energy, since it involves a difference of binding energies. Thus, rewriting the Hamiltonian of (1) by using (3) one obtains a decoupling of the different charge operators in the ionized core, and the dipole operators in the neutral rest of the cluster. The ionization potentials depends on the expectation values $\langle Q_l \rangle$ ($l \in$ ionic core) and $\langle \mathbf{P}_k \rangle$ ($k \in$ rest of the cluster), which have to be calculated self-consistently. The physics underlying the calculation can be understood as follows. A given charge distribution $\{\langle Q_l \rangle\}$ in the ionic core induces a dipole distribution $\{\langle \mathbf{P}_k \rangle\}$ in the rest of the cluster. The potential due to this dipole distribution affects the wave function of the hole and consequently produces a rearrangement of the charge distribution and so on. We have calculated vertical ionization potentials for Xe_n, Kr_n and Hg_n clusters for different sizes m and shapes of the ionic core. We use compact cluster structures, which maximize the total coordination number, and thus are expected to be the most stable in van der Waals clusters. For Xe_n and Kr_n clusters, best agreement is obtained with the experimental results [6, 7] by assuming a compact ionic core of four atoms, as is shown in Fig. 1. Note, that the agreement with experiment is better for Xe_n clusters than for Kr_n clusters. Furthermore, it is remarkable that for Xe_n clusters both the experimental and our calculated $I_p(n)$ show a change in the size dependence at $n = 13$ (indicated with an arrow in Fig. 1a. By analyzing the charge distribution of Xe_n^+ clusters as a function of n, we observe that the hole tends to localize more and more for increasing cluster size. For $n = 10$ almost 80% of the positive charge lies at the atom

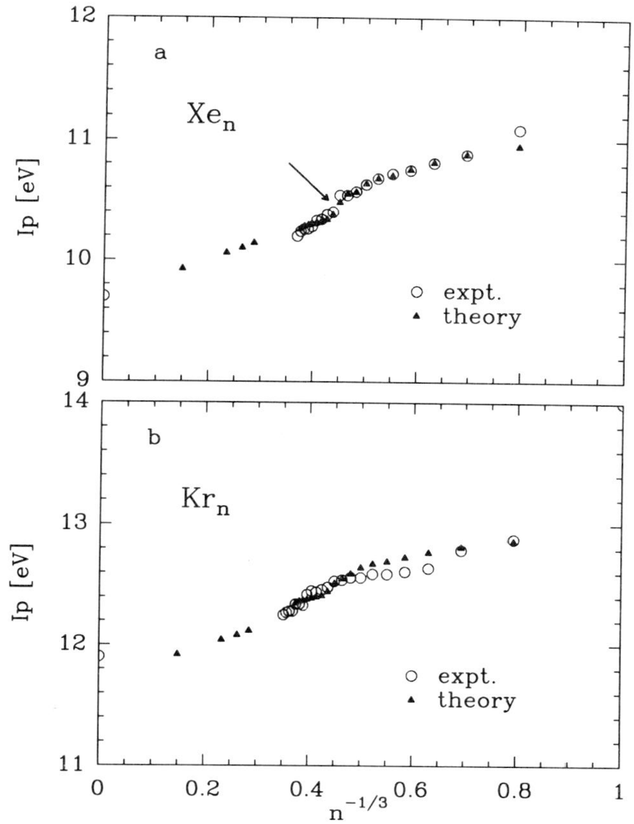

Fig. 1. Results for the size dependence of the ionization potential of **a** Xe_n, and **b** Kr_n clusters, obtained by assuming a compact ionized subcluster of size $m = 4$. The corresponding experimental results of [7] are also given for sake of comparison. In **a** the *arrow* indicates a change in the size dependence of the calculated and measured ionization potentials of Xe_n clusters (see text)

with highest coordination, and the rest of the charge is distributed within the other 3 atoms belonging to the ionic core. The compact structure of $n = 13$ has one atom at the center of the cluster and 12 surface atoms surrounding it. The first coordination shell is complete. Thus, for $10 \leqq n \leqq 13$ the decrease of the ionization potential is essentially governed by the increase of the polarization energy of the atoms surrounding a localized hole. For clusters larger than 13 atoms, neutral atoms occupy the second coordination shell around the central (positively charged) atom. Then, the gain in polarization energy for $n > 13$ is much smaller than for $n \leqq 13$ because the central charge is screened by the first coordination shell. Thus, we can explain the change in the size dependence of the ionization potential of Xe_n clusters as arising from a saturation of the polarization energy due to the completion of the first coordination shell. This effect is more evident in Xe_n clusters due to the fact that Xe has a larger atomic polarizability [8] α ($\alpha(\mathrm{Xe}) = 4.04$ Å^3, $\alpha(\mathrm{Kr}) = 2.484$ Å^3). If our physical picture is correct, this effect should be also observable in the ionization potential of Hg_n clusters, since the atomic polarizability of Hg is also large [8] ($\alpha(\mathrm{Hg}) = 5.7$ Å^3). Experimental results for $I_p(n)$ of Hg_n clusters, obtained by Haberland et al. [9] by electron impact ionization, do exhibit such a behaviour, as is shown in Fig. 2. We present also calculations by assuming

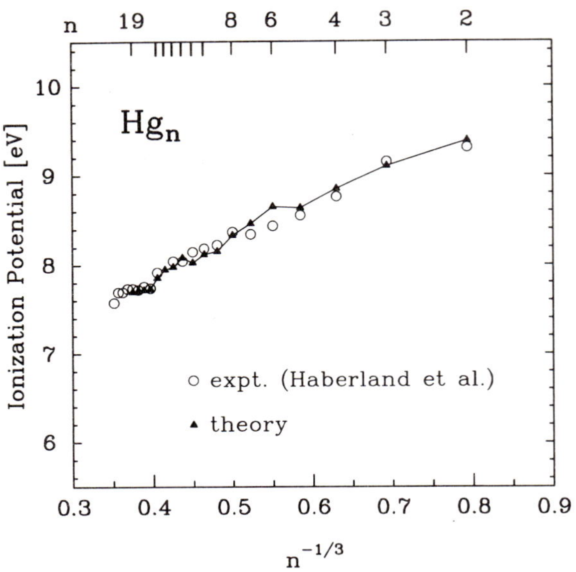

Fig. 2. Calculated size dependence of the ionization potential of Hg_n clusters, obtained by using a ionized trimer core ($m=3$, triangular shape). Results are compared with those obtained experimentally by electron impact ionization (see [7])

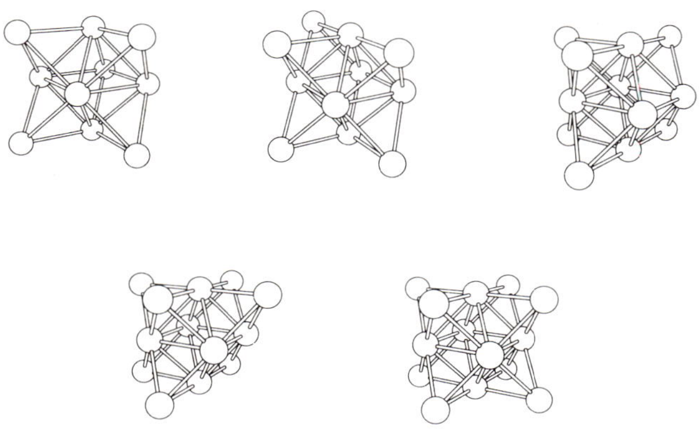

Fig. 3. Illustration of the cluster structures for which the calculated $I_p(n)$ of Hg_n clusters yield best agreement with experiment for $10 \leq n \leq 15$. Since the n atoms are at the surface, the ionization cross section might be large than in the case of compact structures

a compact ionic core of three atoms. It is remarkable that $I_p(n)$ shows a 'plateau' for $16 \leq n \leq 20$. It is a goal of our theory to describe this. However, for Hg_n clusters the saturation of the polarization energy seems to occur only at $n=16$ and not at $n=13$ as it would be if compact structures were assumed. This means other cluster structures, different from the compact ones may be important for the ionization potential of Hg_n clusters with $10 \leq n \leq 15$. In Fig. 3 the structures are shown, which yield best agreement with experiment for $10 \leq n \leq 15$. Note, that these structures have a total coordination number which is smaller than that of the (*fcc* or icosahedral) compact structures, and are therefore less stable. However, they seem to have a much larger ionization cross section than the compact ones. Note, that in the structures shown in Fig. 3, the n atoms of the cluster belong to the surface. The difference between the total coordination numbers of these structures and the compact ones increases with the cluster size. This means that for increasing cluster size the compact structures must begin to dominate. At $n=16$, the saturation in the polarization energy has been reached, indicating that the compact structures govern the dependence of the $I_p(n)$ from this size on. For clusters larger than 20 atoms, the ionic core model fails to describe the further size dependence of $I_p(n)$ for Hg_n clusters. This is due to the transition from van der Waals to covalent bonding. Larger clusters are covalent, and therefore the assumption that the atoms surrounding a charge remain neutral is no longer valid. In the weak covalently bonded Hg_n clusters, i.e., near the transition, a small interatomic charge transfer must be more favourable than local polarization. Itinerancy of the electrons must become more important. However, as long as the covalent bonding is weak, the polarization energy must still dominate the behaviour of $I_p(n)$. Therefore, we expect that the hole remains still localized and that the covalent bonds surrounding this partially localized hole become polarized through interatomic charge transfer. In this way the system can gain polarization energy. Since the polarizability of a dimer is expected to be larger than that of an atom, the gain in polarization energy should be larger than in the van der Waals region. This could explain the relative abrupt decrease of $I_p(n)$ of the covalent clusters. For further increase of the cluster size, the screening of the charge fluctuations must increase, the covalent bonds become stronger, and finally a transition to metallic bonding occurs when the *sp* gap vanishes. Thus, $I_p(n)$ must converge to the behaviour predicted by the classical spherical droplet model [10]. In order to describe the change from atomic polarization to bond polarization and later to metallic behaviour, our theory for Hg_n^+ clusters should be extended in order to describe the change of the bond character, as it has been done for the neutral clusters [1], where one obtains an abrupt change in the binding energy when covalency occurs.

We thank Dr. G. Pastor for useful discussions. This work was supported by the Deutsche Forschungsgemeinschaft through SFB 337.

References

1. Garcia, M.E., Pastor, G.M., Bennemann, K.H.: Phys. Rev. Lett. **67**, 1142 (1991)
2. Garcia, M.E.: Ph.D. Thesis, Freie Universität Berlin (1992)
3. Garcia, M.E., Pastor, G.M., Bennemann, K.H.: Phys. Rev. B**48**, 8388 (1993)
4. Haberland, H.: Surf. Sci. **156**, 305 (1985)
5. Amarouche, M., Durand, G., Malrieu, J.P.: J. Chem. Phys. **88**, 1010 (1988); Kuntz, P.J., Valldorf, J.: Z. Phys. D**8**, 195 (1988); Böhmer, H.U. et al.: Z. Phys. D**11**, 239 (1989)
6. Kamke, W., de Vries, J., Krauss, J., Kaiser, E., Kamke, B., Hertel, I.V.: Z. Phys. D**14**, 339 (1989)
7. Ganteför, G., Bröker, G., Holub-Krappe, E., Ding, A.: J. Chem. Phys. **91**, 7972 (1989)
8. Handbook of chemistry and physics, 66th edn. Boca Raton: CRC Press 1985
9. Haberland, H., Kornemeier, H., Langosch, H., Oschwald, M., Tanner, G.: J. Chem. Soc. Faraday Trans. **86**, 2473 (1990). Photoionization experiments have been performed by Rademann, K., Kaiser, B., Even, U., Hensel, F.: Phys. Rev. Lett. **59**, 2319 (1987)
10. Pastor, G.M., Stampfli, P., Bennemann, K.H.: Phys. Scr. **38**, 623 (1988); Europhys. Lett. **7**, 419 (1988)

Fission barriers for Na_N^{2+} cluster dissociation

F. Garcias[1], J.A. Alonso[2], M. Barranco[3], J.M. López[2], A. Mañanes[4], and J. Németh[5]

[1]Departament de Física. Universitat de les Illes Balears. E-07071 Palma de Mallorca. Spain.
[2]Departamento de Física Teórica, Universidad de Valladolid. E-47011 Valladolid. Spain.
[3]Departament ECM. Facultat de Física. Diagonal, 647. E-08028 Barcelona. Spain.
[4]Departamento de Física Moderna. Universidad de Cantabria. E-39005 Santander. Spain.
[5]Institute of Theoretical Physics. Eötvös University. H-1088 Budapest. Hungary.

Received 28 June 1993; Revised 11 March 1994

Abstract. The competition experimentally observed between asymmetric fission and neutral monomer evaporation as dissociation channels of excited doubly charged sodium clusters has been investigated by means of an axially symmetric, fully selfconsistent Kohn-Sham method. Ellipsoidal equilibrium configurations for parent and daughter clusters have been considered using a deformed jellium model.

PACS: 36.40.+d, 31.15.+q, 31.20.S, 31.90.+s

Since the experimental discovery of Coulomb explo- sion phenomenon, the study of stability of metal clusters in charged states has motivated an important experimental and theoretical research. Basically, the fundamental open questions are referred to the number of atoms for which a multiply charged cluster is stable, the possible dissociation channels for unstable clusters and the charge and mass distribution of fragments.

it is well established that multiply charged metal clusters M_N^{q+} can be observed in mass spectra only if the number of atoms N exceeds a critical size N_c that depends on the metal species M and on the charge state q. For instance, the critical sizes for doubly charged alkali clusters are 27 for Na [1], and 19 for K [2] and Cs [3]. Triply charged K and Na clusters are observable above $N_c \sim 70$ [2,4] and very recently, Na_N^{8+} clusters have been identified above $N \sim 500$ [4]. These critical sizes are experimentally determined studying charged clusters produced by multi-step ionization of hot larger clusters that lose a sizeable part of their excitation energy mainly evaporating neutral atoms. This causes the ionized clusters M_N^{q+} to shrink up to sizes $N \sim N_c$. For cluster sizes around N_c, monomer evaporation competes with another dissociation channel, asymmetric fission, in which two charged fragments of very different sizes are emitted. Electron shell effects are manifested, with preferential emission of magic fragments such as K_3^+ [5].

Old theoretical studies of cluster stability were based on pure energetic criteria which only involved the energies of the initial and final states [6]. Different dissociation channels can easily studied and it is possible to provide the correct trends for neutral and singly charged aggregates. However, stable multiply charged clusters have been detected below (theoretically or experimentally) predicted N_c values. This occurs if they are created in a multi-step ionization process starting from cold neutral clusters [1], and indicates that cluster fragmentation is a barrier-controlled process. For neutral or singly charged clusters the dissociation channels are given by the energy balance because of the absence of barrier. For multiply charged clusters there exists a finite energy barrier against fragmentation, although the total energy is reduced after dissociation. These facts have prompted the interest of theoreticians to obtain these barriers.

Let us recall the model commonly used to study the dissociation problem in the case of doubly charged alkali clusters [1,7]. The heat of monomer evaporation ΔH_e is a positive quantity defined as:

$$\Delta H_e = E(M_{N-1}^+) + E(M) - E(M_N^{2+}) \tag{1}$$

where $E(M_N^{q+})$ is the total energy of the M_N^{q+} cluster. The heat of fission ΔH_f is calculated as:

$$\Delta H_f = E(M_{N-p}^+) + E(M_p^+) - E(M_N^{2+}) \tag{2}$$

Different fission channels can be regarded: monomer emission ($p = 1$), charged trimer emission ($p = 3$), symmetric fission ($p = N/2$), etc. ΔH_e and ΔH_f are very sensitive to shell effects.

The heat of fission is related to the fission F_m and fusion B_m maxima of the corresponding barriers by the relationship:

$$F_m = B_m + \Delta H_f \tag{3}$$

Fission is favored over monomer evaporation when F_m is smaller than ΔH_e, and the other way around if it is larger. If the fission barrier height vanishies, the cluster is absolutely unstable against fission.

Assymetric fission and its competition with monomer evaporation in Na_N^{2+} clusters has been addressed in some detail within an Extended Thomas-Fermi (TF) method and the spherical jellium model [7,8]. The agreement on the value of N_c between theory and experiment was qualitative, obtaining $N_c = 36$ instead of 27 for Na. Inclusion of other fission channels, like symmetric fragmentation [9] or Na_3^+ emmision [8], leads to a reduction of the theoretical critical number. As a general trend, the latter processes are hindered by a barrier higher than that of completely asymmetric fission, thus indicating that monomer emission is the preferred fission channel within the TF model.

In those works [7-9], the cluster fragmentation has been parametrized in terms of the distance between the centers of two jellium spheres representing the fragments, starting from the touching configuration. A more detailed description of the barrier requires a sequence of jellium shapes continuously connecting the initial configuration with the final one corresponding to two separated fragments. Several shape parametrizations of this kind exist [10]. If the goal is the calculation of the barrier height and not the barrier itself, one can take advantage of the fact that the maximum of the barrier (i.e., the saddle point) corresponds to a configuration in which the emerging fragments are already formed [11,7,8], and slightly tied up by the electron cloud (see figure 4 below). Fission and fusion barriers only differ at shorter distances, which is completely irrelevant for the problem at hand.

The semiclassical TF method cannot account for shell structure effects, which must be included to provide the correct picture for studying dissociation channels and critical sizes for stability of alkali clusters. The aim of this work is the study of the role played by shell effects on the fission barrier height. We present some results obtained from an axially symmetric, fully selfconsistent Kohn-Sham (KS) calculation.

Figure 1 shows the fission barrier for $Na_{42}^{2+} \rightarrow 2Na_{21}^+$ obtained from the KS approximation (lower curve) in comparison with the result from the previous TF model (upper curve). The vertical line at $s \sim 22a.u.$ corresponds to the touching distance of the spherical fragments, that is, the sum of their radii. Both barriers are referred to the initial spherical configuration of Na_{42}^{2+}. Horizontal lines at right indicate the energy of the final state in each case. Shell structure confers the daughter clusters a special stability and ΔH_f becomes negative in the KS method. One can see that, when shell effects are taken into account, the fission barrier height reduces drastically, from $F_m(\text{TF}) = 3.1\text{eV}$ to $F_m(\text{KS}) = 0.4\text{eV}$.

The $1/s$ curve in figure 1 (dashed lines) is a pure Coulomb barrier obtained from the classical expression

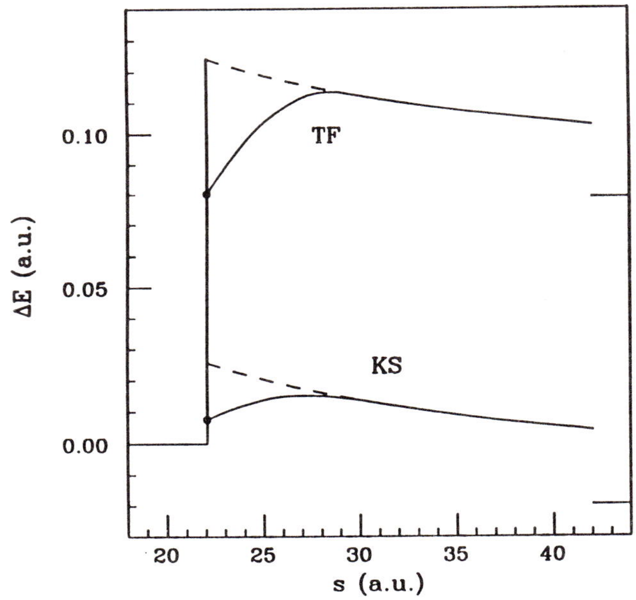

Fig. 1. Fission barrier in atomic units ($a.u.$) for symmetric dissociation of Na_{42}^{2+} obtained from the TF model (upper solid curve) and from the KS appoximation (lower solid curve)

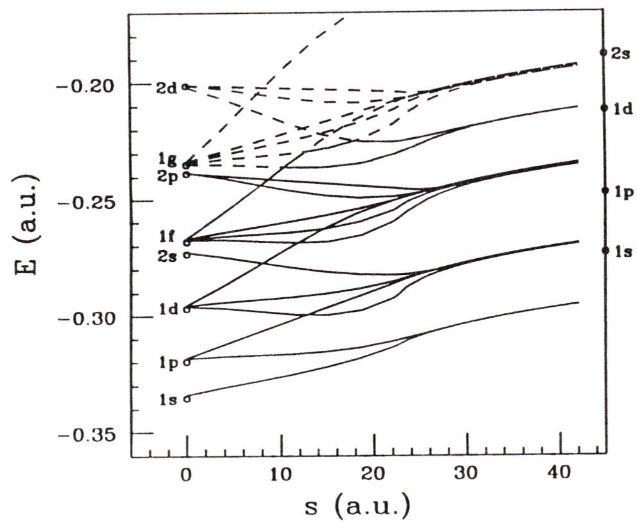

Fig. 2. KS single-particle energies as a function of fragment separation for symmetric fission of Na_{42}^{2+}. Dots at $s = 0$ are the spherical single-electron levels of Na_{42}^{2+} and those at right correspond to the spherical configuration of Na_{21}^+. Dashed curves represent empty levels

$$F_m^{cl}(a.u.) = \frac{1}{r_s[(N-p)^{1/3} + p^{1/3}] + \delta} + \Delta H_f \qquad (4)$$

We have used $r_s = 4a.u.$ for sodium and $\delta = 0$. The δ-parameter serves to take care of the electron spill-out. As figure 1 indicates (see also figure 3 below), a judicious choice of the δ-value may allow for a fair and very quick estimate of F_m. This is the basis of some approximate calculations that work quite well in reproducing the experimental critical sizes [5,9].

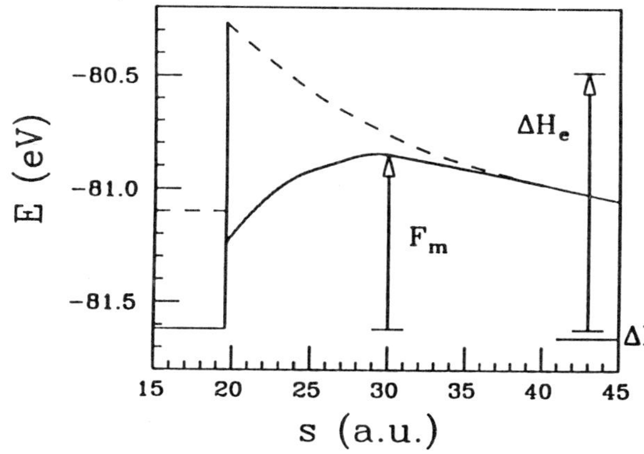

Fig. 3. KS fission barrier for $Na_{44}^{2+} \rightarrow Na_{41}^{+} + Na_{3}^{+}$

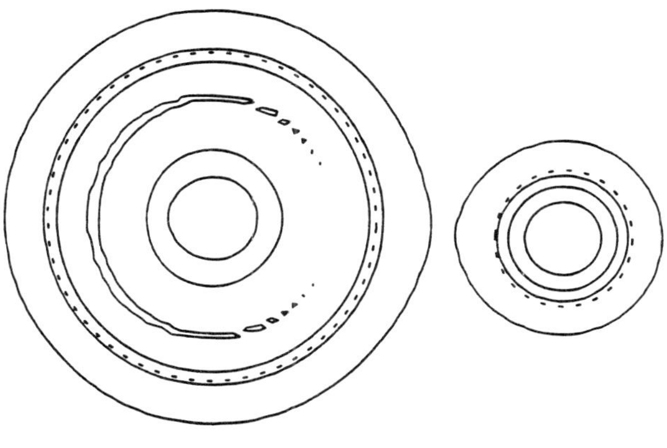

Fig. 4. KS equidensity lines for $Na_{44}^{2+} \rightarrow Na_{41}^{+}+Na_{3}^{+}$ at the saddle configuration. From outside inwards, solid curves correspond to electron densities $n_e = 0.5, 1$ and $2 \times 10^{-3} a.u.^{-3}$. The innermost ones correspond all to $n_e = 4 \times 10^{-3} a.u.^{-3}$. Dashed circles represent the jellium spheres

Figure 2 displays the KS single-particle energies in the case of $Na_{42}^{2+} \rightarrow 2Na_{21}^{+}$. One can see how the single-particle levels evolve from the spherical configuration of the parent cluster to the spherical configuration of two identical fragments. Beyond the saddle point ($s \sim 27a.u.$) the single-particle spectrum of Na_{21}^{+} can be clearly identified.

Within the KS approximation, the symmetric fission of Na_{42}^{2+} is favored over monomer evaporation and completely asymmetric fission: this model gives F_m $(Na_{21}^{+}) = 0.4$eV, $\Delta H_e = 0.9$eV and $F_m(Na^{+}) = 1.2$eV, respectively. Unfortunately it is impossible to detect symmetric fission with current experimental methods [1].

Shell effects cause some clusters to be tightly bound (spherical magic clusters), and to adopt deformed equilibrium configurations to many others. Even in the jellium model, the calculation of quantities entering eqs. (1-3) is a formidable task when the parent clusters and/or final products are not magic. In this case, to obtain the energy of each cluster one has to carry out a deformed calculation in which the positive background adopts shapes whose equilibrium deformation has to be determined selfconsistently [12].

Figure 3 shows the fission barrier and heats corresponding to $Na_{44}^{2+} \rightarrow Na_{41}^{+} + Na_{3}^{+}$. The vertical line at $s \sim 20a.u.$ indicates the touching configuration of fragments. The horizontal dashed line at -81.1eV represents the energy of Na_{44}^{2+} treated as a spherical cluster, whereas the horizontal solid line at ~ -81.6eV indicates its energy when ellipsoidal deformations are considered. In this case ΔH_f is negative and very small. For Na_{44}^{2+} the deformed KS model yields $F_m(Na_3^{+}) = 0.7$eV, $F_m(Na^{+}) \sim 1.6$eV and $\Delta H_e = 1.1$eV, thus predicting charged trimer emission as most favorable dissociation channel.

Figure 4 displays several equidensity lines for the saddle configuration in figure 3. Polarization effects in the electron density, which are clearly visible at the touching configuration [7], are almost negligible at the maximum of the barrier.

As a final example, let us comment that for the dissociation $Na_{24}^{2+} \rightarrow Na_{21}^{+} + Na_{3}^{+}$, the ellipsoidal initial configuration has an energy 0.3eV higher than the saddle point. Consequently, Na_{24}^{2+} will spontaneously fission.

To conclude, much effort still has to be dedicated to disentangle the different dissociation channels of multiply charged metal clusters. From the theoretical point of view, the combination of the KS method and the deformed jellium model is useful in situations where more fundamental calculations [11] are prohibitive, and may help find sound approximations that make possible the study of the de-excitation of highly-charged clusters [4].

We wish to thank Maria Garcies for the technical help with figures. This work has been supported in part by the DGICYT (Spain), grants PB 89-0332 and PB 89-0352-C02-01. It was presented at the 2nd International Conference on Atomic and Nuclear Clusters (Santorini, June 28 - July 2, 1993). Additional financial support from the Vice-Rectorat d'Investigació of the Univeritat de les Illes Balears is gratefully acknowledged.

References

1 Bréchignac, C., Cahuzac, Ph., Carlier, F., and de Frutos, M.: Phys. Rev. Lett. **64**, 2893 (1990)
2 Bréchignac, C., Cahuzac, Ph., Carlier, F., and Leygnier, J.: Phys. Rev. Lett. **63**, 1368 (1989)
3 Martin, T.P.: J. Chem. Phys. **81**, 4426 (1984)
4 Martin, T.P., Näher, U., Göhlich, H., and Lange, T.: Chem. Phys. Lett. **196**, 113 (1992)
5 Bréchignac, C., Cahuzac, P., Carlier, F., Leygnier, J., and Sarfati, A.: Phys. Rev. B **44**, 11386 (1991)
6 Iñiguez, M.P., Alonso, J.A., Aller, M.A., and Balbás, L.C.: Phys. Rev. B **34**, 2152 (1986)
7 Garcias, F., Alonso, J.A., López, J.M., and Barranco, M.: Phys. Rev. B **43**, 9459 (1991)
8 López, J.M., Alonso, J.A., Garcias, F., and Barranco, M.: Ann. Physik **1**, 270 (1992)
9 Barranco, M., Alonso, J.A., Garcias, F., and López, J.M.: Clustering Phenomena in Atoms and Nuclei, eds. Brenner, M., Lönnroth, T., and Malik, F.B. (Springer-Verlag, Berlin, 1992) pp. 305

10 Hasse, R.W., and Myers, W.D.: Geometrical Relationships of macroscopic nuclear physics (Springer-Verlag, Berlin, 1988)

11 Barnett, R.N., Landman, U., and Rajagopal, G.: Phys. Rev. Lett. **67**, 3058 (1991)

12 Ekardt, W., and Penzar, Z.: Phys. Rev. B **38**, 4273 (1988)

This article was processed using Springer-Verlag TEX Z.Physik D macro package 1.0 and the AMS fonts, developed by the American Mathematical Society.

Theoretical calculations of the relative stability of Li_6 isomers

G. Gardet, F. Rogemond and H. Chermette

Institut de Physique Nucléaire de Lyon, IN2P3-CNRS et Université Claude Bernard, 43, Bd du 11 Novembre 1918, F-69622 Villeurbanne Cedex, France

Received 28 June 1993

Abstract. The DFT approach has been used to study the relative stability of Li_6 clusters. The three stable geometries, C_{2v}, C_{5v}, C_{3v}, have been obtained at the LSD level of approximation. For these geometries, we have also calculated the energies with different "non local" functionals and evaluate absorption spectra. Most of the results indicate the C_{2v} structure as the more stable.

PACS: 36.40.+d, 31.50.+w

The stability of different conformers of the Li_6 cluster has been the subject of recent studies, both at experiment and theorical level. In this last case, *ab initio* accurate calculations have been performed by Koutecký *et al* [1-2]. These calculations showed that the 3 conformers of Li_6 are quasidegenerate, the CI expansion leading in their best calculation to the order of decreasing stability: $C_{2v} > C_{5v} > C_{3v}$. However, in order to achieve an agreement with experimental data, it is necessary to perform high level CI calculations, which are known to be quite expensive in computational resources, since they scale (at least) as n^5, n being the number of electrons. This precludes the use of such a method for significantly bigger clusters.

On the other hand, the DFT has proven to be able to provide valuable results about electronic structure relative energies and related properties of molecules or clusters, at least when the so-called non local density functionals are used. The purpose of this study is to investigate the accuracy and predicting power of the DFT in the relative stability of Li_6 clusters, for which experimental optical data related to their electronic structure are available [2].

In this work, we optimized the geometries of the conformers at the LDA level, and for these geometries, we calculated the electronic structure and energies at the NLSD level. We calculated also the absorption spectrum in 1.5 to 3.0 eV visible range.

We checked that the structure obtained are true minima, and we obtained for them the energies reported in Table 1, for the different levels of approximation in the calculation. We used the Vosko-Wilk-Nusair (VWN) fonctional [3] at the LSD level, and at the NLSD level, combinations of Perdew-Wang 86 (PW86) [4], Becke (B) [5], Perdew-Wang 91 (PW91) [6] for exchange and Perdew 86 (P86) [7], PW91, Lee-Yang-Parr (LYP) [8] for correlation. For sake of comparison, best results obtained by Koutecký *et al.* [1] are also reported. All the results reported here are singlets, which are found slighter more stable than corresponding triplets, in accord with Koutecky's results [1].

One sees that the C_{2v} structure is slightly more stable than others at the LSD level, as well as PW91-PW91 and B-P86, and in accord to Koutecky's results. On the contrary, the PW86-P86 and PW91-LYP functionals lead to a different order.

The difference in energies is in all cases significantly small so that it may be necessary to perform geometry optimizations at the NLSD level of approximations. We already obtained the B-P86 case which leads to the same trend as the B-P86 energies at the local geometries. The geometrical structures (bond lengths, angles) are, however much closer to the Koutecky's ones when the geometry optimization is performed at the NLSD level. This is certainly one case where the so-called non local corrections are necessary to be taken into account at the self-consistent field level. This is not really surprising because of the diffuse nature of the bonding orbitals in such a compound. Work is in progress for other functionals combinations as well as for the two other geometries reported here (C_{5v} , C_{3v}).

For the first time transition intensities have been calculated within the DFT approach. The excited configurations used here are not true states, and cannot describe well-resolved transitions, but only the optical absorption broad bands (excited configurations): the ensemble of related states are equivalent to the ensemble of true states, which are linear combinations of them, and one can argue that for transition energies significantly

Table 1. Energies of Li_6-clusters (milliHartrees)

	C_{2v}	C_{5v}-C_{2v}	C_{5v}	C_{3v}-C_{5v}	C_{3v}
VWN	-44 244.61	2.11	-44 242.50	6.76	-44 235.74
PW86-P86	-45 133.63	-0.67	-45 134.30	4.16	-45 130.14
PW91-PW91	-45 007.12	0.68	-45 006.44	5.63	-45 000.81
B-P86	-45 039.81	0.08	-45 039.73	4.31	-45 035.42
PW91-LYP	-44 980.48	-3.37	-44 983.85	2.14	-44 981.72
Hartree-Fock	-44 656.69	2.77	-44 653.92	-6.43	-44 660.35
CI	-44 757.56	3.17	-44 754.39	0.13	-44 754.26

larger than the related bandwith, the calculated intensities are, at the first order, a good approximation of more exact intensities calculated from states. Transition energies and intensities have been calculated at the LSD level, Fig.1, and the results indicate a larger agreement with the experimental spectrum for the C_{2v} structure, compared to the C_{5v} one. This was also the conclusion obtained by Koutecký *et al.*

However the small difference found between our transition energies and Koutecky's ones can be related to the difference in the corresponding structures: some bond lengths being significantly larger in our (LSD) structures than in Koutecky's. Work using geometries optimized at the non-local spin density approximation level is in progress.

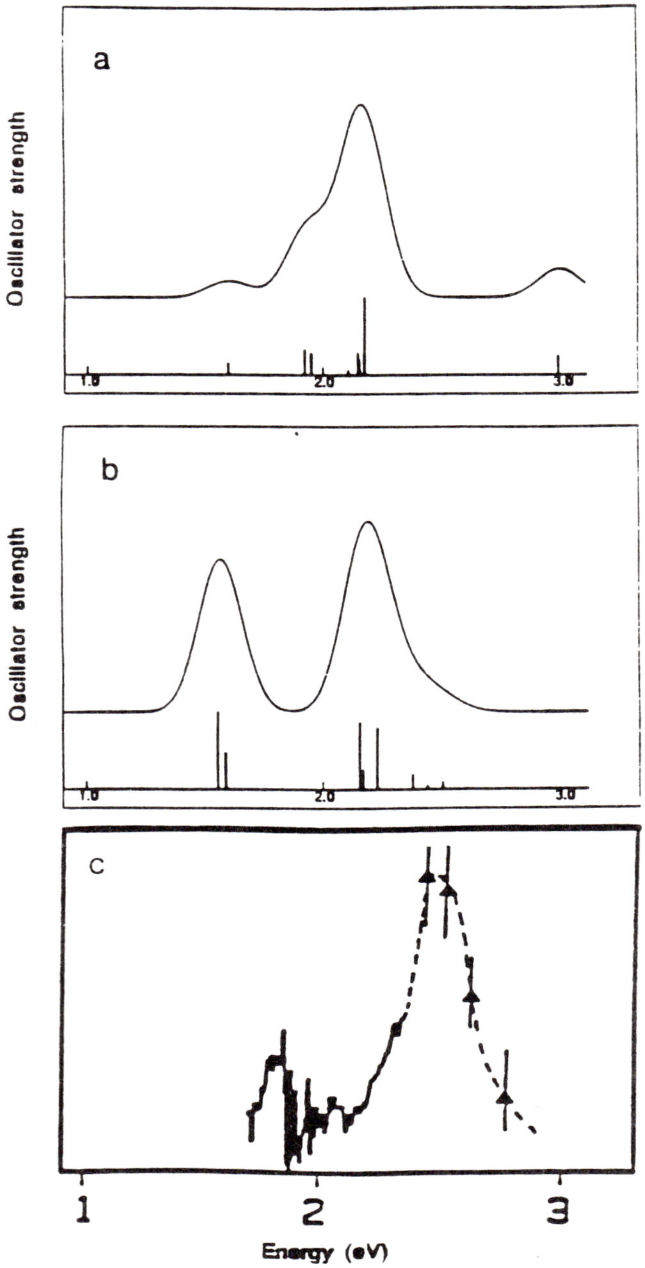

Fig. 1. a) Theoretical visible absorption spectrum: C_{2v} geometry b) Theoretical visible absorption spectrum: C_{5v} geometry c) Experimental visible absorption spectrum [2]

References

1 Koutecký J., Boustani I., Bonačić-Koutecký V., J. Quantum Chem. **38**, 149 (1990)
2 Dugourd Ph., Blanc J., Bonačić-Koutecký V., Broyer M., Chevaleyre J., Koutecký J., Pittner J., Wolf J.P., Wöste L., Phys. Lett. **67**, 2638 (1991)
3 Vosko S.H., Wilk L., Nusair M., Can. J. Phys. **58**, 1200 (1980)
4 Perdew J.P., Wang Y., Phys. Rev. **B33**, 8800 (1986)
5 Becke A.D., Phys. Rev. **A38**, 3098 (1988)
6 Perdew J.P., in Electronic Structure of Solids (1991), edited by P. Ziesche and H. Eschrig (Akademic Verlag, Berlin)
7 Perdew J.P., Phys. Rev. **B33**, 8822 (1986)
8 Lee C., Yang W., Parr R.G., Phys. Rev. **B37**, 785 (1988)

This article was processed using Springer-Verlag TEX Z.Physik A macro package 1.0 and the AMS fonts, developed by the American Mathematical Society.

Van der Waals adducts: spectroscopy and reactivity of styrenes clustered with various molecules

A. Giardini Guidoni, A. Mele, S. Piccirillo, M. Coreno, M. Snels*, R. Teghil**

Università degli Studi di Roma "La Sapienza", Dipartmento di Chimica, P. le A. Moro, 5 I-00185 Rome, Italy

Received 28 June 1993

Abstract. Mass selective one color resonant enhanced two photon ionization has been used to study spectroscopy and reactivity of small clusters of 4-fluorostyrene with several solvent molecules ($C_2H_5NH_2$, $(C_2H_5)_2NH$ and CH_3OH). It has been demonstrated that dissociative proton- (DPT) and electron transfer (DET) as well as fragmentation processes occur in small cluster ions.

In particular, it has been shown that the ion chemistry of 4-fluorostyrene- $C_2H_5NH_2$ and 4-fluorostyrene- $(C_2H_5)_2NH$ clusters is dominated by DPT and DET respectively.

PACS: 36.40.+d

1. Introduction

In the past years studies of molecular complexes of atomic compounds with reactive and nonreactive moietieshave received much attention [1-3]. The progress in this field leads to a better understanding of condensation and nucleation phenomena, energetics of inter- and intramolecular modes and kinetics of reactive processes.

Spectroscopic studies provide information on the perturbation of the electronic states of the solute induced by the solvent. In these complexes the solute solvent interactions are of various type, ranging from hydrogen bonding to dipole and dispersive forces. The evolution of the cluster absorption spectrum as a function of increasing number of solvent molecules can be related to a coalescence process leading to the condensed phase. Valuable information on the solute solvent intracluster reactions can be obtained through mass spectrometric identification of cluster ions [4,5].

Studies in our laboratory have been addressed up to now to spectroscopy of styrene (STY) and 4-fluorostyrene (4FSTY) clustered with rare gas atoms and N_2, CH_4, CCL_4 molecules [6,7]. In the wavelength spectra of these adducts a red shift of the $0^0_0 S_0 \rightarrow S_1$ electronic transition with respect to the pure solute has been observed. The presence of intracluster van der Waals (vdW) stretching and bending frequencies, together with internal vibronic modes of the solute molecule, is evident in the spectra. Model calculations for rare gas atoms [8] are in agreement with the observed vdW modes.

Recently our attention has been focused on chemically relevant solvents. Spectroscopy and reactivity studies of 4FSTY adducts with amines [$C_2H_5NH_2 = MEA$, $(C_2H_5)_2NH = DEA$] and CH_3OH are here reported. The occurrence of intracluster charge and proton transfer for some of these solvents is discussed.

2. Experimental

The experimental apparatus has been described in reference [6] and will be here only summarized. The molecular clusters are formed by supersonic expansion of He seeded with styrenes and solvent molecules. Radiation from a frequency doubled commercial dye laser crosses the skimmed beam and the clusters are ionized via a one color (1+1) R2PI process. The ions produced are mass selected and detected in a linear time of flight mass spectrometer. Wavelenght spectra are measured by positioning boxcar gates on the mass peaks of interest and recording their intensity as the wavelenght of the UV light is scanned.

3. Results and discussion

Spectroscopy. The spectra of 4FSTY clusteres with MEA, DEA and CH_3OH in the frequency range 34100 - 34900 cm^{-1} show broad absorption bands as well as some sharp peaks. In the 4-FSTY-MEA spectrum (Fig. 1a) the sharp peak, red shifted by $-37\ cm^{-1}$ with respect to the 0^0_0 electronic origin of 4FSTY, is attributed to the 0^0_0 electronic origin of this

* Instituto Materiali Speciali, C. N. R., Tito Scalo (Pz) Italy

** Università della Basilicata, Via N. Sauro, Potenza, Italy

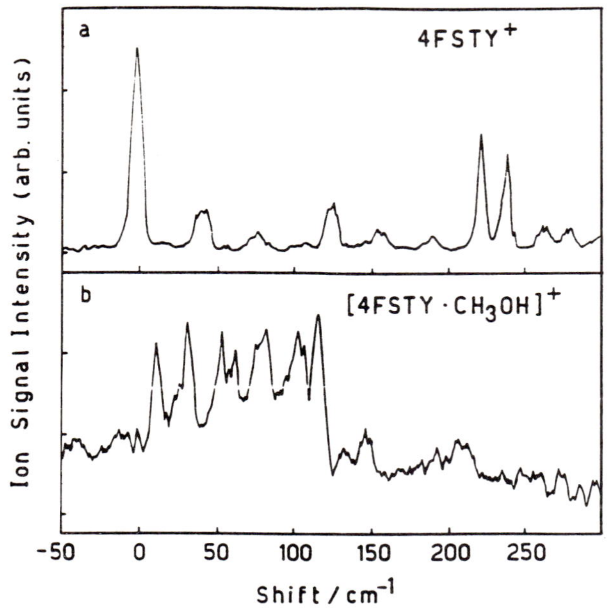

Fig. 1. Wavelength dependence spectrum of: $(4FSTY\text{-}MEA)^+$ (a), $MEAH^+$ (b), $[4FSTY\text{-}(MEA)_2]^+$ (c)

cluster. Other, blue shifted transitions are assigned to vanderWaals vibrations in the 4FSTY-MEA complex and slightly perturbed intramolecular vibrations of the 4FSTY molecule. A broad featureless band is observed in the $(4FSTY\text{-}(MEA)_2)^+$ wavelenght spectrum (Fig. 1c). The absence of sharp features in this spectrum might be explained by the high density of vanderWaals vibrational modes.

The spectrum recorded while monitoring the $(4FSTY\text{-}CH_3OH)^+$ ion signal (Fig. 2b), in the wavelenght range 34250 - 34600 cm^{-1}, shows several sharp peaks blue shifted with respect to the 4FSTY 0^0_0 electronic origin. The first intense peak is at $\Delta\nu = +11 cm^{-1}$ and can be tentatively assigned to the 0^0_0 electronic origin of the 4FSTY-CH_3OH cluster. The blue shift can be explained by a hydrogen bonding interaction of the CH_3OH with the π electron cloud of 4FSTY, which is weakened upon electronic excitation ($\pi \rightarrow \pi^*$). The assignment of the other features is complicated by the fact that fragmentation of larger $4FSTY(CH_3OH)_n$ clusters contributes to the $(4FSTY\text{-}CH_3OH)^+$ signal as has been demonstrated in [5,7].

Reactivity The REMPI mass spectra obtained from mixtures of 4FSTY and solvent molecules S (S = MEA, DEA, CH_3OH) seeded in He, are dominated by $[(4FSTY)_nS_m]^+$ ions (m=1-5, n=1-3). Weaker sequences S_m^+ and protonated S_mH^+ (m=1-5) indicate the occurrence of intracluster ion chemistry involving dissociative proton- (DPT) and electron transfer (DET) as well as fragmentation processes.

The following scheme represents several possible reactions:

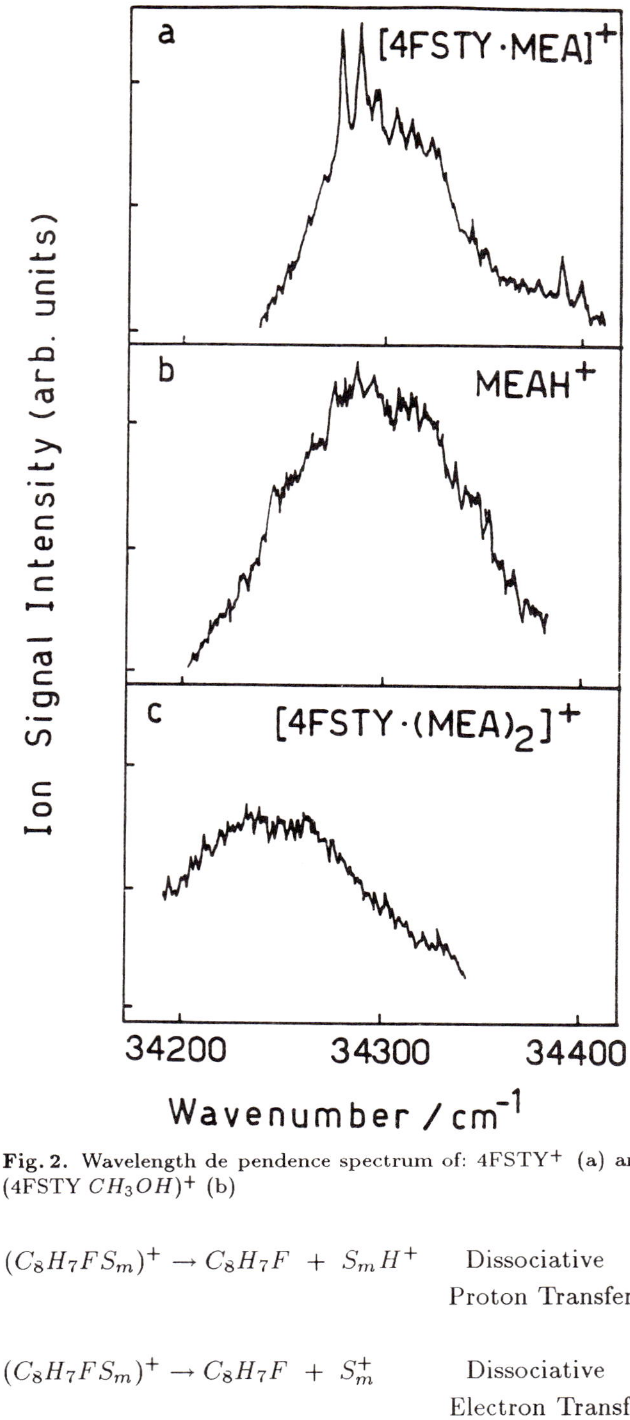

Fig. 2. Wavelength de pendence spectrum of: $4FSTY^+$ (a) and $(4FSTY\ CH_3OH)^+$ (b)

$$(C_8H_7FS_m)^+ \rightarrow C_8H_7F + S_mH^+ \quad \text{Dissociative Proton Transfer}$$

$$(C_8H_7FS_m)^+ \rightarrow C_8H_7F + S_m^+ \quad \text{Dissociative Electron Transfer}$$

In the TOF mass spectrum of 4FSTY-MEA complexes [6], very intense $(MEA)_mH^+$ peaks are present, while the intensity of $(MEA)_m^+$ ions is several orders of magnitude lower. This is in good agreement with literature data [4] which predict an exothermic (0.35 eV) DPT reaction and an endothermic (0.35 eV) DET for the 4FSTY-MEA cluster. The dominance of the DPT reaction can also be demonstrated from the wavelenght spectra, which are very similar for $[4FSTY\ MEA]^+$ and $MEAH^+$ (Fig. 1a,b). Fragmentation of larger clusters may also contribute to the $MEAH^+$ signal.

The red shoulder in the $MEAH^+$ wavelength spectrum (Fig. 1b) might be originated by the fragmentation of $[4FSTY\ (MEA)_2]^+$ (Fig. 1c) or larger clusters.

In the R2PI TOF mass spectrum of 4FSTY and DEA mixtures [6] $(DEA)_n^+$ cluster ions are at variance much more intense than the $(DEA)_nH^+$ ions. The DPT and DET reactions are both exothermic by 0.15 and 0.3 eV respectively for the $(4FSTY\text{-}DEA)^+$ ion [7]. It has been shown that the Dissociative Electron Transfer process, if energetically allowd, always dominates other competing processes [4,5]. From our mass spectra one can conclude that DET processes are predominant in the chemistry of 4STY-DEA ions.

In the mass spectrum of 4FSTY-methanol clusters, $[(4FSTY)_n(CH_3OH)_m]^+$ and $(CH_3OH)_pH^+$ ions with $p \geq 2$ are present. From thermochemical data available [9,10,11] we can reasonably assume that for the $[4FSTY(CH_3OH)_m]^+$ cluster ion, DET is endothermic for m up to 3 and DPT is exothermic from n higher than 2. These energetics constraints can explain the absence of the protonated methanol $(CH_3OH_2)^+$ ion.

4. Conclusion

A rich ion chemistry has been revealed by studing the dynamics of small mixed clusters by multiphoton ionization spectroscopy with tunable lasers. It is important to recognize that the products of intracluster reactions can be completely different from the usual gas phase ion chemistry. The structure of the molecular complexes, together with the presence of solvent molecules can account for sometimes unexpected reactions. Data here reported show that R2PI is an useful tool in understanding the chemical behaviour of ionized molecular clusters. It appears that competing reactions are possible and that their occurrence can be rationalized in terms of the energetics involved. In particular the different behaviour of the 4FSTY-MEA_n and 4FSTY-DEA_n of 4FSTY-$(CH_3OH)_n$ adducts indicates that the energetic threshold plays an important role in determinating the efficiency of a given reaction process.

References

1 Bernstein, E.R.: in "Atomic and molecular clusters", ed. Bernstein, E.R., Elsevier 1990, p.551

2 Even, U., Ben Hovin, N., and Jortner, J., Chem. Phys. Lett. **150**, 138 (1990)

3 Leutwyler, S., and Bosiger, J.: Chem. Rev. **90**, 488 (1990)

4 Brutchy, B., Eggert, J., Janes, C., and Baumgartel, H.: J. Phys. Chem. **95**, 5041 (1991)

5 Keesee, R.G., and Castleman, A.W.: in "Atomic and molecular clusters", ed. Bernstein, E.R., Elsevier 1990, p.507

6 Giardini-Guidoni, A., Piccirillo, S., Coreno, M., Snels, M., Morone, A., and Teghil, R.: J. Appl. Surf. Sci. **69**, 340 (1993)

7 Piccirillo, S., Coreno, M., Giardini-Guidoni, A., Pizzella, G., Snels, M., and Teghil, R.: J. Mol. Struct. **293**, 197 (1993)

8 Consalvo, D., van der Avoird, A., Piccirillo, S., Coreno, M., Giardini-Guidoni, A., Mele, A., and Snels, M.: J. Chem. Phys. december 1993

9 Lias, S.G., Liebman, J.F., and Levin, R.D.: J. Phys. Chem. Ref. Data **13**, 695 (1984)

10 Keesee, R.G., and Castleman, A.W.: J. Phys. Chem. Ref. Data **15**, 1011 (1986)

11 Brutschy, B.: J. Phys. Chem. **94**, 8637 (1990)

This article was processed using Springer-Verlag TEX Z.Physik D macro package 1.0 and the AMS fonts, developed by the American Mathematical Society.

Ground state systematics of spheroidally deformed sodium clusters in the jellium model

[a]T. Hirschmann, [a]M. Brack, [b]J. Meyer

[a]Institut für Theoretische Physik der Universität, D-93040 Regensburg, Germany; [b]Institut de Physique Nucléaire, Université Lyon-I, F-69622, Villeurbanne, France

Received 28 June 1993

Abstract. We solve the Kohn-Sham equations self-consistently for spheroidal sodium clusters with $8 \leq Z \leq 40$. We use a smooth fermi-like jellium density to simulate the influence of the ions in the surface region and obtain similar results as Ekardt and Penzar, but slightly different regions of prolate-to-oblate transitions. We present the systematics of potential energy curves and of quadrupole and hexadecupole moments.

PACS: 36.40; 35.20.Wg; 71.45.Nt

The spheroidal jellium model of Ekardt and Penzar [1] was able to improve the systematics of simple metal clusters. They used the standard jellium model with a rectangular jellium distribution. However, modifications of the shape or the radial surface profile of the jellium density has a non-negligible effect on the total Kohn-Sham potential. It was shown by Rubio et al [2] that a surface diffuseness of the jellium density of about one a.u. minimizes the total energy of spherical clusters. Although no microscopic legitimation can be given to this procedure, it improves static polarizabilities, the positions of dipole resonances, ionization potentials and electron affinities of spherical clusters [2,3]. We therefore employ a simple Fermi-function type parametrization of the jellium density ρ_j with a diffuseness $\alpha = 1.0\ a.u.$:

$$\rho_j(z,\varrho) = \frac{\rho_0}{1+\exp\left[\left(\sqrt{z^2q^{-4/3}+\varrho^2q^{2/3}}-R_0\right)\alpha^{-1}\right]} \quad (1)$$

where z, ϱ are cylindrical coordinates and q is the axis ratio of the spheroid. The parameter δ used by Ekardt and Penzar [1] and in the figures below is related (in lowest order of δ) to q by $q = 1 + \delta$. R_0 is fixed to the number of ions. By construction, the volume within the half-density surface is approximately constant. We solve the Kohn-Sham (KS) equations by expanding the wavefunctions in an optimized harmonic-oscillator basis.

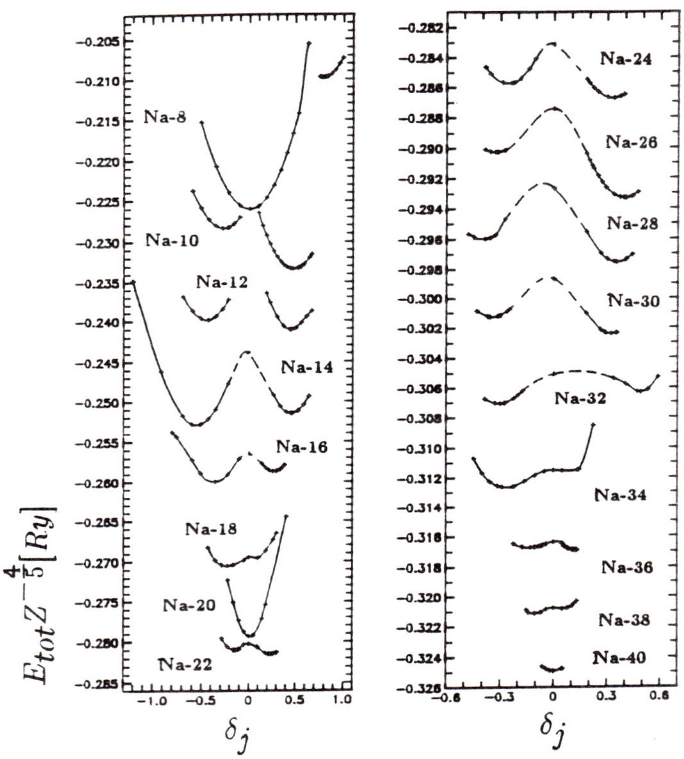

Fig. 1. Deformation energies $E_{tot}(\delta)$ for sodium clusters in scaled units of Ry $Z^{-\frac{4}{5}}$

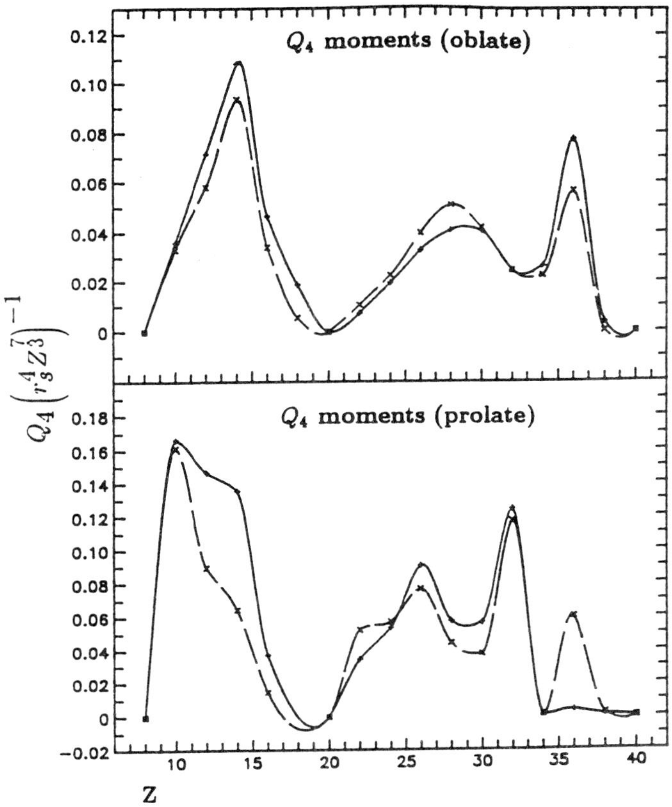

Fig. 2. Hexadecupole moments of electrons $Q_4^{(e)}$ (continuous lines) and of jellium $Q_4^{(j)}$ (dashed lines) for both oblate isomers (upper figure) and prolate isomers (lower figure). The scaled units are $r_s Z^{-\frac{7}{3}}$

In accordance with the Born-Oppenheimer hypothesis, the total cluster energy is minimized with respect to the electronic degrees of freedom for a fixed shape (i.e. deformation δ) of the jellium density. Varying δ allows to obtain the potential energy surfaces (PES) discussed below.

In Fig. 1 we show deformation energies (PES) for sodium clusters with $8 \leq Z \leq 40$. Na-8, Na-20 and Na-40 reveal the well-known shell closings. The deformation strength $|\delta|$ is maximum for the half-filled 1d and 1f-shells, respectively. Oblate-to-prolate transitions occur between Na-12/14 and Na-30/32, respectively. The small difference in energy, however, makes it difficult to judge the particle number at which the transitions take place, since the isomeric energy difference slightly depends on details of the jellium density. Na-36 slightly favours a prolate deformation with two oblate neighbours. From the near-degeneracy we conclude that Na-36 would be triaxially deformed. The situation is similar to the case of Na-16, which also follows a spheroidal shell closing. Na-16 has been found triaxial , with a negative quadrupole moment [4].

We emphasize that the ground-state deformations found from the curves in Fig. 1 agree with those of Ekard and Penzar [1] within the accuracy of their determination. This shows that the diffuseness of the jellium density has a very little effect on the ground-state deformation systematics discussed here

The systematics of oblate and prolate ground-state shapes which we find is qualitatively well confirmed by the experimental analysis of the dipole resonances in the photoabsorption cross sections measured by the Copenhagen group [3]. Since in our calculations there are no restrictions on the electronic degrees of freedom except those which arise from the built-in symmetries of the Hamiltonian, we can obtain a criterion for the quality of our jellium deformation by comparing the electronic quadrupole and hexadecupole moments $Q_2^{(e)}, Q_4^{(e)}$ with those of the jellium density $Q_2^{(j)}, Q_4^{(j)}$. We find only slight differences of less than 10% between $Q_2^{(e)}, Q_2^{(j)}$, but the hexadecupole moments differ quite appreciably in some regions, as seen in Fig. 2 where we plot $Q_4^{(e)}, Q_4^{(j)}$ versus Z, both for the oblate and the prolate minima. The explicit variation of higher multipolarities of the clusters leads to a pronounced isomerism. Hereby the jellium model can be taken to reflect the geometry of the ionic distribution through their lowest multipole moments and thus is partially reconciled with the more elaborate MD or ab initio methods.

Acknowledgement. Extensive discussions with P.-G. Reinhard, S. M. Reimann and M.Seidl have been very valuable. This work has been partially supported by the Commission of the European Communities.

1 Ekardt, W., Penzar, Z.: Phys. Rev. B **38**, 4273 (1988) and Penzar, Z., Ekardt, W.: Z. Phys. D–Atoms, Molecules and Clusters **19**, 69 (1990)

2 Rubio, A., Balbás, L. C., Alonso, J. A., Z. Phys D **19**, 93 (1991)

3 Reinhard, P.-G., Weisgerber, S., Genzken, O., Brack, M., in: *Nuclear Physics Concepts in the Study of Atomic Cluster Physics* (Springer, Berlin, 1992), p. 254; and Genzken, O., Brack, M., Bresele, H., Villing, G., Reinhard, P.-G.: Preprint Regensburg Univ. (1993)

4 Reinhard, P.-G.: (private communication, 1992) using the code of Lauritsch, G., Reinhard, P.-G., Meyer, J., Brack, M., Phys. Lett. A **160**, 179 (1991)

5 Borggreen, J., Chowdhury, P., Kabaïli, Lundsberg-Nielsen, L., Lützenkirchen, K., Nielsen, M. B., Pedersen, J., Rasmussen, H. D.: NBI preprint (submitted to Phys. Rev. B); Pedersen, J. et al, these Proceedings

This article was processed using Springer-Verlag TEX Z.Physik D macro package 1.0 and the AMS fonts, developed by the American Mathematical Society.

Parametrization of the effective potential in sodium clusters

B.A. Kotsos and M.E. Grypeos

Department of Theoretical Physics, Aristotle University of Thessaloniki, GR 54006, Greece

Received 28 June 1993

Abstract. The effective radial electronic potentials for neutral sodium clusters determined by the local density approximation and the jellium model are parametrized by means of (symmetrized) Woods-Saxon and "Wine-Bottle" symmetrized Woods-Saxon potentials. The potential parameters are determined by various least-squares fitting procedures. Particular attention is paid to the dependence of the radius parameter R on the particle number N and it is realized that for relatively small values of N, complex expressions of R as a function of N, are more appropriate than the standard one $R = r_0 N^{1/3}$. It is also found that improved results in these cases are obtained with an expression, of the form $R = r_0 N^{1/3} + b$, which is still very simple.

PACS: 36.40.+d

1. Introduction

The knowledge of the mean field potentials for electrons in metal clusters is needed in various calculations, as for example, in the case of single-particle level densities, electronic binding energies etc [1,2]. In practice, it is sometimes convenient to use parametrized forms of these potentials, as in ref. [2] where the shell and supershell structure was studied for Na clusters.

In this paper we investigate in some detail a parametrization procedure of the effective radial electronic potential $V_{eff}(r)$ in the spherical and homogeneous jellium model as this is determined using the local-density approximation. The density functional theory of Hohenberg, Kohn and Sham [3], has been used repeatedly in studies of metallic clusters, as for example, in refs. [4-8]. In our investigation, we consider the effective potentials of neutral sodium clusters determined systematically by Ekardt [7] in his spherical jellium background model (SJBM) study of work function of metal clusters, of the self-consistent determination of the charge density and self-consistent effective one-particle potential.

Ekardt's potentials have been parametrized in ref. [2] by a spherical Woods-Saxon (W-S) potential

$$V_{WS}(r) = -\frac{V_0}{1 + e^{(r-R)/\alpha}}, \quad 0 \leq r \leq \infty \tag{1}$$

with $V_0 = 6eV, R = r_0 N^{1/3}, r_0 = 2.25\mathring{A}$ and $\alpha = 0.74\mathring{A}$. Nishioka et al. [2] have also considered another type of potential obtained in ref. [9] through a semiclassical approximation, applying the Khon-Sham density functional method to a positive jellium background. The inner part of this potential is slightly shallower and the outer part is deeper than the Woods-Saxon potential. This potential of "wine-bottle" shape has also been used in ref. [2] for the study of supershells.

The aim of this paper is to consider Ekardt's effective potentials for neutral sodium clusters and to discuss a parametrization procedure, by means of potentials of symmetrized Woods-Saxon shape and of "wine-bottle symmetrized Woods-Saxon" shape and determine their parameters through least squares fits. Emphasis is given, in particular, to the dependence of the potential radius R on the particle number N of the cluster. These potentials are given and discussed in section 2 and 3, respectively, while in the last section numerical results are given and commented upon.

2. The symmetrized Woods-Saxon potential

The symmetrized Woods-Saxon potential we are using is the following:

$$V_{SF}(r) = -V_0 f_{SF}(r) = -V_0 \frac{sinh(R/\alpha)}{cosh(r/\alpha) + cosh(R/\alpha)} =$$

$$= -V_0[[1 + exp\frac{(r-R)}{\alpha}]^{-1} + [1+exp\frac{(-r-R)}{\alpha}]^{-1} - 1], \quad 0 \leq r \leq \infty \tag{2}$$

The form factor of this potential has the shape of the symmetrized Fermi distribution which has been used to describe succesfully the densities of nuclei for a wide range of mass numbers [10]. The potential (2) has also been recently used in analyses of hypernuclei [11] and also as a nuclear single particle potential [12]. It is very closely related to the so-called "cosh" potential used in ref. [13] as a cluster-core potential in Nuclear Physics.

The potential (2), which unlike the Woods-Saxon potential has zero slope at the origin, resembles somehow the Gaussian one for small values of R/α, while for $R/\alpha \gg 1$ becomes very close to the Woods-Saxon one.

One feature of the potential (2) is that its volume integral is given exactly by the following simple analytic expression:

$$|4\pi \int_0^\infty V_{SF}(r) r^2 dr| = \frac{4\pi}{3} V_0 R^3 [1 + (\frac{\pi\alpha}{R})^2] \qquad (3)$$

The same holds for its m.s.radius:

$$< r^2 >_{SF} = \frac{3}{5} R^2 [1 + \frac{7}{3} (\frac{\pi\alpha}{R})^2] \qquad (4)$$

More generally, the n-th moment of the potential may be given analytically [12]. In the case of the Woods-Saxon potential there are additional exponential terms, and thus its expression becomes more complex.

Since the condition $R/\alpha \gg 1$ is well satisfied, both for nuclei and atomic clusters, except for the very light ones, the symmetrized Woods-Saxon potential gives almost identical results to the Woods-Saxon's one, even for small values of the mass number and particle number N, respectively. The present analysis corroborates also this fact.

3. The "wine-bottle" symmetrized Woods-Saxon potential

Another parametrization of the effective potential, which seems to be rather interesting, is the following one which is of "wine-bottle" shape

$$V_{WB}(r) = -V_0 (1 + \frac{\omega r^2}{R^2}) \frac{sinh(R/\alpha)}{cosh(r/\alpha) + cosh(R/\alpha)}, \quad 0 \le r \le \infty \qquad (5)$$

for suitable choices of the parameter ω.

The volume integral of the potential (5) may be obtained analytically though its expression is a little more complex than that of the potential (2). The result of integration leads to:

$$|4\pi \int_0^\infty V_{WB}(r) r^2 dr| = \frac{4\pi V_0}{3} R^3 [(1 + \frac{3\omega}{5}) + (1 + 2\omega)(\frac{\pi\alpha}{R})^2 + \frac{7}{5}\omega(\frac{\pi\alpha}{R})^4] \qquad (6)$$

It is seen immediately that for $\omega = 0$ this expression goes over to (3), as it should be the case. Analytic expressions for the various moments of the potential may be also derived.

We note that the potentials (2) or (5) should be also able to roughly reproduce the effective potential in the inhomogeneous jellium model [14] and the potential (5) of the modified jellium potential of ref. [15]. We also note that the substitution of a polynomial of a higher degree (or a suitable function of r) for $(1 + \omega \frac{r^2}{R^2})$ in (5), is likely to reproduce the wavy character of the local equivalent potential of ref. [16] (see fig. 2, of this reference) for individual clusters. Finally, we may point out that it would be of interest to consider also the possibility of parametrizations of the form (2) or (5) of the average potential for the atom in the cluster, in the spirit of ref. [17].

4. Numerical results and discussion

In this section we give some of the results of our numerical calculations. We considered Ekardt's effective local mean potentials for the valence electrons in neutral sodium clusters with particle numbers N = 8, 18, 20, 34, 40, 58, 68, 90, 106, 132, 138, 168, 186, 198 and we performed "an overall least squares fit" of the symmetrized Woods-Saxon potential to the potential values of the above clusters by treating V_0, α and the parameter r_0 in $R = r_0 N^{1/3}$ as adjustable parameters. Depending on the size of each cluster, about 15-30 potential values, covering the interval where the potential differs practically from zero have been estimated from the corresponding figures of ref. [7]. The best fit values are: $V_0 = 6.05eV$, $r_0 = 2.34\AA$, $\alpha = 0.79\AA$.

The same procedure was repeated by adopting as R a fairly complex function of N which results from the observation that, to a good approximation, the volume integral of the potential, for each cluster varies with N as cN+d where c and d are constants. In this case (in which there is an additional adjustable parameter) a marked improvement in the quality of the fitting, compared to the one in the previous case, is observed. A very similar quality in the fitting is obtained if R is taken to be of the form

$$R = r_0 N^{1/3} + b \qquad (7)$$

The best fit values in the latter case are: $V_0 = 6.03eV$, $r_0 = 2.10\AA$, $b = 1.09\AA$, $\alpha = 0.78\AA$.

Another least squares fitting procedure we attempted was to consider the average values of the potential depths and the diffuseness parameter α ($V_0 = 6.05eV$, $\alpha = 0.79\AA$) and determine the parameters in R by least suares fitting of the expression of $< r^2 >_{SF}$ to the corresponding values obtained by fitting of the symmetrized W-S potential to the values of $V_{eff}(r)$ for individual clusters. The best fit values in this case are: $r_0 = 2.16\AA$, $b = 0.795\AA$.

It should be noted that the best fit values depend on the clusters considered. Thus, if among the values of N only those with N$\geq$ 90 are included in the fitting,

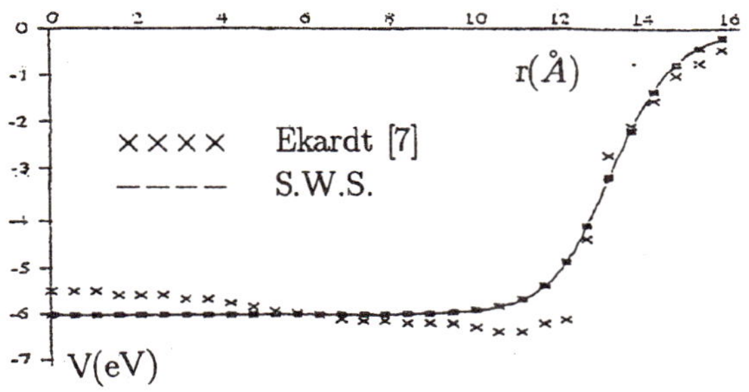

Fig. 1. The $V_{eff}(r)$ of Ekardt [7] for N=198 (denoted by (×) points) and the corresponding symmetrized Woods-Saxon one (solid line) plotted with the best fit values of the "overall fitting" and $R = r_0 N^{1/3} + b$.

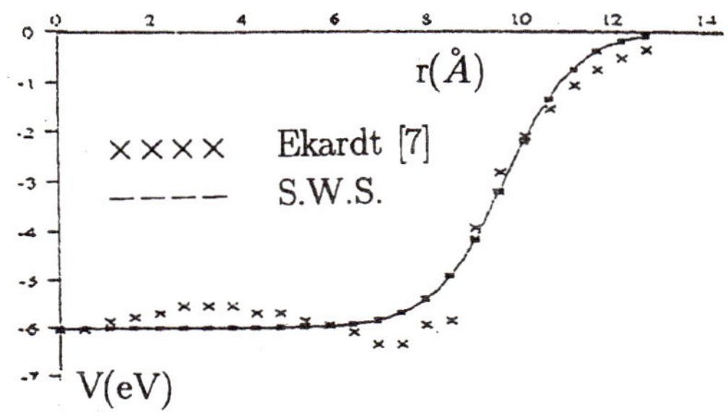

Fig. 2. The $V_{eff}(r)$ of Ekardt [7] for N=68 (denoted by (×) points) and the corresponding symmetrized Woods-Saxon one (solid line) plotted with the best fit values of the "overall fitting" and $R = r_0 N^{1/3} + b$.

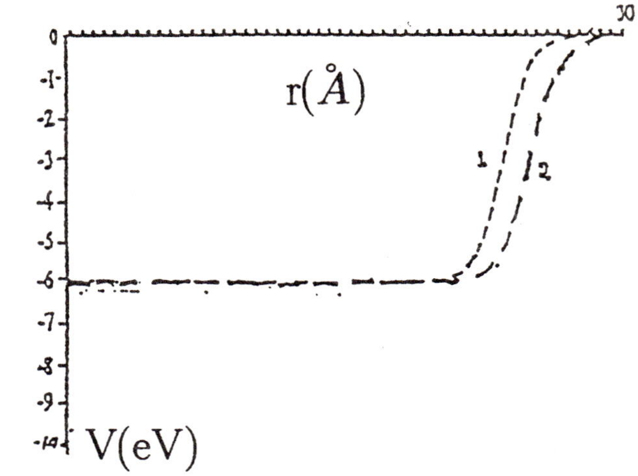

Fig. 3. The Woods-Saxon potential (No 1) with the parameters of ref. [2] for the cluster with N=1000 and the corresponding symmetrized Woods-Saxon potential of this work (No 2) using the expression $R = r_0 N^{1/3} + b$ and the best fit values: $V_0 = 6.014eV$, $r_0 = 2.28\AA$ and $b = 0.114\AA$, $\alpha = 0.87\AA$.

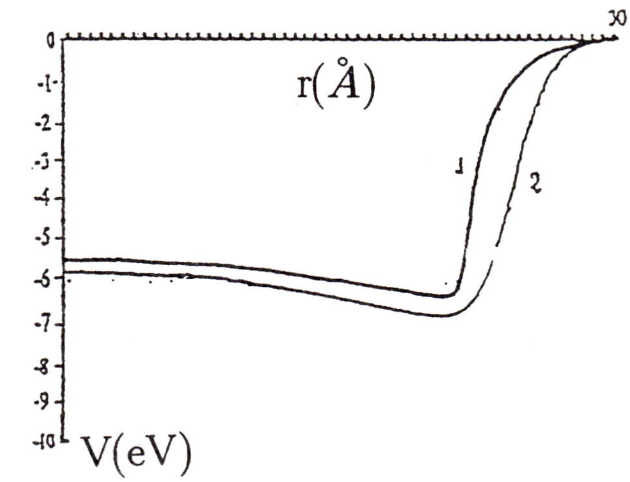

Fig. 4. The wine-bottle-potential (No 1) with N=1000 obtained with the method of ref. [10] and used in ref. [2] and the "Wine-bottle" symmetrized Woods-Saxon potential (No 2) of this work using the expression $R = r_0 N^{1/3} + b$ and the best fit values: $V_0 = 6.03eV$, $r_0 = 1.97\AA$ and $b = 1.64\AA$, $\alpha = 0.86\AA$, $\omega = 0.102$.

the best fit values in the first of the above mentioned fitting procedures with expression (7) for R become: $V_0 = 6.01eV$, $r_0 = 2.28\AA$, $b = 0.114\AA$ and $\alpha = 0.87\AA$. The improvement in the quality of fitting, compared to the one in which the expression $R = r_0 N^{1/3}$ is used, becomes now much smaller.

Least-squares fits analogous to the previously mentioned ones have also been carried out with the wine-bottle potential and the best fit values were obtained. For example in the case of N$\geq$ 90, mentioned just previously we find $V_0 = 6.03eV$, $r_0 = 1.97\AA$, $b = 1.64\AA$, $\alpha = 0.864\AA$, $\omega = 0.102$.

From our analysis we have realized that for some clusters the fitting of $V_{eff}(r)$ by a (symmetrized) Woods-Saxon potential is fairly satisfactory (see e.g. fig. 1), while for others, this is not the case (see fig. 2). All these figures were obtained with the above mentioned "overall fitting" and expression (7) for R. In certain cases the fitting is improved if the "Wine-bottle" symmetrized Woods-Saxon potential is used. Different choices of ω for the individual clusters may, however, be necessary in many cases.

In fig. 3 and 4 we have plotted the W-S potential and the wine-bottle potential (denoted by 1 in each figure) of ref. [10] used by Nishioka et al. [2] (see their fig. 1) together with the symmetrized Woods-Saxon and Wine-Bottle symmetrized Woods-Saxon potential, respectively, obtained in this work. It is seen that there is a marked difference in the surface region. Finally in fig. 5 the variation of the radius of the (symmetrized) Woods-Saxon potential with respect to the $N^{1/3}$ is shown. The best fit values used are those of the above mentioned "overall least squares fitting" and are given after expression (7). The functional dependence of R on $N^{1/3}$ is apparent.

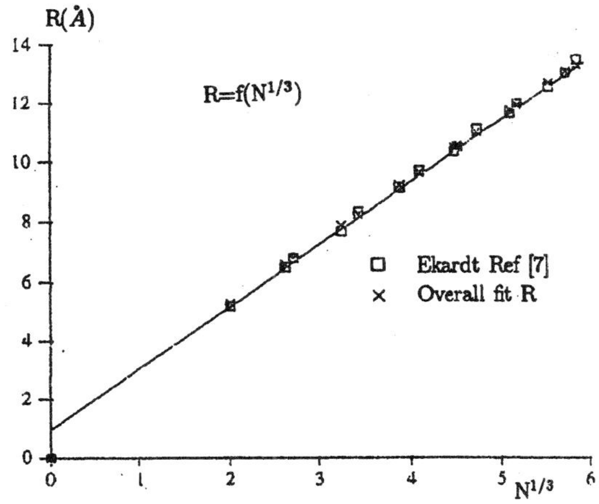

Fig. 5. The variation of the radius R of the Symmetrized Woods-Saxon potential with respect to cubic root of the particle number N. The values of R obtained by performing least squares fitting of $V_{SF}(r)$ to the $V_{eff}(r)$ of Ekardt for individual clusters are denoted by □, while those values obtained by an "overall least squares fitting" using the expression $R = r_0 N^{1/3} + b$ are denoted by ×.

We are indebted to Drs G.S. Anagnostatos and T.P. Martins for useful discussions or correspondence.

References

1 He Heer, A., Knight, W.D., Chou, M.Y., and Cohen, M.L.: in Solid State Phys., ed. by Ehrenreich, H., Seitz, F., and Turnbull, D., (Acad., N.Y. 1987) Vol 40, p.93

2 Nishioka, H., Hansen, K., and Mottelson, B.R.: Phys. Rev. B **42**, 9377 (1990)

3 Hohenberg. P., and Kohn, W.: Phys. Rev. B **136**, 864 (1964); Kohn, W., and Sham, L.J.: Phys. Rev. A **140**, 1133 (1965); Gummarson, O., and Lundgvist, B.I.: Phys. Rev. B **13**, 4274 (1976)

4 Martins, J.L., Car, R., and Buttet, J.: Surf. Sci. **106**, 265 (1981)

5 Snider, D.R., and Sorbello, R.S.: Solid State Communications **47**, 385 (1983)

6 Beck, D.E.: Solid State Commun. **49**, 381 (1984)

7 Ekardt, W.: Phys. Rev. B **29**, 1558 (1984)

8 Chou, M.Y., Cleland, A., and Cohen, M.L.: Solid State Commun. **52**, 645 (1984)

9 Brack, M.: Phys. Rev. B **39**, 3533 (1989)

10 Eldyshev, Yu V., Lukyanov, V.,and Pol, Yu S.: Sov. J. Nucl. Phys. **16**, 282 (1973); Burov, V.V., Eldyshev, Yu V., Lukyanov, V.,and Pol, Yu S.: JINR preprint E4, 8029 Dubna (1974)

11 Lalazissis, G.A., Grypeos, M.E., and Massen, S.E.: J. Phys. G Nucl. Part. Phys. **15**, 303 (1989)

12 Grypeos, M.E., Lalazissis, G.A., Massen, S.E., and Panos, C.P.: J. Phys. G Nucl. Part. Phys. **17**, 1093 (1991)

13 Buck, B., and Pilt, A.: Nucl. Phys. A **280**, 133 (1977)

14 Lange, T., Göhlich, H., Bergmann, T., and Martin, T.P.: Z. Phys. D, Atoms, Molecules and Clusters **19**, 113 (1991); Lange, T.: Doctoral Thesis (Stuttgart, 1992)

15 Yannouleas, C., and Broglia, R.A.: Europhysics Letters **15**, 843 (1991)

16 Puente, A., Casa, M., and Krivine, H.: Exchange energy: A non-local self-consistent quantal and semiclassical approach "Condensed matter theory", Vol. 8 (in press)

17 Anagnostatos, G.S.: Z. Phys. D, Atoms Molecules and Clusters **19**, 121 (1991)

This article was processed using Springer-Verlag TEX Z.Physik D macro package 1.0 and the AMS fonts, developed by the American Mathematical Society.

Z. Phys. A 349, 335–336 (1994)

Elastic $\alpha+{}^{28}$Si scattering – thick target measurements and results

P. Manngård

Department of Physics, Åbo Akademi University, SF-20500 Turku, Finland

Received: 27 October 1993

Abstract. Excitation functions for elastic α-particle scattering from a thick target have been measured. The formalism developed for elemental analysis has been adopted to calculate cross sections. Results from $\alpha+{}^{28}$Si elastic scattering, $E_\alpha=6.5$–19 MeV, are presented. Spins have been deduced from angular distributions and are shown to form a rotational band with odd and even parities. This result is compared to an old GCM phase shift prediction taken from literature.

PACS: 25.55.Ci

1. Introduction

At low energies excitation functions are measured by changing, in small steps, the energy of the particles bombarding a thin target. At higher energies the cross section decreases and the measuring time must be increased. Also, the energy interval to be covered is usually very large as compared to the small energy step. This is particularly true when α-particle scattering is used to investigate a presumed α-cluster structure. We have developed a fast method to obtain the excitation function by direct measurement of particles scattered from a thick target. Details of the method are described elsewhere [1] and only the main concepts are presented in this paper.

2. Outline of the thick target method

In material analysis backscattering is used to determine the elemental composition (the number of target nuclei) of a sample using a *known* interaction (Coulomb). Our method uses the same formalism [2], but the number of target nuclei is known and the unknown parameter is the cross section. In a thin target experiment with a well-defined beam energy E_0 the measured spectrum displays a peak at energy $K\cdot E_0$, where K is the kinematic factor. When a thick target is used, the particle looses energy as it penetrates into the target, scatters backward off a nucleus at depth n and looses energy again on its way out. Thus, its final energy, as measured in the detector, depends on the depth where the scattering occured. The measured spectrum is a continuous distribution with maximum energy $K\cdot E_0$ for particles scattered at the surface and with the number of counts in each channel proportional to the cross section. The energy scale is stretched due to the energy loss on the way out. To obtain the excitation function one needs the correspondence between the energy in the detector and the energy before scattering for different depths. Since the stopping cross section is known, these values can be obtained in a step by step calculation and the measured spectrum converted to an excitation function.

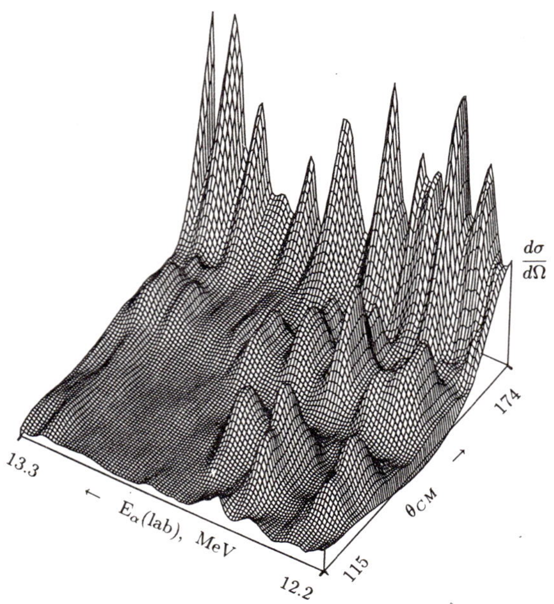

Fig. 1. Elastic scattering cross section at 21 angles larger than 107° and $E_\alpha=12.2$–13.3 MeV. Spline functions have been used for the graphics

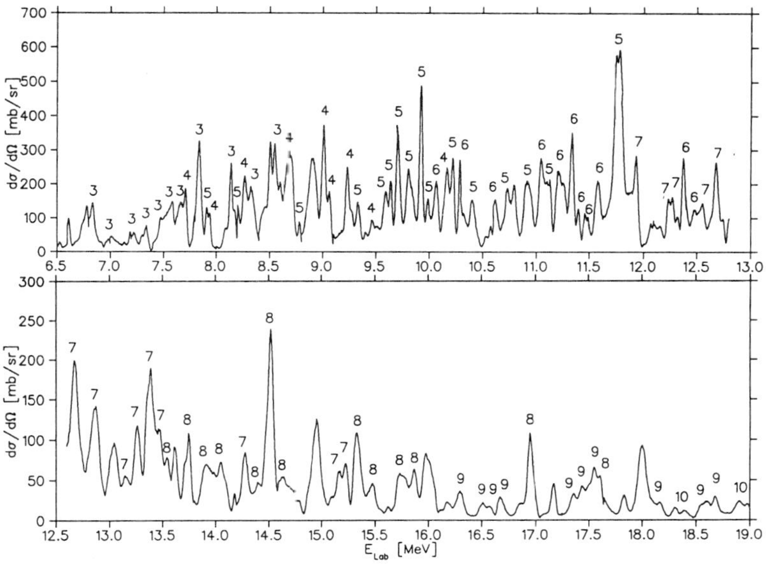

Fig. 2. The elastic scattering excitation function at 173°. Numbers denote the l-values as obtained from angular distributions fitted by squared Legendre polynomials

3. Measurements and results

The thick target measurements were carried out at the Åbo Akademi Accelerator Laboratory ($E_\alpha = 12.5 - 19$ MeV) with the energy step 400 keV and at the The Svedberg tandemaccelerator in Uppsala, Sweden ($E_\alpha = 6.5$–12.7 MeV) with the energy step 300 keV down to $E_\alpha = 8.8$ MeV and 200 keV for the lower energies. The step size is limited because inelastic scattering (at the surface) to the first excited state overlaps with elastic scattering at a certain depth and due to energy straggling affecting the resolution [1]. The setup consisted of 7 Si surface barrier detectors spaced 10° apart, which then were rotated twice by 3°. Thus, we obtained 21 pieces of excitation functions covering the step length for each beam energy. After joining these pieces together the excitation functions were merged into $E \times \theta$-matrices. Such a matrix is shown in Fig. 1. Note that, it is possible to obtain angular distributions in arbitrary energy intervals from this kind of data. This procedure was performed for the structures seen in the excitation function at 173° shown in Fig. 2. The resulting angular distributions were fitted by squared Legendre polynomials to obtain the dominating partial wave (numbers close to the peaks in Fig. 2). Each partial wave is bunched into a certain energy interval with l-values increasing with α-energy or excitation energy in ^{32}S. For each l-value a mean excitation energy is obtained using the peak area as a weight factor. In Fig. 3 we have plotted this result versus $l(l+1)$ together with a GCM phase shift prediction by Langanke et al. [3]. Except for the predicted odd-even splitting the agreement is fairly good. For the moment of inertia we obtain $4.85 \cdot \hbar^2 (\text{MeV})^{-1}$, corresponding to a radius 7.6 fm for a rigid $\alpha + {}^{28}Si$ rotator. This is roughly the location of the top of the effective potential barrier.

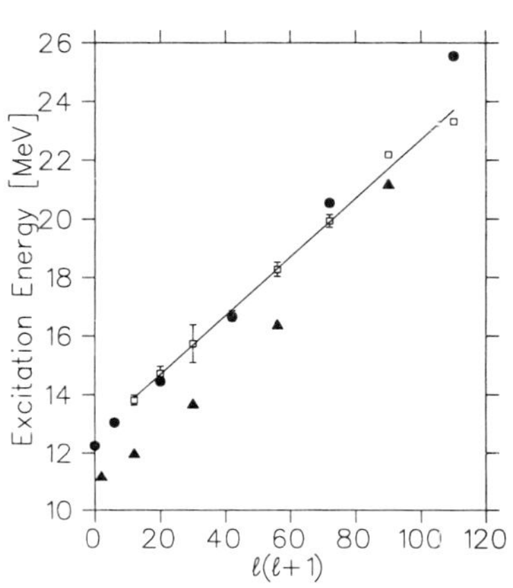

Fig. 3. *Open squares* with error bars represent excitation energy in ^{32}S of the centroids of the barrier resonances. The *solid line* is a fit to the centroids. The same for a GCM phase shift calculation [3] is given by *solid circles* (even l) and *triangles* (odd l)

4. Conclusions

The thick target method allows for fast and detailed measurements of excitation functions in fairly large energy steps. The method is ideally suited for the energy interval $E_\alpha \sim 15$–30 MeV that has so far been rather poorly covered. Such detailed data would facilitate the extension of the optical model description towards lower energies. In the case of ^{28}Si, the method has been used to identify the molecular-type barrier resonances predicted by Langanke et al.. The reason(s) for the fragmentation of these resonances still remains to be clarified.

The author is indepted to his co-workers M. Brenner, V. Goldberg, T. Lönnroth, A. Pakhomov, V. Pankratov and K.-M. Källman in particular for preparing the graphs. Support from the Magnus Ehrnrooth foundation is gratefully acknowledged.

References

1. Källman, K.-M., Goldberg, V.Z., Lönnroth, T., Manngård, P., Pakhomov, A.E., Pankratov, V.V.: Nucl. Instrum. Methods A**338** (1994) (submitted)
2. Chu, W.-K., Mayer, J.W., Nicolet, M.-A.: Backscattering spectrometry. New York: Academic Press 1978
3. Langanke, K., Stademann, R., Frekers, D.: Phys. Rev. C**29**, 40 (1984)

Z. Phys. A 349, 297–299 (1994)

Weak coupling in the upper part of the *sd*-shell and α-clustering in ^{38}Ar

I. Miyamoto[1], **S. Ohkubo**[1,2]

[1] Department of Applied Science, Kochi Women's University Kochi, 780 Japan
[2] Nuclear Physics Laboratory, University of Oxford, Oxford OX1 3RH, UK

Received: 27 October 1993

Abstract. It is suggested that weak coupling picture for the α-cluster states holds not only in the ^{44}Ti$\sim^{40}$Ca region but also in the upper part of the *sd*-shell region. To confirm this experimentally, the α-clustering aspects of ^{38}Ar is investigated.

PACS: 25.55.Ci; 21.60.Gx; 27.30.+t

Alpha-cluster picture has been successful in light nuclei up to ^{20}Ne in the *sd*-shell region. Recent observations [1] of a theoretically predicted parity doublet $K=0^-$ band [2] with an α-cluster structure both in ^{44}Ti and ^{40}Ca have shown that the α-cluster structure is important even in the *fp*-shell region. However, α-clustering aspects in the *upper part* of the *sd*-shell remain unclear. In this communication we suggest that the weak coupling picture for α-cluster states holds not only in the beginning of the *fp*-shell region of ^{44}Ti$\sim^{40}$Ca but also across the shell closure. To confirm this, attention is drawn to the importance of the observation of a well-developed α-cluster $K=0^-$ band in ^{38}Ar.

The α-hole (in the $d_{3/2}$ orbit) interaction in the *fp*-shell region estimated by using the experimental binding energies [4] is repulsive and weak at about 0.48 MeV, which supports weak coupling in the region. However, we see in Fig. 1 that weak coupling of the energy levels seems to hold not only in the beginning of the *fp*-shell but also in the upper part of the *sd*-shell across the shell closure. In ^{38}Ar not only does the rotational $K=0^+$ band built on the first excited 0^+ state [5] resemble that in ^{40}Ca, ^{42}Ca and ^{44}Ti but also the $B(E2)$ values between the states of the band are similarly very much enhanced. This $K=0^+$ band can be considered to be a $4p-6h$ intruder state, while in ^{36}Ar the 0_2^+, wich is enhanced in the α-transfer reactions [6], can be considered to be a $4p-8h$ intruder state.

Correspondence to: S. Ohkubo

One of the direct ways to confirm that weak coupling based on the α-cluster picture persists across the shell closure is to observe the α-cluster $K=0^-$ band in ^{38}Ar which is a parity-doublet partner of the $K=0^+$ band. To know how the band is enhanced in the α-transfer reactions in spite of the fragmentation [1], α-widths as well as location of the energy levels are needed.

α-particle scattering from nuclei in this region does not show a pronounced backward angle anomaly (BAA) which is typically seen in α-scattering from ^{40}Ca, because of increased absorption. This is one of the reasons why little attention has been paid to studies of α-clustering in this region. However, we notice that angular distributions at backward angles still retain a characteristic feature peculiar to the BAA. For the $\alpha+^{34}$S system the angular distribution of α-particle scattering at $E_\alpha=18$ MeV [7] shows a characteristic feature at backward angles more or less similar to $\alpha+^{36}$Ar scattering.

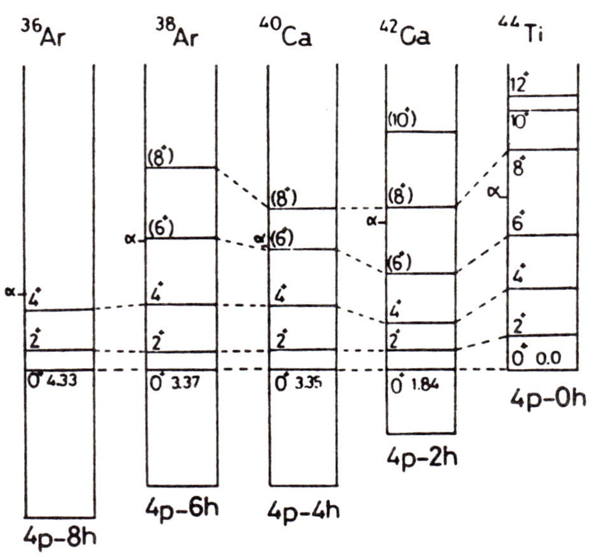

Fig. 1. Weak coupling feature of the experimental energy levels of the $K=0^+$ bands

Although no systematic experimental angular distributions at backward angles are available in a wide energy region, this does not prevent us from getting an intercluster potential for the $\alpha+{}^{34}$S system from an analysis of the angular distribution, if we start from a potential checked with nuclei in this region.

We analyse the experimental data of the lowest energy of $E_\alpha = 18$ MeV available by starting from a potential obtained for $\alpha+{}^{36}$Ar [3]. Although a model-independent potential is recommended to get a perfect agreement of the calculated angular distributions with the data, we retain the same form factor as [3], since it is sufficient to get a unified description of bound and scattering states. We searched for potential parameters which fit the experimental data biasing the experimental data at large angles by using a visual fitting procedure. During the search all the potential parametes are varied except diffuseness. The potential parameters obtained to fit the data are $V = 190.3$ MeV, $r = 1.359$ fm, $a = 1.21$ fm, $W = 7.86$ MeV, $r_i = 1.97$ fm, $a_i = 1.00$ fm and $r_{\mathrm{coul}} = 1.3$ fm: The volume integral per nucleon pair for the real part of the potential, 361 MeVfm^3, is consistent with others in this mass region.

By decomposing the scattering wave into its internal and barrier wave components based on a semiclassical theory [8], we see that the backward rise is still dominated by the internal wave in spite of increased absorption. This means that the inner region of the real part of the obtained optical potential is reliable, allowing us to discuss the structure. Using this real potential, we investigate α-clustering aspects of the $\alpha+{}^{34}$S system. We impose the Wildermuth condition to take into account the Pauli principle. Since the core ^{34}S is not a closed nucleus, the number of the forbidden states should be l-dependent. However, here we simply take $N = 2n + l \geqq 12$ based on the picture that four particles (two neutrons and two protons) in the $d_{3/2}$ orbit are elevated to the fp-shell orbit [3].

The calculated absolute binding energy with respect to the α-threshold of the band head of the first Pauli-allowed band ($N = 12$) falls within a few MeV of the head of the experimental $K = 0^+$ band. In Fig. 3 the calculated results with a potential normalized by a factor 1.024, which is chosen so that the band head 0^+ state coincides with the experimental one, correspond well with the experiment. The $K = 0^+$ band in ^{38}Ar is found to have a more developed α-cluster character than that in ^{40}Ca and ^{44}Ti. Calculated $B(E2)$ values between the states of the band, $2^+ \to 0^+$ (13 w.u.), $4^+ \to 2^+$ (17), $6^+ \to 4^+$ (16), $8^+ \to 6^+$ (13), $10^+ \to 8^+$ (8.9) and $12^+ \to 10^+$ (4.3), correspond to the experimental values [5] $4^+ \to 2^+$ (30^{+10}_{-6} w.u.), $6^+ \to 4^+$ (35^{+30}_{-14}), $8^+ \to 6^+$ ($20 \sim 36$), respectively. The fact that the present potential reproduces not only the energy levels of the $K = 0^+$ band but also the $B(E2)$ values shows that the potential is good for discussing the properties of the $K = 0^-$ band.

In Fig. 3 the $K = 0^-$ band is located about 2.5 MeV above the α-threshold. The calculated α-decay widths and corresponding dimensionless α-reduced widths at channel radius $a = 7.15$ fm are 1^- (0.21 keV, 27%), 3^- (1.89 keV, 25%), 5^- (9.53 keV, 20%), 7^- (18.87 keV, 14%), 9^- (17.16 keV, 6.7%), 11^- (7.34 keV, 2.1%), 13^- (1.17 keV, 0.3%), which are much larger than those of the ^{44}Ti. This suggests the band can be well observed in the α-transfer reactions even if it is fragmented by a coupling with shell-like states. It will confirm that the weak coupling picture persists in the upper part of the sd-shell. An experiment to observe this band (as well as the $K = 0^+$ band) in the ^{34}S(^{7}Li, t)^{38}Ar reaction is in progress [9]. Weak coupling might further persist for α-cluster states in lighter nuclei such as ^{32}S and ^{28}Si.

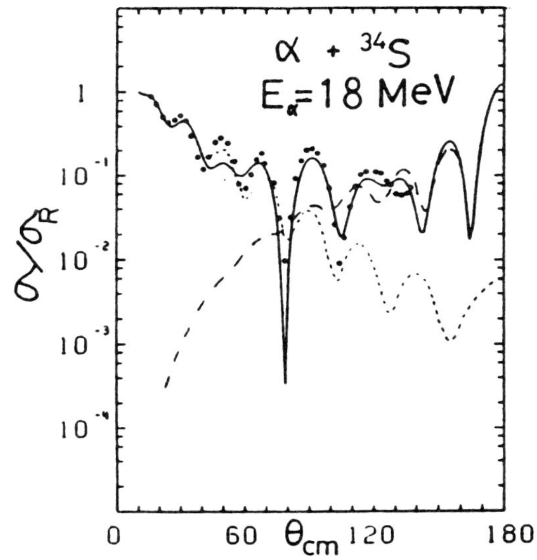

Fig. 2. Experimental angular distribution (*circles*) for $\alpha+{}^{34}$S scattering is compared with the calculated one (*solid lines*), which is decomposed into its internal (*dashed lines*) and barrier (*dotted lines*) wave components

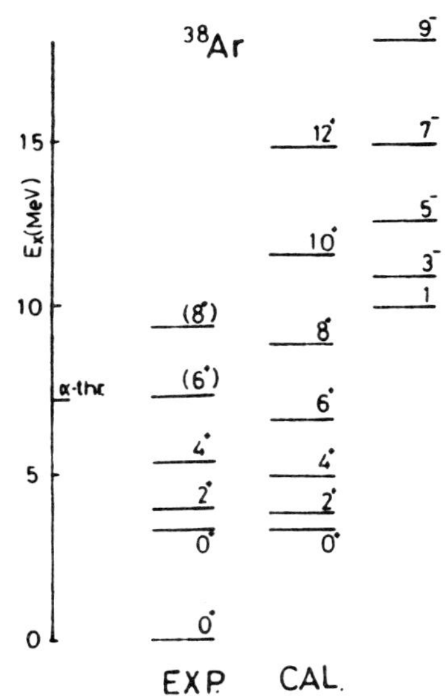

Fig. 3. Experimental and calculated energy levels in ^{38}Ar

References

1. Yamaya T. et al.: Phys. Rev. C **41**, 2421 (1990); Yamaya T. et al.: Phys. Lett. B **306**, 1 (1993)
2. Michel F., Reidemeister G., Ohkubo S.: Phys. Rev. Lett. **57**, 1215 (1986); Phys. Rev. C **37**, 292 (1988); Ohkubo S.: Phys. Rev. C **38**, 2377 (1988)
3. Ohkubo S., Umehara K.: Prog. Theor. Phys. **80**, 598 (1988); Reidemeister G., Ohkubo S., Michel F.: Phys. Rev. C **41**, 63 (1990)
4. Arima A., Horiuchi H., Sebe T.: Phys. Lett. **24 B**, 129 (1967)
5. Betz P. et al.: Z. Phys. A **271**, 195 (1974)
6. Becker A. et al.: Nucl. Phys. A **412**, 159 (1984)
7. Leighton L.G., Wolf A.C.: Nucl. Phys. A **151**, 71 (1977)
8. Brink D.M., Takigawa N.: Nucl. Phys. A **279**, 159 (1977)
9. Ogino K. et al.: Private communications

Z. Phys. A 349, 339–340 (1994)

Properties of ^{8}Be and ^{12}C deduced from the folding–potential model

P. Mohr[1], H. Abele[1], V. Kölle[1], G. Staudt[1], H. Oberhummer[2], H. Krauss[2]

[1] Physikalisches Institut, Universität Tübingen, D-72076 Tübingen, Germany
[2] Institut für Kernphysik, TU Wien, A-1040 Vienna, Austria

Received: 27 October 1993

Abtract. The α–α differential cross sections are analyzed in the optical model using a double–folded potential. With the knowledge of this potential bound and resonance–state properties of α–cluster states in ^{8}Be and ^{12}C as well as astrophysical S–factors of ^{4}He$(\alpha,\gamma)^8$Be and ^{8}Be$(\alpha,\gamma)^{12}$C are calculated. Γ_γ–widths and B(E2)–values are deduced.

PACS: 25.55.Ci; 25.70.Ef; 24.10.Ht

In order to determine reaction rates for the triple–alpha process [1], we have calculated in the potential model the α+α differential cross section as well as the bound and resonance–state properties of α–cluster states in ^{8}Be and ^{12}C. The potentials are calculated using the folding procedure [2, 3].

In this approach the nuclear densities are derived from nuclear charge distributions [4] and folded with an energy and density dependent NN–interaction $v_{\rm eff}$ [2]. By means of a normalization factor λ the depth of the potential can be adjusted to elastic scattering data and to bound and resonant state energies of nuclear cluster states. In this work the folding potentials were determined using the computer code DFOLD [5]. The imaginary part of the optical potential can be neglected at the low energies considered.

Using the folding procedure for the α+α scattering potential we fitted the experimental differential cross section for all projectile energies below the break–up threshold ($E_{\rm lab} \approx$ 7–35 MeV) using a double–folded potential with a volume integral per nucleon pair $J_R/(A_P * A_T) =$ 273.4 MeVfm^3 and a renormalization factor λ = 1.656. An excellent agreement between the calculated and experimental cross section has been found [6]. With the same potential we reproduce the resonance energies and widths of the ^{8}Be ground–state: $E_{\rm R}^{\rm th}$ = 92.1 keV and of the 2^+– and 4^+–state in ^{8}Be (see Table 1). Especially for the ground state the calculated values agree excellently with the experimental data. This result confirms that this nucleus can be well described in a two-cluster approach [7]. This marked α–cluster structure is also reflected by the values of the α–particle spectroscopic–amplitudes for these ^{8}Be–states [8]: S(g.s.)=1.48, S(2^+)=1.46, and S(4^+)=1.39.

The s–wave function for the relative motion of the two α–clusters in the ^{8}Be ground state can be folded with the nucleon densities of both α–particles to calculate the ^{8}Be density [6]. In our calculation we used an upper cut-off radius of 20 fm. Averaging over 4π we obtain the nucleon density distribution. The result is shown in Fig. 1. With the knowledge of the ^{8}Be density distribution we are able to calculate the optical potential for the α–^{8}Be system using the folding procedure again. Since no experimental phase shifts for α–^{8}Be are available, we have to fit the normalization factors λ of the folding potentials to the bound and resonance–state energies of ^{12}C.

The ground state (0_1^+), the first (2_1^+) and second (0_2^+) excited, and the (2_2^+) state in ^{12}C are those states which are of interest in the triple–alpha process (see Table 2). With respect to a $\alpha\otimes^8$Be clustering in ^{12}C, the 0_1^+ and 2_1^+ bound states belong to the ground state band with a harmonic oscillator quantum number Q=4, whereas for the 0_2^+– and 2_2^+–state we assume that they can be interpreted as members of a band with Q=6. We assume the broad 10.3 MeV level in ^{12}C to be the 2_2^+ state. The adopted spin assignment of this level is 0^+ [10]. However, an assignment of 2^+ for this state is strongly suggested by experimental results [11] as well as by microscopic calculations [12].

Using the double–folded α–^{8}Be potential we have fitted the energies of these ^{12}C states. These states have quite a different $\alpha\otimes^8$Be(g.s.) cluster structure as can be seen from the values of the spectroscopic factors: S(g.s.)=0.557 [8], S(2_1^+)=0.154 [13], and S(^{12}C, 0_2^+)=1.8 [14]. Therefore, the α–^{8}Be folding potentials ($J_R/(A_P * A_T)$ = 378.6 MeVfm^3) have to be renormalized separately to the ground and excited states in ^{12}C. The normalization factors λ for the potential obtained in this way are listed in Table 2. Using the optical model we also calculated the alpha widths Γ_α of the resonant 0_2^+ and 2_2^+ states. The good agreement between the experimental and the calculated α–width for the 0_2^+–state confirms the assumption that this state has a marked $\alpha\otimes^8$Be(g.s.)–structure.

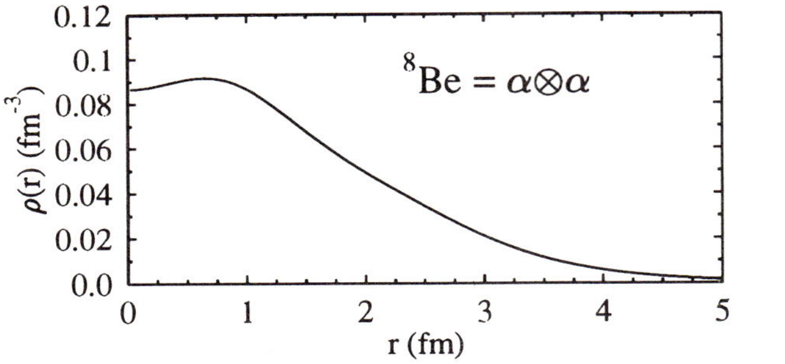

Fig. 1. Nucleon density distribution of ^{8}Be calculated in the potential model

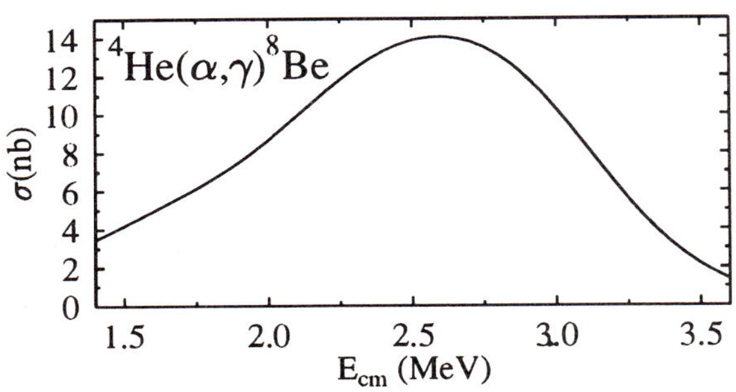

Fig. 2. Capture cross section of ^{4}He$(\alpha,\gamma_0)^{8}$Be near the 2^{+} resonance

Table 1. Relevant data of resonant states in ^{8}Be. Experimental data are taken from [9]

J^{Π}	E^{calc} [MeV]	E^{exp} [MeV]	$\Gamma_\alpha^{\text{calc}}$ [eV]	$\Gamma_\alpha^{\text{exp}}$[eV]
0^+	0.0921	0.09212(5)	6.86	6.8 ± 1.7
2^+	2.78	3.13(3)	$1.3 \cdot 10^6$	$1.5(2) \cdot 10^6$
4^+	11.02	11.5(3)	$\gtrsim 3 \cdot 10^6$	$\approx 3.5 \cdot 10^6$

Table 2. Relevant data of bound and resonant states in ^{12}C. Experimental data are taken from [10]

J^{Π}	E [MeV]	λ	$\Gamma_\alpha^{\text{calc}}$ [eV]	$\Gamma_\alpha^{\text{exp}}$[eV]
0_1^+	0.0	1.144	—	—
2_1^+	4.437	1.010	—	—
0_2^+	7.654	1.286	7.5	8.3±1.0
2_2^+	10.3	1.152	$\gtrsim 2.0 \cdot 10^6$	$(3.0\pm0.7)\cdot10^6$

Table 3. Γ_γ–widths and B(E2)–values

	E_f [MeV]	J_f^π	E_i [MeV]	J_i^π	Γ_γ[meV]	B(E2)[W.u.]
^{8}Be	0.0	0^+	3.04	2^+	8.4	75
^{12}C	4.437	2_1^+	7.654	0_2^+	4.1	8.8
	7.654	0_2^+	10.3	2_2^+	17	100

Furthermore, we calculated the cross sections and astrophysical S-factors of the reactions ^{4}He$(\alpha,\gamma_0)^{8}$Be and ^{8}Be$(\alpha,\gamma_1)^{12}$C in the direct capture model [3, 15]. These cross sections are essentially determined by the overlap of the scattering wave functions in the entrance channel, the bound state wave functions in the exit channel, and the electric quadrupole operator. The wave functions were calculated using the α–α and α–^{8}Be potential, respectively. In order to compute the reaction ^{4}He$(\alpha,\gamma_0)^{8}$Be the resonance energy of the ^{8}Be ground state was lowered by about 100 keV. This means that this state was assumed to be weakly bound. For the description of the bound states the α–particle spectroscopic factors given above are used. In Fig. 2 the cross section of the resonant E2–capture into the ^{8}Be ground state is shown. Excellent agreement is found with the result of a calculation in a bremsstrahlung model [16]. Approximating the $\sigma(E)$ cross sections near the resonance energy by a Breit-Wigner parametrization, Γ_γ–widths and B(E2)–values can be deduced. These quantities are given in Table 3.

Acknowledgement. We want to thank the the DFG (project Sta290/2), and the Austrian Science Foundation (FWF) (project P8806-PHY).

References

1 Oberhummer, H., Krauss, H., Grün, K., Rauscher, T., Abele, H., Mohr, P., Staudt, G.: Z. Phys. A (in press)
2 Kobos, A.M., Brown, B.A., Lindsay, R., Satchler, G.R.: Nucl. Phys. **A425**, 205 (1984)
3 Oberhummer, H., Staudt G.: In: Nuclei in the Cosmos. H. Oberhummer(ed.) p. 29. Berlin Heidelberg New York: Springer 1991
4 De Vries, H., Jager, C. W., de Vries, C.: At. Data Nucl. Data Tables **36**, 495 (1987)
5 Abele, H.: computer code DFOLD, Univ. Tübingen (unpublished)
6 Abele, H.: Ph.D. thesis, Univ. Tübingen 1992
7 Wildermuth, K., Tang, Y.C.: A Unified Theory of the Nucleus. Braunschweig: Vieweg & Son 1977
8 Kurath, D.: Phys. Rev. **C7**, 1390 (1973)
9 Ajzenberg–Selove, F.: Nucl. Phys. **A490**, 1 (1988)
10 Ajzenberg–Selove, F.: Nucl. Phys. **A506**, 1 (1990)
11 Jacquot, C., Satamoto, Y., Jung, M., Girardio, L.: Nucl. Phys. **A201**, 247 (1973)
12 Fujiwara, Y., et al.: Prog. Theor. Phys. Suppl. **68**, 29 (1980)
13 Kwasniewicz, E., Jarczyk, L.: J. Phys. G **12**, 697 (1986)
14 Fukushima, Y., Kamimura, N.: J. Phys. Soc. Jpn **44**, 225 (1978)
15 Mohr, P., Abele, H., Kölle V., Staudt, G., Krauss, H., Oberhummer, H., Denker A., Hammer J. W., Wolf, G.: Phys. Rev. C (in press)
16 Langanke, K.-H., Rolfs, C.: Phys. Rev. C **33**, 790 (1986)

This article was processed using Springer-Verlag TEX Z.Physik A macro package 1.0 and the AMS fonts, developed by the American Mathematical Society.

Similarities in shell structure calculation of atomic clusters and nuclei

H. Müller, Ch. Opitz and M Müller

Institut für Physikalische Chemie der Friedrich-Sciller-Universität Jena, Germany

Received 28 June 1993

Abstract. We apply simple potential models used by nuclear physicists in the early thirties to describe shell structure of large alkali metal clusters up to values of the angular momentum quantum number $l = 17$. We develop more sophisticated analytic models using suitable perturbation operators in order to calculate not only correct magic numbers MN but also proper bunching of subshells up to 1432 electrons. From the calculated aufbau principle we derive a periodic table of alkali metal clusters.

PACS: 36.40.+d; 31.90.+s; 82.70.-y

1. Introduction

In the early thirties nuclear physicists applied a number of "auxiliary potentials": spherical well (infinite walls, GAMOW 1933; finite walls, MARGENAU 1934) and spherical oscillator (HEISENBERG 1935) which introduce various shell models to explain magic numbers MN of stability for atomic nuclei.

Similar models have been used in the literature to successfully describe shell structure in metal clusters [1-9].

Very recent experimental results for various metal clusters, particularly of alkali metals Na_N [10,11], Cs_N [12,13], Li_N [14] give considerable evidence for shell structure in the electronic states up to about 1500 electrons.

We apply here the above mentioned simple models to these large alkali metal clusters and describe their electronic shell structure up to values of the angular momentum quantum number $l = 17$ (chapter 2). In chapter 3 we search for more sophisticated analytic models using correlation between the one-electron states E_{nl} of various models with and without degenerate subshells as a guide and perturbation theory as a practical technique.

2. Standard sphere models

Infinite walls : Within the one-electron states E_{nl} (infinite) ($n = 1, 2, \ldots, 7$; $l = 0, 1, \ldots, 17$) of a 1500 valence electron alkali metal cluster a possible closed subshell configuration could be found for each experimentally observed step but a corresponding grouping of subshells is observed only in some cases.

Finite walls : The comparison of E_{nl} (finite) and E_{nl} (infinite) shows clearly (i) a general depression of all levels E_{nl} (finite) but also (ii) the complete conservation of sequence and degeneration of all subshells at least up to 1500 electrons i.e., all improvements of the infinite model discussed below are also possible for the finite model.

3. Modified analytic models

3.1. Modification by perturbation

As an unperturbed potential we use the infinite sphere model.

The influence of (i) a coulombic or (ii) a harmonic perturbation respectively appears in different sequences of states, in slightly changed magic numbers MN and in improvement of bunching subshells.

A piecewise constant perturbation can be optimized with respect to its parameters in order to get (i) not only correct magic numbers MN but also (ii) a proper bunching of subshells i.e., narrow shells with large spacing in between.

A first result of optimization has been achieved with

$$\hat{V}_{opt}^{perturb}[h^2/2mR^2] = \begin{cases} -18 & 0.25 \le r[R] \le 0.40 \\ +18 & 0.75 \le r[R] \le 0.90 \\ 0 & otherwise \end{cases} \quad (1)$$

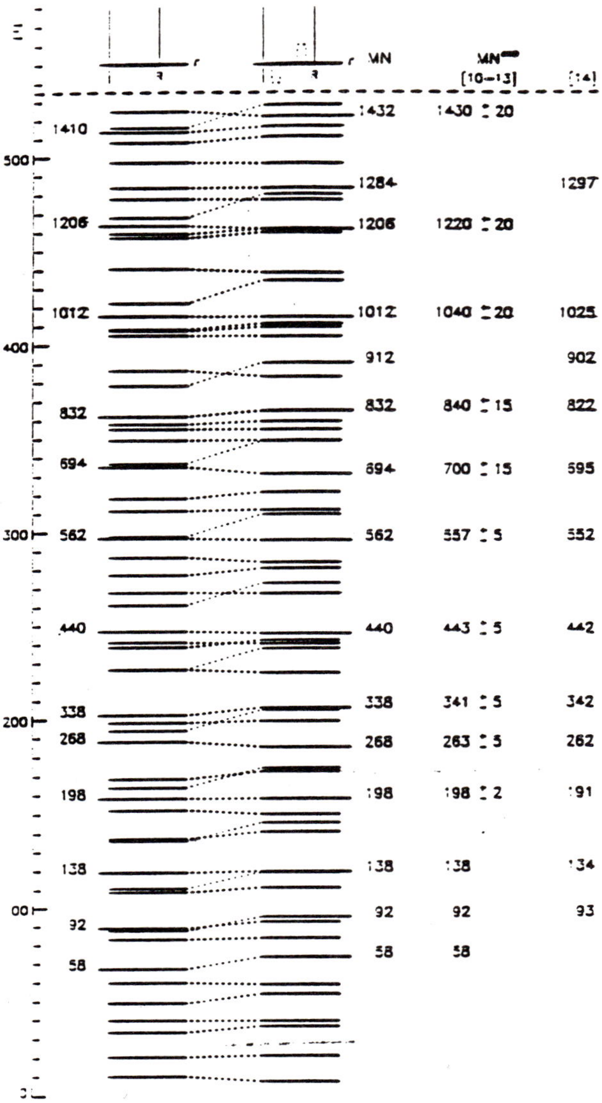

Fig. 1. One-electron states $E = E_{nl}[h^2/2mR_2]$ of a 1500-electron alkali metal cluster calculated by using an optimized piecewise constant perturbation (1) including calculated and measured magic numbers MN^{exp}. Our energy unit depends on particle size because of the well-known relation $R = r_s N^{1/3}$ (r_s is the Wigner-Seitz radius which amounts e.g. for Li, Na, and Cs 1.72 Å, 2.08 Å, and 2.98 Å). As a result the drawn shell structure is valid independently of cluster size for all alkali metal clusters. From this "universal" shell structure it is easier to conclude more general results (cf. section 3.2.).

This simple perturbation leads to a bunching of subshells which explains the observed magic numbers MN up to 1432 electrons within the experimental error (Fig. 1). In this moment, it does not seem, however, that a potential like (1) could be justified, to our present knowledge, on the basis of a many-body theoretical approach. But eq. (1) is only a first step in optimization of a piecewise constant potential

$$\hat{V}^{perturb} = \begin{cases} \Delta U_{12} & R_1 \leq r \leq R_2 \\ \Delta U_{34} & R_3 \leq r \leq R_4 \\ \vdots & \\ \Delta U_{2k-1,2k} & R_{2k-1} \leq R_{2k} \\ 0 & otherwise \end{cases} \tag{2}$$

By means of eq. (2) which is not treated fully up to now we are able to generate a wide class of potential e.g. starting from Coulomb-potential via Woods-Saxon-, wine-bottle-, soft-wall-potential and so on up to harmonic potential. This is what we call model correlation, and it would be possible that the final result allows a simple intepretation.

3.2. Periodic table of alkali metal clusters (PTC)

We have used the aufbau principle generated from our last modified analytic sphere model
$1s^2\ 1p^6\ 1d^{10}\ 2s^2\ 1f^{14}\ 2p^6\ 1g^{18}\ 2d^{10}\ 3s^2\ 1h^{22}\ 2f^{14}\ 1i^{26}\ 3p^6\ 2g^{18}\ 1j^{30}\ 3d^{10}\ 4s^2\ 2h^{22}\ 1k^{34}\ 3f^{14}\ 4p^6\ 1l^{38}\ 2i^{26}\ 3g^{18}\ 1m^{42}\ 4d^{10}\ 2j^{30}\ 5s^2\ 3h^{22}\ 1n^{46}\ 2k^{34}\ 4f^{14}\ 5p^6\ 1o^{50}\ 3i^{26}\ 2l^{38}\ 4g^{18}\ 1q^{54}\ 5d^{10}\ 6s^2\ 3j^{30}\ 2m^{42}\ 4h^{22}\ 1r^{58}\ 5f^{14}\ 3k^{34}\ 2n^{46}\ 6p^6\ 1t^{62}\ 4i^{26}\ 2o^{50}\ 3l^{38}\ 5g^{18}\ 6d^{10}\ 1u^{66}\ 7s^2\ 4j^{30}\ 2q^{54}\ 3m^{42}\ 5h^{22}\ 1v^{70}\ \ldots$
in order to construct a "periodic table of alkali metal clusters" (PTC) as shown in Fig. 2.

Subshell closing which generates magic numbers MN is given in heavy types together with the corresponding MN as an upper index within our PTC. Magic clusters A_{MN} are most frequently of (ns)-type

$$A_2, A_{20}, A_{198}, A_{440} \ldots$$

(in analogy to the alkaline earth metals, group II elements from the periodic table of elements PTE) and of (np)-type

$$A_8, A_{40}, A_{138}, A_{562} A_{1012} \ldots$$

(in analogy to noble gases, group VIII elements from PTE) but besides A_{18} no cluster analoga to transition elememts exists.

3.3. Comparison of calculation and experiment

If we compare calculated and measured MN for the various models discussed here including two further models used by Martin et al. [10-13], as a final result we find three groups of models

1. models which mainly fail in calculating MN (coulombic potential, spherical oscillator)

2. models which calculate MN with some errors (spherical well with coulombic perturbation, pseudo quantum numbers [10-13], special jellium model [10-13])

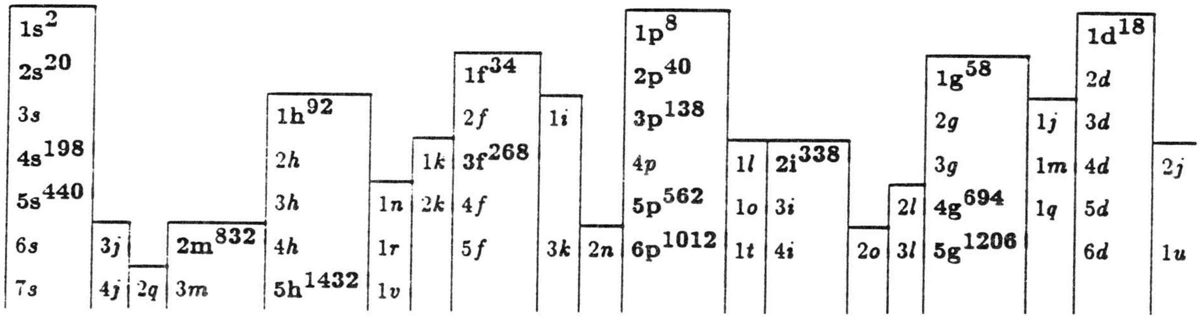

Fig. 2. Periodic table of alkali metal clusters

3. models which calculate possible closed-subshell configurations for each experimentally observed MN

(a) standard sphere models (infinite and finite walls)

(b) modified analytic sphere models (harmonic or piecewise constant perturbation).

4. Conclusion

As we have shown above, a useful model does not only have to calculate correct magic numbers MN (as all members of the 3rd group) but also proper bunching of subshells. This challenge is fulfilled only by models of 3rd group (b) and particularly by analytic sphere model including optimized constant perturbation up to 1432 electrons.

We gratefully acknowledge financial support from the Fonds der Chemischen Industrie and the Deutsche Forschungsgemeinschaft.

References

1 Knight, W.D., Clemenger, K., de Heer, W.A., Saunders, W.A., Chou, M.Y., and Cohen, M.L.: Phys. Rev. Lett. **52**, 2141 (1984)

2 Begemann, W., Dreihofer, S., Meiwes-Broer, K.H., and Lutz, H.O.: Z. Phys. D **3**, 183 (1986)

3 Bergmann, T., Limberger, and H., Martin, T.P.: Phys. Rev. Lett. **60**, 1767 (1988)

4 Martins, J.L., Car, R., and Buttet, J.: Surf. Sci. **106**, 265 (1981)

5 Ekardt, W.: Ber. Bunsen-Ges. Phys. Chem. **88**, 289 (1984)

6 Clemenger, K.: Phys. Rev. B **32**, 1359 (1985)

7 Ishii, Y., Ohniski, S., and Sugano, S.: Phys. Rev. B **33**, 5271 (1986)

8 Bergmann, T., and Martin, T.P.: J. Chem. Phys. **90**, 2848 (1989)

9 Bjørnholm, S., Borggreen, J., Echt., O., Hansen, K., Pederson, J., and Rasmussen, H.D.: Phys. Rev. Lett. **65**, 1627 (1990)

10 Martin, T.P., Bergmann, T., Göhlich, H., and Lange, T.: J. Phys. Chem. **95**, 6421 (1991)

11 Martin, T.P., Bergmann, T., Göhlich, H., and Lange, T. in Pacchioni, G., et al. (ed.): Cluster Models for Surface and Bulk Phenomena, New York, Plenum Press 1992

12 Göhlich, H., Lange, T. Bergmann, T., and Martin, T.P.: Phys. Rev. Lett. **65**, 748 (1990)

13 Martin, T.P., Bergmann, T., Göhlich, H., and Lange, T.: Chem. Phys. Lett. **172**, 209 (1990)

14 Brechignac, C., Cahuzac, Ph., de Frutos, M., Roux, J.-Ph., and Bowen, K.: Physics and Chemistry of Finite Systems: From Clusters to Crystals, ed. by Jena, P., et al., Kluwer Academic Publishers 1992

This article was processed using Springer-Verlag TEX Z.Physik D macro package 1.0 and the AMS fonts, developed by the American Mathematical Society.

Vibrating potential model for giant resonances in deformed metal clusters and atomic nuclei

V.O. Nesterenko *

Bogolubov Theoretical Laboratory, Joint Institute for Nuclear Research, Dubna 141980, Russia

Received: 28 June 1993

Abstract. The vibrating potential model is extended to provide a description of giant resonances in metal clusters and atomic nuclei with any kind of deformation.

PACS: 21.60 Ev; 36.40.+d

Much attention has been paid last time to collective excitations of valence electrons in metal clusters (MC). These excitations are considered as counterparts of giant resonances (GR) in atomic nuclei. (see reviews [1,2] and refs. therein). The vibrating potential model (VPM) is known to be appropriate for investigation of (GR) in both atomic nuclei [3-5] and MC [6]. However, this model encounters considerable difficulties in the case of deformed systems, especially for MC. In this paper the multipole expansion of both single-particle potential and density of valence electrons is used to obtain equations of the VPM for systems with any kind of deformation [7]. The strength function procedure [8] extremely simplifying the calculations is also proposed for the VPM.

The VPM equations coincide with those of the schematic RPA with separable forces. The strength constant of residual forces is determined from the self-consistency condition. Let us consider MC in the most simple external field $f(\mathbf{r}) = r^\lambda(Y_{\lambda\mu}(\Omega) + d \cdot Y^\dagger_{\lambda\mu}(\Omega))$ where the coefficient $d = \pm 1$ is used to assure the hermiticity of the Hamiltonian. Then, the dispersion equation for the collective excitations of valence electrons is [6]:

$$X_t \equiv 2\sum_{kk'} \frac{< k'|Q|k >^2 (\epsilon_k + \epsilon_{k'})}{(\epsilon_k + \epsilon_{k'})^2 - \omega_t^2} = \kappa_0^{-1} \tag{1}$$

where

$$\kappa^{-1} = \int \nabla Q(\mathbf{r}) \cdot \nabla f(\mathbf{r}) n_0(\mathbf{r}) d\mathbf{r} \tag{2}$$

is the inverse strength constant and

$$Q(\mathbf{r}) = \nabla V_0(\mathbf{r}) \cdot \nabla f(\mathbf{r}) + \int \frac{\nabla n_0(\mathbf{r}') \cdot \nabla f(\mathbf{r}')}{|\mathbf{r} - \mathbf{r}'|} d\mathbf{r}' \tag{3}$$

is the operator of residual interaction. Here, $| k >$ and ϵ_k are the single-particle eigenstate and eigenenergy of the static hamiltonian with a single-particle potential $V_0(\mathbf{r})$, ω_t is the root of equation (1), $n_0(\mathbf{r})$ is the static ground state density of valence electrons. The electron-ion interaction (ions are considered in jellium approximation as a uniform positive charged background) and exchange - correlation effects are taken into account through the ground state characteristics $V_0(\mathbf{r})$ and $n_0(\mathbf{r})$. The direct Coulomb interaction between valence electrons is responsible for the second term in (3). If this Coulomb term is neglected we obtain the same equations as for description of isoscalar GR in nuclei. Thus, equations (1)-(3) can be used to study GR in both MC and nuclei.

Equations (1)-(3) take place for both spherical and deformed systems. But in the deformed case they are too general to be convenient for estimations and numerical calculations. One can get more appropriate equations using the multipole expansion of the single-particle potential and ground state density of valence electrons. Then, expression for the operator (3) is written as [7]

$$Q^{(d)}_{\lambda\mu}(\mathbf{r}) = \sum_{LM} Y_{LM}(\Omega)(2\lambda+1)\sqrt{\frac{\lambda(2\lambda-1)}{4\pi(2L+1)}}$$

$$\cdot \sum_{lm}(C^{LM}_{lm\lambda\mu} + d(-1)^\mu C^{LM}_{lm\lambda-\mu})$$

$$[(M^{(1)}_{\lambda Ll}\frac{dV_{lm}(r)}{dr} r^{\lambda-1} - M^{(2)}_{\lambda Ll} V_{lm}(r) r^{\lambda-2})$$

$$-\frac{4\pi}{2L+1} \cdot (M^{(3)}_{\lambda Ll} r^{-(L+1)} \int_0^r n_{lm}(r_1) r_1^{\lambda+L} dr_1$$

$$+M^{(4)}_{\lambda Ll} r^L \int_r^\infty n_{lm}(r_1) r_1^{\lambda-L-1} dr_1)]. \tag{4}$$

Here $C^{LM}_{lm\lambda\mu}$ is the Clebsch-Gordan coefficient, expressions for $M^{(i)}_{\lambda Ll}$ are given in the Appendix, $V_{lm}(r)$ and $n_{lm}(r)$ are the coefficients of multipole expansion of the single-particle potential $V_0(\mathbf{r})$ and electron density $n_0(\mathbf{r})$, respectively. In the case of atomic nuclei, only terms with the coefficients $M^{(1)}_{\lambda Ll}$ and $M^{(2)}_{\lambda Ll}$ take place.

* E-mail: NESTER@LTP03.JINR.DUBNA.SU

For systems with a large number of particles the direct solving of the **RPA** equations takes considerable computer time. This is especially true for deformed systems. For investigation of GR in such systems the strength function method [8] is very useful. Within this method we can avoid finding roots of the secular equation (1) and get information about GR through the strength function

$$b(E\lambda\mu,\omega)=\sum_t B(E\lambda\mu, gr\to\omega_t)\rho(\omega-\omega_t) \quad (5)$$

where $\rho(\omega-\omega_t)=\Delta/[2\pi((\omega-\omega_t)^2+(\Delta/2)^2)]$. Here, $B(E\lambda\mu,gr\to\omega_t)$ is the reduced probability of $E\lambda\mu$ transition from the ground state to the one-phonon state with excitation energy ω_t determined from the secular equation (1). The quantity Δ is an averaging parameter. Following the prescription [8] we obtain

$$b(E\lambda\mu,\omega)=\frac{1}{\pi}\left(Im\left(\frac{\tilde{X}_t^2(z)}{X_t(z)-\kappa^{-1}_{\lambda\mu d}}\right)_{z=\omega+i\Delta/2}\right.$$
$$+\Delta\sum_{k<k'}p^{\lambda\mu}_{kk'}\left(\frac{1}{(\omega-(\epsilon_k+\epsilon_{k'}))^2+(\Delta/2)^2}\right.$$
$$\left.\left.-\frac{1}{(\omega+(\epsilon_k+\epsilon_{k'}))^2+(\Delta/2)^2}\right)\right) \quad (6)$$

where

$$\tilde{X}_t^{\lambda\mu}=2\sum_{kk'}\frac{<k'|Q|k>p^{\lambda\mu}_{kk'}(\epsilon_k+\epsilon_{k'})}{(\epsilon_k+\epsilon_{k'})^2-\omega_t^2} \quad (7)$$

and $p^{\lambda\mu}_{kk'}$ is the single-particle element for the operator of $E\lambda$ transition.

If all collective strength is assumed to be concentrated in one peak (one-pole approximation), we can get the estimation

$$\omega^2_{\lambda\mu}=(\omega^{(0)}_{\lambda\mu})^2+\frac{\int(\nabla Q_{\lambda\mu}(\mathbf{r}))^2 n_0(\mathbf{r})d\mathbf{r}}{\int Q(\mathbf{r})_{\lambda\mu}\nabla f(\mathbf{r})\cdot\nabla n_0(\mathbf{r})d\mathbf{r}} \quad (8)$$

where $\omega_{\lambda\mu}$ and $\omega^{(0)}_{\lambda\mu}$ are the energy of the collective peak and its unperturbed value, respectively. Using a spherical jellium approximation for electron density, the oscillator form $V_0(r)=1/2\cdot\omega_0^2r^2$ for single-particle potential and the estimation $\omega^{(0)}_\lambda=\lambda\omega_0$ for unperturbed energy we obtain the simple expression for excitation energy of $E\lambda$ GR in spherical MC:

$$\omega_\lambda=\sqrt{\frac{\lambda}{2\lambda+1}\omega_p^2+\omega_0^2\lambda(\lambda-1)} \quad (9)$$

where ω_p is the plasma frequency. This expression exactly reproduces the estimations for frequencies of E1 and E2 GR, obtained within the sum rule approach [6,9], but the λ dependence in the second term of (9) is in contradiction with the $\sim(2\lambda+1)(\lambda-1)$ dependence in [9,10]. The possible reason for this discrepancy could be the use for the static potential in (9) of the oscillator form which is not consistent with the ground state density of electrons.

Finally, the equations of the VPM [6] are derived to make them suitable for description of GR in systems (MC and atomic nuclei) with any deformation. Any single-particle potentials and electron (nucleon) densities for which the coefficients of the multipole expansion are known can be used within this model. Other kinds of electric and magnetic GR can be considered in a similar manner by making use of the external fields as transverse or longitudinal combinations of eigenfunctions of the Helmholz equation. It is very promissing to use some new opportunities to study GR, provided by MC, as compared with atomic nuclei. For example, in atomic nuclei a large fragmentation of $E\lambda$ GR of high multipolarity makes their investigation difficult. In MC fragmentation of electric GR decreases drastically if we pass from negative charged to positive charged clusters [11]. So positive charged MC can provide a unique chance for investigation of electric GR of high multipolarity in Fermi systems.

The author is grateful to Profs. E. Lipparini, N.Lo Iudice, Drs. V.Kresin and W.Kleinig for useful discussions.

Appendix

$$M^{(1)}_{\lambda Ll}=A_{\lambda Ll}-B_{\lambda Ll},$$

$$M^{(2)}_{\lambda Ll}=l\cdot A_{\lambda Ll}-(l+1)\cdot B_{\lambda Ll},$$

$$M^{(3)}_{\lambda Ll}=(l+\lambda+L+1)\cdot A_{\lambda Ll}+(l-\lambda-L)\cdot B_{\lambda Ll},$$

$$M^{(4)}_{\lambda Ll}=(l+\lambda-L)\cdot A_{\lambda Ll}+(l-\lambda+L+1)\cdot B_{\lambda Ll},$$

$$A_{\lambda Ll}=\sqrt{(l+1)(2l+3)}\begin{pmatrix} l+1 & \lambda-1 & L\\ \lambda & l & 1\end{pmatrix}\cdot C^{L0}_{l+10\lambda-10},$$

$$B_{\lambda Ll}=\sqrt{l(2l-1)}\begin{pmatrix} l-1 & \lambda-1 & L\\ \lambda & l & 1\end{pmatrix}\cdot C^{L0}_{l-10\lambda-10}.$$

References

1 Kresin V.V.: Phys.Rep. **220**, 1 (1992)
2 Nesterenko V.O.: Sov.J. Part.Nucl. **23**, n.6, 726 (1992)
3 Rowe D.J.: In: Nucl. Collect. Motion. London: Methuen (1970)
4 Lipparini E. and Stringari S.: Nucl.Phys. **A371**, 430 (1981)
5 Aberg S.: Phys.Lett. **157B** 9 (1985)
6 Lipparini E. and Stringari S.: Z.Phys.D **18**, 193 (1991)
7 Nesterenko V.O.: Preprint JINR E4-93-338, Dubna (1993)
8 Malov L.A., Nesterenko V.O. and Soloviev V.G: Teor.Mat.Fiz. **32**, 134 (1977)
9 Serra Ll. et al.: Phys.Rev. **B39**, 8247 (1989)
10 Nix J.R. and Sierk A.J.: Phys.Rev. **C21**, 396 (1980)
11 Yannouleas C., Chem.Phys.Lett. **193** 587 (1992)

This article was processed using Springer-Verlag TEX Z.Physik D macro package 1.0 and the AMS fonts, developed by the American Mathematical Society.

A hollow-cathode plasma sputter source for production of metal cluster beams

M. B. Nielsen[1], J. Borggreen[1], K. Hansen[1], J. Pedersen[1], B. Holst[1], M. Krogh[1], H. D. Rasmussen[1], E. Søndergård[1] and A. Xenoulis[2]

[1] The Niels Bohr Institute, University of Copenhagen, DK-4000 Roskilde, Denmark
[2] N.C.S.R. "Demokritos", Athens, Greece

Received 28 June 1993

Abstract. We have developed a source for the production of clusters of refractory metals by plasma sputtering. One aim is to produce metal clusters with a density of $\sim 10^{12}$ atoms/cm^3 in a size range of 10^2–10^5 atoms per cluster for the purpose of performing X-ray diffraction studies. Net sputtering rates have been measured for a Cu cathode, as a function of pressure, flow velocity, current, and cathode geometry. Computer simulations of the mass loss from the cathode reproduce the experimental results with reasonable agreement.

PACS: 36.40.+d, 52.75.–d, 61.46.+w

1. Introduction

The production of clusters from refractory metals demands the use of either an extremely hot oven or a laser for the evaporation of the material. Alternatively, a primary ion source or an electric discharge may be used to sputter atoms of the material into the gas phase. This type of source has the crucial advantage of a very high yield. While most measurements of cluster properties do not require particularly intense beams, the highest possible cluster beam intensity becomes a determining factor in the investigation of the crystalline structure of freely suspended metal clusters via X-ray or electron diffraction.

Sputtering ion sources have already had some impact in the investigation of cluster properties, see e.g. refs. [1–4]. Here we report on the development of a metal cluster ion source, based on a dc plasma discharge in conjunction with aggregation in an argon gas.

2. Experimental setup

A schematic diagram of the apparatus is shown in Fig. 1. Ar at $\sim$ 1 mbar flows through the hollow cathode at a adjustable rate (1–80 m/s). When a potential difference between anode and cathode of some 300–500 V is applied an electric discharge is ignited. The Ar^+ ions, accelerated towards the inner walls of the cathode, sputter atoms from the cathode into the gas. Typical currents are 100–1000 mA.

The inert gas serves two purposes apart from maintaining the plasma: On the one hand collisions with the small clusters provide cooling to remove the heat of condensation, thus facilitating formation of larger clusters. On the other hand the flowing gas sweeps the atoms and clusters away from the hot plasma region.

The source has been investigated in several geometries. In the one described here, the inert gas enters the apparatus through the cathode and then proceeds through the anode, which is at ground potential.

The cathode is manufactured from the desired cluster metal (designed to allow a minimum of material), and has the geometrical form of a cone. External cooling of the cathode is required, since typically 200–400 watts are deposited on it during the discharge.

It is mounted in an oil cooling unit and the conical shape allows good thermal an electrical contact to the cooling unit which is connected to a HV power supply. The cathode can easily be removed for inspection or replacement.

3. Results and discussion

The rate with which atoms are removed and escape from the cathode depends on two competing processes: sputtering and redeposition. Backscattering of sputtered atoms contribute to redeposition [5], but is here considered not to be significant. The main contribution to redeposition comes from diffusion of atoms back to the wall. The data presented here, of the weight lost from the cathode under various discharge conditions, monitor the combined result of sputtering and redeposition.

Fig. 2 shows that the apparent sputtering rate from a 5.0 mm diameter 40.0 mm long hollow cathode significantly decreases for argon pressures both above and below about 2.0 mbar. The filled circles in the figure represent numerical simulations of the mass loss. The

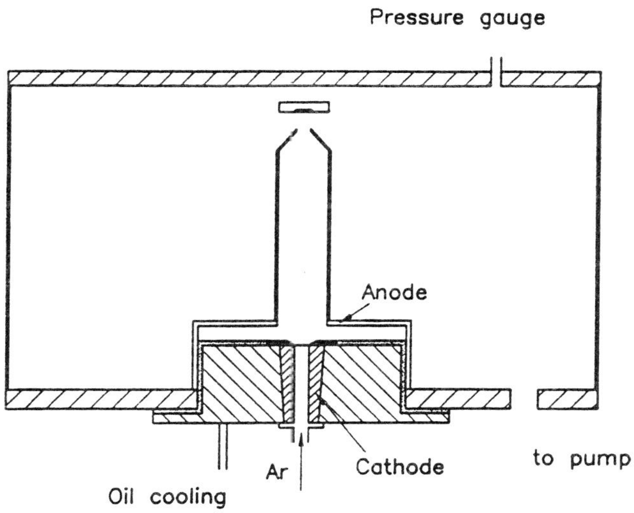

Fig. 1. Schematic of the experimental setup

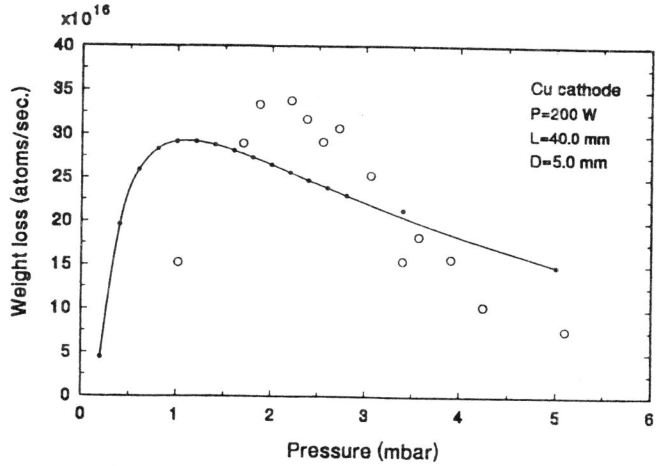

Fig. 2. The dependence of mass loss from the cathode versus pressure. The filled circles represent the numerical result

calculations uses Fick's first law for flux of particles A in a gas of particles B. The total flux is the sum of diffusion and collective flow. The curve can be understood qualitatively as follows.

Cathode atoms leave the surfaces with energies from a few eV up to several hundred eV, but the majority will have energies below 10 eV [6]. When secondary atoms have thermalized within the carrier gas, they will obtain the flow velocity determined by the velocity profile. In the low pressure range the mean free path is so long that diffusion to the opposite cathode wall becomes important. For decreasing pressures this effect will result in decreasing net sputter yields in terms of mass loss from the cathode, see Fig. 2. In the high pressure domain, on the other hand, the mean free path is so short that secondary atoms are confined close to the surface. Due to the low flow velocity close to the wall, the atoms will have a long residence time in the cathode. This leads to a high redeposition probability. The important role of redeposition phenomena is furthermore supported by the considerable reduction observed in the apparent sputtering rate when the flow velocity was decreased.

With an applied current of $\sim$ 500 mA a density of sputtered atoms up to 3×10^{14} atoms/cm^3 is obtained. At this current, the temperature close to the exit of the cathode was measured 800 K. For these conditions the Cu vapor has a partial pressure of 4×10^{-2} mbar, which is far beyond supersaturation. Thus conditions facilitating aggregation are present.

The cluster size distribution has been examined with a transmission electron microscope, which has a resolution of approximately 8 Å. At a distance of $\sim$ 15 cm from the cathode exit samples have been taken to probe the beam. The size distribution was observed to be in the range of 10–100 Å, which correspond to Cu_n with $30 < n < 30 \times 10^3$.

4. Summary

We have presented the state of a newly developed sputter source. The source promises very high beam density and a very high mass distribution. Also, the source shows very stable running conditions and is easy to handle. The principle allows many different materials, which can easily be exchanged within minutes, and promises a broad spectrum of changeable parameters to control the mass distribution. The source at the present stage has a higher density than required, but this is a the exit of the cathode only. The beam must be guided into high vacuum before diffraction is possible. This problem is yet to be solved.

Acknowledgement. The authors thank Dr. Andy Horsewell for performing the TEM studies determining the presence of clusters. This work is supported partly by the Danish Natural Science Foundation, the Thrige Foundation and the SARC Foundation.

References

1. Haberland, H., Karrais, M., Mall, M.: Z. Phys. D **20**, 413 (1991)
2. Leopold, D. G., Ho, J., Lineberger, W. C.: J. Chem. Phys. **86**, 4 (1987)
3. Chow, G. M., Chien, C. L., Edelstein, A. S.: J. Mater. Res. **6**, 9 (1991)
4. Castens, D. H. W., Kozlowski, J. F., Gruen, D. M.: High Temp. Sci.: **4**, 301 (1972)
5. Chapman, B.: Glow Discharge Processes. New York: John Wiley & Sons, 1982
6. Westwood, W. D.: J. Vac. Sci. Technol. **15**, 1 (1978)

This article was processed using Springer-Verlag TEX Z.Physik D macro package 1.0 and the AMS fonts, developed by the American Mathematical Society.

Z. Phys. A 349, 343–344 (1994)

Symmetrized harmonic oscillator SU_3 states for multi-cluster systems

A. Novoselsky[1], J. Katriel[2]

[1] Racah Institute of Physics, The Hebrew University, Jerusalem 91904, Israel
[2] Department of Chemistry, Technion-Israel Institute of Technology, Haifa 32000, Israel

Received: 27 October 1993

Abstract. An algorithm is presented for the construction of single- and multi-cluster harmonic oscillator wave functions that are coupled into well-defined irreducible representations of SU_3 as well as of the symmetric group S_N. The single-cluster harmonic oscillator SU_3 wave functions are constructed recursively, using the SU_3 coefficients of fractional parentage. To construct multi-cluster wave functions with a well-defined permutational symmetry we diagonalize an appropriate set of single-cycle class operators of the symmetric group involving all the constituent particles. The formalism is applicable to the study of multi-cluster systems in nuclear physics where the wave functions are expressed in terms of harmonic oscillator SU_3 states

PACS: 21.60.Gx;21.45.+v;02.20.Df

Harmonic oscillator states that belong to a given irreducible representation (irrep) of the special unitary group in three dimensions, SU_3, have been extensively used in nuclear physics for the last thirty five years. The pioneering work was done by Elliot [1] who in 1958, introduced a collective model describing the rotational structure of light nuclei where the nuclear wave function is simultaneously classified according to its symmetric group irrep and the appropriate SU_3 irrep. This classification scheme has been used in 1975 by Arima and Iachello for describing the low-energy collective states of heavy deformed even-even nuclei in the interacting boson model [2].

In the non-relativistic description of the nucleon as a cluster of three quarks, the state of each quark is specified by its orbital motion, spin, flavor and color. The wave function of a baryon is decomposed into an orbital part, assumed to be totally symmetric, which is coupled to a symmetric spin-isospin SU_4 state and to an antisymmetric singlet SU_3 color state [3]. Harmonic oscillator wave functions are commonly used for describing the orbital state of each quark. Therefore, the baryon orbital wave function is totally symmetric with respect to permutations and at the same time belongs to a well-defined harmonic oscillator SU_3 irrep. Moreover, in this approach, the orbital wave function of the deuteron (two baryons) and the α particle (four baryons) are the totally symmetric harmonic oscillator SU_3 states for two and four-cluster systems, respectively [3].

The extensive use in nuclear and quark physics of harmonic oscillator SU_3 states with well-defined permutational symmetry motivated the search for a general algorithm for constructing these states. An important step was taken by Hecht [4], who constructed the states by using the SU_3 coefficients of fractional parentage (cfps). However, his algorithm is limited to the s-d shell and to simple cases in other shells.

A year ago we developed the first comprehensive method for constructing harmonic oscillator SU_3 states that belong to a well-defined irrep of the symmetric group [5]. The states are obtained recursively by using the SU_3 cfps, which are calculated by diagonalization of the two-cycle class operator of the symmetric group. The eigenvalues uniquely identify the irreps of the symmetric group whereas the eigenvectors are the SU_3 cfps. In practice, it is sufficient to calculate the matrix element of the transposition of the last two particles in every step. These matrix elements are obtained by using the SU_3 Racah coefficients.

Consider a single cluster with N particles in one major harmonic oscillator shell. (The last restriction was imposed for simplicity.) Each particle posseses the SU_3 irrep $(\lambda, 0)$ where $\lambda = 0, 1, 2, \cdots$ is the number of oscillator quanta. The N-particle state belongs to the SU_3 irrep $(\lambda^{(N)}, \mu^{(N)})$ where an additional quantum number $\beta^{(N)}$ is needed in order to take care of possible multiplicity. The sequence of irreps $\Gamma_2, \Gamma_3, \cdots, \Gamma_N$ of the symmetric group-subgroup chain $S_2 \subset S_3 \subset \ldots \subset S_N$ completely determines the permutational symmetry of the wave function. This sequence of irreps is equivalent to a Yamanouchi symbol Y_N. Therefore, the harmonic oscillator SU_3 states that belong to a well defined irrep Γ_N of S_N as well as to a given irrep $(\lambda^{(N)}, \mu^{(N)})$ of SU_3 are

$$|(\lambda, 0)^N; (\lambda^{(N)}, \mu^{(N)}) Y_N \beta^{(N)} \alpha^{(N)} > \quad (1)$$

Correspondence to: A. Novoselsky

where we denote the set of quantum numbers which identify the states by $\alpha_{(N)}$.

Three years ago we introduced a procedure for coupling various clusters of identical particles into an overall well-defined angular momentum and permutational symmetry [6]. This procedure can also be applied for constructing multi-cluster harmonic-oscillator SU_3 states that belong to a well-defined irrep of the symmetric group.

Consider a system of identical particles distributed in several clusters. The states that belong to each cluster are characterized as in (1). We form all the multi-cluster states belonging to a particular total SU_3 irrep by using the SU_3 Wigner coefficients. The set of states generated in this way is used as a basis set within which the class operators of the single-cycle classes of the symmetric group are diagonalized. The eigenstates of these class operators belong to irreps of the symmetric group that are fully specified by the corresponding eigenvalues. Full details of this method are given in [6].

The harmonic-oscillator SU_3 states for multi-cluster systems that are obtained in this procedure are

$$|(\lambda, 0)^N; (Y_{N_1}, Y_{N_2}, \cdots, Y_{N_m})(\lambda^{(N)}, \mu^{(N)}) \Gamma_N \beta^{(N)} \alpha^{(N)} > \tag{2}$$

where the N particles are distributed in m clusters such that there are N_i particles in cluster i and $\sum_{i=1}^{m} N_i = N$. Note that for each cluster only the Yamanouchi symbol Y_{N_i} remains as a good quantum number since we have to sum over all the SU_3 irreps $(\lambda^{(N_i)}, \mu^{(N_i)})$ and the additional quantum numbers $\beta^{(N_i)}$ and $\alpha^{(N_i)}$ of each cluster. Note also that the multi-cluster state (2) is characterized by the irrep Γ_N of the symmetric group and not by the Yamanouchi symbol Y_N.

In conclusion, a method for constructing symmetrized harmonic oscillator SU_3 states for multi-cluster systems is presented. This procedure is easy to implement as a computational scheme and is applicable to various multi-cluster calculations in nuclear physics as well as in non-relativistic quark calculations.

References

1. Elliot, J.P.: Proc. R. Soc. London **245**, 128 (1958); **245**, 562 (1958)
2. Iachello, F., Arima, A., The Interacting Boson Model. Cambridge: Cambridge University Press 1987
3. Oka, M., Yazaki, K.: Baryon-Baryon interaction from quark model viewpoint: in Quarks and Nuclei. Singapore: World Scientific 1984
4. Hecht, K.T.: Nucl. Phys. **62**, 1 (1965)
5. Novoselsky, A., Katriel, J.: Ann. Phys. (N. Y.) **229**, 55 (1994)
6. Katriel, J., Novoselsky, A.: Ann. Phys. (N. Y.) **211**, 1 (1991)

This article was processed using Springer-Verlag TEX Z.Physik A macro package 1.0 and the AMS fonts, developed by the American Mathematical Society.

Z. Phys. A 349, 341–342 (1994)

Search for cluster decay of ^{114}Ba

Yu.Ts.Oganessian, Yu.A.Lazarev, V.L.Mikheev, Yu.A.Muzychka, I.V.Shirokovsky, S.P.Tretyakova, V.K.Utyonkov

Laboratory of Nuclear Reactions, Joint Institute for Nuclear Research, 141980 Dubna, Russia

Received: 27 October 1993

Abstract. A search for the ^{12}C decay of ^{114}Ba produced in the ^{58}Ni(^{58}Ni,2n) reaction was performed by using polycarbonate track detectors. Assuming the total half-life of ^{114}Ba to be $\gtrsim$0.1 s, the lower limit $T_c \geq 10^3$ s for the partial ^{12}C-decay half-life was established.

PACS: : 23.90.+w; 28.85.Ca

1. Introduction

First theoretical predictions of nuclear lifetimes for cluster radioactivities in the new region above closed shells Z=50, N=50 were made by Greiner et al. [1,2]. In their calculations, the most favourable case here is predicted to be the decay of the very neutron-deficient isotope ^{114}Ba by spontaneous ^{12}C emission leading to the daughter nucleus ^{102}Sn. As yet, the isotope ^{114}Ba has not been identified. Ground-state decay properties expected for ^{114}Ba sensitively depend on its unknown mass value for which different mass models [3] yield results varying by more than 2 MeV. This causes large variations of the partial half-life T_c predicted for the ^{12}C decay of ^{114}Ba (Fig.1). Even more dramatic uncertainties arise for the partial α-decay half-life of this nucleus. As for EC(β^+) decay of ^{114}Ba, from systematics [5,6] its partial half-life is expected to be in the range of 0.1 to 1 s. If we assume the total ^{114}Ba half-life to be , say, 0.1 s, then the branching ratio b_c for the ^{12}C decay of ^{114}Ba turns out to be of the order of 10^{-10}. It is clear that experiments on probing such small branching ratios are virtually precluded because of insufficient rates of the ^{114}Ba production by any feasible means.

However, more recent lifetime calculations performed for the ^{12}C decay of ^{114}Ba by Kadmensky et al. [4] have led to T_c values which are 10^6–10^7 times lower compared to estimations of [1,2] (see Fig. 1). This brings predicted branching ratios b_c to a detectable level. It is noteworthy that calculations of [4] and those of [1,2] give identical results for α decay of ^{114}Ba.

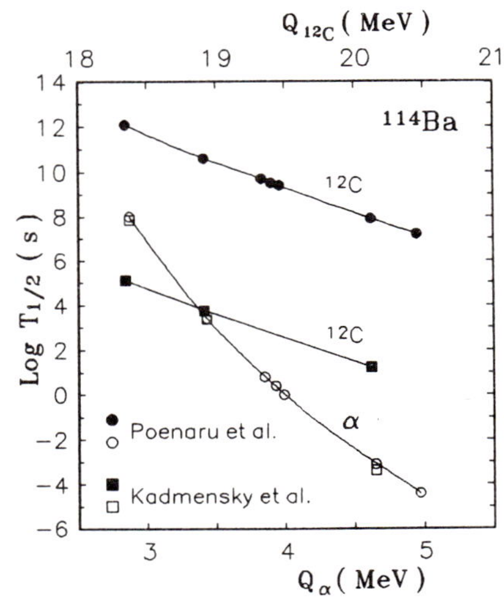

Fig. 1. Partial half-lives for α and ^{12}C decays of ^{114}Ba predicted by Poenaru et al. [2] and Kadmensky et al. [4] as functions of the released energies Q_α and $Q_{^{12}C}$

2. Experimental technique

Guided by the above predictions, we carried out first experiments to search for carbon radioactivity of ^{114}Ba. For the production of ^{114}Ba, we employed the reaction ^{58}Ni(^{58}Ni,2n). The irradiations were performed at the Dubna U400 cyclotron by using the wheel technique outlined in Fig. 2. A beam of ^{58}Ni projectiles with an incident energy of 280 MeV struck tangentially the lateral surface of a cooled copper cylinder onto which 3.5 mg/cm^2 of natural nickel were electrodeposited. This cylindrical target (serving simultaneously as a recoil catcher) rotated at a rate of 10 rev/s which allowed efficient searches for decays with $T_{1/2} \gtrsim 10$ ms. To detect ^{12}C clusters, 200-μm polycarbonate films were arranged all around the cylindrical target except for the beam input zone. The polycarbonate track detectors were calibrated by using beams of ^{12}C projectiles from the U400 cyclotron. The calibration was made in the ^{12}C energy range of 7 to 24 MeV.

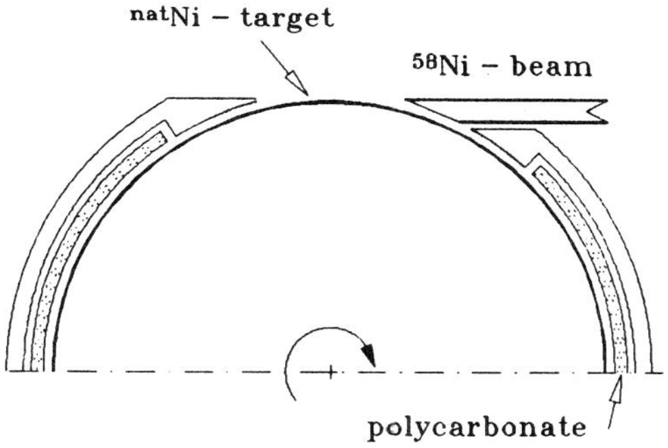

Fig. 2. Scheme of the experiment

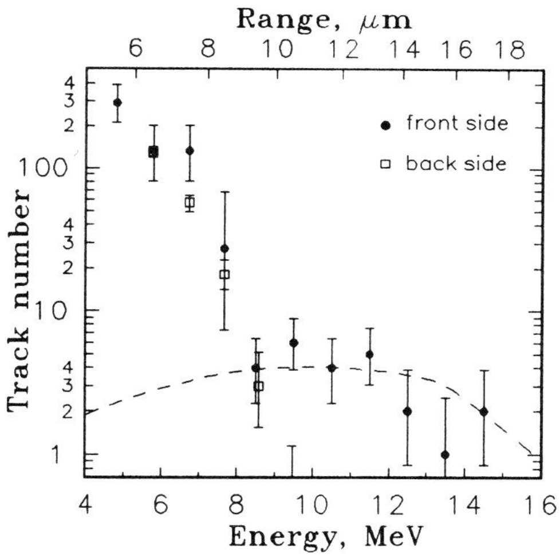

Fig. 3. Range (energy) spectra from ^{12}C track measurements on the front and back sides of the polycarbonate detectors. *Dashed line* shows the calculated energy spectrum of ^{12}C clusters from the decay of ^{114}Ba recoils implanted in the catcher

In searching for the ^{12}C decay of ^{114}Ba, the main background was due to tracks from ^{12}C recoils produced in polycarbonate by fast neutrons accompanying the ^{58}Ni beam. Therefore the back side of the polycarbonate detectors (see Fig. 2) was carefully scanned as well, which allowed us to estimate the background and compare it with the track pattern revealed on the front polycarbonate side viewing the target surface.

3. Results and discussion

The ^{12}C range spectra resulting from the ^{nat}Ni+^{58}Ni bombardment with a total ^{58}Ni beam dose of 3.5×10^{17} are shown in Fig. 3. No tracks with the length larger than 11μm were revealed on the back side of the polycarbonate detectors. On the front side, however, some ten tracks were found to have a length exceeding that of most of the background tracks produced by ^{12}C recoils. From the viewpoint of ranges, these longer tracks might be due to ^{12}C decays of ^{114}Ba. The time distribution of the corresponding decay events would then be roughly uniform within the interval of $\approx$0.1 s. However, taking into account the low number of relevant events as well as the presence of the considerable background, so far we consider it reasonable to set an upper limit of the effect in question.

According to statistical-model calculations, the production cross section of ^{114}Ba in the ^{58}Ni+^{58}Ni reaction is expected to be of the order of 1 μb. Then, assuming for the width of the excitation function a FWHM value of 20 MeV, we obtain an estimate of 2×10^6 for the total number of the ^{114}Ba nuclei produced. The detection efficiency of ^{12}C decays is estimated to be about 0.07. This value includes a factor of 0.5 for detection in 2π geometry, a factor of 0.7 due to losses in the beam input zone, a factor of $\approx$0.4 reflecting the fact that the track length determination is possible only for those ^{12}C clusters which escape the catcher in a certain solid angle, and, finally, a factor of $\approx$0.5 allowing for losses of low-energy ^{12}C clusters in the background range (see Fig. 3). All in all, it follows from our data that $b_c\leq10^{-4}$ for the ^{12}C decay of ^{114}Ba. Assuming the total ^{114}Ba half-life to be $\gtrsim$0.1 s, we obtain the lower limit $T_c\geq10^3$ s which can be compared with the T_c predictions shown in Fig. 1. Obtaining more positive conclusions about the ^{12}C decay of ^{114}Ba represents a challenge for future experiments.

References

1. Greiner W., Ivascu M., Poenaru D., Sandulescu A.: Treatise on Heavy Ion Science, D.A.Bromley (ed.), Vol.**8**, p.641. New York: Plenum Press 1989
2. Poenaru D.N., Schnabel D., Greiner W.: At. Data Nucl. Data Tables **48**, 231 (1991)
3. Haustein P.E. (ed.): At. Data Nucl. Data Tables **39**, 185-393 (1988)
4. Kadmensky S.G., Kurgalin S.D., Mikheev V.L., Furman V.I., Tchuvilsky Yu.M.: Izv. Akad. Nauk (Moskva) Ser. Fiz. **57**, 12 (1993)
5. Takahashi K., Yamada M., Kondoh T.: At. Data Nucl. Data Tables **12**, 101 (1973)
6. Möller P., Randrup J.: Nucl. Phys. **A514**, 1 (1990); Report LBL–27504, Berkeley, 1989

This article was processed by the author using the LaTeX style file *pljour2* from Springer-Verlag.

Z. Phys. A 349, 305–306 (1994)

Cluster preformation in closed- and mid-shell nuclei

D. N. Poenaru[1,2,3], **W. Greiner**[1,2]

[1] Institut für Theoretische Physik der Universität, Robert-Mayer-Strasse 8-10, D - 60325 Frankfurt am Main, Germany
[2] Gesellschaft für Schwerionenforschung, Darmstadt, Germany
[3] Institute of Atomic Physics, Bucharest, Romania

Received: 27 October 1993

Abstract. By using our semiclassical method to estimate the preformation probability of the touching point configuration as the penetrability of the inner part of the barrier, it is shown that this quantity is strongly dependent on the nuclear structure. We are illustrating this dependence by comparing the low probability of cluster preformation in a doubly magic nucleus ^{208}Pb, with much higher values obtained in a heavier parent nucleus leading to ^{208}Pb heavy fragment, for different emitted clusters like ^{4}He, ^{12}C, ^{20}Ne, ^{24}Mg, etc.

PACS: 23.70.+j; 21.10.Jx; 25.85.Ca; 23.60.+e

From more than 150 various cluster decay modes [1] predicted within our analytical superasymmetric fission model (ASAFM), the following radioactivities are already confirmed: ^{14}C, ^{20}O, ^{23}F, $^{24-26}$Ne, ^{28,30}Mg, and ^{32,34}Si. The experimentally determined half-lives [2-4] of the corresponding parent nuclei: ^{221}Fr, $^{222-224,226}$Ra, ^{225}Ac, ^{228,230}Th, ^{231}Pa, $^{232-234}$U, ^{236,238}Pu, and ^{242}Cm, are in good agreement with our estimations.

Strong shell effects are always present: the daughter nucleus is either ^{208}Pb (in eight of these cases) or some of its neighbours (see also [5]). A new island of cluster emitters with daughter nuclei around ^{100}Sn is expected [6] to be observed in the future.

One of the three (model–dependent) quantities, whose product determines the measurable decay constant $\lambda = \nu S P_s$, is the cluster preformation probability, S. According to our semiclassical method [7] it can be calculated as the penetrability of the prescission part of the fission barrier $S_{sc} = \exp(-K_{ov})$, where K_{ov} is the corresponding action integral for overlapping fragments.

We denote $b^2 = (E_{cor} + E^*)/E_b^0$, where E_{cor} is the correction energy (similar to the Strutinsky shell correction), E^* is the excitation energy concentrated in the separation degree of freedom. Other quantities are the separation distances $R_i = R_0 - R_e$ and $R_t = R_e + R_d$, where $R_j = r_0 A_j^{1/3}$ ($j = 0, e, d$; $r_0 = 1.2249 fm$) are the radii of parent, emitted and daughter nuclei, Q is the released energy, and $E_b^0 = E_i - Q$ is the barrier height before correction. The interaction energy at the top of the barrier, in the presence of a nonnegligible angular momentum, $l\hbar$, is given by:

$$E_i = e^2 Z_e Z_d / R_t + \hbar^2 l(l+1)/(2\mu R_t^2) \tag{1}$$

where μ is the reduced mass. We have:

$$\log S_{sc} = -0.09537(E_b^0 A_e A_d/A)^{1/2}(R_t - R_i) \cdot \left[\sqrt{1-b^2} - b^2 \ln \frac{1+\sqrt{1-b^2}}{b}\right] \tag{2}$$

Unlike within approximation leading to the "universal curves" of even-even nuclei, where $\log S = -0.598(A_e - 1)$ depends only on the mass number of the emitted cluster, the nuclear structure effects could be taken into consideration in the above relationship. An example is given in Fig. 1, where we have considered emitted clusters with $Z_e = N_e$. The heavier the cluster, the larger is the difference between the probability to be formed in a nucleus leaving a daughter ^{208}Pb and the preformation in this doubly magic nucleus itself. Even higher cluster preformation probability is obtained for neutron-rich nuclei (^{14}C, ^{20}O, ^{24}Ne, ^{28,30}Mg, ^{34}Si) experimentally observed. An "experimental" value is defined as the ratio $S_{exp} = \lambda_{exp}/(\nu P_s)$. The equation of the universal curves of even-even combinations parent-emitted clusters is expressed as:

$$\log T = 0.598(A_e - 1) - \log P_u - 22.169 \tag{3}$$

where P_u is the penetrability of the external Coulomb barrier, given by

$$-\log P_u = 0.22873(\mu_A Z_d Z_e R_b)^{1/2} \left[\arccos\sqrt{r} - \sqrt{r(1-r)}\right] \tag{4}$$

in which $r = R_t/R_b$, $R_b = 1.43998 Z_d Z_e/Q$, and $\mu_A = A_d A_e/A$. In spite of the rough approximation, $S = S(A_e)$, on which these $\log T = f(-\log P_u)$ straight lines are based, the up-to-now 14 even-even half-life measurements are well reproduced (within a ratio 3.86, or rms=0.587 orders of

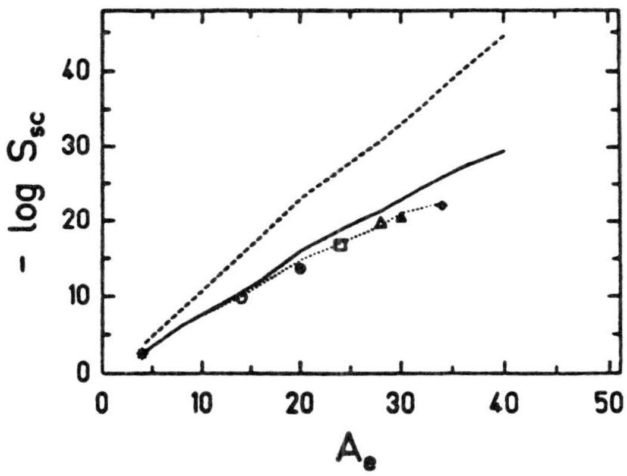

Fig. 1. Preformation probability of even-even emitted clusters with $Z_e = N_e$ from ^{208}Pb (*dashed line*), from various nuclei leading to ^{208}Pb daughter (*full line*), compared to that of neutron-rich emitted nuclei ^{14}C, ^{20}O, ^{24}Ne, ^{28,30}Mg, ^{34}Si with ^{208}Pb heavy fragment (*dotted line*) and the experimental points

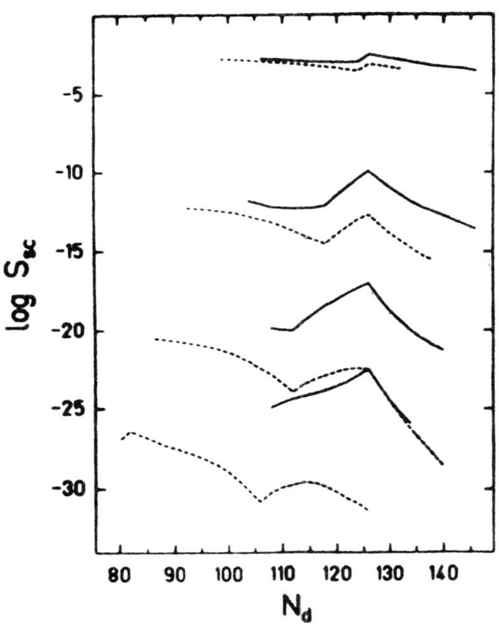

Fig. 2. In descending order from the top to the bottom we present the preformation probability of ^{4}He, ^{14}C, ^{24}Ne and ^{34}Si versus neutron number of the daughter. For each cluster the parent nuclei are either Pb isotopes (*dashed line*) or heavier nuclides leading to heavy fragments which are Pb isotopes (*full lines*)

magnitude). A similar result is obtained with the "hand-pocket" formula presented in [8].

Nevertheless, the fact that a given cluster is preformed with a larger probability in a parent nucleus corresponding to ^{208}Pb daughter than in the doubly magic nucleus ^{208}Pb is also illustrated in Fig. 2, in which we consider few clusters: α-particle; ^{14}C; ^{24}Ne and ^{34}Si. Instead of only one value of S at a given mass number A_e (a horizontal line) independent on Z_e, Z_d, A_d, one can see a whole range, increasing at a constant N_d from $Z = 82$ (dashed line) to $Z = 82 + Z_e$. The maximum value is reached at the magic neutron number of the daughter $N_d = 126$.

Acknowledgement. This work has been supported by Gesellschaft für Schwerionenforschung, Darmstadt, by Kernforschungszentrum Karlsruhe, Germany and by Institute of Atomic Physics, Bucharest, Romania. One of us (DNP) received a donation of computer equipment from Soros Foundation for an Open Society.

References

1. Poenaru, D.N., Greiner, W. (eds) Nuclear Decay Modes (to be published)
2. Price, P.B.: An. Rev. Nucl. Part. Sci. **39**, 19 (1989)
3. Hourani, E.: Chap.6 in [1]
4. Bonetti, R., Guglielmetti, A.: Chap.7 in [1]
5. Broglia, R.A.: Present conference
6. Poenaru, D.N., Greiner, W., Gherghescu, R.:Phys. Rev. C**47**, 2030 (1993)
7. Poenaru, D.N., Greiner, W.: Phys. Scri.**44**, 427 (1991)
8. Blendowske, R., Fliessbach, T., Walliser, H.: Chap.5 in [1]

This article was processed using Springer-Verlag TEX Z.Physik A macro package 1.0 and the AMS fonts, developed by the American Mathematical Society.

Z. Phys. A 349, 349–350 (1994)

The breakup of ^{24}Mg into ^{16}O and ^{8}Be

The Charissa Collaboration: J.S. Pople[1], N.M. Clarke[1], B.R. Fulton[1], M.J. Leddy[1], J.T. Murgatroyd[1], Y. Chan[2], R.G. Stokstad[2], W.D.M. Rae[3], W.N. Catford[4], S.P. Fox[5], G.J. Gyapong[5], D.L. Watson[5], S.J. Bennett[6]

[1] School of Physics and Space Research, University of Birmingham, Edgbaston, Birmingham B15 2TT, UK
[2] Lawrence Berkeley Laboratory, USA
[3] University of Oxford, UK
[4] University of Surrey, UK
[5] University of York, UK
[6] Wayne State University, USA

Received: 27 October 1993

Abstract. The breakup of ^{24}Mg into ^{16}O and ^{8}Be fragments has been studied using the reactions ^{12}C(^{24}Mg, ^{16}O^{8}Be) ^{12}C and ^{12}C(^{20}Ne, ^{16}O^{8}Be) ^{8}Be. In the latter case, discrete states are observed near 24–28 MeV of excitation in ^{24}Mg and the yield from this reaction is an order of magnitude greater than that of the former. This implies the excited configuration populated in ^{24}Mg is favoured by the transfer of an alpha-particle and would therefore suggest an association with a 4-particle 4-hole configuration. This suggests a link with the octupole stabilised deformed minimum which appears in Nilsson-Strutinsky calculations of the potential energy surface in ^{24}Mg, and also with the $\alpha - {}^{16}\mathrm{O} - \alpha$ structure predicted in cranked cluster model calculations. In the excitation spectrum no states appear above 31 MeV indicating a possible band termination in disagreement with recent results using the ^{16}O + ^{12}C reaction. These results are discussed in terms of the Harvey model.

PACS: 24.30.-v;25.70.-z;27.30.+t

1. Introduction

The symmetric fission of ^{24}Mg into two ^{12}C fragments was observed by Fulton et al. (1986) in the scattering of 170 MeV ^{24}Mg from a ^{12}C target. The three body kinematics were reconstructed to yield the spectrum of the ejectile excitation energy E_x (Fig. 1).

The E_x spectrum shows a number of sharp peaks between 20–27 MeV in ^{24}Mg, indicating breakup from specific states. At this excitation, the statistical model predicts > 1000 states per MeV, so it is surprising to observe any structure in the spectrum or decays by emission of clusters, rather than n, p or α emission. Measurements of the spins (Freer et al. 1990) indicated a rotational band (inset Fig. 1) with the bandhead at 20.5 MeV and a rotational spacing of 75 keV, implying a large deformation of ^{24}Mg with a moment of inertia comparable to that for two touching ^{12}C nuclei (Erb and Bromley 1981).

The cranked cluster model (CCM) (Marsh and Rae 1986) shows a number of stable configurations for ^{24}Mg which can be associated with the secondary minima in the potential energy surface in the Nilsson-Strutinsky (NS) calculations (Leander and Larsson 1975). The oscillator model of Harvey (1975) predicts that the mass-asymmetric 3:1 prolate minimum with $\epsilon = 1.0, \epsilon_3 = 0.3, \gamma = 0°$ in the NS calculations (corresponding to the $\alpha - {}^{16}\mathrm{O} - \alpha$ shape in the CCM) may be associated with the ^{12}C + ^{12}C barrier resonances (Fulton et al. 1986) and is allowed to breakup into either ^{12}C + ^{12}C or ^{16}O + ^{8}Be (Fulton and Rae 1990) with all nuclei in their ground states. For identical bosons, only even spin states can be excited, but for ^{16}O + ^{8}Be breakup, odd spin states may also be observed. The prolate minimum in the NS calculations has a wavefunction with a four-particle four-hole configuration, suggesting it can be excited by an alpha-transfer reaction.

2. Experiments

Two experiments have been performed; one at the NSF, Daresbury using the ^{12}C(^{24}Mg,^{16}O^{8}Be)^{12}C reaction at 170 MeV; the other at the 88 inch Cyclotron at the Lawrence Berkeley Laboratory, California using the ^{12}C(^{20}Ne,^{16}O^{8}Be) ^{8}Be reaction at 160 MeV. The energies and angles of the ^{16}O ions are measured by three telescopes, consisting of 12 mm square position sensitive silicon ΔE detectors (22 mm) backed by 300 μm E detectors, 120 mm from the target. The ^{8}Be detector is a 50 mm square silicon detector 1000 mm thick, with 16 independent position sensitive strips each 3 mm high. Signals from both ends of the strips yield energy and position of the incident particles. A front foil of 175 mg/cm^2 Mylar prevents damage from heavy ions. The resolution in energy and horizontal position are typically 300 keV and 0.2 mm, respectively. Beams of up to 50 enA were incident on carbon targets of thickness 400 mg/cm^2. The ^{16}O(^{8}Be) detector angles were 12° or 15°(21°).

Correspondence to: Dr N.M. Clarke, School of Physics and Space Research, University of Birmingham, Edgbaston, Birmingham B15 2TT, UK

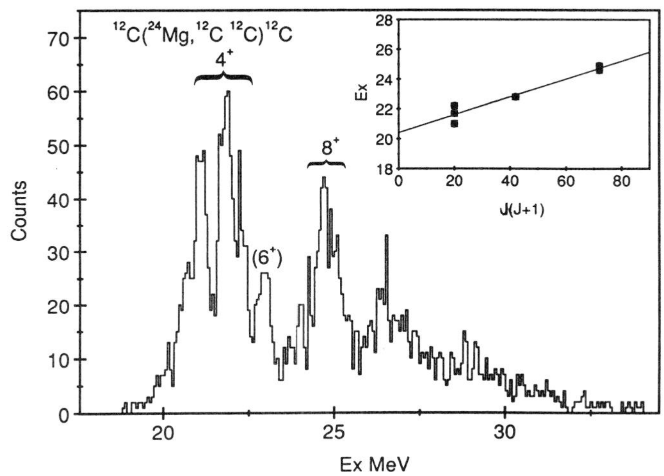

Fig. 1. Spectrum of breakup states in ^{24}Mg and the proposed rotational band (inset)

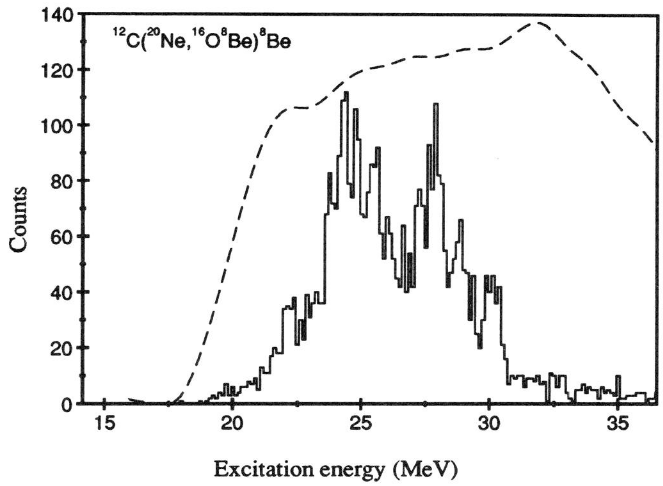

Fig. 2. Spectrum of ^{24}Mg breakup states into ^{16}O + ^{8}Be

The relative energy of the ^{8}Be is obtained from the energy and angles of the α-particles, correcting for energy losses in the Mylar foil. By placing a software window on the ^{8}Be(g.s.) peak at 92 keV, and demanding a coincidence with an ^{16}O ion in a telescope, the total energy in the system, E_{tot} can be obtained (Harvey 1975). A peak in the E_{tot} spectrum at the correct Q-value indicates that the ^{16}O and two ^{8}Be ejectiles are all in their ground states. A window on this peak enables reconstruction of the relative energy of the ^{16}O and ^{8}Be to obtain the excitation spectrum of ^{24}Mg (Fulton and Rae 1990).

3. Results

Our results indicate that the yield from the alpha transfer reaction $^{12}C(^{20}Ne,^{16}O\ ^{8}Be)^{8}Be$ is 10 times greater than from the $^{12}C(^{24}Mg,^{16}O\ ^{8}Be)^{12}C$ reaction. This suggest that the configuration in ^{24}Mg which decays by $^{16}O + {}^{8}Be$ emission corresponds to the 3:1 octupole stabilised deformed minimum which appears as the $\alpha - {}^{16}O - \alpha$ structure in the CCM; this is unique in having the $4p - 4h$ structure, and would be favoured by the alpha transfer reaction.

Figure 2 shows a preliminary E_x spectrum for the $^{12}C(^{20}Ne,^{16}O\ ^{8}Be)^{8}Be$ reaction, indicating the presence of discrete states near 24, 25 and 26 MeV of excitation which are also seen in the $^{12}C(^{24}Mg,^{16}O\ ^{8}Be)^{12}C$ reaction and in Fig. 1. The detection efficiency profile is shown as the dashed line in Fig. 2 and is high even at E_x = 35 MeV, so the lack of states in this region suggest that the spectrum terminates for other reasons. The CCM calculations indicate a limiting angular momentum of J = 12 for the $\alpha - {}^{16}O - \alpha$ band. If this band corresponds to that populated in $^{24}Mg \rightarrow {}^{12}C + {}^{12}C$ breakup, then the inset of Fig. 1 would imply a band termination near E_x = 30 MeV for J = 12, in good agreement with these observations. A recent experiment (Vinciguerra et al. 1993) using the $^{16}O + {}^{12}C$ reaction found J = 14 states in ^{24}Mg near 35, 36 MeV which decay by $^{12}C + {}^{12}C$ emission and appear to be evidence against a band termination. However, the Harvey model implies that the ^{28}Si configuration populated by $^{16}O + {}^{12}C$ *cannot* breakup along the z axis into an alpha plus the 3:1 prolate minimum in ^{24}Mg, but *can* produce a configuration similar to the ^{24}Mg ground state which can decay by $^{12}C + {}^{12}C$. This further implies that the $^{16}O + {}^{12}C$ reaction is populating a different band of states from those produced by alpha transfer reactions and which breakup into $^{16}O + {}^{8}Be$.

References

K.A. Erb, D.A. Bromley: Phys. Rev. **C23**,2781(1981)

M. Freer et al.: Nuclear Structure and Reaction Dynamics, IOP Conference no. 109. R.R. Betts, J.J. Kolata(eds.) Bristol, Philadelphia 1990

B.R. Fulton, S.J. Bennett, C.A. Ogilvie, J.S. Lilley, D.W. Banes, W.D.M. Rae, S.C. Allcock, R.R. Betts, A.E. Smith: Phys. Lett. **181B**, 233(1986)

B.R. Fulton, W.D.M. Rae: J. Phys. G **16**, 333(1990)

M. Harvey, Proc. 2nd Conf. on Clustering Phenomena in Nuclei, College Park, 1975 USDERA report ORO-4856-26 p 549

G. Leander, S.E. Larsson: Nucl. Phys. **A239**, 93 (1975)

S. Marsh, W.D.M. Rae: Phys. Lett. **180B**, 185(1986)

D. Vinciguerra, E. Costanzo, P. Figuera, M. Lattuada, S. Pirrone, S. Romano, M. Zadro: Proc. XXXI International Winter Meeting on Nuclear Physics, Bormio, Italy, January 1993 (to be published)

Variational approach to metal cluster dynamics

João da Providência Jr. and Raphael de Haro Jr.*

Departamento de Física, Universidade de Coimbra, P-3000 Coimbra, Portugal

Received 28 June 1993

Abstract. A fluid dynamical model, which has been previously applied to study the collective modes in atomic nuclei, is here applied to investigate the vibrational modes of the valence electrons in a metal cluster.

PACS: 36.40.+d

The semiclassical method which we are considering has been applied to describe collective modes in atomic nuclei [1,2], in good agreement with microscopic calculations based on selfconsistent HF plus RPA. Recently the same variational method has also been applied [3] to obtain the bulk plasmon dispersion in a metal.

The equilibrium state is represented by the Slater determinant $|\phi_0>$ which minimizes the energy. We consider the quantum mechanical lagrangian $L = ih < \phi|\dot{\phi} > - < \phi|H|\phi >$ with $|\phi >= exp(i\hat{S}/h)|\phi_0 >$, where $\hat{S}$ is a hermitian one-body operator. For small amplitude deviations from the equilibrium state the lagrangian L reduces to

$$L^{(2)} = \frac{i}{2h} < \phi_0|[\hat{S}, \dot{\hat{S}}]|\phi_0 > - \frac{1}{2h^2} < \phi_0|[\hat{S}, [H, \hat{S}]]|\phi_0 > \tag{1}$$

In the classical limit we replace commutators by Poisson brackets and operators $\hat{\mathrm{A}}$ by their Wigner transforms $\mathrm{A}(\mathbf{x}, \mathbf{p})$. From the energy $E =< \phi|H|\phi >$ we obtain, in the classical limit the energy functional E[f]. The Wigner transform of the one-body density matrix $\hat{\rho}$ associated with the Slater determinant $|\phi >$ is the time dependent distribution function $f(\mathbf{x}, \mathbf{p}, t)$ which is related to the equilibrium distribution function f_0 by the canonical transformation

$$f = f_0 + \{f_0, S\} + \frac{1}{2}\{\{f_0, S\}, S\} + \ldots \tag{2}$$

where S is Wigner transform of the generator $\hat{\mathrm{S}}$.

Our method consists in expanding the generator of the fluctuations S in powers of the momentum,

$$S(\mathbf{x}, \mathbf{p}, t) = \chi(\mathbf{x}, t) + \mathbf{p} \cdot \mathbf{s}(\mathbf{x}, t) \tag{3}$$

and in using the variational principle to obtain an approximate solution. We derive, from the quantal lagrangian (1), the following semiclassical lagrangian,

$$L^{(2)} = \int d^3x [\frac{n_0}{2}(\mathbf{s} \cdot \nabla\dot{\chi} - \dot{\mathbf{s}} \cdot \nabla\chi) - \frac{n_0}{2m}(\nabla\chi) \cdot (\nabla\chi)] - E^{(2)}[\mathbf{s}] \tag{4}$$

where

$$E^{(2)}[\mathbf{s}] = \int d^3x [\frac{\tau_0}{3}((\nabla \cdot \mathbf{s})^2 + \frac{1}{2}(\partial_\alpha \mathbf{s}_\beta + \partial_\beta \mathbf{s}_\alpha)^2) + \sum_\sigma \alpha_\sigma \frac{\sigma(\sigma - 1)}{2} n_0^\sigma (\nabla \cdot \mathbf{s})^2] + \frac{e^2}{2} \int d^3x_1 d^3x_2 \frac{(\nabla \cdot (n_0\mathbf{s}))_1 (\nabla \cdot (n_0\mathbf{s}))_2}{|\mathbf{x}_1 - \mathbf{x}_2|} \tag{5}$$

where n_0 and τ_0 are respectively the equilibrium density and the equilibrium kinetic energy density, which are directly obtained from the equilibrium distribution function f_0. The constants α_σ $(\sigma = 1, 2, 3)$ represent effective parameters, associated with a one-body delta interaction, which describes the effect of the more conventional pseudo-potential, and with a two-body δ interaction (simulating exchange effects) and with a three-body δ interaction (simulating correlation effects). These parameters are adjusted in order to reproduce the properties of the homogeneous electron gas (density, energy and bulk modulus).

* Permanent address: Instituto de Física, Universidade Federal do Rio de Janeiro, 21945 - Ilha do Fundão, Rio de Janeiro, Brasil

Following the method of ref. [4] we make an expansion of the dynamical fields χ and $\mathbf{s}$ in multipoles and, for each multipolarity, we express the radial dependence by a polynomial in (r/R), $\mathbf{s} = \nabla[\sum_{k=k_{min}}^{k_{max}} b_k(t)(r/R)^k Y_{l0}]$, $\chi = \sum_{k=k_{min}}^{k_{max}} a_k(t)(r/R)^k Y_{l0}$. Inserting expressions for $\mathbf{s}$ and χ into eqs. (5) and (6) we obtain the following lagrangian

$$L^{(2)} = \frac{1}{2}\sum_{kq}\{A_{kq}[(b_k\dot{a}_q - \dot{b}_k a_q) - \frac{1}{m}a_k a_q] - B_{kq}b_k b_q\} \tag{6}$$

where A_{kq} and B_{kq} are symmetric real matrices. Requiring the action integral to be stationary for arbitrary variations of the variables a_q and b_q, and assuming harmonic time dependence for the variables a_k and b_k ($\ddot{b}_k = -\omega^2 b_k$), we obtain the following eigenvalue equation

$$\omega^2 m \sum_k A_{qk} b_k = \sum_k B_{qk} b_k \tag{7}$$

Solving the eigenvalue eq. (7) we obtain the normal modes for each value of l, which are characterized by the eigenfrequencies ω_j and the eigenvectors $(\chi^{(j)}, \mathbf{s}^{(j)})$. These eigenmodes fulfill an orthogonality relation and additionally they satisfy the energy wheighted sum rule (m_1) and the cubic energy wheighted sum rule (m_3). The eigenvalue equation (7) which we have obtained is equivalent to eq. (30) of ref. [5] which was derived from a multidimensional extension of the sum rule approach.

In the following table we present some results for a sodium cluster of 92 atoms.

l_i^π	$\hbar\omega_i$ [eV]	m_1 [%]	m_3 [%]
0_1^+	6.0666	97.662	96.568
1_1^-	3.4191	100	100
2_1^+	3.8806	99.947	99.751
3_1^-	4.2232	99.735	98.832

For the evaluation of m_1 and m_3 we have considered the operator r^2 for $l = 0$ and the operator $r^l\, Y_{l0}$ for $l > 0$. We have considered the following parameters

α_1 [eV $\AA^{-3}$]	α_2 [eV $\AA^{-6}$]	α_3 [eV $\AA^{-9}$]
-5.8348	-133.17	1642.9

E/N [eV]	n_0 [$\AA^{-3}$]	B [eV]
-6.2669	2.5435×10^{-2}	4.3372×10^{-2}

where B is the bulk modulus.

References

1 Andō, K., Nishizaki, S.: Prog. Theor. Phys. **68**, 1196 (1982)
2 da Providência, J.P.: Prog. Theor. Phys. **75**, 862 (1986)
3 da Providência, Jr.J., Barberán, N.: Phys. Rev. **B45**, 6935 (1992)
4 Nishizaki, S., Andō, K.: Prog. Theor. Phys. **73**, 889 (1985)
5 Brack, M.: Phys. Rev. **B39**, 3533 (1989); Reinhard, P.-G., Brack, M., and Genzken, O.: Phys. Rev. **A41**, 5568 (1990)

This article was processed using Springer-Verlag TEX Z.Physik D macro package 1.0 and the AMS fonts, developed by the American Mathematical Society.

Dipole excitation of Na clusters with a non-local energy density functional

A. Puente[1], Ll. Serra[1,2], and M. Casas[1]

[1]Departament de Física, Universitat de les Illes Balears, E-07071 Palma de Mallorca, Spain
[2]Dipartimento de Fisica, Università di Milano, Via Celoria 16, I-20133 Milano, Italy

Received 28 June 1993; Revised 11 March 1994

Abstract. The elementary dipole excitations of the ionized clusters Na_9^+, Na_{21}^+ and Na_{41}^+ are ivestigated by solving the equations of the Random-Phase Approximation. The ground and excited states are described using the jellium model for the ionic background and a non-local energy density functional for the valence electrons. Non-local effects are specifically analyzed. The excitation energies thus obtained approach better than those of the Local Density Approximation both the full Hartree-Fock and the experimental results.

PACS: 36.40.+d, 31.20.-d, 71.45.-d

1. Introduction

After the discovery by Knight et al. [1] of quantum shell structure in small drops of sodium, the study of metal clusters has been the object of a great deal of both experimental and theoretical investigations [2]. The first theoretical efforts were focused on the description of the ground state properties. Microscopic ab initio calculations have been carried out for clusters with a small number of particles [3] and, for bigger sizes, approaches based on the Density-Functional theory and the jellium model have been more successful [4]. It is now well established that the shell structure in simple metal clusters can be understood from the quantized motion of the valence electrons in the mean field created by their mutual interactions and the positive ionic background (usually described in the jellium model) [5].

The investigations of the dynamical properties have revealed the existence of a dipole resonance [6] that has been interpreted as a collective electronic excitation (surface plasmon). Within a simple picture, it can be explained as the result of the coherent motion of the delocalized valence electrons with respect to the ionic background.

The theoretical effort in the description of the static and dynamical properties of the alkali-metal clusters is usually centered in the Kohn-Sham (KS) formalism, with the Local Density Approximation (LDA) fir the electronic exchange and correlation energies [7,8]. The limitations of the local approximation have been largely studied and there are several good models that go beyond the LDA, using parametrized pair-correlation functions [9] or introducing non local gradient corrections to the exchange-correlation energy [10]. Nevertheless some of these corrections diverge at large distance when we try to incorporate them into a variational selfconsistent scheme. The limitations of the LDA approach could have some important consequences in systems like the alkali clusters, where one of the biggest contributions to the binding energy is precisely the Coulomb exchange. Indeed, within the KS+LDA model more than 70% of the total binding energy of the cluster Na_{92}^+ is due to the exchange term.

The ground state and the dipole excitation of the closed shell Na clusters have been recently studied using a non-local HF central potential [11] which takes into account, in an exact way, the exchange associated to the Coulomb interaction. However, in this framework the correlation energy is not included and the method can not be applied to very big clusters, for which a semiclassical formalism is more appropriate.

In this paper we present an analysis of the dipole response properties of Na charged clusters described with a non-local energy density functional (NLEDF), that gives a better description of the ground state properties of spherical Na clusters and can be applied in a semiclassical formalism for big clusters [12]. We have first obtained the single particle wave functions describing the KS ground state, and then solved the RPA equations with the residual interaction between electrons induced by our NLEDF. A comparison with both the standard LDA predictions and the experimental values is done. In section 2 we present a brief reminder of our NLEDF. Section 3 explains the RPA formalism that we have used and the corresponding residual interaction. Finally, in section 4 we present and discuss our results and draw some conclusions.

2. Non-local Energy Density Functional: Ground State

As it shown in ref. [12], starting from the HF exchange energy of a spin saturated system, and expanding the angular average of the first order density matrix [13], we have constructed a NLEDF and its corresponding variational one body potential without problems of divergence at large distances. In the EDF formalism the total binding energy of an interacting electron system is given by

$$E[n(\mathbf{r})] = T[n(\mathbf{r})] + \int v(\mathbf{r})n(\mathbf{r})d^3r \\ \frac{1}{2}\int\int \frac{n(\mathbf{r}_1)n(\mathbf{r}_2)}{|\mathbf{r}_1-\mathbf{r}_2|}d^3\mathbf{r}_1 d^3\mathbf{r}_2 + E_{xc}[n(\mathbf{r})] \tag{1}$$

where $T[n(\mathbf{r})]$ is the kinetic energy of the non-interacting system of density $n(\mathbf{r})$. The ionic background is modelled as a sperical jellium with a Wigner Seitz radius of 4 a.u. for Na clusters. As it is usual, the exchange correlation term $E_{xc}[n(\mathbf{r})]$ is split into exchange and correlation part. We shall focus in the exchange part, which for a spin saturated system is exactly written as

$$E_x = -\frac{1}{2\nu}\int\int \frac{n^2(\mathbf{r}_1,\mathbf{r}_2)}{|\mathbf{r}_1-\mathbf{r}_2|}d^3\mathbf{r}_1 d^3\mathbf{r}_2 \tag{2}$$

where $n(\mathbf{r}_1,\mathbf{r}_2)$ is the first order density matrix and ν is the degeneracy factor.

Using the standard Density Matrix Expansion (DME) [13] up to second order we obtain

$$E_x = \int d^3r(-\frac{9\pi}{2\nu k^2}n^2 - \frac{35\pi}{6\nu k^4}n[\frac{3}{5}k^2 n - (\tau - \frac{1}{4}\nabla^2 n)]) \tag{3}$$

The next approximation is to choose the value of the "dummy" parameter k, which appears due to the truncation of the expansion. Taking $k = k_F(\mathbf{r}) = (6\pi^2 n/\nu)^{1/3}$ as in the standard DME, the next term of the expansion becomes divergent [11]. However, choosing k constant (around the Fermi momentum k_F computed at the saturation density) and using for the exchange energy the expansion (3), it is possible to derive from (1) a selfconsistent one body potential well behaved at large distances. In the deriviation of the single particle equations an effective mass $m^*(\mathbf{r})$ appears due to the inclusion of the non local terms of (3)

$$\frac{h^2}{2m^*} = \frac{h^2}{2m} + \frac{35\pi n}{6\nu k^4} \tag{4}$$

The mean value of m^* is $\sim 0.5m$ at the interior of the sodium cluster going to m at the surface. This big reduction of the effective mass in the interior of the cluster is showing the importance of the non local terms in the electron-electron interaction. As it is shown in [12], the NLEDF seems to approach better the ground state properties of spherical Na clusters than the LDA. In particular, their first ionization potential is closer to the experimental values than the result obtained within LDA.

3. RPA formalism and residual interaction

In the RPA the elementary excitations of the system are described as the result of correlated particle-hole (ph) transitions in the independent particle scheme of the ground state. The amplitudes of a particle excited state on the basis elements spanning the discrete space of ph transitions are obtained by diagonalizing the Hamiltonian $H = H_0 + V$, the sum of the KS Hamiltonian H_0 and the residual ph interaction V, solving the RPA equations

$$\begin{pmatrix} A & B \\ -B^* & -A^* \end{pmatrix}\begin{pmatrix} X^{(n)} \\ Y^{(n)} \end{pmatrix} = E_n \begin{pmatrix} X^{(n)} \\ Y^{(n)} \end{pmatrix} \tag{5}$$

where matrices A and B are written in terms of angular-momentum-coupled matrix elements of the residual interaction [8].

The residual ph interaction is obtained from the energy density functional of the cluster. This two-body interaction in our approximation is given by

$$v(1,2) = \frac{1}{|\mathbf{r}_1-\mathbf{r}_2|} + \frac{d^2\varepsilon_{cor}}{dn^2}\delta(\mathbf{r}_{12}) + v_{ex}(1,2) \tag{6}$$

where the first two contributions are the well known terms coming from the direct Coulomb and correlation energy (in the LDA approximation) [7], and $v_{ex}(1,2)$ is the new contribution associated to our NLEDF for the exchange energy (3). Following a similar procedure to that of phenomenological Skyrme interactions for nuclei or He drops [14], one can easily obtain

$$v_{ex}(1,2) = \frac{35\pi}{6k^4}(\frac{1}{2}\overleftarrow{k_r}^2\delta(\mathbf{r}_{12}) + \frac{1}{2}\delta(\mathbf{r}_{12})\overrightarrow{k_r}^2 + \overleftarrow{k_r}\delta(\mathbf{r}_{12})\overrightarrow{k_r}) - \frac{8\pi}{k^2}\delta(\mathbf{r}_{12}) \tag{7}$$

where $\overrightarrow{k_r}(\overleftarrow{k_r})$ is the relative momentum acting to the right (left) in the matrix element. The NLEDF induces an effective contact interaction which depends on the relative velocities of the electrons and in this way, simulates to some extent finite range effects of the exchange interaction.

The strength function for the dipole excitation from the ground state $|0>$ to the set of excited states $|n>$ (with excitation energies E_n is defined as

$$S(E) = \sum_n \delta(E-E_n)|<n|Q|0>|^2 \tag{8}$$

where $Q = \sum_i r_i Y_{10}(\Omega_i)$. The transition matrix element entering the strength function are obtained within RPA in terms of the solutions of eq. (5) and the set of single particle states u_i as

$$< n|Q|0 >= \sum_{ph}(X_{ph}^{(n)} - Y_{ph}^{(n)}) < u_p||Q||u_h > \quad (9)$$

The solution of (5) requires the knowledge of all the particle-hole states. Above certain energy the excited single particle states are in the continuum and, in order to solve the single particle equations with the selfconsistent potential, we have enclosed the system in a spherical box of a radius about 3 times larger than the root mean square radius of the cluster. We require all the radial wave functions to vanish at this distance, obtaining in this way a discrete spectrum of states that replaces the continuum. We have checked that the results of our calculations are not sensitive to the precise distance where one puts the box. Finally we note that we have included enough particle-hole configutarions so that the Thomas-Reiche-Khun and cubic energy weighted sum rules [15] are satisfied with an accuracy of $\simeq 0.3\%$ and 3% respectively.

4. Results

We have obtained the solutions of the RPA equations (5) for the dipole excitation. In figures 1,2 and 3 we compare the free and RPA strength functions for Na_9^+, Na_{21}^+ and Na_{41}^+ obtained using the NLEDF model (upper plots), with the LDA results (lower plots). In the calculations presented here, we have assumed $k = 0.35$ $a.u.$, which is a value close to the Fermi momentum ($k_F = 0.48$ $a.u.$) and, as explained in section (2), yields a well behaved potential at large distances.

Quite remarkably, there is a large difference between LDA and NLEDF in the free strength functions. The big reduction of the effective mass in NLEDF causes a lower level density, as compared to LDA, and consequently, the ph transitions are placed at higher energies. In fact, the free intergrated strength is much higher in NLEDF than in LDA, as can be immediately noticed from the figures.

Despite this discrepancy between the two models in the free response, the RPA results do not change very much. This is due to the fact that within NLEDF, and contrary to the LDA, the residual interaction (6) becomes attractive, and this shifts the free strength to lower energies, in overall agreement with the LDA values. The RPA excitation energies thus obtained always lie much below the particle emission threshold (7.98, 5.51 and 4.96 eV for the ionized clusters Na_9^+, Na_{21}^+ and Na_{41}^+ respectively), and approach better than the LDA the full HF results [11]. Of course, the NLEDF residual interaction (6) is able to restore the Thomas-Reiche-Kuhn sm rule in the RPA strength.

For small clusters the dipole oscillation is shared by few states (a single peak exhausts 81% of the cubic energy weighted sum rule in the case of Na_9^+). When the cluster size grows the splitting into several states (Landau damping) also grows. However, the strength obtained using the NLEDF model remains more concetrated than in the LDA approach. Contrary

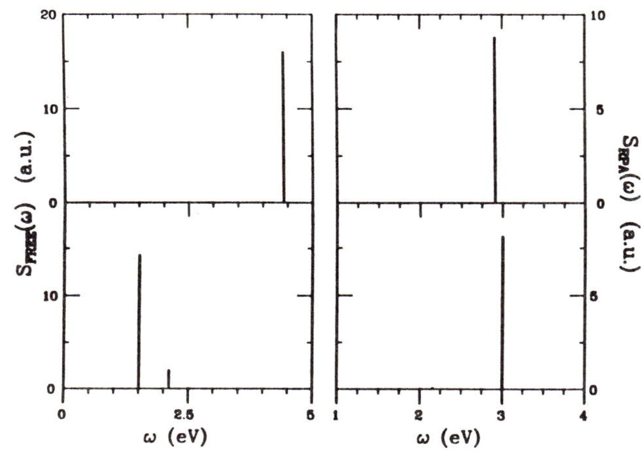

Fig. 1. Free and RPA strength function obtained using the NLEDF (upper plots) and the LDA approach (lower plots) for Na_9^+

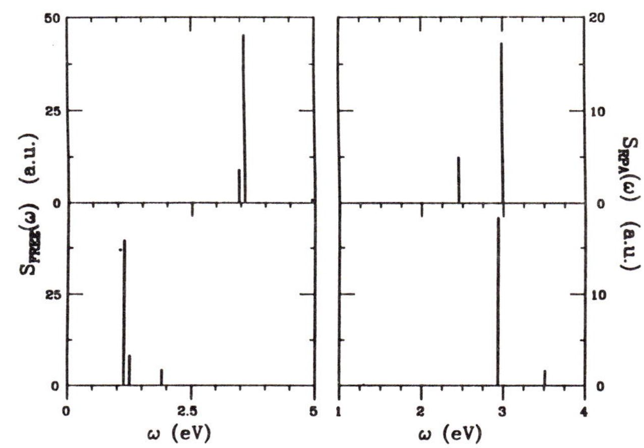

Fig. 2. Same as Fig. 1 for Na_{21}^+

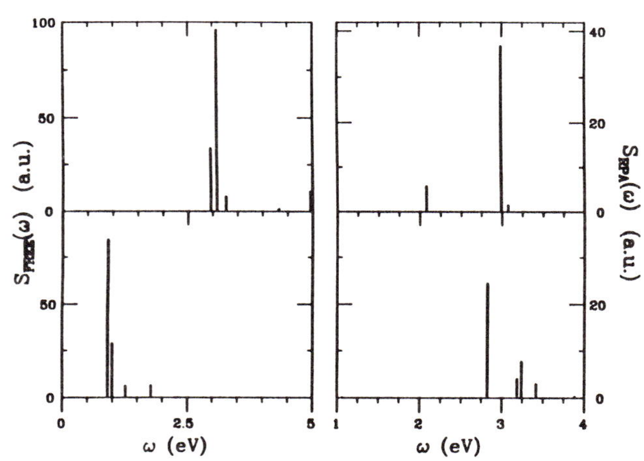

Fig. 3. Same as Fig. 1 for Na_{41}^+

to the local approach, in Na_{21}^+ and Na_{41}^+ there is an RPA state, with sizeable strength, at an energy lower than the most intense traqnsition lying at around 3 eV. This fragmentation in two peaks seems to be more in agreement with recent experimental results for Na_{21}^+ [16] which also yield the most collective state at higher energy. In fact, our predictions for the mean dipole excitation energy (2.94 eV for Na_9^+, 2.91 eV for Na_{21}^+ and 2.93 eV for Na_{41}^+) are more compatible with the experimental results of ref. [16] (2.62 eV for Na_9^+, and 2.65-2.66 eV for Na_{21}^+) than the LDA prediction (3.07 eV for Na_9^+, 3.,07 eV for Na_{21}^+ and 3.08 eV for Na_{41}^+). Although the mean excitation energy given by NLEDF is lower than the LDA one for all the clusters analyzed, our results still have a blue shift of about 10% in comparison with the experimental value. This shows that the ionic geometry can play some role in the determination of the dynamical properties, specially for small drops.

As a main conclusion, starting from the DME one can obtain a NLEDF and its corresponding selfconsistent residual interaction, which is attractive and shifts the mean dipole excitation energy of the free response to lower energies, thus approaching our results to the experimental values.

It is a pleasure to acknowledge useful discussions with Nguyen Van Giai and E. Lipparini. This work has been supported by DGICYT (Spain), grant (PS 90-0212). One of us (Li.S.) thanks the Commission of the European Communities for financial support (Community training project 920027).

References

1 Knight, W.D., Clemenger, K., de Heer, W.A., Saunders, W.A., Chou, M.Y., and Cohen, M.L.: Phys. Rev. Lett. **52**, 2141 (1984)

2 Z. Phys. D -"Small Particle and Inorganic Clusters" **19** and **20**, (1991)

3 Bonačić-Koutecký, V., Fantucci, P., and Koutecký, J.: Chem. Rev. **91**, 1035 (1991)

4 Ekardt, W.: Phys. Rev. B **29**, 1558 (1984)

5 de Heer, W.A., Knight, W.D., Chou, M.Y., and Cohen, M.L.: Solid State Phys. **40**, 93 (1987)

6 Bréchignac, C., Cahuzac, P., Carlier, F., and Leygnier, J.: Phys. Rev. Lett. **63**, 1368 (1989); Pollak, S., Wang, C.R.C., and Kappes, M.M.: J. Chem. Phys. **94**, 2496 (1991)

7 Ekardt, W.: Phys. Rev. B **31**, 6360 (1985)

8 Yannouleas, C., Broglia, R.A., Brack, M., and Bortignon, P.F.: Phys. Rev. Lett. **63**, 255 (1989)

9 Rubio, A., Balbás, L.C., Serra, Ll., and Barranco, M.: Phys. Rev. B **42**, 10950 (1990)

10 Langreth, D.C., and Mehl, M.J.: Phys. Rev. B **28**, 1809 (1983)

11 Guet, C. and Johnson, W.R.: Phys. Rev. B **45**, 11283 (1992)

12 Puente, A., and Casas, M.: Z. Phys. D **26**, S131 (1993)

13 Negele, J., and Vautherin, D.: Phys. Rev. C **5**, 1472 (1972)

14 Stringari, S.: Phys. Lett. **106A**, 2267 (1984)

15 Serra, Ll., Garcias, F., Barranco, M., Navarro, J., Balbás, C., and Mañanes, A.: Phys. Rev. B **39**, 8247 (1989)

16 Bréchignac, C., Cahuzac, Ch., Carlier, F., de Frutos, M., and Leygnier, J.: Chem. Phys. Letters **189**, 28 (1992)

This article was processed using Springer-Verlag TEX Z.Physik D macro package 1.0 and the AMS fonts, developed by the American Mathematical Society.

Experimental investigation on carbon cluster structure

F. Roussel, M. Perdrix and P. Pradel

Service des Photons Atomes et Molécules, C.E.A. Centre d'Etudes de Saclay, 91191 Gif sur Yvette CEDEX France

Received 28 June 1993

Abstract. Photofragmentation of carbon cluster ions, produced by a pulsed classical laser ablation source, have been investigated by measuring the depletion of the isolated cluster ion packet as a function of the photodissociation laser fluence. The experiment has been performed for C_{16}^+ to C_{36}^+ irradiated at first by the second harmonic of a YAG laser, and in a second step by the third harmonic. The comparison between the depletion curves obtained from the two experiments leads us to speculate on the presence of at least two or three valence isomers displaying different stabilities.

PACS: 36.40.+d

1. Introduction

The increasing number of publications on clusters reflects the great interest taken by the scientific community in this subject. In particular, carbon clusters have been a major focus for research in such diverse fields as catalysis and combustion. Their study is also of astrophysical interest. While the physical and electronic structures of small carbon clusters (with the atom number n<10) are reported in the literature [1,2,3], structures of carbon cluster with intermediate sizes remain in the speculative state, especially for those situated in the C_{20}^+ - C_{32}^+ region. These clusters are formed in states that are less stable than those of the other clusters, when they are produced without cooling [4,5,6]. In a supersonic cluster source [3,7] an important role of the carrier gas is the cooling of the growing clusters. Thus their fragmentation rate is slowed [2].

According to theoretical investigations [1], the reactivity data [8] suggests that the reactive small cluster ions (n<10) are linear in structure, whereas the unreactive larger cluster with 10<n<20 present a monocyclic ring structure with no reactive site.

Moreover, O'Brien et al. [3], Geusic et al. [7] and Radi et al. [4] have found the same C_n^+ dissociation pattern for larger clusters: clusters smaller than C_{32}^+ dissociate by C_3 loss whereas clusters with even numbers of C atoms larger than C_{32}^+ lose only C_2 when they fragment. This abrupt change in the fragmentation pattern was proposed to be due to a structural change from two-dimensional mono or polycyclic rings to a three dimensional cage structure. This proposal lead us to perform a comparative photofragmentation experiment on these clusters: the decrease of the parent cluster intensity versus the irradiation fluence, reflects their stabilities and therefore their structures.

2. Experimental

We have constructed a new apparatus working according to the principles described by Smalley and co-workers [9]. The carbon clusters are generated by a pulsed vaporization laser focussed onto a rotated target rod in the mist of a helium gas pulse. After the molecular beam has been skimmed twice, the cluster ions are extracted and focussed by means of a Wiley and McLaren device [10], before entering the field free part of the first time-of-flight (T.O.F.). Following the mass gate is the interaction zone in which cluster photofragmentation is studied. The dissociation laser irradiates at right angle the selected ion packet.

In order to perform fluence dependence studies, the laser pulse energy was controlled by rotating a half wave plate before a Glan-polarizer. The laser was specially modified to have a good homogeneity for the output laser beam.

The remaining parent cluster ions and the positively charged fragments are mass analysed in a high resolution T.O.F. mass spectrometer (reflectron). Ions leaving from the reflectron, pass through the last field free region before detection with dual microchannel plates.

3. Results and discussion

The depletion dependence of C_{21}^+ and C_{28}^+ versus the irradiation laser fluence is shown in Figure 1. The irradiation light was the 2nd harmonic of a Yag laser ($\lambda_2 = 532nm; h\nu_2 = 2.33eV$). Each dot represents the experimental data point resulting from the ratio laser on to laser off signal. The relative fluctuation is ± 10%. The depletion curves present two or three parts separated by thresholds W_i. These different parts cannot be the consequence of the opening of a new fragmentation channel, corresponding to the same cluster structure because, from another photofragmentation experiment, Geusic and co-workers [7] have shown that most of the fragments are produced for approximately 2 $mJcm^{-2}$. A more realistic explanation supported by the curve fragmentation pattern, is the existence of several valence isomers with different stabilities, in the same cluster packet.

For a cluster packet consisting of several valence isomers, the ion parent depletion can be written as:

$$I^+ = I_0^+ \; [\; N_1 exp - \sigma_1 W/h\nu S + N_2 exp - \sigma_2 (W - W_1)/h\nu S + \cdots\cdots] \tag{1}$$

where we impose the conditions:

$$\begin{aligned} N_1 + N_2 + \ldots &= 1 \\ W - W_1 &= 0 \quad \text{if} \quad W < W_1 \\ W - W_i &= 0 \quad \text{if} \quad W < W_i \end{aligned}$$

N_1, N_2,... being the population densities for the first, second ... isomer, σ_1, σ_2,... the total photofragmentation cross-section for the population N_1, N_2,... according to $h\nu$; W_1, W_2,... are the thresholds from which the depletion of the following isomer begins.

The solid curve in figure 1 represents the best computer fit through the experimental data, according to equation 1, using the Simplex method. An obvious change in the law leading the photofragmentation process is shown. For C_{21}^+, above the threshold W_1, at least a second valence isomer is dissociate. Likewise, the depletion pattern of C_{28}^+ suggests that at least three valence isomers with different stabilities constitute the C_{28}^+ cluster packet.

These results can be compared with those obtained from mobility experiments performed by von Helden et al. [11]. Under their experimental conditions, they have shown that the ionic clusters such as $C_{11}^+ < C_n^+ < C_{20}^+$ present only a stable monocyclic ring structure, whereas clusters having between 21 and 28 carbon atoms present both mono and bicyclic ring structure possibilities. Concerning larger clusters from 29 and 41 atoms, mobility experiments conclude that these clusters present at least three valence structure possibilities. C_{21}^+ in figure 1 shows two valence isomers, in agreement with [11], which could be the mono and bicyclic ring structures supposed in the work of von Helden et al.. But, we have found three possible valence structures for C_{28}^+, instead of two for the mobility experiment. This

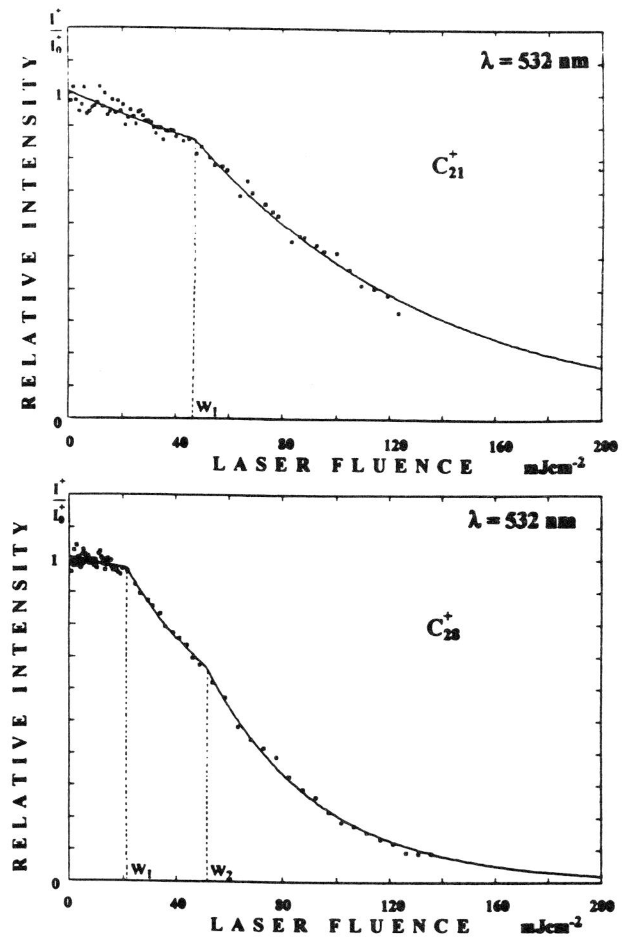

Fig. 1. Depletion laser fluence dependence of C_{21}^+ and C_{28}^+ with 532 nm irradiation light. A change in the depletion law appears for W_1 and W_2 fluences. This is interpreted by the presence of more stable valence isomers which begins to fragment only above W_1 or W_2. The solid curve is the best computer fit through the experimental data.

difference can be explained according to the following assumption: it seems possible that multiple bicyclic ring structures are simultaneously present in the cluster packet, but not resolved in the ion chromatography experiment [12]. In this case, two different bicyclic structures with a close mobilities, indiscernable by "the arrival time distribution method" [12] could present a different stability with respect to a radiation excitation and then would be distinguish in a fluence dependence experiment.

From the fit of the experimental data in figure 1, we cannot directly use the parameters such as the isomer population as well as the photodissociation cross-sections, because the solution of the equation 1 is far from the unity. However, the location of the different thresholds are known with relatively good accuracy because they are observed directly on the curves. So, another photofragmentation experiment has been performed using another irradiation light (the third harmonic of a Yag laser, $\lambda_3 = 355nm, h\nu_3 = 3.53eV$) in order to observe the depletion behaviour and the corresponding threshold positions. We observe that the dissociation pattern is the same for the two wavelengths for each cluster: as the function of the laser fluence. It seems that the photodissociation of these cluster ions would not be direct, but would be the consequence of the internal conversion of the energy brought by the photon, whatever its wavelength, via an excited electronic state [7]. The next step in the investigation, will be the measurement of the fragments produced by the different isomers, in order to make progress in the understanding of the valence isomer population belonging to these cluster packets.

References

1 Raghavachari, K., and Binkley, J.S.: J. Chem. Phys. **87**, 2191 (1987)

2 Bernholc, J., and Phillips, J.C.: J. Chem. Phys. **85**, 3258 (1986)

3 O'Brien, S.C., Heath, J.R., Curl, R.F., and Smalley, R.E.: J. Chem. Phys. **88**, 220 (1988)

4 Radi, P.P., Bunn, T.L., Kemper, P.R., Molchan, M.E., and Bowers, M.T.: J. Chem. Phys. **88**, 2809 (1988)

5 O'Keefe, A., M.Ross, M., and Baronavski, A.P.: Chem. Phys. Letters **130**, 17 (1986)

6 Pradel, P., Monchicourt, P., Laucagne, J.J., Perdrix, M., and Watel, G.: Chem. Phys. Letters **158**, 412 (1989)

7 Geusic, M.E., Jarrold, M.F., McIlrath, T.J., Freeman, R.R., and Brown, W.L.: J. Chem. Phys. **86**, 3862 (1987)

8 McElvany, S.W., Creasy, W.R., and O'Keefe, A.J.: J. Chem. Phys. **85**, 632 (1986); McElvany, S.W., Dunlap, B.I., and O'Keefe, A.J.: J. Chem. Phys. **86**, 715 (1987); McElvany, S.W.: J. Chem. Phys. **89**, 2063 (1988)

9 Brucat, P.J., Zheng, L.S., Pettiette, C.L., Yang, S., and Smalley, R.E.: J. Chem. Phys. **84**, 3078 (1986)

10 Wiley, W.C., and McLaren, I.H.: Rev. Sci. Instrum. **25**, 1150 (1955)

11 von Helden, G., Hsu, M.T., Kemper, P.R., and Bowers, M.T.: J. Chem. Phys. **95**, 3835 (1991)

12 von Helden, G., Hsu, M.T., Gotts, N.G., Kemper, P.R., and Bowers, M.T.: Chem. Phys. Letters **204**, 15 (1993)

This article was processed using Springer-Verlag TEX Z.Physik D macro package 1.0 and the AMS fonts, developed by the American Mathematical Society.

Z. Phys. A 349, 361–362 (1994)

Coexistence of α-clustering and shell structure in ^{40}Ca

Toshimi Sakuda [1], **Shigeo Ohkubo** [2,3]

[1] Department of Physics, Miyazaki University, Miyazaki, 889-21 Japan
[2] Nuclear Physics Laboratory, University of Oxford, Oxford OX1 3RH, UK
[3] Department of Applied Science, Kochi Women's University, Kochi, 780 Japan

Received: 27 October 1993

Abstract. The energy spectrum and α-spectroscopic factors of ^{40}Ca are calculated by using an $\alpha+{}^{36}$Ar orthogonality condition model. Overall agreement is obtained with the experimental data. The model successfully produces the parity-doublet $K^{\pi}=0^{+}$ and 0^{-} $\alpha+{}^{36}$Ar cluster bands. It is shown that the coexistence and interference of α-cluster states and shell-model states are indispensable to understand the structure of ^{40}Ca.

PACS: 21.60.Gx; 27.40.+z

It is well established that α-cluster structure is present in light nuclei of mass number $A \leqq 24$. There have been several studies [1–4] with a view of the extension of α-cluster spectroscopy in the *fp*-shell region. Recently, the α-transfer reactions to ^{44}Ti [5] and ^{40}Ca [6] have been carried out and have disclosed the presence of the parity-doublet of α+core cluster bands. These observations strongly suggest that the α-cluster structure may persist throughout the *fp*-shell region. In particular, the ^{40}Ca is the analog of ^{16}O in the *fp* shell and is of considerable interest.

In order to investigate the real existence of the α-cluster states in ^{40}Ca, the microscopic $\alpha+{}^{36}$Ar orthogonality condition model (OCM) [7] is used. An obvious advantage of this model is to provide a unified description of the cluster states and important shell-model states. The model space is generated as a product of the internal wave functions and the relative ones:

$$\frac{\mathscr{A}}{\sqrt{\binom{40}{4}}}\{\phi(\alpha)[\phi_L({}^{36}\mathrm{Ar})\,Y_l(\hat{\mathbf{r}})]_J\,R_{Nl}(r)\}, \tag{1}$$

where $R_{Nl}(r)$ is a radial harmonic oscillator wave function with N oscillator quanta of the relative motion. The internal wave functions $\phi(\alpha)$ and $\phi_L({}^{36}\mathrm{Ar})$ are assumed to be $(0s)^4$ and $(sd)^{-4}$ $(\lambda\mu)=(08)$ configurations, respectively. The present model space includes important shell-model states such as the sd-closed $(\lambda\mu)=(00)$ configuration, the (21) and (32) of $1p-1h$ states, and the (12,8) of $4p-4h$ states.

The total Hamiltonian of the system is written as

$$H=H(\alpha)+H({}^{36}\mathrm{Ar})+T_r+V_{\alpha-\mathrm{Ar}}. \tag{2}$$

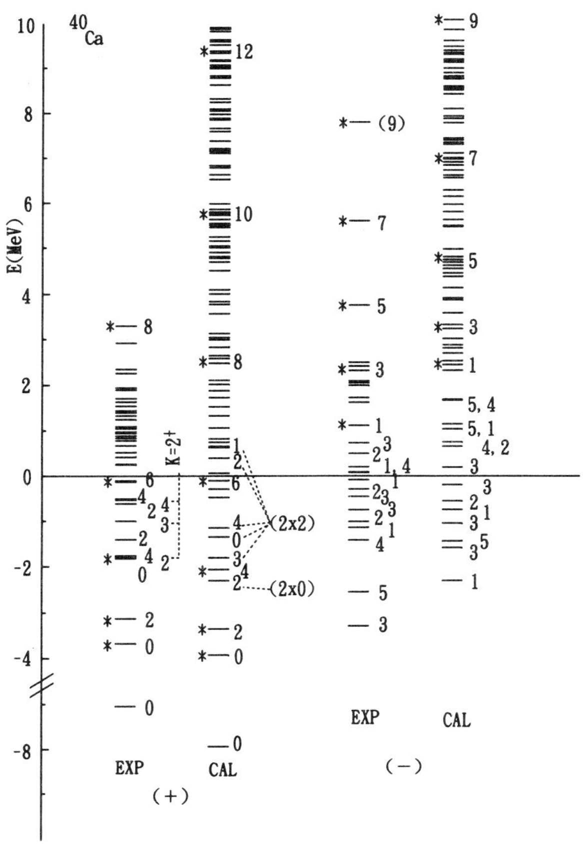

Fig. 1. Calculated and experimental energy spectra of ^{40}Ca. The energy scale is measured from the α-threshold. The states marked by an *asterisk* are the members of $\alpha+{}^{36}$Ar cluster bands. The multiplets of weak coupling $^{36}\mathrm{Ar}(2^{+})\times l$ band are indicated by *dotted*

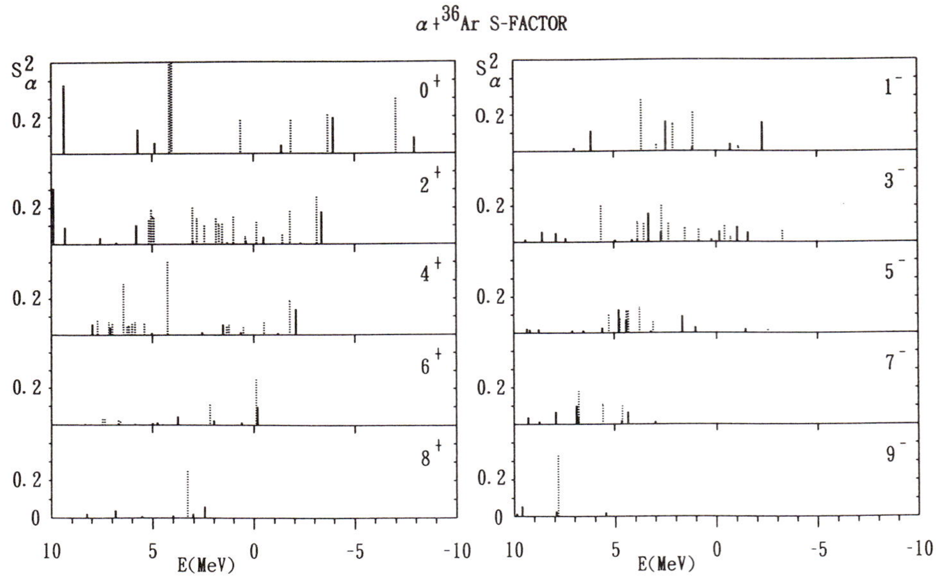

Fig. 2. Calculated (*full line*) and experimental (*dashed line*) α-spectroscopic factors of ^{40}Ca. Experimental values are from [6]

The inter-cluster potential $V_{\alpha-\mathrm{Ar}}$ is constructed by folding a two-nucleon interaction and can be expressed in operator form as

$$V_{\alpha-\mathrm{Ar}} = V^c + V^L \mathbf{L}^2 + V^{T2}[T_2, Y_2]_0 + V^{T2'}[T_2', Y_2]_0 + V^{T4}[T_4, Y_4]_0 \qquad (3)$$

where

$$T_{\lambda\mu} = \sum_{i\in\mathrm{Ar}} y_{\lambda\mu}(\xi_i), \qquad T'_{\lambda\mu} = \sum_{i\in\mathrm{Ar}} \xi_i^2\, y_{2\mu}(\xi_i), \qquad (4)$$

and $\mathbf{L}$ is the angular momentum of ^{36}Ar. The effective two-nucleon interaction used in this calculation is HNY force with the depth parameter $V_0(^3E) = -490$ MeV. The depths V^{T2} and $V^{T2'}$ are multiplied by factors 1.35 and 0.65, respectively, so as to give a good agreement for low-lying states energies.

The calculated and experimental energy spectra are shown in Fig. 1, where energies are measured from the $\alpha + {}^{36}$Ar threshold. In the experimental spectra, the states marked by an asterisk are strongly populated in the α-transfer reactions [6]. In the calculated spectra, the marked states belong to $K^\pi = 0^\pm$ $\alpha + {}^{36}$Ar cluster bands. The calculation gives good correspondence to the both of the experimental $K^\pi = 0^\pm$ cluster bands. The identification of the parity-doublet bands gives a strong support to the establishment of the $\alpha + {}^{36}$Ar cluster model of ^{40}Ca.

One would expect α-cluster states based on the first excited 2^+(1.97 MeV) state of ^{36}Ar around 2 MeV below the α-threshold. We show some of the weak coupling $(2^+) \times l$ band by dotted lines. The observed 2_2^+, 3_1^+ and 4_2^+ states, which are commonly interpreted as $4p-4h$ $K^\pi = 2^+$ band, are good candidates for the members. Further support for this identification comes from the strong $E2$ transitions of the levels. The other members of the weak coupling multiplet are less clear. The observed 0_3^+, 2_3^+ and 4_3^+ states are usually classified as the $8p-8h$ band [9], which are outside the framework of the present model. The low-lying negative-partiy states are in reasonable agreement with the experiments. These are dominated by mixtures of $1p-1h$ and $3p-3h$ configurations.

In Fig. 2, the calculated S_α^2 factors are compared with the experimental data. It is characteristic that α-cluster strenght is shared over many levels and some levels other than α-cluster levels possess appreciable S_α^2 factors. We can, however, identify the cluster states of the parity-doublet bands, which have noticeably large S_α^2 factors. The fragmentation of the α-cluster strength is the result of the coupling of $\alpha + {}^{36}$Ar cluster states with shell-model states.

To summarize, the existence of the parity-doublet $\alpha + {}^{36}$Ar cluster bands has been confirmed. It has been shown that the coexistence and interference of α-cluster states and shell-model states play very important roles in explaining the energy spectra and S_α^2 factors in ^{40}Ca. The α+core cluster model should certainly be equally appropriate for other nuclei in the *fp*-shell region as well.

References

1. Michel, F., Reidemeister, G., Ohkubo, S.: Phys. Rev. Lett. **57**, 1215 (1986); Phys. Rev. C**37**, 292 (1988); C**41**, 63 (1990)
2. Ohkubo, S.: Phys. Rev. C**38**, 2377 (1988)
3. Merchant, A.C.: Phys. Rev. C**36**, 778 (1987); C**37**, 414 (1988)
4. Pal, K.F., Lovas, R.G.: Phys. Lett. **96**B, 19 (1980)
5. Yamaya, T., et al.: Phys. Rev. C**42**, 1935 (1990); C**41**, 2421 (1990)
6. Yamaya, T., et al.: Phys. Lett. **306**B, 1 (1993); Private communication
7. Ogawa, T., Suzuki, Y., Ikeda, K.: Prog. Theor. Phys. **57**, 1072 (1977)
8. Fortune, H.T., et al.: Phys. Rev. C**19**, 756 (1979)
9. Gerace, W.J., Green, A.M.: Nucl. Phys. A**93**, 110 (1967); A123, 241 (1969)

Phonon density of states in nano-clusters

W. Schommers[1], C. Syros[2], C. Politis[3,4,5], and N. Athanasiou[6]

[1]Kernforschungszentrum Karlsruhe, Institut für Materialforschung I, P.O.Box 3640, 76021 Karlsruhe, Germany
[2]University of Patras, Laboratory of Nuclear Technology, 26110 Patras, Greece
[3]University of Patras, 26110 Patras, School of Engineering, Greece
[4]Kernforschungszentrum Karlsruhe, Institut für Nucleare Festkörperphysik, P.O.Box 3640, 76021 Karlsruhe, Germany
[5] University of California, Department of Electrical and Computer Engineering, San Diego, La Jolla, California 92093-0407, U.S.A.
[6] Commission of the European Communities; Joint Research Center, Institute for Transuranium Elements, P.O.Box 2340, 76125 Karlsruhe, Germany

Received 28 June 1993

Abstract. The phonon density of states for nano-clusters has been studied theoretically using molecular dynamics; in this way anharmonic effects could be fully cosidered. Comparisons are made with results obtained from bulk calculations. The cluster-properties have been studied for several particle numbers and temperatures, respectively.

It turned out that the properties of clusters are sensitive to small variations in the interactions between the particles. Therefore, in the calculation of cluster-properties the precise knowledge of the interaction is necessary. In our study we have chosen krypton nano-systems because for this substance a reliable interaction potential is available. First estimations have also been performed for metal (Al, Pb) systems.

PACS: 61.25.Bi, 61.43.Bn, 63.20.-e, 68.35.Ja

1. Introduction

The properties of clusters are in general quite different from the corresponding bulk states. This is mainly due to the fact that with decreasing number of atoms a relative increase of the number of surface atoms is accompanied; in the case of nano-clusters the structure and dynamics is already dominated by the surface atoms.

These cluster-effects are essentially influenced by *anharmonicities*, which are distinctly larger at the surface than in the bulk: The surface particles are, in general, less bonded than the bulk particles and, therefore, the mean-square amplitudes of the particles at the surface are significantly larger than in the bulk, a fact leading to (enhanced) anharmonicities.

The harmonic approximation should be valid for surface vibrations of *semi-infinite* noble-gas crystals [1] below one-sixth of the melting temperature T_m (as compared to about 1/3 T_m for the bulk). The melting temperature of krypton is 116 K. Thus, in the case of krypton usual lattice-dynamics calculations for the *phonon density of states* should only be valid for T=19 K. For T>19 K we have to consider anharmonic effects. This is the situation for *semi-infinite crystals*; however, the *anharmonic effects* are still larger in the case of *clusters* where the number of surface particles are larger than in the case of semi-infinite crystals.

2. The model

Molecular dynamics (MD) calculations (see, for example, [2]) are important in studing classical many-particle systems with strong anharmonicities, since anharmonicity is treated without approximations. Therefore, we have calculated the phonon density of states on the basis of MD calculations, and we are not restricted to low temperatures as in the case of the usual lattice-dynamics calculations.

In order to carry out the first-principles calculation of cluster properties, it is necessary to have an accurate knowledge of the atomic interactions. This requirement is nearly satisfied in the case of noble-gas solids. In contrast to metals, the potential functions of noble-gas solids do not depend on the density, and they are the same at the surfaces and in the bulk of the crystal. At free metal surfaces the local background electron density may be changed from its average bulk value and produces concomitant changes in the potential functions. Such difficulties do not arise at noble-gas surfaces and, therefore, we want to restrict our study to noble-gas (krypton) surfaces. First estimations have been performed on metal systems (Al, Pb).

We have chosen a krypton system because for this substance a reliable pair potential is available. The krypton potential of Barker et al. [3] should be more realistic than the more familiar Lennard-Jones [6-12] potential because it undoubtedly correlates a much wider range of experimental data [4-7].

For a classical cluster-system consisting of N atoms, Hamilton's equations were solved by iteration. The time step used in the calculation was 10^{-14}sec. We did the calculations for 8500 time steps (0.85×10^{-10}sec). The magnitudes of the initial velocities of all atoms were chosen to be equal, but Maxwell's distribution was reached after 120 time steps (1.2×10^{-12}sec). The initial directions of the velocities were assumed to be randomly distributed, so that the velocity averaged over all atoms is zero and remains zero for all time steps.

3. Results for the phonon density of states

3.1. Nano-clusters

The phonon density of states (PDOS) can be calculated from the velocity autocorrelation function (VAF)

$$\psi(t) = \frac{< \boldsymbol{\nu}(0) \cdot \boldsymbol{\nu}(t) >}{< \boldsymbol{\nu}^2(0) >} \tag{1}$$

where $\boldsymbol{\nu}(t)$ is the velocity at time t for one atom of the ensemble and the brackets $< \ldots >$ denots a statistical average.

The Fourier transform of $\psi(t)$ yields a frequency spectrum $f(\omega)$ which is, in the case of the *harmonic* solid, the frequency spectrum of the normal modes. Thus, for an *anharmonic* system we can define a generalized PDOS by means of the Fourier trnasform of $\psi(t)$:

$$f(\omega) = \frac{2}{\pi} \int_0^{\infty} \psi(t') cos(\omega t') dt' \tag{2}$$

where $f(\omega)$ is normalized to unity:

$$\int_0^{\infty} f(\omega) d\omega = 1 \tag{3}$$

With the velocities $\boldsymbol{\nu}_i(t)$, $i = 1, \ldots, N$, calculated on the basis of molecular dynamics, $\psi(t)$ can be determined in the following way:

$$\psi(t) = \frac{1}{N_\tau} \frac{1}{N} \sum_{j=1}^{N_\tau} \sum_{l=1}^{N} \frac{\boldsymbol{\nu}_i(\tau_j) \cdot \boldsymbol{\nu}_i(t + \tau_j)}{\boldsymbol{\nu}_i^2(\tau_j)} \tag{4}$$

$\psi(t)$ is independent of the initial time τ_j (Liouville theorem). The number of initial N_τ was chosen to be 200. Then we obtain, for example, in the case of i=108, a statistical error of 0.7%.

We did the calculations for several particle numbers and temperatures, respectively. For example, Fig. 1 shows the PDOS for N=500 (corresponding to a box size of $\sim 3nm$) and T=37K. Fig. 2 shows the PDOS for N=500 and T=85K, and in Fig. 3 is the PDOS given for N=108 (corresponding to a box size of $\sim 2nm$) and T=91K.

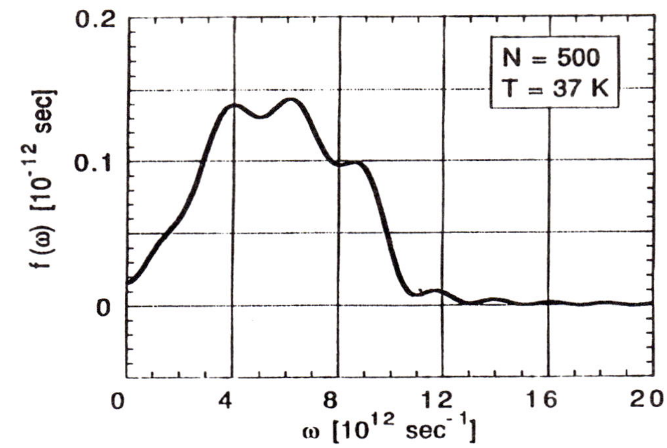

Fig. 1. Phonon density of states for a nano-cluster consisting of N=500 particles at T=37K

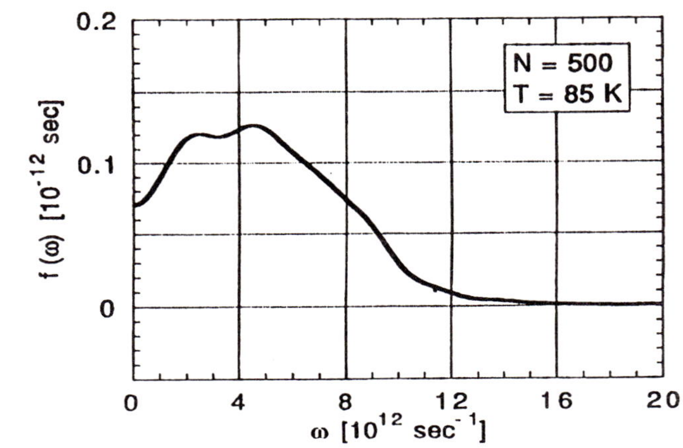

Fig. 2. Phonon density of states for a nano-cluster consisting of N=500 particles at T=85K

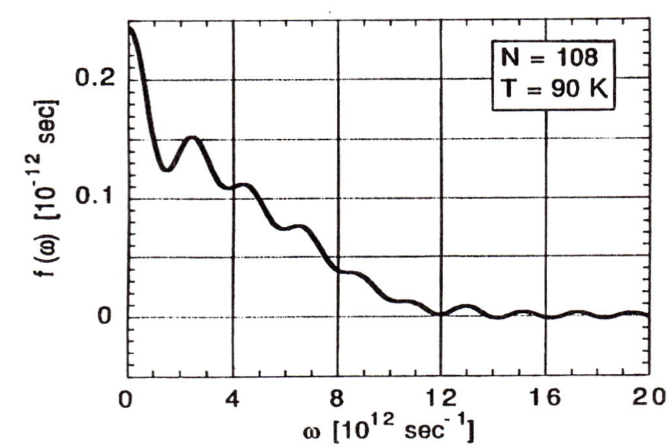

Fig. 3. Phonon density of states for a nano-cluster consisting of N=500 particles at T=90K

We observe by comparison that with decreasing particle number and increasing temperature the PDOS is shifted to low frequencies. More information can be obtained when we compare our cluster-results with the corresponding bulk data. Therefore, we have also performed molecular dynamics calculations for the bulk.

3.2. Phonon density of states for the bulk

The bulk calculations have been performed for N=500 particles. To avoid boundary effects, periodic boundary conditions were imposed on the system.

We did calculations for two temperatures: 7K and 102K. The melting temperature of krypton is 116K. The results are given in Fig. 4. The arrows in Fig. 4a indicate the positions of the peaks in $f(\omega)$ which have been obtained on the basis of a three-nearest-neighbour force-constant fit to experimental dispersion curves [8]. The positions of the peaks obtained from our calculations are the same. The absolute values of $f(\omega)$ differ slightly from each other. The numerical uncertainties in our calculation cannot explain these differences. Quantum effects (not considered in our calculations) should also contribute but are expected to be small in the case of krypton [9,10]. The numerical errors in the force constant fit (in particular in connection with the sampling process) are not quoted in [8].

4. Discussion

When we compare the cluster-results (Fig 1-3) with the corresponding results in the bulk, we recognize that the curves are strongly shifted to low frequencies; this effect cannot be explained exclusively by a temperature effect but this effect is relatively small in comparison to the size-effect.

Such an effect must have consequencies for other properties of the materials, since - as is well known - many other properties can be expressed in terms of phonons.

First estimations for the PDOS on metal-clusters (Al, Pb) showed that the effects, observed for krypton, are of the same order in metals.

Such a phonon-softening effect can influence the superconducting transition temperature T_c because such an effect can essentially enhance the electron-phonon interaction strenght (see, for example, [10])

$$\lambda = 2 \int_0^\infty \alpha^2(\omega) \frac{f(\omega)}{\omega} d\omega \tag{5}$$

(note that ω appears in the denominator), provided $\alpha^2(\omega)$ in the Eliashberg function $\alpha^2(\omega)f(\omega)$ does not compensate this effect.

One of the authors (W. Sch.) would like to thank Prof. Eliashberg for interesting discussions.

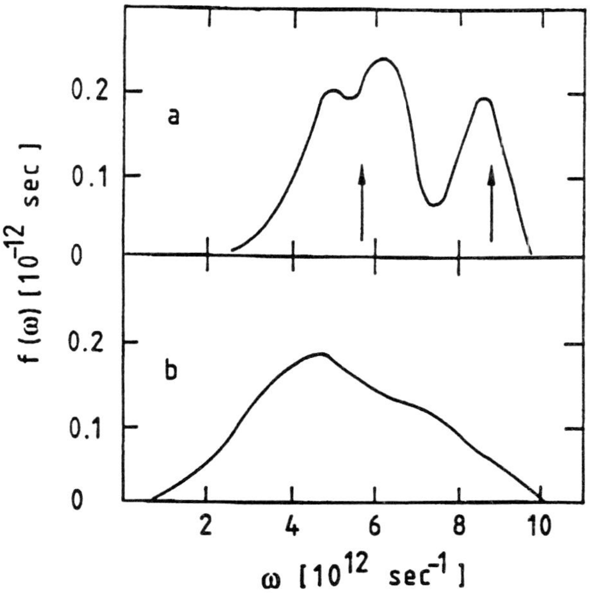

Fig. 4. Phonon density of states for the bulk (N=500). a. T=7K; b. T=102K. (For comparison, the melting temperature is 116K.)

References

1 Schommers, W.: Phys. Rev. B **32**, 6845 (1985)
2 Schommers, W.: in Structure and Dynamics of Surfaces I, ed. by Schommers, W., and von Blanckenhagen, P., Topics in Current Physics, Vol. 41, Springer-Verlag, Berlin, Heidelberg, 1986
3 Barker, J.A. et al.: J. Chem. Phys. **61**, 3081 (1974)
4 Egelstaff, P.A., Salacuse, J.J., Schommers, W., and Ram, J.: Phys. Rev. A **30**, 374 (1984)
5 Salacuse, J.J., Schommers, W., and Egelstaff, P.A.: Phys. Rev. A **34**, 1516 (1986)
6 Romahn, U., Göbel, H., Schommers, W., and von Blanckenhagen, P.: Modern Phys. Lett. B **5**, 1873 (1971)
7 Schommers, W., von Blanckenhagen, P., and Syros, C.: Modern Phys. Lett. B **6**, 23 (1992)
8 Skalyo, J., Endok, Y., and Shirane, G.: Phys. Rev. B **9**, 1797 (1974)
9 Rare Gas Solids, Vol. 1, ed. by Klein, M.L., and Venables, J.A., (Academic, 1976)
10 Barron, T.H.K., and Klein, M.L.: Proc. Phys. Soc. **85**, 833 (1965)
11 Allen, P.B., and Dynes, R.C.: Phys. Rev. B **12**, 905 (1975)

This article was processed using Springer-Verlag TEX Z.Physik D macro package 1.0 and the AMS fonts, developed by the American Mathematical Society.

Charge transfer and radical ion formation in fullerene solution

Y. Tai, T. Osaki, M. Tawaza, S. Tanemura, K. Inukai, S. Sakakibara, K. Ishiguro[1], Y. Sawaki[1], Y. Saito[2], H. Shinohara[2], and H. Nagashima[3].

GIRIN: The Government Industrial Research Institute, Nagoya, Kita-ku, Nagoya 462, Japan
[1]Faculty of Engineering, Nagoya University, Chikusa-ku, Nagoya 464, Japan
[2]Faculty of Engineering, Mie University, Kamihama-cho, Tsu 514, Japan
[3]School of Material Science, Toyohashi University of Technology, Tenpaku-cho, Toyohashi 441, Japan

Received 28 June 1993

Abstract. Electron transfer kinetics in C_{70}/donor system was investigated by means of laser flash spectrometer in benzonitrile and in acetonitrile-benzene (1:4 in volume) solution. Rehm-Weller semi-empirical equation could be fitted successfully for the quenching rate of triplet C_{70} by various donors when the diffusion-controlled quenching rate was corrected according to the difference in viscosity between acetonitrile and the solvents studied. Not only the correlation of electron transfer rate to the free energy difference between initial and final state but also the rate constant data were fairly similar to those reported for C_{60}/donor system by Foote et al. These coincidences can be attributed to the very small difference in triplet energy and reduction potential of these molecules.

Free radical ion yield in C_{60} or C_{70}/donor system in benzonitrile was estimated from the absorbance of donor$^{+\cdot}$ compared with that in anthraquinone (AQ)/donor system whose radical ion yield (0.88) is known. The obtained value for each fullerene was significantly lower than that value.

PACS: 82.30.Fi

1. Introduction

Carbon fullerenes have been attracting much attention in every field of science because of their beautiful structures. Extensive studies have been carried out to reveal some of their important characteristics like superconductivity in alkali metal doped C_{60}. There also have been interesting results on fullerenes in photochemistry. Energy transfer from triplet C_{60} to oxygen resulted to the production of singlet O_2 with high efficiency [1]. Foote et al. observed the electron transfer from various electron donors to triplet C_{60} in solution [2].

In this report, we present the results for the following two experiments. We measured the quenching rate for triplet excited C_{70} by various electron donors using laser flash spectrometer. The quenching rate for C_{60} has already been reported to reveal that it was controlled by the free energy change in electron transfer through Rehm-Weller semi-empirical relation [2]. C_{70} has almost the same reduction potential as that of C_{60} [3] whereas the structural difference makes each photochemical property somewhat distinctive. It is therefore attractive to examine how such structural factors affect the electron transfer in solution.

The second we point out herein is that the free radical ion yield in fullerene/donor system was significantly low compared with that in AQ/donor system. The free radical ion yield for an organic acceptor like AQ, being quenched its triplet state, was reported to be very high in polar solvent like acetonitrile [4]. However, C_{60}, having degenerated LUMO [5], may wear different ST mixing property from those of organic acceptors.

2. Experimental

The method for the kq measurement has been described previously [6].

Free radical ion yield in benzonitrile was estimated as follows. The absorption intensity at λ_{max} of donor cation in AQ/donor system was measured about 1.5μs after the laser pulse [7]. Then the absorption intensity of cations was measured in C_{60} or C_{70}/donor solution whose the optical density of the fullerene at the excitation wavelength (355nm) was controlled to be equal to the AQ/donor solution. The free radical ion yield in AQ/donor system has already been reported [4]. Accordingly the free ion yield in the fullerene/donor system was estimated through the comparison of the absorption intensity of donor cations in each solution. The donors used are dimethylaniline, tetramethyl-p-phenilenediamine, and triphenylamine.

3. Result and discussion

3.1. Quenching rate for triplet C_{70}

Closed circles in Fig. 1 indicate the kq values in benzonitrile plotted against the free energy change (ΔG_0) in electron transfer. ΔG_0 was estimated using eq. (1) [8].

$$\Delta G_0 = [E(D^{+\cdot}/D) - E(A^{-\cdot}/A) - \Delta E_{0,0} - e_0^2/a\varepsilon] \quad (1)$$

$E(D^{+\cdot}/D)$ indicates the energy required for the oxidation of a donor. $E(A^{-\cdot}/A)$ is the reduction energy for C_{70} (−0.51V) [9]. $\Delta E_{0,0}$ and $e_0^2/a\varepsilon$ are the excitation energy for C_{70} to the triplet state (1.53V) and the polarization energy of the solvent (0.085V in benzonitrile), respectively [2]. The quenching rate remained almost the same value ($\sim 5 \times 10^9 M^{-1}s^{-1}$) in $\Delta G_0 < -20kJmol^{-1}$, whereas it abruptly decreased with the increase in the free energy difference in $\Delta G_0 > -20kJmol^{-1}$.

Rehm and Weller suggested a semi-empirical relation between ΔG_0 and kq for charge separation reaction in acetonitrile [8]. The solid line in Fig. 1 is the calculated curve using Rehm-Weller equation. In the region $\Delta G_0 > -20kJmol^{-1}$, it reproduced the experimental data, whereas the experimental kq values were apparently lower than those calculated in $\Delta G_0 < -20kJmol^{-1}$. In this region of ΔG_0, the rate constant is considered to be governed by diffusion of each ion. Thus we applied the diffusion rate $5.6 \times 10^9 M^{-1}s^{-1}$, estimated from the Einstein-Stokes relation, to Rehm-Weller equation (dotted line) to fit the experimental data in the whole ΔG_0 region. Such a correlation of electron transfer rate to the free energy difference and almost the same rate constant values in each C_{70}/donor case have already reported in C_{60}/donor system in benzonitrile [2]. Our result inferred that C_{70}, having practically the same excitation and reduction energies as C_{60}, offers similar rate correlation with C_{60} in being quenched by donors because the thermal factors are decisive of those reactions.

Open circles in Fig. 1 indicate the rate constants obtained using acetonitrile-benzene (1:4) as a solvent. The mixed solvent has smaller viscosity than benzonitrile, which fact means that the quenching rate in each C_{70}/donor case should be enhanced [6]. The obtained data, however, showed smaller rates than in benzonitrile. The polar effect of the solvent causes only slight change in ΔG_0 ($\sim 1kJmol^{-1}$) and is considered to have no practical influence to the rate constants although the dielectric constant of the mixed solvent is smaller than that of benzonitrile. The reason for the small rate constants in the mixed solvent is not understood but it might be effective if the encounter distance of the ions (a in eq. (1)) changed according to the solvent polarity.

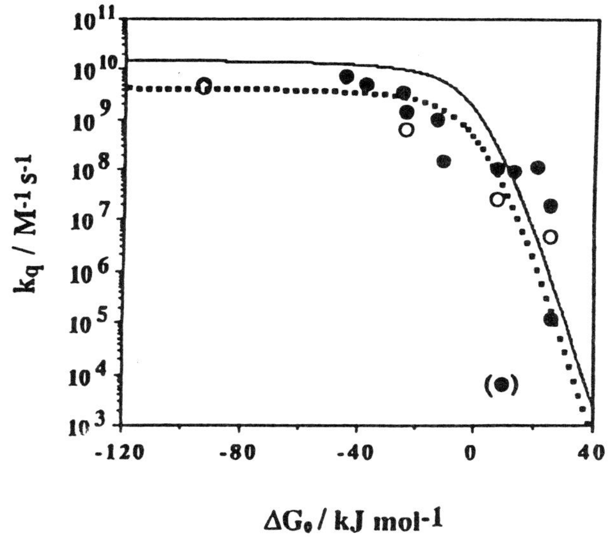

Fig. 1. Correlation of quenching rate constants (kq) for triplet C_{70} by donors with free energy change (ΔG_0) in electron transfer; • : kq in benzonitrile, o : kq in acetonitrile-benzene (1:4). The solid line was calculated from the Rehm-Weller equation, and the dotted line was calculated using the diffusion rate constant of $5.6 \times 10^9 M^{-1}s^{-1}$ in benzonitrile. For the ΔG_0 value of pyrrole (i.e., (•)), see ref. [10]

3.2. Free ion yield

For all cases of C_{60} or C_{70}/donor used, the absorbance of amine cations were suppressed to 1/2~1/5 compared with that in AQ/donor system. Although the transient absorption of amine cations in raw spectra was not very clear because of the overlap with those of anions, the spectra of cations obtained though the extraction of the absorption of anions was quite similar to those reported previously. The small absorbance of amine cation implies that the probability for ion pairs to recombine themselves is relatively high even though they should be produced as triplet pairs. Dinse et al. measured the spin relaxation rate for C_{60} anion in toluene by means of pulsed esr combined with a visible laser [11]. The obtained data were in the order of $10^6 s^{-1}$ around room temperature. It is thus hardly considered that ST mixing becomes very high compared with the case of AQ/donor if there is no specific interaction within fullerene$^{-\cdot}$ and donor$^{+\cdot}$. We are now planning to perform pico-second flash photolysis to investigate the nature of electron transfer events more precisely.

References

1 Terazima, M., Hirota, N., Shinohara, H., and Saito, Y.: J. Phys. Chem. **95**, 9080 (1991)

2 Arbogast, J.W., Foote, C.S., and Kao, M.: J. Am. Chem. Soc. **114**, 2277 (1991)

3 Allemand, P.M., Koch, A., Wudl, F., Rubin, Y., Diederich, F.N., Alvarez, M.M., and Whetten, R.L.: J. Am. Chem. Soc. **113**, 1050 (1991)

4 Haselbach, E., Vauthey, E., and Suppan, P.: Tetrahedron **44**, 7335 (1988)

5 Saito, S., and Oshiyama, A.: Phys. Rev. Lett. **66**, 2637 (1991)

6 Osaki, T., Tai, Y., Tazawa, M., Tanemura, S., Inukai, K., Ishiguro, K., Sawaki, Y., Saito, Y., Shinohara, H., and Nagashima, H.: Chem. Lett. 789 (1993)

7 Early events relating to charge transfer are already completed at this time range in the sample solutions used. In this region, radical ions were terminated by diffusion controlled encounter to the opposite charges.

8 Rehm, D., and Weller, A.: Isr. J. Chem. **8**, 259 (1970)

9 The reduction potential for C_{70} is practically the same as that for C_{60} [3], hence, −0.51V in acetonitrile is used,

10 The oxidation potential of pyrrole was observed as 1.2V vs. SCE by a rotating ring-disk electrode; however, the accurate oxidation potential could not be obtained because of the conducting polymeric film formed on the electrode surface. See, Diaz, A.F., Martinez, A., and Kanazawa, K.K.: J. Electroanal. Chem. **130**, 181 (1981)

11 Ruebsam, M., Dinse, K.P., Plueschau, M., Fink, J., Kraetschmer, W., Fostiropoulos, K., and Taliani, C.: J. Am. Chem. Soc. **114**, 10059 (1992)

This article was processed using Springer-Verlag TEX Z.Physik D macro package 1.0 and the AMS fonts, developed by the American Mathematical Society.

Z. Phys. D 31, 287–289 (1994)

On the electronic shell structure of metallic spheres and circular discs

B. Tatievski, P. Stampfli, K.H. Bennemann

Institut für Theoretische Physik, Freie Universität Berlin, Arnimallee 14, D-14195 Berlin, Germany

Received: 28 June 1993 / Final version: 20 May 1994

Abstract. The electronic shell structure of spherical and circular potential wells of finite depth is examined as an extension of the work of Balian and Bloch. We obtain analytical expressions for the electronic density of states, which are in good agreement with numerical results. The position of the maxima and minima in the density of states depends strongly on the depth of the potential via the quantummechanical scattering phases. For circular discs the oscillating structure in the density of states shows a different supershell pattern than for spheres. This is a consequence of the difference in the effective quantummechanical multiplicity of closed paths, which determine the oscillating structure.

PACS: 36.40; 33.10

I. Introduction

The mass spectra of large metallic clusters [1] have a periodic structure extending to very large sizes of several thousand atoms and which results from electronic shells [2, 3]. The semiclassical theory by Balian and Bloch [2] has been used in a simplified model by Stampfli and Bennemann [4] to obtain analytical expressions for the ionization potential, photoyield and cohesive energy of these clusters, which are in good agreement with experiment [1] and numerical calculation [3]. In this work we discuss in more detail the electronic shell structure of spherical and of circular potential wells.

II. Theory

Balian and Bloch [2] obtained the electronic density of states from a Green's function, for which they used a multiple scattering expansion and a semiclassical approach. Oscillating structures in the density of states result from closed scattering circuits in the vicinity of stationary classical paths being regular polygons and star polygons. The resulting contribution to the density of states is (using atomic units with $\hbar=e=m=1$)

$$\Delta\rho_s(k)=\operatorname{Im}\sum_{\substack{p>2t\\ t\geqq 1}}\frac{4(-1)^t(\cos\varphi_{p,t})(\sin\varphi_{p,t})^{3/2}R^2(kR)^{1/2}}{\pi^{1/2}p^{1/2}}\cdot\exp(i\Phi_{p,t})-\operatorname{Im}\sum_{t\geqq 1}\frac{R^2}{\pi t}\exp[2it\delta_{2t,t}+4ikRt] \tag{1}$$

where

$$\Phi_{p,t}=p(\delta_{p,t}-\pi/2)+2kRp\sin\varphi_{p,t}+3\pi/4 \tag{2}$$

is the total phase of a wave propagating along a closed polygonal path with p corners and t cycles around the center of the sphere ($p=nt$ for a regular polygon). R is the radius of the sphere and $\varphi_{p,t}=\pi t/p$ is one half of the angle subtended by one side of the polygon. Each reflection at the surface contributes a phase of $\exp(i\delta_{p,t})=-1$ for an infinitely deep potential well. Proceeding similarly as Balian and Bloch we obtain the structure in the density of states of a circular disc, which may be an acoustic membrane, or a microlithographically prepared layer as

$$\Delta\rho_c(k)=\operatorname{Im}\sum_{\substack{p\geqq 2t\\ t\geqq 1}}\frac{2\alpha_{p,t}(\sin\varphi_{p,t})^{3/2}R^2}{\pi^{1/2}p^{1/2}(kR)^{1/2}}\cdot\exp(i\Phi_{p,t}), \tag{3}$$

where $\alpha_{p,t}=2$ for $p>2t$ and $\alpha_{2t,t}=1$. Note that the electronic shell structure of spheres and circular discs has already been discussed by Berry and Tabor [5] using Einstein-Brillouin-Keller theory; however no explicite equations have been given.

For a finite potential step at the surface we propose to use (1–3) together with a quantummechanical evaluation of the scattering phase $\delta_{p,t}$. The smallest distance

This paper was originally submitted in connection with the 2nd. Int. Conference on Atomic and Nuclear Clusters held in Santorini from 28. June–2. July 1993 and is published here as a regular article after an independent refereeing procedure according to the standards of Z. Phys. D

from the center to the sides of the polygon is $d = R\cos\varphi_{p,t}$ and thus the angular momentum of a wave ψ traveling along this path is given by $\hbar l = kR\cos\varphi_{p,t}$. Assuming, that the potential is approximately constant we put

$$\psi_l(r,\vartheta,\varphi) \cong [h_l^{(1)}(kr)/h_l^{(1)}(kR) + e^{i\delta_{p,t}} h_l^{(2)} kr)/h_l^{(2)}(kR)]\, Y_{l,m}(\vartheta,\varphi) \tag{4}$$

inside the sphere and

$$\psi_l(r,\varphi) \cong [H_l^{(1)}(kr)/H_l^{(1)}(kR) + e^{i\delta_{p,t}} H_l^{(2)}(kr)/H_l^{(2)}(kR)] \exp(\pm il\varphi) \tag{5}$$

inside the circular disc ($0 \ll r < R$). The scattering phase then follows from the usual boundary conditions at the surface.

III. Results

We use in our calculation parameters that are typical for clusters of sodium atoms [6]. The free electron radius is $r_s = 2$ Å, the Fermi-energy is $E_F = 3.24$ eV, the threshold of the photoelectric effect is $\phi = 2.35$ eV and the Fermi-wavevector is $k_F = 0.92 \cdot 10^8$ cm^{-1}. Thus the electronic potential inside the metal is lower than the vacuum level by $V_0 = \phi + E_F = 5.59$ eV.

Results are shown in Fig. 1 for the shell structure in the electronic density of states ρ of a sphere of radius $R = 17$ Å. For simplicity we consider a single discontinuous potential step at the surface of the cluster because the scattering phases $\delta_{p,t}$ and the electronic wavefunctions can then be calculated analytically. The electronic potential is then $V(r) = 0$ for $r < R$ and $V(r) = V_0$ for $r > R$. Note that more realistic surface potentials could be used using (5), which requires numerical calculation. In order to obtain a smooth curve we used a constant imaginary part $\mathrm{Im}(kR) = 0.3$ for the wavevector in the semiclassical approximation (1–3, shown as solid lines) and a corresponding broadening of the energy levels in the numerical calculation (dotted lines). The semiclassical result (solid line) agrees very well with the numerical results (dotted line), except for very small k, for which the semiclassical approximation is not valid, and for large k near the continuum of free states. The density of states for a sphere (Fig. 1) has rather regular oscillations (shells), which are modulated by a slowly varying factor [2–4] giving rise to supershells. Overall, the oscillation is quite well approximated by triangular ($p = 3$ and $t = 1$) and square ($p = 4$ and $t = 1$) orbits as presumed in our previous work [4]. In addition, the pentagon ($p = 5$ and $t = 1$) makes as small contribution.

Note, the finite potential well gives a similar shell structure as an infinitely deep potential well [2]. However, the scattering phases $-\pi < \delta_{p,t}(k) < 0$, which are obtained from quantum mechanics, now have different values resulting in entirely different shell positions. This can be clearly seen in Fig. 2, where the dashed line shows the result for an infinite potential step ($V_0 \to \infty$) at the cluster surface and the solid line is the same as in Fig. 1. The positions of the maxima (electronic shells) of the two curves are strongly different. In some cases ($k/k_F \simeq 0.7$) the oscillations are rougly opposite in phase such that respective minima coincide with maxima. This suggests that the Einstein-Brillouin-Keller (EBK) theory [5] and the related theory of Gutzwiller fail for finite potential barriers. These theories only take into account the potential in the region that can be reached by a classical particle. Thus it is neglected that the quantum-mechanical energy eigenvalues also depend on the potential in the classically forbidden region. This would result in $\delta_{p,t}(k) \equiv -\pi$ independently of the height of the potential step and the same shell structure (dashed line in Fig. 2)

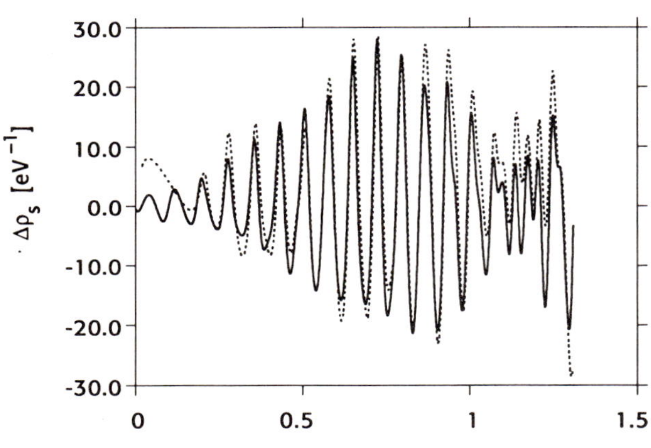

Fig. 1. Electronic shell structure for a sphere with a finite potential step V_0 at the surface. The result of the semiclassical approximation (1) is shown as a *solid line* and the result of a numerical calculation is shown as a *dotted line*. Note, parameters appropriate for Na-clusters are used, thus $V_0 = \phi + E_F = 5.59$ eV and the continuum of free states begins at $k/k_F = 1.3$. The radius of the sphere is $R = 17$ Å, corresponding to 600 sodium atoms

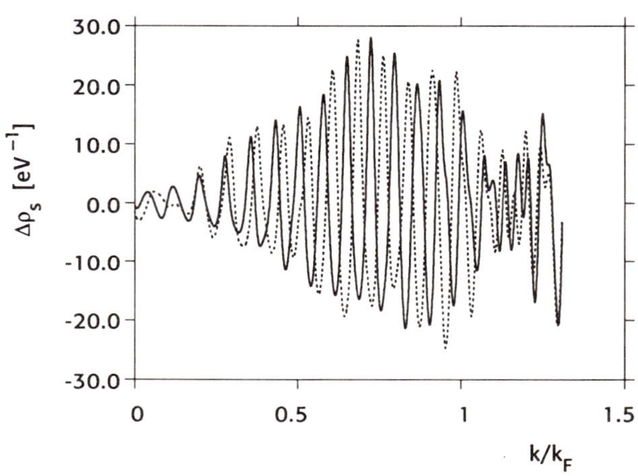

Fig. 2. Electronic shell structure for a sphere with radius $R = 17$ Å. The *solid line* shows the same result as in Fig. 1. For comparison the *dashed line* shows the result obtained for an infinite potential step $V_0 \to \infty$ at the surface. Note the difference in the position of the shells (maxima of $\Delta\rho_s$)

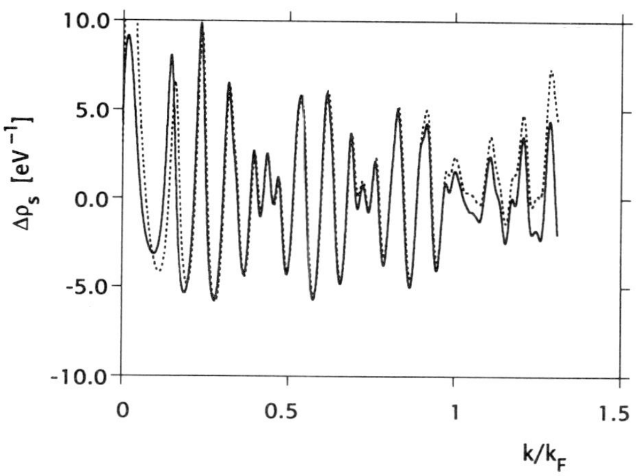

Fig. 3. Electronic shell structure for a circular disc with a finite potential step V_0 at the surface. The result of the semiclassical approximation (3) is shown as a *solid line* and the result of a numerical calculation is shown as a *dotted line*. The same parameters V_0 and R are used as in Fig. 1

would then be obtained for any finite or infinite potential step. Clearly, the results of the EBK theory are then in severe disagreement with more accurate calculations, see Figs. 1 and 2.

In contrast to the sphere, the electronic shell structure of a circular disc (Fig. 3) results in oscillations with a more rapidly varying amplitude (supershell pattern) because the linear orbit ($p=2$ and $t=1$), which can be neglected for the sphere, is now of similar importance as the triangular and square orbits (see 3). The larger difference between the length of the linear orbit and the triangular and square orbits results in a more rapid beating pattern, containing only about two to three shells per supershell. Note, the importance of each orbit is in part determined by its number of distinct orientations (multiplicity) [5]. From the quantization of angular momentum follows that the multiplicity of the paths is $(2l+1)$ (see (4), $m=-l, -l+1, \ldots, l$) for a sphere. This strongly suppresses the linear orbit ($l=0$) in comparison to the other orbits ($l \gg 1$) and results in the semiclassical factor $(\cos \varphi_{p,t}) \propto l$ in (1). In contrast, a circular disc only allows $m = \pm l$, see (5), because the path has to lie inside its plane. Thus the linear orbit only counts once ($m=0$) and the other orbits count twice, which only slightly reduces the effect of the linear orbit (see the factor $\alpha_{p,t}$ in (3)).

A direct comparison with experiments [1] and previous numerical calculations [3] for the electronic shell structure of metallic spheres requires a numerical evaluation of the scattering phases $\delta_{p,t}$ for a realistic potential $V(r)$, which is left to future work. Bot note that the present (1) has exactly the form proposed previously by Stampfli and Bennemann [4]. In addition, the factor $R^2(kR)^{(1/2)}$ and the predominance of the triangle and square orbits too have been conjectured correctly. The resulting shell structure [4] in the ionization potential and in the cohesive energy was in good agreement with experiments and other calculations. Thus we expect that the present Eq. (1) should be in similar or better agreement.

IV. Conclusions

In conclusion, we discuss the shell structure in the electronic density of states of spheres and circular discs with a finite potential step at their surface. Good agreement is obtained between the results of a semiclassical theory and numerical calculations. The supershell pattern of circular discs has a much shorter beating period than for spheres. Quantum mechanics has an important role in the semiclassical theory because it determines the scattering phases at the surface and the effective multiplicity of paths. Thus the electronic shell structure has more in common with electromagnetic resonances than with classical particles moving along mechanical paths.

References

1. Martin, T.P., Bergmann, T., Gölich, H., Lange, T.: Z. Phys. D **19**, 25 (1991); Pedersen, J., Bjornholm, S., Borggreen, J., Hansen, K., Martin, T.P., Rasmussen, H.D.: Nature **353**, 733 (1991)
2. Balian, R., Bloch, C.: Ann. Phys. **69**, 76 (1971)
3. Nishioka, H.: Z. Phys. D **19**, 19 (1991); Nishioka, H., Hansen, K., Mottelson, B.R.: Phys. Rev. B **42**, 9377 (1990)
4. Stampfli, P., Bennemann, K.H.: Phys. Rev. Lett. **69**, 3471 (1992); Z. Phys. D **25**, 87 (1992)
5. Berry, M.V., Tabor, M.: Proc. R. Soc. London A **349**, 101 (1976)
6. Ashcroft, N.W., Mermin, N.D.: Solid State Physics. New York: Holt, Rinehardt and Winston 1976

Z. Phys. A 349, 345–346 (1994)

Absolute alpha-decay width of ^{212}Po in a mixed shell-and-cluster model

K. Varga[1], **R. G. Lovas**[1], **R. J. Liotta**[2]

[1] Institute of Nuclear Research of the Hungarian Academy of Sciences, Debrecen, P. O. Box 51, H–4001, Hungary
[2] Physics at Frescati, Royal Institute of Technology, Frescativägen 24, S–10405 Stockholm, Sweden

Received: 27 October 1993

Abstract. We have reproduced the absolute width of the α decay of the ground state of ^{212}Po in a model in which the shell model is combined with a ^{208}Pb+α cluster model, and found that the amount of core+α clustering in the parent state is ~30%.

PACS: 23.60.+e; 21.60.Gx; 21.10.Jx; 27.80.+w

1. Introduction

Due to technical difficulties in describing heavy nuclei in a large enough domain of the configuration space, until recently there existed no microscopically well-founded model of α decay capable of reproducing absolute decay widths (see e.g. [1]). Here we report on an approach devised to be such, and we shall show its application to the transition ^{212}Po(g.s.)→^{208}Pb(g.s.)+α. For details see [2].

2. Model

^{212}Po is a typical shell-model nucleus, but the shell-model wave function is only correct in the nuclear interior, while decay properties depend on an asymptotic region. That is why we complement a shell-model-like wave function with a cluster-model-like term: $\Psi = \Psi^{\text{sh.}} + \Psi^{\text{cl.}}$. In the shell model an explicit treatment of core nucleons can be avoided because, in the heavy-core limit, the valence orbits are orthogonal to the core orbits. The cluster-model term can be written as

$$\Psi^{\text{cl.}} = \mathcal{A}\{\Phi_c(\xi_c)\psi_\alpha^{\text{cl.}}(\xi_\alpha, \mathbf{R})\} \equiv \mathcal{A}\{\Phi_c(\xi_c)\Phi_\alpha(\xi_\alpha)\phi(\mathbf{R})\},$$

where $\mathcal{A}$ is an antisymmetrizer, while Φ and ϕ are intrinsic and relative wave functions, respectively. To apply a similar approximation to $\Psi^{\text{cl.}}$, the product $\Phi_\alpha\phi$ has to be expressed in terms of single-particle (s.p.) functions. If Φ_α is built up from 0s oscillator states $\varphi^{(\alpha)}(\mathbf{x}) = (\alpha/\pi)^{3/4}\exp(-\frac{1}{2}\alpha x^2)$, and $\phi(\mathbf{R})$ is expanded as

$$\phi(\mathbf{R}) = \sum_k f_k \int d\hat{\mathbf{s}}_k Y_{LM}(\hat{\mathbf{s}}_k)\varphi^{(4\alpha)}(\mathbf{R} - \mathbf{s}_k),$$

then $\Phi_\alpha\phi$ will become a combination of such terms. Although the s.p. states involved, $\varphi^{(\alpha)}(\mathbf{x}-\mathbf{s})$, are not orthogonal to the core orbits, the antisymmetrizer makes the contibutions, to Ψ, of the non-orthogonal components vanish. Therefore, $\varphi^{(\alpha)}(\mathbf{x}-\mathbf{s}_k)$ can be replaced by $\psi_k(\mathbf{x}) = P\varphi^{(\alpha)}(\mathbf{x}-\mathbf{s}_k)$, where P projects off the s.p. orbits filled by core nucleons, and $\Phi_\alpha\phi$ can be replaced by

$$\psi^{\text{cl.}}(\xi_\alpha, \mathbf{R}) = \sum_k f_k \int d\hat{\mathbf{s}}_k Y_{LM}(\hat{\mathbf{s}}_k)\psi_k(\mathbf{x}_1)...\psi_k(\mathbf{x}_4)\chi,$$

with χ being the spin-isospin state of α. With this substitution, the whole function $\psi = \psi^{\text{sh.}} + \psi^{\text{cl.}}$ in $\Psi = \mathcal{A}\{\Phi_c\psi\}$ is built up from s.p. orbits orthogonal to those in Φ_c, which allows the problem to reduce to a four-nucleon problem.

To solve $H\psi = E\psi$ with the ansatz $\psi = \psi^{\text{sh.}} + \psi^{\text{cl.}}$, the coefficients f_k, along with those of the shell-model term, are treated as variational parameters. The variational problem then reduces to the diagonalization of

$$H = \sum_{i=1}^{4}(T_i + U_i) + \sum_{i<j}^{4} V(i,j),$$

where T_i is the kinetic energy, U_i is a realistic s.p. potential of Woods–Saxon shape with a Coulomb term, and $V(i,j)$ is a residual interaction.

Contrary to [3], we avoid a bias for α formation by using a large shell-model basis, and, in contrast with [4], we do not presuppose the existence of a model α-cluster state near the physical state. We fix all parameters to independent experimental data, and include the interaction between protons and neutrons as well.

The α formation is characterized by the amplitude $g(r) = \langle\psi_r|\psi\rangle$, where $\psi_r = r\int d\hat{\mathbf{r}} Y^*_{LM}(\hat{\mathbf{r}})\Phi_\alpha\delta(\mathbf{r}-\mathbf{R})$, and by the spectroscopic factor $S = \langle g|g\rangle$ or the amount of clustering $\mathcal{S} = \langle g|\hat{A}^{-1}|g\rangle$ [5], where $\hat{A}$ is the integral operator of

Correspondence to: R.G. Lovas

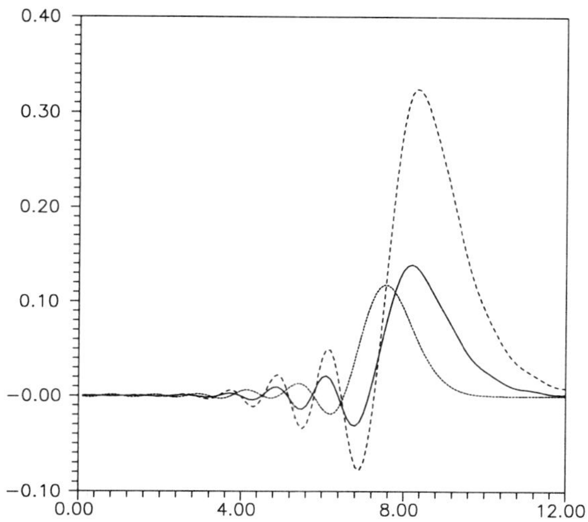

Fig. 1. Formation amplitudes in the pure shell model (multiplied by 10, *dotted line*), in the pure cluster model (*dashed line*) and in the combined model (*solid line*) with the V1 force

Table 1. ^{208}Pb+α spectroscopic factors S and amounts of clustering $\mathcal{S}$ in the three models

Force		Shell m.	Cluster m.	Mixed m.
V1	S	1.5×10^{-4}	0.15	0.025
	$\mathcal{S}$	3.7×10^{-2}	1	0.30
B1	S	1.7×10^{-4}	0.19	0.023
	$\mathcal{S}$	4.2×10^{-2}	1	0.23

kernel $\langle\psi_r|\psi_{r'}\rangle$. To calculate the decay width, Γ, we adopt the standard formula $\Gamma = 2P_L(\hbar^2/2M_\alpha a_L)g^2(a_L)$, where M_α is the (reduced) mass, P_L is the Coulomb penetration factor, and a_L is the channel radius.

3. Results

We used s.p. potentials consistent with the nucleon densities of ^{208}Pb and reproducing the empirical s.p. levels. The effective forces Volkov 1 (V1) [6] and Brink-Boeker B1 [7] reproduce the yrast levels of the core+two-nucleon subsystems very well in corresponding shell models. The decay energy is also reproduced excellently by these forces provided cluster-model states spanning the interval $0 \leq s \leq 20$ fm are included.

The formation amplitudes produced by the shell model, the cluster model and the mixed model with the V1 force are displayed in Fig. 1. The curves show the enormous importance of the inclusion of the cluster states. The resulting width, 1.5×10^{-15} and 2.1×10^{-15} MeV, with V1 and B1, respectively, agrees with the experimental value (Γ=1.52×10^{-15} MeV) very well.

The spectroscopic factor and the amount of clustering are given in Table 1. The value of S is 20 times higher than the best shell-model estimate [8]. The amount of clustering, $\mathcal{S}$=0.23–0.30, is also substantially larger than previous microscopic estimates ($\sim$0.01, [9]). At the same time the ratio $S/\mathcal{S}$=12, which is a measure of the Pauli effects, is much smaller than ever assessed ($\sim$300, [9]).

4. Summary

We have reproduced the absolute α-decay width of ^{212}Po in a microscopic model containing no free parameters. Since the best shell-model calculations, which underrate Γ by an order of magnitude, were only able to allow for correlations between like nucleons properly, the present improvement can be interpreted as a result of a treatment of all correlations on the same footing. In view of the agreement of the width with experiment, the surprisingly high value obtained for the amount of α clustering can be considered realistic.

This result is consistent with the good performance of the extreme cluster model [10], which assumes $\mathcal{S}$=1. The extreme model describes a whole range of cases of cluster decay, including the α decay of ^{212}Po, and its success helps to view our results over a more general background. One can thus infer that the high amount of clustering must be a general property of the decaying states of heavy nuclei.

Acknowledgement. This work was supported by OTKA, Hungary (grants No. 3010 and F4348), and by NFR, Sweden.

References

1 Arima, A.: Proc. Int. Symp. on Developments of Nuclear Cluster Dynamics, p. 338. Sapporo 1988, (eds.) Akaishi, Y. et al. Singapore: World Scientific 1989
2 Varga, K., Lovas, R.G., Liotta, R.J.: Nucl. Phys. **A550**, 421 (1992)
3 Steinmayer, T., Sünkel, W., Wildermuth, K.: Phys. Lett. **125B**, 437 (1983)
4 Okabe, S.: J. Phys. Soc. Jpn. S **58**, 516 (1989)
5 Beck, R., Dickmann, F., Lovas, R.G.: Ann. Phys. (N.Y.) **173**, 1 (1987)
6 Volkov, A.B.: Nucl. Phys. **74**, 33 (1965)
7 Brink, D.M., Boeker, E.: Nucl. Phys. **A91**, 1 (1967)
8 Tonozuka, I., Arima, A.: Nucl. Phys. **A323**, 45 (1979)
9 Blendowske, R., Fliessbach, T., Walliser, H.: Z. Phys. A **339**, 121 (1991)
10 Buck, B., Merchant, A.C., Perez, S.M.: Phys. Rev. Lett. **65**, 2975 (1990); J. Phys. G **17**, L91 (1991)

Z. Phys. A 349, 347–348 (1994)

Microscopic multicluster description of neutron halos

K. Varga[1], R. G. Lovas[1], Y. Suzuki[2]

[1] Institute of Nuclear Research of the Hungarian Academy of Sciences, Debrecen, P. O. Box 51, H–4001 Hungary
[2] Physics Department, Niigata University, Niigata, 950–21 Japan

Received: 27 October 1993

Abstract. We give a unified description for the ground states of ^{6}He and ^{8}He in a model comprising an α cluster and single-neutron clusters arranged into different configurations. Stochastic methods for selecting the the terms of the trial function yield excellent numerical convergence both for the energy and wave function. The ^{6}He and ^{8}He energies are reproduced by the same effective force very well.

PACS: 21.60.Gx; 27.20.+n

1. Introduction

Neutron-halo nuclei are currently in the focus of intense interest [1]. Their description requires a unified approach with very flexible neutron orbits and correct treatment of the centre-of-mass motion and of the Pauli principle. The approaches used hitherto, such as the cluster-orbital shell model [2] and the few-body approaches [3], run into difficulties in going beyond a passive-core-plus-two-nucleon approximation.

In this contribution we present a fully microscopic approach, which is extremely flexible. The model is based on a linear variational method, and it is made feasible by stochastic sampling in setting up the basis. We use the three-cluster system ^{6}He=α+n+n as a testing ground, and present results for the five-cluster system ^{8}He=α+n+n+n+n as well. In Sect. 2 we outline the model. In Sect. 3 we explain how we generate the basis and summmarize the results.

2. Model

Our model is a generalization of that outlined in Ref. [4]. It employs a trial function, which is a sum over various cluster arrangements μ, each associated with a particular set of intercluster Jacobi coordinates $\{\rho_1^\mu, ..., \rho_n^\mu\}$. For α+n+n, there are two non-trivially different arrangements, viz. μ=(αn)n and α(nn), while for α+n+n+n+n there are nine: $\{[(\alpha n)n]n\}n$, $\{[\alpha(nn)]n\}n$, $[(\alpha n)(nn)]n$, $\{\alpha[(nn)n]\}n$, $[(\alpha n)n](nn)$, $[\alpha(nn)](nn)$, $(\alpha n)[(nn)n]$, $\alpha\{[(nn)n]n\}$ and $\alpha[(nn)(nn)]$. The coupling of the nucleon spins to S, and the orbital angular momenta l_i ($i = 1, ..., n$) to L follow the pattern of the Jacobi coordinates.

Correspondence to: K. Varga

The trial function is

$$\Psi_{(S,L)JM} = \sum_{\mu} \sum_{l_1...l_n} \Psi^{\mu}_{[S,(l_1...l_n)L]JM},$$

where

$$\Psi^{\mu}_{[S,(l_1...l_n)L]JM} = \mathcal{A}\left\{\left[\Phi_S \mathcal{X}^{\mu}_{(l_1...l_n)L}(\rho_1^\mu, ..., \rho_n^\mu)\right]_{JM}\right\},$$

where $\mathcal{A}$ is an antisymmetrizer, Φ_{SM_S} is a vector-coupled product of the cluster intrinsic wave functions and

$$\mathcal{X}^{\mu}_{(l_1...l_n)LA} = \sum_{K} D^{\mu}_{K,l_1...l_n} \left[\Gamma^{\mu}_{1k_1l_1} ... \Gamma^{\mu}_{nk_nl_n}\right]_{LA}$$

describes the relative motion, with $D^{\mu}_{K,l_1...l_n}$ ($K = \{k_1, ..., k_n\}$) being the variational parameters, and

$$\Gamma^{\mu}_{iklm}(\rho_i^\mu) \sim (\rho_i^\mu)^l \exp[-\nu^{\mu}_{ik}(\rho_i^\mu)^2] Y_{lm}(\hat{\rho}_i^\mu).$$

The α particle wave function is a 0s oscillator shell-model state of parameter $\nu = m\omega/2\hbar$.

We chose the central effective interaction of Ref. [5]. All matrix elements are calculated analytically.

3. Results

For ^{6}He appreciable contributions only come from the subspaces defined by $\mu_{l_1l_2}$ = $[\alpha(nn)]_{00}$ ("T$_{00}$ subspace"), by $[(\alpha n)n]_{00}$ ("Y$_{00}$"), and by $[(\alpha n)n]_{11}$ ("Y$_{11}$"). In Table 1 we give the energies and rms point matter radii produced by calculations with nearly complete bases in these subspaces and their unions.

The energies obtained in the individual subspaces are very close to each other, and yet, when they are combined, the binding gets stronger moderately. This behaviour is typical of strongly overlapping subspaces [4], in which case the united basis is likely to be largely redundant.

Table 1. Energies and rms radii of ^{6}He with almost complete bases in different model spaces

Subspace	T_{00}	Y_{00}	Y_{11}	$\{T_{00},Y_{00}\}$	$\{T_{00},Y_{11}\}$	$\{Y_{00},Y_{11}\}$	$\{T_{00},Y_{00},Y_{11}\}$	Experiment
E (MeV)	−0.395	−0.387	−0.435	−0.944	−1.015	−0.605	−1.016	−0.975
r (fm)	2.455	2.441	2.451	2.411	2.443	2.469	2.442	2.4 [a]

[a] 2.48±0.03 fm, [6]; 2.33±0.04 fm, [7]

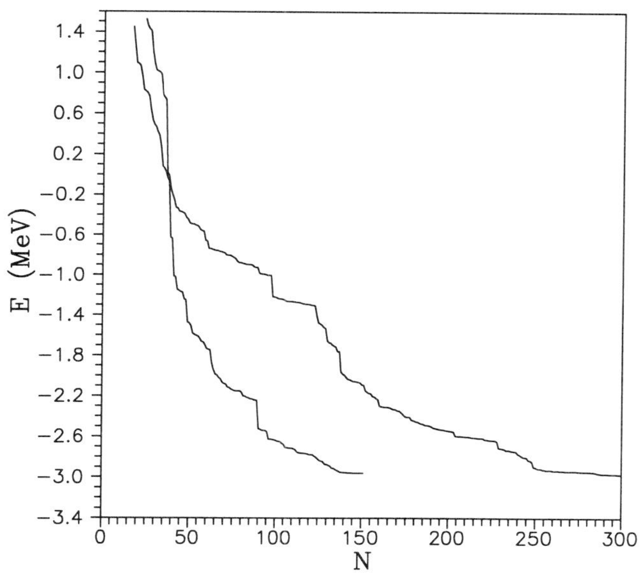

Fig. 1. Convergence of the ^{8}He energy with a (preselected) random basis (*flatter curve*) and with a basis constructed with utility tests and simulated annealing (*steeper curve*) in the subspace $\{[\alpha(nn)]n\}n$ restricted by all orbital momenta set to zero

We constructed less redundant bases by stochastic sampling. We enlarged the basis step by step by selected random elements. They were qualified for admittance if their contribution was either the best among 10 randomly chosen candidates ("preselected basis") or was larger than 0.005 MeV ("utility-tested basis"). Since "utility testing" might terminate itself too soon, we allowed temporary lapses in the rigour of admittance ("simulated annealing").

Whichever selection prescription was used, the energies as well as radii calculated with random bases were always found to converge to the "exact" energy, and, at the expense of ~20 keV loss of binding, it was possible to reduce the basis dimension by a factor of 3–6.

This enabled us to make reliable calculations for the five-cluster system as well. Figure 1 shows that for that case utility testing converges more rapidly. A summary of the calculations for ^{8}He is given in Table 2. The configuration $\{[\alpha(nn)]n\}n+[\alpha(nn)](nn)+\{[(\alpha n)n]n\}n$ seems to be adequate; not even the inclusion of configuration $\{[(\alpha n)n]n\}n'$, which is the best by itself, could deepen the binding appreciably.

We managed to reproduce the g.s. energies of ^{6}He and ^{8}He simultaneously with an accuracy of 100 keV. The calculated matter radii would be slightly larger with a force that reproduces the empirical α-particle radius.

Table 2. Energies and rms radii of ^{8}He in different model spaces. All l_i are zero except in $\{[(\alpha n)n]n\}n'$, where $l_1 = l_2 = 1$

Subspace	N	E (MeV)	r (fm)
$\{[(\alpha n)n]n\}n$	153	−2.509	2.341
$\{[(\alpha n)n]n\}n'$	130	−3.033	2.322
$\{[\alpha(nn)]n\}n$	161	−2.961	2.305
$[\alpha(nn)](nn)$	163	−2.676	2.278
$\alpha[(nn)(nn)]$	176	−1.167	2.188
$(\alpha n)[(nn)n]$	220	>0	
$\alpha\{[(nn)n]n\}$	157	−1.368	2.196
$\{[\alpha(nn)]n\}n+[\alpha(nn)](nn)$	211	−3.292	2.306
$\{[\alpha(nn)]n\}n+\alpha\{[(nn)n]n\}$	223	−3.027	2.225
$\{[\alpha(nn)]n\}n+[\alpha(nn)](nn)$ $+\{[(\alpha n)n]n\}n$	303	−3.395	2.271
$\{[\alpha(nn)]n\}n+[\alpha(nn)](nn)$ $+\alpha\{[(nn)n]n\}$	311	−3.321	2.317
Experiment		−3.112	2.50 [a]

[a] [2,3]

Acknowledgement. This work was supported by the OTKA grants No. 3010 and F4348 (Hungary) and by a Grant-in-Aid for Scientific Research (No. 03640265) of the Ministry of Education, Science and Culture (Japan).

References

1. Ikeda, K., Suzuki, Y. (eds.): Proc. Int. Symp. on Structure and reactions of unstable nuclei (Niigata, 1991). Singapore: World Scientific 1991
2. Suzuki, Y., Ikeda, K.: Phys. Rev. **C38**, 410 (1988)
3. Zhukov, M.V., Danilin, B.V., Fedorov, D.V., Bang, J.M., Thompson, I.J., Vaagen, J.S.: Phys. Reports (Nordita preprint, 92/90 N)
4. Csótó, A., Lovas, R.G.: Phys. Rev. **46**, 576 (1992)
5. Thompson, D.R., LeMere, M., Tang, Y.C.: Nucl. Phys. **A286**, 53 (1977)
6. Tanihata, I., Kobayashi, T., Yamakawa, O., Shimoura, S., Ekuni, K., Sugimoto, K., Takahashi, N., Shimoda, T., Sato, H.: Phys. Lett. **B206**, 592 (1988)
7. Tanihata, I., Hirata, D., Kobayashi, T., Shimoura, S., Sugimoto, K., Toki H.: Phys. Lett. **B289**, 261 (1992)

This article was processed using Springer-Verlag TEX Z.Physik A macro package 1.0 and the AMS fonts, developed by the American Mathematical Society.

Atomic shell structures observed in photoionization spectra of nickel and cobalt clusters

J.L. Vialle, M. Pellarin, B. Baguenard, J. Lermé and M. Broyer

Laboratoire de Spectrométrie Ionique et Moléculaire (URA n°171), CNRS et Université Lyon I, Bât. 205 43 Bd du 11 Novembre 1918 - 69622 Villeurbanne Cedex, France

Received 28 June 1993

Abstract. Nickel and cobalt clusters have been studied by near threshold laser-photoionization and time-of-flight mass spectrometry. In the size domain from 50 up to 800 atoms, the mass distributions of the photoionized products look very similar for nickel and cobalt clusters. In both cases a regular structure is observed which is periodic on a $N^{1/3}$ scale. It is found to be consistent with the filling of successive icosahedral shells of atoms. The recurring details of this structure agree with the so-called umbrellas model.

PACS: 36.40.+d, 33.80.Eh, 82.80.Ms

The observation of intensity anomalies in the mass spectra of laser photoionized clusters provides information about their relative stability and then about their structure. For the alkali clusters, the electronic shells, as predicted by the spherical jellium background model, have been observed as a general rule, at least for not too large sizes (N≤2000) [1]. However, for larger sizes, the occurrence of successive shells of atoms becomes the dominant effect [2]. In the case of trivalent metals the situation is very different depending on whether one considers aluminum or gallium and indium. The main effects observed in indium and gallium clusters are in agreement with pure electronic shell structure [3,4], while the mass spectra of aluminum clusters exhibit a periodic pattern linked to successive shells of atoms [5,6]. The melting point of the element in the bulk seems to be a crucial parameter to settle what is dominant between electronic shell structure and geometric shells of atoms [4,7]. Following this idea we may expect that if any structure appears in cluster mass spectra of transition metals, it will correspond to shells of atoms and provide us information on the geometry of these clusters. We present here the results of an experimental study on nickel and cobalt clusters, in the size range N=50-800.

The metallic clusters are produced by laser vaporization technique. They are photoionized by the U.V. photons of a frequency-doubled tunable dye laser and the ionized products are analyzed in a time-of-flight mass spectrometer perpendicular to the incoming cluster beam. The U.V. photon energy is chosen close above the ionization threshold of the clusters. It has to be slightly decreased as the size of the studied clusters increases, in order to follow the general trend of the cluster ionization threshold. A more complete description of our experimental set up has already been published [5].

Mass spectra of cobalt and nickel clusters are displayed in the upper part of Fig. 1. They clearly exhibit strong similarities. In particular a shell structure is observed in both spectra for the sequence of sizes indicated by dotted lines in the figure: N=55, 147, 309, 561. The photon energy has been chosen to provide marked effects in the center of the mass range (147≤N≤561). A slight blue shift of the laser wavelength would enhance the shell effects observed for the smaller clusters (N<147), a red shift would do the same for teh larger ones (N>561). This sequence of numbers is well known. It corresponds to the filling of successive atomic shells, the outer shape of the cluster being a cuboctahedron or an icosahedron. For both polyhedra, the number N_K of atoms in a cluster with K shells is given by the same formula [8]:

$$N_K = 1/3(10K^3 - 15K^2 + 11K - 3)$$

On the inside of every shell (for instance from N=147 to N=309), one can see a series of several sharp irregularities. They are small but highly reproducible. This subshell pattern looks very similar from shell to shell as well as from nickel to cobalt (see Fig. 1, lower part). It permits to settle the question of the cluster shape. We found indeed that this pattern is in good agreement with the icosahedral geometry and the filling sequence proposed by Martin et al. for calcium and magnesium clusters [9,10]. Atoms successively cover the twenty triangular faces of the icosahedron. The stability is enhanced each time a face is fully covered, moreover a greater stability is obtained each time a set

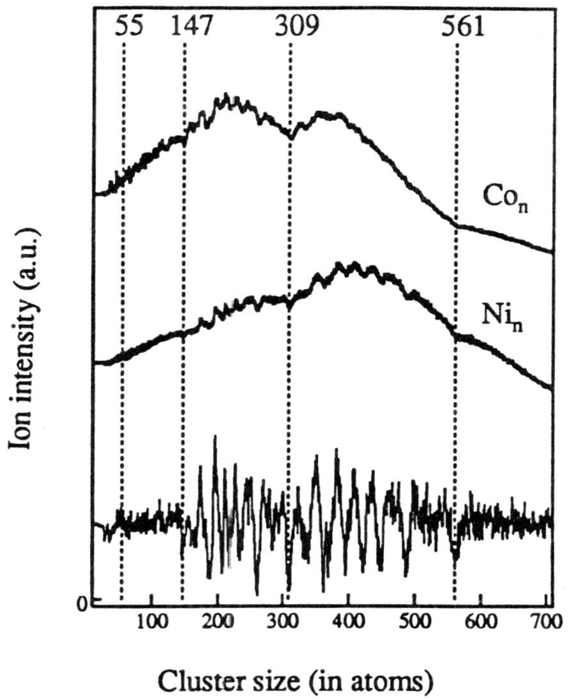

Fig. 1. Mass spectra of cobalt and nickel clusters. Characteristic sizes corresponding to completed shells of atoms are located with dotted lines. The lower spectrum is obtained by subtracting from the untreated nickel spectrum (just above) a sufficiently smoothed bell-shaped envelope. Notice the great similarity between the subshell patterns from shell to shell.

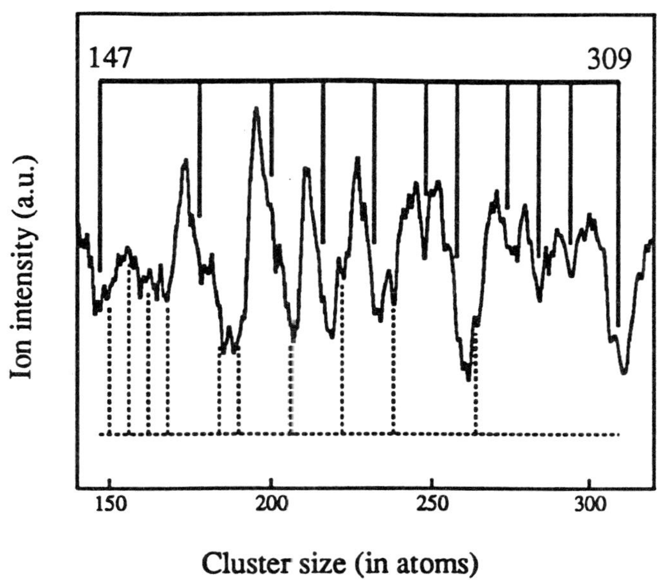

Fig. 2. Comparison of the experimental subshell pattern (N=147-309) with the predictions of the umbrellas model. Full lines correspond to completed umbrellas, dotted lines to completed intermediate triangular faces.

of five adjacent triangular faces, forming a pentagonal umbrella, is completed. In our experiments an enhanced stability appears as a dip in the ion signal. An experimental subshell pattern is compared with the predictions of the umbrellas model in Fig. 2. The agreement is rather good, indicating that the correct cluster shape is the icosahedral one.

We have identified icosahedral atomic shell structure in nickel and cobalt clusters up to 800 atoms. Our results confirm and extend to larger sizes those previously obtained by Riley et al [11,12] in chemical reactivity experiments on small nickel and cobalt clusters. It would be interesting to extend these experiments towards larger sizes (N≈2000) to see how far this structure continues [13].

References

1 Pedersen, J., Bjørnholm, S., Borggreen, J., Hansen, K., Martin, T.P., and Rasmussen, H.D.: Nature **353**, 733 (1991); Bréchignac, C., Cahuzac, Ph., Carlier, F., de Frutos, M., and Roux, J.Ph.: Phys. Rev. B **47**, 2271 (1993)

2 Martin, T.P., Bergmann, T., Göhlich, H., and Lange, T.: Chem. Phys. Lett. **172**, 209 (1990)

3 Baguenard, B., Pellarin, M., Bordas, C., Lermé, J., Vialle, J.L., and Broyer, M.: Chem. Phys. Lett. **205**, 13 (1993)

4 Broyer, M.: invited paper to this conference. To be published in the same issue; Pellarin, M., Baguenard, B., Bordas, C., Broyer, M., Lermé, J., and Vialle, J.L.: submitted to Phys. Rev.

5 Martin, T.P., Näher, U., and Schaber, H.: Chem. Phys. Lett. **199**, 470 (1992)

6 Pellarin, M., Baguenard, B., Broyer, M., Lermé, J., Vialle, J.L., and Perez, A.: J. Chem. Phys. **98**, 944 (1993)

7 Stampfli, P., and Bennemann, K.H.: Phys. Rev. Lett. **69**, 3471 (1992)

8 McKay, A.L.: Acta Cryst. **15**, 916 (1962)

9 Martin, T.P., Näher, U., Bergmann, T., Göhlich, H., and Lange, T.: Chem. Phys. Lett. **183**, 119 (1991)

10 Martin, T.P., Näher, U., Bergmann, T., Göhlich, H., and Lange, T.: Chem. Phys. Lett. **176**, 343 (1991) and J. Phys. Chem. **95**, 6421 (1991)

11 Parks, E.K., Winter, B.J., Klots, T.D., and Riley, S.J.: J. Chem. Phys. **94**, 1882 (1991)

12 Winter, B.J., Klots, T.D., Parks, E.K., and Riley, S.J.: Z. Phys. D **19**, 375 (1991)

13 Cleveland, C.L., and Landman, U.: J. Chem. Phys. **94**, 7376 (1991)

This article was processed using Springer-Verlag TEX Z.Physik D macro package 1.0 and the AMS fonts, developed by the American Mathematical Society.

Z. Phys. D 31, 299–301 (1994)

Molecular dynamics study of $A_{18}B$ Lennard-Jones clusters

Q. Wang, M.P. Iñiguez, J.A. Alonso

Departamento de Fisica Teórica, Universidad de Valladolid, E-47011 Valladolid, Spain

Received: 28 June 1993 / Final version: 11 May 1994

Abstract. Low temperature structures, melting and evaporation temperatures of $A_{18}B$ clusters with Lennard-Jones interactions have been studied using Molecular Dynamics simulations for different values of the parameters representing (a) the size ratio between A and B atoms, and (b) the ratio between the strengths of the $A-B$ and $A-A$ interactions.

PACS: 36.40; 61.20.J

1. Introduction

The special characteristics of clusters of type $A_n B$, where A is an inert gas atom and B is a selected impurity, have been useful in the experimental research on the dynamical properties and melting behaviour of rare gas clusters [1–3]. Infrared spectroscopy [3] results on impurity solvation are in good agreement with molecular dynamics simulations [4, 5] using Lennard-Jones (L-J) interatomic potentials. Using constant energy Molecular Dynamics (MD) we study in this paper the melting transition, the evaporation and the low-temperature structures of mixed $A_{18}B$ clusters with L-J interactions for different sets of potential parameters. Some L-J clusters containing one impurity ($A_{12}B$, $A_{13}B$ and $A_{55}B$) have been studied by Garzón and coworkers [5]. Our interest in $A_{18}B$ comes from the fact that L-J A_{19} clusters have a double icosahedral structure at low temperatures and have an enhanced stability with respect to neighbor sizes. We want to study how the substitution of a matrix atom by an impurity of different size and interaction strength affects the properties of the cluster.

This paper was originally submitted in connection with the 2nd. Int. Conference on Atoic and Nuclear Clusters held in Santorini from 28. June–2. July 1993 and is published here as a regular article after an independent refereeing procedure according to the standards of Z. Phys. D

2. Method

The interactions between equal and unequal atoms (host-host and host-impurity interactions respectively) are assumed to have the following form:

$$V_{ij}(r) = 4\varepsilon_{ij}\left[\left(\frac{\sigma_{ij}}{r}\right)^{12} - \left(\frac{\sigma_{ij}}{r}\right)^{6}\right] \tag{1}$$

where $i, j = A, B$. This Lennard-Jones potential is known to describe well the interaction between inert gas atoms. We have considered different sets of parameters representing different radii and interaction strength for impurity and host atoms, namely $\sigma_{AB}/\sigma_{AA} = 0.75$, 1.0, 1.25, and $\varepsilon_{AB}/\varepsilon_{AA} = 0.7$, 1.0, 2.0. For those potentials the constant energy MD equations have been integrated using Verlet's algorithm [6] with a time step of $0.01\,\tau_A$. This guarantees conservation of the total energy to within 0.01%. This quantity $\tau_A = \left(\frac{m_A \sigma_{AA}^2}{\varepsilon_{AA}}\right)^{1/2}$ is the vibrational period for a dimer of host atoms that, for instance, is equal to 2.2 ps for argon. m_A in the above formula is the atom mass. The calculations start from an initially non-rotating and non-translating structure close to a double icosahedron in which the impurity atom B is at a substitutional position. The stabilities of those structures have been studied from the behaviour of the total energy versus temperature (caloric curve). The heating was performed by scaling the velocities with a factor $f = 1.1$ and each time this is made the system is allowed to propagate in the phase space. Around 3000 time steps are necessary to equilibrate the system, after which a number of additional time steps are used to average the thermodynamic quantities (the averages are indicated by the brackets $\langle\ \rangle$ in the text). So, the cluster temperature is obtained from the time averaged kinetic energy

$$T = \frac{2\langle KE\rangle}{(3N-6)\,k_B} \tag{2}$$

where N is the number of atoms in the cluster and k_B is Boltzmann's constant. Another interesting quantity is the average square atomic displacement δ

$$\delta = \frac{1}{N} \sum_{i=1}^{N} \langle |\mathbf{r}_i(t) - \mathbf{r}_i(t_0)|^2 \rangle \tag{3}$$

over a trajectory ranging from times t_0 to t. The number of time steps used to calculate the thermodynamic averages in (2) and (3) is sufficiently large to keep the temperature fluctuations below 0.005 (reduced units). Typically we have used 10000 steps in the solid region and up to a maximum of 60000 for higher temperatures.

Once the temperature increases above a given value T_e, evaporation of an atom occurs. The evaporation process requires a statistical study for each temperature [7] and we speak of an evaporation probability on a given time. As a rough approximation we can identify T_e as the minimum temperature for which a cluster evaporates one atom within 60000 time steps. Although the T_e values are crude we can extract from them some correlations with the potential parameters that will be displayed in the next section. Once evaporation takes place we stop the heating process. Then, starting from temperatures below and near T_e, the clusters are cooled down by using a factor $f = 0.9$ to scale the velocities, and a new caloric curve is obtained. It is well known [8] that this caloric curve can be different from the previous one obtained by heating. The final cooled clusters do not possess in some cases the icosahedral structures that were initially heated up.

The melting temperature T_m is evaluated as that temperature at which the average square atomic displacement δ suffers a sudden increase with increasing T from very low values. Other descriptions of melting are based on vibrations [9], bond angle changes [10] or isomeric transformations [11].

3. Results

The type of evaporating atom and the corresponding evaporation temperatures T_e defined in the previous section are listed in the last column of Table 1. The impurity

Table 1. Melting (T_m) and evaporation (T_e) temperatures for different potential parameters. Temperatures are in reduced units of ε_{AA}/k_B. Along with T_e the atom that evaporates first (A or B) is given

σ_{AB}/σ_{AA}	$\varepsilon_{AB}/\varepsilon_{AA}$	T_m	T_e
0.75	0.7	0.12	0.22 *B*
0.75	1.0	0.12	0.26 *B*
0.75	2.0	0.12	0.26 *A*
1	0.7	0.22	0.26 *B*
1	1.0	0.24	0.32 *A*
1	2.0	0.25	0.33 *A*
1.25	0.7	0.17	0.23 *B*
1.25	1.0	0.21	0.26 *A*
1.25	2.0	0.05	0.35 *A*

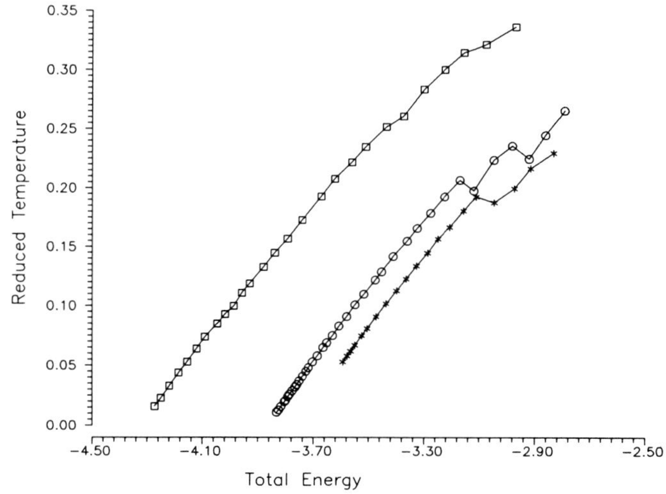

Fig. 1. Total energy per atom as a function of temperature for the case $\sigma_{AB}/\sigma_{AA} = 1.25$. We use reduced units of ε_{AA} and ε_{AA}/k_B respectively. The cases $\varepsilon_{AB}/\varepsilon_{AA} = 0.7$, 1 and 2 are represented by *stars*, *circles* and *squares* respectively

evaporates first for the case of a weakly interacting impurity ($\varepsilon_{AB}/\varepsilon_{AA} < 1$). The opposite occurs when the *A-B* is stronger than the *A-A* interaction: in such a case one host atom evaporates first. These conclusions are independent of the size ratio σ_{AB}/σ_{AA}. Finally, when $\varepsilon_{AB} = \varepsilon_{AA}$, then the smaller atom is the one that evaporates first. For a given value of the size ratio σ_{AB}/σ_{AA}, the evaporation temperature correlates with $\varepsilon_{AB}/\varepsilon_{AA}$. In unimolecular evaporative processes without barrier, the prefered channel is such that, after evaporation, the cluster remains with the highest possible binding energy [12]. This is the lowest energy channel. For a given ratio σ_{AB}/σ_{AA}, the binding energy of $A_{18}B$ correlates with the ratio $\varepsilon_{AB}/\varepsilon_{AA}$. It is evident, due to the nature of the potential (1) that the same has to occur for $A_{17}B$, so the type of evaporated atom from $A_{18}B$ (see Table 1) corresponds in all cases to the lowest energy channel.

Figure 1 shows some caloric curves obtained in the cooling process starting from high temperatures near T_e. For a fixed value of σ_{AB}/σ_{AA}, lower (that is more negative) total energies are found for higher interaction strength ratios $\varepsilon_{AB}/\varepsilon_{AA}$. It is evident that the fluctuations are more important in the transition region. Solid and liquid forms coexist in this region [13] resulting in a bimodal distribution of temperatures which broadens the uncertainty of the averaged values. Probably, the length of our trajectories is not large enough compared to the transit time between the two forms and this may be the reason for obtaining a non-smooth caloric curve in the transition region. On the other hand the caloric curves are very smooth both in the solid and the liquid regions, which indicates that the uncertainties are smaller in these regions.

The influence of σ_{AB}/σ_{AA} and $\varepsilon_{AB}/\varepsilon_{AA}$ on the low temperature structures is now described. The structure for $\sigma_{AB} = \sigma_{AA}$ is the double icosahedron independently of $\varepsilon_{AB}/\varepsilon_{AA}$. If $\sigma_{AB} > \sigma_{AA}$ the structure is a slightly distorted

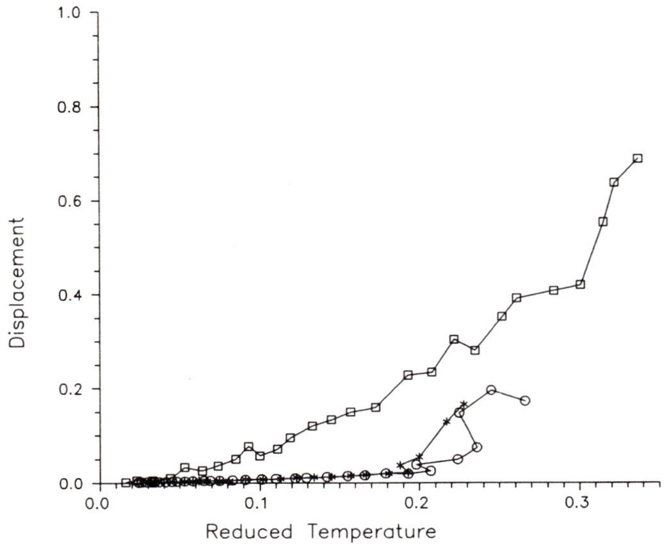

Fig. 2. Average square atomic displacement as a function of temperature in reduced units of $(2^{1/6}\sigma_{AA})^2$ and ε_{AA}/k_B respectively, for the case $\sigma_{AB}/\sigma_{AA}=1.25$. $\varepsilon_{AB}/\varepsilon_{AA}=0.7$, 1 and 2 are represented by *stars, circles* and *squares* respectively

double icosahedron except when $\varepsilon_{AB}/\varepsilon_{AA}$ has the largest value studied. In such a case, with $\varepsilon_{AB}/\varepsilon_{AA}=2$, the structure is a nearly spherical shell with the impurity inside. This is the case in which the largest atomic reorganization occurs. On the other hand for the case $\sigma_{AB}<\sigma_{AA}$ the low temperature structures are more complex. In this case, as the (small) impurity is inside the cluster, the size mismatch between impurity and host atoms destabilizes the double icosahedron structure, still keeping a host atom in the internal region. By comparing the total energies of the low temperature icosahedra for $\sigma_{AB}/\sigma_{AA}=1$ (no size mismatch) we have obtained the obvious result that the most stable cluster occurs for $\varepsilon_{AB}/\varepsilon_{AA}=2$. In this case the impurity atom is in the interior of the cluster, and consequently it has maximum coordination of 12. The case nearest in stability occurs for $\varepsilon_{AB}=\varepsilon_{AA}$ (the pure A_{19} cluster), followed by $\varepsilon_{AB}/\varepsilon_{AA}=0.7$. In the last case the impurity is at the cluster surface.

The melting temperatures T_m are given in Table 1. These have been extracted from the analysis of the calculated average square atomic displacements δ (see Eq. (3)), which have been plotted as a function of temperatures in Fig. 2. Melting temperatures are highest for $\sigma_{AB}/\sigma_{AA}=1$. Melting in clusters is usually understood as the transition from a rigid to a nonrigid structure. Atomic size mismatch introduces elastic tensions that undermine the rigidity of the clusters. This is already evident in the low temperature structures of clusters with $\sigma_{AB}\neq\sigma_{AA}$. So it is understandable that those clusters lose their rigidity at a lower temperature than clusters with atoms of equal size, in which the elastic tensions are minimized. A correlation between T_m and $\varepsilon_{AB}/\varepsilon_{AA}$ appears to exist for the clusters formed by atoms of equal size. However no such correlation occurs when the impurity is smaller than the host atoms. In this case the melting temperature is independent of the ratio $\varepsilon_{AB}/\varepsilon_{AA}$. The large structural distortion induced at low temperature by the small impurity causes also a the large reduction of T_m with respect to the $\sigma_{AB}/\sigma_{AA}=1$ case.

Finally, when the impurity is the large atom an interesting phenomenon occurs (see Fig. 2). First of all, the melting temperatures are below those of the case $\sigma_{AB}/\sigma_{AA}=1$ for the reasons already explained above, that is, size mismatch. Furthermore, when the impurity-host interaction is strongest ($\varepsilon_{AB}/\varepsilon_{AA}=2$) we find a very low melting temperature. This drastic decrease of T_m with respect to all other cases is motivated by the special low temperature structure of the cluster. The large size of the impurity and its strong interaction with the host atoms lead to a spherical structure at low temperature. Rearrangements of this spherical shell preserving the central position of the impurity are easy. This kind of "surface" melting is responsible of the low T_m.

Work supported by DGICYT (Grant PB89-0653). One of us (Q.W.) is grateful to Ministerio de Educación y Ciencia of Spain for a postdoctoral Giant. We also acknowledge discussions with I. Garzón.

References

1. Bosiger, J., Leutwyler, S.: Phys. Rev. Lett. **59**, 1895 (1987)
2. Hahn, M., Whetten, R.L.: Phys. Rev. Lett. **61**, 1190 (1988)
3. Gu, X.J., Levandier, D.J., Zhang, B., Scoles, G., Zhuang, D,: J. Chem. Phys. **93**, 4898 (1990)
4. Perera, L., Amar, F.G.: J. Chem. Phys. **93**, 4884 (1990)
5. Garzón, I.L., Long, X.P., Kawai, R., Weare, J.H.: Z. Phys. D **12**, 81 (1989); Garzón, I.L., Long, X.P., Kawai, R., Weare, J.H.: Chem. Phys. Lett. **158**, 525 (1989)
6. Verlet, L.: Phys. Rev. **159**, 98 (1964)
7. Rey, C., Gallego, L.J., Iñiguez, M.P., Alonso, J.A.: Physica B **179**, 273 (1992)
8. Briant, C.L., Burton, J.J.: J. Chem. Phys. **63**, 2045 (1975)
9. Jellinek, J., Beck, T.L., Berry, R.S.: J. Chem. Phys. **84**, 2783 (1986)
10. Blaisten-Barojas, E., Garzón, I.L., Avalos-Borja, M.: Phys. Rev. B **36**, 8447 (1987)
11. Garzón, I.L., Avalos-Borja, M., Blaisten-Barojas, E.: Phys. Rev. B **40**, 4749 (1989)
12. López, M.J., Jellinek, J.: Phys. Rev. A: (to be published)
13. Beck, T.L., Jellinek, J., Berry, R.S.: J. Chem. Phys. **87**, 545 (1987)

Z. Phys. A 349, 359–360 (1994)

The competition between cluster and constituent multiparticle emission as a means to identify the reaction mechanism

A.C. Xenoulis, A.E. Aravantinos, E. Adamides

N.C.S.R. Demokritos, Athens, Greece

Received: 27 October 1993

Abstract. Since the competition between cluster and constituent multiparticle emission producing the same residual nucleus in a nuclear reaction depends on the mechanism of the reaction, the pn over d ratio in the $^{16}O(^{6}Li, pn/d)$ reaction, measured between 10 and 19 MeV, is used to identify its mechanism as a function of energy.

PACS: 25.70.Gh; 25.70.Hi

1. Introduction

Certain nuclear reactions, such as $(^{6}Li, d)$ and $(^{7}Li, t)$, constitute a testing ground of our understanding of the energy evolution of a reaction mechanism, mainly because the cluster-like α configuration of the projectile permits the coexistence of compound-nucleus evaporation and direct α-transfer mechanisms even at relatively very low bombarding energies. Furthermore, it seems that the energy region for the transition from the one mechanism to the other is rather narrow.

For instance, γ-ray angular distribution measurements in the reaction $^{16}O(^{6}Li, d)$ lead to the conclusion that at 15 MeV the mechanism is predominantly compound nucleus [1]. Deuteron angular distribution measurements at 20 MeV, however, indicated at least coexistence of the two mechanisms [2], while at 24 MeV [3] and certainly at 38 MeV [2] the reaction is predominantly direct.

Apparently, the energy interval between 15 and 20 MeV of bombarding energy is the reaction mechanism's transitional region. Since it may be very important to understand how and why that transition occurs, in the present work we have attempted to map in detail the evolution of the reaction mechanisms utilizing measurements of the competition between pn and d emission.

2. Method and results

The competition between cluster and multiparticle emission, the so called isoproduct ratio, such as between pn and d or $p2n$, dn and t emission populating the same A-2 and A-3 residual nucleus, respectively, is expected to depend strongly on the mechanism of the reaction.

This is so because while in the compound-nucleus mechanism this competition depends on the phase space, and thus mainly on the residual excitation energy [4] available to each competing exit channel, in the direct α transfer d or t cluster emission dominates since multiparticle emission can be induced only by second order effects. It has been shown, furthermore, in cases of evaporation from a compound nucleus that Hauser-Feshbach statistical-model calculations can adequately describe the experimental pn over d competition in about 30 different nuclear reactions [4–6].

These results suggest that, in cases of ambiguous reaction mechanism, a comparison between experimentally measured and statistically predicted emission competition can serve to identify the involved mechanisms as well as to quantify their evolution as a function of energy. This will be demonstrated for the case of the $^{16}O(^{6}Li, pn/d)$ reaction.

The experimental method, utilizing charged particle-γ coincidence techniques, has been extensively described previously [7]. Presently, the intensity of the 1630 keV, first excited to ground state, transition in ^{20}Ne, in coincidence with protons and deuterons has been used. Thus it is important to notice that the experimental data represent the pn over d competition in the production of certainly the 1630-keV state but also of all other excited states in ^{20}Ne which cascade to that state. This is an important difference to bear in mind when the present results are to be compared with previous results where individual states had been measured [1–3].

The pn over d competition associated with the ^{16}O $+ ^{6}Li$ reaction is demonstrated in Fig. 1, where the measured ratio, shown with the data points, is compared

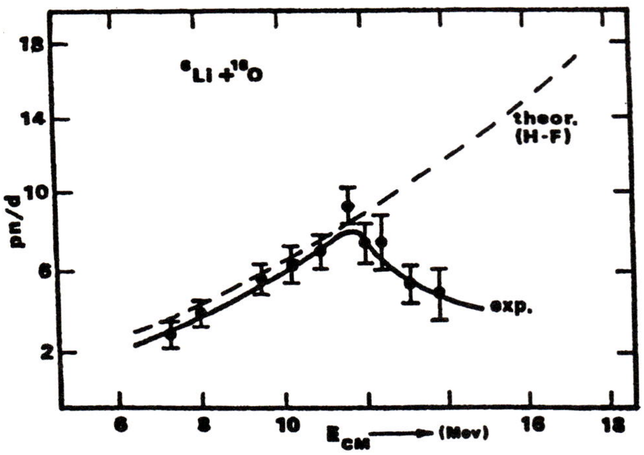

Fig. 1. Comparison between experimentally measured (*data points*) and statistically predicted (*broken line*) pn/d competition in the reaction $^{16}O(^6Li, pn/d)$ ^{20}Ne. The *solid line* through the data points simply serves to guide the eye

with the relevant statistical-model predictions. It is seen that between E_{cm} 7.3 and 11.6 MeV, or correspondigly E_{lab} 10 and 16 MeV, an excellent agreement between theory and experiment is observed. Thus, it may be concluded that between 10 and 16 MeV both exit channels in the reaction $^{16}O(^6Li, pn/d)$ proceed essentially via the compound-nucleus mechanism, substantiating and extending the previous similar evidence [1] obtained at 15 MeV.

Above that energy, however, a sudden decrease in the σ_{pn}/σ_d ratio occurs, which is seen [8] to be due to the increasing flow of deuterons, while the pn emission still follows the statistical-model predictions. That decrease in the competition ratio, therefore, should be due to the introduction and the gradually increasing contribution from the direct α-transfer mechanism. That interpretation is, furthermore, supported by the observation that above 17 MeV the coincident intensity of the 4.97 MeV 2^- unnatural-parity state becomes negligible [8].

The contribution of each mechanism can be estimated in Fig. 1 from the deviation of the experimental points from the statistical-model curve. Thus, it is presently seen that at $E_{lab} = 19$ MeV the $^{16}(^6Li, d)$ reaction is proceeding by about 44% compound-nucleus and by 56% direct α-transfer mechanism. That conclusion is in perfect accord with the previously mentioned, although rather qualitative, results in the vicinity of that energy region [2, 3].

We have not yet been able to understand the reason that at exactly $E_{^6Li} = 16.5$ MeV the reaction ceases to be exclusively compound nucleus. Relevant optimum Q-value calculations we have performed do not indicate any noticable effect that may enhance the direct α-transfer contribution at 16.5 MeV or even above that energy.

References

1. Balamuth, D.P.: Phys. Rev. C **3**, 1565 (1971)
2. Anantaraman, N.: Nucl. Phys. A **313**, 445 (1971)
3. House, L.J., Kemper, K.W.: Phys. Rev. C **17**, 79 (1978)
4. Xenoulis, A.C. et al.: Phys. Lett. **106** B, 461 (1981)
5. Gazis, E.N. et al.: Phys. Rev. C **34**, 872 (1986)
6. Aravantinos, A.E., Xenoulis, A.C.: Phys. Rev. C **35**, 1746 (1987)
7. Xenoulis, A.C. et al.: Phys. Lett. **90** B, 224 (1980)
8. Xenoulis, A.C. et al.: (to be published)

Z. Phys. A 349, 363–365 (1994)

Core-excited α-cluster structure in ^{44}Ti

Taiichi Yamada [1], **Shigeo Ohkubo** [2,3]

[1] Laboratory of Physics, Kantou Gakuin University, Yokohama, 236 Japan
[2] Nuclear Physics Laboratory, University of Oxford, Oxford OX1 3RH, UK
[3] Department of Applied Science, Kochi Women's University, Kochi, 780 Japan

Received: 27 October 1993

Abstract. With the use of a schematic ^{40}Ca* + α model, the low-lying positive $K=0_2^+$ and 2^+ bands in ^{44}Ti, which start from the band-head energy $E_x = 1.90$ MeV and 2.88 MeV, respectively, are shown to be well interpreted by an α-cluster structure with an excited ^{40}Ca core, i.e., ^{36}Ar + α + α structure.

PACS: 21.10.−k; 21.60.Gx; 27.40.+z

1. Introduction

A recent ^{40}Ca(^{6}Li, d)^{44}Ti experiment [1] gave clear evidence of a low-lying negative parity band ($K^\pi = 0^-$) in ^{44}Ti above the α-threshold previously predicted by the α-cluster theory [2]. This means that the α-cluster correlation persists in the ground band ($K^\pi = 0^+$) of ^{44}Ti as well as in the $K^\pi = 0^-$ band. Therefore, the α-cluster viewpoint is indispensable in understanding the low-lying structure of ^{44}Ti. On the other hand, a ^{40}Ca(α, γ)^{44}Ti experiment [3] showed that in the low-lying energy region of ^{44}Ti there appear to be two positive excited bands, a $K^\pi = 0_2^+$ band ($0^+, 2^+, 4^+$) starting from 1.90 MeV and a $K^\pi = 2^+$ band ($2^+, 3^+, 4^+$) starting from 2.88 MeV. In the shell model they are considered to be core-excited states which consist of 8 particles in the *fp*-shell and 4 holes in the *sd*-shell.

In this paper we show that the $K^\pi = 0_2^+$ and 2^+ bands can be interpreted to have α-cluster structure with the core excitation characterized by the dominant configuration of ^{40}Ca*($0^+, 2^+, 4^+$) + α. Here ^{40}Ca*(I^π) ($I^\pi = 0^+, 2^+, 4^+$) denotes a state of the excited $K^\pi = 0^+$ band in ^{40}Ca, whose structure has been well interpreted by the ^{36}Ar + α cluster model [4].

2. The core-excited α-cluster model

In the low-lying energy region below $E_x = 5.3$ MeV for ^{44}Ti, there appear 12 well-defined positive parity states [5] except the ground band, where nine positive parity states, 0^+ states (1.90, 4.84 MeV), 2^+ states (2.53, 2.89, 4.10 MeV), 3^+ state (3.42 MeV) and 4^+ states (3.37, 3.98, 5.31 MeV) are strongly populated in the α-transfer reactions. Therefore, these can be tentatively assigned in terms of the basic configurations of ^{40}Ca*(^{36}Ar + α) + α as

$$J^\pi = 0^+: [^{40}\mathrm{Ca}^*(0^+) \otimes S_\alpha] + [^{40}\mathrm{Ca}^*(2^+) \otimes D_\alpha]$$
$$J^\pi = 1^+: [^{40}\mathrm{Ca}^*(2^+) \otimes D_\alpha]$$
$$J^\pi = 2^+: [^{40}\mathrm{Ca}^*(2^+) \otimes S_\alpha] + [^{40}\mathrm{Ca}^*(0^+) \otimes D_\alpha] + [^{40}\mathrm{Ca}^*(2^+) \otimes D_\alpha]$$
$$J^\pi = 3^+: [^{40}\mathrm{Ca}^*(2^+) \otimes D_\alpha]$$
$$J^\pi = 4^+: [^{40}\mathrm{Ca}^*(4^+) \otimes S_\alpha] + [^{40}\mathrm{Ca}^*(2^+) \otimes D_\alpha] + [^{40}\mathrm{Ca}^*(4^+) \otimes D_\alpha]$$

The Hamiltonian of the ^{40}Ca* + α system with respect to the ^{40}Ca*(0^+) + α threshold energy is given as

$$H = [T_R + h(^{40}\mathrm{Ca}^*) - \varepsilon(^{40}\mathrm{Ca}^*(0^+))]\,\delta_{c,c'} + V_{c,c'}(R), \tag{1}$$
$$c = [^{40}\mathrm{Ca}^*(I^\pi) \otimes L_\alpha]_J, \tag{2}$$

where L_α denotes the orbital angular momentum referring to the relative coordinate $\mathbf{R}$ between an α-particle and a ^{40}Ca* core and T_R denotes the relative kinetic energy with respect to the relative coordinate $\mathbf{R}$, and $h(^{40}\mathrm{Ca}^*)$ and $\varepsilon(^{40}\mathrm{Ca}^*(0^+))$ are the intrinsic Hamiltonian of the ^{40}Ca* nucleus and the intrinsic energy of the ^{40}Ca*(0^+) state, respectively. The inter-cluster potential $V_{c,c'}$ which is derived by folding the nucleon-nucleon interactions with the wave functions of ^{40}Ca*(I^π) nucleus and α-particle is expanded in the following multipole series:

$$V_{c,c'}(R) = (-)^J \sqrt{(2I+1)(2L+1)(2I'+1)(2L'+1)} \cdot \sum_\lambda (L0L'0|\lambda 0)\, W(ILI'L'; J\lambda)\, V^{(\lambda)}_{L,L'}(R), \tag{3}$$

where the multipole transition potential, $V^{(\lambda)}_{L,L'}(R)$, is symmetric for exchange for L and L'. In general $V^{(\lambda)}_{L,L'}(R)$

Correspondence to: S. Ohkubo

with λ larger than 2 is very small, so only $\lambda=0$ is taken into account for the diagonal part and $\lambda=2$ for the non-diagonal part.

The Hamiltonian matrix elements for two configuration of $J^\pi=0^+$ are given as

$$\langle[^{40}\mathrm{Ca}^*(0^+)\otimes S_\alpha]_{J=0^+}|H|^{40}\mathrm{Ca}^*(0^+)\otimes S_\alpha]_{J=0^+}\rangle = E_0\,, \tag{4}$$

$$\langle[^{40}\mathrm{Ca}^*(2^+)\otimes D_\alpha]_{J=0^+}|H|^{40}\mathrm{Ca}^*(2^+)\otimes D_\alpha]_{J=0^+}\rangle = E_0+0.55+\varepsilon_D\,, \tag{5}$$

$$\langle[^{40}\mathrm{Ca}^*(0^+)\otimes S_\alpha]_{J=0^+}|H|^{40}\mathrm{Ca}^*(2^+)\otimes D_\alpha]_{J=0^+}\rangle = \sqrt{5}\,V_1\,. \tag{6}$$

where the value of 0.55 MeV in Eq. (5) denotes the energy difference between the ^{40}Ca*(0^+) and ^{40}Ca*(2^+) states. Similar Hamiltonian matrix elements are also obtained for $J=2^+$ and $J=4^+$ states. We diagonalize the Hamiltonian matrix elements with four parameters E_0 (energy of $[^{40}\mathrm{Ca}^*(0^+)\otimes S_\alpha]$, V_1, V_2 (coupling strength among different basic configurations) and ε_D (excitation energy of α-particle from S state to D state with respect to the ^{40}Ca* core). The parameters are determined so as to reproduce the energies of the 1.90 MeV and 4.84 MeV 0^+ states and energies of the 2.53 MeV, 2.89 MeV and 4.10 MeV 2^+ states.

3. Results and discussion

The obtained energy spectra are shown in Fig. 1 and are good agreement with the experimental data below $E_x=5.3$ MeV. In the present model, there appear the 1^+ state which is characterized by the main configuration of $[^{40}\mathrm{Ca}^*(2^+)\otimes D_\alpha]$. Experimentally, it has not been observed, but it might correspond to one of the levels below $E_x=5$ MeV whose spin and parity are not assigned. Since the ^{40}Ca*(2^+) state has prolate deformation, it is considered that the unnatural parity 1^+ state appears at energy slightly higher than the $J^\pi=3^+$ state. It should be noted that the present α-cluster model with core excitation could reproduce the experimental energies of the $8p-4h$ states without performing a laborious calculation like the conventional shell model. In order to investigate quantitatively the validity of the parameter values determined above, we evaluate E_0 and ε_D by solving the ^{40}Ca*$(0^+)-\alpha$ potential problem in which the inter-cluster potential is derived within the framework of the $^{36}\mathrm{Ar}+\alpha+\alpha$ model. In the range of about 10% reduction of the potential strength, we could reproduce the experimental energy of $J^\pi=0_2^+$ ($E_x=1.90$ MeV). Taking into account the reduction, E_0 and ε_D become -5.6 and 0.46 MeV, respectively, which are in good agreement with the parameter values determined phenomenologically ($E_0=-5.6$ MeV, $\varepsilon_D=0.5$ MeV).

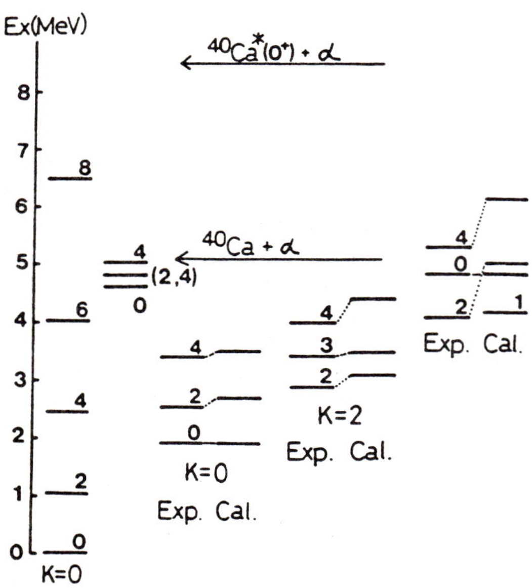

Fig. 1. Comparison of the experimental positive parity energy spectra of ^{44}Ti below $E_x=5.3$ MeV with the calculated ones

With the use of the wave functions obtained by the schematic procedure, we calculate the $B(E2)$ values and compare them with the experimental data. The $E2$-transition operator can be effectively given as

$$M(E2)=\frac{e}{2}\,[Q(^{40}\mathrm{Ca})+\tfrac{40}{11}\,Q(\mathbf{R})]\,, \tag{7}$$

where $Q(^{40}\mathrm{Ca})$ is the quadrupole operator for ^{40}Ca core part and $Q(\mathbf{R})$ for the relative coordinate $\mathbf{R}$. Table 1 shows the calculated $B(E2)$-transition rates. They are in gratifying agreement with the experiment, especially if one takes into account that no effective charge has been introduced in the present calculation. The discrepancy in the intra-band $B(E2)$ values, about factor two, is due to the fact that the reduced $E2$-transition matrix elements for ^{40}Ca employed here, for which we don't take into account the contribution of the $B(E2)$ from the ^{36}Ar core part itself which is fairly large because of large deformation.

In conclusion, with the use of a schematic $^{40}\mathrm{Ca}^*+\alpha$ model, we have shown that the low-lying $K^\pi=0_2^+$ and 2^+ band states in ^{44}Ti can be interpreted to have an α-cluster structure with the excited ^{40}Ca core, namely $^{36}\mathrm{Ar}+\alpha+\alpha$ structure. Together with these states, nine of the twelve positive parity states except the ground band states below $E_x=5.3$ MeV could be described by the $^{40}\mathrm{Ca}^*+\alpha$ model and the calculated energy spectra and the $B(E2)$ transition rates are in good agreement with the experimental data.

Table 1. Comparison of the calculated $B(E2)$ values with the experimental data [3]

	$B(E2)$ Weisskopf unit				
$J_i\to J_f$	cal.	exp.	$J_i\to J_f$	cal.	exp.
$2_2\to 0_2$	16.6	24	$2_2\to 0_1$	0.44	0.17
$4_2\to 2_2$	17.2	29	$2_3\to 0_1$	1.03	0.71
$3\ \to 2_3$	9.0	<64	$2_2\to 2_1$	0.19	6.7
$4_3\to 2$	2.2	<85	$2_3\to 2_1$	0.82	<3.2
$4_3\to 2_3$	7.1	19	$4_2\to 2_1$	1.61	<2
			$4_3\to 2_1$	0.06	0.24

References

1. Yamaya, T. et al.: Phys. Rev. C **41**, 2421 (1990); Yamaya, T., Ohkubo, S., Okabe, S., Fujiwara, M.: Phys. Rev. C **47**, 2389 (1993)
2. Michel, F., Reidemeister, G., Ohkubo, S.: Phys. Rev. Lett. **57**, 1215 (1986); Phys. Rev. C **37**, 292 (1988); Ohkubo, S.: Phys. Rev. C **38**, 2377 (1988)
3. Simpson, J.J., Dixon, W.R., Storey, R.S.: Phys. Rev. Lett. **31**, 946 (1973)
4. Ohkubo, S., Umehara, K.: Prog. Theor. Phys. **80**, 598 (1988)
5. Table of Isotopes, Lederer, C.M., Shirley, V.S. (eds.), p. 1. New York: Wiley 1978

Laser induced - ion enhanced binary nucleation in vapors

V. Zafiropulos[1], C. Fotakis[1], J.A.D. Stockdale[2]

[1] FORTH - Institute for Electronic Structure and Laser, P.O. Box 1527, GR 71110, Heraklion, Crete, Greece and University of Crete
[2] Martin Marietta Energy Systems, P.O. Box 2003, Oak Ridge, TN 37831-7274, U.S.A.

Received 28 June 1993

Abstract. The formation of positively charged clusters consisting of 10^{11}-10^{12} molecules and 10^{2}-10^{4} unit charges has been observed in mixtures of triophenol vapor with a buffer gas when irradiated by a single 30 ns UV laser pulse. The results indicate a means of controlling the size and the charge of the micron size clusters formed, by varying certain parameters such as the laser intensity and the pressure of the buffer gas. Here we address the fundamental assumptions of a model which describes this nucleation process and predicts the experimental trends.

PACS: 33.80.Eh, 34.20.-b, 34.70.+e, 34.90.+q, 36.40.+d, 51.20.+d

1. Introduction

Recently, laser-induced clustering has received considerable attention due to its potential use for isotope separation [1,2]. production of finely divided powders for catalysts and ceramic materials [2,3] and production of metal clusters [4]. The possibility of initially creating and later controlling the shape and size of such clusters/particles was demonstrated by Tam et al. [5a] and Tanaka et al. [5b] respectively. The latter were able to grow $(CsH)_n$ crystalline particles of four different shapes and various sizes just by varying the experimental parameters.

We have reported [6] results on laser-induced clustering in thiophenol vapor using either single pulses of a KrF excimer laser or many pulses of an excimer pumped dye laser. Charged micron size particles were produced and it was shown that the size and charge can be controlled by varying certain parameters such as the laser pulse energy, the buffer gas pressure and by using the appropriate buffer gas. A qualitative summary of possible mechanisms that may be responsible for the experimental results was also presented. The need of understanding the growth mechanism, of explaining the experimental trends and in general of modelling this kind of nucleation process is obviously of great importance. Here we address the fundamental aspects of a model for that kind of laser-induced nucleation which is described in Ref. [6]. As a first step, this model explains only qualitatively the above process. The quantitative analysis of the laser induced nucleation is a difficult task due to the many unknown parameters and the inherent complexity of the process. A quantitative development of the present model is under way.

2. Summary of observations

The experimental arrangement and the results have been described in detail elsewhere [6]. In brief, a cell containing 1 Torr of thiophenol (PhSH) vapor and a buffer gas was crossed by a KrF excimer laser beam. After the interaction with one laser pulse, positively charged micron size clusters were formed basically within the cylindrical volume defined by the trace of the excimer laser beam. These clusters fell at a terminal velocity towards a probe HeNe CW laser beam (parallel to the excimer laser beam) under the action of gravity or an added applied electric field, and their size and charge were determined [6] by measuring the elapsed time between the excimer laser pulse and the light scattering due to the passage of the clusters through the probe beam.

The results can be summarized as follows: (a) By varying the energy of the laser pulse and keeping the other parameters constant, the size and the charge of the clusters increases while the quantity N/q remains constant, where N is the average number of molecules per cluster and q is the average charge in positive unit charges per cluster. (b) The magnitude of N/q strongly depends on the buffer gas used, e.g. it increases by about two orders of magnitude when switching from N_2 to He. (c) For each buffer gas there exists an optimum pressure for maximum growth and charge accumulation. (d) The clustering threshold was found to occur at 298 nm, which is the onset wavelength of the UV absorption band of

PhSH. (e) The role of the low energy electrons produced during the laser pulse ($PhSH^* + h\nu \rightarrow PhSH^+ + e^-$) was investigated by introducing 1 Torr of CCl_4 (high dissociative electron attachment cross section [7]) into the cell and it was observed that the charge dropped by $\sim 15\%$.

3. Model and comparison with experimental trends

Clustering is initiated by the laser pulse which results in the formation of long lived triplet state molecules and molecular ions. Actually the KrF laser pulse excites the PhSH molecules [8] to the second or third electronically excited π-singlet states PhSH$^*(S_2, S_3)$ which normally undergo intramolecular relaxation to the S_1 state within 1 ps. Under the experimental conditions of this work the above excitation is saturated since even the lowest energy fluence used (17 MW/cm^2) is higher than the saturation intensity for the one photon absorption (3.3 MW/cm^2). The PhSH$^*(S_1)$ molecules mostly [9] decay through intersystem crossing to the triplet manifold PhSH$^*(T_1)$ within a few ns. The lifetime of the triplet state in the gas phase is expected to be at least of the order of the lifetime in the liquid phase, which is 7.6ms [10]. During the 30ns duration of the laser pulse, excited molecules may absorb a second photon producing molecular ions PhSH$^+$. Energetically, the second photon absorption is strongly favored since the addition of the 5eV photon energy to the energy of the singlet state (4.4eV [9]) coincides with the second vertical ionization potential of PhSH (9.4eV [11]). Thus at the end of the 30ns laser pulse there exists a vapor mixture consisting of PhSH(S_0), PhSH$^*(T_1)$, PhSH$^+$ and buffer gas. The latter prevents the rapid diffusion of the excited species out of the volume where they are produced. Due to the increased intermolecular forces in the above mixture [6a] and the favorable free energy of mixing in the liquid state, the gas phase becomes supersaturated relative to liquid mixtures and ion-enhanced binary nucleation takes place.

The first step for the nucleation is the formation of dimers (PhSH)$_2$ resulting from the collisions between a PhSH$^*(T_1)$ and another PhSH$^*(T_1)$ or a PhSH(S_0) since it is well known that the average interaction energy between two molecules with permanent dipole moments is always negative (attractive). On the contrary, the average interaction energy between one ion and one molecule is positive, thus excluding the formation of (PhSH)$_2^+$. The excited dimers are quenched by collisions with the buffer gas which exists in excess. Other possible initiating processes, such as production of PhSSPh*, are not favored [8].

Due to the supersaturation of the gas phase the nucleation continues with the formation of trimers, tetramers, etc. The existing ions are attached to the growing clusters with a slow rate since they are only a very small fraction of the supersaturated gas mixture. As the nucleation progresses, the volume of the supersaturated vapor expands following the laws of diffusion and the supersaturation progressively decreases. As a result the quantity ΔG_{max} [12] increases making it difficult for new nuclei to reach the critical radius and thus supporting the nuclei which have grown larger than the critical radius during the early stages following the laser pulse to grow to micron sizes within a few tens of ms [6a].

Focussing more on the role of the ions it can be understood that although they do not greatly contribute to the supersaturation of the gas mixture (the experimental results [6a] show an ion to neutral ratio ranging from $1 : 4 \times 10^7$ to $1 : 4 \times 10^9$) they do play a key role in controlling the experimental trends. For instance, a high ionic concetration due to a high laser intensity would increase the surface tension of the growing nuclei although it will not considerably alter the supersaturation (saturated one-photon absorption in our case). This would result (see eq.IX-7 of Ref. [12]) to an increased ΔG_{max} producing fewer particles but grown to larger sizes and this is exactly what is observed [6a], i.e. as the laser intensity increases the size and the charge of the clusters increase.

The rate at which the ions and the excited molecules diffuse out of their formation region dictates the behavior of the supersaturation of the gas phase and together with the concetration of ions they dictate the final size of the particles formed. For instance the diffusion of PhSH$^*(T_1)$ through He is about three times faster than through N_2 and, therefore, during nucleation the average supersaturation is considerably lower when He is used as buffer gas. This means that in the presence of He there are fewer particles but they grow to larger sizes, as is experimentally observed (Fig. 2 of Ref. 6a) The quantity N/q, described in the previous section, is another example of the important role of diffusion in the nucleation process. In the presence of N_2 the ions form short lived complexes of the type PhSH$^+(N_2)_n$ which diffuse much slower than the excited PhSH$^*(T_1)$ molecules. The result would be an unequal increase of the volume containing the ions and the volume containing the excited molecules. Assuming the octahedral structure PhSH$^+(N_2)_6$ it is estimated that the diffusion rate of PhSH$^*(T_1)$ is three times faster than that of PhSH$^+(N_2)_6$. Thus the inner volume containing both ions and excited molecules will have a high concetration of ions and the quantity N/q is expected to attain its lowest possible value. On the other hand in the presence of He, complexes like PhSH^+He_n are expected to be both unstable and lighter than PhSH$^+(N_2)_n$ producing a low concetration of ions in the nucleation volume. As a result N/q will reach its highest possible value, which is exactly what the experimental results show (Fig. 4 of Ref. [6a]).

At this point we should refer to the quenching of the excited PhSH$^*(S_1)$ by the buffer gas which affects the growth mechanism. As the pressure of the buffer gas increases the quenching of PhSH$^*(S_1)$ increases [8] and the concetration of the PhSH$^*(T_1)$ in the vapor mixture decreases lowering the supersaturation. At the same time the increased buffer gas pressure tends to prevent diffusion, keeping the gas mixture in a smaller volume corresponding to a tendency of increasing the supersaturation. By starting from low and going to higher pressures the results arising from the PhSH$^*(S_1)$ quenching predominate and the supersaturation reaches the lowest value (optimum growth). For higher pressures the decrease of the diffusion prevails and the supersaturation starts rising. By solving the rate equations and making some reasonable assumptions it can be shown that the optimum growth in the presence of Ar occurs at a lower pressure than in the presence of N_2 (Fig. 5 of Ref. [6a]).

Finally, it is worthwhile to look at the role of the low energy electrons in the nucleation process. A possible use of these electrons is their attachment to PhSH$^*(T_1, S_0)$ molecules [8] and through the harpooning effect [6a,7] their contributions to the nucleation [6a]. Under the present experimental conditions this mechanism is not favored since the addition of CCl_4 vapor into the nucleation cell helps in capturing electrons which otherwise move quickly out of the nucleation volume. The negative ions CCl_4^- formed either dissociate into Cl^- and CH_3 [7] or transfer their electron to a growing positively charged cluster, in both cases reducing the final charge of the clusters.

We wish to acknowledge support from the NATO Collaborative Research Grant #0423/86. We would also like to express our gratitude to Professor S. Leach, Dr. G. Wilemski and Dr. C. Kalpouzos for the helpful discussions with them.

References

1 Borsella, E., Catoni, F., and Freddi, G.: J. Chem. Phys. **73**, 316 (1980)

2 Borsella, E., Fantoni, R., and Caneve, L.: reports RT/TIB/87/25 and RT/TIB/87/26, ENEA Centro Ricerche Energia Frascati CP 65 65-00044, Frascati, Rome, Italy (1987) and references therein

3 Kizaki, Y., Kandori, T., and Fujiyama, Y.: Jap. Appl. Phys. **24**, 800 (1985); Suyama, Y., Marra, R.A., Haggerty, J.S., and Bower, H.K.: Am. Ceram. Soc. Bull. **64**, 1356 (1985)

4 Allegrini, M., Bicchi, P., Dattrino, D., and Moi, L.: Opt. Commun. **49**, 39 (1984)

5 (a) Tam, A., Moe, G., and Happer, W.: Phys. Rev. Lett. **35**, 1630 (1975); Tanaka, T., Mitsui, T., Sugiyama, K., Kitano, M., and Yabuzaki, T.: Phys. Rev. Lett. **63**, 1390 (1989)

6 (a) Zafiropulos, V., Kollia, Z., Fotakis, C., and Stockdale, J.A.D.: J. Chem. Phys. **98**, 5079 (1993); (b) J. Chem Phys. **94**, 2374 (1991)

7 Christophorou, L.G., and Stockdale, J.A.D.: J. Chem. Phys. **48**, 1956 (1968)

8 Pinnudawage, L.A., Christophorou, L.G., and Hunter, S.R.: J. Chem. Phys. **90**, 6275 (1989)

9 Russell, P.G.: J. Phys. Chem. **79**, 1347 (1975)

10 Perkampus, H.H., and Vollbrecht, H.R.: Spectrochim. Acta A **27**, 2173 (1971)

11 Frost, D.C., Herring, F.G., Katrib, A., McDowell, C.A., and McLean, R.A.N.: J. Phys. Chem. **76**, 1030 (1972)

12 Adamson, A.W.: Physical Chemistry of Surfaces. 5th Edn. New York, Chichester, Brisbane, Toronto, Singapore: John Wiley & Sons 1989, chapter IX

This article was processed using Springer-Verlag TEX Z.Physik D macro package 1.0 and the AMS fonts, developed by the American Mathematical Society.

Z. Phys. A 349, 367–370 (1994)

Nuclear clusters in dissipative medium

V. Zagrebaev

Department of Theoretical Physics, Chuvash State University, Cheboksary, 428900 Russia

Received: 27 October 1993

Abstract. An exact solution of channel coupling equations describing the relative motion of two heavy clusters was analyzed without additional (commonly used) simplifications concerning a restricted number of the channels and low excitation energy in comparison with the total energy. It was found that excitation of even very simple quantum system with only one degree of freedom may reveal the dissipative features at slow collisions if the spectra of this system is sufficiently random. An excitation of such a system at sub-barrier tunnelling increases significantly (by several orders) the transmission probability, in spite of the large dissipation of relative motion kinetic energy occurs. At deep sub-barrier energies the soft degrees of freedom are excited with the most probability. Dynamics of wave packet in dissipative environment was also analyzed and a new effect of wave packet sticking to a potential wall due to its tangling and time delay inside the inelastic channels was found.

PACS: 25.70; 25.85; 24.75.+i

A strong channel coupling is inherent in interaction of nuclear clusters with atomic nuclei and with each other due to intense excitation of collective and particle degrees of freedom. This leads to appearance of dissipative phenomena in such processes as fusion of the clusters with atomic nuclei, cluster radioactivity, heavy cluster transfer reactions, nucleus fragmentation into the clusters, formation of nuclear quasi-molecular states. Meanwhile a conservation of the quantum effects – coherent nucleon motion inside a cluster, subbarrier tunneling of the clusters, formation of such light fragments as α-particles, the Pauli exclusion principle – forces us to consider these dissipative phenomena within a quantum channel coupling approach. In connection with it, numerous fundamental questions arise before us. What are the main reasons and mechanisms of nuclear viscosity: one-body excitations inside a time-depended mean field [1], two-nucleon collisions [2], chaotic motion of nucleons in asymmetric deformed nucleus [3], excitations of collective degrees of freedom [4], nucleon exchange [5], chaotization of coupling matrix elements [6], ...? If there are several mechanisms, then what are their relative contributions depending on mass and energy of interacted nuclei? Do the dissipative processes survive at high energies? Is it possible at all to distinguish correctly the dissipative and potential forces in heavy ion experiments? At last, how can we consider the friction forces within quantum theory of collisions?

Using the multi-dimensional and multi-channel semiclassical approximation [7] parallel with exact solution of stationary and non-stationary Shrödinger coupling equations we examine the quantum problem of formation and relative motion of heavy clusters with intrinsic degrees of freedom and try to answer the following questions:

(1) May a simple quantum system of only one degree of freedom display a dissipative response?
(2) What is the mechanism of subbarrier dissipative tunnelling?
(3) What is the dynamics of wave packet motion in dissipative media?

1. Model

The total Hamiltonian of nucleus-nucleus relative motion is given as

$$H(\mathbf{r}, \xi) = T(\mathbf{r}) + U(r) + H_A(\xi) + V(\mathbf{r}, \xi), \tag{1}$$

and

$$H_A \phi_\nu(\xi) = \varepsilon_\nu \phi_\nu(\xi). \tag{1a}$$

A number of internal degrees of freedom ξ is usually very large ($=4[A_1+A_2]-6$). However, not all of them are coupled with relative motion in equal manner. Moreover, on initial reaction stage some of them (e.g., surface vibrations, rotation states of deformed nuclei, one-particle excitations near the Fermi surface) seem to be excited with a most probability.

Let the relative motion of two interacted nuclei is coupled with only one internal degree of freedom ξ. Total wave function can be decomposed over the eigenstates of Hamiltonian $H_A(\xi)$:

$$\Psi(\mathbf{r}, \xi, t)=\sum_{v} \psi_v(\mathbf{r}, t)\, \phi_v(\xi). \tag{2}$$

The channel wave functions $\psi_v(\mathbf{r}, t)$ satisfy a set of coupled equations with the matrix elements $V_{v\mu}(\mathbf{r}) = \langle\phi_v|\, V(\mathbf{r}, \xi)\,|\phi_\mu\rangle$. In a case of strong coupling (processes with a large dissipation of kinetic energy) we can not get an exact solution of these equations. So, we use here a simplified model of a straight motion along x-axis. In this case we have the following set of coupled equations

$$i\hbar \frac{\partial \psi_v(x, t)}{\partial t}=\left[-\frac{\hbar^2}{2m}\frac{\partial^2}{\partial x^2}+U(x)+\varepsilon_v\right]\psi_v(x, t) + \sum_{\mu} V_{v\mu}(x)\, \psi_\mu(x, t). \tag{3}$$

The matrix elements were chosen in the following form

$$V_{v\mu}(r)= V_0 \cdot f(x) \cdot V(v, \mu), \tag{4}$$

where $V(v, \mu)= V(\mu, v)$, and

$$f(x)=\mathrm{e}^{-\left(\frac{x-R_V}{a}\right)^2}. \tag{5}$$

For linear coupling of relative motion with harmonic oscillator

$$V(v, \mu)=\sqrt{v}\,\delta_{\mu, v-1}+\sqrt{v+1}\,\delta_{\mu, v+1} \qquad (v, \mu=0, 1, 2, \ldots),$$

and $\varepsilon_v=\varepsilon_{v-1}+\Delta\varepsilon_v$, $\Delta\varepsilon_v=\hbar\omega$.

2. Viscous response of a simple quantum system

An excitation of classical harmonic oscillator with a frequency ω by heavy projectile moving by with a velocity v and interacting with it in linear manner (for example, $V(x, \xi)= V_0 \cdot \xi \cdot \mathrm{e}^{-\left(\frac{x}{\Delta x}\right)^2}$), is given by a simple formula [8]:

$$E_{\mathrm{loss}}=\frac{1}{2M}\left|\int_{-\infty}^{\infty} F(t)\,\mathrm{e}^{-\mathrm{i}\omega t}\,\mathrm{d}t\right|^2 =\pi\frac{V_0^2}{8M}\tau^2\, \mathrm{e}^{\left[-\frac{1}{8}(\omega\tau)^2\right]} \tag{6}$$

Here M is inertial parameter of oscillator, and $\tau=\frac{2\Delta x}{v}$ is time of flight (in peripheral nuclear collisions $\Delta x \approx\sqrt{b a/2}$, where b is impact parameter, and a is diffuseness of nuclear forces). Maximal energy loss occurs when time of flight is approximately equal to a period of internal vibrations: $\tau_{\mathrm{opt}}=\frac{\sqrt{8}}{\omega}=\frac{\sqrt{8}}{2\pi}T\approx\frac{1}{2}T$. In slow collisions ($\tau\gg T$) the energy loss of a projectile is exponentially small (absolute elastic response), whereas in fast collisions a target absorbs always a non-zero energy due to transfer of linear momentum in a sudden impact but not due to a real dissipative process. Therefore, we can speak at all about elastic or viscous response of a system only at **slow** influence upon it. At excitation of linear oscillator with an energy $\Delta\varepsilon=\hbar\omega$ collisions with initial

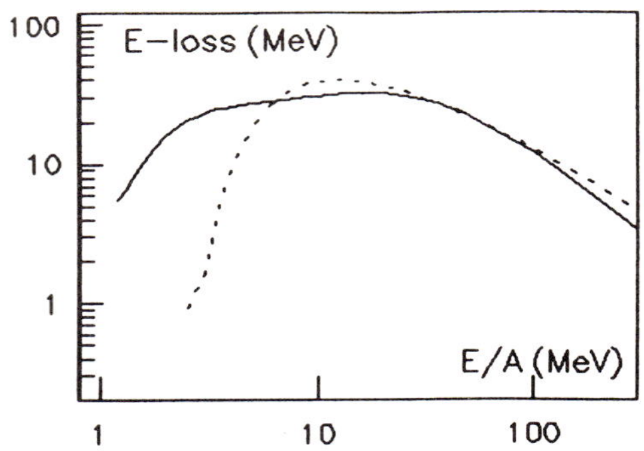

Fig. 1. Energy loss of a heavy projectile at excitation of linear (*dashed line*) and 'chaotic' oscillator (*solid line*). $U(x)=0$, $V_0=$ 10 MeV, $\Delta\varepsilon=10$ MeV

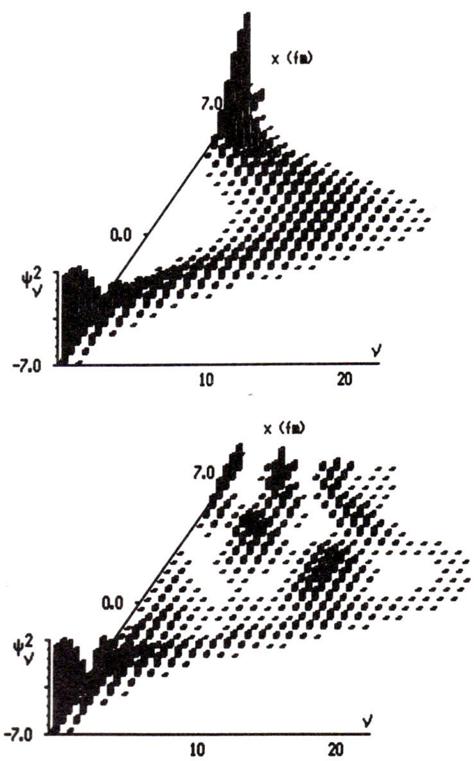

Fig. 2. Distribution over the channels at excitation of linear (*up*) and non-linear (*down*) oscillator by a slow projectile moving in positive direction of x-axis. $E/A=2.5$ MeV, $\Delta\varepsilon=20$ MeV. $V_{v\mu}$ are regular in both cases, $V_0=15$ MeV

energies $\frac{E-V_C^B}{A}<\frac{2m_N}{\hbar^2}\left(\frac{\Delta x\,\Delta\varepsilon}{2\pi}\right)^2$, (where V_C^B is a Coulomb barrier, and m_N is nucleon mass) may be treated as slow ones meaning $\tau<T=2\pi/\omega$.

In Fig. 1 the dissipative energy is shown depending on initial energy of the projectile interacted with a simple quantum system of only one degree of freedom and the corresponding channel wave functions at different values of x are shown in Fig. 2. We see that a quantum linear oscillator displays also an absolutely elastic response even at strong channel coupling. In order to clear up the main response features of non-linear oscillator, we solved the same equation (3) in two cases: (a) with completely chaotic matrix elements $V_{\nu\mu}$, (b) with completely chaotic spectra ε_ν at a fixed $\langle\Delta\varepsilon_\nu\rangle=\Delta\varepsilon$. It was found that just the chaotization of energy spectra (not the matrix elements!) leads to the viscous response of a simple quantum system with only one degree of freedom at slow influence upon it from an incident particle (see Fig. 1 and Fig. 2).

3. Dissipative tunnelling

The problem of subbarrier dissipative transmission is mainly motivated by nuclear heavy-ion fusion-fission reactions and heavy-fragment radioactivity. Channel coupling approach seems to be the most consistent one to solve this problem [9]. Feynman's path integration with inclusion of classically forbidden trajectories is another possibility (see, for example, [10]). However, if there is strong channel coupling, one failed to get physically clear results in both cases without additional simplifications.

We analyzed an exact solution of (3) taking into account **all** the channels which can be reached at a given value of coupling interaction (see Fig. 2) and a total excitation energy $\varepsilon_\nu^{\max}$ was allowed to be comparable (or even more) with initial relative motion kinetic energy. Potential barrier $U(x)$ was chosen in Gauss form to avoid the Coulomb scattering problems – Fig. 3.

The total transmission probability is shown in Fig. 4. Significant enhancement of barrier penetrability at subbarrier energies due to channel coupling is wonderful and well known fact. Physical interpretation of this phenomenon is not clear completely. Generally used explanation is the following one. If we neglect the excitation energy ($\varepsilon_\nu=0$) and choose separable coupling matrix elements (4), then (3) can be decoupled and a set of barriers of different heights arises. This barrier distribution decreases transmission probability at above-barrier energies and increases it at sub-barrier ones (see, for example, [11]). One more reason for subbarrier penetrability enhancement is an appearance of effective advanced forces (independent on sign of V_0) induced by channel coupling in barrier region [10].

In our view, an additional study of this question is required. In particular, we found an unusual effect of preferable transmission not in entrance ($\nu=0$) but in high excited channels ($\nu\neq0$) with significant dissipation of relative motion kinetic energy – Fig. 5. Now we can give only 'mathematical' explanation of this effect. The waves in channels $\mu\neq0$ arise only under the barrier due to coupling with $\psi_{\nu<\mu}$ and have quite different (increasing) behavior in comparison with the wave in entrance channel – Fig. 6. We have found also that the effect of transmission probability enhancement due to coupling channel is strongly sensitive to the change of coupling interaction radius. If the excitation of intrinsic degrees of freedom occurs behind or before the barrier then penetration probability sharply decreases – Fig. 7. The results shown in Figs. 5–7 are qualitatively the same both for excitation of linear and non-linear oscillator.

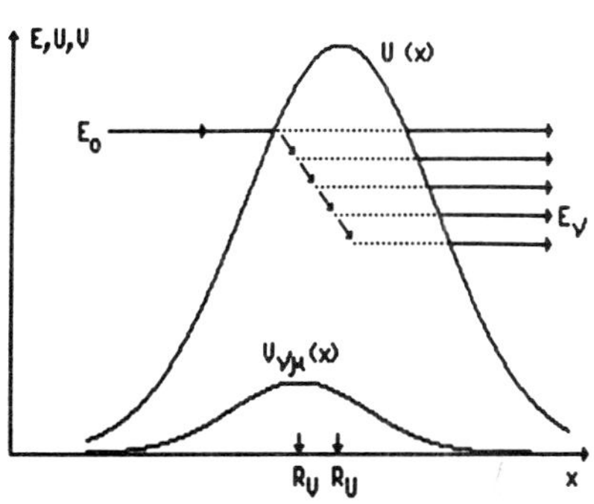

Fig. 3. Schematic view of subbarrier tunnelling with channel coupling

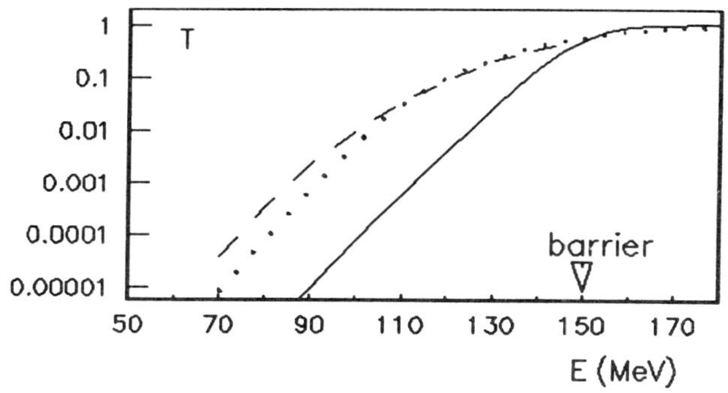

Fig. 4. Barrier penetrability without channel coupling (*solid line*), with excitation of linear (*dots*), and 'chaotic' oscillator (*dashed line*). $U_0=150$ MeV, $V_0=20$ MeV, $\Delta\varepsilon=10$ MeV

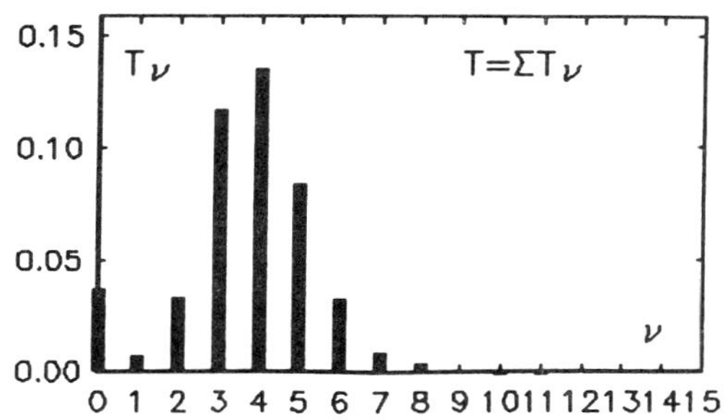

Fig. 5. Barrier penetrability in different channels at $E=140$ MeV, $V_0=15$ MeV

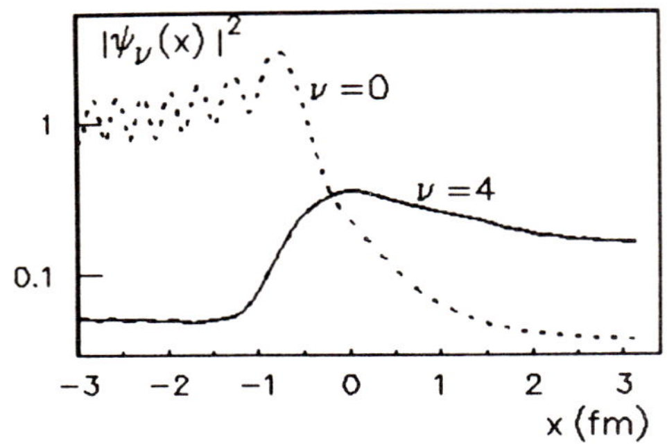

Fig. 6. Amplitudes of the channel wave functions in the channels $\nu=0$ and $\nu=4$. Plane wave of unit amplitude with an energy of 140 MeV come from the left in the channel $\nu=0$, $V_0=15$ MeV

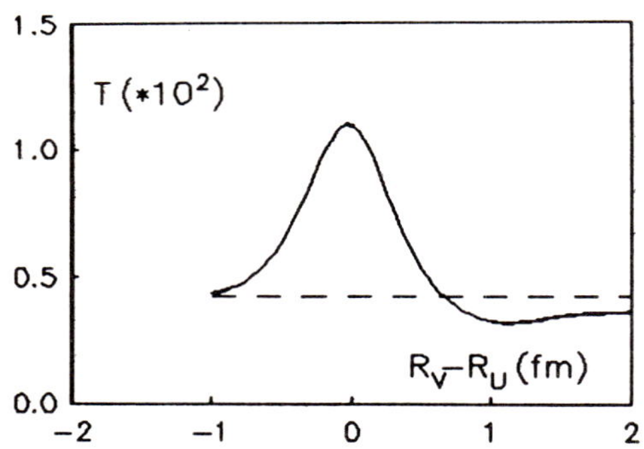

Fig. 7. Barrier penetrability depending on the radius of coupling interaction $V_{\nu\mu}(r)$ at $E=120$ MeV and $V_0=10$ MeV (*solid line*). *Dashed line* shows the corresponding value without channel coupling. $a_V=a_U=0.7$ fm

4. Summary

We have analyzed an exact solution of channel coupling equations without additional (commonly used) simplifications concerning a restricted number of the channels and low excitation energy in comparison with an energy of relative motion. It was found that excitation of even very simple quantum system with only one degree of freedom may reveal the dissipative features at slow collisions if the spectra of this system is sufficiently random. An excitation of such a system at subbarrier tunnelling increases significantly (by several orders) the transmission probability, in spite of the large dissipation of relative motion kinetic energy. Penetration probability depends strongly not only on the value of coupling interaction V_0 but also on relation between its radius and radius of the barrier. Maximal penetrability is observed at $R_V = R_U$. Calculations show that at the deep sub-barrier energies the soft degrees of freedom corresponding to the small values of $\Delta\,\varepsilon$ are excited with the most probability. We analyzed also the dynamics of wave packet in dissipative environment and found a new effect of wave packet sticking to a potential wall due to its tangling and time delay inside the inelastic channels. Such an effect can be displayed experimentally, for example, in formation of short-living nuclear molecules at sub-barrier energies.

References

1. Gross, D.M.E.: Nucl. Phys. A**240**, 472 (1975); Hofmann, H., Siemens, P.J.: Nucl. Phys. A**257**, 165 (1976)
2. Mantzouranis, G., Pauli, H.C.: Phys. Rev. C**22**, 1550 (1980); Pal, S., Granguly, N.K.: Nucl. Phys. A**370**, 175 (1981)
3. Swiatecki, W.J.: Nucl. Phys. A**488**, 375c (1988)
4. Broglia, R.A., et al.: Phys. Rev. Lett. **41**, 25 (1978); Catara, F., Lombardo, U.: Nucl. Phys. A**455**, 158 (1986)
5. Randrup, J.: Ann. Phys. **112**, 356 (1978); Nucl. Phys. A**327**, 490 (1979)
6. Nemes, M.C., Weidenmüller, H.A.: Phys. Rev. C**24**, 450 (1981)
7. Zagrebaev, V.I.: Ann. Phys. **197**, 33 (1990)
8. Landau, L.D., Lifshiz, E.M.: Mechanics p. 83. Moscow: Nauka 1965
9. Dasso, C.H., Landowne, S., Winther, A.: Nucl. Phys. A**432**, 495 (1985)
10. Fröbrich, P., Lipperheide, R.L., Möhring, K.: Z. Phys. B**78**, 325 (1990)
11. Rowley, N.: Nucl. Phys. A**538**, 205c (1992)

Monopole and quadrupole incompressibility of the hydrogen atom using deformed oscillation trial wave functions

L. Zamick

Department of Physics and Astronomy, Rutgers University, Piscataway, New Jersey 08855, USA

Received 28 June 1993

Abstract. We use deformed harmonic oscillator trial wave functions for the hydrogen atom. For the cases considered 1S and 2P it is shown that it does not pay to deform, but one can calculate the quadrupole incompressibility and compare it with that of the monopole. The virial theorem (E=-E_k) is satisfied with deformed oscillator trial wave functions when the minimum energy is obtained.

PACS: 31.20.-d

1. Introduction

In the beginning of his book "The Harmonic Oscillator in Modern Physics: from Atoms to Quarks" M. Moshinsky [1] discusses the use of harmonic oscillator trial wave functions to describe the hydrogen atom. He shows for example that using a single oscillator wave function Ψ_{1S} one gets 84% of the binding energy. The situation is not improved by admixing Ψ_{2S} but then when higher n values are allowed one gets a steady convergence towards the exact answer.

In this work we show that with a bit more effort one can obtain expressions for the energy when deformed oscillator wave functions are used. One can address the question of whether it pays to deform the hydrogen atom and one can obtain the quadrupole incompressibilities.

2. The spherical oscillator

Before discussing the deformed oscillator trial wave function we present simpler results for the spherical oscillator. Many of the results have been discussed by Moshinsky [1].

The spherical oscillator hamiltonian is

$$H_0 = \frac{p^2}{2m} + \frac{1}{2}m\omega^2 r^2 \tag{1}$$

The wave functions are written in terms of the oscillator length parameter $b = (h/m\omega)^{1/2}$ e.g.

$$\Psi_{1s} = N exp(\frac{-r^2}{2b^2})Y_{00} \qquad N^2 = \frac{4}{\sqrt{\pi}} \tag{2}$$

The Hamiltonian for the hydrogen-like atom is of course

$$H_h = \frac{p^2}{2m} - \frac{Ze^2}{r} \qquad 3$$

The trial energy is

$$E_t \quad = \quad <\Psi_{1s} H_h \Psi_{1s}> \quad = \quad \frac{3}{4}\frac{h^2}{mb^2} - \frac{2Ze^2}{\sqrt{\pi}b} \tag{4}$$

The condition for a minimum dE/db=0 yields $b = \frac{1}{Z}(3\sqrt{\pi}/4)\alpha_0$ where $\alpha_0 = h^2/(me^2)$ is the Bohr radius. [It should be noted that the exact mean root square radius for hydrogen is $\sqrt{3}/\alpha_0$ whereas the trial wave function result is $\sqrt{27\pi/32}\alpha_0 = 1.699\alpha_0$.]

We express the energy in units of mZ^2e^4/h^2

$$E_{trial} \quad = \quad \text{kinetic} + \text{potential} \quad = \quad \frac{4}{3\pi} - \frac{8}{3\pi} \quad = -\frac{4}{3\pi}$$

Note that the virial theorem is satisfied with the trial wavefunction. The exact result is $(-\frac{1}{2})$. The ratio $\frac{E_t}{E_{exact}} = 0.84$. We can get an expression for the incompressibility

$$K_t \quad = \quad b^2\frac{d^2E}{db^2} \quad = \quad \frac{9}{2}\frac{h^2}{mb^2} - \frac{4}{\sqrt{\pi}}\frac{Ze^2}{b} \tag{5}$$

We see that $K_t = 2|E_t|$ i.e. the incompressibility is twice the binding energy.

We note that if one uses the exact wave function $\Psi = Ne^{-r/\alpha}$ one obtains the same result except that E_t is replaced by the exact energy for the exact wavefunction

$$E = \frac{h^2}{2m\alpha^2} - \frac{Ze^2}{\alpha} \qquad (6)$$

The condition $\frac{dE}{d\alpha} = 0$ yields $a = \alpha_0/Z$ and $K = a^2\frac{d^2E}{d\alpha^2} = Z^2e^4m/h^2 = 2|E_{1s}|$.

3. Deformed trial wave functions and the quadrupole incompressibility

We now attempt to obtain a deformed solution for the 1S state of hydrogen. We assume axial symmetry and introduce two oscillator length parameters b_z and $b_x(= b_y)$. Our trial wave function is

$$\Psi = Ne^{-(x^2+y^2)/2b_x^2}e^{-z^2/2b_z^2}$$
$$N^2 = \frac{1}{b_0^3\pi\sqrt{\pi}} \qquad b_0^3 = b_xb_yb_z \qquad (7)$$

We introduce a deformatiom parameter $y = b_0/b_z$. For the case of a prolate deformation we have $y < 1$; for an oblate deformation we have $y > 1$.

The kinetic energy is

$$\frac{h\omega_x}{2}(1/2+1/2) + \frac{h\omega_z}{2}(1/2) = \frac{h^2}{mb_0^2}[\frac{1}{2y} + y^2/4] \qquad (8)$$

The potential energy is

$$-Ze^2N^2\int e^{-r^2cos^2\theta/b_z^2}e^{-r^2sin^2\theta/b_x^2}\frac{1}{r}r^2drd(cos\theta)d\phi \quad (9)$$

After integrating r and ϕ we have

$$-\pi Ze^2N^2b_x^2\int_{-1}^{+1}\frac{d\mu}{[1-\mu^2F^2]} \quad \text{where } F = \sqrt{1-y^3} \ (10)$$

The expression for the total energy is

$$\text{PROLATE } E = \frac{h^2}{mb_0^2}(\frac{1}{2y}+\frac{y^2}{4}) - \frac{Ze^2}{b_0\sqrt{\pi}}y\frac{1}{F}ln(\frac{1+F}{1-F}) (11)$$

$$\text{OBLATE } E = \frac{h^2}{mb_0^2}(\frac{1}{2y}+\frac{y^2}{4}) - \frac{2Ze^2}{b_0\sqrt{\pi}}y\frac{tan^{-1}(F)}{F} \quad (12)$$

We write b_0 in the following way

$$b_0 = b(spherical)q = \frac{1}{Z}3\frac{\sqrt{\pi}}{4}\alpha_0 q \qquad (13)$$

For q=1 one of course gets the spherical result.
The energy can be written as

$$E = mZ^2e^4/h^2\frac{4}{3\pi}\epsilon \qquad (14)$$

$$\epsilon = \frac{4}{3q^2}[\frac{1}{2y}+\frac{y^2}{4}] - \frac{y}{q}\frac{1}{F}ln(\frac{1+F}{1-F}) \quad \text{prolate} \qquad (15)$$
$$= \frac{4}{3q^2}[\frac{1}{2y}+\frac{y^2}{4}] - \frac{2y}{q}\frac{tan^{-1}(F)}{F} \quad \text{oblate} \qquad (16)$$

Note that in the limit $q \longrightarrow 1$ $y \longrightarrow 1$ ϵ approaches -1.

We can now ask whether it pays to deform? The answer is no. If we search in the parameter space of y and q we find that the solution is $\epsilon = -1$ corresponding to $y = 1$ $q = 1$.

To make the above point more clear we expand the expression for E in powers of a deformation parameter ξ defined by

$$\xi = 1 - y \quad \text{prolate}$$

$$\xi = y - 1 \quad \text{oblate}$$

We find in both the prolate and oblate cases for $q = 1$ $\epsilon = -1 + 7/5\xi^2$. Thus the lowest solution at least for small deformation ion corresponds to $\xi = 0$ (or $y = 1$). We can obtain an expression for the quadrupole incompressibility

$$K_Q = b_z^2\frac{d^2E}{db_z^2} = \frac{14}{5}|E_{1s}| = \frac{d^2E}{d\xi^2} \qquad (17)$$

where $|E_{1s}|$ is the calculated binding energy of the hydrogen atom. Recall that the monopole incompressibility K_M is $= 2|E_{1s}|$.

4. The 2P orbit

Do we get deformed solutions for orbits which do not have angular momentum L=0? We consider the 2P orbit. Our trial solution for M = 0 is

$$\Psi = N \ z/b_z \ e^{-(x^2+y^2)/2b_x^2}e^{-z^2/2b_z^2} \ N^2 = \frac{2}{\pi\sqrt{\pi}b_0^3} \ (18)$$

In the spherical limit the result is

$$E = 5/4\frac{h^2}{mb^2} - 4/(3\sqrt{\pi})Ze^2/b \qquad (19)$$

and the minimum condition $\frac{dE}{db} = 0$ yields

$$b = \frac{h^2}{mZe^2}\frac{15}{8}\sqrt{\pi} \qquad (20)$$

In units of $\frac{mZ^2e^4}{h^2}$ the energy is $-\frac{16}{45\pi} = -0.1132$. This should be compared with the exact result of $-1/8 = -0.125$.

When we allow for deformation, the expression for the energy is

$$E = \frac{mZ^2e^4}{h^2}\frac{16}{45\pi}\epsilon \tag{21}$$

PROLATE $\epsilon = \frac{4}{5q^2}[\frac{1}{2y} + \frac{3}{4}y^2] - \frac{3y^4}{q}[\frac{1}{F^2(1-F^2)} - \frac{1}{2F^3}ln(\frac{1+F}{1-F})]$ (22)

OBLATE $\epsilon = \frac{4}{5q^2}[\frac{1}{2y} + \frac{3}{4}y^2] - \frac{3y^4}{qF^3}[tan^{-1}(F) - \frac{F}{(1+F^2)}]$ (23)

Direct calculations show that even in this case the lowest solution is the spherical one $y = 1$, $q = 1$.

This work was supported in part by th U.S. Department of Energy, DE-FG05-86 ER 40299.

References

1 Moshinsky, M.: The Harmonic Oscillator in Modern Physics: From Atoms to Quarks, Gordon and Breach Science Publishers, New York, 1969.

This article was processed using Springer-Verlag TEX Z.Physik D macro package 1.0 and the AMS fonts, developed by the American Mathematical Society.

Conference Programme[1]

Monday, 28 June	Morning Session – Chairman: W. von Oertzen

A Survey of Alpha-Clustering
P.E. Hodgson

Oscillations in Finite Quantal Fermi Systems
R.A. Broglia

Clustering Aspects of Nuclei with Octupole and Superdeformation
S. Åberg

Supershell Structure in Gallium Clusters: A Fermionic System of More Than 6000 Electrons
M. Broyer

Afternoon Session – Chairman: J. Borggreen

Giant Dipole Resonances in Giant Nuclei: Optical Response and Shapes of Charged Sodium Clusters
J. Pedersen, J. Borggreen, P. Chowdhury, N. KeBaili, L. Lundsberg-Nielsen, K. Lützenkirchen, M.B. Nielsen and H.D. Rasmussen

Giant Resonances in Hot Nuclei: Challenges for the Optical Response of Metal Clusters
A. Bracco, F. Camera

Optical Spectra of Hot and Cold Sodium Cluster Ions
Ch. Ellert, H. Haberland, Th. Reiners, M. Schmidt

RPA in Nuclei and Metal Clusters
P.-G. Reinhard

Optical Response of $\mathrm{Li_xNa_{N-x}}(0 \leq x \leq N = 8, 20)$ Bimetallic Clusters and of $(\mathrm{Na_8})\mathrm{Zn}$ and $(\mathrm{Na_9})\mathrm{X}(\mathrm{X} = \mathrm{F}, \mathrm{Cl}, \mathrm{Br}, \mathrm{I})$ Doped Clusters
by L.C. Balbàs, A. Rubio, M.B. Torres

[1] Footnotes denote changes made during the conference.

High Clustering States in the $^{62}Ni(^{58}Ni, ^{58}Ni)^{62}Ni$ Reaction
U. Abbondanno, G. Vannini, M. Bettiolo, P. Boccaccio, R.A. Ricci, L. Vannucci, M. Bruno, M. D'Agostino, P.M. Milazzo, N. Cindro

---

Tuesday, 29 June — **Morning Session – Chairman: G.M. Temmer**

Nuclear Cluster Molecules
B.R. Fulton

Collisional Processes Involving Atomic Cluster Ions
T. Kondow

Fragmentation of Gold Clusters Stored in a Penning Trap
St. Becker, G. Dietrich, H.-U. Hasse, N. Klisch, H.-J. Kluge, D. Kreisle, St. Krückeberg, M. Lindinger, K. Lützenkirchen, L. Schweikhard, H. Weidele and J. Ziegler

Cluster-Cluster and Atom-Cluster Collisions
R. Schmidt

Alpha-Particles and Heavy Clusters in Binary Nuclear Molecules
M.Brenner

Search for Missing Decay Strength of High-Spin Molecular States in ^{48}Cr
R.W. Zurmühle, C. Lee, D.P. Benton, J.T. Murgatroyd, S.P. Barrow, Y. Miao, M.G. Wimer, Z. Liu, A.H. Wuosmaa, M. Freer, R.R. Betts, B.R. Fulton, A.St.J. Murphy

Electronic Structure and Properties of Designer Clusters and Cluster-Assembled Crystals
P. Jena, S.N. Khanna

Afternoon Session – Chairman: P. Manngard

Nuclear Clustering Aspects in Astrophysics
S. Kubono

Alpha Clustering and the Stellar Nucleosynthesis of Carbon
H. Oberhummer, K. Grün, H. Krauss, H. Abele, G. Staudt

The Van Der Waals Interaction Between Finite Fermi Systems: Application to Small Metal Particles
J.M. Pacheco, W. Ekardt

Alpha Chain States in 4N-Nuclei
A.C. Merchant, W.D.M. Rae

Many-Particle Decays of α-Chain Structures in ^{24}Mg
A.H. Wuosmaa

Linear α-Chain States in Nuclei Calculated with Skyrme Interactions
L. Zamick, D.C. Zheng

Wednesday, 30 June — Afternoon Session – Chairman: A.C. Xenoulis

---

Energetics of the Fission Process
F. Gönnenwein

The Fission of Highly Charged Metal Clusters
U. Näher, T.P. Martin

Multi-dimensional Tunneling and Nuclear Fission Process
A. Iwamoto

High-Accuracy Coupled-Cluster Calculations of Potentials Between Atoms or Small Molecular Groups
U. Kaldor

Nonlinear Approach in the α-Cluster Model
K.A. Gridnev

Spectroscopy of Microclusters in Superfluid Helium
J.H.M. Beijersbergen, Qin Hui, M. Takami

Thursday, 1 July — Morning Session – Chairman: J.A. Alonso

New Response Theory for Finite Many-Electron-Systems Beyond the TDLDA
W. Ekardt

Effects of the Cluster Surface on the Electronic Shell Structure: Faceting, Roughness and Softness
J. Mansikka-aho, M. Manninen, E. Hammarén

Electron Diffraction on Free Small Cu Particles. Anomalous Structures and Lattice Contraction
D. Reinhard, P. Berthoud, D. Ugarte, B.D. Hall, R. Monot

Very Neutron-Rich Exotic Nuclei
F. Pougheon

Clustering Aspects of Light Exotic Nuclei
J.S. Vaagen

Neutron-Rich Nuclei Studied with AMD
H. Horiuchi, Y. Kanada-En'yo, A. Ono

Ultraviolet Laser Facility
C. Fotakis

Afternoon Session – Chairman: V.K. Lukyanov[2]

Ionic Hydrogen Clusters: Structure and Fragmentation
B. Farizon, M. Farizon, M.J. Gaillard, E. Gerlic, S. Ouaskit

Statistical Multifragmentation
D.H.E. Gross

Cluster Emission from Hot Nuclei
J.S. Edwards, S. Picozzi, Z. Shehadeh, F.B. Mailk, I. Reichstein

The Island of Thira and the Aegean Civilization
Ch. Doumas

Friday, 2 July **Morning Session - Chairman: M.E. Grypeos**

Probing the Structure and Dynamics of Cage-Like Clusters: From Water to Met-Cars
A.W. Castleman, Jr.

The Cubic Structures of Water Clusters
S.C. Farantos

Shell Structure: A Mesoscopic Property of Nuclear and Atomic Many Particle Systems
V. Strutinski[3]

Parity Doublets in C_{60}
R. Friedberg, T.D. Lee, and H.C. Ren

Synthesis of C_{60} Atomic Clusters and Magnetization Studies of Superconducting Rb_3C_{60}
C. Politis

Valence Change in Rare Earth Clusters
C. Brechignac, Ph. Cahuzac[4]

---

[2] Unable to attend, replaced by Prof. V.Z. Goldberg.
[3] Died en route to the conference. His time in the program was devoted to speeches given in his honor, focusing upon his life and his outstanding contributions to the field, which made him a pioneer in nuclear physics. Speakers were Prof. V. Pugatch, from his home institute, and Prof. R.A. Broglia, collaborator.
[4] Speech topic was changed to fission.

Dissociative Ion Capture by Neutral Rare Gas Clusters
P. Scheier, M. Lezius, B. Schiestl, M. Foltin, T.D. Märk

Observation of α-Cluster Bands in fp-Shell Nuclei
T. Yamaya, M. Saito, M. Fujiwara, T. Itahashi, K. Katori, T. Suehiro, S. Kato, S. Hatori, S. Ohkubo

---

Afternoon Session – Chairman: S. Ohkubo

Recent Results on Cluster Radioactivity
A.A. Ogloblin[5]

Nuclear Structure Effects in Cluster Radioactivity
D.N. Poenaru, W. Greiner

Spontaneous Emission of ^{14}C Clusters from $A = 221$ Nuclei
R. Bonetti, C. Chiesa, A. Guglielmetti, C. Migliovino, P. Monti

^{14}C Cluster Emission: A Tool for the Study of Nuclear Structures
M. Hussonnois, G. Ardisson

Self-organisation of Clusters on Surfaces
M.R.C. Hunt, R.E. Palmer

Concluding Remarks
Walter Greiner

[5] Unable to attend.